The Handbook of Continuous Crystallization

The Handbook of Continuous Crystallization

Edited by

Nima Yazdanpanah
Massachusetts Institute of Technology, USA
Email: nima@mit.edu

and

Zoltan K. Nagy
Purdue University, USA
Email: zknagy@purdue.edu

ROYAL SOCIETY OF CHEMISTRY

Print ISBN: 978-1-78801-214-0
PDF ISBN: 978-1-78801-358-1
EPUB ISBN: 978-1-83916-131-5

A catalogue record for this book is available from the British Library

The Royal Society of Chemistry is a charity, registered in England and Wales, Number 207890, and a company incorporated in England by Royal Charter (Registered No. RC000524), registered office: Burlington House, Piccadilly, London W1J 0BA, UK, Telephone: +44 (0) 20 7437 8656.

Visit our website at www.rsc.org/books

Printed in the United Kingdom by CPI Group (UK) Ltd, Croydon, CR0 4YY, UK

Preface

Continuous crystallization is a common technique for most bulk chemical manufacturing processes. The traditional applications have been in fine chemicals, fertilizers, and salts and sugars, and it has recently been adopted in the pharmaceutical industry. The continuous nature of the process includes continuous feed addition and product removal, as well as continuous solid formation (nucleation, crystallization, and precipitation). The classic crystallization process (mainly batch) and other continuous manufacturing processes have been traditionally taught in chemical engineering courses. This book is intended to gather the most important aspects of the continuous crystallization process, and to cover many important concepts and applications.

This book highlights the fundamental concepts, applications, challenges, and aspects related to modeling and control of continuous crystallization processes. It should serve as a resource for practicing chemical engineers as well as for chemists, analysts, technologists, and operations and management team members. The book is also a useful resource for graduate students working in the field. It intends to bridge the gap between theory and practice with a comprehensive overview of the continuous crystallization processes.

Each chapter can be used as a stand-alone source of information for the particular topic in this domain, as well as integrated with other chapters to provide a unique reference for interested readers:

Chapter 1 covers the continuous nucleation, which is the first step in the continuous crystallization, by explaining common techniques for primary and secondary nucleation, heterogeneous and homogeneous crystallization, and crystal growth and breakage.

The Handbook of Continuous Crystallization
Edited by Nima Yazdanpanah and Zoltan K. Nagy

Published by the Royal Society of Chemistry, www.rsc.org

Chapter 2 is on the mathematical modeling of the continuous crystallization, which spans from the molecular front level to process scale, with equations and models for crystal growth, population balance, and equipment modeling.

Chapter 3 is on a particular method in continuous processing, Oscillatory Baffled Reactor, which has been recently adopted as Oscillatory Baffled Crystallizer, for applications in pharma and fine chemicals.

Chapter 4 covers the process control and techniques to keep the process under control, efficient, and optimized. The higher-level view in this chapter holistically considers the continuous crystallization process as a unique unit operation.

Chapter 5 is on a particular plug flow method for continuous crystallization, in which the slug flow in the tubular crystallizer creates small individual micro crystallizers for specific applications. Fundamental analysis is also provided for design strategies, from both fluid flow and heat transfer perspectives.

Chapter 6 explains the applications of continuous crystallization in bulk chemical and traditional industries at large scale of equipment and process. The mode of operation and criteria for design and applications in multipurpose plants are covered.

Chapter 7 is on process intensification and running the continuous crystallization at a very small scale and challenging fashion. The downscaling and process intensification has its own extreme challenges and limitations with Multiphysics that this chapter covers.

Chapter 8 covers the continuous membrane crystallization for unique applications at different scales. The combination of membrane processing and crystallization has a variety of advantages and challenges that are explained in this chapter.

Chapter 9 is on instrumentation and utilization of process analytical technologies (PAT tools) for continuous crystallization. Different sensors and probes, integration, data collection, monitoring, and model-free control for solid and liquid phases are covered.

Chapter 10 explains a niche application for continuous protein crystallization. The difference between batch and continuous operating modes and a discussion on how and why a continuous process could be employed for protein crystallization is covered here with examples and case studies.

Chapter 11 covers continuous melt crystallization at different scales and applications. The coverage spans from fundamental melt crystallization to equipment design to different techniques and methods.

Chapter 12, on Continuous Crystallization of Isomers and Chiral Compounds, provides a holistic overview of the process of crystallization of isomers and chiral compounds with numerous examples, case studies, and technical discussions. This chapter describes the process for attaining pure enantiomers through enantioselective crystallization in continuous mode.

Chapter 13 explains process integration, downstream processes, and continuous filtration and drying for continuous crystallization processes at different scales. The chapter also covers some common isolation process problems, their potential causes and solution strategies.

Chapter 14 is on Eutectic Freeze Crystallization (EFC) for treating saline streams to recover both water and dissolved salts. Thermodynamic modeling is used to predict the eutectic temperature, composition, ice and salt recoveries for binary, ternary and multicomponent streams.

Chapter 15 provides an Economic Analysis of Continuous Crystallization with technoeconomic analyses of different crystallization configurations, process modeling and optimization with three extensive case studies. The chapter highlights the effect of a variety of process considerations for crystallizer design.

Chapter 16 explains mechanistic modeling tools for digital design and operation of continuous crystallization processes. Flowsheet modeling techniques are demonstrated with a couple of case studies, for workflow development, risk management through sensitivity analysis, batch to continuous conversion, and process control.

There are many people to thank who made this work possible. First, we would like to thank all the contributors to this book. We also would like to thank the Royal Society of Chemistry for the outstanding support and help during the preparation of this work.

Nima Yazdanpanah
Zoltan K. Nagy

Contents

The Handbook of Continuous Crystallization
Edited by Nima Yazdanpanah and Zoltan K. Nagy

Published by the Royal Society of Chemistry, www.rsc.org

CHAPTER 1

Nucleation and Crystal Growth in Continuous Crystallization†

JOHN MCGINTY[a], NIMA YAZDANPANAH*[b], CHRIS PRICE[a], JOOP H. TER HORST[a] AND JAN SEFCIK[a]

[a]EPSRC Centre for Innovative Manufacturing in Continuous Manufacturing and Crystallisation, University of Strathclyde, Glasgow, UK; [b]Department of Chemical Engineering, Massachusetts Institute of Technology, USA
*E-mail: nima@mit.edu

1.1 Introduction

Continuous crystallization has been a common process in the fine chemicals and petrochemical industries for decades. The advantage of continuous crystallization led other industries to employ the technology on the manufacturing scale. Continuous crystallization processes need smaller process equipment leading to substantial reduction in capital and operating costs. While batch processes can demonstrate significant batch-to-batch variability in product quality, continuous crystallization processes tend to give the continuously created crystals the same process experience, irrespective of their time in production, and therefore are potentially more consistent. Additionally, using a recycle, it is feasible to achieve higher yields in continuous processes. Continuous crystallization also tends to yield lower impurity inclusion in the products due to lower accumulation of impurity in the mother liquor by continuous fresh feed addition.

---

†Electronic supplementary information (ESI) available. Colour version of Figure. 1.1. See DOI: 10.1039/9781788013581

The Handbook of Continuous Crystallization
Edited by Nima Yazdanpanah and Zoltan K. Nagy

Published by the Royal Society of Chemistry, www.rsc.org

The pharmaceutical industry is beginning to adopt continuous manufacturing.[1] Although crystallization has been identified as one of the bottlenecks for full adoption of continuous manufacturing as it is considerably slower than the upstream continuous synthesis,[2–4] a number of principal drivers for switching from batch to continuous crystallization have been identified.[5]

Continuous crystallization is a collection of sub-processes such as solution feeding, supersaturation generation, heat transfer, evaporation, primary nucleation, secondary nucleation, crystal growth, agglomeration and particle suspension. While it sets the rate of formation of new crystalline particles over which the crystallizing mass is distributed in a continuous crystallization process, the sub-process of crystal nucleation is poorly understood and controlled.[6] Once a supersaturation is created crystallization can commence. Crystallization is a collection of the subprocesses of crystal primary nucleation, crystal growth, secondary nucleation and agglomeration, which are all governed by the prevailing supersaturation as well as other parameters. The rates of these subprocesses determine the crystalline product quality.

In order to achieve enhanced control over the crystal nucleation and growth in a continuous crystallization processes, a higher level of nucleation understanding and control is needed. Nucleation can be circumvented by seeding which is especially useful during start-up to minimize the peak of supersaturation associated with conventional unseeded crystallization and along with this reduce the risk of encrustation forming, as once formed this ultimately limits the duration of a continuous crystallization process. Seed suspension can also be continuously fed into a crystallizer operating at a steady state as an additional input. Continuous nucleators also have been proposed as a workaround technique, as well as *in situ* milling, or the inherent *in situ* secondary nucleation in mixed suspension mixed product removal (MSMPR) due to particle–particle or particle–impeller attrition.

1.2 Crystal Nucleation

Industrial crystallization involves the formation of a particulate crystalline phase from a thermodynamically metastable solution.[7,8] A continuous crystallization process will have a clear (particle free) undersaturated solution as an input and a slightly supersaturated suspension as an output. The product crystals will need to be generated in the crystallizer by creating the supersaturation driving force for crystallization using an external action.

One of the ways to define the driving force for crystallization is by the supersaturation ratio S:

$$S = C/C^* \tag{1.1}$$

The supersaturation ratio S is defined by the concentration C and solubility C^* at the current value of the parameters being adjusted to generate supersaturation (temperature, solvent mixture composition, pH *etc*). [Note: concentration can have various units (*e.g.*, mole fraction or mg per mL solvent), which will result in different values for S and therefore it is important that it is clearly specified which units are used]. The supersaturation can be increased by, for instance, a concentration increase through solvent evaporation or a solubility decrease by decreasing the temperature. Crystal growth would reduce the solution concentration and thus the supersaturation.

If the concentration exceeds the solubility, the supersaturation ratio $S > 1$, the solution is supersaturated and any crystals present can grow. If the concentration is lower than the solubility ($S < 1$) the solution is undersaturated and any crystals present will tend to dissolve. At thermodynamic equilibrium the solution is saturated, concentration and solubility are equal ($S = 1$), any crystals present will be maintained in equilibrium with the flux of molecules arriving and leaving the collective crystal surface being in balance. Since the supersaturation ratio drives the crystallization process, the solubility of a compound is a crucial parameter in the crystallization process design. For instance, a strongly increasing solubility with temperature and a sufficiently small solubility at a low temperature direct the preferred supersaturation generation method towards cooling. In addition, the difference between the inlet concentration and the end point solubility is strongly associated with the yield and productivity of a crystallization process.

Within an industrial crystallization process, crystals can be formed from an initially clear solution (primary nucleation) or due to the presence of parent crystals (secondary nucleation). In turn, primary nucleation generally is divided into homogeneous and heterogeneous nucleation. In a supersaturated solution new crystals can be formed in the absence of crystalline solids of the same substance, which is termed primary nucleation, or in the presence of crystalline solids of the same substance, which is termed secondary nucleation. Primary and secondary nucleation will be discussed in respectively Sections 1.2.1 and 1.2.2. Both primary and secondary nucleation as well as crystal growth kinetics vary widely under thermodynamically metastable conditions. However, the nucleation rate varies over many orders of magnitude while growth rate has more gentle increase with increasing supersaturation. During heterogeneous primary nucleation, the crystals form at surfaces such as dust particles, crystallizer wall, air–solution interface or deliberately added template particles. Homogeneous primary nucleation takes place in the absence of heterogeneous particles in a clear solution. It is important to note that in the laboratory and more so in large-scale processes on an industrial scale, the presence of many different heterogeneous particles or surfaces is impossible to avoid. Despite their importance, usually no information is available on the amount and kind of heterogeneous particles that are finally responsible for the occurrence of heterogeneous nucleation.

Thus, while an unseeded batch cooling crystallization process usually relies on primary nucleation to provide the crystals, during a continuous crystallization process the omnipresent crystals continuously generate more crystals through secondary nucleation. Only in extreme cases are there indications that homogeneous nucleation is the dominant nucleation mechanism. Introduction of crystals into a crystallization process is based on either nucleation or seeding. Seeding relies on addition of previously formed crystals while nucleation implies birth of new crystals. The nucleation rate expresses the number of new crystals that are generated per unit of time per unit solution volume at a given composition and temperature. Nucleation events could be evenly distributed across the bulk fluid volume. However, it may more often be the case that locally extreme conditions (supersaturation, fluid dynamics, mixing points) lead to local nucleation events. While the resulting suspension is distributed over the entire crystallizer, the generation of crystals through nucleation can be highly localized.

1.2.1 Primary Nucleation

In supersaturated solutions, the nucleation rate varies highly non-linearly with supersaturation. Classical Nucleation Theory describes the supersaturation dependent nucleation rate J as a function of a supersaturation dependent nucleation barrier $B/\ln^2 S$:[9,10]

$$J = AS\exp(-B/\ln^2 S) \tag{1.2}$$

where A and B are constants. The nucleation barrier is very large for supersaturated solutions close to the solubility line resulting in a negligible nucleation rate and prolonged lifetimes of the metastable solutions. At very high supersaturations, far away from the solubility line, the energy barrier for nucleation vanishes and spinodal decomposition takes over from nucleation.

The values of the nucleation rate constants A and B depend on the primary nucleation rate mechanism taking place.[6] Primary nucleation can take place in the bulk volume of a particle free solution (homogeneous primary nucleation) or at interfaces (heterogeneous primary nucleation) due to the presence of crystallizer wall, solution–air interface and suspended foreign particles such as dust particles. Heterogeneous particles or surfaces promoting heterogeneous nucleation are characterized by B-values that are much lower than those in the case of homogeneous nucleation from a clear solution: the enhanced heterogeneous nucleation is due to the reduction of the nucleation barrier. This, even though the A-value for heterogeneous nucleation is orders of magnitude lower than that for homogeneous nucleation. Since there will always be interfaces or particles present in industrial solutions to nucleate onto, often heterogeneous nucleation is assumed to be the dominant primary nucleation mechanism.

However, if nucleation events are heterogeneous and are related to external fluid interfaces, such as vessel walls or fluid–air interface with a defined surface area, it may be appropriate to express nucleation rate as number of crystals generated per unit surface area per unit time. Furthermore, if nucleation events are related to some other localized environment, such as a region of high shear (*e.g.*, due to pump or agitator), inlet stream mixing point or external field impact (*e.g.*, by ultrasonic transducer), it may be appropriate to express nucleation rate simply as number of crystals generated per unit time within the given local volume.

Dependencies of the nucleation rate on solution composition and temperature within the metastable zone vary widely from system to system. There are currently no reliable tools able to quantitatively (and often even qualitatively) predict how nucleation rates depend on solution composition and temperature, and even the fundamental mechanisms (homogeneous or heterogeneous; single step or multistep) are strongly debated in scientific literature.[6,8,9] Until there is major progress in predictive computational tools in this area, information about nucleation and growth rates needs to be obtained experimentally.

Experimental measurements of crystal nucleation rates are challenging because nuclei are small so that they have to grow into a detectable range, and nucleation and growth are inextricably linked. In order to measure the nucleation rate, it is possible to count the increase in the number of detectable particles assuming that the change of number of particles is only due to nucleation,[3,11,12] or it is possible to determine induction times from which primary nucleation kinetics can be estimated using various assumptions (isothermal/polythermal, constant/variable nucleation rate, single/multiple nuclei, growth time to detection, *etc.*).[13,14]

Fouling and encrustation can be related to (heterogeneous) nucleation and agglomeration of crystals in regions with high local supersaturations such as cooling surfaces, mixing points and contact lines of boiling liquid, air and crystallizer wall. Fouling needs to be monitored[15] and mitigated when operating continuous crystallization processes as it can compromise the steady state operation as well as product quality attributes.

Primary nucleation in stagnant fluid in absence of external fields is typically addressed in textbooks: but what about effects such as mixing, shear, pressure and electromagnetic fields? Effects of pressure waves (ultrasound), high power lasers and fluid shear on nucleation have been observed: can we understand and use them to develop better crystallization processes?

1.2.1.1 Mixing-induced Supersaturation

The effect of local concentration gradient can be important when supersaturation is mixing-induced, *e.g.*, in antisolvent or reactive crystallization, where two fluid streams need to be mixed to obtain required solution composition. If the mixing process is much slower than the nucleation process at the final solution composition, it can be expected that nucleation would proceed within the mixing region before mixing is completed and local composition heterogeneity in the mixing region would significantly influence resulting nucleation outcome.[11,16,17] On the other hand, if the mixing process is much faster than nucleation, it can be expected that nucleation would only proceed once mixing is complete and local concentration gradients would not be significant. The significance of mixing can thus be assessed by comparison of characteristic times of mixing τ_m[18–20] and of nucleation τ_n and by analogy with chemical reactions, we can define a dimensional ratio of the two timescales τ_m/τ_n, similar to the Damköhler number.[21,22] While mixing time scales have been investigated and quantified across a range of mixing conditions,[23–25] it is less straightforward to quantify nucleation time scales (*e.g.*, inverse of induction time couples both nucleation and growth time scales and is dependent on relevant volume where nucleation occurs as well as observation method).

1.2.1.2 Shear

Heterogeneity of flow environment is inherent in vessels, pipes and pumps. Uniform flow field can be achieved in some idealized rheometry environments such as Couette (in the gap between two cylinders with inner cylinder rotating[26] or cone-and-plate cells). In pressure driven laminar flow in pipes the shear rate varies from zero in the center to the maximum value at the pipe wall. Ranges of shear rates experienced by fluids can vary widely depending on the nature of agitation, agitator shape, size and movement, vessel geometry and resulting flow regimes (*e.g.*, stirred tanks, Taylor–Couette flows, oscillatory flows).[27,28]

The effect of local flow environment on nucleation has been repeatedly noted in previous literature but in early reports it was unclear whether effects were due to primary or secondary nucleation (Figures 1.1 and 1.2).[29–33]

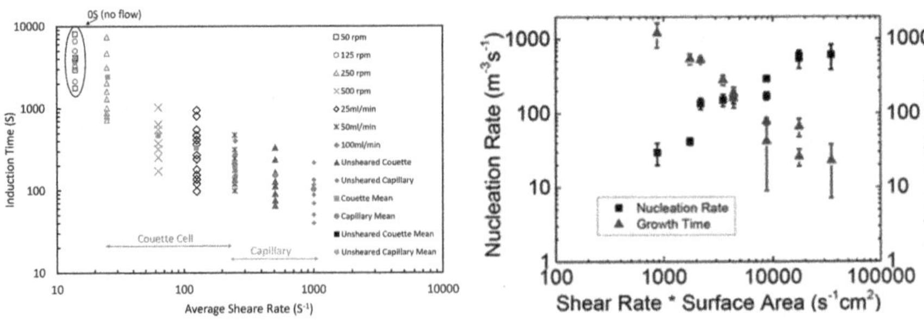

Figure 1.1 Induction time (s), nucleation rates J, and growth times t_g estimated from a model fit as a function of the product of the average shear rate and surface area. Adapted from ref. 34 with permission from American Chemical Society, Copyright 2015, and ref. 35, https://pubs.acs.org/doi/abs/10.1021/acs.cgd.5b01042, with permission from American Chemical Society, Copyright 2016.

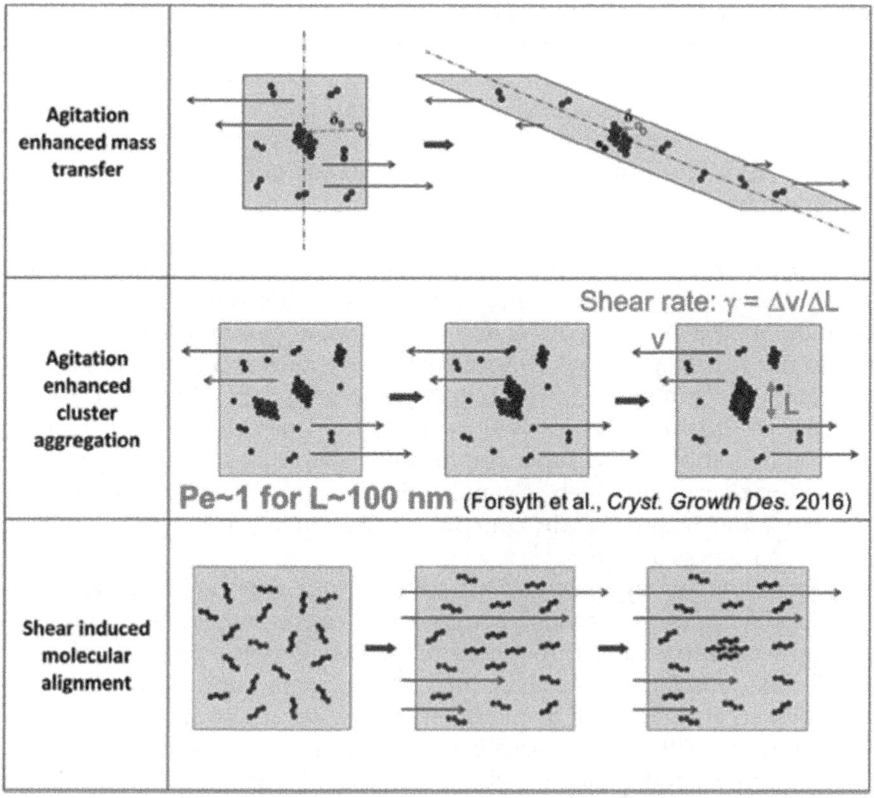

Figure 1.2 Effect of shear and fluid micro-mixing on nucleation.

1.2.1.3 *External Fields (Ultrasound, Laser, Electromagnetic)*

The principal benefit of using external fields, such as ultrasound or laser, to control nucleation lies in avoiding the problems typically encountered with conventional seeding; inconsistent seed attributes and seed history, incomplete seed wetting and dispersion resulting in polycrystalline particle formation, seed dissolution due to delayed seed addition, *etc.* Well-controlled nucleation offers the possibility of exquisite control of product particle size. These aspects are particularly important in continuous crystallization where seeding is an essential element in the start-up process required to minimize the peak supersaturation attained and so to delay the onset of encrustation. The benefit of externally induced nucleation in starting up a continuous crystallization is that it can be initiated at supersaturation levels lower than the metastable zone limit for spontaneous nucleation and potentially at levels lower than would be considered robust for conventional seed addition.

In a classical continuous crystallization, following successful initiation *via* seed addition, the next major operational challenge is to manage the secondary nucleation rate such that the available surface area for crystal growth is commensurate with achieving almost complete desupersaturation within the target residence time whilst also producing crystals with the required size distribution. Typical strategies for commodity materials manufacture revolve around making large particles which are easy to isolate by filtration and washing. This is usually achieved by actively suppressing nucleation and operating at modest supersaturations with extended residence times.

1.2.1.3.1 Ultrasound-induced Nucleation. Sonocrystallization is the application of ultrasound to influence crystallization processes. The most common approach may be more accurately termed "sononucleation" where ultrasound is used to trigger controlled primary nucleation. Insonation can also make a distinctive contribution by generating new particles through breakage of existing crystals (see next section on secondary nucleation) allowing operation at low supersaturation levels whilst maintaining a large population of crystals which are appropriately small for direct formulation. This can allow breaking of the constraints of conventional continuous crystallization process design.

The mechanism by which ultrasound triggers nucleation is not fully understood, however, it is widely accepted that cavitation plays a central role. A sound wave propagates through a solution as alternating periods of compression and rarefaction. Acoustic intensity, the amplitude of the wave, can be expressed in microns of displacement of the source of the sound. When the amplitude of the sound wave is sufficiently large cavitation occurs and bubbles form from the release of dissolved gas and evaporated solvent vapor. These bubbles shrink during the compression phase and then they expand again during the subsequent rarefaction phase as the sound wave propagates (this repeating oscillation is known as stable cavitation). If the amplitude of the sound wave is large enough bubbles of solvent vapor form during the rarefaction. These bubbles can coalesce and transient cavitation may occur. Under appropriate conditions bubbles increase in size cycle by cycle until they reach a critical size, perhaps 100–200 μm, when they collapse catastrophically. This transient cavitation phenomenon is linked with triggering nucleation.

Much of the literature relates to aqueous solutions where the range over which transient cavitation is possible is limited to about 15–20 cm from the acoustic source, due to the shielding effect of cavitation bubbles reducing the intensity of insonation further from the acoustic source. This results in a design constraint for large scale operation. Tubular geometries with multiple acoustic sources are required to insonate large volumes with a maximum duct diameter of around 40 cm.

There are a number of reported applications with APIs in organic solvents for example, however the essential physical data required for modelling *i.e.*, velocity of sound and solution density as a function of temperature and concentration and the cavitation threshold, are rarely available. A consequence of the lack of fundamental measurements linking the underlying physics of ultrasound with measured intensity maps is that the approaches to process development are structured but empirically based.

Early pharmaceutical applications of ultrasound in crystallization include Pfizer's patent to reduce the crystal size of procaine penicillin.[36] Principe and Skauen (1962) report the use of ultrasound in the preparation of microcrystalline particles of the hormone progesterone as an alternative to size reduction by milling.[37] They report the effect of intensity on particle size and size distribution, reporting smaller particles formed at higher intensities. Wyeth's Hem (1967) investigated mechanisms by which ultrasound might be effective in producing small uniform crystals and suggested that the beneficial effects of ultrasound on crystallization are linked to cavitation arising from the passage of ultrasound thorough the solution.[38] He was the first to suggest that cavitation bubbles act as nucleation sites.

Three key examples indicate the benefits of controlled ultrasound induced nucleation.

Howard Anderson *et al.* (1995) obtained US5471001A, the first patent for continuous sonocrystallization. It relates to evaporative crystallization of adipic acid from aqueous solution and significantly was applied at manufacturing scale.[39] Whilst the patent makes no explicit reference to nucleation, the claimed benefits are consistent with manipulation of the nucleation rate allowing the process to operate at reduced supersaturation to deliver crystals with improved powder flow characteristics.

Michael Midler secured US3892539A in 1971 for Merck & Co. Inc., which relates to the use of ultrasound to break up large crystals at the base of a continuous fluidized bed crystallizer effectively both managing the upper particle size and generating new nuclei at a controlled rate whilst operating at a modest supersaturation.[40] The technology was applied to facilitate continuous resolution of an API.

Whilst not typically a continuous process, a particularly attractive application for the pharmaceutical industry is reliably triggering nucleation in a sterile environment.[41] Insonation substantially reduces the induction time even at modest levels of supersaturation generated by addition of an antisolvent. The resulting controlled nucleation was shown to deliver more consistent particles than in the equivalent unseeded process and since conventional seeding is not favored in sterile manufacture, due to the risk of introducing biological contamination, this is clearly beneficial.

1.2.1.3.2 Laser-induced Nucleation. Application of continuous wave[42,43] or pulsed lasers[44,45] can dramatically shorten induction times in a wide range of solutions. In principle, this provides an intriguing opportunity for accurate spatial and temporal control of nucleation in both batch and continuous systems. Furthermore, laser-induced nucleation can lead to different polymorphs being nucleated compared to identical solutions in the absence of lasers.[46]

As laser-induced nucleation has been reported for systems which were not significantly absorbing light at the laser wavelengths used, this phenomenon does not necessarily involve photochemical effects.

For pulsed lasers, it has been established that a certain threshold laser power is needed to induce nucleation[47,48] and that the probability of nucleation can further increase with increasing laser power. It has been also observed that the probability distribution of induction time in laser irradiated glycine solutions shows bi-exponential distribution, where a certain fraction of samples undergo fast laser-induced nucleation while the rest undergo much slower spontaneous nucleation.

However, the mechanism of laser-induced nucleation in the absence of a photochemical effect is not yet clear. Several mechanisms have been suggested, such as polarization of clusters and cavitation or bubble formation, *e.g.*, caused by heating of nanoparticulate impurities.[42–47]

1.2.1.3.3 Effects of Electric or Magnetic Fields on Nucleation. A theoretical description of the effect of an electric field on the homogeneous nucleation rate preceded experimental work and concluded that depending on the ratio of dielectric constants between solution and solid the nucleation rate will decrease or increase.[49] One of the first experimental reports on the combination of electric field and crystallization showed that the electric field enabled the crystallization of an enzyme.[50] Later on, it was shown that crystals of lysozyme formed with a preferred orientation on the electrode, probably because of the field's influence on the nucleation process.[51] A decrease in induction time of the protein BPTI was for instance observed in the presence of an electric field assumed to occur due to electromigration.[52] Another interesting effect is that particle motion is induced in a suspension in an isolator solvent after which the particles collect at a specific electrode.[53] This effect was used to separate a mixture of crystals that collect at different electrodes under the influence of an electric field.

Electric fields are nowadays known to be able to locally enhance or inhibit nucleation, although the mechanism seems not yet clear. Despite its potential, as far as we know, localized electric field induced nucleation has not yet been applied to continuous crystallization processes.

1.2.2 Secondary Nucleation

Secondary nucleation is believed to be the predominant source of nuclei in the vast majority of crystallization processes. Secondary nucleation also could cause polymorphism "whether contact secondary nuclei originate from parent crystals *via* microattrition or from semiordered solute clusters at the interface of parent crystals" (Figures 1.4 and 1.5),[54,55] however, detailed discussion of this topic is not in the scope of this chapter. The importance of secondary nucleation can be explained with two examples of seeding and particle attrition:

1.2.2.1 Seeded Crystallization

In order to keep the crystal number constant, primary or secondary nucleation needs to be prevented, therefore a zero (or negligible) rate of secondary nucleation is required. Typically, this is difficult to achieve in suspension crystallizers because of the preponderance of secondary nucleation. In continuous stirred tank (CST) crystallization, the secondary nucleation is typically required for steady supply of new crystals, so the rate of secondary nucleation needs to be controlled. This should be the main focus of the process design such that the population density can be maintained at a modest supersaturation consistent with faceted growth, impurity rejection and delaying the onset of encrustation. The capability to manipulate CST is dominated by the ability to manipulate the secondary nucleation rate and allow the system to readjust to a new steady state through growth. Sucrose represents a special case where the seed crystals are added as very fine particles and because of the high solution viscosity, crystal collisions are relatively rare and sufficiently gentle that very few secondary nuclei form. This is why commercial granular sucrose has a tightly controlled particle size typically between 355 μm and 500 μm depending on the seed load selected.

Secondary nucleation has been defined[56] as: "nucleation which takes place only because of the prior presence of crystals of the material being crystallized". Therefore, secondary nucleation of a solid phase of a substance occurs due to presence of particle(s) of the same substance *vs.* heterogeneous primary nucleation which occurs due to presence of other interfaces.

There are two distinct ways to form new crystals in the presence of pre-existing crystals. In the first one, small (abrasion) or large pieces (fracture) can break off from existing crystals and the resulting crystal fragments become new crystals. It can be argued that this should not be called nucleation at all, as there is no formation of a new solid phase domain, just mechanical division of an existing solid phase domain already present. This typically happens due to relatively high energy collisions of crystals with impellers, vessel walls, each other, or due to high energy turbulence eddies or cavitation caused by impellers or external fields, such as ultrasound. Mechanisms of formation of crystal fragments are related to fracture and abrasion mechanics of parent crystals.

In the second distinct way to form new crystals in the presence of pre-existing crystals, the existing crystals keep their structural integrity but induce formation of new crystals through contact with the surrounding supersaturated fluid phase, resulting in formation of new solid phase domains within the fluid phase. This typically happens under conditions of relatively low energy interactions with other solids or fluids, such as gentle tapping, sliding across surfaces or sedimenting. Such mechanisms of secondary nucleation are much less understood.

Despite decades of study and discussion on secondary nucleation, the mechanism and underlying physics are still in the center of scientists' debates, from "Although secondary nucleation appears to be major source of nuclei in industrial crystallization, little is known about the mechanism by which such nuclei are produced",[57] and some 35 years later...."Despite years of studies, the mechanism of contact secondary nucleation has not been resolved".[55] Although various complex models for secondary nucleation can be proposed, it is experimentally challenging to determine which mechanism occurs in reality, and this might also differ from compound to compound. However, the main secondary nucleation scenarios can be categorized as:

- Parent crystal present in "stagnant" solution
- Parent crystal subjected to fluid shear
- Parent crystal subjected to collision with impellers or vessel walls
- Parent crystal subjected to collision with others crystals

1.2.2.2 Attrition, Fragmentation, Breakage

These can be seen as mechanical separation of preformed crystals (or their pieces) from larger crystals or aggregates/agglomerates, typically due to collisions with impellers, vessel walls or with other crystals or due to fluid action on crystal (*e.g.*, fluid shear or turbulent eddies). This can be also seen as crystal breakage or de-aggregation/de-agglomeration. The collision-based processes can be studied separately from crystal growth, *i.e.*, in saturated solutions or non-solvents.

Small pieces of larger crystals are broken off due to mechanical collisions (*e.g.*, with impeller) and these small crystalline pieces appear to be new particles. This effect can be influenced by suitably designing the crystallizer vessel, agitation and suspension density.[58–63] Attrition rate is a function of the number of collisions and the energy impact of those collisions as well as system-specific properties such as the material's hardness, solution viscosity and density difference between solution and crystalline material. The collision rate with the stirrer can be increased by, for instance, increasing the stirrer speed or changing the stirrer design. Another way to increase the collision rate is by increasing the suspension density so that there are more crystals. Also the crystal size plays a role: above a certain size the impulse of the particle originating from the density difference between solution and solid becomes too large for the fluid to drag the particle with it and the probability of collisions drastically increases.[11] However, the effect of attrition can be influenced by suitably designing the crystallizer vessel, agitation and suspension density.[64,65]

At high stirring speeds, macroabrasion of crystals results in fragments that serve as nucleation sites. As opposed to contact nucleation, which involves microabrasion of the crystals, this phenomenon results in the rounding of edges and corners of crystals. This process is referred to as collision or attrition breeding. Attrition causes small particles of acetaminophen crystals that have already been formed on the surface of the excipient to break and enter the solution, thus causing secondary nucleation (acting as seeds). A special case occurs at very high supersaturation, where there is very fast dendritic/needle-like growth from the surface of crystals, and pieces of crystalline material can be easily broken off by fluid flow or mechanical action.

The effect of mixing on secondary nucleation has been reported in several studies.[11,66] An analogy of relation mixing and nucleation rate with reaction Damköhler number has been proposed for visualization of the effect of mixing severity.[12,67] The seed (fragment) generation is a proportion of the collision energy and frequency of collision in the size range of crystal length. Three different types of collision can result in attrition:

- Crystal–crystal impact: a function of both the local micromixing environment and the overall macromixing circulation.
- Crystal–impeller: a function of the impeller speed, the shape of the blade, and the construction material.
- Crystal–wall impact: a function of eddy turbulence, particle velocity and shape, and crystallizer design.

The critical mixing factors have been identified as impeller type and speed and their influence on local turbulence and overall circulation. Particle damage is a function of the shear produced by the agitator, which is nominally proportional to the agitator tip velocity. The attrition and fragmentation can be formed by contact of the crystals with a pump impeller (circulation line), the stirrer blade (traditional pitch blade impeller), or due to impact of the slurry on vessel walls (radial *vs.* vertical circulation in the liquid phase). Fluidized bed crystallizers have been used as an alternative approach to provide high mass transfer and mixing will minimize particle–particle and particle–wall attrition.[68]

Key effects in mechanically induced attrition and breakage in stirred tanks can be listed as:

- Supersaturation
- Crystal concentration or mass (magma density)
- Crystal mechanical properties
- Individual crystal size (or mass), aspect ratio, which also has an anisotropic aspect based on slice energy in the lattice and preponderance and nature of defects agitation/flow
- turnover time (s)
- impeller rotation rate (s^{-1})
- impeller tip velocity (ms^{-1})
- power input or energy dissipation rate (local *vs.* overall intensity of mixing W/L)
- impeller geometry (marine propellers or profiled blades *vs.* pitched flat blades with sharp edges) and mechanical properties (soft surface coating)

- vessel geometry and flow patterns. For instance, classically proportioned stirred vessel with baffles *vs.* bespoke draft tube baffle units widely used in continuous suspension crystallizers. Generally avoid conical base vessels for crystallization; these are encountered in many pilot plants because of the wide dynamic volume range and low minimum stir volume but require very intense agitation to prevent crystals settling in the base of the cone.[16,69]
- Contact secondary nucleation: crystal–wall contact[57,70]
- Secondary nucleation due to fluid shear[71,72]
- Secondary nucleation threshold[73]

Ottens *et al.*[74] developed a mechanistic description of the mechanical interaction of the crystals with the crystallizer component in which the nucleation rate of the crystals is assumed to be proportional to the collision energy and frequency of collision. Evans *et al.*[75] used the same basic approach and considered additional collision mechanisms, such as crystal–impeller collisions due to turbulence and crystal–crystal collisions induced by gravitational force. Instead of discussing the detailed model, a simpler nucleation rate model is used here:

$$B = K_{\mathrm{E}}(S)\left(K_{\mathrm{c-i}}\frac{N_{\mathrm{P}}}{P_0}k_{\mathrm{v}}\rho_{\mathrm{c}}P_{\mathrm{susp}}m_3 + K_{\mathrm{c-c}}\rho_{\mathrm{sl}}\varphi_{\mathrm{c}}\varepsilon^{5/4}L_{50}{}^4m_0\right) \tag{1.3}$$

where N_{P} is the Newton number of the impeller, P_0 is the power input of the stirrer, ρ is the density, P_{susp} is the minimum power required to suspend particles in the vessel, φ_{c} is the volume fraction of crystals, ε is the dissipated power by the impeller per unit mass of suspension, K_{E} is the number of nuclei per collision, $K_{\mathrm{c-i}}$ is the crystal–impeller collision constant, $K_{\mathrm{c-c}}$ is the crystal–crystal collision constant, and

$$m_3 = \int_{L_{\mathrm{c-i}}}^{\infty} L^3 n(L)\mathrm{d}L$$

$$m_0 = \int_{L_{\mathrm{c-c}}}^{\infty} n(L)\mathrm{d}L$$

where $L_{\mathrm{c-i}}$ and $L_{\mathrm{c-c}}$ are lower integration boundaries of the moments.
van Beusichem[76] developed a secondary nucleation model that considers both crystal–impeller and crystal–crystal collisions. For the crystal–impeller collisions, Ottens[74] assumed that the collision frequency, ω_{L}, of a crystal with the impeller is size independent and proportional to the circulation time, t_{c}:

$$\omega_{\mathrm{L}} \propto \frac{1}{t_{\mathrm{c}}} = \frac{K_{\mathrm{i}}N_{\mathrm{i}}D_i{}^3}{V_{\mathrm{c}}} \tag{1.4}$$

where t_c is the circulation time (the time interval between two subsequent passages of a crystal through the impeller area), V_c is the total volume of the crystallizer, K_i is the impeller discharge coefficient, N_i is the impeller speed, and D_i is the impeller diameter.

Assuming that only the collisions with the tip of the impeller blade are effective, the impact energy E_L is proportional to the mass and the tip speed of the impeller:

$$E_L \propto m_L (ND)^2 \tag{1.5}$$

Then, the overall nucleation rate B is proportional to crystal–impeller collisions, $K_{c\text{–}i}$:

$$B \propto K_{c-i} \int_{L_{c-i}}^{\infty} \frac{KN^3D^5}{V_c} m_L n_L \,\mathrm{d}L \tag{1.6}$$

The power dissipated by the impeller per unit mass of suspension, ε, is defined as

$$\varepsilon = \frac{P_0 N^3 D^5}{V_c} \tag{1.7}$$

By $m_L = \rho_c K_V L L^3$, equation 1.6 can be written as

$$B \propto K_{c-i} \frac{K_i}{N_P} K_V \rho_c \varepsilon \int_{L_{c-i}}^{\infty} L^3 n_L \,\mathrm{d}L \tag{1.8}$$

A detailed study of the model was performed by van Beusichem.[76]

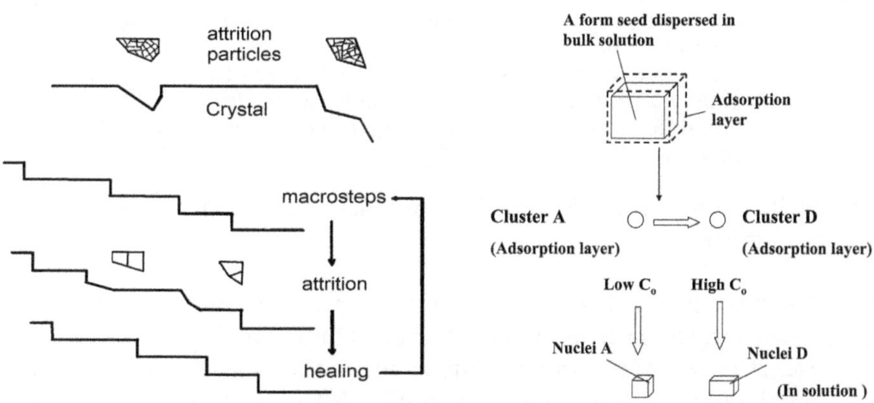

Figure 1.3 Attrition *vs.* contact and shear-secondary nucleation mechanism. Reproduced from ref. 77 with permission from Elsevier, Copyright 1994 and from ref. 78 with permission from American Chemical Society, Copyright 2016.

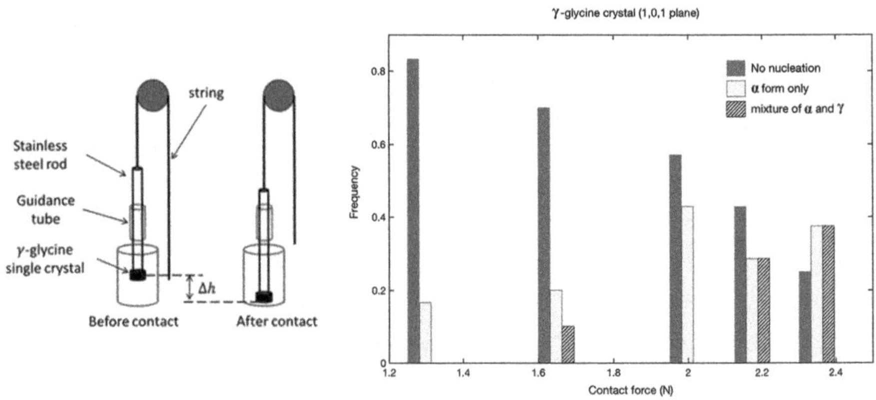

Figure 1.4 Secondary nucleation effect on polymorphism. Adapted from ref. 55 with permission from American Chemical Society, Copyright 2014.

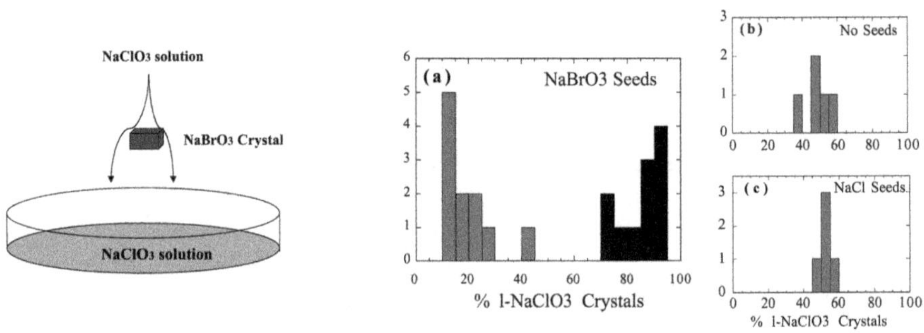

Figure 1.5 Chirally selective crystallization of $NaClO_3$ using seeds of $NaBrO_3$ crystals. (a) Histograms of percentage of L-$NaClO_3$ crystals obtained with L- and D-$NaBrO_3$ seeds, shown by light and dark bars, respectively. (b) Histogram of percentage of L crystals obtained with no seed. (c) Histogram of percentage of L crystals obtained with achiral NaCl seed. Reproduced from ref. 79, http://dx.doi.org/10.1103/PhysRevLett.84.4405 with permission from American Physical Society, Copyright 2000.

Quantification of secondary nucleation kinetics can be performed at various conditions:

1. Batch seeding with single or multiple crystals by counting numbers of newly formed particles (either total or as function of time)
2. Steady state MSMPR experiments by measuring steady state crystal size distribution (CSD) and analyzing number density data (assuming absence of primary nucleation, agglomeration and breakage plus ensuring that the outlet stream is fully representative of the crystallizer contents)
3. Other arrangements (unsteady state/batch/plug flow) by measuring CSD and fitting with population balance models to estimate secondary nucleation kinetics (assuming certain functional expressions)

1.3 Continuous Crystallization

A continuous crystallization process will have a clear undersaturated solution as a feed flow and a crystal suspension as an outgoing flow. The method with which a supersaturated solution with $S > 1$ is created in a crystallizer defines the crystallization method used.

By decreasing the temperature of a solution generally the solubility of a crystalline solid decreases and if it is changed to below the solution concentration *cooling crystallization* can take place. A continuous cooling crystallization would have a hot, concentrated but undersaturated feed and a cold suspension as an outflow representative of the crystallizer bulk composition and temperature.

In case of *evaporative crystallization* the concentration is increased by solvent evaporation. Since solvent evaporation rate is fast at the boiling point of a solution, evaporative crystallization usually takes place close to the three-phase equilibrium point between crystals, solution and solvent vapor. This point can be shifted for process optimization since the boiling temperature decreases with decreasing pressure. A continuous evaporative crystallization process would have a hot undersaturated feed, a suspension outflow taken from the base of the crystallizer and an additional evaporated solvent flow. One process challenge is to ensure sufficient height difference between the boiling surface and the outlet to ensure sufficient hydrostatic head to suppress boiling in the outlet stream.

In case of *antisolvent crystallization* an antisolvent is mixed with the solution. While the addition of antisolvent decreases the overall concentration, the solubility in the mixed solvent is also decreased. If the solubility decrease due to the antisolvent is larger than the concentration decrease due to dilution, supersaturation is created and crystallization can occur. A continuous antisolvent crystallization would have a solution and an antisolvent feed flow and a suspension outflow.

For *reactive crystallization* two solutions each containing one of the reagents are mixed. The reagents react to form a solute with a lower solubility so that the solute concentration is higher than the solubility and crystallization occurs. A reactive crystallization could be performed for instance by adding a high pH solution to a low pH solution of a compound: the change in pH upon mixing the solutions reduces the solubility of the compound and crystallization can occur. A continuous reactive crystallization would have two solution feeds and a suspension outflow. Reactive crystallization is a combination of chemical reaction and crystallization. The reaction rate and solute generation rate (reaction product) defines the solute concentration, and at the solution temperature, the supersaturation ratio would be determined. Although the reactive crystallization normally runs at steady state, any temperature change or disturbance in any reagent flow (concentration) could disturb the crystallization phenomena, including the nucleation rate and crystal growth. Yazdanpanah *et al.*

demonstrated a dynamic continuous reactive crystallization case, where a temperature shift in reactor temperature had a nonlinear effect on reaction rate and solubility.[80] In this case, the higher temperature increases the reaction rate, hence the solute concentration, by generating more solute from the reaction. However, the solubility of the solute increases as the temperature ramps up. Therefore, the supersaturation ratio decreases at higher temperature. The dynamic effect and nonlinearity were modelled in Aspen Plus and bi-direction effects of temperature on crystal size, nucleation rate, and steady-state crystal chord length were demonstrated. The transition time (unsteady-state period) and amount of residence time required for reaching the steady state, are important aspects in controlling the process and defining control strategies. Another crucial aspect of the reactive crystallization in a dynamic case is to monitor and model impurity inclusion and the effect of the concentration of unreacted reagents on the crystal's purity, shape, and yield.[80,81] Continuous reactive crystallization also could be challenging with concentration PAT tools if a change in concentration of reagents affects the reflective index of the solution or generates water that shifts the Raman/FTIR base line or creates noise and significant peaks that mask the solution/solute characteristic peaks. Further discussion on PAT tools are provided in Chapter 9.

Both antisolvent and reactive crystallization use mixing-induced supersaturation to enable crystallization. Here by mixing we mean bringing together two fluids resulting in a molecularly homogeneous/isotropic solution mixture on a molecular scale, while agitation refers to the process of keeping contents of a vessel from developing segregation in terms of composition, temperature or dispersion homogeneity (*e.g.*, suspension of solids). Depending on the reaction rate magnitude of difference to crystallization rate, the Damköhler number and mixing *vs.* reaction rate study could be useful in designing an efficient continuous reactive crystallization. Some intense mixing, such as impinging jet mixing and adding the reagent/antisolvent inside an *in situ* high shear mixer have been proposed to overcome these challenges.[82,83]

Assuming that agglomeration and breakage ("mass fracture") are negligible, the CSD developing in a steady state is then a function of the prevailing nucleation rate B and growth rate G in the crystallizer as well of the residence time $\tau = V_{cr}/\varphi$. Under the further assumptions that the crystallizer is perfectly mixed (composition, temperature and solids uniformly distributed within the crystallizer volume, the feed stream instantly mixed with the crystallizer volume and the outlet stream identical to the crystallizer contents; *i.e.*, MSMPR conditions), growth rate G of the equally shaped crystals is independent of crystal size and constant, the number-based size distribution $n(L)$ can be shown to be function of only the nucleation rate B in the crystallizer, the growth rate G of the crystals and the residence time τ:

$$n(L) = (B/G)\exp(-L/G\tau) \tag{1.9}$$

Interestingly, for a continuous crystallization process under MSMPR conditions, valuable kinetic data therefore can be extracted from a measured number-based CSD of a product by fitting a straight line to the log-normal CSD plot of $\ln(n)$ against crystal size L. The slope and intercept then are equal to respectively $(-G\tau)^{-1}$ and $\ln(B/G)$. It is important to note that the nucleation rate determined in this way is an average over the entire crystallizer volume: if the nucleation rate is locally occurring, the volume in which this occurs has to be known in order to determine the true nucleation rate. The relation for the number-based CSD $n(L)$ is also valid for other crystallization methods under MSMPR conditions.

If, for instance, the secondary nucleation rate due to attrition with the stirrer is the dominant nucleation mechanism, the stirrer speed influences the attrition. After an increase in the stirrer speed for a suspension in a steady state continuous process, the nucleation rate B would increase and with it an increased total crystal surface area would be available for growth. This would decrease the prevailing average supersaturation in the crystallizer and decrease the growth rate G. The newly developing CSD $n(L)$ would be characterized by a larger intercept $\ln(B/G)$ and a more negative slope $(-G\tau)^{-1}$: the average number-based crystal size would decrease.

Although MSMPR conditions often do not hold, the model is useful and helps understanding of continuous crystallization processes in agitated vessels. Generally, increasing the value of B/G and decreasing the value of $G\tau$, *i.e.*, increasing the nucleation rate B and decreasing the growth rate G as well as the residence time τ, will result in smaller crystals. These are general criteria for design of any seed generation/nucleator units.

1.3.1 Crystalline Product Quality Attributes

All continuous crystallization processes result in a suspension flow with a certain suspension density of the crystalline product having a certain solid form, CSD, shape and purity. These represent the crystalline product quality attributes, which determine performance in downstream processing steps (isolation, formulation) and of the final product.

Control of crystal nucleation in a continuous crystallization process is crucial to control the final product quality attributes. In continuously seeded crystallization processes for instance, crystal nucleation has to be avoided to allow control over the particle size, however, continuously seeded, continuous crystallizers are quite rare. Moreover, in unseeded continuous crystallization the (secondary) nucleation process has to continuously produce small crystals with a constant rate to remain in the steady state allowing control over the product size. The nucleation process thus directly influences the resulting crystal size distribution. It can furthermore cause significant issues with solid form control if nucleation of undesired polymorphs occurs. Fouling and encrustation are other significant issues in the continuous crystallization which are covered in Chapter 3.

The kind of nucleation occurring depends strongly on the continuous crystallization process configuration chosen. A simple continuous crystallization concept is to use a single continuous agitated vessel with a continuous feed solution and suspension outflow, such as a CST, or MSMPR crystallizer. The various crystallization methods can be performed by either maintaining the crystallizer at a lower temperature than the feed solution, evaporating solvent in the crystallizer or adding an additional antisolvent or solution feed to the crystallizer. It is also possible to combine various crystallization methods, for instance, an antisolvent crystallization or a pH shift with a cooling crystallization, if the product still has a substantial solubility at the higher temperature after the antisolvent addition or pH shift, respectively.

Growth is a bulk average (driven by bulk composition and temperature), while nucleation is a local phenomenon (*e.g.*, mixing zone, impeller, vessel wall, liquid/air interface, temperature heterogeneity, external field) occurring at the locations with highest supersaturation. During mixing of the feed and crystallizer suspension locations with a relatively high concentration and low temperature exist at the feed inlet and/or cooling surfaces. In evaporative crystallization, locally increased concentrations occur in the boiling zone due to the selective evaporation of the solvent. These locations therefore are associated with a high supersaturation at which primary nucleation could occur and be determined. Secondary nucleation can occur for instance due to large crystals colliding with the stirrer so that small crystal fragments (secondary nuclei) are created.

The encrustation in continuous cooling crystallization (MSMPR or oscillatory baffled crystallizer (OBC)) mostly happens at the heat transfer interface that provides a high local supersaturation ratio, which initiates intense and uncontrolled local nucleation. The continuous nucleation on the heat transfer surface is deteriorative to the process stability and longevity of the runs and there have been some control strategies proposed to avoid the rapid nucleation on the surface.[84,85]

In CST type continuous cooling and evaporative crystallization processes new particles are typically generated through secondary nucleation due to collisions of particles with an agitator. However, there are other types of agitated vessels where secondary nucleation may not be dominant (see below). In case of mixing induced crystallization processes such as antisolvent and reactive crystallization, particles often are generated by primary nucleation due to concentration heterogeneities introduced by the mixing process.

Another continuous crystallization concept is a *plug flow crystallizer* (PFC). A PFC is a long tube in which the crystals are allowed to grow along the length of the tube while they are taken along with the solution flowing through the tube. Use of a PFC allows all crystals to have the same residence time in the tube if the back mixing is sufficiently small. In principle, this allows a tight control over the CSD. In a plug flow type continuous crystallization processes there are 2 routes towards generating particles in the process.

First, primary nucleation can be generated locally at the start of the tube. Second, continuous seeding can be used to prevent the nucleation process altogether.

1.3.2 Continuous Heterogeneous Crystallization on Excipient Surfaces

The direct crystallization of an API on a crystalline excipient surface is called "heteroepitaxy." The heterosurface orders prenucleation aggregates, so nucleation becomes energetically favorable.[86,87] In the crystallization process detailed in this example here (Figure. 1.6), the API (acetaminophen) nucleates and grows on an excipient surface (D-mannitol). The API-excipient system selection, induction time measurement, and molecular interaction modeling have been studied.[86–89] The dynamic conditions of

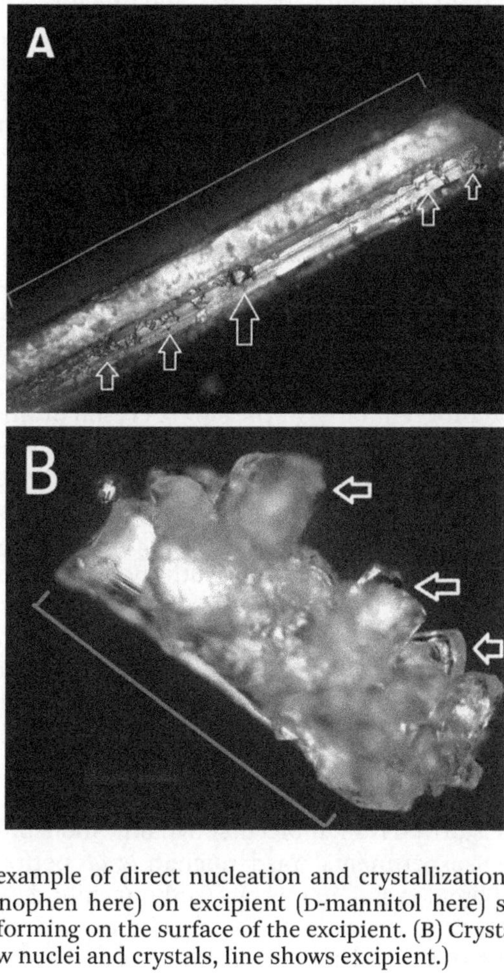

Figure 1.6 An example of direct nucleation and crystallization of a solute (acetaminophen here) on excipient (D-mannitol here) surface. (A) Nuclei are forming on the surface of the excipient. (B) Crystal growth. (Arrows show nuclei and crystals, line shows excipient.)

continuous crystallization and the effects of all transport phenomena and kinetic effects have significant impact on the process, which have been demonstrated and addressed. Continuous crystallization and process stability, spontaneous nucleation (bulk nucleation) at high supersaturation ratio, secondary nucleation and mixing, particle–impeller attrition, and heterogeneous nucleation rate are critical aspects to be controlled and monitored. Understanding the mechanisms driving epitaxy at a molecular level is critical for controlling epitaxial nucleation and growth of crystals. Of particular interest for this concept, the induction time, preferential nucleation rate, and final properties of the composite are highly affected by the mechanisms driving epitaxy and proper API-excipient selection. Therefore, understanding the mechanisms controlling epitaxial ordering is fundamental for controlling the final properties of the crystalline material.

Numerous studies investigated epitaxial ordering on crystalline and other highly ordered surfaces to understand the effect of lattice matching, functional group matching (surface functionality), and interaction energy.[86] The explanations rely on a partial or total matching between the two opposing lattice planes, which significantly lowers the nucleation free energy barrier and eventually promotes nucleation. Alternately, molecular dynamics modelling could demonstrate that molecular functionality and chain orientation, in such a manner as to utilize as many hydrogen bonding groups as possible to stabilize the prenucleation aggregate of crystalline substrates, are dominant in promoting heterogeneous nucleation of a model API.

Figure. 1.7 demonstrates the surface chemistry and molecular interaction (bond matching) between the solute crystal and excipient (crystalline or polymeric). The hydrogen bond propensity and bond formation potential between the active groups on the exposed surface of excipient and also chemical bonds at different faces of the solute crystal promote preferential nucleation and crystallization. In this way, the solute molecules prefer to interact and form local nuclei on a certain side of the excipient, in competition with other surfaces (*i.e.* mixer blade, crystallizer glass wall...) or homogeneous bulk nucleation. Therefore, the nucleation and crystal growth on the surface of the excipient benefit from both the heterogenous nucleation and surface Gibbs free energy and also chemical bond formation and surface chemistry.

Heterogeneous nucleation by definition can occur on any foreign particle, such as dust, which is the distinction between homogeneous and heterogeneous nucleation. The presence of excipient crystals in the crystallization solution enhances the nucleation and yields the epitaxy. As mentioned previously, the difference in excipient selection is based on the matching ranking matrix, which is driven from molecular dynamic modeling and induction time measurement experiments. Yazdanpanah *et al.* performed an experiment with two different excipient crystals (sodium chloride and D-mannitol) at the same supersaturation ratio of solute solution (acetaminophen), and demonstrated by molecular dynamic calculations and experimental results

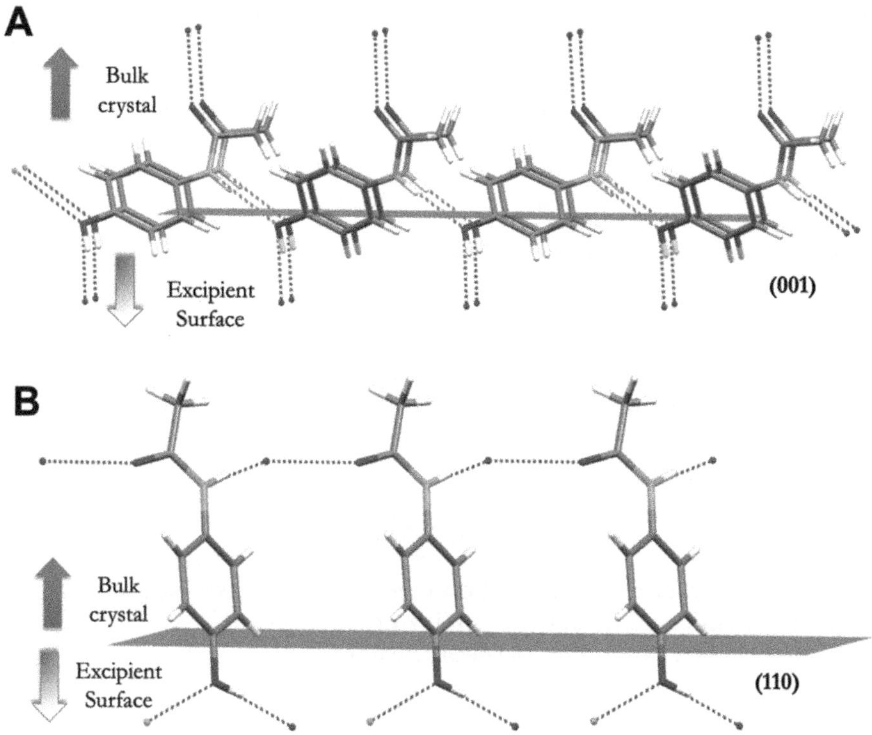

Figure 1.7 Analysis of surface chemistry between solute crystal (acetaminophen here) and excipient. (A) face 001 (B) face 110.

that under the same process conditions, the acetaminophen preferred the D-Mannitol surface for nucleation. In a high supersaturation solution, the solute crystals formed on the surface of the D-mannitol excipient, but in the presence of sodium chloride, bulk (homogeneous) nucleation takes place and the excipient was ineffective.[68]

Population balance modeling can be used to unravel the kinetics of nucleation and growth. In the modeling of the system, the following assumptions could be made. Heterogeneous nucleation is the dominant nucleation event, and primary nucleation is assumed to be absent in the system. The nucleation rate increases with increasing surface area of the excipient, and *vice versa*. The active surface of the excipient available for nucleation is assumed to be constant. The growth rate is proportional to the surface area of acetaminophen that is in contact with the solution. Aggregation and breakage of API crystals are absent in the system.

The dynamics of the size distribution of crystals of API growing over the excipient surface are described by

$$\frac{\partial n(\nu,t)}{\partial t}=-\frac{n(\nu,t)}{\tau}-\frac{\partial}{\partial \nu}\left(\frac{\mathrm{d}\nu}{\mathrm{d}t}n(\nu,t)\right)+\dot{N}\delta(\nu-\nu_{\mathrm{c}}) \tag{1.10}$$

where t is the time, v is the volume, and $n(v,t)$ is the time-dependent volume-based population balance density function, with $n(v,t)\mathrm{d}v$ representing the number of particles in the volume range of v to $v + \mathrm{d}v$ per unit volume of the reactor. The volume growth rate (G) of any particle is defined as

$$G = \frac{\mathrm{d}v}{\mathrm{d}t} = k_{g1}\mathrm{e}^{\frac{-k_{g2}}{RT}}\sigma^{g}k_{\mathrm{a}}\left(\frac{v}{k_{\mathrm{v}}}\right)^{\frac{2}{3}} \tag{1.11}$$

where the relative supersaturation is defined by

$$\sigma = \frac{C - C_{\mathrm{eq}}}{C_{\mathrm{eq}}} \tag{1.12}$$

T is the absolute temperature, R is the universal gas constant, k_{g1} and k_{g2} are the growth rate constants, g is the growth rate order, and k_{a} is the shape factor of the particle used to calculate the surface area of particles.

The nucleation rate expression is

$$\dot{N}_{\mathrm{heterogen}} = k_{b1}\mathrm{e}^{\frac{-k_{b2}}{RT}}\sigma^{b}Ar \tag{1.13}$$

where b is the order of the nucleation rate and Ar is the surface area of the excipient per unit volume of the crystallizing solution.

The mass balance of API particles can be written as

$$\frac{\mathrm{d}C}{\mathrm{d}t} = \frac{(C_{\mathrm{i}} - C)}{\tau} - \frac{\dot{N}n_{\mathrm{c}}}{Ar_{\mathrm{v}}} - \frac{k_{\mathrm{a}}\left(\frac{1}{k_{\mathrm{v}}}\right)^{\frac{2}{3}}k_{g}S^{g}}{V_{\mathrm{m}}}\int_{0}^{\infty} n(v,t)v^{2/3}\,\mathrm{d}v \tag{1.14}$$

where $k_{g} = k_{g1}\mathrm{e}^{\frac{-k_{g2}}{RT}}$, C and C_{i} are the concentrations of solute in the reactor and in the inlet slurry, respectively, and V_{m} is the specific volume of the solute.

To simplify the analysis, the moments of the population density function are solved here. The jth moment M_{j} of the population density function is defined by

$$M_{j} = \int_{0}^{\infty} n(v,t)v^{j}\,\mathrm{d}v \tag{1.15}$$

The 0th moment M_0 represents the total number of solute crystals present in the system, and M_1 is the total volume of solute crystallized. As $k_{\mathrm{a}}\left(\frac{1}{k_{\mathrm{v}}}\right)^{\frac{2}{3}}M_{\frac{2}{3}}$ is the total surface area of crystals, the rate equation for the concentration of solute can be written as

$$\frac{\mathrm{d}C}{\mathrm{d}t} = \frac{(C_{\mathrm{i}} - C)}{\tau} - \frac{\dot{N}n_{\mathrm{c}}}{Ar_{\mathrm{v}}} - \frac{k_{\mathrm{a}}\left(\frac{1}{k_{\mathrm{v}}}\right)^{\frac{2}{3}}k_{g}\sigma^{g}M_{\frac{2}{3}}}{V_{\mathrm{m}}} \tag{1.16}$$

Multiplying eqn (1.16) with v^i and integrating over the interval of 0 to ∞ results in the generalized rate equation of the moments:

$$\frac{\mathrm{d}M_j}{\mathrm{d}t} = -\frac{M_j}{\tau} + \dot{N}v_c^j + jk_a\left(\frac{1}{k_v}\right)^{\frac{2}{3}} k_g \sigma^g M_{j-\frac{1}{3}} \tag{1.17}$$

The dynamics of the crystallization process can be obtained by numerical solution of the set of ordinary differential equations:

$$\frac{\mathrm{d}v}{\mathrm{d}t} = k_{g1}\mathrm{e}^{\frac{-k_{g2}}{RT}} \sigma^g k_a \left(\frac{v}{k_v}\right)^{\frac{2}{3}}$$

$$\frac{\mathrm{d}C}{\mathrm{d}t} = \frac{(C_i - C)}{\tau} - \frac{\dot{N}n_c}{Ar_v} - \frac{k_a\left(\frac{1}{k_v}\right)^{\frac{2}{3}} k_g \sigma^g M_{\frac{2}{3}}}{V_m}$$

$$\frac{\mathrm{d}M_0}{\mathrm{d}t} = -\frac{M_0}{\tau} + \dot{N}$$

$$\frac{\mathrm{d}M_1}{\mathrm{d}t} = -\frac{M_1}{\tau} + k_a\left(\frac{1}{k_v}\right)^{\frac{2}{3}} k_g \sigma^g M_{\frac{2}{3}} + \dot{N}v_c$$

$$\frac{\mathrm{d}M_{2/3}}{\mathrm{d}t} = -\frac{M_{2/3}}{\tau} + \frac{2}{3}k_a\left(\frac{1}{k_v}\right)^{\frac{2}{3}} k_g \sigma^g M_{\frac{1}{3}} + \dot{N}v_c^{2/3}$$

$$\frac{\mathrm{d}M_{1/3}}{\mathrm{d}t} = -\frac{M_{1/3}}{\tau} + \frac{1}{3}k_a\left(\frac{1}{k_v}\right)^{\frac{2}{3}} k_g \sigma^g M_0 + \dot{N}v_c^{1/3}$$

$$\dot{N} = k_{b1}\mathrm{e}^{\frac{-k_{b2}}{RT}} \sigma^b Ar$$

The shape factor of solute can be obtained from the literature or by image analysis techniques and microscope images. Seeded batch growth studies help in obtaining the growth rate parameters explicitly. Batch data of heterogeneous nucleation and growth experiments can be used to finally fit the nucleation rate parameters, which is a safe assumption for the continuous crystallization.

The previous method was for direct nucleation and crystallization on the surface of the excipient crystalline particles. Yazdanpanah *et al.* proposed a new technique for direct crystallization on the surface of an amorphous polymeric film (continuous medium) in an end-to-end continuous fashion.[90] A schematic of the process is shown in Figure. 1.8. The process consists of three sections: (1) excipient film preparation by melt or solution deposition on the polymer on a back-layer surface, (2) a crystallization pool for direct nucleation and

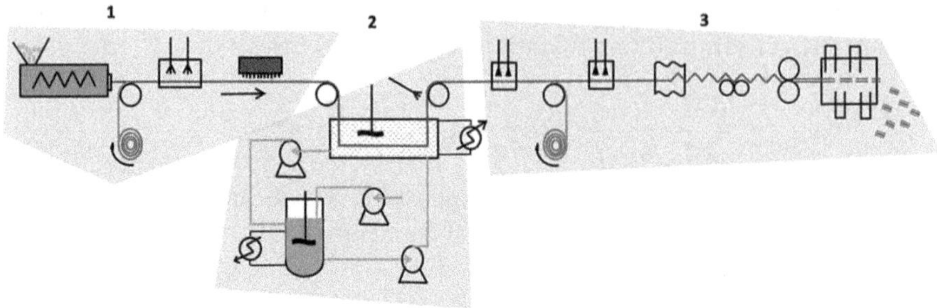

Figure 1.8 Continuous direct nucleation and crystallization on surface of polymer film process line scheme. Section 1: Excipient film preparation, Section 2: Continuous direct crystallization, Section 3: Roll-to-roll continuous product processing.

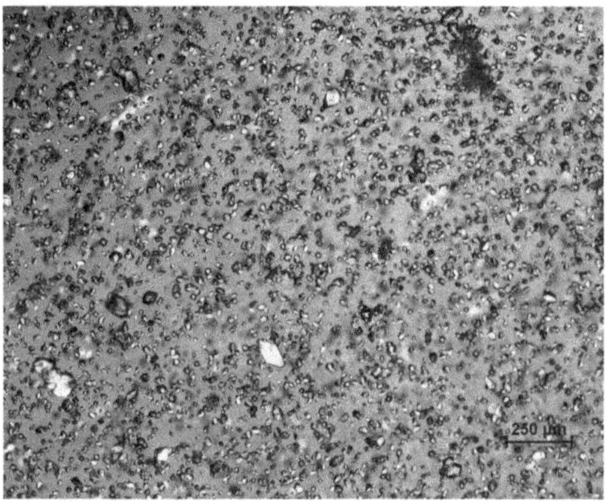

Figure 1.9 Deposited crystals on the surface of the polymeric excipient film from a continuous process.

crystallization (heterogeneous) on the surface of the film travelling through the solution, and (3) the processing section ("downstream"). Figure. 1.9 shows the film surface after Section 2 where crystals formed on the surface of the excipient film. The control strategy for tuning the amount of crystal deposition on the surface area has been proposed by controlling the supersaturation ratio of the crystallization solution and also by residence time (linear velocity of the film through travels through the pool).

1.3.3 Agitated Vessel Type Crystallization Process

Agitation in crystallization vessels can be conveniently achieved using impellers in CSTs, but there are alternatives including airlift/draft tube/fluidized bed crystallizers, OBC, and Taylor–Couette crystallizers (TCC).[26,29] A typical CST configuration is shown in see Figure. 1.10.

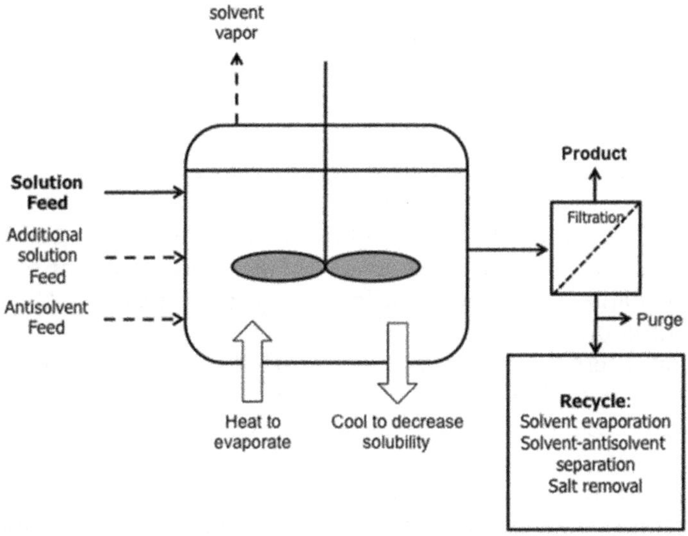

Figure 1.10 A CST Crystallizer can be used for continuous cooling, evaporative, antisolvent and reactive crystallization.

In a continuous cooling crystallization for instance the hot solution feed at temperature with concentration and flow rate enters the crystallizer and is mixed up with a suspension which has a specific steady state solution concentration and suspension density of crystals in the solution at temperature. The outflow with rate contains a suspension having a solution concentration and a suspension density at temperature. The volume change of a solution upon crystallization may be negligible, meaning that for a constant suspension volume in the crystallizer, the feed and outflow rate are equal. For a continuous cooling crystallization in steady state operation, temperatures, concentrations and suspension densities in crystallizer and outflow are often assumed equal. Then in steady state the CSD and other product quality aspects in the crystallizer suspension and outflow suspension are equal.

The crystallizer suspension has a temperature and concentration lower than those of the feed. Due to the temperature decrease the low solubility induces a supersaturation ratio $S_{cr} = C_{cr}/C_{cr}^*$ used for crystal growth and the other subprocesses of crystallization. Crystal growth is an average bulk phenomenon, determined by the average supersaturation S_{cr} in the crystallizer solution through which the crystals travel.

The developing CSD in a continuous cooling crystallization process is determined, among other things, by the local nucleation rate, which results in a number-based rate B of crystal formation in the suspension volume, usually given in units of $m^{-3}s^{-1}$. The crystals formed have a chance to end up in the outflow: while some crystals might end up in the outflow right after they formed, some others may take a long time to be removed from the process.

The formed crystals thus all experience a different residence time in the crystallizer and therefore a different growth time. In steady state condition, the CSD in the suspension and in the outflow therefore consist of crystals of various sizes.

In a continuous crystallization process the nucleation rate strongly affects final product quality aspects such as the CSD. Since in steady state a suspension of crystals is present there is a high likelihood that one of the secondary nucleation processes is the dominant nucleation mechanism. However, due to their high local supersaturation values upon mixing, in antisolvent and reactive crystallization primary nucleation mechanisms might be dominant in the mixing zones.

Continuous crystallizers can easily be used in a cascade configuration in which the outgoing suspension of one crystallizer is the feed for the subsequent crystallizer. This can increase yield and average product size by for instance decreasing temperature or increasing antisolvent fraction (Figure. 1.11) in each cascade crystallizer while keeping relatively mild supersaturation conditions to avoid crystal nucleation and promote crystal growth in the subsequent crystallizers. Note that for a cascade of (well-mixed) agitated vessels the overall residence time distribution becomes narrower as the number of vessels increases, ultimately approaching a plug flow process.

There are also large, more complex units, which have been routinely used for industrial crystallization of some commodity chemicals (see Chapter 6).

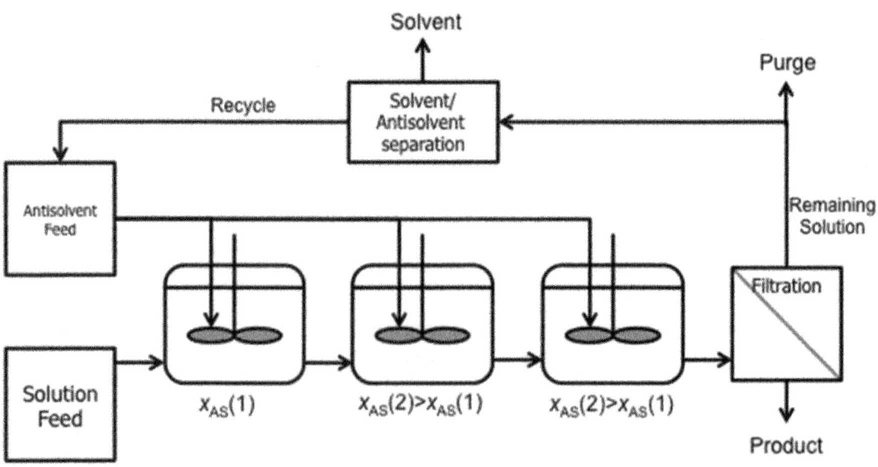

Figure 1.11 A cascade of 3 CST crystallizers for a continuous antisolvent crystallization process with a recycle of the antisolvent. Each subsequent crystallizer operates at a higher antisolvent fraction to enforce a driving force for crystallization in order to increase the suspension density through nucleation and growth. The filtration step continuously separates the crystal product from the remaining solution. After the solvent/antisolvent separation the antisolvent is recycled back into the process. A purge stream enables impurities to leave the system.

1.3.4 Plug Flow Type Crystallization Process

There are several approaches available to provide all crystals with the same (or very similar) residence time which is the hallmark of a plug flow process. These include: (1) turbulent flow conditions, (2) oscillatory flow conditions, or (3) segmented/slug flow.

In order to achieve turbulent rather than laminar flow conditions in a simple long straight tube, the flow rate would need to be very large, resulting in small residence times. For crystals with a typical fast growth rate of 10^{-7} m s^{-1}, the residence time needed to make 100 μm crystals is 1000 seconds. For a 1 cm^2 section area of a 10 m long tube ($V = 0.001\ m^3 = 1l$) this indicates a flow rate of $\varphi = V_{cr}/\tau = 1$ mL s^{-1} which is much too low to induce turbulence in the tube. Therefore, either the PFC should be very long or the flow rate φ should be decoupled from the mixing in the tube.

This decoupling of flow rate and turbulence induction can be achieved by using an OBC (see Chapter 3). An OBC is a PFC equipped with periodically spaced orifice baffles creating well-mixed conditions in the individual cells between the baffles due to an oscillatory flow motion with specific frequency and amplitude superimposed on the net flow φ through the OBC (Figure. 3.5). The oscillatory flow in combination with the baffles give a good radial mixing to allow mass transfer between solution and crystals while the slow net flow φ gives a plug flow-like behavior in the OBC determining the average residence time of the crystals. Since the oscillation cannot achieve perfect plug flow some back-mixing of crystals will occur resulting in a narrow yet substantial residence time distribution of the crystals.

In the case of a PFC in the absence of seeding, primary nucleation is required to generate the initial distribution of crystals that grow out throughout their residence time while travelling through the length of the crystallizer. Details of primary and secondary nucleation, locality and also (non-ideal) backmixing are discussed in Chapter 3.

1.4 Continuous Seeding and Nucleators

1.4.1 Continuous Seeding

It is often taken for granted that primary nucleation processes cannot be well controlled. Therefore, seeding is commonly used in industry to get improved crystallization processes and product control. Perhaps one could see the wide application of seeding as a symptom of our inability to understand and control crystal nucleation.

The generation of seeding suspensions by dry or wet milling is nowadays routinely performed in industry. However, there are many challenges related to these seeding suspensions. Ultimately, to seed a continuous crystallization process a suspension of seed crystals of desired solid form, particle size and number density is needed. It should be able to grow well under suitable conditions, while avoiding secondary nucleation by providing mild crystallization conditions, to achieve required performance of purification and particle product attributes.

A suspension of seed crystals can be added as a feed so that the continuous crystallizer is used to grow the seeds. In this case it is important to prevent any kind of nucleation in order to control the final product size. Therefore, relatively mild supersaturation conditions will have to be used. In case of a continuous cooling crystallization in a plug flow crystallizer, for instance, the temperature of the suspension will have to be reduced along the length of the crystallizer. This can be done in a way equivalent to cooling profiles used in batch cooling crystallization processes. The seed suspension feed stock will have to be made in a nucleation or seeding unit. One way of doing this is by using a continuous nucleator, or to introduce seed crystals to invoke crystal nucleation in the clear solution right at the start of the plug flow crystallizer. This can be achieved, for example, by mixing with an antisolvent or by performing reactive crystallization to create a relatively high supersaturation. While a high supersaturation is needed for the local nucleation to take place, a mild supersaturation is needed to create the conditions under which only growth occurs. Therefore, in the remainder of the crystallizer, it is difficult to control the supersaturation to only grow crystals. Since the mixing in an industrial scale crystallizer has some intrinsic locality, the mixing time could be lower than the reaction time (the Damköhler number), and scaling issues might arise. It is therefore preferable to locally induce nucleation by other means, for instance by external energies such as ultrasound, laser light and electric fields.

1.4.2 Decoupling Nucleation and Growth in Continuous Crystallization

As discussed above, in the context of continuous crystallization, there are several general scenarios of how nucleation can be employed to obtain the required crystalline product quality attributes (see Figure. 1.12):

1. secondary nucleation, *e.g.*, under mechanical impact, typically in MSMPR.
2. primary nucleation under conditions where supersaturation is mixing-induced, *e.g.*, in antisolvent or reactive crystallization, in MSMPR or PFC.
3. primary nucleation under conditions where the driving force for nucleation is localized and induced by shear or external fields, *e.g.*, ultrasound or laser, typically in PFC but could also be MSMPR bypass.

There are 2 approaches to follow in order to get a crystal product with a narrow size distribution: (1) seeding, or (2) nucleation zone/nucleator approach.

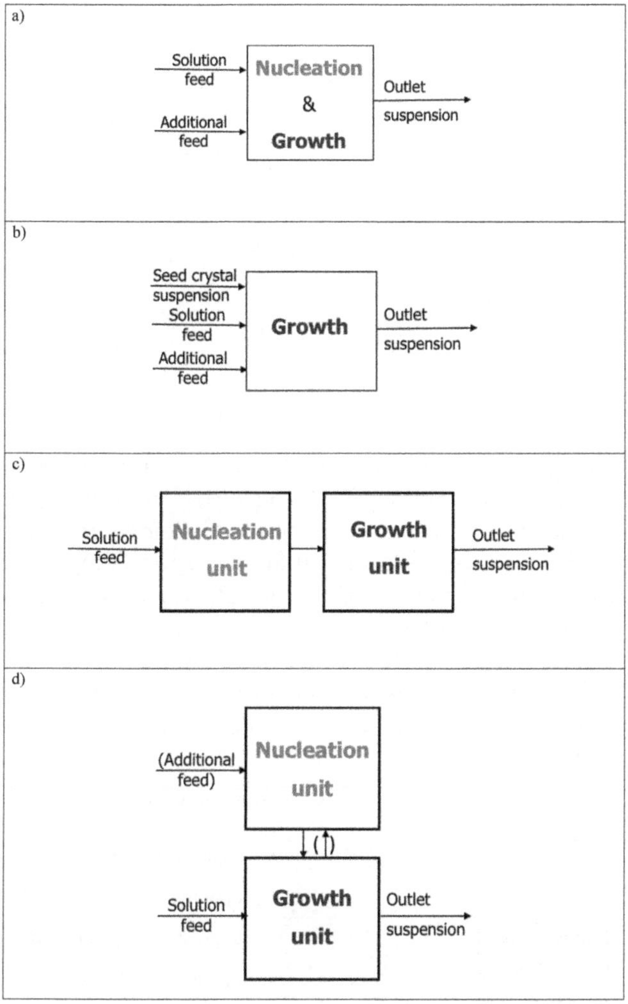

Figure 1.12 Approaches to decoupling nucleation and growth in continuous crystallization. Deconstructing continuous crystallization: continuous nucleation unit (nucleator) for continuous production of seeding suspensions followed by continuous growth unit for continuous seeded crystallization. (a) typical (unseeded) continuous crystallizer configuration. (b) typical seeded continuous crystallizer configuration. (c) a series configuration with a single solution feed; the first unit (nucleator) is dedicated to obtaining a continuous seeding suspension stream which is fed into the second unit to perform seeded crystallization. (d) Continuous seeded crystallization (growth unit) where the seed crystal suspension is generated in a dedicated continuous nucleation unit (nucleator) which may have an independent feed and/or a recycle from the growth unit.

1.4.3 Continuous Nucleators

There have been various techniques proposed for continuous nucleators to employ in continuous crystallization applications, which some of them are listed below. The key concept for almost all of the techniques is to induce local high supersaturation and local nucleation enhancement by mixing, shear or external fields. However, each of them has some limitations and niche applications that are explained in this chapter (see Figure. 1.13).

- static mixers[91–94]
- capillary jet[83,95–98]
- high shear wet mill[99–102]
- secondary nucleator by tapping[54,55,103,104]
- ultrasound[105–108]
- laser[43–48]

To locally trigger primary nucleation in a continuous crystallization process the solution should typically be either highly supersaturated, undergo high shear or be exposed to external fields such as ultrasound or laser light. The local increase in supersaturation is most directly generated by the rapid mixing of solutions or solution and antisolvent. High shear is most commonly applied with a wet mill either upstream or inside an MSMPR. In addition to inducing nucleation (which is grinding here) the wet mill also acts as a size reduction tool. The most prevalent and diverse external field applied to nucleation is ultrasound. Ultrasound

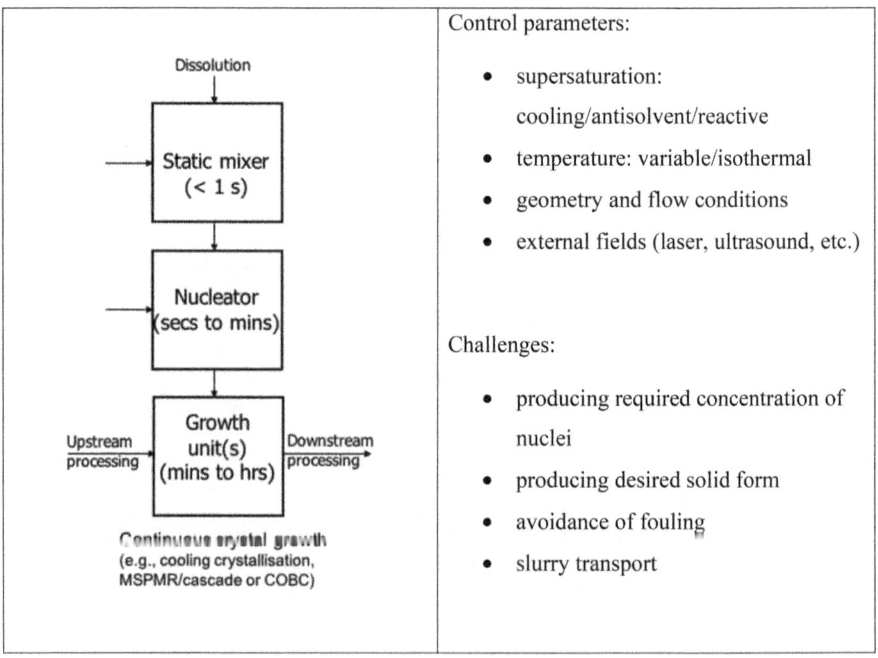

Figure 1.13 Example of continuous nucleator concept.

can either be applied to a low supersaturated solution where nucleation would not otherwise take place (which can be considered as increase of supersaturation locally) or to a highly supersaturated solution to promote nucleation further and better control particle size. However, the multi-effects of grinding and particle breakage by ultrasound, combination of secondary nucleation, high supersaturation, cavitation and so on could initiate some complexity. In addition to primary nucleation there has been specific research into triggering continuous secondary nucleation in a controllable manner by means of contact secondary nucleation where the generated secondary nuclei are directly fed into a tubular crystallizer. In this way the secondary nucleation process and the subsequent crystal growth process are decoupled. The differences in operating windows for the different subprocesses in crystallization and the need for specific conditions for local nucleation have been framed before.[109]

The continuous nucleation process can be undertaken in a variety of equipment utilizing a variety of crystallization modes. Rapid mixing is typically employed in a range of static mixers including standard T and Y shaped mixers, Rushton-type mixers, co-axial mixers, radial mixers and pipes with inserts. Cooling and simple solution addition are usually employed in either a standard tubular device, a glass channel, an OBC or a cascade of MSMPR Crystallizers of one or more stages. Each piece of equipment has a set of advantages and disadvantages with regards to mixing intensity, capacity, temperature control and heat transfer characteristics. This equipment can be used in conjunction with novel technologies such as the wet mill and ultrasound. Furthermore, the crystallization modes can be combined with novel operation strategies including slug flow and periodic flow.

1.4.4 Supersaturation Control by Rapid or Non-rapid Mixing

One method of generating local high supersaturation is to rapidly mix solutions in a static mixer in an effort to reach a target supersaturation level instantaneously. The rapid mixing should result in a crystallization process which is dominated by primary nucleation with minimal crystal growth which leads to the creation of a seed suspension with a narrow particle size distribution (PSD) and a small mean particle size. Many studies which utilize rapid mixing strategies have applied these to antisolvent,[91,110–114] reactive[92–94] and cooling[111,115,116] crystallizations with a specific desire to control PSD and/or mean particle size. Some of this research[91–94] has compared continuous mixing processes in static mixers with conventional batch processes and shown that the more controlled mixing environment in the static mixer does result in a narrower PSD with a smaller mean particle size. Furthermore, when using a static mixer the PSD and mean particle size can be controlled by changing the initial supersaturation and the flow velocity.[91,93,94,110,111,113] Sometimes ultrasound is applied even to rapidly mixed, highly supersaturated solutions in an effort to further decrease particle size due to enhanced primary nucleation rates.[92,112] Sonication may also help reduce the extent of fouling and blockages in a tubular device. With rapid mixing there is the issue of what happens to the seed suspension immediately after it is created. An attractive answer to this is to employ slug flow in the tubular device to allow for the seeds to grow to a larger size while maintaining a narrow PSD.[115]

In addition to controlling particle size, other studies[117–119] have used rapid mixing to demonstrate the control of solid form in antisolvent and reactive crystallization processes. The polymorphic outcome of a single component system was shown to be controlled *via* the solution concentration and antisolvent mass fraction.[119] It has also been demonstrated that a solid solution can be produced continuously by means of a rapid mixing process.[118] Being able to produce these solid solutions rapidly and at steady state for extended periods of time is a marked improvement over the small batch crystallizations which were previously performed.

Some studies have shown high supersaturation generation with non-rapid cooling or simple addition leads to a crystallization process in which primary nucleation is still dominant but crystal growth plays a larger role. This has been shown for antisolvent and cooling crystallizations in standard tubular devices[120,121] and in OBCs.[122–124] It was demonstrated that for a continuous crystallization process in a standard tubular device it is crucial that the flow rate, residence time and temperature profile are well controlled so that seeds are continuously produced from the crystallizer rather than a clear solution.[121] This essentially comes down to ensuring the supersaturation profile along the tubular crystallizer is suitable for the crystallization process being performed. Supersaturation profile control for an antisolvent crystallization process has also been shown in a standard tubular device with inserts.[120] In this process the crystal size distribution was controlled by varying the number of antisolvent addition points along the length of the tube showing that the final PSD depends on the relative nucleation and growth rates of the compound being crystallized. OBCs have the advantage over standard tubular devices that they provide good mass and heat transfer even at low flow rates due to the fluid oscillation. This advantage has been applied successfully to continuous seed generation *via* antisolvent crystallization where it has been shown that crystal size can be decreased by increasing the initial supersaturation level.[122,123] Continuous seed generation *via* cooling crystallization has also been demonstrated in the OBC where increasing initial supersaturation and mixing intensity reduces particle size due to the increased nucleation rate.[124]

1.4.5 Ultrasound Induced Nucleation

By using ultrasound (whether in the form of a bath, probe or transducer) primary nucleation can be triggered at relatively low supersaturation levels. Many studies have demonstrated this phenomenon for cooling crystallization processes in standard tubular devices,[125–127] glass channels[128,129] and

OBCs.[130] In the standard tubular devices ultrasound assisted continuous seed generation is combined with slug flow in order to better control the PSD after the initial nucleation step. This has been demonstrated by first applying ultrasound during cooling to generate the seed suspension followed by the addition of air slugs to create the slug flow.[125,126] However, this has also been demonstrated by firstly combining a solution with immiscible liquid slugs to create the slug flow followed by the application of ultrasound during cooling to generate the seeds within the slugs.[127] With either procedure the result is that the ultrasound directly produces small seeds with a narrow PSD and the slug flow allows for a narrow PSD to be maintained as the crystals grow because the slugs minimize the axial dispersion in the tube. In the work which uses glass channels, continuous seed generation was performed by firstly cooling the clear solution inside a length of tubing to generate supersaturation before exposing this solution to ultrasound inside the glass channel by means of an ultrasonic transducer.[128,129] These studies show that the mean particle size can be controlled with both supersaturation and ultrasonic power, with increasing supersaturation and ultrasonic power leading to a smaller mean particle size. Inserting an ultrasound probe in a OBC demonstrated that ultrasound assisted continuous seed generation could be implemented reliably at a larger scale.[130] It was shown that using sonication allowed for the mean particle size to be controlled with a narrower PSD than in the equivalent batch process. In addition, the product had consistent quality for 12 hours with no fouling or agglomeration issues being observed.

1.4.6 Fully Continuous Crystallization in an MSMPR Cascade

Continuous seed generation can also be the first step in a cascade of MSMPRs of one or more stages. Studies have demonstrated this for processes which use any combination of antisolvent,[131–134] reactive[135] and cooling[132,133,135–141] crystallization. Some studies have looked at single stage MSMPR configurations which utilize continuous flow with a recycle loop in an effort to better control the product quality and minimize waste.[132,140] It was shown that the recycle ratio and crystallizer temperature could be used to control purity and yield while the use of additives could control the particle size and shape. Other work has investigated single stage MSMPRs without recycle loops. One study looked at feeding solution and antisolvent into the MSMPR where nucleation takes place followed by continuously feeding the seed suspension into an agglomeration vessel.[131] Conditions were found that allowed the nucleation and agglomeration stages to be decoupled. Another study looked at feeding solution through a membrane into an MSMPR filled with well stirred antisolvent.[134] By using the membrane the particle size was controlled by the solution concentration, the antisolvent/solvent volume ratio and the type of antisolvent. These studies show that antisolvent crystallization is typically employed to continuously generate seeds in a single stage MSMPR, but ultrasound assisted cooling crystallization has also been employed.[138] It was demonstrated that utilizing ultrasound led to a shorter time until steady state with a lower supersaturation level being reached at steady state. In addition, sonication resulted in significantly smaller particle sizes, reduced agglomeration and improved crystal habit.

Continuous cooling crystallization is usually employed in a cascade of MSMPRs rather than a single MSMPR because residence time is increased and the flow becomes closer to plug flow. These processes commonly involve continuous seed generation in the first MSMPR stage followed by crystal growth.[136,137,139,141] Some studies show that process yield and purity can be controlled by changing the temperature and residence time of each stage of the MSMPR as this changes the supersaturation profile.[136,137] In addition, the mean particle size can be controlled by changing the number of stages in the MSMPR cascade. Another study shows that the PSD can be controlled in a two stage MSMPR cascade by implementing a nucleation control strategy where particle chord count information from the FBRM (see Chapter 9) is used to implement heating or cooling rates to maintain particle chord counts in a desired setpoint range during the crystallization process.[141] This allows for greater control over nucleation and growth in an MSMPR cascade without adding additional stages. In addition to controlling particle size in an MSMPR cascade, a study has demonstrated that the polymorphic outcome of a co-crystallization can be controlled when continuous periodic flow is utilized.[139] It was found that the polymorphic outcome of the co-crystallization depended on the feeding regime employed for combining the two components of the co-crystal in the first MSMPR in addition to the initial operating temperature.

A cascade of MSMPRs can also be utilized to employ a combination of crystallization modes. This is typically achieved by combining cooling crystallization with either antisolvent or reactive crystallization.[133,135] In the combined cooling and antisolvent process it was shown that adding antisolvent in a later stage lead to the entire process operating at a lower supersaturation level which resulted in crystals which were less agglomerated with better crystallinity.[133] On the other hand the purity and yield of the crystals were unaffected by the antisolvent addition. In the combined cooling and reactive process it was demonstrated that the reactant molar ratio and temperature of each stage must be well controlled in order to maximize the yield of the reaction as well as the yield of the crystallization.[135] In addition, the temperature profile had to be well controlled so that supersaturation was not generated too rapidly and a high crystal purity was obtained.

1.4.7 Continuous MSMPR Cascade with Batch Crystallization Start Up

Sometimes it is not possible to achieve fully continuous seed generation within a series of one or more MSMPRs. In these cases[142–146] a crystal seed suspension is generated in batch mode before continuous flow is implemented.

This has been demonstrated using a single stage MSMPR with a recycle loop in an effort to minimize waste and obtain a system where nucleation and growth kinetics could be estimated.[145] In this work the nucleation rate increased with increasing supersaturation and suspension density. It has also been shown that a two stage MSMPR cascade with a recycle loop can be used to better control the PSD utilizing kinetic parameters obtained from the single stage MSMPR process.[146] By changing the temperature profile across the two MSMPR stages the mean particle size can be controlled where a steeper temperature decrease and a shorter residence time results in smaller particles.

With regards to work that did not use recycle loops, one study looked at using antisolvent addition to generate a seed suspension in batch before implementing continuous flow in a single stage MSMPR and how this compares with the equivalent batch process.[143] It was found that the MSMPR crystallizer was able to access PSDs that were both smaller and larger compared with batch. Another study looked at cooling crystallization to generate the seed suspension in batch so that steady state operation was reached sooner after continuous flow began.[144] In this work it was demonstrated that the nucleation rate increased with increasing suspension density and this depends on the temperature profile. As well as having an MSMPR in the form of a stirred tank reactor there has been research into using a TCC as an MSMPR where continuous cooling crystallization takes place after the initial seed suspension is created in batch mode.[142] The TCC operates by imposing the crystallization under a Taylor vortex flow induced in the cylinder gap by rotating the inner cylinder where the rotation speed can be altered in addition to the solution feeding strategy (number of feeding ports used). It was demonstrated that the PSD was controlled by the solution concentration, feeding strategy and the rotation speed of the inner cylinder as these factors controlled the supersaturation profile which determine primary and secondary nucleation rates.

1.4.8 High Shear Wet Mill in MSMPR Configuration

A wet mill can be used in an MSMPR configuration for the purpose of delivering high shear to a solution leading to continuous seed generation. The wet mill can either be placed inside the MSMPR[100,147,148] or upstream of the MSMPR.[147,148] The studies have demonstrated that whether the wet mill is used inside or upstream of the MSMPR, it acts both as a continuous seed generator and a size reduction tool in a continuous crystallization process. A clear conclusion from using the wet mill is that increasing the rotor speed decreases the mean particle size, as would be expected.[100] This decrease in particle size is due to increased secondary nucleation in addition to increased primary nucleation depending on the process. Furthermore, using wet milling has been shown to significantly enhance the yield of a process especially when used upstream of the MSMPR in the form of a high shear nucleator due to the increased nucleation rates.[147] Using the wet mill upstream has also

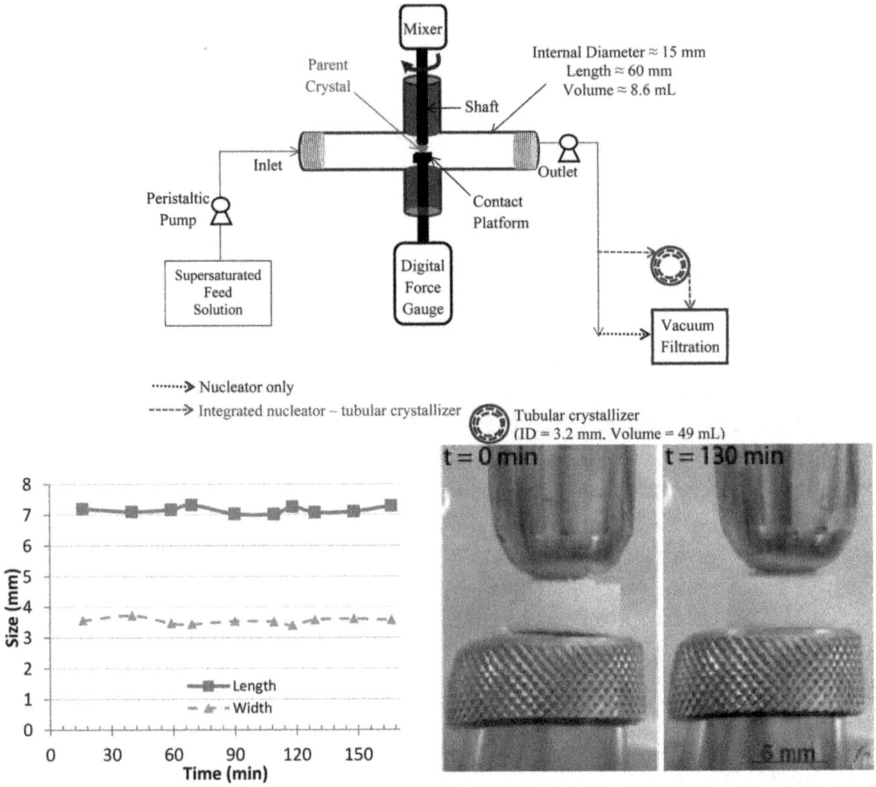

Figure 1.14 Example of continuous contact nucleator. Reproduced from ref. 149 with permission from American Chemical Society, Copyright 2013.

been shown to reduce the time required to reach steady state. As with the standard MSMPR cascade, the upstream wet mill feeding to an MSMPR can be combined with a nucleation control strategy where particle chord count information from the FBRM is used to implement heating or cooling rates to maintain particle chord counts in a desired setpoint range during the crystallization process.[148] This additional control in the MSMPR allows for the PSD to be more finely tuned.

1.4.9 Secondary Nucleators

In addition to devices for inducing primary nucleation there has also been research into devices for inducing secondary nucleation from a parent crystal or tablet. Recent studies have demonstrated devices which use contact secondary nucleation as a means of creating seed crystals for continuous tubular crystallizers.[149,150] Firstly it was demonstrated that secondary nuclei could be continuously generated in such a fashion that the size of the nuclei could be controlled by the supersaturation of the feed solution and the residence time (by changing flow rate),[149] see Figure. 1.14. As this secondary nucleation process is decoupled from the growth stage the PSD of the final crystal product can be well controlled by tailoring the subsequent tubular crystallizer to allow for the required growth. Building on this work it was then shown that for the same type of experimental setup the secondary nucleation rate could be controlled by changing the contact force, contact frequency and contact area.[150] It was also shown that a feature of contact secondary nucleation is the generation of nuclei with a narrow PSD with the contact force, frequency and area not having a significant effect on width. This feature clearly makes it very attractive to implement this technique in a continuous crystallization process.

Abbreviations

API	active pharmaceutical ingredient
CSD	crystal size distribution
CST	continuous stirred tank
MSMPR	mixed suspension mixed product removal
OBC	oscillatory baffled crystallizer
PFC	plug flow crystallizer
TCC	Taylor–Couette crystallizer

Roman Symbols

A	constant in supersaturation dependent nucleation rate equation
Ar	surface area of the excipient per unit volume of the crystallizing solution
B	nucleation rate
B	order of the nucleation rate
C	concentration
C^*	solubility
D_i	impeller diameter
E_L	impact energy
G	crystal growth rate
g	growth rate order
J	supersaturation dependent nucleation rate
k_a	shape factor of the particle used to calculate the surface area of particles
$K_\mathrm{c-c}$	crystal–crystal collision constant
$K_\mathrm{c-i}$	crystal–impeller collision constant
K_E	number of nuclei per collision
k_{g1} and k_{g2}	growth rate constants
K_i	impeller discharge coefficient
$L_\mathrm{c-c}$	lower integration boundaries of the moments for crystal–crystal collision
$L_\mathrm{c-i}$	lower integration boundaries of the moments for crystal–impeller collision
M_j	population density function

$N(v,t)$	time-dependent volume-based population balance density function
$n(L)$	number-based size distribution
N_i	impeller speed
N_P	Newton number of the impeller
P_0	power input of the stirrer
P_{susp}	minimum power required to suspend particles in the vessel
R	gas constant
S	supersaturation ratio
S_{cr}	average supersaturation
T	temperature
t_c	circulation time
V_c	total volume of the crystallizer
V_m	specific volume of the solute

Greek Symbols

ε	dissipated power by the impeller per unit mass of suspension
ρ	density
φ_c	volume fraction of crystals
ω_L	collision frequency
σ	relative supersaturation
τ	residence time
τ_m	time of mixing
τ_n	time of nucleation
φ	flow rate

References

1. K. Plumb, Continuous Processing in the Pharmaceutical Industry: Changing the Mind Set, *Chem. Eng. Res. Des.*, 2005, **83**(6), 730–738.
2. I. R. Baxendale, R. D. Braatz, B. K. Hodnett, K. F. Jensen, M. D. Johnson and P. Sharratt, *et al.*, Achieving Continuous Manufacturing: Technologies and Approaches for Synthesis, Workup, and Isolation of Drug Substance. May 20–21, 2014 Continuous Manufacturing Symposium, *J. Pharm. Sci.*, 2015, **104**(3), 781–791.
3. J. H. ter Horst, C. Schmidt and J. Ulrich, 32-Fundamentals of Industrial Crystallization, in *Handbook of Crystal Growth*, ed. P. Rudolph, Elsevier, Boston, 2nd edn, 2015, pp. 1317–1349.
4. C. J. Brown, T. McGlone, S. Yerdelen, V. Srirambhatla, F. Mabbott and R. Gurung, *et al.*, Enabling precision manufacturing of active pharmaceutical ingredients: workflow for seeded cooling continuous crystallisations, *Mol. Syst. Des. Eng.*, 2018, **3**(3), 518–549.
5. J. Chen, B. Sarma, J. M. B. Evans and A. S. Myerson, Pharmaceutical Crystallization, *Cryst. Growth Des.*, 2011, **11**(4), 887–895.

6. R. J. Davey, S. L. M. Schroeder and J. H. ter Horst, Nucleation of Organic Crystals—A Molecular Perspective, *Angew. Chem., Int. Ed.*, 2013, **52**(8), 2166–2179.
7. J. W. Mullin, *Crystallization*, 2001.
8. A. Myerson, *Handbook of Industrial Crystallization*, Butterworth-Heinemann, 2002.
9. D. Kashchiev, *Nucleation: Basic Theory With Applications*, Butterworth Heinemann, Oxford, Boston, 2000.
10. D. Kashchiev and G. M. van Rosmalen, Review: Nucleation in solutions revisited, *Cryst. Res. Technol.*, 2003, **38**(7–8), 555–574.
11. R. Ó. Meadhra, H. J. M. Kramer and G. M. Van Rosmalen, Model for secondary nucleation in a suspension crystallizer, *AIChE J.*, 1996, **42**(4), 973–982.
12. A. Chianese and H. J. Kramer, *Industrial Crystallization Process Monitoring and Control*, Wiley, 2012.
13. S. A. Kulkarni, S. S. Kadam, H. Meekes, A. I. Stankiewicz and J. H. ter Horst, Crystal Nucleation Kinetics from Induction Times and Metastable Zone Widths, *Cryst. Growth Des.*, 2013, **13**(6), 2435–2440.
14. R. R. E. Steendam, L. Keshavarz, M. A. R. Blijlevens, B. de Souza, D. M. Croker and P. J. Frawley, Effects of Scale-Up on the Mechanism and Kinetics of Crystal Nucleation, *Cryst. Growth Des.*, 2018, **18**(9), 5547–5555.
15. A. Koswara and Z. K. Nagy, Anti-Fouling Control of Plug-Flow Crystallization via Heating and Cooling Cycle, *IFAC-PapersOnLine*, 2015, **48**(8), 193–198.
16. L. G. Bauer, R. W. Rousseau and W. L. McCabe, Influence of crystal size on the rate of contact nucleation in stirred-tank crystallizers, *AIChE J.*, 1974, **20**(4), 653–659.
17. A. M. Neumann, S. K. Bermingham, H. J. Kramer and G. M. Van Rosmalen, Modeling the Attrition Process in a 22 Liter Draft Tube Crystallizer, *Mixing and Crystallization*, Springer, 2000, pp. 267–279.
18. C. J. Callahan and X.-W. Ni, An investigation into the effect of mixing on the secondary nucleation of sodium chlorate in a stirred tank and an oscillatory baffled crystallizer, *CrystEngComm*, 2014, **16**(4), 690–697.
19. J. M. Schall, J. S. Mandur, R. D. Braatz and A. S. Myerson, Nucleation and Growth Kinetics for Combined Cooling and Antisolvent Crystallization in a Mixed-Suspension, Mixed-Product Removal System: Estimating Solvent Dependency, *Cryst. Growth Des.*, 2018, **18**(3), 1560–1570.
20. D. O'Grady, M. Barrett, E. Casey and B. Glennon, The Effect of Mixing on the Metastable Zone Width and Nucleation Kinetics in the Anti-Solvent Crystallization of Benzoic Acid, *Chem. Eng. Res. Des.*, 2007, **85**(7), 945–952.
21. V. V. Kelkar and K. M. Ng, Design of reactive crystallization systems incorporating kinetics and mass-transfer effects, *AIChE J.*, 1999, **45**(1), 69–81.

22. N. S. Tavare, Mixing, reaction and precipitation: Environment micro-mixing models in continuous crystallizers—I. premixed feeds, *Comput. Chem. Eng.*, 1992, **16**(10), 923–936.
23. N. S. Tavare, Mixing in Continuous Crystallizers, *AIChE J.*, 1986, 705–732.
24. Y. J. Yi and A. S. Myerson, Laboratory Scale Batch Crystallization and the Role of Vessel Size, *Chem. Eng. Res. Des.*, 2006, **84**(8), 721–728.
25. P. C. Chen, C. C. Chen, M. H. Fun, O. Y. Liao, J. J. Jiang and Y. S. Wang, *et al.*, Mixing and Crystallization Kinetics in Gas-liquid Reactive Crystallization, *Chem. Eng. Technol.*, 2004, **27**(5), 519–528.
26. C. D. Andereck, S. S. Liu and H. L. Swinney, Flow regimes in a circular Couette system with independently rotating cylinders, *J. Fluid Mech.*, 1986, **164**, 155–183.
27. D. L. Marchisio, M. Soos, J. Sefcik and M. Morbidelli, Role of turbulent shear rate distribution in aggregation and breakage processes, *AIChE J.*, 2006, **52**(1), 158–173.
28. A. S. Moussa, M. Soos, J. Sefcik and M. Morbidelli, Effect of Solid Volume Fraction on Aggregation and Breakage in Colloidal Suspensions in Batch and Continuous Stirred Tanks, *Langmuir*, 2007, **23**(4), 1664–1673.
29. J. Liu and Å. C. Rasmuson, Influence of Agitation and Fluid Shear on Primary Nucleation in Solution, *Cryst. Growth Des.*, 2013, **13**(10), 4385–4394.
30. R. A. Granberg, C. Ducreux, S. Gracin and Å. C. Rasmuson, Primary nucleation of paracetamol in acetone–water mixtures, *Chem. Eng. Sci.*, 2001, **56**(7), 2305–2313.
31. J. Liu, M. Svärd and Å. C. Rasmuson, Influence of Agitation and Fluid Shear on Nucleation of m-Hydroxybenzoic Acid Polymorphs, *Cryst. Growth Des.*, 2014, **14**(11), 5521–5531.
32. J. Liu, M. Svärd and Å. C. Rasmuson, Influence of Agitation on Primary Nucleation in Stirred Tank Crystallizers, *Cryst. Growth Des.*, 2015, **15**(9), 4177–4184.
33. F. L. Nordström, M. Svärd, B. Malmberg and Å. C. Rasmuson, Influence of Solution Thermal and Structural History on the Nucleation of m-Hydroxybenzoic Acid Polymorphs, *Cryst. Growth Des.*, 2012, **12**(9), 4340–4348.
34. C. Forsyth, P. A. Mulheran, C. Forsyth, M. D. Haw, I. S. Burns and J. Sefcik, Influence of Controlled Fluid Shear on Nucleation Rates in Glycine Aqueous Solutions, *Cryst. Growth Des.*, 2015, **15**(1), 94–102.
35. C. Forsyth, I. S. Burns, P. A. Mulheran and J. Sefcik, Scaling of Glycine Nucleation Kinetics with Shear Rate and Glass–Liquid Interfacial Area, *Cryst. Growth Des.*, 2016, **16**(1), 136–144.
36. R. R. Umbdenstock, Preparation of Procaine Penicillin, *US Pat.*, US298185A, 1952.
37. J. R. Principe and D. M. Skauen, Preparation of microcrystalline progesterone using ultrasound, *J. Pharm. Sci.*, 1962, **51**(4), 389–390.

38. S. L. Hem, D. M. Skauen and H. M. Beal, Mechanism of Crystallization of Hydrocortisone by Ultrasonic Irradiation, *J. Pharm. Sci.*, 1967, **56**(2), 229–233.
39. H. W. Anderson, J. B. Carberry, H. F. Staunton and B. C. Sutradhar, Crystallization of adipic acid, *US Pat.*, US5471001A, 1995.
40. J. M. Midler, Process for Production of Crystals in Fluidized Bed Crystallizers, *US Pat.*, US3892539A, 1971.
41. C. J. Price, Ultrasound the key to better crystals for the pharmaceutical industry, *Pharm. Technol. Eur.*, 1997, (October), 78–86.
42. S. Kozyukhin, Y. Vorobyov, P. Lazarenko and M. Presniakov, Isothermal and CW laser crystallization of amorphous Ge2Sb2Te5 thin films, *J. Non-Cryst. Solids*, 2018, **480**, 51–56.
43. T. Uwada, S. Fujii, T. Sugiyama, A. Usman, A. Miura and H. Masuhara, *et al.*, Glycine Crystallization in Solution by CW Laser-Induced Microbubble on Gold Thin Film Surface, *ACS Appl. Mater. Interfaces*, 2012, **4**(3), 1158–1163.
44. I. S. Lee, J. M. B. Evans, D. Erdemir, A. Y. Lee, B. A. Garetz and A. S. Myerson, Nonphotochemical Laser Induced Nucleation of Hen Egg White Lysozyme Crystals, *Cryst. Growth Des.*, 2008, **8**(12), 4255–4261.
45. Y. Hosokawa, H. Adachi, M. Yoshimura, Y. Mori, T. Sasaki and H. Masuhara, Femtosecond Laser-Induced Crystallization of 4-(Dimeth ylamino)-N-methyl-4-stilbazolium Tosylate, *Cryst. Growth Des.*, 2005, **5**(3), 861–863.
46. T. Sugiyama and H. Masuhara, Laser-Induced Crystallization and Crystal Growth, *Chem. – Asian J.*, 2011, **6**(11), 2878–2889.
47. A. J. Alexander and P. J. Camp, Non-photochemical laser-induced nucleation, *J. Chem. Phys.*, 2019, **150**(4), 040901.
48. H. Y. Yoshikawa, R. Murai, H. Adachi, S. Sugiyama, M. Maruyama and Y. Takahashi, *et al.*, Laser ablation for protein crystal nucleation and seeding, *Chem. Soc. Rev.*, 2014, **43**(7), 2147–2158.
49. D. Kashchiev, Nucleation in external electric field, *J. Cryst. Growth*, 1972, **13–14**, 128–130.
50. C. C. Chin, J. B. Dence and J. C. Warren, Crystallization of human placental estradiol 17beta-dehydrogenase. A new method for crystallizing labile enzymes, *J. Biol. Chem.*, 1976, **251**(12), 3700–3705.
51. C. N. Nanev and A. Penkova, Nucleation of lysozyme crystals under external electric and ultrasonic fields, *J. Cryst. Growth*, 2001, **232**(1), 285–293.
52. E. Revalor, Z. Hammadi, J.-P. Astier, R. Grossier, E. Garcia and C. Hoff, *et al.*, Usual and unusual crystallization from solution, *J. Cryst. Growth*, 2010, **312**(7), 939–946.
53. W. W. Li, N. Radacsi, H. J. M. Kramer, A. E. D. M. van der Heijden and J. H. ter Horst, Solid Separation from a Mixed Suspension through Electric-Field-Enhanced Crystallization, *Angew. Chem., Int. Ed.*, 2016, **55**(52), 16088–16091.

54. S. Y. Wong, Y. Cui and A. S. Myerson, Contact Secondary Nucleation as a Means of Creating Seeds for Continuous Tubular Crystallizers, *Cryst. Growth Des.*, 2013, **13**(6), 2514–2521.
55. Y. Cui and A. S. Myerson, Experimental Evaluation of Contact Secondary Nucleation Mechanisms, *Cryst. Growth Des.*, 2014, **14**(10), 5152–5157.
56. G. D. Botsaris, Secondary Nucleation — A Review, in *Industrial Crystallization*, ed. J. W. Mullin, Springer US, Boston, MA, 1976, pp. 3–22.
57. J. Garside, I. T. Rusli and M. A. Larson, Origin and size distribution of secondary nuclei, *AIChE J.*, 1979, **25**(1), 57–64.
58. P. Ayazi Shamlou, S. Stavrinides, N. Titchener-Hooker and M. Hoare, Turbulent breakage of protein precipitates in mechanically stirred bioreactors, *Bioprocess Eng.*, 1996, **14**(5), 237–243.
59. P. Ayazi Shamlou, A. G. Jones and K. Djamarani, Hydrodynamics of secondary nucleation in suspension crystallization, *Chem. Eng. Sci.*, 1990, **45**(5), 1405–1416.
60. G. M. Crawley, F. Gruy and M. Cournil, In-line study of crystal fragmentation in a stirred reactor, *Chem. Eng. Sci.*, 1996, **51**(20), 4537–4550.
61. P. A. Shamlou and N. Titchener-Hooker, Turbulent aggregation and breakup of particles in liquids in stirred vessels, *Process. Solid-Liq. Suspensions*, 1993, **1**, 1–25.
62. A. Chianese, F. Di Berardino and A. G. Jones, On the effect of secondary nucleation on the crystal size distribution from a seeded batch crystallizer, *Chem. Eng. Sci.*, 1993, **48**(3), 551–560.
63. S. K. Bermingham, P. J. T. Verheijen and H. J. M. Kramer, Optimal Design of Solution Crystallization Processes with Rigorous Models, *Chem. Eng. Res. Des.*, 2003, **81**(8), 893–903.
64. K. C. Kee and C. D. Rielly, Measurement of Particle Impact Frequencies and Velocities on Impeller Blades in a Mixing Tank, *Chem. Eng. Res. Des.*, 2004, **82**(9), 1237–1249.
65. S. M. Reeves and P. J. Hill, Mechanisms Influencing Crystal Breakage Experiments in Stirred Vessels, *Cryst. Growth Des.*, 2012, **12**(6), 2748–2758.
66. M. Liiri, T. Koiranen and J. Aittamaa, Secondary nucleation due to crystal–impeller and crystal–vessel collisions by population balances in CFD-modelling, *J. Cryst. Growth*, 2002, **237–239**(Part 3), 2188–2193.
67. E. L. Paul, V. A. Atiemo-Obeng and S. M. Kresta, *Handbook of Industrial Mixing: Science and Practice*, Wiley, 2004.
68. N. Yazdanpanah, C. J. Testa, S. R. K. Perala, K. D. Jensen, R. D. Braatz and A. S. Myerson, *et al.*, Continuous Heterogeneous Crystallization on Excipient Surfaces, *Cryst. Growth Des.*, 2017, **17**(6), 3321–3330.
69. M. Yokota, E. Takezawa, T. Takakusaki, A. Sato, H. Takahashi and N. Kubota, Scale-up effect on the rate of contact nucleation caused by collisions of crystals with an impeller, *Chem. Eng. Sci.*, 1999, **54**(17), 3831–3838.
70. B. H. Shanks and K. A. Berglund, Contact nucleation from aqueous sucrose solutions, *AIChE J.*, 1985, **31**(1), 152–154.

71. C. Y. Sung, J. Estrin and G. R. Youngquist, Secondary nucleation of magnesium sulfate by fluid shear, *AIChE J.*, 1973, **19**(5), 957–962.
72. M.-L. Wang, H.-T. Huang and J. Estrin, Secondary nucleation of citric acid due to fluid forces in a Couette flow crystallizer, *AIChE J.*, 1981, **27**(2), 312–315.
73. S. Srisa-nga, A. E. Flood and E. T. White, The Secondary Nucleation Threshold and Crystal Growth of α-Glucose Monohydrate in Aqueous Solution, *Cryst. Growth Des.*, 2006, **6**(3), 795–801.
74. E. Ottens and E. de Jong, A model for secondary nucleation in a stirred vessel cooling crystallizer, *Cryst. Res. Technol.*, 1974, **9**(8), 873–886.
75. T. Evans, A. Sarofim and G. Margolis, Models of secondary nucleation attributable to crystal-crystallizer and crystal-crystal collisions, *AIChE J.*, 1974, **20**(5), 959–966.
76. R. van Beusichem, *Secondary Nucleation Attributable to the Combined Effects of Crystal–Impeller and Crystal–Crystal Collisions*, Delft, The Netherlands, 2006.
77. A. E. D. M. van der Heijden, J. P. van der Eerden and G. M. van Rosmalen, The secondary nucleation rate: a physical model, *Chem. Eng. Sci.*, 1994, **49**(18), 3103–3113.
78. M. Kitamura and Y. Hayashi, Secondary Nucleation Behavior and the Mechanism in Antisolvent Crystallization of Thiazole Derivative Polymorphs, *Ind. Eng. Chem. Res.*, 2016, **55**(5), 1413–1418.
79. T. Buhse, D. Durand, D. Kondepudi, J. Laudadio and S. Spilker, Chiral Symmetry Breaking in Crystallization: The Role of Convection, *Phys. Rev. Lett.*, 2000, **84**(19), 4405–4408.
80. N. Yazdanpanah, T. O'connor and C. Cruz, Dynamic Modeling of a Continuous Reactive Crystallization Process, *AIChE Annual Meeting 2018*, 10/30/2018, AIChE, Pittsburgh, PA, USA, 2018.
81. L. Derdour, J. M. Reckamp and C. Pink, Development of a reactive slurry salt crystallization to improve solid properties and process performance and scalability, *Chem. Eng. Res. Des.*, 2017, **121**, 207–218.
82. W. J. Liu, C. Y. Ma, J. J. Liu, Y. Zhang and X. Z. Wang, Continuous reactive crystallization of pharmaceuticals using impinging jet mixers, *AIChE J.*, 2017, **63**(3), 967–974.
83. M. Jiang, Y.-E. D. Li, H.-H. Tung and R. D. Braatz, Effect of jet velocity on crystal size distribution from antisolvent and cooling crystallizations in a dual impinging jet mixer, *Chem. Eng. Process.*, 2015, **97**, 242–247.
84. D. Acevedo, X. Yang, Y. C. Liu, T. F. O'Connor, A. Koswara and Z. K. Nagy, *et al.*, Encrustation in Continuous Pharmaceutical Crystallization Processes—A Review, *Org. Process Res. Dev.*, 2019, **23**(6), 1134–1142.
85. A. Majumder and Z. K. Nagy, Dynamic Modeling of Encrust Formation and Mitigation Strategy in a Continuous Plug Flow Crystallizer, *Cryst. Growth Des.*, 2015, **15**(3), 1129–1140.
86. K. Chadwick, J. Chen, A. S. Myerson and B. L. Trout, Toward the Rational Design of Crystalline Surfaces for Heteroepitaxy: Role of Molecular Functionality, *Cryst. Growth Des.*, 2012, **12**(3), 1159–1166.

87. K. Chadwick, A. Myerson and B. Trout, Polymorphic control by heterogeneous nucleation-A new method for selecting crystalline substrates, *CrystEngComm*, 2011, **13**(22), 6625–6627.
88. S. Chunsrivirot, Y. Diao and B. L. Trout, Binding Affinity of a Small Molecule to an Amorphous Polymer in a Solvent. Part 1: Free Energy of Binding to a Binding Site, *Langmuir*, 2011, **27**(20), 12381–12395.
89. Y. Diao, A. S. Myerson, T. A. Hatton and B. L. Trout, Surface Design for Controlled Crystallization: The Role of Surface Chemistry and Nanoscale Pores in Heterogeneous Nucleation, *Langmuir*, 2011, **27**(9), 5324–5334.
90. N. Yazdanpanah, A. Myerson, B. Trout, J. Stojakovic and K. D. Jensen, Continuous Heterogeneous Crystallization and the New Method of Making Tablets, *AIChE Annual Meeting 2017*, 11/1/2017, AIChE, Minniapolis, MN, USA, 2017.
91. S. Ferguson, G. Morris, H. X. Hao, M. Barrett and B. Glennon, In-situ monitoring and characterization of plug flow crystallizers, *Chem. Eng. Sci.*, 2012, **77**, 105–111.
92. M. Furuta, K. Mukai, D. Cork and K. Mae, Continuous crystallization using a sonicated tubular system for controlling particle size in an API manufacturing process, *Chem. Eng. Process.*, 2016, **102**, 210–218.
93. M. Stahl, B. L. Alund and A. C. Rasmuson, Reaction crystallization kinetics of benzoic add, *AIChE J.*, 2001, **47**(7), 1544–1560.
94. A. F. Blandin, D. Mangin, V. Nallet, J. P. Klein and J. M. Bossoutrot, Kinetics identification of salicylic acid precipitation through experiments in a batch stirred vessel and a T-mixer, *Chem. Eng. J.*, 2001, **81**(1–3), 91–100.
95. H. Zheng, R. K. Smith, Y.-w Jun, C. Kisielowski, U. Dahmen and A. P. Alivisatos, Observation of Single Colloidal Platinum Nanocrystal Growth Trajectories, *Science*, 2009, **324**(5932), 1309.
96. N. T. K. Thanh, N. Maclean and S. Mahiddine, Mechanisms of Nucleation and Growth of Nanoparticles in Solution, *Chem. Rev.*, 2014, **114**(15), 7610–7630.
97. S. Roy and T. Ando, Solidification behavior of highly supercooled polycrystalline silicon droplets, *Mater. Sci. Semicond. Process.*, 2012, **15**(6), 722–730.
98. M. Jiang, C. Gu and R. D. Braatz, Understanding temperature-induced primary nucleation in dual impinging jet mixers, *Chem. Eng. Process.*, 2015, **97**, 187–194.
99. B. Ahmed, C. J. Brown, T. McGlone, D. L. Bowering, J. Sefcik and A. J. Florence, Engineering of acetaminophen particle attributes using a wet milling crystallisation platform, *Int. J. Pharm.*, 2019, **554**, 201–211.
100. D. Acevedo, V. K. Kamaraju, B. Glennon and Z. K. Nagy, Modeling and Characterization of an in Situ Wet Mill Operation, *Org. Process Res. Dev.*, 2017, **21**(7), 1069–1079.

101. C. V. Luciani, E. W. Conder and K. D. Seibert, Modeling-Aided Scale-Up of High-Shear Rotor–Stator Wet Milling for Pharmaceutical Applications, *Org. Process Res. Dev.*, 2015, **19**(5), 582–589.
102. T. Kamahara, M. Takasuga, H. H. Tung, K. Hanaki, T. Fukunaka and B. Izzo, *et al.*, Generation of Fine Pharmaceutical Particles via Controlled Secondary Nucleation under High Shear Environment during Crystallization – Process Development and Scale-up, *Org. Process Res. Dev.*, 2007, **11**(4), 699–703.
103. R. W. Rousseau, W. L. McCabe and C. Y. Tai, The stability of nuclei generated by contact nucleation, *AIChE J.*, 1975, **21**(5), 1017–1019.
104. C. Y. Tai, W. L. McCabe and R. W. Rousseau, Contact nucleation of various crystal types, *AIChE J.*, 1975, **21**(2), 351–358.
105. A. Vancleef, S. Seurs, J. Jordens, T. Van Gerven, L. C. J. Thomassen and L. Braeken, Reducing the Induction Time Using Ultrasound and High-Shear Mixing in a Continuous Crystallization Process, *Crystals*, 2018, **8**(8), 326.
106. T. T. H. Nguyen, A. Khan, L. M. Bruce, C. Forbes, R. L. O'Leary and C. J. Price, The Effect of Ultrasound on the Crystallisation of Paracetamol in the Presence of Structurally Similar Impurities, *Crystals*, 2017, **7**(10), 294.
107. S. Gracin, M. Uusi-Penttilä and Å. C. Rasmuson, Influence of Ultrasound on the Nucleation of Polymorphs of p-Aminobenzoic Acid, *Cryst. Growth Des.*, 2005, **5**(5), 1787–1794.
108. S. Kaur Bhangu, M. Ashokkumar and J. Lee, Ultrasound Assisted Crystallization of Paracetamol: Crystal Size Distribution and Polymorph Control, *Cryst. Growth Des.*, 2016, **16**(4), 1934–1941.
109. R. Lakerveld, H. J. M. Kramer, A. I. Stankiewicz and J. Grievink, Application of generic principles of process intensification to solution crystallization enabled by a task-based design approach, *Chem. Eng. Process.*, 2010, **49**(9), 979–991.
110. Y. Zhao, V. K. Kamaraju, G. Y. Hou, G. Power, P. Donnellan and B. Glennon, Kinetic identification and experimental validation of continuous plug flow crystallisation, *Chem. Eng. Sci.*, 2015, **133**, 106–115.
111. M. Jiang, Y. E. D. Li, H. H. Tung and R. D. Braatz, Effect of jet velocity on crystal size distribution from antisolvent and cooling crystallizations in a dual impinging jet mixer, *Chem. Eng. Process.*, 2015, **97**, 242–247.
112. C. Beck, S. V. Dalvi and R. N. Dave, Controlled liquid antisolvent precipitation using a rapid mixing device, *Chem. Eng. Sci.*, 2010, **65**(21), 5669–5675.
113. C. Lindenberg and M. Mazzotti, Continuous Precipitation of L-Asparagine Monohydrate in a Micromixer: Estimation of Nucleation and Growth Kinetics, *AIChE J.*, 2011, **57**(4), 942–950.
114. N. E. B. Briggs, U. Schacht, V. Raval, T. McGlone, J. Sefcik and A. J. Florence, Seeded Crystallization of beta-L-Glutamic Acid in a Continuous Oscillatory Baffled Crystallizer, *Org. Process Res. Dev.*, 2015, **19**(12), 1903–1911.

115. M. Jiang, Z. L. Zhu, E. Jimenez, C. D. Papageorgiou, J. Waetzig and A. Hardy, *et al.*, Continuous-Flow Tubular Crystallization in Slugs Spontaneously Induced by Hydrodynamics, *Cryst. Growth Des.*, 2014, **14**(2), 851–860.
116. M. Jiang, M. H. Wong, Z. L. Zhu, J. Q. Zhang, L. F. Zhou and K. Wang, *et al.*, Towards achieving a flattop crystal size distribution by continuous seeding and controlled growth, *Chem. Eng. Sci.*, 2012, **77**, 2–9.
117. P. MacFhionnghaile, V. Svoboda, J. McGinty, A. Nordon and J. Sefcik, Crystallization Diagram for Antisolvent Crystallization of Lactose: Using Design of Experiments To Investigate Continuous Mixing-Induced Supersaturation, *Cryst. Growth Des.*, 2017, **17**(5), 2611–2621.
118. S. A. Raza, U. Schacht, V. Svoboda, D. P. Edwards, A. J. Florence and C. R. Pulham, *et al.*, Rapid Continuous Antisolvent Crystallization of Multicomponent Systems, *Cryst. Growth Des.*, 2018, **18**(1), 210–218.
119. V. Svoboda, P. MacFhionnghaile, J. McGinty, L. E. Connor, I. D. H. Oswald and J. Sefcik, Continuous Cocrystallization of Benzoic Acid and Isonicotinamide by Mixing-Induced Supersaturation: Exploring Opportunities between Reactive and Antisolvent Crystallization Concepts, *Cryst. Growth Des.*, 2017, **17**(4), 1902–1909.
120. A. J. Alvarez and A. S. Myerson, Continuous Plug Flow Crystallization of Pharmaceutical Compounds, *Cryst. Growth Des.*, 2010, **10**(5), 2219–2228.
121. L. Hohmann, R. Gorny, O. Klaas, J. Ahlert, K. Wohlgemuth and N. Kockmann, Design of a Continuous Tubular Cooling Crystallizer for Process Development on Lab-Scale, *Chem. Eng. Technol.*, 2016, **39**(7), 1268–1280.
122. C. J. Brown, J. A. Adelakun and X. W. Ni, Characterization and modelling of antisolvent crystallization of salicylic acid in a continuous oscillatory baffled crystallizer, *Chem. Eng. Process.*, 2015, **97**, 180–186.
123. R. Pena, J. A. Oliva, C. L. Burcham, D. J. Jarmer and Z. K. Nagy, Process Intensification through Continuous Spherical Crystallization Using an Oscillatory Flow Baffled Crystallizer, *Cryst. Growth Des.*, 2017, **17**(9), 4776–4784.
124. P. Cruz, F. Rocha and A. Ferreira, Effect of operating conditions on batch and continuous paracetamol crystallization in an oscillatory flow mesoreactor, *CrystEngComm*, 2016, **18**(47), 9113–9121.
125. R. J. P. Eder, S. Schrank, M. O. Besenhard, E. Roblegg, H. Gruber-Woelfler and J. G. Khinast, Continuous Sonocrystallization of Acetylsalicylic Acid (ASA): Control of Crystal Size, *Cryst. Growth Des.*, 2012, **12**(10), 4733–4738.
126. M. Jiang, C. D. Papageorgiou, J. Waetzig, A. Hardy, M. Langston and R. D. Braatz, Indirect Ultrasonication in Continuous Slug-Flow Crystallization, *Cryst. Growth Des.*, 2015, **15**(5), 2486–2492.
127. D. Rossi, R. Jamshidi, N. Saffari, S. Kuhn, A. Gavriilidis and L. Mazzei, Continuous-Flow Sonocrystallization in Droplet-Based Microfluidics, *Cryst. Growth Des.*, 2015, **15**(11), 5519–5529.

128. R. Jamshidi, D. Rossi, N. Saffari, A. Gavriilidis and L. Mazzei, Investigation of the Effect of Ultrasound Parameters on Continuous Sonocrystallization in a Millifluidic Device, *Cryst. Growth Des.*, 2016, **16**(8), 4607–4619.
129. J. Jordens, E. Canini, B. Gielen, T. Van Gerven and L. Braeken, Ultrasound Assisted Particle Size Control by Continuous Seed Generation and Batch Growth, *Crystals*, 2017, **7**(7), 20.
130. H. Siddique, C. J. Brown, I. Houson and A. J. Florence, Establishment of a Continuous Sonocrystallization Process for Lactose in an Oscillatory Baffled Crystallizer, *Org. Process Res. Dev.*, 2015, **19**(12), 1871–1881.
131. R. Pena and Z. K. Nagy, Process Intensification through Continuous Spherical Crystallization Using a Two-Stage Mixed Suspension Mixed Product Removal (MSMPR) System, *Cryst. Growth Des.*, 2015, **15**(9), 4225–4236.
132. S. Y. Wong, A. P. Tatusko, B. L. Trout and A. S. Myerson, Development of Continuous Crystallization Processes Using a Single-Stage Mixed-Suspension, Mixed-Product Removal Crystallizer with Recycle, *Cryst. Growth Des.*, 2012, **12**(11), 5701–5707.
133. H. T. Zhang, J. Quon, A. J. Alvarez, J. Evans, A. S. Myerson and B. Trout, Development of Continuous Anti-Solvent/Cooling Crystallization Process using Cascaded Mixed Suspension, Mixed Product Removal Crystallizers, *Org. Process Res. Dev.*, 2012, **16**(5), 915–924.
134. R. Othman, G. T. Vladisavljevic, E. Simone, Z. K. Nagy and R. G. Holdich, Preparation of Microcrystals of Piroxicam Monohydrate by Antisolvent Precipitation via Microfabricated Metallic Membranes with Ordered Pore Arrays, *Cryst. Growth Des.*, 2017, **17**(12), 6692–6702.
135. J. L. Quon, H. Zhang, A. Alvarez, J. Evans, A. S. Myerson and B. L. Trout, Continuous Crystallization of Aliskiren Hemifumarate, *Cryst. Growth Des.*, 2012, **12**(6), 3036–3044.
136. A. J. Alvarez, A. Singh and A. S. Myerson, Crystallization of Cyclosporine in a Multistage Continuous MSMPR Crystallizer, *Cryst. Growth Des.*, 2011, **11**(10), 4392–4400.
137. J. C. Li, T. T. C. Lai, B. L. Trout and A. S. Myerson, Continuous Crystallization of Cyclosporine: Effect of Operating Conditions on Yield and Purity, *Cryst. Growth Des.*, 2017, **17**(3), 1000–1007.
138. O. Narducci, A. G. Jones and E. Kougoulos, Continuous crystallization of adipic acid with ultrasound, *Chem. Eng. Sci.*, 2011, **66**(6), 1069–1076.
139. K. A. Powell, G. Bartolini, K. E. Wittering, A. N. Saleemi, C. C. Wilson and C. D. Rielly, *et al.*, Toward Continuous Crystallization of Urea-Barbituric Acid: A Polymorphic Co-Crystal System, *Cryst. Growth Des.*, 2015, **15**(10), 4821–4836.
140. K. A. Powell, A. N. Saleemi, C. D. Rielly and Z. K. Nagy, Monitoring Continuous Crystallization of Paracetamol in the Presence of an Additive Using an Integrated PAT Array and Multivariate Methods, *Org. Process Res. Dev.*, 2016, **20**(3), 626–636.

141. Y. Yang, L. C. Song and Z. K. Nagy, Automated Direct Nucleation Control in Continuous Mixed Suspension Mixed Product Removal Cooling Crystallization, *Cryst. Growth Des.*, 2015, **15**(12), 5839–5848.
142. A. T. Nguyen and W. S. Kim, Influence of feeding mode on cooling crystallization of L-lysine in Couette-Taylor crystallizer, *Korean J. Chem. Eng.*, 2017, **34**(7), 2002–2010.
143. S. Ferguson, G. Morris, H. X. Hao, M. Barrett and B. Glennon, Characterization of the anti-solvent batch, plug flow and MSMPR crystallization of benzoic acid, *Chem. Eng. Sci.*, 2013, **104**, 44–54.
144. G. Y. Hou, G. Power, M. Barrett, B. Glennon, G. Morris and Y. Zhao, Development and Characterization of a Single Stage Mixed-Suspension, Mixed-Product-Removal Crystallization Process with a Novel Transfer Unit, *Cryst. Growth Des.*, 2014, **14**(4), 1782–1793.
145. G. Morris, G. Power, S. Ferguson, M. Barrett, G. Y. Hou and B. Glennon, Estimation of Nucleation and Growth Kinetics of Benzoic Acid by Population Balance Modeling of a Continuous Cooling Mixed Suspension, Mixed Product Removal Crystallizer, *Org. Process Res. Dev.*, 2015, **19**(12), 1891–1902.
146. G. Power, G. Hou, V. K. Kamaraju, G. Morris, Y. Zhao and B. Glennon, Design and optimization of a multistage continuous cooling mixed suspension, mixed product removal crystallizer, *Chem. Eng. Sci.*, 2015, **133**, 125–139.
147. Y. Yang, L. C. Song, T. Y. Gao and Z. K. Nagy, Integrated Upstream and Downstream Application of Wet Milling with Continuous Mixed Suspension Mixed Product Removal Crystallization, *Cryst. Growth Des.*, 2015, **15**(12), 5879–5885.
148. Y. Yang, L. C. Song, Y. Q. Zhang and Z. K. Nagy, Application of Wet Milling-Based Automated Direct Nucleation Control in Continuous Cooling Crystallization Processes, *Ind. Eng. Chem. Res.*, 2016, **55**(17), 4987–4996.
149. S. Y. Wong, Y. Q. Cui and A. S. Myerson, Contact Secondary Nucleation as a Means of Creating Seeds for Continuous Tubular Crystallizers, *Cryst. Growth Des.*, 2013, **13**(6), 2514–2521.
150. Y. Q. Cui, J. J. Jaramillo, T. Stelzer and A. S. Myerson, Statistical Design of Experiment on Contact Secondary Nucleation as a Means of Creating Seed Crystals for Continuous Tubular Crystallizers, *Org. Process Res. Dev.*, 2015, **19**(9), 1101–1108.

CHAPTER 2

Fundamentals of Population Balance Based Crystallization Process Modeling

BOTOND SZILAGYI[a], ANIRUDDHA MAJUMDER[b] AND ZOLTAN K. NAGY*[a,c]

[a]Davidson School of Chemical Engineering, Purdue University, West Lafayette, IN 47907-2100, USA; [b]School of Engineering, University of Aberdeen, Aberdeen AB24 3UE, UK; [c]Department of Chemical Engineering, Loughborough University, Loughborough, LE11 3TU, UK
*E-mail: zknagy@purdue.edu

2.1 Introduction

Crystallization modeling and its pertinent areas, including the model based analysis, optimization, design and control, have been intensively investigated in the past decades from the academic perspective, starting from the famous work of Hulburt and Katz,[1] with the recent increasing attention from the pharmaceutical and food industries. In addition to the process analytical technology (PAT) based crystallizer analysis, the model based investigation and *in silico* design of crystallization systems is a major trend in the crystallization community.[2] The principal crystallization process development objectives are generally related to a combination of crystal purity, size, shape, polymorphic form and yield. Since the population balance model (PBM) framework is able to consider all these properties,[3,4] it has become an

The Handbook of Continuous Crystallization
Edited by Nima Yazdanpanah and Zoltan K. Nagy

Published by the Royal Society of Chemistry, www.rsc.org

enabling technology for high performance crystallization process development and design. This chapter aims to give a brief but comprehensive overview of PBM based description of crystallization processes.

The chapter starts with an introduction describing the modeling of fundamental crystallization mechanisms, followed by a brief description of the modeling of mixed suspension mixed product removal (MSMPR) crystallizers and plug flow crystallizers (PFC). A high-level description of the most commonly used numerical solution techniques is presented, and at the end of the chapter demonstrative examples are given for advanced population balance-based crystallization models, which include topics such as modeling of solvent mediated polymorphic transformation, preferential crystallization and growth rate dispersion.

2.2 Modeling of Fundamental Crystallization Mechanisms

Crystallization is a dispersed phase process where the crystals or the solid phase are dispersed in solution as continuous phase. The various subprocesses that can be present in the crystal phase are growth, nucleation, dissolution, aggregation and breakages as shown in Figure 2.1. While nucleation, growth and dissolution involve mass transfer between the continuous and dispersed phases, aggregation and breakage are purely due to the crystal–crystal interactions. The changes in the crystal phase resulting from the aforementioned processes are tracked by the population balance equations (PBEs). On the other hand, the changes in the continuous phases (*e.g.*, concentration and temperature) are tracked by mass and energy balance equations. These changes that take place due to mass transfer and

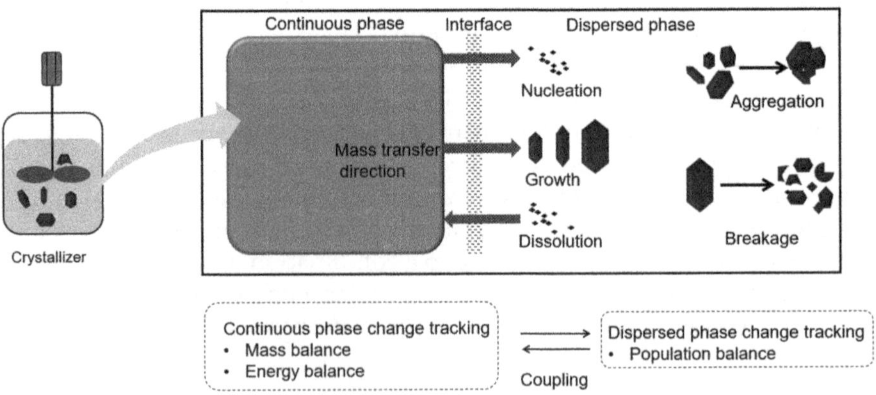

Figure 2.1 Schematic diagram showing the various subprocesses present in crystallization processes. Population balance modeling tracks the changes in the dispersed phase (*i.e.*, crystals) while mass and energy balance equations track the changes in the continuous phase (*i.e.*, solution). These two sets of equations need to be coupled for describing the crystallization process.[5]

heat transfer to the crystals and the surroundings can affect the crystal nucleation, growth and dissolution processes. This is because the driving force of the crystallization is supersaturation which is a function of concentration and temperature.

This section is aimed to provide a brief insight into the modeling of fundamental crystallization mechanisms, namely, the nucleation, growth, agglomeration and breakage of crystals. As usually all crystallization sub-processes are lumped into the same PBE, the adequate description of individual mechanisms is paramount.

2.2.1 The Supersaturation

The concepts of solubility and supersaturation are vital in developing and characterizing the behavior of crystallization systems. The solubility, also called saturation concentration (c_s), is defined as the amount of a substance (solute) that can be dissolved in a given amount of solvent at given set of thermodynamic conditions (temperature, pressure *etc.*). Under certain conditions the actual solute concentration (c) might exceed the solubility, in which case the solution is said to be supersaturated. Supersaturated solution is metastable, and given enough time, the solution will come back to its saturation concentration by crystallizing additional solute. If the actual concentration is lower than the solubility, the solution is undersaturated. Table 2.1 summarizes the widely used expressions for the supersaturation.

There are four methods that are broadly used to generate supersaturation:[6]

- *Cooling*. The solubility often increases with the temperature, therefore, supersaturation can be generated by cooling a high temperature, highly concentrated solution.
- *Evaporation*. The solute concentration can be increased by removing the solvent from the system *via* evaporation. Therefore, the solute can be concentrated by evaporation until reaching and passing the solubility.
- *Antisolvent addition*. The solubility depends on solvent nature; the solubility can significantly differ in different solvents. In this case, the solvent in which the solubility is higher is called "solvent", whereas the low solubility solvent is called "antisolvent". Preparing a highly concentrated solution in the "solvent" and then adding antisolvent to it leads to solubility decrease, hence, generates supersaturation.

Table 2.1 Widely used supersaturation expressions.

Mechanism	Model-equation	Remark
Supersaturation ratio	$S = \frac{c}{c_s}$	dimensionless. $S = 1$ indicates saturated solution
Absolute supersaturation	$\Delta c = c_s - c$	expressed in units of concentration
Relative supersaturation	$\sigma = \frac{c - c_s}{c_s} = S - 1$	dimensionless. $\sigma = 0$ indicates saturated solution

- *Chemical reaction.* Chemical reaction can be used, in which the reagents are well soluble, but the product is lowly soluble in the applied solvent system. Often fast, ionic reactions are employed, which leads to extremely high supersaturations compared to the achievable supersaturation domains by other methods. This process is also called precipitation.

The crystallization process can be conveniently described in the phase diagram. Here we use the $c - T$ phase diagram for demonstration purposes, which can easily be adapted to antisolvent crystallization. Phase diagrams are not generally used for precipitation, due to high supersaturation and extremely fast kinetics, and evaporative crystallization, where the temperature is not an adjustable parameter as the boiling point is thermodynamically linked to the solvent boiling point, concentration and pressure. In the $c - T$ phase diagram (see Figure 2.2.) there are two curves: the solubility curve and the nucleation curve. The solubility curve is determined by thermodynamics and is a function of temperature. At the solubility curve the solution is in saturation equilibrium. The nucleation curve, which is also called the metastable limit, is thought of as a region where the nucleation rate increases rapidly rather than a sharp boundary (in contrast with the solubility line). These two curves divide the phase diagram in three important zones, namely the stable zone, under the solubility line; labile zone, over the metastable limit where both spontaneous nucleation and growth occurs, and the metastable zone. The metastable zone lies between the two curves, where the existing crystals are growing and there is no nucleation.

2.2.2 Nucleation

Nucleation is the formation of a solid crystalline phase. Nucleation mechanisms are commonly lumped into one of two categories: primary and secondary nucleation.[6]

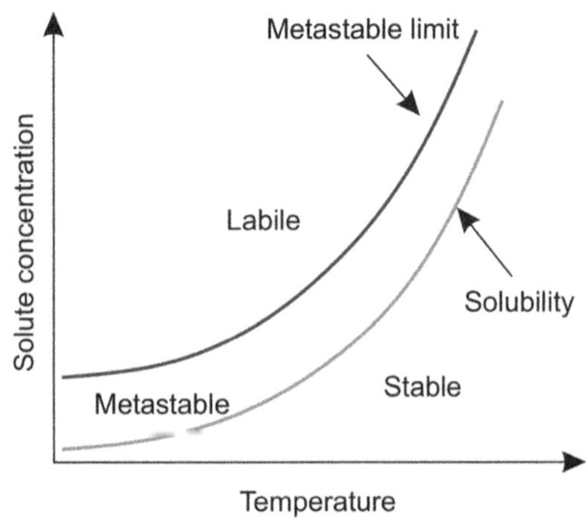

Figure 2.2 $c - T$ phase diagram with the solubility and metastable limit.

- Primary nucleation is the formation of a solid phase from a clear liquid. Primary nucleation is usually categorized as homogeneous nucleation and heterogeneous nucleation.
    - Homogeneous nucleation occurs in the pure bulk solution. It is determined by the formation of stable nuclei in a supersaturated solution, which means molecules of solute come close together to form the nuclei.[7]
    - Heterogeneous nucleation is induced by foreign surfaces such as impurities present in the solution and can become significant at much lower supersaturation levels than homogeneous primary nucleation.

The first and most famous theory to describe nucleation processes is the classical nucleation theory (CNT). CNT is based on the condensation of a vapor from a liquid, but it may be extended to crystallization from melts and solutions. The theory considers that the overall excess free energy (ΔG) between the small spherical particle (with r radius) and the solution is given by the sum of the surface excess free energy (ΔG_S) and the volume excess free energy (ΔG_V). The former is often called "surface term", and the latter "volume term". The surface term is the excess free energy between the surface of a particle and the bulk of a particle and it is a positive quantity, which is proportional to the particle surface. The volume term is the excess free energy between a very large particle ($r \rightarrow \infty$) and the bulk solution, which is a negative quantity proportional to the particle volume. The overall free energy of a particle with radius r then can be written as:

$$\Delta G = \Delta G_v + \Delta G_s = 4\pi r^2 \sigma_s + \frac{4}{3}\pi r^3 \Delta G_v \tag{2.1}$$

where σ_S (N m^{-1}) is the interfacial tension and ΔG_V (J m^{-3}) is the free energy change of transformation per unit volume. Taking the derivative of ΔG with respect to the radius, eqn (2.1) becomes:

$$\frac{\mathrm{d}(\Delta G)}{\mathrm{d}r} = 8\pi r \sigma_s + 4\pi r^2 \Delta G_v \tag{2.2}$$

Taking into account that the second term of eqn (2.2) is a negative quantity, it is obvious that eqn (2.2) has a maximum. This maximum is the critical radius (r_c) of the *nuclei*: below this size, the free energy of the cluster decreases if the radius decreases (*i.e.* the cluster dissolves), and over this radius the free energy decreases with the radius, so the cluster reduces its free energy by growth. Therefore, r_c is considered to be the smallest stable *nuclei* size under the given set of thermodynamic conditions.

$$r_c = -\frac{2\sigma_s}{\Delta G_v} \tag{2.3}$$

The critical activation energy, *i.e.* the activation energy belonging to r_c, can be expressed from eqn (2.1) and (2.3) (Figure 2.3):

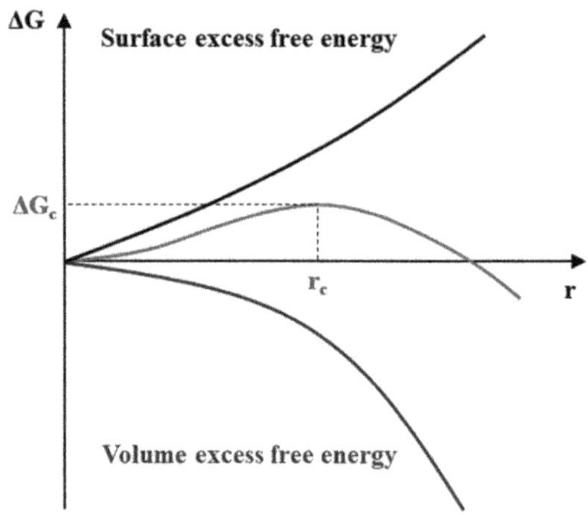

Figure 2.3 Graphical representation of the terms appearing in the CNT.

$$\Delta G_c = -\frac{16\pi\sigma_s^3}{3(\Delta G_v)^2} \tag{2.4}$$

There are numerous nucleation rate models that go beyond the assumptions of CNT. A good example is the two step nucleation model. According to the two-step mechanism, the crystalline nucleus appears inside pre-existing metastable clusters of size several hundred nanometers, which consist of dense liquid and are suspended in the solution.[8]

- Secondary nucleation is generated by the existing crystals and can be induced by:
  - contact of crystals with an external surface (impeller, wall of the vessel or another crystal), which generates fracture/attrition
  - continuous removal of dendrites due to free energy driving force
  - fluid shear
  - initial breeding or dust breeding

Secondary nucleation is generally more easily controlled than primary nucleation and is the most dominant mechanism in most industrial crystallization processes.[9] Attrition can be induced by agitation or pumping and can generate significant secondary nucleation in industrial crystallization systems. The greater the intensity of agitation the greater the rate of secondary nucleation. The modeling of nucleation is highly complex due to the variety of mechanisms, collectively termed as nucleation, and the high uncertainty associated to these processes. Table 2.2 summarizes some commonly used empirical relationships to capture nucleation kinetics for particular process conditions.

Table 2.2 Summary of commonly used empirical and semi-empirical nucleation rate models.

Mechanism	Model-equation	Remark
Homogeneous nucleation	$B = k_b \exp\left[\dfrac{-16\pi\sigma_s{}^3 v^2}{3k^3T^3\ln^2(S)}\right]$	σ_s: interfacial tension; v: molecular volume; k: Boltzmann constant
Heterogeneous nucleation	$B = k_b \exp\left[\dfrac{-16\pi\sigma_s{}^3 v^2 f(\varphi)}{3k^3T^3\ln^2(S)}\right]$	Extra factor $f(\varphi)$ is to consider foreign surface effects
Two step nucleation	$B = \dfrac{k_2 C_1 T \exp\left(-\dfrac{\Delta G_2^*}{k_B T}\right)}{\eta(C_1,T)\left[1+\dfrac{U_1}{U_0}\exp\left(\dfrac{\Delta G_C^0}{k_B T}\right)\right]}$	Two step nucleation rate model. See the literature for details[8]
Primary nucleation[a]	$B = k_b\sigma^b$	Commonly used nucleation expression
Secondary nucleation[a]	$B = k_b\sigma^b\mu_2$	μ_2: second moment (see eqn (2.46))
	$B = k_b\sigma^b\mu_3$	μ_3: third moment (see eqn (2.46))
	$B = k_b\sigma^b\mu_3^j \exp\left(-\dfrac{E_b}{RT}\right)$	E_b: activation energy; R: gas constant
	$B = k_b\sigma^b\mu_3^j N^k$	N: impeller stirring rate

[a]Both relative and absolute supersaturation can be used.

2.2.3 Growth and Dissolution

The newly born *nuclei* and existing crystals in the supersaturated solution grow with time. From the mass transfer point of view, three major successive steps are required for the crystal growth process to happen, which are illustrated in Figure 2.4:

- convective mass transport of solute molecules from the bulk solution to the boundary layer of the crystal, which in a well-mixed system is usually significantly faster than the
- conductive (diffusive) mass transport through the boundary layer to the crystal surface and
- surface integration of the molecule to the crystal lattice.

The diffusion step can be described by the diffusion equation, which, in terms of notations of Figure 2.4 has the form:

$$\dot{m}_{c,d} = \frac{dm_{c,d}}{dt} = DA\frac{c - c_i}{\delta} = k_d A(c - c_i) \tag{2.5}$$

where D (m^2 s^{-1}) is the diffusion coefficient and A (m^2) is the crystal surface. The surface integration step is characterized by:

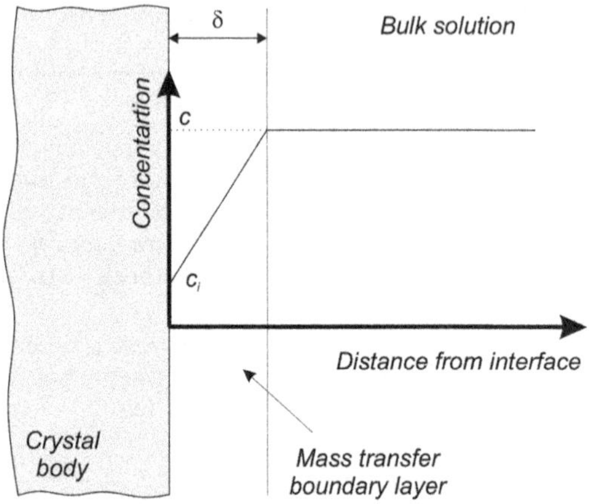

Figure 2.4 Growing crystal – solution interface.

$$\dot{m}_{c,i} = \frac{dm_{c,i}}{dt} = k_i A(c_i - c_s)^i \tag{2.6}$$

where k_i (m^{3i} $kg^{i-1}s^{-1}$) is the rate constant of surface integration and i is the supersaturation exponent. According to the equations, neither the diffusion step nor the surface reaction step proceeds as long as the solution is not supersaturated. The main issue with the practical application of eqn (2.5) and (2.6) is the difficulty of the measurement of c_i. To overcome this limitation, the overall growth rate is expressed as a function of bulk concentration and solubility:

$$\dot{m}_c = \frac{dm_c}{dt} = k_G A(c - c_s)^g \tag{2.7}$$

where k_G (m^{3g} $kg^{g-1}s^{-1}$) is an overall crystal growth rate coefficient and g (–) is the overall supersaturation exponent. In the overall growth process the slower mechanism will control the overall rate. If the growth rate is limited by mass transfer through the laminar film then the growth is said to be diffusion controlled. In this case $g = 1$. If the slowest step is the surface integration, the growth is surface integration controlled, when usually but not necessarily $g > 1$.

Eqn (2.7) describes the variation of crystal mass due to growth, however, it is often more convenient to express the growth rate in terms of the linear surface displacement velocity than in terms of rate of change of crystal mass. In this case the crystal mass and surface should be expressed as a function of linear crystal size. Assuming spherical shape with L diameter and ρ_c crystal density, eqn (2.7) can be reformulated as:

$$\frac{d\left(\rho_c L^3 \frac{\pi}{6}\right)}{dt} = k_G L^2 \pi (c - c_s)^g \Rightarrow 3L^2 \rho_c \frac{\pi}{6}\frac{dL}{dt} = k_G \pi L^2 (c - c_s)^g$$

$$\Rightarrow G = \frac{dL}{dt} = k_{g,c}(c - c_s)^g \tag{2.8}$$

Here, G (µm s^{-1}) is the linear growth rate, expressed in units of velocity and $k_{g,c}$ (µm s^{-1} × m^{3g} kg^{g-1}) denotes a combined growth rate constant, which is obtained by combining the constants of the previous equation (k_G,0.5ρ_c) into one. Eqn (2.8) is a simplified semi-empirical formula derived based on the considerations of the two-step (diffusion-integration) growth rate model.

It's worth noticing that the surface integration step has multiple mechanisms, including spiral growth, typically dominant at lower supersaturations, and surface nucleation, which typically occurs at higher supersaturation. This means that different growth rate expressions are required to accurately describe the dominant crystal growth mechanism in different supersaturation domains. There are predictive models that successfully capture these mechanistic changes in the function of supersaturation, which also enables the prediction of crystal shapes.[10]

The reverse process of the crystallization is known as dissolution, which occurs in the undersaturated region. As dissolution proceeds, the concentration of the solute increases. If given enough time at fixed conditions, the solute will eventually dissolve up to the solubility concentration or to complete dissolution. The dissolution follows similar steps to the growth, however, form mechanistic point of view, the surface disintegration step is significantly different.

Numerous empirical and semi-empirical growth and dissolution rate functions were developed and successfully applied, which are summarized in Table 2.3.

It worth noticing that the crystal growth is sensitive to certain impurities present in the system. It was observed that the impurities, sometimes even in ppm concentrations, have a significant growth rate inhibition effect. Based on the widely accepted mechanism for the growth rate inhibition the impurity adsorbs on the surface of the growing crystal and blocks the growing sites of the surface from additional solid deposition. In equilibrium, adsorption

Table 2.3 Summary of commonly used empirical growth rate models.

Mechanism	Model-equation	Remark
Size independent growth[a]	$G = k_g\sigma^g$	Commonly used growth expression
Temperature dependent	$G = k_g\sigma^g \exp\left(-\frac{E_g}{RT}\right)$	Semi-empirical model
Size dependent growth[a]	$G = k_g\sigma^g(1 + \gamma L)^p$	Size dependent model
Power law size dependent[a]	$G = k_g\sigma^g L^p$	Size dependent model
Burton-Cabrera-Frank (BCF)	$G = \frac{k_g}{k_{bcf}} S^2 \tanh\left(\frac{k_{bcf}}{\sigma}\right)$	Includes effects of surface defects
Size independent dissolution[a]	$D = k_d\sigma^g$	Commonly used dissolution expression
Size independent dissolution[a]	$D = k_d c_s \sigma$	Dissolution mechanism controlled by mass transfer

[a]Both relative and absolute supersaturation can be used.

isotherms can be applied to express the equilibrium coverage of the surface. Assuming that the Langmuir isotherm applies, the equilibrium coverage is expressed as:

$$\theta_{eq} = \frac{Kc_i}{1 + Kc_i} \quad (2.9)$$

where K (m^3 kg^{-1}) is the Langmuir constant and c_i (kg m^{-3}) is the impurity concentration. Then, the growth rate inhibition, related to the growth rate in pure media, can be written as:

$$\mathrm{Gi} = \mathrm{G}(1 - \alpha\theta_{eq}) \quad (2.10)$$

where α (−) accounts for the effectiveness of impurity on growth rate inhibition.[11] It is important to highlight that the growth rate inhibition is facet-specific, therefore growth rate inhibitors can be used to selectively block the growth rate of selected crystal facets, enabling efficient crystal shape control.[12]

2.2.4 Modeling Crystal Agglomeration

In certain crystallization systems the unification of crystals, including by flocculation, aggregation and agglomeration, occurs, which can influence the crystal size distribution (CSD), and so has to be taken into account in the model development. From a mechanistic perspective, it is important but difficult to distinguish among these mechanisms. Generally speaking, a flocculate is a group of particles that are connected by weak cohesive forces. Agglomeration is the unification of primary particles by cementation through a crystalline bridge. The formation of this bridge requires crystal growth, therefore, agglomeration occurs in supersaturated solutions and these groups are harder to destroy. The bonding forces in aggregates are stronger than in flocculates, but weaker than in agglomerates. Agglomeration is dominant under the approximately 10 μm size range, including the submicron range, and it is generally negligible for crystals exceeding 50 μm.[13] Therefore, agglomeration is naturally accompanied by high nucleation rates, typically those in reaction crystallization/precipitation.

For isotropic crystals the agglomeration does not conserve the shape, but it conserves the volume. It is a general agglomeration modeling assumption to consider the linear size of the agglomerate as the sphere-equivalent diameter. For the shape of binary agglomerates this might be a crude assumption, but for highly agglomerating systems spherical agglomerates are likely to be formed, as Figure 2.5 presents.

The rate of agglomeration between two crystals (with sizes L and λ) is described by the agglomeration function (β):

$$\beta(L,\lambda) = \alpha_0\beta_0(L,\lambda) \quad (2.11)$$

where α_0 (−) is an agglomeration constant and β_0 (s^{-1}) is the agglomeration kernel.

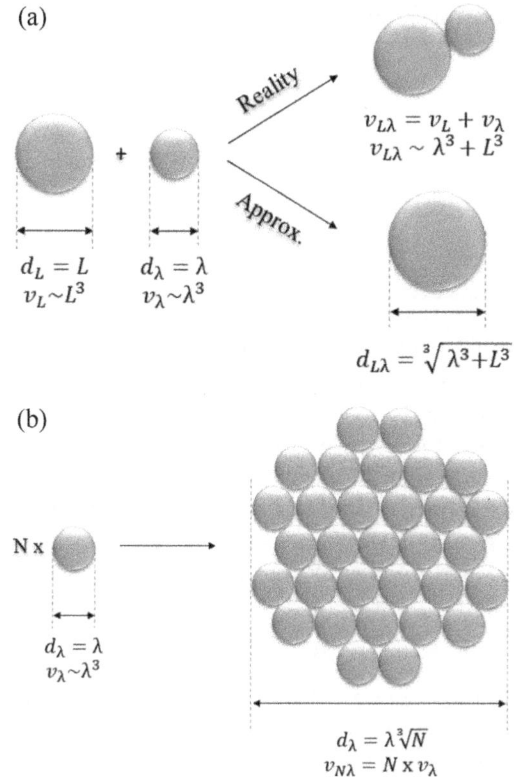

Figure 2.5 (a) Agglomeration of two primary spherical crystals leads to anisotropic agglomerate shape, whose diameter is often approximated with sphere-equivalent diameter; (b) successive agglomeration processes lead to quasi-spherical agglomerates.

The agglomeration constant depends on the hydrodynamic conditions (frequency of collisions), collision efficiencies (the likelihood that the particles will form an agglomerate after the collision) *etc.* For rigorous modeling of the agglomeration constant, hydrodynamic models are applied, which take into account the stirring conditions, drag and lift forces as well as the crystal size and shape. The description of such a model is far beyond the scope of this chapter, and it can be found in the literature.[14] From practical perspective, α_0 is often considered to be constant and it is estimated based on experimental data, coupled with a rationally chosen agglomeration kernel (Table 2.4).

It was observed that for different agglomeration systems the agglomeration rate has different size dependencies. For instance, in the case of un-stirred systems the Brownian motion is the driving force of agglomeration. However, the constant kernel is often used as an approximation of the Brownian

Table 2.4 Commonly used agglomeration kernels.

Name	$\beta_0(L,\lambda)$
Constant	1
Brownian motion	$\frac{(L+\lambda)^2}{L\lambda}$
Sum	$L^3+\lambda^3$
Hydrodynamic	$(L+\lambda)^3$
Differential force	$(L+\lambda)^2\lvert L-\lambda\rvert^2$

kernel for its simplicity. In stirred systems the agglomeration rate is known to be size dependent, which is described by the hydrodynamic kernel, which is often approximated by the sum kernel. In the case of collision due to turbulent inertia effects or velocity gradients due to differential sedimentation, the differential force kernel is applicable.[15]

The death rate of L sized crystals as a result of an agglomeration event with any other sized crystal (λ) can be described as a function of number of crystals (n_L, n_λ) and the agglomeration function:

$$D_{\text{agg}}=\beta(L,\lambda)n_L n_\lambda \tag{2.12}$$

According to the eqn (2.12), the agglomeration is a nonlinear process depending on the square of crystal number. The birth rate of crystals of linear size L by agglomeration of two crystals of volumes λ^3 and $L^3-\lambda^3$ as a function of their number $\left(n_\lambda, n_{(L^3-\lambda^3)^{1/3}}\right)$ can be described as:

$$B_{\text{agg}}=\frac{1}{2}\beta\left(\lambda,\left(L^3-\lambda^3\right)^{1/3}\right)n_\lambda n_{(L^3-\lambda^3)^{1/3}} \tag{2.13}$$

The agglomeration rates can be extended for population of crystals, characterized by a population density function $n(L,t)$.

$$\frac{\partial n(L,t)}{\partial t}=B_{\text{agg}}(L)-D_{\text{agg}}(L) \tag{2.14}$$

This formulation, taking into account the mass conservation, gives rise of integral terms in both functions:

$$D_{\text{agg}}(L)=n(L,t)\int_0^{L_{\max}}\beta(L,\lambda)n(\lambda,t)\mathrm{d}\lambda \tag{2.15}$$

$$B_{\text{agg}}(L)=\frac{L^2}{2}\int_0^L\frac{\beta\left[\left(L^3-\lambda^3\right)^{1/3},\lambda\right]}{\left(L^3-\lambda^3\right)^{2/3}}n\left(\left(L^3-\lambda^3\right)^{1/3},t\right)n(\lambda,t)\mathrm{d}\lambda \tag{2.16}$$

$B_{\text{agg}}(L)$ expresses the rate of birth of L size crystals by agglomeration of λ and $(L^3-\lambda^3)^{1/3}$ sized crystals. The rate of production is proportional with the number of these crystals within the CSD $n((L^3-\lambda^3)^{1/3},t)$ and $n(\lambda,t)$. $D_{agg}(L)$ states that if an L size crystal agglomerates with any crystal (λ), an L size crystal disappears from the system.

2.2.5 Modeling Crystal Breakage

Fragmentation of crystals might occur in certain crystallization systems, and is a collective name for all particle disintegration mechanisms, including the breakage, attrition, de-agglomeration *etc.* While the breakage can be neglected in some systems, it might play key role in the formation of CSD in others, like the binary breakage of rod-like crystals, the attrition (collision induced secondary nucleation), de-agglomeration during dissolution *etc.* The proper modeling of these fragmentation mechanisms requires mechanistic models that involve crystalline properties (shape, density, Young modulus) and solvent properties (density, viscosity) as well as stirring properties (stirring energy, circulation time). Generally speaking, the smaller the specific power input, the smaller the fragmentation rates are, but the presentation of such a complex model is beyond the scope of this subsection and it can be found in the literature.[13,16] Instead, the generally used empirical and semi-empirical breakage functions will be presented here.

The overall rate of breakage of a λ size crystal is described by the so-called selection function:

$$S(\lambda) = b_0 S_0(\lambda) \tag{2.17}$$

Where b_0 is the breakage rate constant, dependent of crystal size, shape and hydrodynamic conditions, but in practice it is often used as a fitting parameter. $S_0(\lambda)$ describes the size dependency of the selection function. Various empirical selection functions were successfully applied, which are summarized in the Table 2.5.[15,17] These functions aim to capture the experimental observation that the larger crystals break up with higher probability. The tangent hyperbolic selection function explicitly involves a critical crystal size (L_{crit}), under which the crystals don't tend to break.

The daughter crystals, formed by the breakage, can have any size between zero and the parent size with the restriction that the volume must be

Table 2.5 Commonly used breakage selection functions.

Name	$S_0(\lambda)$
Constant	1
Power law	λ^k
Exponential	$\exp(k\lambda)$
Tangent hyperbolic	$\frac{1}{2}\left[\tanh\left[k(\lambda - L_{crit})\right]+1\right]$

Table 2.6 Commonly used daughter distribution functions.

Name	$b(L/\lambda)$
Symmetric fragmentation	$2\delta\left(L-\frac{\lambda}{2^{1/3}}\right)$
Parabolic distribution	$\frac{3CL^2}{\lambda^3}+\left(1-\frac{C}{2}\right)\left(\frac{72L^8}{\lambda^9}-\frac{72L^5}{\lambda^6}+\frac{18L^2}{\lambda^3}\right)$
Mass ratio 1:2	$\delta\left(L-\lambda\left(\frac{1}{3}\right)^{1/3}\right)+\delta\left(L-\lambda\left(\frac{2}{3}\right)^{1/3}\right)$
Uniform distribution	$2/\lambda$
Attrition	$2\delta(L-\lambda)$

conserved in the breakage process. There are two limiting daughter distribution functions for binary breakage: the symmetric fragmentation, when two, equal sized crystals are produced, and attrition, when an infinitesimally small crystal is produced, and the other has the size of parent crystal. This kernel is often used to model the collision induced secondary nucleation. It is worth noticing that other distribution functions were also developed and successfully applied to crystallization processes, which are summarized in Table 2.6.[15]

The death rate of λ sized crystals by breakage, as a function of its number n_λ and the selection function $S(\lambda)$, can be calculated as:

$$D_{\text{bre}} = S(\lambda)n_\lambda \tag{2.18}$$

According to eqn (2.18), the breakage, in contrast with agglomeration, is a first order process, showing linear dependency on the particle number. Then, the formation of crystals of size L from breakage of crystals of size λ can be obtained, in the knowledge of the daughter distribution function:

$$B_{\text{bre}} = b(L/\lambda)S(\lambda)n_\lambda \tag{2.19}$$

Naturally, the breakage description can be extended to population of crystals $n(L,t)$:

$$\frac{\partial n(L,t)}{\partial t} = B_{\text{bre}}(L) - D_{\text{bre}}(L) \tag{2.20}$$

With the terms:

$$D_{\text{bre}}(L) = S(L)n(L,t) \tag{2.21}$$

$$B_{\text{bre}}(L) = \int_{\lambda}^{L_{\max}} b(L\backslash\lambda)n(\lambda,t)S(\lambda)\mathrm{d}\lambda \tag{2.22}$$

$B_{\text{bre}}(L)$ function gives the number of crystals of size L produced from the breakage of a $\lambda > L$ crystal, which depends on the rate of breakage of λ crystals

$S(\lambda)$, and on the probability that the λ crystal will form an L size fragment $b(L/\lambda)$. The more λ size crystals in the system, the more L will be produced $n(\lambda,t)$. D_{bre} is expressed in terms of breakage probability of L size crystals ($S(L)$) and their actual number within the population $n(L,t)$.

2.3 Modeling the MSMPR Crystallizer

The continuous tank crystallizer is an important unit operation in food, pharmaceutical and chemical industries. The continuous mixed suspension mixed product removal (MSMPR) crystallizer is a widely used modeling approximation to describe continuous crystallization processes with unclassified product removal. The MSMPR concept is the analogue of the continuous stirrer tank reactor (CSTR) from chemical reactor engineering, with similar modeling assumptions:

- the crystallizer is perfectly mixed in all three (micro, meso and macro) scales,
- the temperature field is homogeneous in the slurry,
- the product removal is unclassified and the product stream has identical thermodynamic properties with the crystallizer slurry.

This section presents and discusses the derivation of a cooling MSMPR model-equation system, although the model can be easily adapted to other types of crystallization process. The scheme of a continuous jacketed MSMPR crystallizer is illustrated in Figure 2.6.

Considering a jacketed crystallizer, which is the typical cooling strategy for this equipment, the state variables to be modeled are the system volume (mass), solute concentration, crystallizer temperature, coolant temperature

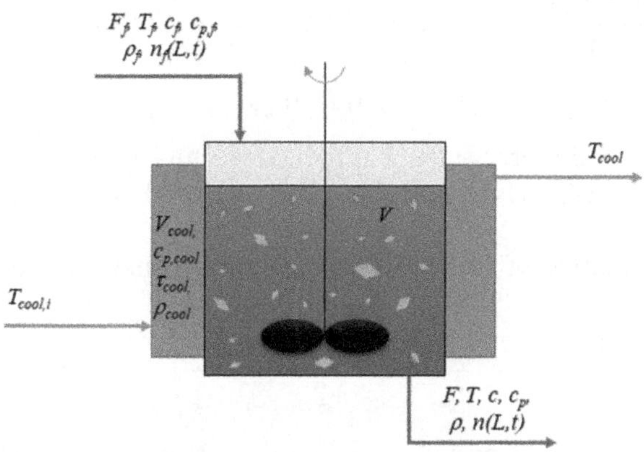

Figure 2.6 Scheme of a continuous jacketed MSMPR crystallizer with the most important process variables.

and the CSD. The crystallizer mass balance, describing the mass accumulation in the MSMPR, can be written as:

$$\frac{d(V\rho)}{dt} = F_f\rho_f - F\rho \tag{2.23}$$

The initial condition $V(t = 0) = V_0$ gives the volume of slurry (m^3) being initially in the crystallizer, subscript *f* stands for the feeding stream, F ($m^3\ s^{-1}$) is the flowrate and ρ ($kg\ m^{-3}$) is the density.

The component balance equation for the crystallizing compound, taking into account the liquid–solid mass transfer generated by nucleation (B – with constant *nuclei* size L_n) and growth (G), as well as the impact of feeding and evacuation streams is:

$$\frac{d(Vc)}{dt} = -V\rho_c k_V \left[BL_n^{\ 3} + 3G\int_0^{L_{max}} L^2 n(L,t)\,dL \right] + F_f c_f - Fc \tag{2.24}$$

The initial condition $c(t=0) = c_0$ is for the initial concentration, ρ_c is the crystal density and k_V is the volumetric shape factor of the crystal, which is a shape-specific constant, used to relate the crystal volume (v_c) to the linear crystal size as:

$$v_c = k_V L^3 \tag{2.25}$$

The crystallization is associated with the release of the latent heat of crystallization (ΔH_c). Hence, the heat of crystallization needs to be included into the energy balance. Due to the homogeneous temperature field MSMPR assumption, it is considered that the heat of crystallization is absorbed by the whole mass of suspension. The energy balance takes into account the impact of feeding and evacuation streams, the heat exchange between the crystallizer slurry and coolant as well as the heat of crystallization:

$$\frac{d(V\rho c_p T)}{dt} = -\Delta H_c V\rho_c k_V \left[BL_n^{\ 3} + 3G\int_0^{L_{max}} L^2 n(L,t)\,dL \right] + F_f\rho_f c_{p,f} T_f - F\rho c_p T - UA(T - T_{cool}) \tag{2.26}$$

where c_p ($J\ kg^{-1}K^{-1}$) denotes the specific heat of the slurry, T (°C) is the temperature and UA ($J\ s^{-1}K^{-1}$) is the product of the overall heat transfer coefficient and heat transfer area.

The energy balance of the coolant, assuming constant coolant volume, takes the form:

$$\frac{dT_{cool}}{dt} = \frac{T_{cool,i} - T_{cool}}{\tau_{cool}} + \frac{UA}{V_{cool}\rho_{cool}c_{p,cool}}(T - T_{cool}) \tag{2.27}$$

subject to the $T(t = 0) = T_0$ and $T_{cool}(t = 0) = T_{cool,0}$ initial conditions, which are the jacket and crystallizer temperatures at the initial moment of time. For the description of particle population, the monovariate population density function (often referred to as CSD) is used $n(L,t)$ (# m^{-1} m^{-3}), which gives the number of crystals within the L, $L + \mathrm{d}L$ size range at t time moment in unit volume of suspension. The CSD is governed by the population balance equation (PBE), which, assuming nucleation with constant nucleon size, growth, breakage and agglomeration mechanisms, takes the form:

$$\frac{\partial[Vn(L,t)]}{\partial t} + V\frac{\partial[Gn(L,t)]}{\partial L} = V\left[B\delta(L-L_n) + B_{\mathrm{agg}} - D_{\mathrm{agg}} + B_{\mathrm{bre}} - D_{\mathrm{bre}}\right] + F_{\mathrm{f}}n_{\mathrm{f}}(L,t) - Fn(L,t) \tag{2.28}$$

with the $n(L,t = 0) = n_0(L)$ initial, and $\lim_{L\to\infty}n(L,t) = 0$ boundary conditions. The initial condition gives the initial distribution in the crystallizer (seeds), whereas the boundary condition states that the crystals have finite size. $n_f(L,t)$ denotes the CSD in the feeding stream: it is 0 if the crystallizer is fed with liquid only, but $n_f(L,t)$ can represent a continuous seeding CSD of an upstream crystallizer. B_{agg} and B_{bre} are birth (source) functions, which describe the rate of production of new crystals of size L by agglomeration and breakage. D_{agg} and D_{bre} stands for death (sink) functions for the rate of consumption of crystals of size L by agglomeration and breakage. The commonly used nucleation, growth agglomeration and breakage functions are described in the first section of this chapter. The set of ordinary and partial differential eqn (2.23)–(2.26) represents the closed model of the continuous cooling MSMPR crystallizer.

There are numerous practical simplifications that can be made to simplify further the model equations.

Assumption 1. *Constant volume, density and specific heat.*

For instance, assuming constant suspension density, specific heat and crystallizer working volume, which is a reasonable approximation of many pharmaceutical crystallization systems, the characteristic differential equations for the crystallizer are greatly simplified:

$$\frac{\mathrm{d}c}{\mathrm{d}t} = -\rho_{\mathrm{c}}k_{\mathrm{V}}\left[BL_n{}^3 + 3G\int_0^{L_{\max}} L^2 n(L,t)\mathrm{d}L\right] + \frac{c_{\mathrm{f}} - c}{\tau} \tag{2.29}$$

$$\frac{\mathrm{d}T}{\mathrm{d}t} = -\frac{\Delta H_{\mathrm{c}}\rho_{\mathrm{c}}k_V}{\rho c_{\mathrm{p}}}\left[BL_n{}^3 + 3G\int_0^{L_{\max}} L^2 n(L,t)\mathrm{d}L\right] + \frac{T_{\mathrm{f}} - T}{\tau} - \frac{UA}{V\rho c_{\mathrm{p}}}(T - T_{\mathrm{cool}}) \tag{2.30}$$

$$\frac{\partial n(L,t)}{\partial t} + \frac{\partial[Gn(L,t)]}{\partial L} = B\delta(L-L_n) + B_{\mathrm{agg}} - D_{\mathrm{agg}} + B_{\mathrm{bre}} - D_{\mathrm{bre}} + \frac{n_{\mathrm{f}}(L,t) - n(L,t)}{\tau} \tag{2.31}$$

where $\tau = V/F$ denotes the mean residence time. For a process simulation, the model-equations are solved simultaneously, with one of the suitable numerical solution methods presented in the fifth subsection of this chapter.

Assumption 2. *Steady-state operation with nucleation and growth.*

Continuous crystallizers are most often operated in steady-state mode. Assuming stable operation, in steady-state all state variables are constant in time ($n(L,t) \rightarrow n(L)$). With negligible agglomeration and breakage, no crystals in the feeding stream, and assuming size independent growth, eqn (2.31) reduces to the classical characteristic MSMPR equation:

$$G\frac{\mathrm{d}n(L)}{\mathrm{d}L} = -\frac{n(L)}{\tau} \tag{2.32}$$

The nucleation is taken into account as an additional, traditional boundary condition of eqn (2.32):

$$\lim_{L\to 0}\frac{\mathrm{d}n(L)}{\mathrm{d}t} = \mathrm{B} \Rightarrow \lim_{L\to 0}\left[\frac{\mathrm{d}n(L)}{\mathrm{d}L}\frac{\mathrm{d}L}{\mathrm{d}t}\right] = Gn(0) \tag{2.33}$$

where $n(0)$ denotes the number of zero sized particles (nuclei). Then, the variables of eqn (2.32) can be separated and an analytical solution for the differential equation can be obtained:

$$n(L) = -n(0)\exp\left(-\frac{L}{G\tau}\right) \tag{2.34}$$

Therefore, the semi logarithmic CSD plot can be used to simultaneously determine the nucleation and growth rates, as Figure 2.7 presents.

A linear population density function is obtained in the MSMPR if and only if the modeling assumption holds for the investigated crystallization system (size independent growth, negligible breakage and agglomeration *etc.*). Practically, the CSD often deviates from the ideal linear form, but this deviation enables the underlying crystallization mechanisms to be inferred.[18] An example is presented in Figure 2.8.

We would like to remind the reader that the model equations were derived for a cooling crystallizer but they are not restricted to systems with that particular method of producing supersaturation.

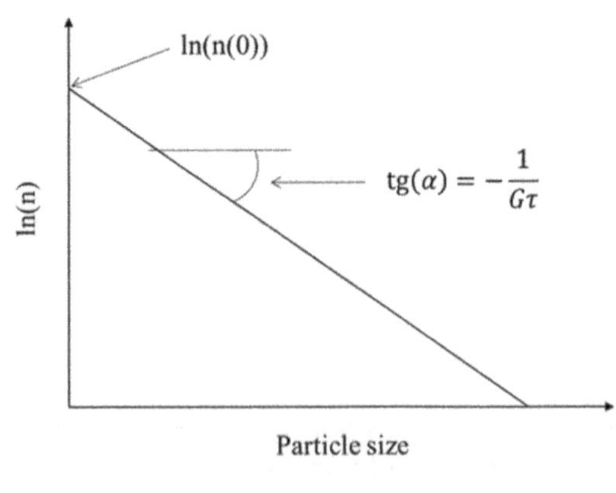

Figure 2.7 Nucleation and growth kinetics determination from the steady-state population density function of MSMPR crystallizers.

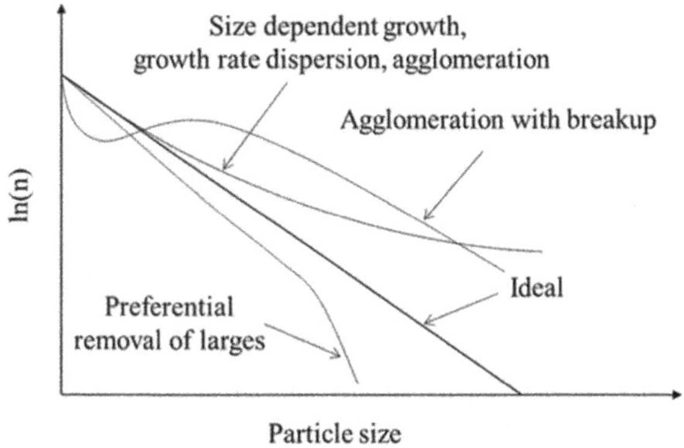

Figure 2.8 Typical deviations from the ideal MSMPR behavior.

2.3.1 MSMPR Crystallizer Configurations

MSMPR crystallizers have gained significant ground in the chemical, food and pharmaceutical industries in the last century due to their simple design and robust operability. However, in the pharmaceutical applications, MSMPR crystallizers often fail to deliver the rigorous CSD requirements, as they inherently produce crystals over a broad size range. In order to overcome this limitation, crystallization systems based on MSMPR cascades were introduced and optimized.[19] The MSMPR cascade design consists of various stages where the operating conditions are chosen to promote different mechanisms (*i.e.* nucleation, dissolution, or growth). In the first MSMPR nucleation occurs, whereas the subsequent crystallizers are designed to promote the crystal growth, and potentially also the dissolution of the fines. Figure 2.9 presents three MSMPR cascades with identical combined volumes.

Each individual MSMPR of the cascade is modeled with the equations presented in the previous section, taking into account that the feeding stream of a crystallizer has identical physical-chemical properties with the corresponding upstream crystallizer.

In order to illustrate the impact of multiple MSMPR stages on the product CSD, a numerical simulation has been carried out for the configurations presented in Figure 2.9. The simulated crystallization system consists of the following solubility and kinetic equations:

$$c_s(T) = \sum_{i=0}^{2} a_i T^i \tag{2.35}$$

$$B = k_b(c - c_s)^b$$

$$G = k_g(c - c_s)^g$$

In the simulations the kinetic and solubility data was taken from the literature,[20] which is listed, together with the process parameters, in Table 2.7.

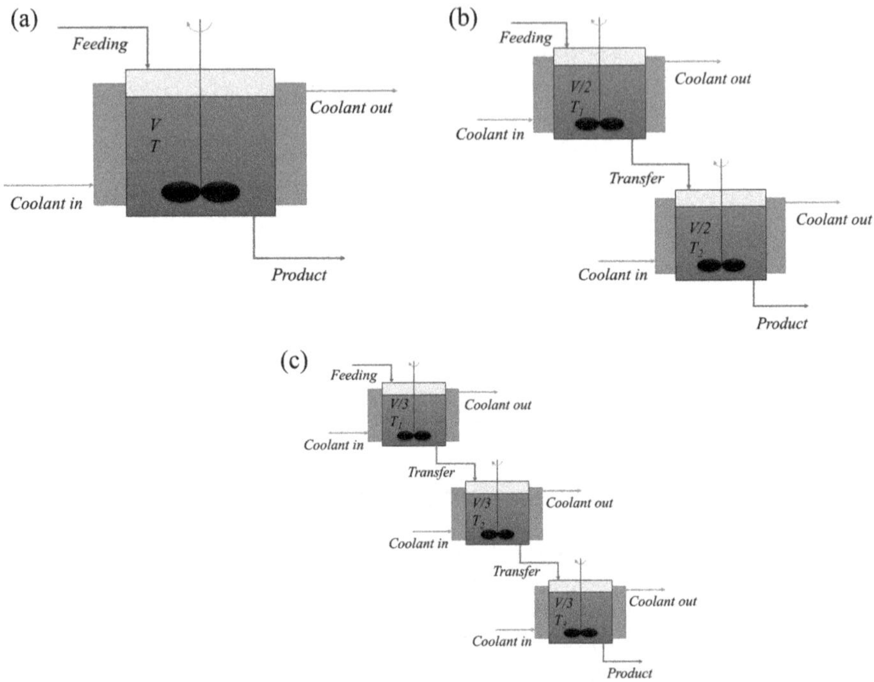

Figure 2.9 Illustration of various MSMPR crystallizer configurations involving (a) one, (b) two and (c) three MSMPR stages with identical combined volumes.

The population and mass balance equations were solved using the HR FVM (see Subsection 2.5), and the steady state CSDs of the product stream of three configurations are plotted in Figure 2.10. According to Figure 2.9 and Table 2.7, the combined volume, mean residence time and yield of all combinations are identical. It worth noting that the one MSMPR produced the expected linear number based density function, but the MSMPR cascades produced a peak in the number based particle density function. The difference between the product CSDs is shown better in the normalized volume based CSD graphs, which demonstrate clearly that the MSMPR cascade produces significantly narrower CSD than a single stage MSMPR.

Practically, the operating conditions as well as the number of required stages are optimized with respect to various quality related, technical and economic objectives.

2.4 Modeling the Tubular Crystallizer

Although MSMPR crystallizers remain the most utilized platform for continuous crystallization, largely due to familiarity in terms of operation and control and availability of equipment, they possess some weaknesses for

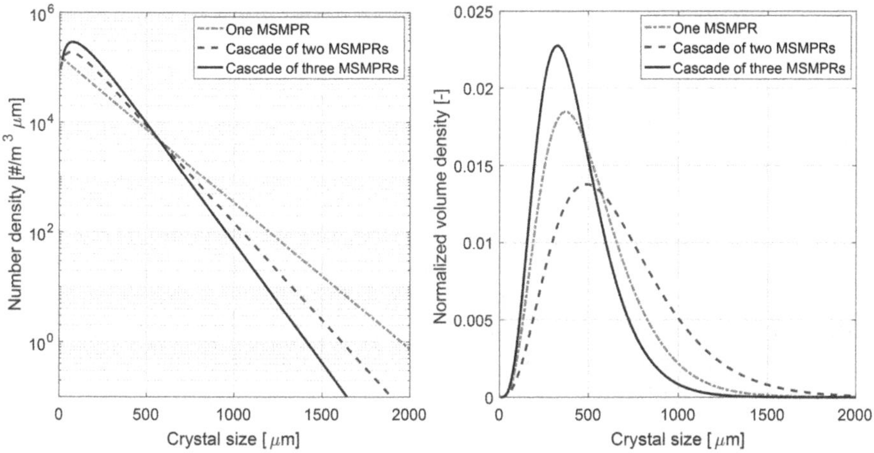

Figure 2.10 Comparison of number and volume based CSDs produced by an MSMPR crystallizer as well as MSMPR cascades consisting of two and three MSMPR units. The combined volume, mean residence time and yield is identical in all configurations.

Table 2.7 Process and kinetic parameters applied in the simulation of MSMPR cascade configurations.

Parameter (unit of measure)	Value
a_0 (g g^{-1})	5.48×10^{-3}
a_1 ((g g^{-1}) °C^{-1})	-1.93×10^{-4}
a_2 ((g g^{-1}) °C^{-2})	7.09×10^{-6}
$\ln(k_b)$ (−)	11.84
b (−)	2.63
$\ln(k_g)$ (−)	−14.90
g (−)	1
ρ_c (kg m^3)	1412
k_V (−)	0.32
τ (min) – 1 MSMPR	60
τ (min) – cascade of 2 MSMPRs	30
τ (min) – cascade of 3 MSMPRs	20
c_f (g g^{-1})	$c_s(T_f)$
T_f(°C)	80
T(°C) – 1 MSMPR	20
T(°C) – cascade of 2 MSMPRs	{50, 20}
T(°C) – cascade of 3 MSMPRs	{60, 40, 20}

the application of crystallization. These include non-uniform temperature control, highly localized shear regions due to agitators, challenges with handling solids at transfer lines, and non-linear scalability.[21] Plug flow crystallizers (PFCs) offer advantages in each of the challenges faced by MSMPR crystallizers and, thus, are promising platforms for applying continuous crystallization. In a PFC, solution is fed in at the inlet and supersaturation is generated by cooling or antisolvent addition as the solution moves through the tube. Due to the generation of supersaturation, crystallization takes place, and the product crystals are collected at the outlet. The key aspects of an ideal PFC are:

- perfect mixing in the radial direction, and
- no mixing in the axial direction.

These two assumptions lead to an identical residence time for all the crystals. In theory, near plug flow conditions can be achieved using a series of MSMPRs with the number of reactors approaching infinite. The drawbacks of this include higher overall running costs. Near plug-flow conditions can be obtained in a tubular type reactor at turbulent flow requiring significantly high flow rates, which can lead to very long reactors for maintaining the required residence time and therefore large capital costs. However, variants of tubular reactors that have provisions for static mixture[22] or oscillatory flow[21,23] can overcome the requirement of a very long tube.

The PBE can be used for modeling of crystallization process in PFC. Here the CSD is a function of both internal coordinates (*e.g.*, crystal size) and external coordinates (*e.g.*, location in physical space). Assuming that the CSD n is a function of one internal coordinate L and one external coordinate x, then the PBE becomes:

$$\frac{\partial}{\partial t}(n(L,x,t))+\frac{\partial}{\partial L}(Gn(L,x,t))+\frac{\partial}{\partial x}(un(L,x,t))=S \tag{2.36}$$

With respect to the boundary and initial conditions:

$$Gn(L,x,t)|_{L=0}=B_0$$
$$n(L,x,t)|_{x=0}=n_{\text{seed}}(L) \tag{2.37}$$

where G (μm s^{-1}) is the growth rate, μ (m s^{-1}) is the mean flow velocity, $n_{\text{seed}}(L)$ (# μm^{-1} m^{-3}) is the seed distribution at the feed, S (# μm^{-1} m^{-3} s^{-1}) is the source term accounting for aggregation and breakages, and B_0 (# m^{-3} s^{-1}) is the nucleation rate. The details of the derivation of the PBE can be found in Appendix A.1. The first term in the left hand side of eqn (2.36) is the rate of change of the CSD along the crystallizer, the second term is the contribution due to crystal growth, the third term is the contribution due to convection resulting from fluid flow, and the term on the right hand side is the source term accounting for aggregation and breakage. If there is significant mixing in the axial direction, it can cause residence time distribution (RTD) of the crystals in the slurry leading to the broadening of the product CSD. In order to capture this phenomenon, a dispersion model that incorporates a term for dispersion in the axial direction can be used. The PBE shown in eqn (2.36) without the source terms for aggregation and breakage can be written as:

$$\frac{\partial}{\partial t}(n)+\frac{\partial}{\partial L}(Gn)+\frac{\partial}{\partial x}(un)=D\frac{\partial^2}{\partial x^2}(n) \tag{2.38}$$

where D (m^2 s^{-1}) is the axial diffusivity. However, it has been found that axial dispersion is often not sufficient to describe the experimentally observed broadening of the product CSD.[22] In such a case, a dispersion term that takes into account the growth rate dispersion can be incorporated as follows.

$$\frac{\partial}{\partial t}(n)+\frac{\partial}{\partial L}(Gn)+\frac{\partial}{\partial x}(un)=D_{\mathrm{G}}\frac{\partial^2}{\partial L^2}(n) \tag{2.39}$$

where D_{G} (μm^2 s^{-1}) is growth diffusivity for the random fluctuations of the growth rate of the crystals, which is discussed in the last section of this chapter.

This PBE needs to be supplemented with appropriate mass and energy balance equations. Assuming that no crystallizer fouling or encrustation takes place and crystal volume is negligible compared to the liquid volume, the mass balance equation can be written as:

$$\frac{\partial}{\partial t}(c)+\frac{\partial}{\partial x}(uc)=-\rho_{\mathrm{c}}k_{\mathrm{V}}\frac{\partial}{\partial t}(\mu_3) \tag{2.40}$$

Where k_{V} (–) is the volume shape factor for the crystals, ρ_{c} (kg m^{-3}) is the crystal density, and μ_3 (μm^3) is the third moment of the CSD which can be considered as a measure of crystal volume. The derivation of the population balance, mass balance and energy balance equations can be found in the Appendices. In eqn (2.40), the first term denotes the rate of change in concentration along the PFC, the second term is the advection term which takes into account the change in concentration due to fluid flow and the term on the right hand side is the depletion term due to crystal growth.

2.4.1 Case Study: PFC With Multiple Feeding Points

In this section, we present a case study on antisolvent crystallization in a tubular crystallizer. Antisolvent is a fluid which, when mixed with the solution of an active pharmaceutical ingredient (API), reduces the solubility of the API in the mixed solvent. Antisolvent crystallization is particularly important for separation of substances which are temperature sensitive. The system considered here is the crystallization of flufenamic acid in ethanol (solvent) and water (antisolvent). The case study is based on the work by Alvarez and Myerson.[22] The PFC consists of four glass reactor modules each of which has a volume of 76 cm^3 (600 mm long, 12.7 mm internal diameter). The crystallizer is maintained at 25°C using a water bath temperature control. The feed solution contains 1.4 mg mL^{-1} of flufenamic acid in ethanol and the feed rate is maintained at 100 mL min^{-1}. One of the objectives of this study is to investigate the effect of number of injection points of the antisolvent on the CSD. Thus, the total antisolvent flowrate 200 mL min^{-1} is distributed equally in 1,2,3 and 4 injection points, *e.g.*, one flow of 200 mL min^{-1} for one injection point, two flows of 100 mL min^{-1} for two injection points and so on (Figure 2.11). The solubility of flufenamic acid, and the nucleation and growth rate equations are described by

$$c(\text{mg mL}^{-1}) = 336.0\exp(-0.108(\text{volume \% antisolvent}))$$

$$G = k_g(c - c_s)^g$$

$$B = k_b(c - c_s)^b \tag{2.41}$$

The experimentally determined kinetic parameters are shown in Table 2.8.

The crystallization process was simulated at steady-state. It has been shown that GRD is necessary to explain the observed broadening of the product CSD. In such case, the PBE in eqn (2.39) becomes:

$$\frac{\partial}{\partial L}(Gn) + \frac{\partial}{\partial x}(un) = D_G \frac{\partial^2}{\partial L^2}(n) \tag{2.42}$$

In dimensionless form, the above equation can be written as.

Table 2.8 Estimated nucleation and crystal growth kinetic parameters and 95% confidence intervals for flufenamic acid at different number of antisolvent addition points.[22]

# Addition points	$k_g(\times 10^{-7}\text{m s}^{-1})$	g	$k_b(\times 10^{8}\#\,\text{m}^{-3}\text{s}^{-1})$	b
1	8.2 ± 0.64	1.1 ± 0.12	1.5 ± 0.12	2.0 ± 0.13
2	9.8 ± 40	1.0 ± 0.02	1.4 ± 0.11	2.0 ± 0.09
3	9.8 ± 1.01	1.1 ± 0.08	1.3 ± 0.15	2.1 ± 0.14
4	9.9 ± 0.58	1.1 ± 0.08	1.3 ± 0.08	2.1 ± 0.09

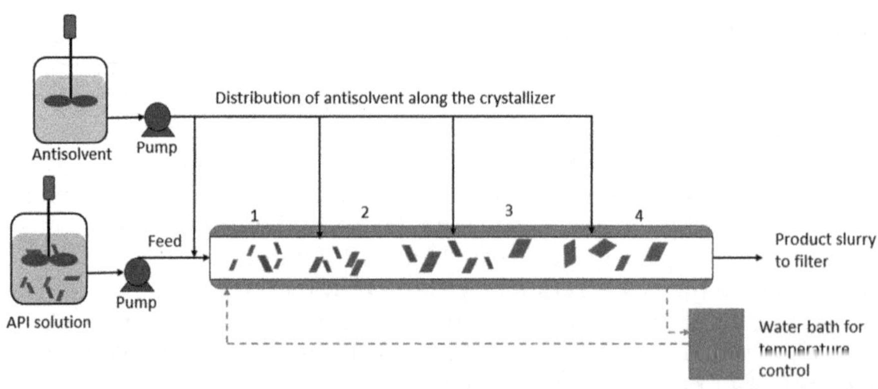

Figure 2.11 Schematic diagram of the PFC with multiple antisolvent addition points.

$$\frac{\partial \tilde{n}}{\partial \tilde{L}} + \frac{\partial \tilde{n}}{\partial \tilde{x}} = \frac{1}{\mathrm{Pe_G}} \frac{\partial^2 \tilde{n}}{\partial^2 \tilde{L}} \quad (2.43)$$

where $\tilde{n}$, $\tilde{L}$ and $\tilde{x}$ are the dimensionless variables for population density, crystal size and axial distance, respectively, defined as

$$\tilde{n} = \frac{n}{n_0}; \tilde{L} = \frac{L}{G_0 \tau}; \tilde{x} = \frac{x}{\bar{x}} \quad (2.44)$$

where n_0 (# µm^{-1} m^{-3}) is the initial nuclei density, G_0 (µm s^{-1}) is the initial crystal growth rate, τ (s) is the residence time and $\bar{x}$ (m) is the effective length of the crystallizer. $\mathrm{Pe_G}$ is the Péclet number for size dispersion resulting from growth rate fluctuation, defined as

$$\mathrm{Pe_G} = \frac{G\bar{L}}{D_G} \quad (2.45)$$

$\mathrm{Pe_G}$ is a dimensionless number which provides information on the relative strength of the crystal growth rate and dispersion due to growth rate fluctuation. If $\mathrm{Pe_G}$ is large, then the relative importance of the dispersion is low and *vice versa*. By fitting the experimentally obtained CSD with the simulation results, the value of $\mathrm{Pe_G}$ is found to be 5. In Figure 2.12 comparison of the

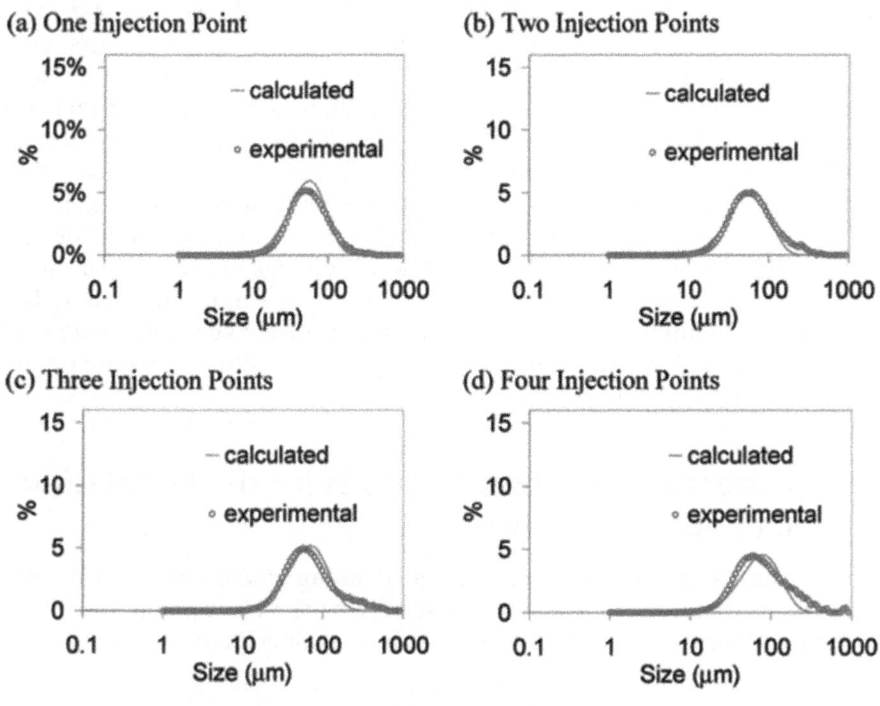

Figure 2.12 Comparison of calculated and experimental crystal size distribution for flufenamic acid using the growth rate dispersion model ($\mathrm{Pe_G}$ = 5) for various injection points ((a)–(d)). Reprinted from ref. 22 with permission of the American Chemical Society, Copyright 2010.

simulation and experimental results are shown for 1–4 injection points for antisolvent. As can be seen, a good agreement is found between the simulation and experimental data. In particular, the location of the peak and width of the final CSD is captured reasonably well. The mean crystal size is found to be 67, 80, 94 and 106 μm for 1, 2, 3 and 4 points of antisolvent addition, respectively. For one injection point where all the antisolvent is added at the inlet, a large number of small crystals is created due to nucleation leading to smaller mean size of the product crystals. For two injection points, 50% of the antisolvent is added in the inlet. This results in lower supersaturation compared to the one injection point and therefore fewer number of small crystals were generated due to nucleation. The remaining 50% antisolvent is added to the next addition point. This addition generates supersaturation which is mostly consumed by growth of the smaller crystals resulting in larger crystals compared to the previous case. For three and four injection points, similar mechanisms are observed by Alvarez and Myerson.[22] Ridder et al.[24] carried out optimization studies for antisolvent crystallization of flufenamic acid using the same kinetic parameters but without considering GRD. They found that equal distribution of the antisolvent among the addition points does not necessarily provide the optimal supersaturation profile along the PFC. In order to maximize the average crystal size and minimize the coefficient of variation (a measure of the variability of crystal size), the optimum antisolvent profile is found to be such that, at the first segment, about 30% of the total antisolvent is added, which generates enough supersaturation so that nucleation occurs. At the second segment almost no antisolvent is added so that the crystals from the first segment can grow in moderate supersaturation without further nucleation. In the subsequent two segments, the remaining 30% and 40% of the total antisolvent is added to facilitate the further growth of the crystals. Su *et al.*[25] investigated the design and optimization of the antisolvent crystallization of paracetamol in acetone with water as antisolvent in a PFC. They arrived at the similar conclusion that there are optimum locations and amount of antisolvent additions that avoid excessive nucleation and promote crystal growth resulting in large average crystal size with less variability.

2.5 Numerical Solution Methods for the Population Balance Equations

Various numerical methods were developed and/or applied to solve the PBE. This subsection aims to give a brief overview of the most popular solution techniques, focusing, but not limiting to the one dimensional PBE eqn (2.31).

2.5.1 Moment Based Methods

The standard method of moments (SMOM) is the most widely used solution technique of the PBEs,[26] which relies on the moment transformation:

$$\mu_k = \int_0^\infty L^k n(L,t)\mathrm{d}L, k = 0,1,2\ldots \tag{2.46}$$

where μ_k (μm^k) denotes the k-th moment of the distribution. The first four moments (k = 0, 1, 2, 3) have physical meaning, being proportional, in row, with the specific number, length, surface area and volume of particles. Superior moments can also be computed, but no physical meaning can be associated to them.

Applying the model transformation rule eqn (2.46) on the PB eqn (2.31).

$$\begin{aligned}&\int_0^\infty L^k \frac{\partial n(L,t)}{\partial t}\mathrm{d}L + \int_0^\infty L^k \frac{\partial[Gn(L,t)]}{\partial L}\mathrm{d}L\\ &= \int_0^\infty L^k B\delta(L-L_n)\mathrm{d}L + \int_0^\infty L^k B_{\mathrm{agg}}\,\mathrm{d}L - \int_0^\infty L^k D_{\mathrm{agg}}\,\mathrm{d}L\\ &+\int_0^\infty L^k B_{\mathrm{bre}}\,\mathrm{d}L - \int_0^\infty L^k D_{\mathrm{bre}}\,\mathrm{d}L + \int_0^\infty L^k \frac{n_f(L,t)-n(L,t)}{\tau}\mathrm{d}L\end{aligned} \tag{2.47}$$

Depending on the form of kinetic functions applied in eqn (2.47), discussed in the first section of this chapter, the integrals might or might not be evaluated,[4] which delimits the applicability of the SMOM for numerous practical problems. For nucleation and growth processes eqn (2.47) simplifies to:

$$\begin{aligned}\frac{\mathrm{d}\mu_k}{\mathrm{d}t} &= \int_0^\infty L^k B\delta(L-L_n)\mathrm{d}L - \int_0^\infty L^k \frac{\partial[Gn(L,t)]}{\partial L}\mathrm{d}L\\ &+\int_0^\infty L^k \frac{n_f(L,t)-n(L,t)}{\tau}\mathrm{d}L\end{aligned} \tag{2.48}$$

Which, assuming size independent growth, applying the moment transformation rule eqn (2.46) leads to the system of moment equations:

$$\frac{\mathrm{d}\mu_k}{\mathrm{d}t} = BL_n{}^k + kG\mu_{k-1} + \frac{\mu_{f,k}-\mu_k}{\tau} \tag{2.49}$$

Eqn (2.49) is an ordinary differential equation, which can be solved simultaneously with the mass and energy balance equations. Although they are averaged quantities, the moments enable the calculation of numerous relevant statistical properties of the particle population, as summarized in Table 2.9.

Table 2.9 gives various mean sizes and distributional properties, but the exact reconstruction of the CSD is not possible. However, various methods have been developed to approximate the original CSD based on its leading moments.[3]

Even though the integrals of eqn (2.47) can be evaluated, the resulting equations system might not be closed. A typical example is the size dependent growth:

$$G(L) = GL^j \tag{2.50}$$

Table 2.9 Statistical properties of the particle population calculated from the moments of the distribution.

Property	Expression in terms of moments
Number based mean size	μ_1/μ_0
Sauter mean diameter	μ_3/μ_2
Volume mean diameter	μ_4/μ_3
Variance (of equivalent normal PDF)	$\sqrt{\frac{\mu_2}{\mu_0}-\left(\frac{\mu_1}{\mu_0}\right)^2}$
Coefficient of variation	$\sqrt{\frac{\mu_0\mu_2}{\mu_1^2}-1}$

In this case, the moment equation becomes:

$$\frac{d\mu_k}{dt}=BL_n^k+\int_0^\infty kL^{k-1}GL^j n(L,t)dL+\frac{\mu_{f,k}-\mu_k}{\tau}=BL_n^k+kG\mu_{k-1+j}+\frac{\mu_{f,k}-\mu_k}{\tau} \tag{2.51}$$

It is obvious that the moment equations can be solved if and only if $j=0$ or 1. However, in the applied side, no such restrictions exist for the value of size exponent. This is the so-called closure problem of the SMOM. The most generic solution to overcome the closure problem is the quadrature approximation of the population density function:

$$\tilde{n}(L,t)=\sum_{i=1}^{I}w_i(t)\delta[L-L_i(t)] \tag{2.52}$$

where i denotes the number of points used in the approximation, w_i is a quadrature weight and L_i is a quadrature abscissa. Applying the quadrature approximation eqn (2.52) to the moment transformation eqn (2.46), the moments can be approximated in terms of quadrature weights and abscissas:

$$\mu_k=\int_0^\infty L^k n(L,t)dL\cong\int_0^\infty L^k\sum_{i=1}^{I}w_i(t)\delta[L-L_i(t)]dL=\sum_{i=1}^{I}w_i(t)L_i^k(t) \tag{2.53}$$

Using the quadrature approximation, the system of moment equations can be approximated, and this is known as the quadrature method of moments (QMOM).[27] Then, using the simplified notations for better readability $w_i(t)\rightarrow w_i$ and $L_i(t)\rightarrow L_i$, eqn (2.51) becomes:

$$\frac{d\mu_k}{dt}\cong BL_n^k+kG\sum_{i=1}^{I}w_iL_i^{k-1+j}+\frac{1}{\tau}\left(\sum_{i=1}^{I}w_{f,i}L_{f,i}^k-\sum_{i=1}^{I}w_iL_i^k\right) \tag{2.54}$$

According to eqn (2.54) it is obvious that the closure problem doesn't exist anymore in the QMOM. Various QMOM implementations were developed[28] and the QMOM was successfully extended to aggregation-breakage problems.[15] Generally speaking, the QMOM provides high accuracy even with reduced number of points (i is generally < 5).

A major common limitation of all moment based methods is that they are not able to accurately calculate the CSD. If accurate CSD calculation is required, another numerical solution technique has to be applied. In addition, the implementation of dissolution is difficult because of the zeroth moment (crystal number): the zeroth moment is decreasing only if there are crystals at the left boundary of the crystal size domain, which can disappear from the system. This information is however not available, since the CSD cannot be exactly reconstructed from the moments.

2.5.2 Method of Characteristics

The method of characteristics (MOCH) provides an elegant way to determine the evolution of the crystal size distribution for crystallization processes.[29] The MOCH for first order PDEs determines lines, called characteristic lines, along which the PDE degenerates into a set of ODEs.[30] The MOCH is demonstrated through the pure size independent growth batch PBE:

$$\frac{\partial n(L,t)}{\partial t}+G\frac{\partial n(L,t)}{\partial L}=0 \tag{2.55}$$

MOCH finds curves in the (L–t) plane, which reduces the PBE to a system of ODEs. The (L–t) plane can be expressed in a parametric form by $L = L(Z)$ and $t = t(Z)$, where the parameter Z gives the measure of distance along the characteristic curve. Then, the population density function can be redefined as:

$$n(L,t) = n(L(Z),t(Z)) \tag{2.56}$$

Applying the chain rule on we can obtain:

$$\frac{\mathrm{d}n(L,t)}{\mathrm{d}Z}=\frac{\mathrm{d}t}{\mathrm{d}Z}\frac{\partial n(L,t)}{\partial t}+\frac{\mathrm{d}L}{\mathrm{d}Z}\frac{\partial n(L,t)}{\partial L} \tag{2.57}$$

Comparing the coefficients of eqn (2.55) and (2.57), it can be concluded that:

$$\frac{\mathrm{d}t}{\mathrm{d}Z}=1\Rightarrow \mathrm{d}t=\mathrm{d}Z \tag{2.58}$$

$$\frac{\mathrm{d}L}{\mathrm{d}Z}=G\Leftrightarrow\frac{\mathrm{d}L}{\mathrm{d}t}=G \tag{2.59}$$

$$\frac{\mathrm{d}n(L,t)}{\mathrm{d}Z}=0\Leftrightarrow\frac{\mathrm{d}n(L,t)}{\mathrm{d}t}=0 \tag{2.60}$$

with initial conditions, corresponding to $t = Z = 0$, $L = L_0$ and $n(L,0) = n_0(L_0)$. To obtain the dynamic evolution of the crystal size distribution, $n(L,0)$, eqn (2.59) and (2.60), with prescribed growth expressions, can be integrated repeatedly for different initial values $[L_0, n_0(L_0)]$.

For the expression of growth rate, the actual concentration is required. The accurate evaluation of the mass balance requires very fine CSD discretization, which degrades the simulation time. MOCH is often combined with the QMOM, where the QMOM provides the accurate concentration profile, which is then used as an input to the MOCH for growth rate calculation.[31] This elegantly and efficiently decouples the accuracy of the method from the calculation time.

The combined MOCH – QMOM is very efficient and accurate for growth-only problems. The implementation of nucleation, however reduces its efficiency. In addition, the implementation of secondary mechanisms (breakage, agglomeration) is possible, but complicated.

2.5.3 Finite Volume Methods

The finite volume method (FVM) is a standard approach for the numerical solution of hyperbolic partial differential equations,[32] which makes it well suited for the solution of PBEs. The FVM relies on the discretization of the crystal size domain and the finite volume approximation of the population density function on the defined grid. The method is demonstrated on the one dimensional PBE with nucleation and growth:

$$\frac{\partial n(L,t)}{\partial t} + G\frac{\partial n(L,t)}{\partial L} = B\delta(L - L_n) \tag{2.61}$$

Denoting with h the size and k the time interval, n_l^m is the approximate (discrete) population density function of the continuous $n(L,t)$, defined as:

$$n_l^m \approx \frac{1}{h}\int_{(l-1)h}^{lh} n(L,t)\mathrm{d}L \tag{2.62}$$

where m and l are integers such that $m \geq 0$ and $N \geq l \geq 1$. N stands for the mesh size (*i.e.* the number of discretization points). Then, the PBE reduces to a system of algebraic equations:

$$n_l^{m+1} = n_l^m - \frac{k}{h}\left(G_l n_l^m - G_{l-1}n_{l-1}^m\right) - \left[\begin{array}{l} \frac{kG_l}{2h}\left(1 - \frac{kG_l}{h}\right)\left(n_{l+1}^m - n_l^m\right)\phi_l \\ -\frac{kG_{l-1}}{2h}\left(1 - \frac{kG_{l-1}}{h}\right)\left(n_l^m - n_{l-1}^m\right)\phi_{l-1} \end{array}\right] + \frac{n_{l,f}^m - n_l^m}{\tau} + \varepsilon_b\frac{k}{h}B \tag{2.63}$$

ε_b is a binary existence variable with values $\{0, 1\}$: $\varepsilon_b = 1$ if $l = 1$ (nucleon size) and is 0 otherwise. The second term in the right hand side of eqn (2.63) is the first order term. The FVM can also be solved using the first order term only, but this is known to lead to numerical diffusion – an artificial broadening – of the CSD. The second order term – third term in the right hand side of eqn (2.63) – is aimed to reduce the numerical diffusion, but, this term leads to numerical oscillations around the sharp fronts of the CSD. The flux limiter function (ϕ_l) adaptively turns the second order term off around the sharp CSD variations, avoiding the numerical oscillation, and keeping the second order term active in the rest of the solution to reduce numerical diffusion. The FVM combined with flux limiter function is also known as the high resolution finite volume method (HR FVM) due to its ability to capture sharp fronts or discontinuity without numerical oscillation. $\phi_l = f(\theta_l)$ is calculated from the ratio of consecutive gradients (θ_l):

$$\theta_l = \frac{f_l^m - f_{l-1}^m}{f_{l+1}^m - f_l^m} \tag{2.64}$$

Although numerous flux limiter functions were developed for different finite volume schemes, for crystallization the Van Leer flux limiter is the most widely used (Gunawan *et al.* 2004):

$$\phi(\theta_l) = \frac{|\theta_l| + \theta_l}{1 + |\theta_l|} \tag{2.65}$$

The presented FVM is a discrete time formulation, which means that in addition to the crystal size, the time is also discretized. In these schemes the time step is recalculated in every iteration to satisfy the Courant–Friedrichs–Lewy (CFL) criterion. The numerical system is stable if CFL ≤ 1.

$$\text{CFL} = \frac{k}{h}\max\{G_l\} \tag{2.66}$$

Practically, the CFL is fixed and k is expressed from eqn (2.66).

The FVM methods are generic PBE solvers as they are able to provide full CSD. The implementation of nucleation, growth and dissolution is straightforward. FVM was successfully applied for the numerical solution of aggregation-breakage problems. On the other hand, the computational expense of the FVM is significantly higher than that of the moment based methods.

2.6 Advanced Crystallization Modeling – Case Studies

In this section, case studies will be discussed for topics that are beyond the standard application of the PBMs in pharmaceutical crystallization process, without the aim of providing a comprehensive description of all possible application areas. The basic PBEs with nucleation and growth will be used for demonstration. Naturally, these PBMs can be combined (extended) with the models presented so far, in terms of crystallization mechanisms, crystallizer type and numerical solution techniques.

2.6.1 Modeling Solvent Mediated Polymorphic Transformation

Numerous APIs have multiple polymorphic forms that can be obtained under similar process conditions, however, it is of primary importance to ensure that the end-product contains only the desired polymorph. Therefore, the mathematical modeling of polymorphic crystallization has great interest in pharmaceutical crystallization.

2.6.1.1 Model Equations for a MSMPR Crystallizer

To keep the simplicity without losing generality, the polymorphic crystallization modeling is presented through the example of the crystallization of a substance with two polymorphic forms (form I and form II). The fundamental difference that is related to the crystal lattice structure is that each polymorphic form has individual solubility. Typical solubility curves of the two polymorphs are presented in Figure 2.13 through the example of ortho-amino benzoic acid in 90% water 10% isopropyl-alcohol, adapted from the literature.[33]

The form I and form II crystals in the suspension can be treated as individuals belonging to two separate populations. In this context, two population density functions can be distinguished within all the solids, one for the form I $n_{\mathrm{I}}(L,t)$ and another for form II $n_{\mathrm{II}}(L,t)$. The population density function of all solids in the crystallizer is represented by the sum $n_{\mathrm{I}}(L,t) + n_{\mathrm{II}}(L,t)$. Given the actual concentration and temperature conditions in the crystallizer, three cases can occur with respect to solubility conditions:

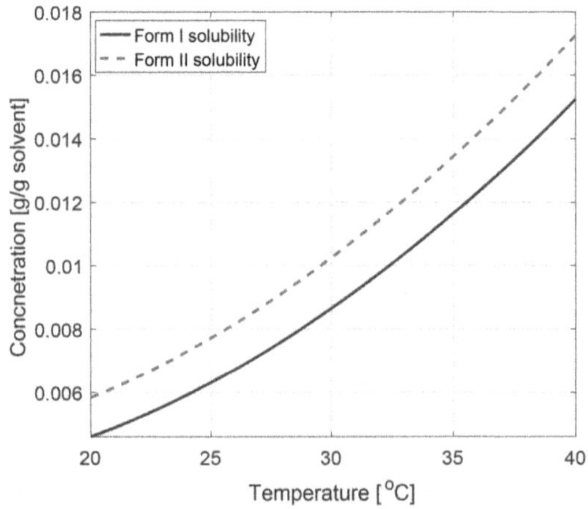

Figure 2.13 Typical polymorphic solubility curves – case of ortho-amino benzoic acid solubility in 90% water – 10% isopropyl alcohol.

- $c < \min\{c_{sol,I}(T), c_{sol,II}(T)\}$ the system is undersaturated for both populations. All crystals are dissolving,
- $c > \max\{c_{sol,I}(T), c_{sol,II}(T)\}$ the system is supersaturated for both populations. All crystals are growing and nucleation might occur for both forms,
- $\min\{c_{sol,I}(T), c_{sol,II}(T)\} < c < \max\{c_{sol,I}(T), c_{sol,II}(T)\}$ the more soluble crystals are dissolving whereas the less soluble are growing.

It should be highlighted under the concentration conditions described by the last bullet-point the more soluble crystals will fully be transformed to the less soluble crystals. This process is called solvent mediated polymorphic transformation. The mathematical model of such a system consists of the PBEs of form I and form II crystals, which are coupled though the mass balance.

The form I crystal PBE takes the form:

$$\frac{\partial n_i(L,t)}{\partial t} + \frac{\partial [G_i n_i(L,t)]}{\partial L} = B_i \delta(L - L_n) + \frac{n_{i,f}(L,t) - n_i(L,t)}{\tau}, \text{if } c > c_{sol,i}(T), i = \text{I}, \text{II} \tag{2.67}$$

$$\frac{\partial n_i(L,t)}{\partial t} - \frac{\partial [D_i n_i(L,t)]}{\partial L} = \frac{n_{i,f}(L,t) - n_i(L,t)}{\tau}, \text{if } c < c_{sol,i}(T), i = \text{I}, \text{II} \tag{2.68}$$

with the corresponding initial and boundary conditions. Here B_i, G_i and D_i denote the nucleation, growth and dissolution rate of form i under the given thermodynamic conditions.

The mass balance equation takes into consideration the impact of crystallization or dissolution of both polymorphs:

$$\frac{dc}{dt} = -\sum_i \rho_{c,i} k_{V,i} R_{V,i} + \frac{c_f - c}{\tau} \tag{2.69}$$

with the corresponding initial condition, where $R_{V,i}$ denotes the impact of crystallization on the liquid concentration. Depending on the solubility conditions, these functions have the forms listed in Table 2.10.

The four PB eqn (2.67) and (2.68) are solved simultaneously with the mass balance eqn (2.69) with a suitable solution method. Since during the solvent

Table 2.10 Impact of crystallization of the two polymorphs on the liquid concentration as a function of concentration conditions.

	$c < \min\{c_{sol,I}, c_{sol,II}\}$	$c > \max\{c_{sol,I}, c_{sol,II}\}$	Otherwise
$R_{V,I}$	$-3D_I \int_0^{L_{max}} L^2 n_I(L,t) dL$	$B_I L_n^3 + 3G_I \int_0^{L_{max}} L^2 n_I(L,t) dL$	$B_I L_n^3 + 3G_I \int_0^{L_{max}} L^2 n_I(L,t) dL$
$R_{V,II}$	$-3D_{II} \int_0^{L_{max}} L^2 n_{II}(L,t) dL$	$B_{II} L_n^3 + 3G_{II} \int_0^{L_{max}} L^2 n_{II}(L,t) dL$	$-3D_{II} \int_0^{L_{max}} L^2 n_{II}(L,t) dL$

mediated polymorphic transformation one population is dissolving, the efficient moment based methods are generally not suitable.

2.6.1.2 Solution Mediated Polymorphic Transformation in a PFC

The cooling crystallization of L-glutamic acid is investigated in a PFC with 25 segments. Each of these segments are 60 cm long and with internal diameter of 12.7 mm. The feed at the inlet contains seeds of both the metastable α-form and the stable β-form. However, the temperature of the first segment is such that the feed concentration lies between the metastable and stable solubility curves so that the metastable form dissolves and stable form grows in size. This solution mediated polymorphic transformation is modeled using steady state population balance equation (PBE) where nucleation, growth and dissolution kinetics are considered to account for the transformation.

At steady state, the PBEs describing the crystallization process for supersaturated and undersaturated regions take the forms:

$$u_x \frac{\partial n_i(x,L)}{\partial x} + \frac{\partial[G_i n_i(x,L)]}{\partial L} = B_i \delta(L - L_n), \text{ if } c > c_{\text{sol},i}(T), i = \alpha, \beta \tag{2.70}$$

$$u_x \frac{\partial n_i(x,L)}{\partial x} - \frac{\partial[D_i n_i(x,L)]}{\partial L} = 0, \text{ if } c < c_{\text{sol},i}(T), i = \alpha, \beta \tag{2.71}$$

with the boundary condition $n_i(0,L) = n_{i,\text{seed}}(L)$. Note that Greek letters (α, β) are used interchangeably with the roman numerals (I, II) to distinguish between polymorphic forms. In this case study, the applied kinetic equations for the α and β forms are written as:

$$B_\alpha = k_{\text{b}\alpha}(S_\alpha - 1)\mu_{\alpha,3}$$

$$B_\beta = k_{\text{b}\beta,1}(S_\beta - 1)\mu_{\alpha,3} + k_{\text{b}\beta,2}(S_\beta - 1)\mu_{\beta,3}$$

$$G_\alpha(t) = k_g(S_\alpha - 1)^{g\alpha}$$

$$G_\beta = k_{g\beta,1}(S_\beta - 1)^{g\beta} \exp\left(-\frac{k_{g\beta,2}}{S_\beta - 1}\right), \tag{2.72}$$

$$D_\alpha = k_{d\alpha}(1 - S_\alpha)$$

where the saturation concentration for both polymorphic forms in (g kg^{-1}) as a function of temperature (in °C) is given by the second order power law model eqn (2.35).

The seed crystal size distribution n_i (# per m^2 kg solvent) is given as:

$$n_i(L,0) = n_{\text{seed},i} = \frac{\lambda_i}{\sqrt{2\pi}\sigma_{i,\text{seed}}} \exp\left(-\frac{(L - \mu_{i,\text{seed}})^2}{2\sigma_{i,\text{seed}}^2}\right); i \in \{\alpha, \beta\}. \tag{2.73}$$

The parameters for the seed distribution are shown in Table 2.11 below

The mass balance equation needs to be considered which accounts for the changes in solution concentration due to growth and dissolution of both polymorphs.

$$\frac{dC}{dx} = -3\frac{10^3}{\rho_{\text{solv}}}\sum_i \rho_{c,i} k_{V,i} G_i \mu_{i,2}, i = \alpha, \beta \tag{2.74}$$

The experimentally determined kinetic parameters are taken from Hermanto *et al.*,[34] which are listed in Table 2.12.

The inlet concentration is taken as C = 20(g per kg solvent) and an exponentially decaying temperature profile is applied along the crystallizer going from 45°C to 33°C. The resulting PBE is solved using HR FVM. The results are summarized in Figure 2.14. It can be seen that although both the metastable α-form and stable β-form are in the feed, the CSD at the outlet indicates that the stable form has grown in size and metastable form has almost disappeared due to dissolution. The disappearance of the α-form can be further explained with the help of Figure 2.14b where the operating curve in the

Table 2.11 Parameters for seed distribution.

	λ_i	$\sigma_{i,\text{seed}}$ (m)	$\mu_{i,\text{seed}}$ (m)
α	2×10^{10}	2×10^{-6}	30×10^{-6}
β	2×10^{10}	4×10^{-6}	50×10^{-6}

Table 2.12 Kinetic parameters used in this case study.

Parameter (unit of measure)	Value
$k_{b,\alpha}$(# m^{-3}s^{-1})	exp(17.233)
$k_{g\alpha,0}$(m s^{-1})	exp(1.878)
g_α (−)	1.859
$E_{g\alpha}$ (J mol^{-1})	10.671
$k_{d,\alpha}$(m s^{-1})	exp(−10.260)
$k_{b\beta,1}$(# m^{-3}s^{-1})	exp(15.801)
$k_{b\beta,2}$(# m^{-3}s^{-1})	exp(20.00)
$k_{g\beta,0}$(m s^{-1})	exp(52.002)
$k_{g\beta,2}$(m s^{-1})	exp(−0.252)
g_β (−)	1.047
$E_{g,\beta}$(J mol)	exp(12.078)
ρ_{solv}(kg m^{-3})	990
$\rho_{c,\alpha}$(kg m^{-3})	1540
$\rho_{c,\beta}$(kg m^{-3})	1540
$k_{V,\alpha}$ (−)	0.480
$k_{V,\beta}$ (−)	0.031
$a_{\alpha,0}$ (g kg^{-1})	4.564
$a_{\alpha,1}$ ((g kg^{-1}) °C^{-1})	0.03032
$a_{\alpha,2}$ ((g kg^{-1}) °C^{-2})	8.437×10^{-3}
$a_{\beta,0}$ (g kg^{-1})	6.622
$a_{\beta,1}$ ((g kg^{-1}) °C^{-1})	−0.1165
$a_{\beta,2}$ ((g kg^{-1}) °C^{-2})	7.644×10^{-3}

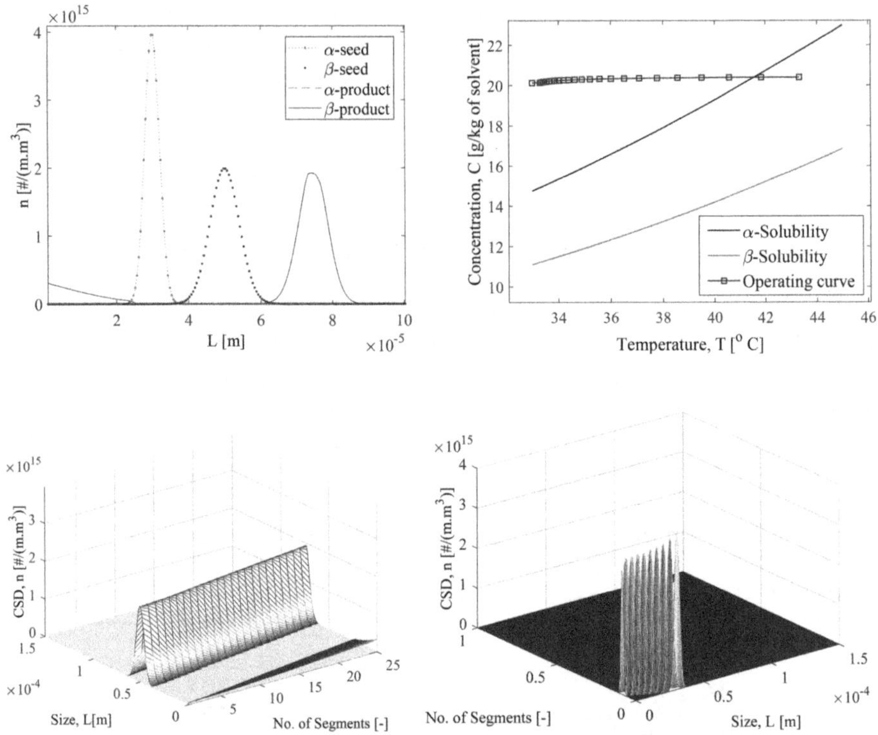

Figure 2.14 Simulation results for the polymorph transformation. (Top left) Seed and product CSD from the PFC, α-form completely dissolved at the outlet. (Top right) Operating curve in the phase diagram where starting point lies in between two solubility curves. (Bottom left) Evolution of the β-form along the PFC. (Bottom right) Dissolution of the metastable α-form in the first segment.

phase diagram is shown. As can be seen, the inlet point lies in between the solubility curves, *i.e.*, supersaturated with respect to the stable form while undersaturated with respect to the metastable form, resulting in dissolution of the metastable crystal form.

2.6.2 Modeling Preferential Crystallization of Enantiomers

Chiral molecules or enantiomers are molecules which are non-superimposable mirror images of each other. Chirality is very common in organic molecules, *e.g.*, amino acids. These molecules have identical physical properties, however, their biological activities can be different. For instance, *S* (−)-fluoxetine shows remarkable therapeutic effects in preventing migraines, while the racemic (equimolar mixture of both enantiomers) drug has no effect.[35] Thus, the pharmaceutical regulatory authority demands that the drugs are manufactured in enantiopure form. The enantiomers are typically found in an equimolar mixture known as a racemic mixture. Due to the identical physical properties, separation of enantiomers from the racemic mixture is a challenging engineering task. There are various means available for resolution of enantiomers that include classical chiral separation techniques such as preferential crystallization,[36–38] chromatography[39,40] and membrane separation.[41] Among these means, preferential crystallization is an attractive and efficient way to separate the racemic mixtures since typically no auxiliaries and reagents other than solvent are needed. However, preferential crystallization can only be used for conglomerate forming systems where the crystal phases consist of either pure D enantiomer or L enantiomer which is up to 10% of the available systems.[36] Crystallization can be carried out in standard equipment readily available in the pharmaceutical industry. There is a good number of process configurations reported in literature involving both batch and MSMPR crystallizers, some of which also include liquid phase exchange between two coupled crystallizers for better productivity and yield.[42,43] Next we compare the performance of the two process configurations – coupled preferential crystallization in batch (CPC) and coupled preferential crystallization in MSMPR (CPC-MSMPR) crystallizers.

The crystallizer configuration considered involves two MSMPR tanks with volume $V = 0.45$ L which are coupled through liquid phase exchange as shown in Figure 2.15. Each MSMPR crystallizer contains saturated slurry with racemic liquid phase at 36°C (T_{sat} = 45°C). Feed slurry with the same conditions is fed in and cooling crystallization is carried out. Product slurry from each MSMPR vessel is filtered to collect the product crystals. However, crystal free liquid phase rich in E_1enantiomer is obtained from the MSMPR 2 at a suitable location and added to the other MSMPR 1 so that the depletion of supersaturation of the preferred enantiomer can be compensated

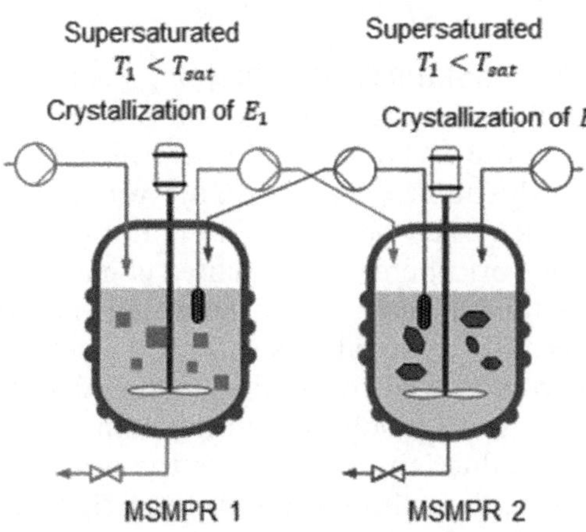

Figure 2.15 Coupled preferential crystallization in MSMPR crystallizers (CPC-MSMPR), where each vessel initially containing supersaturated racemic solution is continuously fed with a slurry containing the seed of the specific enantiomer. Exchange of liquid phase makes use of the depletion of the counter enantiomer in the other vessel.

to some extent. In other words, such exchange of liquid phase mimics the racemization of the liquid phase. This process is based on the principle that the nucleation of the enantiomers is an activated process. As a result, when a crystallizer is seeded with preferred enantiomer, the crystallization of the counter enantiomer does not readily occur at low supersaturations. The performance of the continuous process configuration CPC-MSMPR is compared with the batch CPC configuration in terms of productivity and yield. It is to be noted that the difference in CPC configuration as compared to the CPC-MSMPR is that there are no continuous feed and product withdrawal from the coupled crystallizers.

A population balance based model is developed for this coupled crystallizer system. The PBE with nucleation and growth has the following form

$$\frac{\partial n_{k,j}}{\partial t} + \frac{\partial}{\partial L}\left(G_{k,j} n_{k,j}\right) = \frac{q}{V}\left(n_{k,j,\text{feed}} - n_{k,j}\right); \quad j \in \{\text{MSMPR1}; \text{MSMPR2}\}; k \in \{E_1, E_2\} \tag{2.75}$$

with boundary condition $G_{k,j} n_{k,j}(L,t)|_{L=0} = B_{k,j}$. In eqn (2.75), $n_{k,j}$ is the number density or CSD of component k and in MSMPR j, L is the characteristic size of the crystals, $G_{k,j}$ is the growth rate of the crystals, q is the feed rate, V is the volume of the crystallizer, S is the supersaturation, $B_{k,j}$ is the nucleation rate and L_0 is the nuclei size. The PBE in eqn (2.75) needs to be supplemented by appropriate mass and energy balance equations. Moreover, the liquid phase exchange will affect the enantiomer concentrations as shown in Figure 2.15. The mass balance equation can be written as

$$\frac{\partial m_{k,j,L}}{\partial t} = -3k_{\text{V}}\rho_{\text{c}} \int_0^{L_{\max}} L^2 G_{k,j} n_{k,j} \, \mathrm{d}L + F_{\text{ex}}\left(\rho_{L,j^*} w_{k,L,j^*} - \rho_{L,j} w_{k,L,j}\right) + q\left(\rho_{L,\text{feed}} w_{k,L,j,\text{feed}} - \rho_L w_{k,L,j}\right) \tag{2.76}$$

where $j \neq j^*$ with $j, j^* = \{\text{MSMPR1, MSMPR2}\}$; $m_{k,j,L}$ is the concentration of the enantiomer k in liquid phase, ρ_L is the solution density and F_{ex} is the volumetric flowrate of the exchange liquid per unit mass of the solution. It is to be noted that the liquid phase exchange rate is assumed to be the same for both MSMPR vessels so that there is no net change in volume in the vessel. In this work the energy balance equation is not solved, it is assumed that the temperature in both vessels is maintained at 36 °C by the cooling jacket. All the equations will form a set of highly coupled PDEs which needs to be solved numerically. The quadrature method of moments (QMOM)[44] is used to solve the PBEs, as in this study we are primarily interested in productivity and yield of the process which does not require the CSD information. Simulations were carried out using MATLAB computational software.

The well-studied D/L-threonine in water is taken as the model system in this study. The required rate expression (growth and nucleation) and the kinetic parameters for this system are experimentally determined and verified which

Table 2.13 The simulation parameters used for the CPC-MSMPR and CPC process configurations.

		CPC-MSMPR		CPC	
		Tank 1	Tank 2	Tank 1	Tank 2
Liquid phase	$m_{rac,L}$	97.48 g	97.48 g	97.48 g	97.48 g
	m_{H_2O}	369.83 g	369.83 g	369.83 g	369.83 g
Solid phase	$m_{E1,S}$	2.00 g	—	2.00 g	—
	$m_{E2,S}$	—	2.00 g	—	2.00 g
Temperature	T	36 °C	36 °C	36 °C	36 °C
Exchange rate	F_{ex}	80 mL min^{-1}		80 mL min^{-1}	
Feed rate	q	80 mL min^{-1}	80 mL min^{-1}	80 mL min^{-1}	80 mL min^{-1}

can be found elsewhere.[42,45] The vessel size used in the configurations shown in Figure 2.15 is 0.45 L. The performance of the continuous crystallizer configuration CPC-MSMPR is compared with the batch configuration CPC in terms of productivity and yield. The details of the process conditions used are shown in Table 2.13.

The seed distribution is taken as the log-normal distribution as follows:

$$f_{k,j,\text{seed}} = A_{k,j,\text{seed}} \frac{1}{\sqrt{2\pi}\sigma_{k,j} L} \exp\left(-\frac{1}{2}\left[\ln\left(\frac{L}{L_{\text{mean},k,j}}\right)\frac{1}{\sigma_{k,j}}\right]^2\right) \tag{2.77}$$

where $A_{k,j,\text{seed}}$ (−) is the scaling factor that can be used to adjust the seed mass, $\sigma_{k,j} = 0.34$ and $L_{\text{mean},k,j} = 95$ μm.

For the continuous configurations (*i.e.*, CPC-MSMPR) the productivity is calculated as follows.

$$\text{Prod}_{k,j} = (\dot{m}_{k,S,\text{prod},j} - \dot{m}_{k,S,\text{prod}})\text{ee}_{S,j} \tag{2.78}$$

while for the batch configurations productivity is found as

$$\text{Prod}_{k,j} = \left(\frac{m_{S,k,j} - m_{\text{seed},k,j}}{t}\right)\text{ee}_{S,j} \tag{2.79}$$

where $\dot{m}_{k,S,\text{prod},j}$ (kg s^{-1}) and $\dot{m}_{k,S,\text{prod}}$ (kg s^{-1}) are the solid crystal mass flow rates in the product stream and feed stream, respectively and $\text{ee}_{S,j}$ (−) is the enantiomeric excess used to define the purity and as a weighting factor in the definition of productivity so that the purity of the products is taken into account.

$$\text{ee}_{S,j} = \frac{|m_{S,k,j} - m_{S,k^*,j}|}{m_{S,k,j} + m_{S,k^*,j}} \times 100\% \tag{2.80}$$

In this study the target ee_S is set to be 99%. The productivity and yield of the crystallizer configurations are shown in Figure 2.16. As can be seen, the CPC-MSMPR configuration provides significantly higher productivity compared to the CPC configuration. Due to the operation in batch mode, the productivity for CPC configuration is limited by the consumption of the supersaturation which has no provisions to replenish from feed streams. In contrast, for the CPC-MSMPR crystallizer there is a steady supply of racemic solution to the tanks from feed streams. This allows the CPC-MSMPR configuration to maintain the high level of productivity and purity for about 8 h, indicating that the process has reached a steady state. Thus, the superiority of the continuous configuration in terms of productivity as compared to the batch configurations has been shown by these simulation results for the system considered.

However, as can be seen in Figure 2.16b, this high productivity comes at the cost of low yield. When it comes to yield, the CPC configuration performs better than the CPC-MSMPR configuration. This can be explained by considering the residence time for each configuration. The residence time for the slurry in the CPC-MSMPR configuration is much shorter (5.63 min) as compared to the CPC configuration (8 h) and thus the time available for the seed crystals to grow is much shorter. The shorter residence time for CPC-MSMPR can be attributed to the steady feed flow and product removal from the crystallizer. However, one can choose an appropriate feed flow rate for CPC-MSMPR configuration by prioritizing the trade-off between productivity and yield. The low yield of the CPC-MSMPR configuration can be compensated if there is a provision for recycling the solution obtained from the product stream after filtering the product crystals followed by racemization. The overall yield that can be obtained with recycle is equivalent to that of the

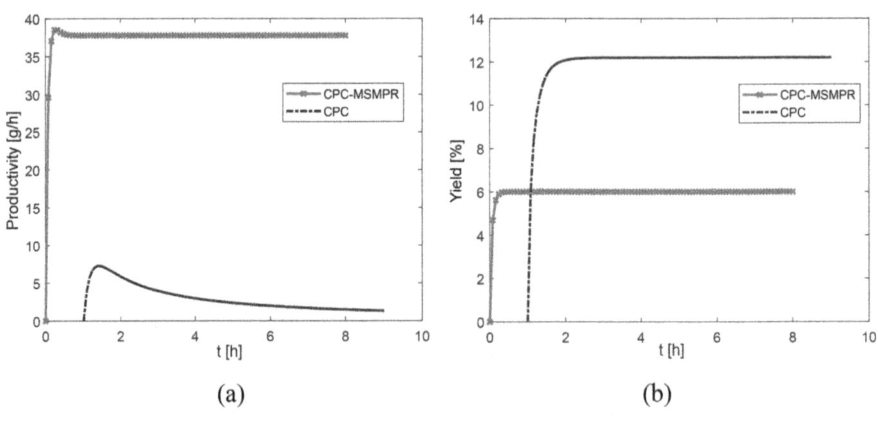

Figure 2.16 (a) Comparison of productivities for CPC and CPC-MSMPR configurations. Process parameters used are seed mass 8 g, feed flow rate 80 mL min^{-1}, exchange rate 80 mL min^{-1}, and racemate mass 70 g. The productivity for CPC-MSMPR is found to be significantly higher than the CPC configuration. (b) The curves showing comparison of yields. The yield of CPC-MSMPR is the lower than the CPC configuration. Adapted from ref. 47, https://doi.org/10.3390/pharmaceutics9040055, under the terms of the CC BY 4.0 license, https://creativecommons.org/licenses/by/4.0/.

equilibrium batch process or even greater.[46] However, the increase of yield by using recycle stream can be limited if there is a build-up of impurities within the system.[46]

2.7 The Growth Rate Dispersion (GRD)

In the traditional crystallization modeling practice, the McCabe's ΔL law is applied, which states that the growth rate of crystals is constant in time and it doesn't depend on the crystal size. An important consequence of this law is that two crystals under identical thermodynamic conditions (temperature, supersaturation *etc.*) exhibit identical growth rate. If this is true, in a seeded batch pure growth crystallization process the width of the CSD should remain constant throughout the growth process. This is a reasonable approximation of growth rate in numerous cases, but for many systems strong deviations were observed from the McCabe's ΔL law. This experimental evidence suggests that some microscopic mechanisms must exist that can cause a fluctuation of the observed growth rate, therefore, there is a growth rate variance in a population of crystals under macroscopically identical conditions.

It seems reasonable to assume that GRD is caused by the interplay of various mechanisms, whose relative importance depends on the crystalline system, the operating conditions and the particle history. From a modeling perspective, there are two general techniques to describe GRD: the random fluctuations (RF) and constant crystal growth (CCG) approach. The CCG assumes that each crystal is intrinsically born with an inherent growth rate, which can differ from crystal to crystal, but it remains constant over the lifetime of the crystal. Practically, in the CCG model the growth rate is often assumed to be some standard probability density function (normal, lognormal, gamma *etc.*). In contrast, RF assumes that the growth rate of crystals fluctuates in time around a given, constant expected value. The spread of growth rates is described by a coefficient of dispersion, that gives rise to a second order differential.

For the illustration of CCG, let $F(G_R)dG_R$ represent the distribution of relative growth rates around a nominal value[1] for the crystals such that $F(G_R)dG_R$ is the fraction of the total number of crystals having a relative growth rate of G_R. Table 2.14 gives the typically applied probability density functions for approximating the growth rate dispersion (Figure 2.17).[48,49]

Then, for the characterization of the crystal population, which also takes into account the growth rate of individual crystals, a bivariate population density function is introduced $n(L,G_R,t)$, which gives the number of crystals in the $(L,L + dL) \times (G_R,G_R + dG_R)$ size and relative growth rate domain in t time moment in unit volume of suspension. The size density function can be extracted from the bivariate population density function, since the joint probability density function is related to the normalized density function, by the integral:

Table 2.14 Commonly used probability density functions for the description of growth rate dispersion.

Name	Probability density function	Notes
Normal	$F(G_R)=\frac{1}{\sqrt{2\pi\sigma}}\exp\left[-\frac{(G_R-\mu)^2}{2\sigma^2}\right]$	Can lead to negative growth rates for high dispersion (σ) values. μ is the mean (nominal) growth rate
Lognormal	$F(G_R)=\frac{1}{G_R\sigma\sqrt{2\pi}}\exp\left[-\frac{(\ln(G_R)-\mu)^2}{\sigma^2}\right]$	Cannot lead to negative growth rates
Gamma	$F(G_R)=\frac{1}{\Gamma(k)m^k}G_R{}^{k-1}\exp\left[-\frac{G_R}{m}\right]$	Γ: Gamma function; m, k are the parameters of the function

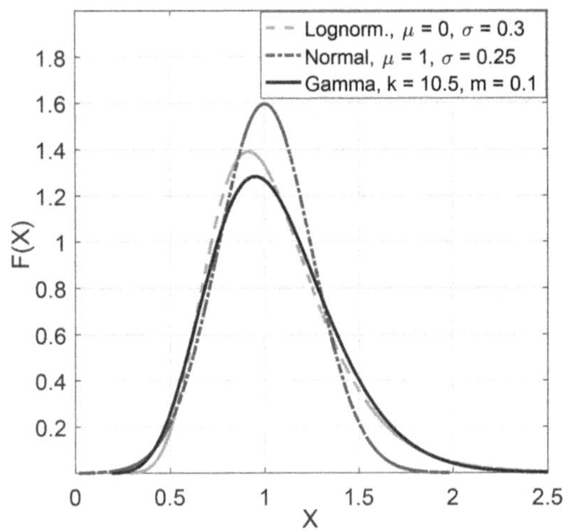

Figure 2.17 Widely applied distribution functions to model growth rate dispersion.

$$n(L,t)=\int_0^\infty n(L,G_R,t)\mathrm{d}G_R \tag{2.81}$$

where $n(L,t)$ becomes the marginal density function of L. With a similar approach, the growth rate dispersion function can be de-convoluted from the 2D CSD as:

$$n(G_R,t)=F(G_R)=\int_0^\infty n(L,G_R,t)\mathrm{d}L \tag{2.82}$$

The PBE for an MSMPR crystallizer with nucleation and growth mechanisms takes the form:

$$\frac{\partial n(L,G_{\mathrm{R}},t)}{\partial t}+G\frac{\partial\left[n(L,G_{\mathrm{R}},t)\right]}{\partial L}$$
$$=B\delta(L-L_n)+\frac{n_f(L,G_{\mathrm{R}},t)-n(L,G_{\mathrm{R}},t)}{\tau} \quad (2.83)$$

Subject to corresponding boundary condition:

$$\lim_{\substack{L\to\infty \\ G_{\mathrm{R}}\to\infty}} n(L,G_{\mathrm{R}},t)=0 \quad (2.84)$$

The energy and mass balance equations for brevity are not repeated here. It was shown in the literature that it is possible to fit the parameters of GRD functions, listed in Table 2.14, to the steady-state CSD collected in MSMPR crystallizers.[48] In the same work it was concluded that there is only a slight difference between the different probability density function performances in the fitting of the experimental data.

The RF model assumes constant growth rate with random fluctuations around the growth rate, noted by F_{g}, which can be modeled as:

$$F_{\mathrm{g}}=\frac{\partial\left[D_{\mathrm{g}}n(L,t)\right]}{\partial L} \quad (2.85)$$

where D_{g} stands for the growth rate diffusivity. The PBE with nucleation, growth and RF mechanisms for an MSMPR crystallizer can be written as:

$$\frac{\partial n(L,t)}{\partial t}+\frac{\partial[\mathrm{Gn}(L,t)]}{\partial L}-\frac{\partial^2\left[D_{\mathrm{g}}n(L,t)\right]}{\partial L^2}$$
$$=B\delta(L-L_n)+\frac{n_f(L,t)-n(L,t)}{\tau} \quad (2.86)$$

As it can be seen, RF gives rise to a second order term in the PBE, which is a diffusive term in the partial differential equation. It must be noted that from a theoretical perspective, RF and CCG approaches are not exclusive, but are rarely combined with each other.

Appendix

A1 Derivation of the Population Balance Equation for Plug Flow Crystallizer

The general balance equation for a quantity in a control volume can be written as follows.

$$\begin{bmatrix}\text{Rate of}\\ \text{accumulation}\end{bmatrix}=\begin{bmatrix}\text{Rate of}\\ \text{entry}\end{bmatrix}-\begin{bmatrix}\text{Rate of}\\ \text{departure}\end{bmatrix}\pm\begin{bmatrix}\text{Rate of}\\ \text{generation/ depletion}\end{bmatrix}$$

We consider the number density function n for the crystals in a PFC which depends on one internal coordinate size L, one external coordinate – axial location along the crystallizer x and time t. Taking number balance at the volume element $A\Delta x\Delta L$

$$A\Delta x\Delta L\left(n|_{t+\Delta t} - n|_t\right) = A\Delta L\Delta t\left(un|_x - un|_{x+\Delta x}\right)$$
$$+A\Delta x\Delta t\left(G^{(k)}n|_L - G^{(k)}n|_{L+\Delta L}\right) + SA\Delta x\Delta L\Delta t$$

If the cross-sectional area of the PFC is constant, then we have

$$\Rightarrow \frac{n|_{t+\Delta t} - n|_t}{\Delta t} = \frac{un|_x - un|_{x+\Delta x}}{\Delta x} + \frac{Gn|_L - Gn|_{L+\Delta L}}{\Delta L} + S$$

In the limit Δt, $\Delta x \to 0$, we have

$$\Rightarrow \frac{\partial n}{\partial t} = -\frac{\partial}{\partial L}(Gn) - \frac{\partial}{\partial x}(un) + S \tag{A1}$$

The source term S due to aggregation and breakage can be set to be zero if these processes are negligible. Then we have

$$\Rightarrow \frac{\partial}{\partial t}(n) + \frac{\partial}{\partial L}(Gn) + \frac{\partial}{\partial x}(un) = 0 \tag{A2}$$

The boundary and initial conditions are:

$$n(L_0, x, t) = \frac{B_0}{G},$$
$$n(L,0,t) = n_{\text{seed}}(L,t),$$
$$n(L,x,0) = n_0(L,x) \tag{A3}$$

A2 Derivation of the Mass Balance Equation for Plug Flow Crystallizer

Taking mass balance for the solute species present in the liquid phase in the volume element $A\Delta x$, as shown in Figure 2.18, we have,

$$(A\Delta xc)\Big|_t^{t+\Delta t} = -(\Delta tuAc)\Big|_x^{x+\Delta x}$$
$$-\Delta tA\Delta x\rho_c \frac{\mathrm{d}}{\mathrm{d}t}\left(\int_{L_0}^{L_{\max}} nL^3\,\mathrm{d}L\right) - \Delta tA\Delta xk_V\rho_c B_0 L_0^3,$$

where on the right hand side of the equation, the contributing terms are due to convection by fluid flow, depletion due crystal growth and nucleation, respectively.

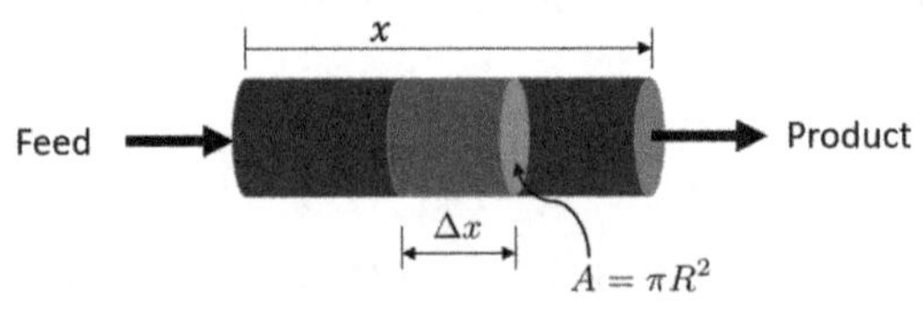

Figure 2.18 Schematic of a plug flow crystallizer showing volume element.

Dividing both side by $A\Delta x$ and in the limit Δt, $\Delta x \rightarrow 0$

$$\Rightarrow \frac{\partial}{\partial t}(c) = -\frac{\partial}{\partial x}(uc) - k_V \rho_c \frac{d\mu_3}{dt} - k_V \rho_c B_0 L_0^3 \tag{A4}$$

where μ_3 is the third moment of CSD. Often the depletion due to nucleation is neglected due to its much smaller contribution as compared to the other terms.

A3 Derivation of the Energy Balance Equation for Plug Flow Crystallizer

We write the energy balance equation in the volume element $A\Delta x$ as shown in Figure 2.18. The contributing terms are the heat transfer due to fluid flow (*i.e.*, convection) and conduction in the x direction, overall heat transfer from the crystallizer to the cooling water and the generation/depletion term due to the heat of crystallization.

$$\rho_{slurry} c_p A\Delta x T\Big|_t^{t+\Delta t} = -uA\Delta t \rho_{slurry} c_p T\Big|_x^{x+\Delta x} + kA\frac{\partial T}{\partial x}\Bigg|_x^{x+\Delta x} + 2\pi R\Delta x\Delta t h(T_{cool} - T) + \left(-\Delta H_c A\Delta x \rho_c k_V \int_0^{L_{max}} nL^3 \, dL\right)\Bigg|_t^{t+\Delta t}$$

where h is the overall heat transfer coefficient, k is the conductivity of the slurry, $c_p$is the heat capacity of the slurry and ΔH_c is the heat of crystallization. Now dividing both sides with $\rho_L c_p A\Delta x\Delta t$ and in the limit Δt, $\Delta x \rightarrow 0$, finally we have

$$\frac{\partial T}{\partial t} = -\frac{\partial}{\partial x}(uT) + \frac{k}{\rho_{slurry} c_p}\frac{\partial^2 T}{\partial x^2} + \frac{2\pi Rh}{\rho_{slurry} c_p A}(T_{cool} - T) + \frac{\rho_c k_V \Delta H_c}{\rho_{slurry} c_p}\frac{\partial \mu_3}{\partial t}$$

Abbreviations

API	active pharmaceutical ingredient
CCG	constant crystal growth
CFL	courant–Friedrichs–Levy
CPC	coupled preferential crystallization
CNT	classical nucleation theory
CSD	crystal size distribution
FVM	finite volume method
GRD	growth rate dispersion
HR	high resolution
MOCH	method of characteristics

MSMPR	mixed suspension mixed product removal
PAT	process analytical technology
PBE	population balance equation
PBM	population balance model
PFC	plug flow crystallizer
QMOM	quadrature method of moments
RF	random fluctuations
SMOM	standard method of moments

Roman Symbols

a_i	parameter of the power-law solubility equation, kg m^{-3}° C^i
A	crystal surface, m^2; heat transfer area, m^2
b	breakage daughter distribution function
b	supersaturation exponent in nucleation rate function
B	nucleation rate, # $m^{-3}s^{-1}$
c	concentration, kg m^{-3}
c_i	concentration at the crystal interface, kg m^{-3}
c_s	solubility, kg m^{-3}
Δc	absolute supersaturation, kg m^{-3}
c_p	specific heat capacity, J $kg^{-1}K^{-1}$
D	diffusion coefficient, m^2 s^{-1}
D_G	growth rate diffusivity, m^2 s^{-1}
ee_S	enantiomeric excess
E_b	activation energy of nucleation, J mol^{-1}
E_g	activation energy of growth, J mol^{-1}
F	feeding flowrate, m^3 s^{-1}
g	supersaturation exponent or the crystal growth
G	crystal growth rate, m s^{-1}, μm s^{-1}
ΔG	free energy, J mol^{-1}
ΔG_v	volume excess free energy, J mol^{-1}
ΔG_s	surface excess free energy, J mol^{-1}
ΔG_c	critical free energy, J mol^{-1} K^{-1}
G_R	growth rate dispersion, internal crystal variable
h	discrete crystal size channel width, m, μm
ΔH_c	heat of crystallization, J kg^{-1}
i	supersaturation exponent or the surface integration step
j	third moment exponent in nucleation rate function
k	Boltzmann constant
	parameter of the tangent hyperbolic breakage selection function
	discrete time step size in the finite volume method, s
k_b	nucleation rate constant, # $m^{-3}s^{-1}$
k_d	rate constant of diffusion, s^{-1}

k_r	rate constant of surface integration, $m^{3i}\ kg^{i-1}s^{-1}$
k_{bcf}	Burton–Cabrera–Frank (BCF) growth rate model parameter
k_G	overall crystal growth rate constant, $m^{3g}\ kg^{g-1}s^{-1}$
k_V	volume shape factor of the crystal
L	crystal size, m, μm
L_i	*i*th quadrature abscissa, m
$\dot{m}$	mass transfer rate, $kg\ m^{-2}\ s^{-1}$
n	crystal number density, #/m^3 μm
N	impeller stirring rate, rot s^{-1}
p	exponent of size dependent growth rate model
Pe_G	Péclet number,
r	crystal radius, m
r_c	critical radius, m
R_V	volume growth rate of crystal population, $m^3\ s^{-1}$
S	supersaturation ratio; breakage selection function, # $m^{-3}\ s^{-1}$
S_0	size dependent breakage kernel
T	temperature, °C, K
u	mean axial velocity, m s^{-1}
U	heat transfer coefficient, W $m^{-2}\ K^{-1}$
v	molecular volume, m^3
v_c	volume of a single crystal, m^3
V	working volume, m^3
w_i	*i*th quadrature weight, # m^{-3}
x	axial coordinate, m

Greek Symbols

α	effectiveness of growth inhibition,
β	agglomeration rate function, # $m^{-3}\ s^{-1}$
β_0	agglomeration kernel, s^{-1}
ϕ_l	flux (slope) limiter function
θ_{eq}	equilibrium coverage of the surface with impurity
θ_l	sharpness of the distribution
σ	relative supersaturation
σ_s	interfacial tension, J m^{-2}
μ_k	moment of the distribution, $m^k\ m^{-3}$
δ	boundary layer thickness, m
γ	parameter of size dependent growth rate model, m^{-1}, μm^{-1}
	crystal size, m, μm
μ_k	*k*th moment of the distribution, $m^k\ m^{-3}$
ρ	density, kg m^{-3}
τ	mean residence time, s

References

1. H. M. Hulburt and S. Katz, Some problems in particle technology : A statistical mechanical formulation, *Chem. Eng. Sci.*, 1964, **19**(8), 555–574.
2. Z. K. Nagy and R. D. Braatz, Advances and New Directions in Crystallization Control, *Annu. Rev. Chem. Biomol. Eng.*, 2012, **3**(1), 55–75.
3. A. Randolph and M. Larson, *Theory of Particulate Processes, Analysis and Techniques of Continuous Crystallization*, Academic Press, Salt Lake City, 1988.
4. D. Ramkrishna, *Population Balances*, Elsevier, 2000, pp. 7–45.
5. A. Gerstlauer, S. Motz, A. Mitrovic and E. D. Gilles, Development, analysis and validation of population models for continuous and batch crystallizers, *Chem. Eng. Sci.*, 2002, **57**(20), 4311–4327.
6. J. W. Mullin and J. Mullin, *Crystallization*, Elsevier, 2001, pp. viii–ix.
7. F. F. Abraham, *Homogeneous Nucleation Theory: The Pretransition Theory of Vapor Condensation*, Academic Press, 1974, (Advances in Theoretical Chemistry: Supplement).
8. P. G. Vekilov, Nucleation, *Cryst. Growth Des.*, 2010, **10**(12), 5007–5019.
9. J. B. Rawlings, S. M. Miller and W. R. Witkowski, Model identification and control of solution crystallization processes: a review, *Ind. Eng. Chem. Res.*, 1993, **32**(7), 1275–1296.
10. M. A. Lovette and M. F. Doherty, Predictive modeling of supersaturation-dependent crystal shapes, *Cryst. Growth Des.*, 2012, **12**(2), 656–669.
11. N. Kubota and J. W. Mullin, A kinetic model for crystal growth from aqueous solution in the presence of impurity, *J. Cryst. Growth*, 1995, **152**(3), 203–208.
12. A. Majumder and Z. K. Nagy, Prediction and control of crystal shape distribution in the presence of crystal growth modifiers, *Chem. Eng. Sci.*, 2013, **101**, 593–602.
13. A. Mersmann, *Crystallization Technology Handbook*, Marcel Dekker Inc., New York, Basel, 2001.
14. D. L. Marchisio and R. O. Fox, *Computational Models for Polydisperse Particulate and Multiphase Systems*, Cambridge University Press, 2010, pp. 1–508.
15. D. L. Marchisio, R. D. Vigil and R. O. Fox, Quadrature method of moments for aggregation-breakage processes, *J. Colloid Interface Sci.*, 2003, **258**(2), 322–334.
16. V. G. Levich, *Physicochemical Hydrodynamics*, Prentice-Hall Inc., 1962.
17. K. Sato, H. Nagai, K. Hasegawa, K. Tomori, H. J. M. Kramer and P. J. Jansens, Two-dimensional population balance model with breakage of high aspect ratio crystals for batch crystallization, *Chem. Eng. Sci.*, 2008, **63**(12), 3271–3278.
18. A. S. Myerson, *Handbook of Industrial Crystallization*, Elsevier, 2002.

19. D. Acevedo, D. J. Jarmer, C. L. Burcham, C. S. Polster and Z. K. Nagy, A continuous multi-stage mixed-suspension mixed-product-removal crystallization system with fines dissolution, *Chem. Eng. Res. Des.*, 2018, **135**, 112–120.
20. E. Temmel, M. Eicke, H. Lorenz and A. Seidel-Morgenstern, A Short-Cut Method for the Quantification of Crystallization Kinetics. 2. Experimental Application, *Cryst. Growth Des.*, 2016, **16**(12), 6756–6768.
21. T. McGlone, N. E. B. Briggs, C. A. Clark, C. J. Brown, J. Sefcik and A. J. Florence, Oscillatory Flow Reactors (OFRs) for Continuous Manufacturing and Crystallization, *Org. Process Res. Dev.*, 2015, **19**(9), 1186–1202.
22. A. J. Alvarez and A. S. Myerson, Continuous Plug Flow Crystallization of Pharmaceutical Compounds, *Cryst. Growth Des.*, 2010, **10**(5), 2219–2228.
23. S. Lawton, G. Steele, P. Shering, L. Zhao, I. Laird and X. W. Ni, Continuous crystallization of pharmaceuticals using a continuous oscillatory baffled crystallizer, *Org. Process Res. Dev.*, 2009, **13**(6), 1357–1363.
24. B. J. Ridder, A. Majumder and Z. K. Nagy, Population Balance Model-Based Multiobjective Optimization of a Multisegment Multiaddition (MSMA) Continuous Plug-Flow Antisolvent Crystallizer, *Ind. Eng. Chem. Res.*, 2014, **53**(11), 4387–4397.
25. Q. Su, B. Benyahia, Z. K. Nagy and C. D. Rielly, Mathematical Modeling, Design, and Optimization of a Multisegment Multiaddition Plug-Flow Crystallizer for Antisolvent Crystallizations, *Org. Process Res. Dev.*, 2015, 1859–1870.
26. A. Randolph and M. Larson, *Theory of Particulate Processes*, Academic Press, Salt Lake City, 1973.
27. R. McGraw, Description of Aerosol Dynamics by the Quadrature Method of Moments, *Aerosol Sci. Technol.*, 1997, **27**(2), 255–265.
28. R. Grosch, H. Briesen, W. Marquardt and M. Wulkow, Generalization and numerical investigation of QMOM, *AIChE J.*, 2007, **53**(1), 207–227.
29. M. J. Hounslow and G. K. Reynolds, Product engineering for crystal size distribution, *AIChE J.*, 2006, **52**(7), 2507–2517.
30. R. J. LeVeque, *Numerical Methods for Conservation Laws*, Mathematics of Computation, 1992, vol. 57, p. 875.
31. E. Aamir, Z. K. Nagy, C. D. Rielly, T. Kleinert and B. Judat, Combined Quadrature Method of Moments and Method of Characteristics Approach for Efficient Solution of Population Balance Models for Dynamic Modeling and Crystal Size Distribution Control of Crystallization Processes, *Ind. Eng. Chem. Res.*, 2009, **48**(18), 8575–8584.
32. R. J. LeVeque, *Finite Volume Methods For Hyperbolic Problems, Cambridge Texts In Applied Mathematics*, Cambridge University Press, Cambridge, UK, New York, 2002, vol. xix, p. 558.

33. E. Simone, A. N. Saleemi, N. Tonnon and Z. K. Nagy, Active polymorphic feedback control of crystallization processes using a combined raman and ATR-UV/Vis spectroscopy approach, *Cryst. Growth Des.*, 2014, **14**(4), 1839–1850.
34. M. W. Hermanto, N. C. Kee, R. B. H. Tan, M.-S. Chiu and R. D. Braatz, Robust Bayesian estimation of kinetics for the polymorphic transformation of L-glutamic acid crystals, *AIChE J.*, 2008, **54**(12), 3248–3259.
35. N. M. Maier, P. Franco and W. Lindner, Separation of enantiomers: needs, challenges, perspectives, *J. Chromatogr. A*, 2001, **906**(1), 3–33.
36. J. Jacques, A. Collet and S. H. Wilen, *Enantiomers, Racemates, and Resolutions*, Wiley, 1981.
37. M. P. Elsner, G. Ziomek and A. Seidel-Morgenstern, Efficient separation of enantiomers by preferential crystallization in two coupled vessels, *AIChE J.*, 2009, **55**(3), 640–649.
38. G. Levilain, M. J. Eicke and A. Seidel-Morgenstern, Efficient Resolution of Enantiomers by Coupling Preferential Crystallization and Dissolution. Part 1: Experimental Proof of Principle, *Cryst. Growth Des.*, 2012, **12**(11), 5396–5401.
39. G. Subramanian, *Chiral Separation Techniques: A Practical Approach*, John Wiley & Sons, 2008.
40. A. Rajendran, G. Paredes and M. Mazzotti, Simulated moving bed chromatography for the separation of enantiomers, *J. Chromatogr. A*, 2009, **1216**(4), 709–738.
41. J. T. F. Keurentjes, L. J. W. M. Nabuurs and E. A. Vegter, Liquid membrane technology for the separation of racemic mixtures, *J. Membr. Sci.*, 1996, **113**(2), 351–360.
42. M. J. Eicke, G. Levilain and A. Seidel-Morgenstern, Efficient Resolution of Enantiomers by Coupling Preferential Crystallization and Dissolution. Part 2: A Parametric Simulation Study to Identify Suitable Process Conditions, *Cryst. Growth Des.*, 2013, **13**(4), 1638–1648.
43. K. Galan, M. J. Eicke, M. P. Elsner, H. Lorenz and A. Seidel-Morgenstern, Continuous Preferential Crystallization of Chiral Molecules in Single and Coupled Mixed-Suspension Mixed-Product-Removal Crystallizers, *Cryst. Growth Des.*, 2015 Apr 1, **15**(4), 1808–1818.
44. D. L. Marchisio, J. T. Pikturna, R. O. Fox, R. D. Vigil and A. A. Barresi, Quadrature method of moments for population-balance equations, *AIChE J.*, 2003, **49**, 1266–1276.
45. M. P. Elsner, G. Ziomek and A. Seidel-Morgenstern, Simultaneous preferential crystallization in a coupled batch operation mode. Part II: Experimental study and model refinement, *Chem. Eng. Sci.*, 2011, **66**(6), 1269–1284.
46. S. Ferguson, F. Ortner, J. Quon, L. Peeva, A. Livingston and B. L. Trout, *et al.*, Use of Continuous MSMPR Crystallization with Integrated Nanofiltration Membrane Recycle for Enhanced Yield and Purity in API Crystallization., *Cryst. Growth Des.*, 2014, **14**(2), 617–627.

47. A. Majumder and Z. Nagy, A Comparative Study of Coupled Preferential Crystallizers for the Efficient Resolution of Conglomerate-Forming Enantiomers, *Pharmaceutics*, 2017 Dec 5, **9**(4), 55.
48. M. A. Larson, E. T. White, K. A. Ramanarayanan and K. A. Berglund, Growth rate dispersion in MSMPR crystallizers, *AIChE J.*, 1985, **31**(1), 90–94.
49. A. M. Zikic, R. I. Ristic and J. N. Sherwood, Three-parameter distribution function fit to growth rate dispersion among small crystals, *J. Cryst. Growth*, 1996, **158**(4), 560–567.

CHAPTER 3

Continuous Crystallisation With Oscillatory Baffled Crystalliser Technology

XIONGWEI NI*[a,b]

[a]NiTech Solutions Ltd, UK; [b]School of Engineering and Physical Sciences, Heriot-Watt University, Edinburgh EH14 4AS, UK
*E-mail: x.ni@nitechsolutions.co.uk, x.ni@hw.ac.uk

3.1 Introduction

Crystallisation is a well-known unit operation for purifying materials. It can be found in the majority of industries from agrochemicals to pharmaceuticals, from food stuff to energetic materials, and over 80% of pharmaceuticals and over 60% of fine/speciality chemicals have crystallisation in their production lines.[1] In chemical engineering and chemistry textbooks,[2–5] the principles and theories of crystallisation are well documented with undergraduate lab projects in crystallisation too.

Mixing is seldom a problem in a lab research environment and can be regarded as uniform; furthermore the specific surface area ($m^2 m^{-3}$) of lab vessels for heat transfer is sufficient to allow any type of controlled or uncontrolled cooling/heating profiles; crystallisation on lab scales is as routine as bread making. When the volume of a crystalliser increases to industrial scale, the specific surface area for heat transfer decreases dramatically as illustrated in Table 3.1 where the specific surface area for a 1 m^3 vessel is reduced by ~93% in comparison to that for a 1 L lab vessel.

The Handbook of Continuous Crystallization
Edited by Nima Yazdanpanah and Zoltan K. Nagy

Published by the Royal Society of Chemistry, www.rsc.org

Table 3.1 Specific surface area calculations.

Diameter (m)	Area (m^2)	Height (m)	Volume (m^3)	Surface area (m^2)	Specific surface area (m^2m^{-3})
0.05	0.0020	0.5	0.001	0.08	80
0.75	0.4418	2.3	1	5.33	5.33
1.51	1.7857	5.6	10	8.44	0.84

At the same time, the mixing worsens, leading to concentration and temperature gradients that would have randomly altered the width of metastable zone within the crystalliser. Spontaneous nucleation would have occurred whenever the metastable zone has been crossed, leading to nucleation dominated crystals that do not grow, which in turn gives wide crystal size distributions. Precisely due to this, prolonged filtration time and lower purity due to solvent entrapment are often the expected outcomes. It is thus the combination of the mixing and temperature control that is problematic in scaling up batch crystallisation processes.

I realized in 2004 that continuous crystallisation was an elegant way forward, as plug flow can deliver the consistent and scale independent mixing environment. At the same time, mass and heat transfer constraints are minimized due to the achievement of plug flow, the two unique elements required by crystallisation, which cannot be attainable *simultaneously* in traditional batch crystallisers. There were however a number of interesting experimental observations that were hard to explain, for example,

- Why were nucleation temperatures higher in the oscillatory baffled crystalliser (OBC) than in stirred tank crystallisers?
- Why/how did OBC allow consistent crystal morphology?
- Why/how did OBC achieve better purities while using faster cooling rates than benchmark cases?
- How did nucleation take place in OBC without seeds, while seeding was essential in benchmarking cases?

3.2 Plug Flow

3.2.1 The Definition

Plug flow is regarded as the "holy grail" of chemical engineering, however the concept of plug flow is generally poorly understood by chemical engineers and chemists. There is no simple accepted concise definition for what constitutes plug flow. In order to appreciate the complex concept of plug flow, it is necessary to introduce velocity profiles in laminar and turbulent tube (pipe) flows. Let us visualise a liquid with a velocity of u flowing from left to right through a longitudinal tubular reactor, Figure 3.1 shows the velocity profiles for laminar, turbulent and plug flows. In laminar flows, the velocity at the centre of the tube (along the dotted line) is equal to that of the incoming flow, u; while the velocity at the wall equals to zero (due to viscosity), giving the

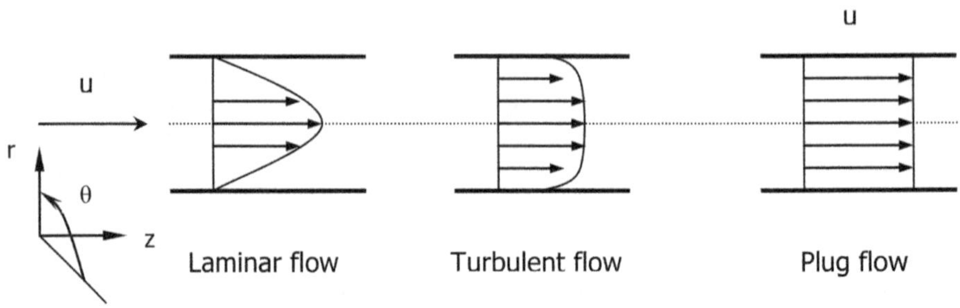

Figure 3.1 Velocity profiles for laminar, turbulent and plug flows.[6]

well-known parabolic velocity profile. Hence there is a velocity gradient (d*u*/d*r*) along the radial direction, consequently the fluid element at the centre exits the tube first and at the wall last, *i.e.* the fluid elements have different residence times in the tube. If we cook soup in a tubular device under laminar flow, the lighter liquid phase in the centre would flow faster and stay shorter in the tube than the heavier solids near the wall, leading to the former being undercooked and the latter overcooked.

In turbulent flows, the laminar parabolic velocity front is significantly flattened, however, there is still a laminar sub-layer remaining in the velocity profile where energies are dissipated and heat transfers are restricted. For a given fluid (ρ and μ) and tube diameter (d), the main criterion that separates the laminar from the turbulent flow is the fluid velocity, generally the net flow Reynolds number, Re_n ($= \rho du/\mu$).

In plug flow, all the velocity components in the tube equal to that of the incoming flow, u, hence there is no velocity gradient in the radial direction, indicating complete mixing across the tube. Because of the velocity profile, all fluid elements travelling through the tube will have an equal residence time, like a 'plug', hence the name. When we use a plug flow device to cook the soup, the mixture of solids and liquid will evenly be cooked. From the above illustrations and descriptions, the definition of the plug flow is a type of flow that satisfies the following criteria:

- The velocity profile in the direction of flow (axial) is flat and $u_z = u$;
- There is no mixing in the axial direction;
- There is complete mixing in the radial direction.

3.2.2 How to Measure Plug Flow

Plug flow is easier to identify than to define; using a tracer is the most widely used methodology. In a tubular reactor, a tracer such as NaCl or KNO_2 with a known concentration and density can be injected at some point along the reactor and conductivity probes placed downstream to the injection point register the concentration of the tracer changing with time (so-called "C" curve or commonly known as the residence time distribution (RTD)), as shown in Figure 3.2. The C-curve spreads out gradually and levels off completely after an infinite length. The shape of the tracer spread is the measure

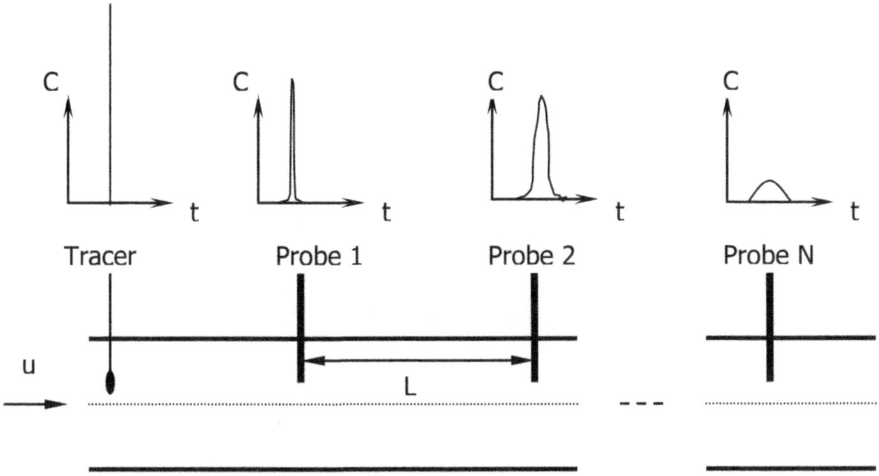

Figure 3.2 Concentration profile of a tracer travelling through a tube.[6]

of the degree of plug flow achievable. For plug flow, the area under the C curve equals to 1, and the width of the curve to 0. This is not possible in the real world, which is why ideal plug flow cannot be achieved.

Using the measured profiles of tracer concentration, the degree of deviation from the true plug flow (*i.e.* the spread) or the axial dispersion coefficient (D) can be quantified. The unit of D is taken as the area spread by tracer per time ($m^2\ s^{-1}$). The lower the value of D, the closer it is to plug flow. Figure 3.3 shows the concentration as a function of a dimensionless time for various axial dispersions. When the dimensionless group of D/uL equals zero, it is a true plug flow; when D/uL approaches infinity, the system behaves as a single continuous stirred tank, where L is the distance (m) between the tracer injection port and the location of the probe in question, which can also be the reactor length.

Another popular method to measure plug flow is to use the Tanks-in-Series model, which views the fluid as flowing through a series of equal-size ideal continuous stirred tank reactors (CSTRs). The number of tanks (N) can be evaluated experimentally from the RTD, Figure 3.4 shows the correlation between the number of tanks and the state of plug flow being achieved. When $N \rightarrow \infty$, it is a true plug flow. When $N = 1$, it is a CSTR system. This confirms that the state of plug flow can never be obtained in a single tank.

3.2.3 How Could Near Plug Flow Be Achieved in the Real World?

Near plug flow can be achieved by using either a number of CSTRs in series or turbulent flow in a tubular reactor. In the former, ideal plug flow would require an infinite number of CSTRs, but this is not feasible. In practice, three to eight CSTRs in series are the typical numbers in most companies I visited. The more reactors are used, the more inventory, capital and energy are needed, not to mention the fact that it is still far from a true plug flow.

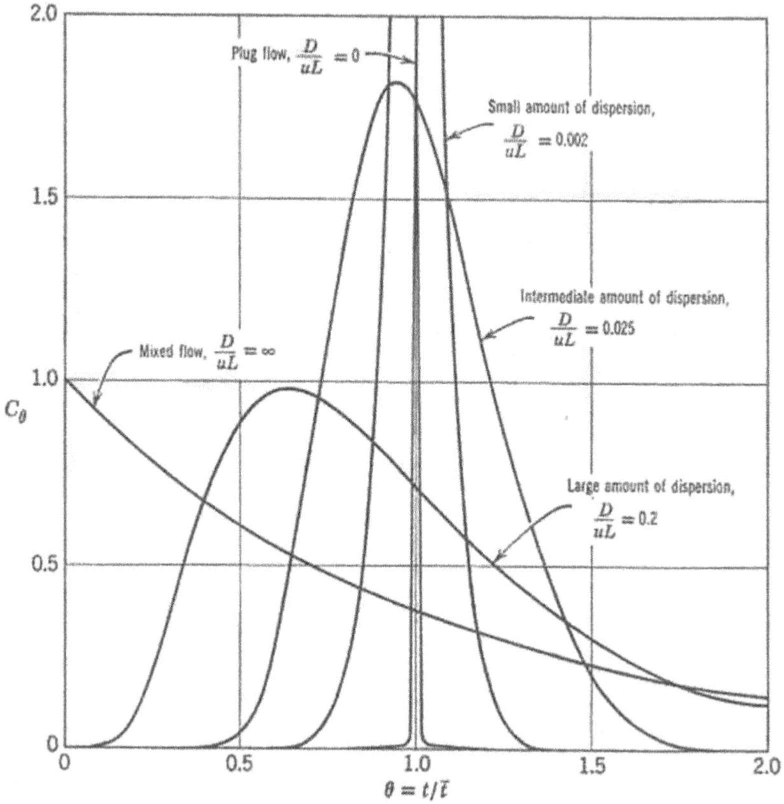

Figure 3.3 Trace concentration as a function of a dimensionless time as predicted by the dispersion model.[6,7]

The alternative is to use a tubular reactor with the main bulk phase flowing through the reactor, while other reactants are added at different stages. While this places lesser demand on energy and cost, the system has a vital drawback in that turbulent flow requires high flow rates at a given tube diameter and this translates as very long pipes for even low to moderate residence times. In reality, only a few reactions with very short reaction times ($\ll$ 5 mins) would be feasible which has limited the adoption of this type of reactor and kept the stirred tank reactor in business as the main workhorse of the chemical industry since the industrial revolution.

In short, fluids in plug flow have uniform velocities in the direction of flow (axial) and perfect mixing in the radial direction. This would provide the consistent fluid mechanical environment for crystallisation. Plug flow cannot be achieved in a batch stirred vessel and can only be realised in flows, hence the use of flow chemistry today.

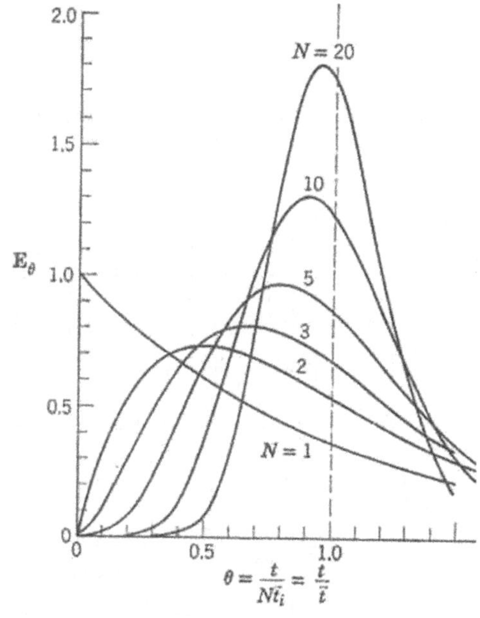

Figure 3.4 Residence time distribution curves for tanks-in-series model.[6,7]

3.3 Continuous Oscillatory Baffled Crystalliser

3.3.1 Principles

Following the above discussions, if plug flow could be achieved under laminar flow, we would have the best of both worlds. NiTech's continuous oscillatory baffled crystalliser (COBC) is one such device as shown in Figure 3.5. It consists of a tube with periodically spaced orifice baffles, superimposed with fluid oscillation. There are three key features for this type of system. Firstly, mixing is achieved by the generation and cessation of eddies. Each baffled cell acts as a CSTR; it is thus possible to have hundreds of "tanks" in series by simply arranging baffle cells in a serpentine fashion (Figure 3.5), leading to plug flow. Secondly, the mixing in the COBC is independent of net flow, so laminar flows can be operated, maintaining much longer residence times that cannot be achieved in turbulent flow systems. Lastly, the COBC allows well controlled mixing from plug flow to chaotic mixing conditions. There are various terminologies to be used in this chapter. In batch set up, OBC denotes oscillatory baffled crystalliser. When reactions are involved, it is an OBR, R stands for reactor. These apply to continuous set ups too with terminologies of COBC and COBR respectively. NiTech DN15 crystalliser is a COBC of 15 mm diameter (Figure 3.6). Figure 3.7 shows the residence time distributions (RTD) in a COBC, and the plug flow characteristics are clearly seen here. It must be noted that the shape of the RTD for plug flow must be in the form

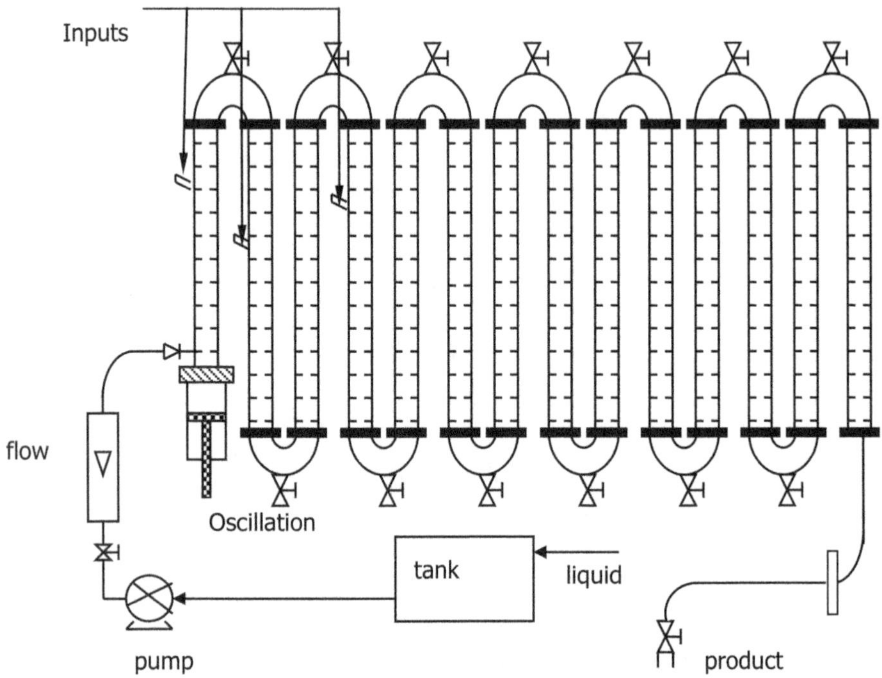

Figure 3.5 Schematic of a COBR (baffles in bends are not shown here, black blocks are the collars for sampling and PAT probes)[11] with permission from Elsevier, Copyright 2001.

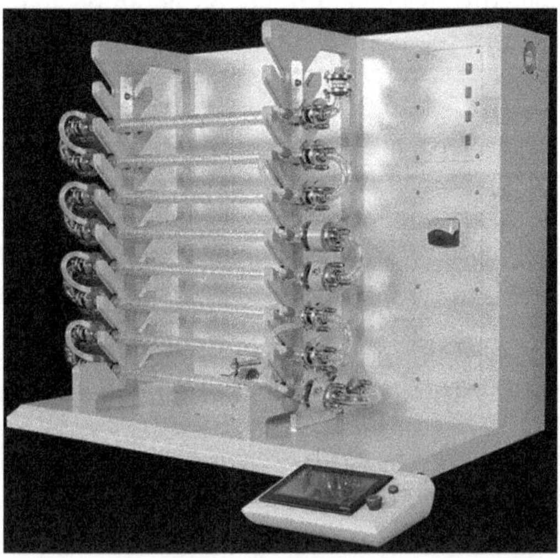

Figure 3.6 A photo of a NiTech DN15 Lite Crystalliser.

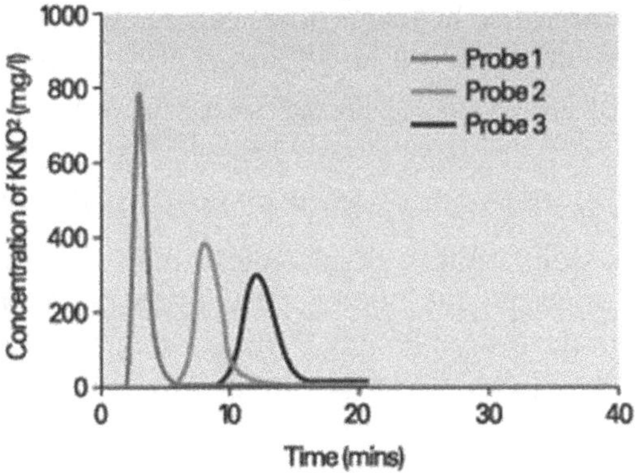

Figure 3.7 RTD profiles in a COBC, $Re_o = 4518$ (6 mm, 3 Hz); $Re_n = 504$ (1.5 L min^{-1}); Probes 1, 2 and 3 were located 3.7, 7.9 and 10.1 m away from the tracer injection port respectively.[12]

of a Gaussian function,[8] and any skewedness or deviation would indicate the existence of laminar flow. From the measured tracer concentration–time profiles in Figure 3.7, the axial dispersion coefficient, *D*, (how narrow the curve) and the number of tanks (*N*) were evaluated using the aforementioned models as 0.0004 $m^2 s^{-1}$ and 550 respectively.

The flow in COBC is laminar and the chaotic nature of the flow is created by eddies that are formed when fluid flows through orifice baffles; the turbulent kinetic energy so generated within each baffled cell is similar to that of turbulent flow.[9] The shear or strain rate in COBC is however lower than that of STR at the same power density,[10] and increases with the increase of oscillation frequency.

3.3.2 Mixing Evaluation in Single Phase

When using a tracer to generate RTDs and then evaluating the axial dispersion coefficient based on RTDs, this is really a single phase dispersion, as the tracer is a miscible liquid. By solving the governing equation of $\frac{\partial C}{\partial t} = D\frac{\partial^2 C}{\partial x^2} - u\frac{\partial C}{\partial x}$ with the help of tracer RTD, axial dispersion in COBC has extensively been studied,[8,13–16] where C is the concentration of tracer (kg m^{-3}) and *u* the net flow velocity (m s^{-1}). The density of the tracer is normally different (heavier) than that of water and such a difference promotes additional dispersion.[17] This density effect on dispersion can be compensated by increasing the density of the bulk fluid so that equal densities were made[8] or can be avoided by using non-intrusive tools, *e.g.* laser induced fluorescence.[18]

How mixing performance in a batch OBC can be quantified is still an open question. Flow visualisation was used initially;[19,20] then local and overall velocity profiles from particle image velocimetry;[21,22] shear or strain rate;[23] ratio of axial to radial velocity components and axial dispersion coefficient[24,25] evaluated from $\frac{\partial C}{\partial t} = D\frac{\partial^2 C}{\partial x^2}$ in absence of flow. Apart from experimental studies, numerical simulations were also performed using dispersion rate,[26] stretch rate,[27] shear strain rate history,[28] axial and radial stretching capacity,[28] RTD,[28,29] asymmetric double time,[30] mixing length and turbulent kinetic energy[9,31,32] and mixing rate[33,34] to provide the key indicators for assessing the mixing.

3.3.3 Mixing Evaluation in Two Phases

3.3.3.1 Liquid–Liquid

When two liquid phases are mixed in an OBC/COBC, one is often the continuous phase (the dominant one), another the discrete phase with generally different densities or/and viscosities from the main phase. A typical example of liquid–liquid mixing consists of an oil phase in an aqueous phase, which is the foundation of the suspension/emulsion polymerisation process. The oil phase (often monomer) must be dispersed as droplets in the aqueous phase as uniformly and small as possible so droplet size distribution is thus one of the key parameters for assessing the performance of mixing/reactor[11,35,36] and for predicting final product properties, *e.g.* particle size distribution.[37] A variant from this is inverse phase polymerisation where an aqueous phase is dispersed as droplets in a denser phase (*e.g.* oil).[38,39] Mixing delivers and maintains the balance of droplet breakage and coalescence, resulting in the onset droplet size distributions in either suspension or emulsion.[40] A wavelet method was used to analyse droplet and particle images[41] and population balance modelling has been applied to evaluate breakage and coalescence rates in this type of system,[42–45] while CFD predictions of oil droplet size distribution are still few and far between. Injecting a dye and analysing the colourimetry allowed the assessment of the mixing performance of two phases.[46] Manninen *et al.* (2013)[47] evaluated the axial dispersion coefficients when viscous non-Newtonian fluid as well as Newtonian fluids of different viscosities were involved in both moving fluid and moving baffle types of OBRs.

3.3.3.2 Solid–Liquid

There have been reports on solid–liquid mixing in OBR/COBR. Mackley *et al.* (1993) studied particle suspension in an OBR,[48] Liu *et al.* (1995) reported their study of carrot particle flow in a COBR[22,49] and Gao *et al.* (1998) investigated particle flocculation in an OBR.[50] Crystallisation is an effective example of solid–liquid mixing during the formation of solid crystals and a number of papers have been published on crystallisation (see Section 3.5), however few consider solid–liquid mixing. From a fluid mechanics viewpoint, solid crystal particles would follow the liquid phase movement, as the fraction of solid or solid loading is typically low (<20w/w%); particle size is small (50–200 s μm) and so is the density difference between particles and solvent/solvent mixture in crystallisation of organic compounds. Kacker *et al.* (2017) studied the RTDs of melamine particles (mean size = 100 μm, density = 1.57 kg m^{-3}, solid loading = 0.5, 1, 5 and 10 w/w%) in water in a NiTech DN15 crystalliser and found that a mixing intensity at 2 mm and 2 Hz was sufficient to obtain plug flow like mixing of all four slurry concentrations tested (Figure 3.8).[46]

When solid loading increases or size becomes larger or the density of particles is higher, terminal velocity of particles of a mean diameter would exceed the fluid velocities generated by both mixing and net flow; particles would then be sinking, affecting mixing and plug flow characteristics as well as crystallisation processes. In these cases, either oscillation velocity (in terms of the product of oscillation amplitude and frequency) or net flow rate or both must be increased to boost the fluid mechanical forces. There are however few papers considering CFD modelling in this respect, with the exception of two PhD projects from CMAC[51,52] where CFD is applied to predict residence time distributions of solid particles of different sizes and their effects on dispersion in COBC with the inclusion of particle interactions.[52–54] More data would not only enable better understanding of the interactions between the two phases, but also assist control, prediction and operation in COBCs.

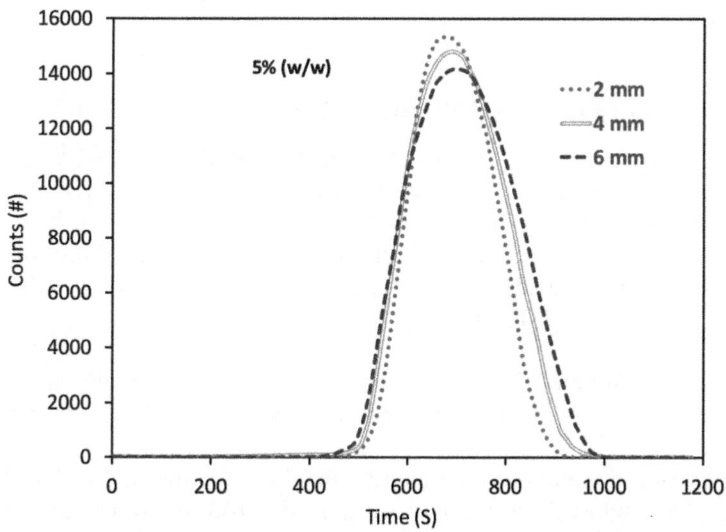

Figure 3.8 Particle count distribution, measured by FBRM at the exit of DN15 crystalliser (f = 2 Hz, flow rate = 100 mL min^{-1}). Adapted from ref. 46 with permission from Elsevier, Copyright 2017.

Residence time distribution of large solid particles in fluid suspensions was studied in a continuous tubular (without baffles) aseptic processing system and it was found that particle interactions promoted an increased mean residence time and dispersion, even for low particle concentrations; mixing particles with different characteristics led to a decrease in both the mean and minimum residence time of each type of particle.[55]

3.3.3.3 *Gas–Liquid*

Investigations of gas–liquid mixing in OBR were for the purpose of evaluating mass transfer performance where liquid is the dominant phase and gas the discrete phase, *e.g.* increase the mass transfer of O_2 or H_2 into liquid as dissolved gases. In OBR, the combination of increased residence time of bubbles, the more uniform bubble size distributions and the smaller bubble sizes together with the increased gas hold up contributed to the enhanced mass transfer rate as well as mass transportation rate in comparison with that from bubble columns and stirred tanks.[56–59] This feature has found applications in hydrogenation,[60,61] oxygenation,[62] hydroformlyation, heterogeneous catalysis,[63] fermentation[64] and so on, where a gas is a reactant. The presence of bubbles and their effect on crystallisation in COBC will be discussed in Section 3.4.

In summary, perfect Gaussian shaped RTDs are obtained when employing the stimulus tracer method; this indicates that the plug flow characteristics have been achieved in COBC. It should be noted that deviations from the Gaussian function occur when either solid or gas is present and appropriate adjustments of oscillation and flow conditions are required in order to maintain the degree of plug flow. In other words, the RTDs of a single phase only service as a guide for cases involving two or more phases.

3.3.4 Moving Fluid *vs.* Moving Baffles

In batch set up, there are two ways of generating oscillation: (a) pulsing fluid at the base or one end of the tube by a piston or bellows arrangement where baffles are stationary, (b) moving a string of baffles at the top by a linear motor, as shown in Figure 3.9. The former is used in all COBCs, and the latter is found in lab OBCs. Eddies are generated similarly in both cases, however, there are some subtle differences:

a) the baffle spacing in the moving baffle set up is usually larger than that in the moving fluid devices for enhanced mixing operation in OBCs;[20]
b) the axial dispersion coefficients in the moving baffle-COBR were 10–17% higher than in the moving fluid COBR due to enhanced shear rate in the former device.[47]

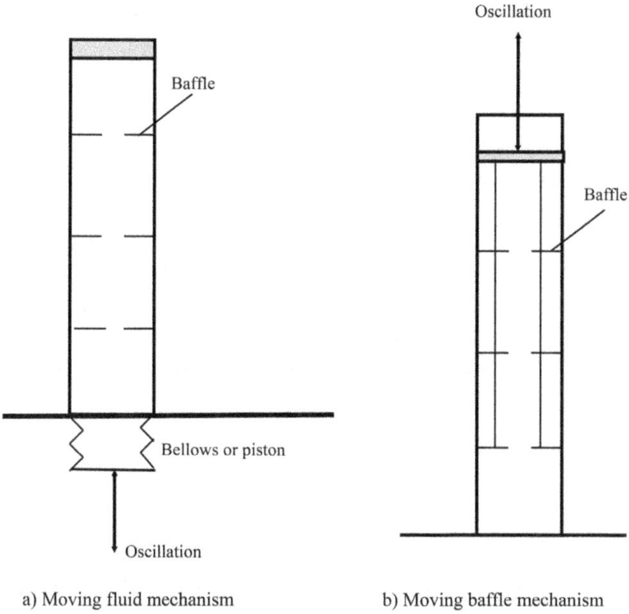

Figure 3.9 Oscillation mechanisms.

For compounds displaying chirality, scraping between the outer edge of the baffles and the inner surface of the wall in the moving baffle OBC can cause homogeneous nucleation of crystals of opposite handedness[65] in seeded and unseeded crystallisation (more details will be given in Section 3.5). In cooling crystallisation of L-glutamic acid, the STC failed to nucleate at the lowest concentration of 20 g kg^{-1}, while the OBC nucleated under the same time period,[16] indicating that scraping in the moving baffle OBC promoted nucleation more readily when compared to the STC system.

In an attempt to smooth the translation of results from batch moving baffle OBCs to COBCs,[16,66–69] a moving fluid OBC was implemented; however, surface renewal in the moving fluid OBC deposits a thin film of fluid onto the surface of column wall on the upwards stroke of oscillation and partial or full evaporation of this film of supersaturated solution leads to local high supersaturation and even more seed-dissimilar nucleation for compounds displaying chirality in the absence of scraping action.[70] It must however be stressed that

i) the effect of surface renewal can be minimized when the height of the stationary fluid in OBC is optimized;[68]
ii) the effect of surface renewal is completely absent in COBCs.

It is rather prejudiced to think one type of OBCs is better than another in terms of supplying process information for continuous operation, as there is no flow in both setups. To add flow onto the batch OBCs seldom works as the flow rate is too low to have any impact. The truth is that essential data for the design and operation of continuous crystallisation in COBCs can be generated from either type of OBC, *e.g.* metastable zone width, nucleation temperature, starting and end temperatures, crystal size distribution, crystal shape and purity, so on. The learnings from this evaluation work have successfully been implemented in COBCs, so much so that all NiTech feasibility works are still being done in moving baffle OBCs. In summary, the moving baffle and moving fluid OBCs are suitable for crystallisation of non-chiral compounds, reactive crystallisation and antisolvent crystallisation as well as reactions.

3.3.5 Scaling Up and Down

3.3.5.1 Scale Up

Plug flow is the epitome of chemical engineering and the uniform radial mixing is the key element within plug flow. What is its scaling up capability? Naturally the uniformity of mixing will utterly be lost when the diameter of the tube increases to that of industrial batch stirred tank crystallisers, with the reappearance of both mixing gradients and the inferior specific surface area as shown in Table 3.1. What is the definite scale up limit for COBC?

Investigations on scaling up OBR were undertaken

a) in mass transfer studies of air to water and air to yeast culture in 25, 50 and 100 mm inner diameter (ID) OBRs,[56,57,59] similar mass transfer coefficients were obtained for similar power densities;
b) in flow visualization and droplet size distribution studies in 50, 213 and 380 mm ID OBRs using silicon oil and water with oil fractions ranging 10 to 36% (v/v), linear correlations were observed using the combination of energy dissipation and oscillatory velocity as the basis;[71,72]
c) in evaluating axial dispersion coefficients in OBRs of 24, 54 and 150 mm ID, the axial dispersion coefficient was a linear function of both the oscillatory and net flow Reynolds number;[73,74]
d) using CFD simulations in 50, 100 and 200 mm ID OBRs by comparing the averaged mean velocity and the averaged mean velocity ratio of axial to radial components in an oscillation cycle, a linear trend was obtained.[75,76]

From the above, together with confidential studies in scaling up COBRs by NiTech funded by the Scottish Enterprise, the characteristics of plug flow with uniform radial mixing can faithfully be scaled up from lab to COBCs < 150 mm diameters. Such a diameter would cater for production rates of both speciality and commodity materials. At the time of composing this chapter, the largest diameter COBR installed in industrial operation is 60 mm ID at a flow rate of 0.115 $m^3\ min^{-1}$.

3.3.5.2 Scale Down

Due to the limited availability of new drug compounds, scaling down COBC becomes necessary, albeit challenging. Some work on meso COBR and meso structures[77] has been reported considering fluid mechanics[51,77–83] and applications[84–91] together with some confidential work on COBCs of 10, 8, 4 and 2 mm ID by NiTech funded by DTI and the Scottish Enterprise.[92] In addition, studies on mixing performance[93–96] and applications[97–99] using hollow micro-capillary tubes were also carried out.

For micro devices it is generally regarded that the diameter/width of the flow channel is smaller than 1 mm; however, the cut-off between meso and macro devices is less clear. In this book chapter, I categorize meso devices as having diameters of tube from 1 to 10 mm and macro COBC having diameters above 10 mm, for the purpose of simplification.

At a scale, the motion of fluids is governed by the Navier–Stokes equation, which is $f = ma$, written for a viscous fluid. It satisfies the conservation of mass, momentum and energy, where mass conservation is included implicitly through the continuity equation. The combined form for incompressible fluids is given as

$$\frac{\partial u}{\partial t} + u \bullet \nabla u = -\frac{1}{\rho}\nabla p + \frac{\mu}{\rho}\nabla^2 u + \frac{F}{\rho} \tag{3.1}$$

where p is the pressure (kN m^{-2}), μ the viscosity (Ns m^{-2}), F the body force (N) and u the flow velocity (m s^{-1}). The terms on the left of eqn (3.1) are the inert force and the terms on the right denote the pressure, viscous and external forces applied to the fluid. Does the flow velocity, u, have a range outside which the eqn (3.1) is invalid?

At macroscopic level, most forms of mixing, *e.g.* oscillating, stirring, turbulent flow, are convective motions, effectively to break and reduce the length scale, allowing diffusion to take place and increasing the local magnitude of mass transfer by diffusion. The upper limit for the flow velocity in the Navier–Stokes equation is equal or lower than the speed of sound.[100]

In small-scale flows, however, the Navier–Stokes description is expected to fail[101,102] when the characteristic hydrodynamic lengthscale approaches the fluid "internal" lengthscale.[103] What is then the fluid "internal" lengthscale? Taking a water molecule for example, it has a size scale of 10^{-10} m. In order to satisfy the continuum equations, we need to define the macroscopic variables, *e.g.* how many molecules are needed to draw a reliable average? 10 is probably too few; 100 seems to be reasonable (10^{-8} m). This means that each "point" in a water continuum is a box with the side about 0.01 μm. To satisfy the macroscopic equations, we need a control volume, say 10 times larger than the "point box", *i.e.* about 0.1 μm. Bear in mind however that this control volume must in turn be much smaller than the computational domain, giving another factor of 10–100 at least, resulting the "internal" lengthscale of 1–10 μm. Thus, the continuum equation cannot be expected to be reliable below 1–10 μm.

In addition, when modelling fluid motions in a microchannel of diameter of 10 μm (10^{-5} m) using CFD codes, if the mesh in the *Y* direction has 100 cells (in the control volume), each cell thus has a scale of 10^{-7} m, which would be smaller than the "internal" lengthscale as discussed earlier. Furthermore, the no-slip boundary condition at the wall commonly fails at a scale of 10^{-7} m.[104,105]

What about meso devices? The flow in meso COBCs is laminar, often in the Stokes (or creeping) flow regime, in order to accommodate the required residence time. There have been debates on why there is no diffusion velocity in the macroscopic Navier–Stokes equations,[106,107] as diffusion is an irreversible flow, while convection is regarded as a reversible event. From a CFD viewpoint, modelling of diffusion velocity is often done with the extended Navier–Stokes equations, where more than one mass conservation equation is solved together with other terms (*e.g.* chemical reactions, density differences) that could create a concentration gradient to act as the driving force.

The relativity of convection and diffusion was captured by a dimensionless number called the Peclet number ($Pe = Lu/D$),[108,109] where L is a characteristic length scale (or reactor system). When the Peclet number is greater than one, the influence of convection exceeds that of diffusion in balancing the overall mass transfer or mass flux, as in macro COBCs. When the Peclet number is less than one, diffusion plays a more significant role in mass transfer than convection, like in micro channels. It is however difficult to quantify the relative contributions of convection and diffusion in meso COBCs. Figure 3.10

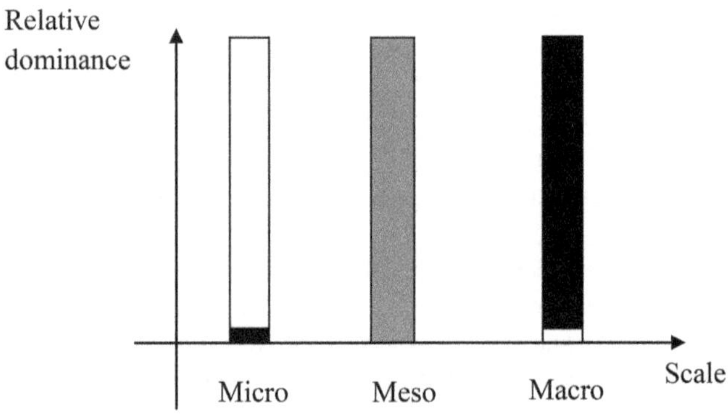

Figure 3.10 Illustration of degree of significance *vs.* scale (Filled rectangles denote convection, empty rectangles denote diffusion, grey colour refers to unknown).

above is what I understood about the relative dominance, *e.g.* convective mixing in macro crystallisers, diffusion mixing in micro channels and somewhere in between in meso devices.

Using the above analysis of the "internal" lengthscale, it is reasonably assumed that the Navier–Stokes equations are still applicable to flows in meso devices, as both the continuum and the no-slip boundary conditions are satisfied. Irrespective of the applicability and validity of the Navier–Stokes equations to diffusion driven flows, how is mixing achieved in a micro/meso channel with laminar or Stokes flows? The fact is that this phenomenon is not new and was demonstrated by Taylor[108–113] some 60 years ago; mixing in small channel devices is achieved in a number of ways:

a) Taylor diffusion allows mass transfer normal to the fluid flow;[110,112]
b) A longer channel or slower flow gives the fluid more time for diffusion;
c) Reversible flows due to oscillation change the directions of flow, shorten the diffusion length and lessen the needed diffusion time;[114]
d) Flow over bends induces vortex shedding with the same above effect;
e) When there are two liquids with different densities or concentrations, a density or concentration gradient is created, enhancing diffusion.[115]

3.3.6 Power Dissipation

How do we compare process performances (*e.g.* mixing, yield, conversion, selectivity, *etc.*) in different types of crystallisers or reactors? Taking the 100 m sprint for example, if you race against Usain Bolt, you will lose every time, because Usain Bolt is an elite athlete and the fastest sprinter in the world, while you are a club runner. We would accept the outcome without asking any questions. If Usain Bolt is racing against you on a motor bike, you will win the race and Usain Bolt would probably get very upset. Is this a fair competition? This leads to the key question: what is the *common* basis for comparison? From the above example, power consumed by the motor bike is more than that by the human being so the sprint time divided by the power consumption would provide a level playing field for comparison of different racing modes; in this case Usain Bolt will win it again.

These exact principles are applicable for comparison of process performance between different devices, *e.g.* OBR and STR. The basis for this is the power density or dissipation rate (W m^{-3}). This is the power dissipated into the liquid, or what liquid is experienced within the reactor. Note that it is not the electrical power of the motor that drives either oscillation or stirring.

Two main models deal with the power density in OBR: a quasi-steady flow model and an acoustic model. The former was initially proposed by Jealous and Johnson (1955)[116] and it has since been applied to the OBR.[117–119] It was defined by the equation:

$$\frac{P}{V}=\frac{2\rho N_{\mathrm{b}}}{3\pi C_{\mathrm{D}}{}^{2}}\left(\frac{1-\alpha^{2}}{\alpha^{2}}\right)x_{0}^{3}\omega^{3}\quad\left(\mathrm{W\,m^{-3}}\right)\tag{3.2}$$

Where ρ is the fluid density (kg m^{-3}), N_b the number of baffles per unit length of reactor (m^{-1}), α the baffle restriction ratio (= $(D_o/D)^2$, where D_o the baffle diameter and D the tube diameter), x_o the oscillation amplitude (m), ω the angular oscillation frequency (= $2\pi f$) and C_D a discharge coefficient for the orifice (normally taken as 0.7,[120–122] although the value depends on the baffle thickness and α[123]). Baird and Stonestreet (1995) suggested that the quasi-steady model was suited for high amplitudes (in the region of 5–30 mm) and low frequency (order of 0.5–2 Hz) oscillation.[119,124]

At higher oscillation frequencies (3 to 14 Hz) and lower amplitudes (1 to 5 mm), the quasi-steady model becomes ineffective at describing the power density. Baird and Stonestreet (1995)[125] proposed the eddy enhancement model, which is defined by:

$$\frac{P}{V}=\frac{1.5\rho x_{0}^{2}\omega^{3}l}{L\alpha}\quad\left(\mathrm{W m^{-3}}\right)\tag{3.3}$$

where L is the baffle spacing (m) and l is the mixing length (m), with the same order of magnitude as the column diameter. The challenge for this model would be the determination of the mixing length. The recent CFD work revealed that the application range originally specified for each model was a myth rather than a fact; both models gave very similar predictions of power density for a wide range of geometric and operating conditions, and can be used interchangeably with high confidence.[126]

The power density of a stirred tank crystalliser is defined in[127]

$$\frac{P}{V}=\frac{P_{0}\rho N_{s}^{3}D_{s}^{5}}{V_{\mathrm{L}}}\quad\left(\mathrm{W\,m^{-3}}\right)\tag{3.4}$$

where N_s is the speed of the stirrer (rps), D_s the diameter of the stirrer (m), V_L the volume of liquid in the STC (m^3) and P_o is the dimensionless power number of the agitator that depends on the type and diameter of blades and the distance with respect to the bottom of the vessel.

It should be stressed that the above equations for evaluating power densities were derived from theories and experimental validations are much needed. Recently a CFD work generated data on pressure drop and phase angle (between pressure and velocity) in a COBR, which were used to validate both power density equations.[126]

3.4 Design and Operation of Continuous Oscillatory Baffled Crystalliser

When cooling crystallisation is carried out in batch crystallisers, be it a lab or production scale, the whole mass of solution is cooled from a starting to an end temperature; this can be regarded as a temporal domain process. The operational principles in continuous crystallisation in COBC are different from the aforementioned in that

a) the mass of crystallisation solution is spread out in the spatial domain in COBC with a very small fraction of it cooled at any given time;
b) while both the start and the end temperatures are the same as that of batch operations, temperature zones are firstly created along the length of COBC;
c) a small fraction of the solution goes through the temperature zones from tropic summer to summer temperature; from summer to autumn, then out of the crystalliser in one pass fashion.

There are new concepts as well as methodologies that can be used to extract real growth rates, steady state characteristics *etc.* in COBC, which will be discussed later.

3.4.1 Linking the Design and Operation With Science

The above statement is common sense, but it is however difficult to follow through in practice in large batch vessels, even though this is routinely achieved in lab scale crystallisation. Figure 3.11 below is a typical diagram showing the metastable zone width (the temperature difference between the supersolubility and solubility curves); the thick black line represents the optimal operational path. The primary aims are to avoid crossing the supersolubility curve where spontaneous nucleation takes place and to operate as close to the solubility curve as possible, since the supersolubility is not well defined and depends on a number of parameters that are scale dependent,[128] such as temperature profile, the rate of generating supersaturation, solution history, impurities and fluid dynamics. When seeds are used (this is the dominant methodology in industry), it is desired to add seeds at the start of the metastable zone.

It reads simply, but this is intrinsically hard to do in industrial crystallisers! Firstly, thermocouples are not long enough and can only be located near the walls of industrial crystallisers, so the temperatures are thus local,

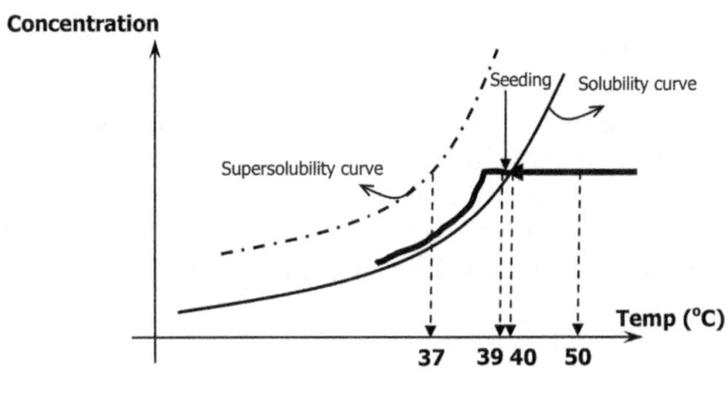

Figure 3.11 Cooling crystallisation profile linking science with design and operation.

not overall, measurements. Secondly, local temperature within industrial crystallisers can vary by a few or over 10 s of degrees depending on scales. Thirdly, some clients have no information/data on supersolubility. What chances of success are there, when the combination of all the above is unknown and changeable? These are the roots of industrial scale crystallisation.

In continuous operation, data for metastable zone width, temperature profile and concentration of solute *etc.* can be identified/measured/traced throughout the operation. As an example of design and operation, I use a challenging batch seeded cooling crystallisation of an API to illustrate the principles. This API had a very narrow metastable zone width of 3 °C, the starting temperature of 50 °C was cooled to 10 °C, while the saturation temperature was 40 °C. Seeds were continuously added at 39 °C (Figure 3.11). With this information and without constraints of heat and mass transfer, the operation was readily designed as

$$\text{Crystallisation time} = \frac{(50-40)\ ^\circ\text{C}}{5\ ^\circ\text{C/min}} + \frac{(40-10)\ ^\circ\text{C}}{1\ ^\circ\text{C/min}} = 32\ \text{min}$$

Cooling from the start temperature in the unsaturated region to the saturation temperature has no effect on crystal properties, and can be done as fast as one wishes, *e.g.* 5 °C min^{-1} here, and this also reduces the overall crystallisation time. This would be unlikely to be feasible in most industrial batches. Different cooling rates can then be selected when cooling from the saturation temperature to the final temperature. In principal, the slower the cooling rate, the larger the crystals, the better the filtration and equally the longer the crystallisation time. With the achievement of plug flow at laminar flow and the associated superior properties of heat and mass transfers, faster cooling rates in COBC yielded similar results to these found in traditional large batches and this makes the selection of cooling rates an easy task.

Once the crystallisation time had been determined, the flow rate was then calculated based on a fixed length of a NiTech DN15 crystalliser and, in turn, the different temperature zones were implemented. For this particular application with a very narrow metastable zone width, four temperature zones were used, *i.e.* 50–40 °C, 40–30 °C, 30–20 °C and 20–10 °C, as shown in Figure 3.12. Generally speaking, the narrower the metastable zone width is, the worse the batch scale up becomes and the more temperature zones it requires in continuous operation. For most APIs and chemicals that I have been involved with, two to three zones were sufficient. Note that the flow rate is common to all temperature zones, this means that the length of DN15 for each temperature zone is different and adjusted by the flow rate and the temperature of coolant in the jackets (shell side).

3.4.1.1 Start-up Process

The start-up procedure involves pumping the solvent through the DN15 and back to the solvent tank, while the feed tank is closed. This process has two purposes: to degas the DN15 crystalliser and to aid the establishment of the temperature zones. With the help of a degas mode included in all NiTech DN15 series now, degas is much quicker. Tube side temperatures at each critical location along the DN15 are continuously measured and the shell side flow rates are automatically adjusted, facilitating the setup of temperature zones. Once the temperature zones have been confirmed, the solvent feed is replaced by the crystallising media; the flow rate in this case was 52 mL min^{-1}, giving a net flow Reynolds number of 110.

3.4.1.2 Operation

The small amount of solution (52 mL per min) was pumped through the crystalliser, went through the regions with temperatures from desert summer to tropical summer (zone 1), where seeds were continuously fed at the precise temperature of 39 °C (Figure 3.12); from tropical summer to summer (zone

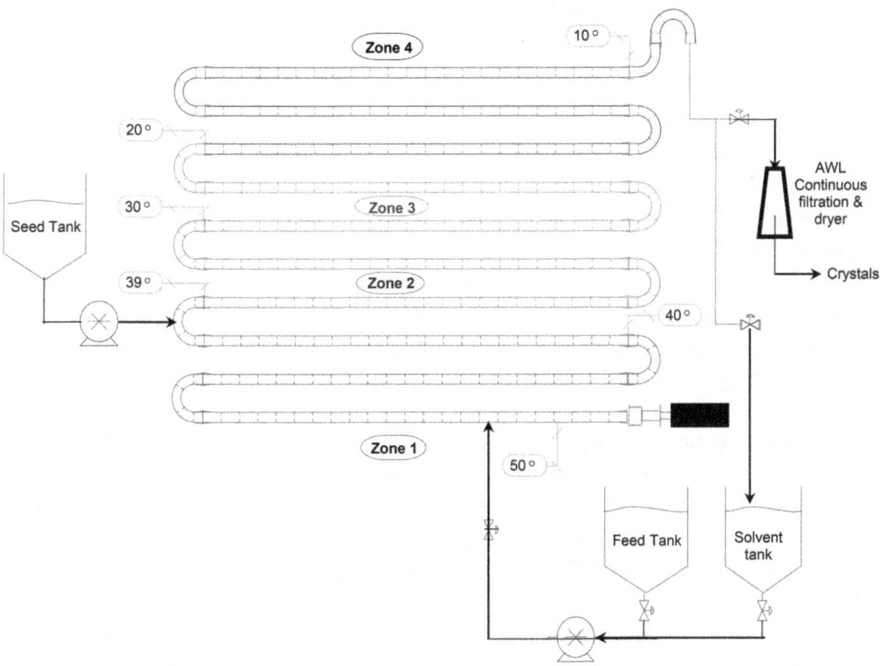

Figure 3.12 Design and operation of a cooling crystallisation in a NiTech DN15 crystalliser.

2), from summer to autumn (zone 3) and from autumn to British summer (zone 4). Crystal particles were well and uniformly suspended along the crystalliser, exited directly into a continuous filtration and dry unit (Alconbury Weston Ltd)[129] and dried crystals were continuously harvested. In this way, the science of crystallisation (Figure 3.11) is closely linked with the design and operation (Figure 3.12), with data-logging continuously. The crystallisation process was run nonstop for 8 hours until the seed tank was emptied; this "cannot be scaled up" industrial batch crystallisation process was made easy here.

3.4.1.3 *Shut Down Process*

The shut-down procedure involves firstly replacing the crystallising feed by the solvent feed and secondly stopping the seeds *after* the solvent has passed through the seeding port. The solvent takes one residence time (RT) to come out of the DN15 and the product crystals can still be collected shortly before reaching this RT; the exit was directed to the waste for a short period, then directed back to the solvent tank after the solvent flow has passed one RT. The temperatures in all zones were then increased to 50 °C to dissolve any residual and make ready for the next run.

3.4.1.4 *The Presence of Bubbles*

The presence of bubbles is a routine event in traditional stirred tank crystallisers; bubbles can come from the solvent tank or through pumping or heating. These bubbles usually rise and coalesce and some would escape from the liquid medium into the free surface while some would remain within. There have been reports of the effect of cavitation on nucleation, for example, using ultrasound, where it influences the primary nucleation process by reducing the induction time and MSZW;[130–133] affects crystal growth[134,135] to a lesser degree than nucleation and improves crystal purity,[136,137] while the underlying mechanisms of this effect are not yet well understood. Ultrasound has also been utilized in continuous crystallisation.[138–141] The reasons for initiating nucleation include the formation of local hot spots due to large energy release on the collapse of cavitation bubbles, creating localised regions of extremely high temperature and pressure,[142,143] or due to the rapid cooling which follows, shockwaves are released from cavitation bubbles.[144,145]

There is a degas mode in NiTech DN15 crystallisers that minimizes the presence of bubbles in the system at the start of operation, however, a few visible bubbles may still be present. It is important to emphasize that the sizes of these bubbles are many orders of magnitude larger than that of the aforementioned minute bubbles caused by ultrasound bombardment and are thus too big to affect nucleation in both seeded and non-seeded crystallisations. Because of the low number of the bubbles, they are discrete, do not coalesce and hence have little effect on mixing. In addition, the bubbles move forward gradually and leave the DN15 crystalliser. As a result, a few visible bubbles in DN15 are as normal as those in industrial batch crystallisers.

It should however be stressed that when bubbles coalesce to form a gas layer at the top of the DN15 crystalliser due to either the boiling point of a particular solvent being lower than the operating temperature or the gas volume fraction exceeding the operational boundary in COBC (Section 3.7), this would severely affect the propagation of the oscillation wave and in turn the mixing, as well as crystallisation. Bear in mind that this is the extreme case, either a new solvent or elevated pressure operation is required (see Section 3.6 for more details).

3.4.1.5 Generic Comments

From the above example and operational principles, the following advantages of continuous crystallisation can be drawn:

a) Fluid mechanical conditions (mixing is one element of these) are consistent once plug flow is achieved;
b) Heat transfer is enhanced, since there is no mixing or concentration gradients;
c) The increased specific surface areas allow any type of cooling profiles to be readily implemented;
d) Crystallising solution goes through the COBC in a one pass fashion and harvesting crystals, such as by filtration and drying, can also be done continuously;
e) There is no accumulation or storage required, avoiding the potential of entrapping impurity;
f) All the above can be obtained in lab, pilot and full scales.

Crystallisation was singled out in 2005 by NiTech as the platform process for NiTech continuous technology; a large number of organic (some inorganic) chemicals, pharmaceuticals, speciality and food compounds have since been tested using NiTech DN15 series, covering a wide range of solubility and crystal shapes; the consistent outcome is that it works and delivers the above benefits when operated correctly (See Section 3.5.4) and within the operational boundary of DN15 (See Section 3.7).

3.5 What Has Been Done?

3.5.1 Cooling Crystallisation

3.5.1.1 Unseeded Cases

Nucleation is the very first step of the crystallisation process when the first nuclei appear from the solution that is supersaturated. Considering the solubility diagram in Figure 3.11, it is certain that nucleation won't happen when the solution is in the unsaturated region, and that spontaneous nucleation will occur when the cooling exceeds the metastable zone width and should principally be avoided. There have however been many successful crystallisations of *inorganic* compounds that utilize spontaneous nucleation as the means of initiating nucleation, *e.g.* pigments, iron oxides, inorganic salts *etc.* In crystallisation of inorganic materials, water is largely used as the sole solvent; dissociation reactions, ionic interactions and pH play a major role in determining the solubility of a particular inorganic species in aqueous solution, as most inorganic materials are ionic.[146] Precipitation is perhaps a better phrase for inorganic crystallisations. The subtle differences between precipitation and crystallisation have been well explained.[2] The former is generally initiated at high supersaturation, resulting in fast nucleation and a large number of very small primary crystals. Agglomeration often occurs soon after nucleation and aging or maturation then follows covering all irreversible changes that take place in a precipitation after its formation.[2] Unlike crystallisation of organic species where a wide range of solvents or/and solvent mixtures can be used, increasing the varieties of interactions between solutes and solvents and complicating the process operation, control and prediction, the formation of hydrates (water molecules constitute part of the crystal structure) results in perhaps the most complications in unseeded crystallisation of inorganics, *e.g.* some compounds form several hydrates with varying water contents. As a result, there are neither issues of morphology nor of size distribution, while yield is the main focus, *e.g.* iron oxides, which were one of the first industrial inorganic compounds I was involved with. Seeded precipitations are occasionally operated, but rather crude methods of seeding are practised, *e.g.* bag loads of product crystals are dumped into crystallisers *via* a manhole.

When nuclei collide with the walls of the crystalliser and with other nuclei, this could lead to breakage of nuclei that could act as seeds for additional nucleation. The perquisite for this to happen would be that a sufficient number of nuclei of reasonably larger sizes would be available in the vessel, however, sizes of nuclei were small from spontaneous nucleation in this case.

It should be stressed that while spontaneous nucleation has been used in both batch and continuous crystallisations, spontaneous nucleation as the means of generating nuclei seldom works in tubular continuous crystallisers for organic compounds, as encrustation occurs, leading to blockage[16,66] (see Section 3.5 for more details).

3.5.1.2 Seeded Cases

Any crystallisation textbook would tell you that within the metastable zone, nucleation may or may not take place, but what does it mean in reality? It is very rare, or we can use the word "certain" again, for homogeneous primary nucleation to take place within the metastable zone. This is the "may not take place" part, while the "may take place" part contains a significant amount of uncertainties where dusts, impurity or any residual left behind from a previous batch could lead to heterogeneous primary nucleation.

What has been overlooked is the primary nucleation caused by variations of local temperature, *i.e.* local supersaturation, within an industrial scale crystalliser, which can vary from a few to 10 s of degrees depending on scales. The added practical difficulty is that it is not feasible to measure such variations in large vessels as most thermocouples can only be located near to the walls of crystallizers. When dealing with organic compounds with narrow metastable zone width, the local temperature variations can readily overshoot the metastable zone, and lead to spontaneous nucleation. Because of the local and random nature of such temperature variations, it is impossible to forecast and prevent these unwanted nucleation moments.

Seeding can readily be adopted in COBC, as illustrated earlier in Section 3.4. Some research works of both public and confidential nature have been undertaken in OBC/COBC,[16,147,148] *e.g.* cooling crystallisation of paracetamol,[149–152] L-glutamic acid,[16,147,153–155] aspirin,[156] a pharmaceutical API,[157] sodium chlorate,[65,69,70,158] adipic acid,[159] urea,[160,161] mannitol,[67] α-lipoic acid : nicotinamide co-crystal,[162] palm oil[163] and cyclopentane hydrate.[164] Seeded cooling crystallisation in NiTech DN15 crystallisers has successfully been running[165] for 8 h,[166] 10 h[147] and 4 weeks in CMAC at the time of writing this chapter.

As the nucleation process is too fast to be captured, nucleation and growth are often lumped together. To be able to separate and control nucleation from growth has been at the heart of crystallisation research worldwide. When tubular crystallisers are used, this presents new possibilities and feasibilities of investigating this subject,[167,168] *e.g.* nucleation was generated by adding two streams in a "Y" shaped tube, which was fed to a continuous device for growth,[66] the challenges were from encrustation in the nucleator due to the unquantified amount of nuclei generated. More research on separating nucleation from growth are needed.

3.5.2 Antisolvent Crystallisation and Seed Generator

3.5.2.1 Antisolvent Crystallisation

Tubular plug flow crystallisers enable multi inputs and hence are well suited for antisolvent crystallisation operation. There are a few of ways to initiate the meeting of solvent and antisolvent, *e.g.* a "Y" shaped tube, an "L" shaped nozzle or a lantern ring within the COBC. The degree of supersaturation within COBC can be adjusted by the flow rates of solvent and antisolvent, which effectively controls the sizes of crystals. Generally speaking, the higher the initial supersaturation, the smaller the crystal size. Supersaturation increases significantly once the two streams meet, then drops sharply; efficient mixing is crucial in order to suspend the mass of solid crystals emerging at the point of contact, which has been the common practical problem in traditional stirred tank crystallisers. The combination of the enhanced mixing that suspends crystal particles along the COBC and the plug flow characteristics that move the generated crystals downstream from the meeting point gives the best of both worlds. As antisolvent crystallisation is normally carried out isothermally, this becomes a relatively simple operation. Some experimental and modelling work on antisolvent crystallisation are in the public domain in COBC[169–171] and other devices.[172] Successful development and control of an antisolvent crystallisation for the production of a pharma API was reported where online FBRM and ATR-IR data were used to control concentration and maintain relative supersaturation.[173] At the time of composing this chapter, a 3 day continuous campaign of antisolvent crystallisation in COBC has been successfully performed.[174]

There are few industrial processes that require both antisolvent and cooling crystallisation steps with essentially two purposes. The first is to separate nucleation from growth as we discussed in the previous section. The second is to rely on antisolvent crystallisation to generate small crystals or crystal agglomerates, then to employ a cooling crystallisation step to either grow crystals or maintain these agglomerates for better filtration.

3.5.2.2 Seed Generator

Seed generation is notoriously time and energy consuming. Generally, the smaller the size and the larger the surface areas, the better the seed production with less seeds needing to be used. Traditionally this is done by either wet or dry milling. The former method entails smashing a solution containing crystals using a high shear stirrer, *e.g.* Silverson homogeniser, in either batch or flow set up. The batch set up resembles a kitchen smoothie machine and foams are often formed when air is entrained into the solution when operating at high rotational speeds; as the majority of the crystals are migrating into the foams, this defeats the purpose of milling. Antifoam agent can be used to alleviate this phenomenon, but the removal of this antifoam brings about more problems. The flow set up necessitates pumping and recycling the solution containing crystals from a tank through a rotor-stator type of high shearing device,[175] or using a turbulent jet.[66] Seeds generated this way often undergo agglomeration, but an ultrasonic device can be used to control and reduce this effect, although lengthening the process.[16]

Dry milling literally involves a ‘pestle and mortar’ type of device to grind crystals, either manually or by machine. This breaks up crystals and the required crystal sizes are obtained by sieving. The grinding process is repeated for oversize crystals, while the fines so generated are waste with hazards. It can often take weeks in order to achieve seed loading for one batch operation. This is not only energy intensive, but also un-scalable.

While the labile region is to be avoided at all costs in cooling crystallisation in COBC, operating antisolvent crystallisation in COBC in the labile region presents an elegant means of generating seeds of small crystal size and large crystal mass, the two essential criteria for seeds. The following is one example. The solvent was a mixture of salicylic acid in isopropyl alcohol (IPA) and the antisolvent was water. By using different flow rates of the two streams, very high initial supersaturation rates were achieved (Table 3.2). Figure 3.13 shows the operational path, *i.e.* it was deeply in the labile region.

Figure 3.14 shows the COBC set up. The system consisted of two straight parallel sections converging into a third straight section with a U-piece, each section was a jacketed glass tube Ø15 mm and 700 mm long with PVDF (polyvinylidene fluoride) baffle inserts fitted throughout. Solvent and antisolvent were held in two 20 L jacketed stirred tanks and pumped into the COBC *via*

Table 3.2 Experimental conditions for seed generation[65,170] with permission from Elsevier, Copyright 2015.

Run	Solvent flow rate (mL min^{-1})	Antisolvent flow rate (mL min^{-1})	Residence time (min)	Initial supersaturation ratio (S)
A	40	40	1.5	5.56
B	40	40	1.5	5.56
C	40	40	1.5	5.56
D	40	60	1.25	8.16
E	40	80	1.0	11.71

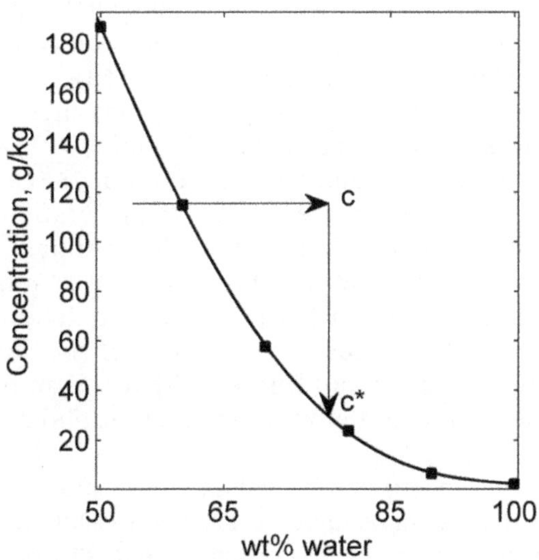

Figure 3.13 Solubility profile of antisolvent crystallisation with a mixture of salicylic acid in IPA as solvent, water as antisolvent.[170]

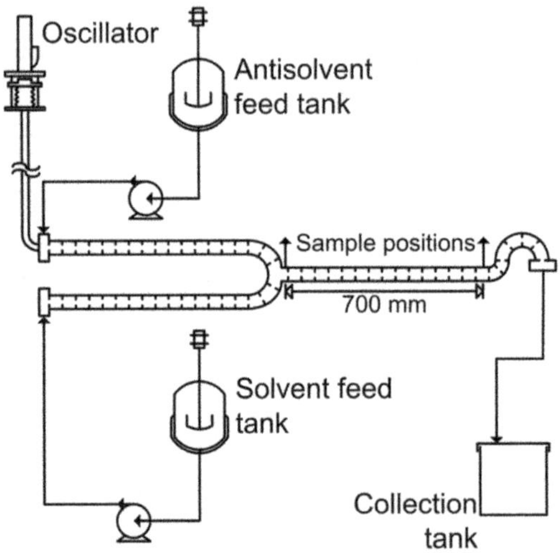

Figure 3.14 Antisolvent crystallisation in COBC as the seed generator[170] with permission from Elsevier, Copyright 2015.

peristaltic pumps (Watson Marlow) through 3.2 mm santoprene tubing. Oscillation was supplied by a PTFE (polytetrafluoroethylene) bellows and linear actuator (Dunkermotoren) configuration to one of the straight inlet sections as shown in Figure 3.14.

Sampling collars were fitted between the outlet of the U-piece (55 mm from the contact point where the two streams meet) and at the end of the third straight outlet (785 mm from the contact point) to allow for process monitoring. Sample solute concentration was measured by FTIR spectroscopy with a wavenumber range of 600–2000 cm^{-1}. Concentration of salicylic acid was measured by following the peak height between 1631 and 1647 cm^{-1}. The crystal size distribution of the solid samples was measured by Malvern Mastersizer S with a measurement range of 0.1–300 μm.

All experiments were undertaken isothermally at 25 °C. To ensure that a steady state operation was achieved, the total duration of the experiments ranged from 50 to 180 times the residence time.[170] Samples were withdrawn from both sample points of the COBC at 15 min intervals and filtered with a 0.45 μm syringe filter (Millipore) to determine the solute concentration *via* FTIR spectroscopy. Samples of the slurry from each sample point were also taken and filtered with a 1.2 μm filter (Whatman GF/C) on a Büchner funnel under vacuum and the solid phase dried overnight at 60 °C.

The top (a) and bottom (b) graphs in Figure 3.15 plot the saturation and particle size *vs.* the number of residence times in the COBC, respectively. It can be seen that the saturation reached a steady state (0.97 ± 0.048) for all cases when solvent and antisolvent flows were equal (Runs A, B & C in Table 3.2); saturation values were very similar at both sampling locations, indicating

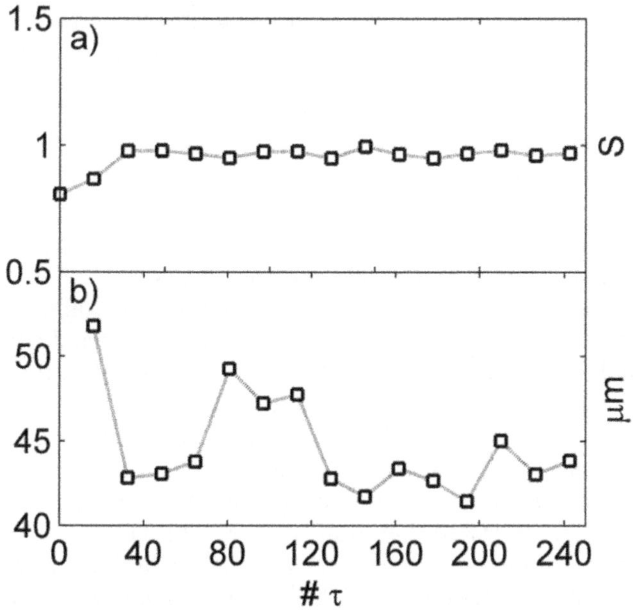

Figure 3.15 Profiles of saturation and crystal sizes in antisolvent crystallisation seed generator[170] with permission from Elsevier, Copyright 2015.

that crystallisation was rapidly completed near the point where the two streams met; saturation became unstable when flow rates were unequal (Runs D & E in Table 3.2).

The steady state of the mean crystal size (45 ± 3.01 μm) was achieved at the 2nd sampling point with equal flow rates of solvent and antisolvent (Runs A, B & C), and was not established when unequal flow rates were employed (Runs D & E). In the extended operation of the steady states, 884 g of seed crystals of uniform size were produced over 6.25 h (or 0.141 kg h^{-1}). These particles are perfect for seeds.

3.5.3 Nucleation by Scraping

Sodium chlorate is an intrinsically achiral inorganic compound, but crystallizes as two enantiomorphic chiral crystals in the cubic space group $P2_13$.[176] The usefulness and the uniqueness of this compound is that the chiral status at the end of crystallisation can retrospectively refer to the nucleation event. In 1898, Kipping and Pope[177] measured the handedness of 3137 $NaClO_3$ crystals from 46 separate crystallisation tests, and showed 50.08% : 49.92% of dextrorotatory (*R*) to laevorotatory (*L*) crystals. This was well within the statistical error of the unbiased 50% : 50% distribution they anticipated. Kondepudi *et al.* repeated the 19th century work of Kipping and Pope and obtained equal numbers of left and right handed crystals in evaporation crystallisation of sodium chlorate[178] through a non-stirred spontaneous nucleation. On the introduction of stirring at 100 rpm, however, >99% crystals displayed a single enantiomorphism. They postulated that this chiral symmetry breaking on stirring was due to secondary nucleation in which a "mother crystal" struck by the stirrer cloned secondary nuclei that have the same chirality as their "mother". They further explained that the depletion of the solute from the solution due to the growth of the primary "Eve" nucleus and the secondary nuclei that were rapidly generated from it, could have reduced the concentration to a level at which the rate of generation of new nuclei of opposite handedness from primary nucleation was effectively supressed. Their hypothetic mechanism was confirmed by the video images of McBride and Carter[179] and supported by the concentration measurements where the solute concentration in the stirred case was much lower with time, due to enhanced secondary nucleation,[180] than that in the non-stirred situation.

Viedma[181] created some innovative experiments in stirred crystallisation of $NaClO_3$ and demonstrated that total symmetry breaking can be achieved by massive instantaneous primary nucleation. This broadened the mechanism proposed by Kondepudi *et al.*[178,180,182] in that the real chiral symmetry breaking occurs in the first steps of the primary nucleation, whereas secondary nucleation only amplifies the symmetry breaking, known as the Viedma ripening. Following on, Viedma[183] showed that complete homochirality and chiral purity can also be attained from an initial system containing glass balls with the presence of both enantiomers in either an equal form or with small enantiomorphic excess. The grinding and abrasive motions between the glass balls and nuclei produced a size distribution of chiral fragments, facilitating a nonlinear autocatalytic phenomenon as well as dynamic dissolution-crystallisation phenomenon. This concept was extended from the simple achiral salts to biologically relevant enantiomeric amino acid molecules and derivatives.[184] Other studies have also demonstrated that chiral symmetry breaking of $NaClO_3$ can be induced by an ultrasonic field,[185,186] cavitation,[187] a temperature gradient under boiling[188] and a heating–cooling cycle.[189,190]

On the effect of stirring speed on the observed enantiomorphic excess (EE), the stirring rate was proportional to the enrichment in the EE; with the increase of the stirring/mixing/grinding rate, the value of EE tended logarithmically towards ±100%,[182] with smaller crystal size[191] and shorter time to achieve total symmetry breaking.[183] The effect of the direction of stirring (clockwise or counter-clockwise) on the other hand had little effect on the EE.[191] Martin *et al.* showed that the chiral symmetry breaking was not attainable when mixing was achieved in flow rather than by mechanical stirring[192] due to there being no collisions between nuclei and mixing objects. When bubbles were introduced into Martin's system, however, both enantiomers were harvested in the product crystals due to the creation of multiple nucleation sites for crystals of both handedness, whereas only a single enantiomer was seen before, without bubbles.[193] By flowing the solution of sodium chlorate over a seed crystal of sodium bromate, this provided heteroepitaxial nucleation leading to product crystals of 100% EE of the seed. In this case, fluid flow was the sole generator of secondary nuclei as collisions between seed crystal and crystalliser were non-existent,[194] a new mechanism to that of stirring.

By using seeded and unseeded crystallisation of $NaClO_3$, the exceptions to the work by Kondepudi *et al.* are demonstrated where mixing does not lead to total symmetry breaking, instead reducing the EE towards racemic state.

3.5.3.1 Experimental Setup and Procedure

A stirred tank crystalliser (STC) (Figure 3.16 left) and an oscillatory baffled crystalliser (OBC) (Figure 3.16 right) were used in the investigations. In order to minimize any difference due to crystalliser geometries, two identical glass vessels were used, while the agitator was changed to suit. The glass vessel was 70 mm in diameter and 1000 mL in volume with jacket. The vessel was sealed with a six-neck glass lid, which allows material to be added and extracted from the vessel, the input of the PAT probes, as well as hosting the shaft of the agitator.

For the STC, the vessel was fitted with a 58 mm diameter, PTFE, twin paddle type impeller, with the PAT probes acting as wall baffles to assist the mixing in the system. The surface area of the impeller is 0.041 m^3. For the OBC, a set of 2 PTFE annular orifice baffles were used with outer diameter of 69 mm and orifice diameter of 33 mm. The baffles were spaced at 105 mm apart using stainless steel spacers and oscillated in the fluid by means of a Copley Motion Ltd linear actuator (Model number STA2504S-55-S-S03X) that was controlled by a PC. The surface area of the orifice baffles is 0.061 m^3, about

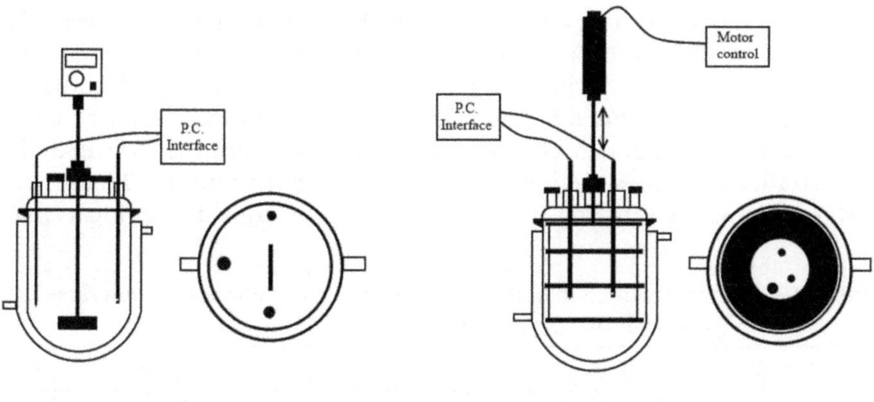

Figure 3.16 The schematics of the STC (left) and the OBC (right). The top view to the right of each system indicates the locations of the FTIR, turbidity and thermocouple[65] with permission from American Chemical Society, Copyright 2012.

1.5 times more than that of the impeller. Mixing conditions for the two crystallisers were selected so that the power dissipation rate was approximately equal, at 120 or 460 W m^{-3}. The determinations of power density for both STC[195] and OBC[119] are given in Section 3.3.6.

3.5.3.2 Seeded Experiments

For each experiment, 369 g $NaClO_3$ was added to the vessel containing 349 mL of distilled water. The solution was held at 40 °C for one hour under agitation to ensure that all solid particles had dissolved. After full dissolution, the solution was filtered using a Whatman glass fibre filter to remove any foreign particles, then cooled to 30 °C ($S = 1.006$, saturation temperature = 31 °C), *i.e.* 1 °C supercooling under controlled conditions. A single seed crystal of known handedness was dipped into the solution for three minutes, then removed. At the same time the agitation was stopped allowing crystals to grow overnight. Product crystals were collected, washed, dried and then analysed using a polarimeter. The left and right handed crystals were then separated, counted and weighed. Each experiment was done three times to establish the error margins within the data. The apparatus was thoroughly cleaned between trials using detergent (Fairy detergent) at 60 °C, rinsed with hot water and allowed to dry prior to use.

Table 3.3 below shows that the same results as Denk and Botsaris (1972)[196] in the stirred tank environment were obtained where 100% chiral symmetry breaking was achieved,[197] but the data in the OBC differed, with the EE of ~94%. While Kondepudi's secondary nucleation mechanism was still a dominant feature in the OBC due to fluid folding and unfolding, as well as collisions between nuclei and the surfaces of baffles, some sort of primary nucleation must also have occurred in order to produce the product crystals of the opposite handedness to the seed.

The cause of the primary nucleation was the scraping actions between the outer edge of the baffles and the inner surfaces of the wall of the column when the baffles are moving up and down the liquid phase,[65,69] similar to scratching the bottom of a beaker containing supersaturated solution using a spatula in undergraduate chemistry labs. By reconfiguring the stirrer in the STC so that it touches the bottom of the vessel when rotating (Figure 3.17 left) – creating the scraping action, and at the same time, by leaving a small

Table 3.3 Summary of EE in both STC and OBC[65] with permission from American Chemical Society, Copyright 2012.

	STC		OBC	
Seed	Left hand	Right hand	Left hand	Right hand
EE (%)	100	100	94.3	94.2
Standard Deviation (SD)	0	0	0.6	1.8

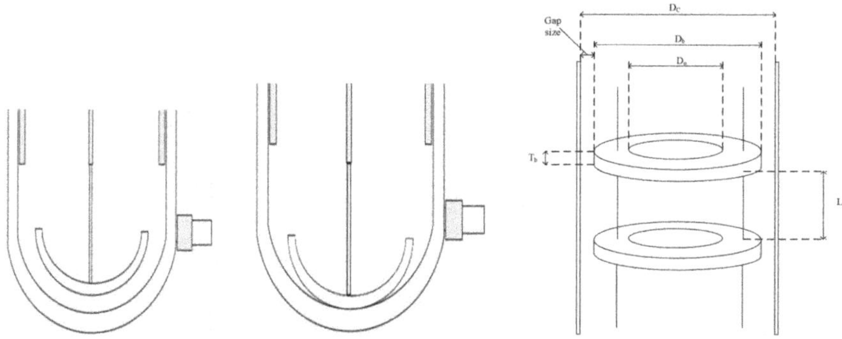

Figure 3.17 Schematic setup of the STC without and with scraping (left) and the OBC with a gap (right)[65] with permission from American Chemical Society, Copyright 2012.

Table 3.4 Summary of EE in the STC with scraping and the OBC without scraping[65] with permission from American Chemical Society, Copyright 2012.

	STC		OBC	
Seed	Left hand	Right hand	Left hand	Right hand
EE (%)	99.2	97.7	98.0	99.4
Standard Deviation (SD)	1.4	2.4	3.5	0.8

gap between the baffle edges and the inner surface of the wall of the OBC (Figure 3.17 right) – removing the scraping action, the EEs were reduced in the STC, while increased in the OBC (Table 3.4).[65,69]

The *first summary* is that scraping in an agitated system prevented chiral symmetry from breaking in seeded crystallisation of $NaClO_3$, as it promoted primary nucleation of the opposite enantiomer in the predominant secondary nucleation environment. Further research reveals that there is a relationship between the EE and the number of scratches as shown in Figure 3.18, where product crystals of seed-similarity decrease exponentially with the number of scratches, approaching 56.6% asymptotically where primary nucleation due to scraping dominates over secondary nucleation by seeding.[158] The 6% or so above the theoretical symmetry of 50% : 50% is the remaining effect of the latter.

The reductions of the EEs (Table 3.4) were amplified when purer $NaClO_3$ was used (Table 3.5). Table 3.6 shows the changes of EE when scraping was added to the STC and removed from the OBC. It can be seen that there was more successful primary nucleation of the opposite handed crystals through scraping for the purer materials than for the bought materials, implying that there was lesser secondary nucleation for the former. A 6 fold reduction in induction time was reported when purified carbamazepine was used.[198] This could be why purer material allowed for more primary nucleation of crystals of opposite handedness. The reason why less impurity has led to less

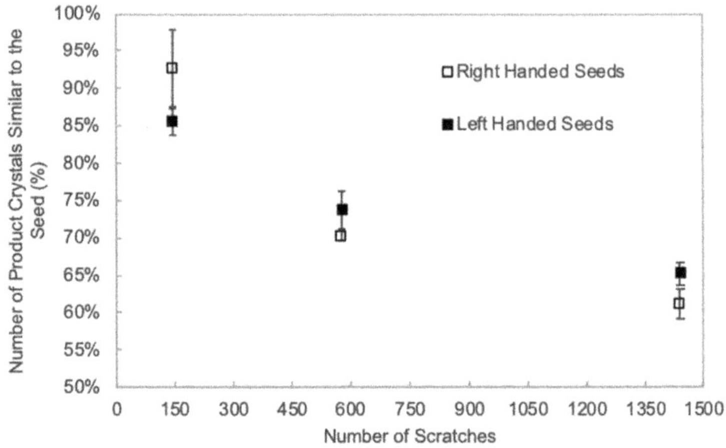

Figure 3.18 The relationship between the EE and the number of scratches.

Table 3.5 Summary of EE in both STC and OBC using purified material[65] with permission from American Chemical Society, Copyright 2012.

	STC		OBC	
Seed	Left hand	Right hand	Left hand	Right hand
EE (%)	100	100	90.9	87.8
Standard Deviation (SD)	0	0	5.9	9.9

Table 3.6 Summary of EE in both STC with scraping and OBC without scraping for purified material[65] with permission from American Chemical Society, Copyright 2012.

	STC		OBC	
Seed	Left hand	Right hand	Left hand	Right hand
EE (%)	82.8	81.7	96.5	81.9
Standard Deviation (SD)	13.8	9.7	2.3	5.6

secondary nucleation is not entirely clear; some previous work suggested that the use of recycled (*i.e.* purified) material showed an analogous effect such as the addition of nucleating agents, *i.e.* crystallisation begins at a higher temperature and a higher crystallisation rate was detected compared to materials without.[199]

The *second summary* is that scraping in a mixed system containing solution of purer solute in seeded crystallisation of $NaClO_3$ prevented further chiral symmetry from breaking.

3.5.3.3 Unseeded Experiments

The procedure for unseeded crystallisation experiments is similar to that for the seeded ones, except that (a) cooling was from 50 to 15 °C without seeds, giving a supercooling of 35 °C; (b) 446.8 g $NaClO_3$ and 400 g of distilled water were used for each experiment; (c) cooling profiles were linear and four cooling rates were used of 0.04, 0.1, 0.2 and 0.4 °C min^{-1}. The fully dissolved and filtered solution was added to the crystalliser and held at 50 °C for a further hour prior to the experiment. A pre-calibrated FTIR spectrometer probe (Shimadzu IR-Prestige 21), a stainless steel turbidity probe (HEL Crystaleyes DMS-2) with a Hastelloy mirror and a stainless steel PT100 thermocouple were inserted in the crystallisers to measure the profiles of solute concentration (peak area), solution turbidity and temperature as a function of time during the experiments. Three runs were undertaken at each cooling rate, ensuring a good repeatability.

The data in Table 3.7 indicate that crystals in the STC have predominantly a single enantiomeric form, while the EEs in the OBC are significantly less than 100%, with both dextro- and laevorotatory crystals. In our experiments, the supercooling was 35 °C and spontaneous nucleation is expected in both crystallisers, as indicated by Denk and Botsaris (1972),[196,197] leading to close to total symmetry breaking as seen in Viedma's work.[181] It must be noted that our experiments differ from Viedma's trials[181] in that

i. our saturated solution was under constant mixing from the start to the finish, while there was no stirring in Viedma's solution when it was cooled from 90 to 20 °C, then sudden and rapid stirring at ~1000 rpm was applied at 20 °C;
ii. our cooling was controlled so that a linear profile was maintained, while natural cooling was used in Viedma's work;
iii. our product crystals were collected and filtered straightaway after each experiment, while Viedma harvested his crystals after two days.

Our data in the STC suggest that controlled cooling at constant mixing contributed to the chiral symmetry breaking, this would be new to Viedma's work. However, is scraping also responsible for the different EEs in the OBC?

Table 3.7 Summary of EE for both the STC and the OBC in unseeded experiments.

	STC				OBC			
Cooling rate (°C min^{-1})	0.4	0.2	0.1	0.04	0.4	0.2	0.1	0.04
EE (%)	99.9	99.5	98.8	96.1	48.8	15.8	31.2	59.4
Standard Deviation (SD)	0.07	0.42	0.7	3.07	13.21	9.94	11.72	17.02

Figures 3.19–3.22 are the corresponding measurements of the solute concentration by the FTIR probe and the percentage transmittance by the turbidity probe for the cooling rates of 0.4 and 0.2 °C min^{-1} respectively with temperature ramps shown on the right hand side of the graphs. At 50 °C and $t = 0$, crystals were fully dissolved in the solution in the vessels, the solute concentrations (peak area) stayed the same in both vessels (Figures 3.19 and 3.21); so did the % transmittance that had the highest values (Figures 3.20 and 3.22), as the solutions were clear of solids. As the cooling took place, the degree of supersaturation was built up, while both the solute concentration and the % transmittance remained the same for a certain time before dropping; the moments of the first noticeable decrease in signals indicate the nucleation (or post-nucleation) events. Both signals continued to decrease as the cooling was proceeded further and levelled off when the temperature was held at 15 °C where most of the crystals had emerged from the solution, denoting the minimum solute concentration and the maximum cloudiness in the crystallisers. The readings of the solute concentrations and the % transmittance started to increase when a heating cycle commenced immediately and reached the maximum when 50 °C was re-attained. Three distinct features can be drawn from these figures:

a) The cooling profiles matched exactly in both crystallisers; for the cooling rate of 0.4 °C min^{-1}, it took about 90 mins to cool down from 50 to 15 °C and about 50 mins to heat it up from 15 back to 50 °C. When the cooling rate was halved at 0.2 °C min^{-1}, the time of cooling was doubled, while the heating slope kept the same. This indicates the robust experimental technique and methodology;
b) The solute concentration started to decrease just a fraction earlier in the OBC than in the STC, *e.g.* 42.5 mins for the OBC *vs.* 47.5 mins for the STC at 0.4 °C min^{-1} as shown in Figure 3.19, likewise 52 mins *vs.* 74.5 mins at 0.2 °C min^{-1} in Figure 3.21. This indicates that nucleation occurred earlier in the OBC. The turbidity measurements support the finding that the drops in the signals (*i.e.* crystal solids appeared) took place earlier in the OBC (at 37.5 mins and 44.5 mins in Figures 3.20 and 3.22 respectively) than in the STC (at 46.5 mins and 74.5 mins in Figures 3.20 and 3.22 respectively);
c) The detection of the nucleation events was generally earlier by the turbidity probe than that by the FTIR probe. This implies that the turbidity probe is more sensitive to the appearance of solids than the FTIR probe to the depletion of solute at the earlier stage of the nucleation.

The earlier arrival of nucleation in the OBC is due to the scraping action. Due to the large supercooling, spontaneous nucleation took place[196] in both the STC and the OBC. According to Viedma's,[181] work to survey unique

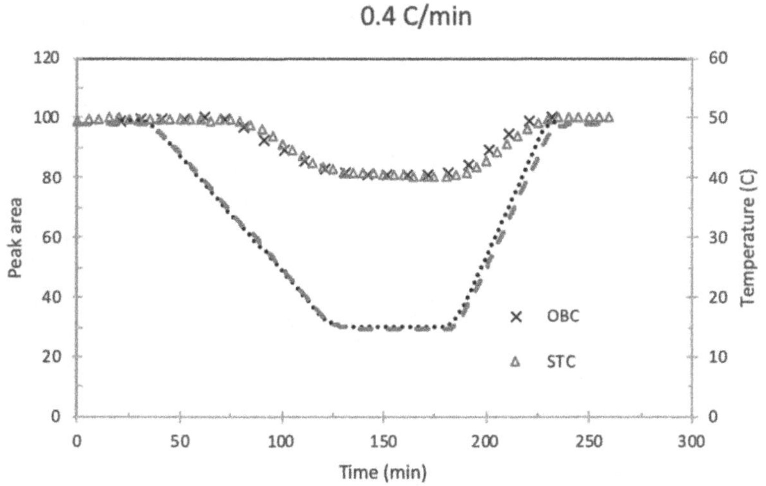

Figure 3.19 Solute concentration (peak area) as a function of time at the cooling rate of 0.4 °C min^{-1} in both OBC and STC with temperature ramps on the right.

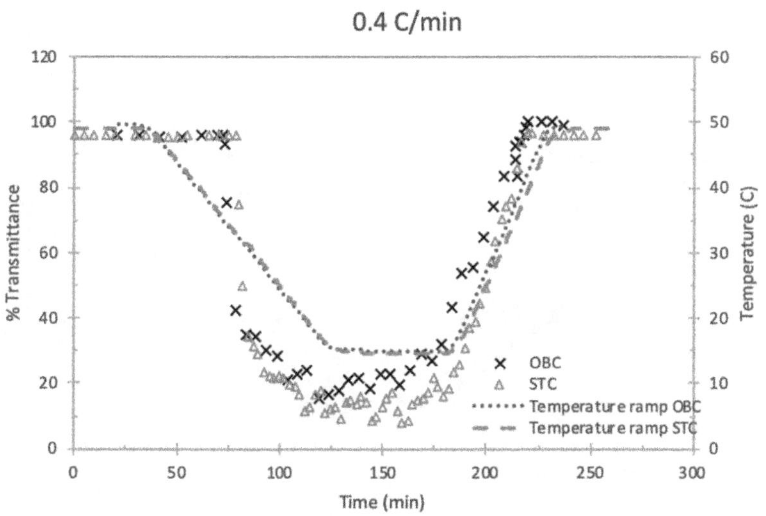

Figure 3.20 Transmittance as a function of time at the cooling rate of 0.4 °C min^{-1} for both OBC and STC with temperature ramps on the right.

chirality after a primary chiral nucleation, processes of chirally autocatalytic secondary nucleation must be present in order to avoid a new primary nucleation of different chirality.[181] The bulk eddy mixing in the OBC creates collisions between nuclei and baffle surfaces (that are 1.5 times larger than that of the stirrer); the surface areas of agitator within crystallisers have been

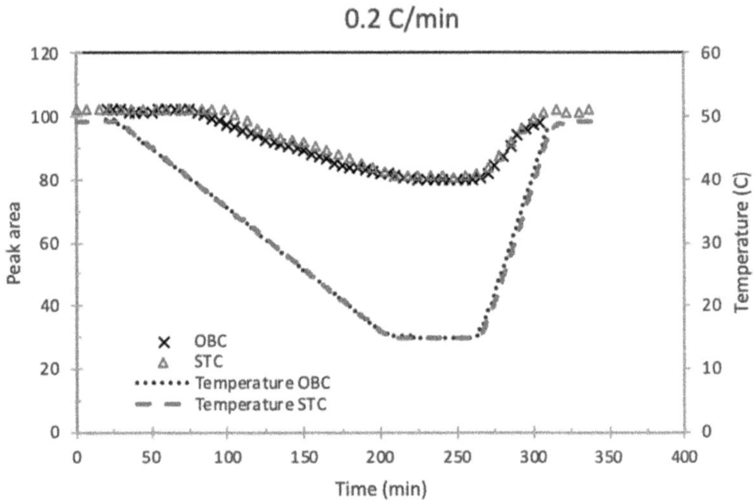

Figure 3.21 Solute concentration (peak area) as a function of time at the cooling rate of 0.2 °C min^{-1} in both OBC and STC with temperature ramps on the right.

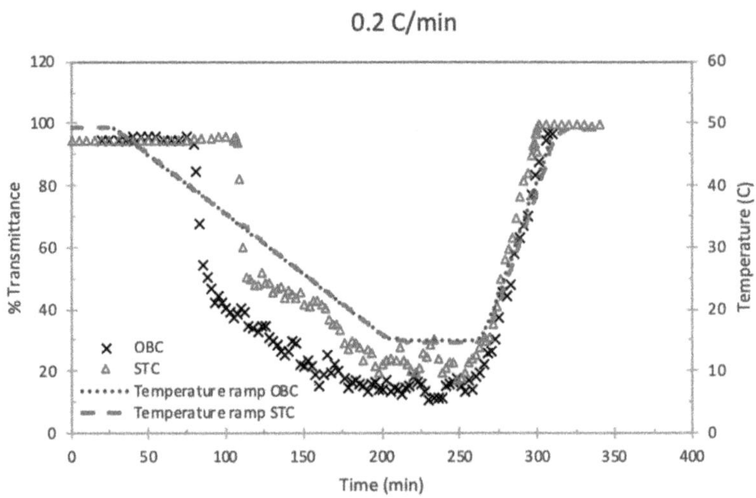

Figure 3.22 Transmittance as a function of time at the cooling rate of 0.2 °C min^{-1} for both OBC and STC with temperature ramps on the right.

known to influence nucleation, *i.e.* the larger the surface areas, the more the nucleation.[200–202] This increased surface area in the OBC would enhance the chirally autocatalytic secondary nucleation and increase the population of crystals of a given handedness. On the other hand, the scraping process can now be referred to as a source of the new primary nucleation, as articulated by Viedma.[181] It is known that the primary nucleation rate is a direct function of temperature and supersaturation ratio.[203] The scraping process at a given frequency would have created local heat due to friction, leading to a small temperature gradient between the scraping and non-scraping parts in the OBC. Consequently, the generation of nuclei of an opposite chirality on scraping would be faster, which competes in the solution with those nuclei from the spontaneous nucleation due to excess supercooling. This could have led to the close to racemic EEs in the OBC as shown in Table 3.7.

The *third summary* is that scraping in the mixed and supersaturated system prevented chiral symmetry from breaking in unseeded crystallisation.

A patent arose from these studies, as the scraping could be utilized as a means to generate primary nucleation when the amount of nuclei could be quantified and controlled. It should be emphasized that the scraping only happens in batch OBRs where the string of baffles moves up and down the column driven by a linear motor. In COBC, baffles are stationary and there is no scraping effect.

3.5.4 Encrustation

The increased specific surface areas in COBC are the unique feature for the enhanced heat transfer processes but are also prone to encrustation in crystallisation when operated incorrectly. Some investigations on fouling were undertaken on other devices.[167,204] Numerical simulation of crystallisation fouling on heat exchanger surfaces was also reported, outlining that fouling layers exhibit complicated structures, which essentially affect flow hydrodynamics, fouling kinetics and heat transfer performance.[205] Fouling resistances, defined as the decrease of efficiency due to fouling, were evaluated for brass, copper, aluminium, bronze and stainless steel surfaces; the energy at the interface of fluid and surface was a key factor in surface crystallisation, besides the surface roughness of the materials.[206]

In terms of cooling crystallisation in NiTech DN15 crystallisers, the following are the culprits for fouling encountered for processes involving industrial compounds: (a) local temperature or local supersaturation; (b) incorrect seeding strategy; (c) insufficient mass of nuclei; (d) suboptimal hardware; (e) oil out; (f) intrinsically sticky compounds.

We shall eliminate the category of compounds that are intrinsically sticky, *e.g.* sucrose. When compounds stick on the walls and impellers of stirred tank crystallisers, these would be amplified in DN15 crystallisers due to the larger specific surface area. We should be confident to say that these are NOT the compounds for continuous crystallisation.

Here Are Some Industrial Challenging Project Examples

3.5.4.1 Case 1 – Due to Local Temperature

In a seeded cooling crystallisation of an API from a pharma company, the seed tank was heated to 50 °C and seeds were pumped into a NiTech DN15 Standard crystalliser *via* an 8 mm pipe. It was anticipated that the temperature of seeds when entering the DN15 would be around about 45 °C, whereas the saturation temperature for this API was reported as 47 °C, *i.e.* the seeds were added at 2 °C supercooling. The DN15 was set up with three temperature zones, 60 to 47 °C, 47 to 30 °C, and 30 to 15 °C and the metastable zone width was unknown. During the trials, heavy encrustation took place within the first two straights of the DN15 crystalliser; the experiments were then stopped.

For seeded crystallisation in general, the amount of seed fed to a NiTech DN15 crystalliser is much smaller than that of the crystallizing solution and a smaller diameter pipe is often used. On examination, the seed pipe was clearly blocked, when no or insufficient seeds were introduced into the DN15, naturally leading to encrustation. This problem is caused by the fact that the amount of heat taken away by the surroundings is often underestimated when the diameter of the tube decreases. Table 3.8 illustrates this point. If the same heat source ($q = UA\Delta T$) from a crystallisation medium and the same heat transfer coefficient (U) are assumed in a crystallisation experiment, the temperature drop over say unit length (ΔT) has an inverse relationship to the cross sectional area (A) available for heat transfer, *e.g.* say the required temperature drop over unit length of a 15 mm diameter pipe is 5 °C, this would then become 20 °C, four times larger, when the diameter of the tube is halved from 15 to 7.5 mm (Table 3.8)! For such a large temperature change, the metastable zone was crossed, leading to spontaneous nucleation within the seed pipe, blocking the pipe. At the same time, encrustation occurred in the DN15 due to lack of seeds.

The remedies for this would come in two ways: (a) the seed pipe must thermally be insulated so that the required temperature drop is maintained; (b) the seed tank is heated up at higher temperatures so that the temperature for seeds to enter the DN15 is what one expects. The former was implemented in this particular case and the problem of encrustation was successfully resolved. During the trials, the metastable zone width for this compound was identified as 7 °C.

De facto, similar blockages happened within the feed pipes from the feed tank into the DN15 due to precisely the same reasons for some APIs with very narrow metastable zones, so the same remedies apply.

3.5.4.2 Case 2 – Due to Incorrect Seeding

A number of repeated encrustation cases happened for seeded cooling crystallisations of both pharmaceutical and speciality chemicals in NiTech DN15 crystallisers. A pharmaceutical company tested one of their APIs in a

Table 3.8 Heat transfer illustrations.

Diameter (m)		Area (m^2)	ΔT (°C)
	$q = UA\Delta T$		$\Delta T = q/UA$
0.015		0.000176715	5
0.0075		4.41786×10^{-5}	20

NiTech DN15 crystalliser, as the filtration time was too long, and the impurity levels were often high and variable. In plant batch operation, seeds were obtained by wet milling product crystals from previous batches and were then pumped into a stirred tank crystalliser to initiate nucleation after the temperature within the crystalliser had dropped from 80 to 60 °C. Note that the decision of when exactly to add the seeds was entirely controlled by plant operators. The amount of seeds added was *estimated* at about 10–15% w/w with the mean size about 200–300 μm. The batch was further cooled to 20 °C over 8 h. The solution was pumped out to a storage tank for filtration, while the batch crystalliser was washed and ready for next batch. The filtration was slow and took 8–12 h.

Applying the same recipe in a NiTech DN15 crystalliser, encrustation was seen and some of the baffled tubes in DN15 were blocked. On examining and calculating the mass of the seeds based on the size, the likely cause for encrustation was the incorrect combination of the two. For this specific case, the seed mass was increased to about 25% due to the large size of the seeds and because milling seeds to smaller sizes was then not an option and encrustation disappeared completely in all subsequent runs.

It is clear that successful seeding requires BOTH correct seed size (μm) and correct seed mass (kg), while the exact figures for the two parameters would be different for different compounds. By assuming that (a) no nucleation takes place (nucleation is totally suppressed by seeds); (b) crystals do not agglomerate or break; (c) there is no change in crystal shape during the entire process and (d) the growth is on seed particles alone, the product crystal size can be estimated by the McCabe's ΔL law as

$$C_S = \frac{W_S}{Y_{theo}} \tag{3.5}$$

where C_S is the seed loading (%), W_S the mass of inoculum seeds (g), Y_{theo} the theoretical mass yield,[207,208] ($= (C_{sat} - C_{final}) \times W_{sol}$) with W_{sol} being the mass of solvent (g) and C_{sat} and C_{final} the saturation and final concentration (g solute g solvent^{-1}) respectively.

The seed mass, W_S, is evaluated from $W_P = W_S + Y_{theo}$, where W_P is the mass of crystals produced (g). Based on the above assumptions of an unagglomerated growth only model, the number of seed crystals (N) can be assumed to be conserved and the crystal morphology (represented by the shape factor (F) = the ratio of surface and volume shape factors) does not change; for particles with a characteristic size of L_S for seeds (μm), L_P for the product crystals (μm) and ρ_S for the density of solid crystal (g mL^{-1}), the ratio between the mass of product crystals and the mass of seeds takes the form:

$$\frac{W_P}{W_S} = \frac{FN\rho_S L_P^3}{FN\rho_S L_S^3} = \left(\frac{L_P}{L_S}\right)^3 \tag{3.6}$$

The left side of the eqn (3.6) undergoes further simplifications using the eqn (3.5) and becomes:

$$\frac{L_{\mathrm{P}}}{L_{\mathrm{S}}} = \left(\frac{(1+C_{\mathrm{S}})}{C_{\mathrm{S}}}\right)^{1/3} \tag{3.7}$$

When the characteristic size of seed crystals, L_{S}, and the seed loading, C_{S}, are known, the ideal characteristic size of the product crystals, L_{P}, can then be calculated. The relationship of the ideal normalized product size ($L_{\mathrm{P}}/L_{\mathrm{S}}$) can graphically be represented as a function of the seed loading/mass, as shown in Figure 3.23. The solid curve represents the effect of seed loading on the normalized product size; a decrease in the normalized product size is seen with increased seed loading. This is because the theoretical mass of crystals produced could consist of either numerous small crystals or fewer large ones; each seed crystal grows to a smaller final size than if there were fewer seeds at the same supersaturation, thus the increase of the seed loading leads to the increase of the number of crystals.

The seed response graph (Figure 3.23) is also a very useful means for assessing seeding strategy, as the seed response curve itself represents the growth-dominated mechanism. Above the curve is associated with normalized product sizes being larger than the ideal values and in order to maintain such crystal growth this region is facilitated by non-encrustation; below the curve is the area where the final normalized product sizes are smaller than $L_{\mathrm{P}}/L_{\mathrm{S}}$. Two phenomena can contribute to this, either the occurrence of primary nucleation leads to a massive number of small crystals without control or there are insufficient seed crystal surfaces for the solute material to grow

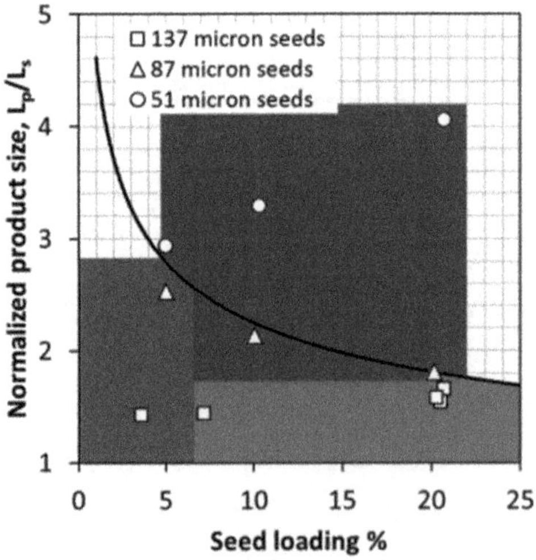

Figure 3.23 Traffic light map for operation. Left – severe blockage, bottom – minor blockage, top – no blockage.

on and solute molecules move to the reactor walls, leading to severe encrustation and blockage when seeds of larger sizes are combined with smaller seed loading. Based on the above principles, when operating in the green zone as shown in Figure 3.23, formulated for a specific organic chemical compound, encrustation in DN15 has totally been avoided.

3.5.4.3 Case 3 – Due to Insufficient Nuclei

One of the largest chemical companies conducted continuous crystallisation trials for one of its fish oil products in a NiTech DN15 crystalliser; the starting and the end temperatures were 10 and −25 °C respectively, with a nucleation temperature of 3 °C. No seeds were used; spontaneous nucleation was the main means of generating nucleation. Two temperature zones were set up, 10 to 0 °C, and 0 to −25 °C. The cooling rates for the first and second zones were 2.5 and 0.75 °C min^{-1} respectively, for the purposes of generating supersaturation in zone 1 and growing crystals in zone 2. Cloudiness of the solution in DN15, the indication of nucleation, was noticed from 3 to 6 °C. Encrustation was observed close to the end of the zone 1 and the start of the zone 2. Following the same principles as the aforementioned seeding strategy, the combination of size and mass of nuclei is critically important for successful operation. When spontaneous nucleation takes place, nuclei are generated, and the size of nuclei is very small (not measured in this case) which is an advantage, meeting one of the two criteria; the amount of nuclei was however unknown and difficult to determine. The encrustation would suggest that the mass of nuclei was not sufficient according to the above principles. The two temperature zones were replaced by three as 10 to 5 °C, 5 to −10 °C and −10 to −25 °C, with cooling rates of 7.5, 7.5 and 0.75 °C min^{-1} respectively, on the hypothesis that more nuclei (*i.e.* more mass) would appear when sufficiently high supersaturation was generated quickly in zone 2. Previously the zone 1 was cooled from 10 to 0 °C at the rate of 2.5 °C min^{-1}, *i.e.* over 4 mins; the zone 2 was now from 5 to −10 °C at a rate of 7.5 °C min^{-1}, *i.e.* over 2 mins. Encrustation was still seen in zone 2 in further trials. One additional observation noted that the degree of blockage was decreased with the increase of oscillation. This hinted that there may be too many nuclei generated which has an adverse effect on the local mixing, causing blockage. By trial and error with a number of cooling rates in zone 2, encrustation was noticeably reduced and finally removed at one cooling rate that seemed to generate the correct mass of nuclei, enabling trouble-free runs thereafter! This is a practical dilemma; on one hand, some of the nuclei would stick on the lowest energy place, *e.g.* walls of the device, leading to encrustation, when the mass of nuclei from spontaneous nucleation is insufficient and cannot complete the nucleation process. On the other hand, when too many nuclei emerge from solution at once, local mixing is compromised with particle sedimentation and agglomeration that would also lead to surface encrustation. What would be the optimal loading of nuclei is compound dependent, for example, encrustation occurred in the DN15 a couple of days later when a new batch of starting materials of different compositions was used. Further adjustments of cooling rates by "trial and error" were done in order to achieve the same result and this indicates that the generation of the correct mass of nuclei is sensitive to the composition of the starting materials. Modelling work is much needed here to guide experimentation or trial and error would be the only solution.

3.5.4.4 Case 4 – Due to Suboptimal Hardware

As I have explained, mixing in COBC is created by the generation and cession of eddies when fluid flows through orifice baffles, propagation of oscillatory motion is achieved by the incompressibility of fluid/solvent/solvent mixtures within the COBC, while the sizes of vortices and the degrees of interactions between eddies are governed by a combination of parameters. These include baffle spacing (the distance between a pair of baffles), baffle diameter (the restriction), baffle thickness (the effective distance for eddy shedding), oscillation frequency and amplitude. In order to accommodate COBC in labs, the straight sections of baffled tubes are connected by jacketed bends, forming a serpentine shape. In Mark I of NiTech DN15 crystallisers, the baffles were not well formed within bends, leading to uneven baffle spacing, as there was a learning curve for the glass manufacturer. When the baffle spacing is larger than the optimal, eddies may be disconnected, affecting both local and overall mixing, leading to particle sedimentation. In addition, the collars that connect straight baffled tubes (or straights) and bends contained no baffle in the Mark I version; this effectively means that the baffle spacing was unintentionally doubled at the joints and consequently heavy sedimentation was seen.[16,147] Furthermore, the initial design of collars for inserting PAT probes also had some issues, *e.g.* the probe tip with respect to tube surface and the angle of probe led to some dead zones at the probe location, causing sedimentation and in some cases primary nucleation. I am glad to report that these suboptimal glassware designs have subsequently been improved from Mark II onwards, with very close to even baffle spacing at bends (Figure 3.24); with a baffle manufactured in every collar to maintain the regular baffle spacing at joints; with better design and plug-in PAT connection, and encrustation due to suboptimal hardware is no longer observed and reported.

In one case a seeded cooling crystallisation of an API was investigated in a NiTech DN15 crystalliser by a global pharmaceutical company, with appropriate seeding strategy (size and mass) and procedures and the appropriate cooling rates. Crystallisation was run for a number of hours and some build-up was observed in one straight baffled tube, which turned into encrustation gradually. We examined every aspect of the nucleation theories and practices, revisited seed preparation, seeding method and temperature, recalibrated the flow rates of the feed and the seeds, as well as the cooling rates, but the build-up and encrustation was still seen at the *same* baffled tube. After having exhausted all possible approaches, tube 3 was swapped with the last tube.

Figure 3.24 A photo showing even baffles around the bends.

The effects of chemically modified glass surfaces on size, morphology and number of crystals formed in crystallisation of four proteins (lysozyme, thaumatin, catalase and ferritin) were clearly demonstrated,[209] and the effects of surface chemistry and morphology of polymeric substrates on nucleation behaviour showed a reduction in nucleation induction time for the crystallisation of aspirin.[210] The impact of materials of construction on nucleation were also reported.[153,211] The exchange of the tubes was the 'game changing' solution for this case and encrustation free operations were accomplished thereafter. A few of days later, the truth was unveiled. A surface defect was clearly visible when this piece of tube was broken up. Since it was initially located at the nucleation zone, the rough edges of the damaged surface would have made it easy for nuclei or crystals to stick on, or for some minute crystal residual of previous runs to be retained and to act as seeds. Once this piece of tube had been moved further downstream, the nucleation process was fully completed, and crystals grew in size; the effect of crystals hanging on the rough surface was minimized. Well, this is just an unexpected case.

3.5.4.5 Case 5 – Due to Recycle

In a chemical company in Europe, a NiTech DN15 crystalliser was used for a cooling crystallisation of an inorganic compound. The solution was fed into DN15 at 60 °C, through two temperature zones (60 to 40 and 40 to 20 °C) and the exit stream from the DN15 was directed back to the feed tank. No seeds were used; spontaneous nucleation was the means of generating nucleation. The start of the cloudiness of the solution was observed and varied from 47 to 35 °C, while the nucleation temperature was supplied as 40 °C. The generation of supersaturation rates was tested by applying different cooling rates and trouble free crystallisation was running for twice the residence time but encrustation was then observed initially in zone 1, later zone 2.

On the examination of the operational conditions, all the appropriate procedures and protocols were followed; would connecting the exit stream back to the feed tank have any influence on encrustation? It turned out that this was the main factor causing the encrustation and when the exit stream was fed to a separate tank, the encrustation issue was resolved. What were the explanations? Crystallisation is a purification process and the solid after crystallisation would be purer than that at the start. Directing the exit stream back to the feed tank acted effectively to alter the purity of the starting materials, even though the temperature of the feed tank was sufficiently high to dissolve the incoming colder stream. It is known that purer materials nucleate later than less pure materials as impurity drives heterogeneous primary nucleation. This means that some would be nucleated at higher temperatures, others at lower ones; the range of nucleation temperatures observed supports this line of thinking. The further postulation, following this line of inquiry, was that nuclei generated at higher temperatures would not be sufficient in mass to complete the nucleation, leading to encrustation, in comparison to nuclei generated at the same temperature in the earlier residence time.

3.5.4.6 Case 6 – Due to Oil out

A cooling crystallisation of an organic compound was undertaken in a NiTech DN15 crystalliser at a lab of a chemical company in Europe; the starting temperature was 75 °C and was cooled to 15 °C. The nucleation temperature was given as about 55 °C from lab batch experiments. Two temperature zones were set up: 75 to 55 °C and 55 to 15 °C at a cooling rate of 5 and 1 °C min^{-1} respectively. Just before reaching the nucleation temperature, the solution became cloudy and the formation of a second liquid phase, *i.e.* oiling out, was noticed. Nucleation took place after oiling out and some oil was stuck on the crystal solids, leading to some encrustation in DN15 and lower crystal purity. Generally, the larger the oil droplet size, the smaller the crystal size. Oscillation conditions seemed to control the crystal size. The trend of crystal size changing with mixing was similar to the work by Takasuga and Ooshima (2014)[212] who explained that the generation of small crystals was due to the fact that the primary and secondary nucleation frequencies increased with the droplet size, while the generation of large crystals was due to the growth of a small amount of crystals generated in small droplets through the consumption of API in the remaining crystal-vacant droplets.

Oiling out has been observed during cooling crystallisation of complex molecules like pharmaceuticals,[213] proteins[214–216] or polymers from solutions, and normally results in slow crystal growth, uncontrollable morphology and low purity in pharmaceutical crystallisation.[217] The molecular forces between solute and solvent and between solute and impurities are deemed as the essential reasons,[217] with the following specific indicators:

a) Melting temperature of solvent/solvent mixture is equal or lower than its operating temperature
   Remedy: lowering crystallisation temperature below the melting point;
b) Too many impurities or there are nucleation inhibitors
   Remedy: removing impurities and/or improving purity of raw materials;
c) Poor solvent selection[218]
   Remedy: select suitable solvent[214] and select preferred solvent ratios;[218]
d) Generation of supersaturation is too fast and too large[219] or there is local high supersaturation
   Remedy: lower supersaturation, facilitating slower nucleation or use seeds[220,221] whenever necessary;
e) Initial concentration of crystallisation is too high[217,221]
   Remedy: screening initial concentration of crystallisation;
f) Altering crystallisation method
   Remedy: using anti-solvent,[222] evaporation, freeze dry[223] or regional melting.

It was reported that better and uniform mixing were used to avoid bonding and wrapping between solid crystals and the oil phase[212] that lowered purity of crystals.[212,217] In crystallisation of vanillin, de Albuquerque and Mazzotti (2014) showed that the vanillin and water mixture displayed a liquid–liquid equilibrium that interacts with vanillin's solid–liquid equilibrium. When the system entered the metastable liquid–liquid equilibrium region, by varying either composition or temperature, the phenomenon of oiling out could occur. By selecting operating points outside the metastable liquid–liquid equilibrium, oiling out was consistently avoided.[219] Lu *et al.* (2012)[217] showed that the initial concentration had a significant effect on oiling out; oiling out did not take place and crystallisation products possessed higher purity at low initial concentrations (*e.g.* 12 mg mL^{-1}). At moderate initial concentration (*e.g.* 25 mg mL^{-1}) small quantities of oil droplets formed and coalesced into larger droplets that subsequently crystallized into product crystals with lower purity. When high initial concentrations (*e.g.* 31 mg mL^{-1}) were employed, a large amount of oil droplets formed a stable oil phase, leading to no crystal products.[217]

Based on the above discussions, the initial concentration (e), the cooling rate that drove the build-up of supersaturation (d) and seeding (d) were the operational parameters available for investigation. Lowering the initial concentration would compromise the yield, which was not what the company wanted.

- The cooling rate was lowered from 1 to 0.75 °C min^{-1} in an attempt to slow the generation of supersaturation and oiling out was noticeably reduced. A further reduction of the cooling rate to 0.5 °C min^{-1} prevented oiling out from occurring all together.
- Seeding was employed at the cooling rate of 1 °C min^{-1}; the addition of seeds in the undersaturated region worked better.

In summary, encrustation occurred in NiTech DN15 crystallisers when operated incorrectly which can be avoided entirely when appropriate operational and seeding strategies are applied.

3.5.5 PAT Implementation

In NiTech crystallisers, the saturated solution becomes supersaturated when flowing through the pre-established temperature zones. The increased specific heat transfer areas provide ample sites for PAT tools that can readily be placed anywhere along the length of the COBC. This is the unique advantage in comparison to industrial batch operation where it is impossible to implement any lab scale PAT tools in scale up vessels, preventing the science learnt in labs from realisation.

Stability of continuous crystallisation and steady states of process parameters in continuous crystallisation, such as solute concentration, crystal size and supersaturation, can be evaluated experimentally with the help of PAT probes. In crystallisation of palm oil melt, two FTIRs and two PVMs were placed at two strategic locations of a known distance along the length of the DN15 crystalliser; the analysis of solute concentrations from FTIRs and particle size from PVMs at two locations *at the same time* provides *spatial* characteristics, *e.g.* the rate of decrease in solute concentration, the rate of supersaturation depletion, the rate of crystal growth and the stability of operation, while the analysis of data at two time intervals at *a specific location* gives the *temporal* characteristics. This allows the determination of the degree of stability of crystal size and solute concentration, the stability of supersaturation *etc.*[224,225] Crystallisation kinetics (growth) was directly calculated from experimental data based on the changes of solute concentration from FTIR as a function of time and equally, experimental data can also be fed into a population balance model to determine crystallisation kinetics. Evolution of crystal particle size distributions (PSD) in a seeded cooling crystallisation of L-glutamic acid was investigated and showed that a constant PSD or steady state was obtained after the 2nd residence time in a seeded crystallisation, while PSD varied and steady state was not achievable in an unseeded case.[147]

FBRM and a UV-NIR spectrophotometer were employed to monitor particle counts and absorbance spectra of methylene blue dye in a COBC to determine residence time distribution of dispersed solid and liquid phase.[46] Other types of PAT tools have also been used in COBC, for example, immersion Raman, non-invasive Raman and THz Raman.

PAT tools have been seen in other types of continuous crystallisation, for example, Powell *et al.* (2016)[226] employed a combination of three components, (a) an integrated PAT array based on attenuated total reflectance ultraviolet/visible (ATR-UV/vis) spectroscopy and focused beam reflectance measurement (FBRM), (b) a crystallisation process informatics software system (CryPRINS), and (c) an ancillary PAT array based on Raman spectroscopy and particle vision microscopy (PVM) to monitor crystallisation and determine when steady state was achieved. This PAT array was also applied to a case-study for the selective crystallisation of 1 : 1 and 3 : 2 co-crystals of *p*-toluenesulfonamide/triphenylphosphine oxide.[227] *In situ* imaging and chord length distribution measurements were integrated for estimation of particle size and shape.[228] *In situ* monitoring of stirring effects on polymorphic transformations during cooling crystallisation of carbamazepine was carried out.[198] Image-based monitoring was also used for early detection of fouling in crystallisation processes.[167] Real-time image processing based online feedback control system was employed for cooling batch crystallisation[229] and for the investigation of the evolution of crystal size and shape during temperature cycling and in the presence of a polymeric additive.[230] Mathematical modelling and/or experimental validation were reported in MSMPR crystallisers for cooling crystallisation,[231] for antisolvent crystallisations,[232–234] for modelling of artefacts in estimations of particle size of needle-like particles from laser diffraction measurements,[235] for fines removal by optimizing spatial temperature profiles with controlled dissolution,[236] and for nucleation kinetics of butyl paraben.[237]

3.6 What Are the Opportunities and Challenges?

While new applications for OBC/COBC are emerging, *e.g.* sonocrystallisers,[238] microwave reactors, reactions incorporating dynamic separation, continuous hydrogenation and functional carbon nanotubes,[239] the following are some opportunities in crystallisation for COBC.

3.6.1 Reactive Crystallisation

Reactive crystallisation has been run as two separate unit operations in traditional batch operation but a tubular continuous reactor/crystalliser presents the unique opportunity of treating it as a single continuous process. One example is the reactive crystallisation of paracetamol,[240] where the solubility for crystallisation was first optimized, from which suitable reagent concentrations for the reaction part were retrospectively determined. The effects of water content and reaction temperature on reaction kinetics and mechanism as well as product purity were jointly investigated using chromatographic methods. Paracetamol form I particles were produced with the presence of water, while 4′-acetoxyacetanilide was identified in the reaction in the absence of water.[240]

Crystallisation kinetics for different solvent compositions, cooling and mixing rates were then evaluated and linked with reaction kinetics that can be used to influence crystal properties.[241] Using the optimized process conditions, crystal particles with a narrow size distribution and a high purity (99%) were successfully obtained. These will be the starting conditions for a COBC performing two unit operations in a one pass fashion.[225]

3.6.2 Co-crystallisation

Co-crystallisation is notoriously difficult in operation as well as in science. Yu *et al.* (2010)[242] established a complete temperature-dependent solute-solute-solvent phase diagram of the caffeine-glutaric acid-acetonitrile system in the temperature range of 10–35 °C using ATR-FTIR; the eutectic points in the phase diagram were located through solution-mediated phase transfer. The operating region of cooling co-crystallisation was prescribed according to the stoichiometry of the co-crystal and the stability zone of the co-crystal at the initial and final temperatures. Operation outside the prescribed operating region resulted in the production of impure co-crystals or single crystals. This work demonstrated how critical it is to control temperature in cooling co-crystallisation. The better heat transfer and mixing characteristics in COBC offer opportunities for better temperature controls for this type of cooling co-crystallisation. There are some reports in co-crystallisation[243,244] in a COBC.[162]

3.6.3 Crystallisation of Energetic Materials

Energetic materials are often obtained *via* a crystallisation process in traditional batch crystallisers; their crystal size and purity are as critical as those of pharmaceuticals.[245–247] As the materials are highly exothermic, explosive and sensitive to external environment, this has severely restricted experimentation, while many modelling studies have been undertaken.[248,249] Thus, safe handling of a large quantity of such materials is paramount.[250] Continuous crystallisers such as COBC have a much smaller inventory and volume in comparison to their traditional counterpart. The superior heat transfer rate and enhanced specific surface areas in COBC allow better temperature controls and enable safer processing of energetic materials with consistent crystal properties. Co-crystallisation of energetic materials has also gained significant attention, as it is a viable tool to make explosives less sensitive without losing their detonation power. Some work has been reported on this aspect,[251,252] while other methods were also employed to process energetic materials from sol–gel[253] and more recently additive manufacture.[254]

3.6.4 Pressurized Crystallisation

When the boiling point of a solvent or solvent mixture is lower than the operational temperature in crystallisation, the solvent boils off and gas bubbles coalesce and form a gas layer within the NiTech DN15 crystalliser. This is detrimental not only to crystallisation due to local high supersaturation *via* evaporation, leading to spontaneous nucleation, but also to the propagation of fluid oscillation, weakening the mixing within the DN15 crystallisers, leading to sedimentation. While solvent screening and reselection are one of the long term goals for this type of crystallisation processes, crystallisation at an elevated pressure is an immediate and viable operational option. At an elevated pressure, the solvent would remain as a liquid form, enabling normal cooling crystallisation processes. How many bars would be required depends on the pressure-temperature phase diagrams of compounds in question. In terms of operation in NiTech DN15 crystallisers, this can simply be done by placing a pressure regulator at the exit. The DN15 crystallisers made of glass can withstand 2–3 bars operational pressures, with the oscillation system being capable of delivering fluid oscillation up to 10 bars; NiTech crystallisers made of stainless steels or alloy (*e.g.* Hastelloy C-22) are for higher pressures with a magnetic oscillation system.

3.6.5 Solvent Swap

There are two types of solvent swap processes. The first one has a mixture of two solvents, one being a lower boiling solvent with good solubility of the solute, the other being a higher boiling solvent with poor solubility of the solute. Evaporation of either solvent achieves the goal of solvent swap. The second type has one solvent that is replaced by a different solvent after either a reaction or crystallisation step. Irrespective of which type, a traditional solvent swap process generally involves three unit operations: a reactor/crystalliser, a distillation column and a crystalliser, which was well illustrated, discussed and demonstrated by Papadakis *et al.* in their research.[255] Elgue *et al.* articulated the economic and environmental criteria for solvent replacement.[256] The amount of time required to perform a solvent swap on a scale can be frustrating, *e.g.* evaporating methanol (or water) is often atrocious on a large scale because of the high enthalpy of evaporation.

Using tubular crystallisers presents a viable alternative to the traditional operation, with some obvious advantages, such as much smaller volumes for evaporation, shorter process times, as well as enabling dynamic separation. For the first type of solvent swap operation, this would involve a continuous crystalliser, a dynamic separator that is operated vertically and a continuous crystalliser. The dynamic separator allows continuous separation of the evaporated solvent, while it accommodates the continuous flow of the second solvent into a second tubular crystalliser. All are operated continuously.

For the second type of solvent swap, it would involve a tubular crystalliser, an AWL continuous filtration/dry unit and a second tubular crystalliser. The continuous filtration unit separates the solute from the first solvent, the second solvent can be used as the washing medium. The filtered solute is then ejected into the dissolution chambers containing the second solvent, which is then fed into the second tubular crystalliser.

3.7 Operational Boundary

According to the laws of wave propagation, the propagation of fluid oscillation in incompressible fluid is at the speed of sound. When solids are present however, this effect would be reduced depending on the solid loading (%w/w), size and density, as solids act as the barrier for propagation. The presence of gaseous substance would dampen the propagation severely due to its compressibility. Like every flow chemistry device in the world, COBC has its operational boundaries, for instance,

- Specific density of particles should generally be lower than 1.5, *e.g.* either copper acetate or lead is too heavy. This would however depend on particle sizes and the combination of heavy and large particles is not for COBC;
- Particle concentration should generally be lower than 25% w/w; this would also depend on the density and mean size of solid particles;
- Viscosity of the main liquid phase should generally be lower than 300 cP at room temperature;
- Small amounts of discrete gas bubbles (~10% v/v) can be accommodated in COBCs, since the size of bubbles is too large to affect nucleation and the amount of gas is too little to affect mixing. Bubbles can also be used when gas is a reactant, *e.g.* hydrogenation, in a vertical set up.

Within these boundaries, successful trials of pharmaceuticals, fine/speciality chemicals and food compounds have been carried out. Full scale industrial operation using COBC has successfully been running since 2007 at a throughput of 0.06 $m^3\ min^{-1}$ (now 0.115 $m^3\ min^{-1}$ in the second installation in 2016).

Roman Symbols

A	area of tube (m^2)
C	trace concentration ($kg\ m^{-3}$)
C_D	discharge coefficient for an orifice (= 0.7 normally)

C_{final}	final concentration (g solute g solvent^{-1})
C_S	Seed loading (%)
C_{sat}	saturation concentration (g solute g solvent^{-1})
d	tube/column diameter (m)
D	axial dispersion coefficient (m^2 s^{-1})
D_o	orifice diameter (m)
D_s	impeller diameter (m)
f	oscillation frequency (Hz)
F	body force (N)
F	shape factor constant (= ratio of surface and volume shape factors)
L	distance between probes, or baffle spacing (m)
L_P	characteristic size of product crystal (m)
L_S	characteristic size of seeds (m)
N	number of tanks
N	number of seed crystals
N_b	number of baffles per unit length (m^{-1})
N_s	rotational speed of a stirrer (rev s^{-1})
P	pressure force (N m^{-3})
Pe	Peclet number
P_o	power number of a stirrer
P/V	power density (W m^{-3})
q	heat generation or release (J)
r	radius (m)
Re_n	net flow Reynolds number
S	initial supersaturation ratio
t	time (mins)
u	axial velocity (m s^{-1})
U	overall heat transfer coefficient (J m^{-2}K^{-1})
V_L	volume of liquid (m^{-3})
W_P	mass of product crystal (kg)
W_S	mass of seed crystal (kg)
W_{sol}	mass of solvent (kg)
x_o	oscillation amplitude (m)
Y_{theo}	theoretical mass yield (%)

Greek Symbols

α	baffle restriction ratio
μ	dynamic viscosity (kg m^{-1}s^{-1})
ρ	density (kg m^{-3})
ρ_S	density of seed crystal (kg m^{-3})
ω	angular oscillation frequency (s^{-1})
ΔT	temperature difference (°C)

Acknowledgements

I am grateful for the unresolved problems of scaling up batch crystallisers that gave me the inspiration of originating and pioneering continuous crystallisation in COBC. I would like to thank Scottish Enterprise for having granted me the SMART Award Scotland to pursuit my innovation. Also to thank my NiTech colleagues, my research students/postdocs at Heriot-Watt University and my collaborators for walking the road with me. To the doubters, ironically you drove me to succeed every bit as much as the early adopters if the truth be known. Finally, I dedicate this chapter to Gordon Burgess, Jonathan Clarke and Neil Todd who believed in me at the very start and invested the seeding funds; to Ian Laird and the late Lihua Zhao for your assiduity, energy, expertise and motivation that have enriched NiTech. We continue to share our friendship in life as well as on golf courses.

References

1. N. Variankaval, A. S. Cote and M. F. Doherty, From form to function: Crystallization of active pharmaceutical ingredients, *AIChE J.*, 2008, **54**, 1682–1688.
2. J. W. Mullin, *Crystallisation*, Butterworth-Heinemann, 2001.
3. W. L. McCabe, J. C. Smith and P. Harriott, *Unit Operations of Chemical Engineering*, McGraw-Hill, 5th edn, 1993.
4. J. H. Harker, J. R. Backhurst and J. F. Richardson, *Chemical Engineering*, 5th edn, 2003.
5. A. G. Jones, *Crystallisation Process Systems*, Butterworth Heinemann, London, 2002.
6. X. Ni, Unwrapping the myth about plug flow, *Chem. Eng.*, 2006, (779), 26–28.
7. O. Levenspiel, *Chemical Reaction Engineering*, John Wiley & Sons, New York, 3rd edn, 1999.
8. M. R. Mackley and X. Ni, Mixing and dispersion in a baffled tube for steady laminar and pulsatile flow, *Chem. Eng. Sci.*, 1991, **46**(12), 3139–3151.
9. X. Ni, H. Jian and A. W. Fitch, Evaluation of turbulent integral length scale in an oscillatory baffled column using large-eddy simulation and digital particle image velocimetry, *Trans. Inst. Chem. Eng.*, 2003, **81**(A8), 842–853.
10. X. Ni, J. A. Cosgrove, A. D. Arnott, C. A. Greated and R. H. Cumming, On the measurement of strain rate in an oscillatory baffled column using particle image velocimetry, *Chem. Eng. Sci.*, 2000, **55**, 3195–3208.
11. N. E. Pereira and X. Ni, Droplet size distribution in a continuous oscillatory baffled reactor, *Chem. Eng. Sci.*, 2001, **56**, 735–739.
12. Ltd NS. Mixing excellence UK2006.

13. M. R. Mackley and X. Ni, Experimental fluid dispersion measurements in periodic baffled tube arrays, *Chem. Eng. Sci.*, 1993, **48**(18), 3293–3305.
14. Y. Sommer de Gelicourt, On single phase axial dispersion in oscillatory baffled columns, MPhil thesis, Heriot-Watt University, 2000.
15. X. Ni, Y. Sommer de Gelicourt, M. H. I. Baird and N. V. Rama Rao, Scale-up of single phase axial dispersion in batch and continuous oscillatory baffled tubes, *Can. J. Chem. Eng.*, 2001, **79**(3), 444–448.
16. N. E. B. Briggs, *Polymorph Control of Pharmaceuticals Within a Continuous Oscillatory Baffled Crystalliser*, Strathclyde University, Glasgow, Scotland, 2015.
17. X. Ni, Y. Sommer de Gelicourt, J. Neil and T. Howes, On the effect of tracer density on axial dispersion in a batch oscillatory baffle column, *Chem. Eng. Sci.*, 2002, **85**, 17–25.
18. A. W. Fitch and X. Ni, On the Determination of Axial Dispersion Coefficient in a Batch Oscillatory Baffled Column using Laser Induced Fluorescence, *Chem. Eng. J.*, 2003, **92**, 243–253.
19. C. R. Brunold, J. C. B. Hunns, M. R. Mackley and J. W. Thompson, Experimental observations on flow patterns and energy losses for oscillatory flow in ducts containing sharp edges, *Chem. Eng. Sci.*, 1989, **44**, 1227–1244.
20. P. Gough, X. Ni and K. C. Symes, Experimental flow visualisation in a modified pulsed baffled reactor, *J. Chem. Technol. Biotechnol.*, 1997, **69**(3), 321–328.
21. X. Ni, S. Liu, M. J. Joye, P. Grewal and C. A. Greated, A study of velocity vector profiles and strain rate distributions for laminar and oscillatory flows in a baffled tube using particle image velocimetry, *J. Flow Visual. Image Process.*, 1995, **2**(2), 135–147.
22. S. Liu, D. J. Mason, X. Ni and C. A. Greated, Particle image velocimetry study of oscillatory flow in baffled tubes, *Sensors and Their Applications VIII*, 1997, pp. 335–340.
23. X. Ni, J. A. Cosgrove, A. D. Arnott, C. A. Greated and R. H. Cumming, On the measurement of strain rate in an oscillatory baffled column using particle image velocimetry, *Chem. Eng. Sci.*, 2000, **55**(16), 3195–3208.
24. A. W. Fitch, *Characterisation of Flow in an Oscillatory Baffled Column Using Digital Particle Image Velocimetry and Laser Induced Fluorescence*, Edinburgh, Heriot-Watt University, 2003.
25. A. W. Fitch and X. Ni, Using non-intrusive laser induced fluorescence in characterisation of mixing in an oscillatory baffled column, *J. Chem. Technol. Biotechnol.*, 2003, **78**, 326–331.
26. T. Howes, M. R. Mackley and E. P. L. Roberts, The simulation of chaotic mixing and dispersion for periodic flows in baffled channels, *Chem. Eng. Sci.*, 1991, **46**, 1669–1677.
27. E. P. L. Roberts and M. R. Mackley, The simulation of stretch rates for the quantitative prediction and mapping of mixing within a channel flow, *Chem. Eng. Sci.*, 1995, **50**(23), 3727–3746.

28. A. Mazubert, D. F. Fletcher, M. Poux and J. Aubin, Hydrodynamics and mixing in continuous oscillatory flow reactors—Part II: Characterisation methods, *Chem. Eng. Process.*, 2016, **102**, 102–116.
29. D. González-Juárez, J. P. Solano, R. Herrero-Martín and A. P. Harvey, Residence time distribution in multiorifice baffled tubes: A numerical study, *Chem. Eng. Res. Des.*, 2017, **118**, 259–269.
30. E. P. L. Roberts and M. R. Mackley, The development of asymmetry and period doubling for oscillatory flow in baffled channels, *J. Fluid Mech.*, 1996, **328**, 19–48.
31. X. Ni, H. Jian and A. W. Fitch, CFD modelling of flow patterns in an oscillatory baffled column, *Chem. Eng. Sci.*, 2002, **57**, 2849–2862.
32. H. Jian and X. Ni, On modelling turbulent flow in an oscillatory baffled column – RANS model or large eddy simulation, *J. Chem. Technol. Biotechnol.*, 2003, **78**, 220–226.
33. M. R. Mackley and R. M. C. Neves Saraiva, Quantitative description of fluid mixing using Lagrangian- and concentration-based numerical approaches, *Chem. Eng. Sci.*, 1999, **54**, 159–170.
34. The effect of geometry on mixing rates for oscillatory flow in tubes with periodically spaced baffles, in *15th IMACS World Congress on Scientific Computation Modelling and Applied Mathematics*, ed. R. M. C. Neves Saraiva and M. R. Mackley, Berlin, Germany, 1997 August 24–29.
35. X. Ni, Y. Zhang and I. Mustafa, An Investigation of Droplet Size and Size Distribution in Methylmethacrylate Suspensions in a Batch Oscillatory-Baffled Reactor, *Chem. Eng. Sci.*, 1998, **53**, 2903–2919.
36. Y. Zhang, X. Ni and I. Mustafa, A study of oil-water dispersion in a pulsed baffled reactor, *J. Chem. Technol. Biotechnol.*, 1996, **66**, 305–311.
37. X. Ni, Y. Zhang and I. Mustafa, Correction of Polymer Particle Size with Droplet Size in Suspension Polymerisation of Methylmethacrylate in a Batch Oscillatory Baffled Reactor, *Chem. Eng. Sci.*, 1999, **54**, 841–850.
38. X. Ni, D. C. Bennett, K. C. Symes and B. D. Grey, On the inverse phase suspension polymerisation of acrylamide in a batch oscillatory baffled reactor, *J. Appl. Polym. Sci.*, 2000, **76**, 1669–1676.
39. X. Ni, J. C. Johnstone, K. C. Symes, B. D. Grey and D. C. Bennett, Suspension polymerisation of acrylamide in an oscillatory baffled reactor: from drops to particles, *AIChE J.*, 2001, **47**(8), 1746–1757.
40. G. Vilar, R. A. Williams, M. Wang and R. J. Tweedie, On line analysis of structure of dispersions in an oscillatory baffled reactor using electrical impedance tomography, *Chem. Eng. J.*, 2008, **141**, 58–66.
41. J. Chen and X. Z. Wang, A wavelet method for analysis of droplet and particle images for monitoring heterogeneous processes, *Chem. Eng. Commun.*, 2005, **192**(4), 499–515.
42. M. J. Hounslow and X. Ni, Population balance modelling of droplet coalescence and break-up in an oscillatory baffled reactor, *Chem. Eng. Sci.*, 2004, **59**, 819–828.

43. X. Ni, D. Mignard, B. Saye, J. C. Johnstone and N. E. Pereira, On the evaluation of droplet breakage and coalescence rates in an oscillatory baffled reactor, *Chem. Eng. Sci.*, 2002, **57**, 2101–2114.
44. D. Mignard, L. Amin and X. Ni, Population balance modelling droplets in an oscillatory baffled reactor using direct measurement of breakage rate constants, *J. Chem. Technol. Biotechnol.*, 2003, **78**, 364–369.
45. D. Mignard, L. Amin and X. Ni, Modelling of droplet breakage probabilities in an oscillatory baffled reactor, *Chem. Eng. Sci.*, 2004, **59**, 2189–2200.
46. R. Kacker, S. I. Regensburg and H. J. M. Kramer, Residence time distribution of dispersed liquid and solid phase in a continuous oscillatory flow baffled crystallizer, *Chem. Eng. J.*, 2017, **317**, 413–423.
47. M. Manninen, E. Gorshkova, K. Immonen and X. Ni, Evaluation of axial dispersion and mixing performance in oscillatory baffled reactors using CFD, *J. Chem. Technol. Biotechnol.*, 2013, **88**, 553–562.
48. M. R. Mackley, K. B. Smith and N. P. Wise, The mixing and separation of particle suspensions using oscillatory flow in baffled tubes, *Trans. Inst. Chem. Eng.*, 1993, **71**(A), 649–656.
49. S. Liu, X. Ni, C. A. Greated and P. J. Fryer, Measurements of velocities of single particles for steady and oscillatory flows in plain and baffled tubes, *Trans. Inst. Chem. Eng.*, 1995, **73**(A), 727–732.
50. S. Gao, X. Ni, R. H. Cumming, C. A. Greated and P. I. Norman, Experimental investigation of particle flocculation in a batch oscillatory baffled reactor, *Sep. Sci. Technol.*, 1998, **33**(14), 68–72.
51. Characterisation of axial dispersion in a Meso-scale Oscillatory Baffled Crystalliser using a Numerical Approach, in *Proceedings of the 27th European Symposium on Computer Aided Process Engineering – ESCAPE 27*, ed. E. N. Kimuli, I. I. Onyemelukwe, B. Benyahia and C. D. Rielly, Barcelona, Spain, 2017 October 1st–5th, 2017.
52. G. Jimeno Miller, Y. C. Lee and X. Ni, *Characterization of Solid–Liquid Mixing in a Continuous Oscillatory Baffled Crystallizer Using CFD*, Glasgow, Scotland, 2015.
53. G. Jimeno Miller, Y. C. Lee and X. Ni, Smoothed particle hydrodynamics – a new approach for modelling flow in oscillatory baffled reactors, *Comput. Chem. Eng.*, 2019, **124**, 14–27.
54. G. Jimeno Miller, Y. C. Lee and X. Ni, The effect of particle size on flow in a continuous oscillatory baffled reactor using CFD, *Comput. Chem. Eng.*, 2019, submitted.
55. P. N. Baptista, F. A. R. Oliveira, J. Sannervik and J. C. Oliveira, The effect of mixing particles with different characteristics on the residence time distribution of particles in two-phase flow in a tubular system, *J. Food Eng.*, 1996, **29**, 361–373.
56. M. R. Hewgill, M. R. Mackley, A. B. Pandit and S. S. Pannu, Enhancement of gas–liquid mass transfer using oscillatory flow in a baffled tube, *Chem. Eng. Sci.*, 1993, **48**(4), 799–809.

57. X. Ni, S. Gao, R. H. Cumming and D. W. Pritchard, A comparative study of mass transfer in yeast for a batch pulsed baffled bioreactor and a stirred tank fermenter, *Chem. Eng. Sci.*, 1995, **50**(13), 2127–2136.
58. X. Ni, S. Gao and L. Santangeli, On the effect of surfactant on mass transfer to water-glycerol solutions in a pulsed baffled reactor, *J. Chem. Technol. Biotechnol.*, 1997, **69**, 247–253.
59. X. Ni and S. Gao, Mass transfer characteristics of a pilot pulsed baffled reactor, *J. Chem. Technol. Biotechnol.*, 1996, **65**, 65–71.
60. F. Navarro Fuentes, M. A. Keane and X. Ni, An investigation in continuous catalytic hydrogenation, *10th Wold Congress of Chemical Engineering*, Barcelona, Spain, 2017.
61. F. Navarro Fuentes, M. A. Keane and X. Ni, A comparative evaluation of hydrogenation of 3-butyn-2-ol over Pd/Al2O3 in an oscillatory baffled reactor and a commercial PARR reactor, *Org. Process Res. Dev.*, 2018, **23**(1), 38–44.
62. A study of pulsed-baffled membrane oxygenation, in *1st European Congress on Chemical Engineering*, ed. X. Ni, R. Cousins, P. Neilson and I. Airdie, Florence, 1997.
63. V. C. Eze, A. N. Phan, C. Pirez, A. P. Harvey, A. F. Lee and K. Wilson, Heterogeneous catalysis in an oscillatory baffled flow reactor, *Catal.: Sci. Technol.*, 2013, **3**(9), 2373–2379.
64. H. K. Gaidhani, An experimental investigation of pullulan fermentation in a batch oscillatory baffled fermenter, PhD thesis, Heriot-Watt University, 2004.
65. C. J. Callahan and X. Ni, Probing into nucleation mechanisms of cooling crystallization of sodium chlorate in a stirred tank crystallizer and an oscillatory baffled crystallizer, *Cryst. Growth Des.*, 2012, **12**, 2525–2532.
66. U. Schacht, *Control of Nucleation in Continuous Crystallisation Processes*, Strathclyde, Glasgow, 2014.
67. L. Palmer, *Investigation of Batch and Continuous Crystallization Processes Using Non-invasive Raman and Acoustic Emission Spectrometries*, Strathclyde, Glasgow, Scotland, 2013.
68. C. J. Callahan and X. Ni, An investigation into the effect of mixing on the secondary nucleation of sodium chlorate in a moving fluid and moving baffled crystallizer, *Can. J. Chem. Eng.*, 2014, **92**, 1920–1925.
69. C. J. Callahan, *The Influence of Hydrodynamic Environment on the Nucleation Mechanism of a Chiral Crystallization*, Heriot-Watt University, Edinburgh, Scotland, 2014.
70. C. J. Callahan and X. Ni, An investigation into the effect of mixing on the secondary nucleation of sodium chlorate in a stirred tank and oscillatory baffled crystallizer, *CrystEngComm*, 2013, **16**(4), 690–697.
71. G. Nelson, A scale up study in suspension polymerisation of methylmethacrylate in a pilot oscillatory baffled reactor, PhD thesis, Heriot-Watt University, 2001.

72. Experimental measurement of droplet size distribution of a MMA suspension in a batch oscillatory baffled column of 0.21m diameter, in *Proceedings of the 10th European Conference on Mixing*, ed. G. Nelson and X. Ni, Delft, Netherlands, 2000.
73. K. B. Smith, The scale-up of oscillatory flow mixing, Ph.D. thesis, University of Cambridge, 1999.
74. K. B. Smith and M. R. Mackley, An experimental investigation into the scale up of oscillatory flow mixing in baffled tubes, *Trans. Inst. Chem. Eng., Part A*, 2006, **A11**, 1001–1011.
75. H. Jian and X. Ni, A numerical study on the scale up behaviour in oscillatory baffled columns, *Chem. Eng. Res. Des.*, 2005, **83**, 1163–1170.
76. H. Jian, *Understanding Unsteadiness and Turbulence in Two Chemical Engineering Systems*, Heriot-Watt University, Edinburgh, 2002.
77. K. Robertson, P. B. Flandrin, A. R. Klapwijk and C. C. Wilson, Design and Evaluation of a Mesoscale Segmented Flow Reactor, *Cryst. Growth Des.*, 2016, **16**(8), 4759–4764.
78. A. N. Phan, A. Harvey and J. Lavender, Characterisation of fluid mixing in novel designs of mesoscale oscillatory baffled reactors operating at low flow rates (0.3–0.6 mL/min), *Chem. Eng. Process.*, 2011, **50**(3), 254–263.
79. A. N. Phan and A. P. Harvey, Development and evaluation of novel designs of continuous mesoscale oscillatory baffled reactors, *Chem. Eng. J.*, 2010, **159**(1–3), 212–219.
80. J. R. McDonough, A. N. Phan and A. P. Harvey, Rapid process development using oscillatory baffled mesoreactors – a state of the art review, *Chem. Eng. J.*, 2015, **265**, 110–121.
81. N. Reis, A. A. Vicente, J. A. Teixeira and M. R. Mackley, Residence times and mixing of a novel continuous oscillatory flow screening reactor, *Chem. Eng. Sci.*, 2004, **59**, 4967–4974.
82. N. Reis, A. P. Harvey, M. R. Mackley, A. A. Vicente and J. A. Teixeira, Fluid mechanics and design aspects of a novel oscillatory flow screening mesoreactor, *Trans. Inst. Chem. Eng., Part A*, 2005, **83**(A4), 357–371.
83. M. Zheng and M. R. Mackley, The axial dispersion performance of an oscillatory flow meso reactor with relevance to continuous flow operation, *Chem. Eng. Sci.*, 2008, **63**(7), 1788–1799.
84. L. N. Ejim, S. Yerdelen, T. McGlone, I. Onyemelukwe, B. Johnston, A. J. Florence and M. Reis Nuno, A factorial approach to understanding the effect of inner geometry of baffled meso-scale tubes on solids suspension and axial dispersion in continuous, oscillatory liquid–solid plug flows, *Chem. Eng. J.*, 2017, **308**, 669–682.
85. A. N. Phan, A. P. Harvey and M. Rawcliffe, Continuous screening of base-catalyzed biodiesel production using new designs of mesoscale oscillatory baffled reactors, *Fuel Process. Technol.*, 2011, **92**(8), 1560–1567.
86. A. N. Phan, A. P. Harvey and V. Eze, Rapid production of biodiesel in mesoscale oscillatory baffled reactors, *Chem. Eng. Technol.*, 2012, **35**(7), 1214–1220.

87. F. R. M. Rasdi, A. N. Phan and A. P. Harvey, Rapid determination of reaction order and rate constants of an imine synthesis reaction using a mesoscale oscillatory baffled reactor, *Chem. Eng. J.*, 2013, **222**, 282–291.
88. V. C. Eze, J. C. Fisher, A. N. Phan and A. P. Harvey, Intensification of carboxylic acid esterification using a solid catalyst in a mesoscale oscillatory baffled reactor platform, *Chem. Eng. J.*, 2017, **322**, 205–214.
89. A. Jawor-Baczynska, J. Sefcik and B. D. Moore, 250 nm glycine-rich nanodroplets are formed on dissolution of glycine crystals but are too small to provide productive nucleation sites, *Cryst. Growth Des.*, 2013, **13**(2), 470–478.
90. A. Jawor-Baczynska, B. D. Moore, H. S. Lee, A. V. McCormick and J. Sefcik, Population and size distribution of solute-rich mesospecies within mesostructured aqueous amino acid solutions, *Faraday Discuss.*, 2013, **167**, 425–440.
91. A. Jawor-Baczynska, B. D. Moore, J. Sefcik and B. D. Moore, Effect of mixing, concentration and temperature on the formation of mesostructured solutions and their role in the nucleation of DL-valine crystals, *Faraday Discuss.*, 2015, **179**, 141–154.
92. NiTech, Mixing and dispersion in DN2, DN4, DN6 and DN8 COBC, 2009.
93. B. Hallmark, M. R. Mackley and F. Gadala-Maria, Hollow microcapillary arrays in thin plastic films, *Adv. Eng. Mater.*, 2005, **7**(6), 545–547.
94. C. H. Hornung, M. R. Mackley, I. R. Baxendale and S. V. Ley, A Microcapillary Flow Disc (MFD) Reactor for Organic Synthesis, *Org. Process Res. Dev.*, 2007, **11**, 399–405.
95. C. H. Hornung and M. R. Mackley, The measurement and characterisation of residence time distributions for laminar liquid flow in plastic microcapillary arrays, *Chem. Eng. Sci.*, 2009, **64**, 3889–3902.
96. C. Parmar, D. Walker, C. H. Hornung and M. R. Mackley, The experimental observation and modelling of microdroplet formation within a plastic microcapillary array, *Chem. Eng. Sci.*, 2009, **64**(22), 4758–4764.
97. C. H. Hornung, B. Hallmark, M. Baumann, I. R. Baxendale, S. V. Ley and P. Hester, *et al.*, Multiple microcapillary reactor for organic synthesis, *Ind. Eng. Chem. Res.*, 2010, **49**(10), 4576–4582.
98. C. H. Hornung, B. Hallmark and M. R. Mackley, A palladium wall coated microcapillary reactor for use in continuous flow transfer hydrogenation, *Adv. Synth. Catal.*, 2010, **352**(10), 1736–1745.
99. D. Russo, D. Spasiano, M. Vaccaro, K. H. Cochran, S. D. Richardson and R. Andreozzi, *et al.*, Investigation on the removal of the major cocaine metabolite (benzoylecgonine) in water matrices by UV254/H2O2 process by using a flow microcapillary film array photoreactor as an efficient experimental tool, *Water Res.*, 2016, **89**, 375–383.

100. L. Quartapelle, *Numerical Solution of the Incompressible Navier–Stokes Equations*, Springer, Basel, AG, 1993.
101. G. E. Karniadakis and A. Beskok, *Microflows: Fundamentals and Simulation*, Springer, New York, 2001.
102. M. Gad-el-Hak, Flow physics, in *Handbook of MEMS*, ed. M. Gad-el-Hak, CRC, Boca Raton, FL, USA, 2002.
103. N. G. Hadjiconstantinou, The limits of Navier–Stokes theory and kinetics extensions for describing small-scale gaseous hydrodynamics, *Phys. Fluids*, 2006, **18**, 111301–111318.
104. E. Lauga, M. P. Brenner and H. A. Stone, *Microfluidics: The No-Slip Boundary Condition*, ed. J. Foss, C. Tropea and A. Yarin, Springer, New York, 2005.
105. G. Nagayama, T. Matsumoto, K. Fukushima and T. Tsuruta, Scale effect of slip boundary condition at solid–liquid interface, *Sci. Rep.*, 2017, 7, 43125.
106. Transport Properties of Molecular-Cluster Mixtures of Gases, in *Proceedings of the 9th Asian Thermophysical Properties Conference*, ed. L. I. Kurlapov, China, Beijing, 2010.
107. P. G. Lemarie-Rieusset, *The Navier–Stokes Problem in the 21st Century*, CRC Press, Taylor & Francis Group, Florida, USA, 2016.
108. G. I. Taylor, Dispersion of soluble matter in solvent flowing slowly through a tube, *Proc. R. Soc.*, 1953, **A219**, 186–203.
109. G. I. Taylor, Conditions under which dispersion of a solute in a stream of solvent can be used to measure molecular diffusion, *Proc. R. Soc.*, 1954, **A225**, 473–477.
110. G. I. Taylor, The dispersion of matter in turbulent flow through a pipe, *Proc. R. Soc.*, 1953, **A219**, 446–468.
111. R. Aris, On the dispersion of a solute by diffusion, convection and exchange between phases, *Proc. R. Soc.*, 1956, **A252**, 538–550.
112. R. Aris, Notes on the diffusion-type model for the longitudinal mixing of fluids in flow, *Chem. Eng. Sci.*, 1959, **9**, 266–267.
113. R. Aris, On the dispersion of a solute in pulsating flow through a tube, *Proc. R. Soc.*, 1960, **A259**, 370–376.
114. R. Aris, On the dispersion of a solute in a fluid flowing through a tube, *Proc. R. Soc.*, 1956, **A235**, 67–77.
115. X. Ni, Y. Sommer de Gélicourt, J. Neil and T. Howes, On the effect of tracer density on axial dispersion in a batch oscillatory baffled column, *Chem. Eng. J.*, 2002, **85**(1), 17–25.
116. A. C. Jealous and H. F. Johnson, Power Requirements for Pulse Generation in Pulse Columns, *Ind. Eng. Chem.*, 1955, **47**(6), 1159–1166.
117. M. R. Mackley, Process innovation using oscillatory flow within baffled tubes, *Trans. Inst. Chem. Eng.*, 1991, **69**(A), 197–199.
118. M. R. Mackley and P. Stonestreet, Heat transfer and associated energy dissipation for oscillatory flow in baffled tubes, *Chem. Eng. Sci.*, 1995, **50**(14), 2211–2224.

119. M. H. I. Baird and P. Stonestreet, Energy dissipation in oscillatory flow within a baffled tube, *Trans. Inst. Chem. Eng.*, 1995, **73**(5), 503–511.
120. A. Liao, *Characterisation of Crystallisation of L-Glutamic Acid in a Batch Oscillatory Baffled Crystalliser*, Heriot-Watt University, Edinburgh, 2008.
121. G. Nelson, *Scale-up Study of Suspension Polymerisation in an Oscillatory Baffled Reactor*, Heriot-Watt University, Edinburgh, 2001.
122. X. Ni and A. Liao, Effects of cooling rate and solution concentration on solution crystallization of L-glutamic acid in an oscillatory baffled crystallizer, *Cryst. Growth Des.*, 2008, **8**(8), 2875–2881.
123. R. L. Daugherty, J. B. Franzini and E. J. Finnemore, *Fluid Mechanics With Engineering Aspects*, McGraw-Hill Book Co., Singapore, 1989.
124. M. H. I. Baird and N. V. R. Rao, Power dissipation and flow patterns in reciprocating baffle-plate columns, *Can. J. Chem. Eng.*, 1995, **73**(4), 417–425.
125. M. H. I. Baird and P. Stonestreet, Energy dissipation in oscillatory flow within a baffled tube, *Trans. Inst. Chem. Eng.*, 1995, **73**(A), T503–T511.
126. G. Jimeno Miller and X. Ni, On the evaluation of power density models for oscillatory baffled crystallizer using CFD, *Chem. Eng. Process.*, 2018, **134**, 153–162.
127. R. K. Sinnott, Equipment Selection, Specification and Design, in *Coulson and Richardson's Chemical Engineering Volume 6-Chemical Engineering Design*, ed. R. K. Sinnott, Butterworth-Heinemann, Oxford, 3rd edn, 1999, pp. 471–473.
128. J. Ulrich and C. Strege, Some aspects of the importance of metastable zone width and nucleation in industrial crystallizers, *J. Cryst. Growth*, 2002, **237–239**(3), 2130–2135.
129. AWL, Continuous filtration, http://www.a-w-l.co.uk/continuous-filtration.
130. R. Chow, R. Blindt, R. Chivers and M. Povey, A tudy on the primary and secondary nucleation of ice by power ultrasound, *Ultrasonics*, 2005, **43**, 227–230.
131. J. Dodds, F. Espitalier, O. Louisnard, R. Grossier, R. David and M. Hassoun, *et al.*, The Effect of Ultrasound on Crystallisation-Precipitation Processes: Some Examples and a New Segregation Model, *Part. Part. Syst. Charact.*, 2007, **24**, 18–28.
132. C. Virone, H. J. M. Kramer, G. M. van Rosmalen, A. H. Stoop and T. W. Bakker, Primary nucleation induced by ultrasonic cavitation, *J. Cryst. Growth*, 2006, **294**, 9–15.
133. N. Lyczko, F. Espitalier, O. Louisnard and J. Schwartzentruber, Effect of ultrasound on the induction time and the metastable zone widths of potassium sulphate, *Chem. Eng. J.*, 2002, **86**, 233–241.
134. S. R. Patel and Z. V. P. Murthy, Effect of process parameters on crystal size and morphology of lactose in ultrasound-assisted crystallization, *Cryst. Res. Technol.*, 2011, **46**, 243–248.

135. G. Ruecroft, D. Hipkiss, T. Ly, N. Maxted and P. W. Cains, Sonocrystallization: The Use of Ultrasound for Improved Industrial Crystallization, *Org. Process Res. Dev.*, 2005, **9**, 923–932.
136. C. J. Price, Take some solid steps to improve crystallization, *Chem. Eng. Process.*, 1997, **93**, 34–43.
137. T. T. H. Nguyen, A. Khan, L. M. Bruce, C. Forbes, R. L. O. O'Leary and C. J. Price, The Effect of Ultrasound on the Crystallisation of Paracetamol in the Presence of Structurally Similar Impurities, *Crystals*, 2017, **7**(294), 1–24.
138. O. Narducci, A. G. Jones and E. Kougoulos, Continuous crystallization of adipic acid with ultrasound, *Chem. Eng. Sci.*, 2011, **66**, 1069–1076.
139. M. Jiang, C. D. Papageorgiou, J. Waetzig, A. Hardy, M. Langston and R. D. Braatz, Indirect ultrasonication in continuous slug-flow crystallization, *Cryst. Growth Des.*, 2015, **15**, 2486–2492.
140. D. Rossi, R. Jamshidi, N. Saffari, S. Kuhn, A. Gavriilidis and L. Mazzei, Continuous-flow sonocrystallization in droplet-based microfluidics, *Cryst. Growth Des.*, 2015, **15**, 5519–5529.
141. H. Siddique, C. J. Brown, I. Houson and A. J. Florence, Establishment of a Continuous Sonocrystallization Process for Lactose in an Oscillatory Baffled Crystallizer, *Org. Process Res. Dev.*, 2015, **19**, 1871–1881.
142. K. S. Suslick, The chemical effects of ultrasound, *Sci. Am.*, 1989, **260**, 80–86.
143. R. E. Apfel, Sonic effervescence: A tutorial on acoustic cavitation, *J. Acoust. Soc. Am.*, 1997, **101**, 1227–1237.
144. W. Lauterborn and C.-D. Ohl, Cavitation bubble dynamics, *Ultrason. Sonochem.*, 1997, **4**, 65–75.
145. W. Lauterborn, T. Kurz, R. Mettin and C. Ohl, Experimental and theoretical bubble dynamics, *Adv. Chem. Phys.*, 1999, **110**, 295–380.
146. A. S. Myerson, *Handbook of Industrial Crystallisations*, Butterworth-Heinemann, Boston, 2002.
147. N. E. B. Briggs, U. Schacht, V. Raval, T. McGlone, J. Sefcik and A. J. Florence, Seeded crystallization of β-L-glutamic acid in a continuous oscillatory baffled crystallizer, *Org. Process Res. Dev.*, 2015, **19**, 1903–1911.
148. S. Y. Wong, Y. Cui and A. S. Myerson, Contact Secondary Nucleation as a Means of Creating Seeds for Continuous Tubular Crystallizers, *Cryst. Growth Des.*, 2013, **13**, 2514–2521.
149. C. M. Chew, R. I. Ristic, R. D. Dennehy and J. De Yoreo, Crystallization of paracetamol under oscillatory flow mixing conditions, *Cryst. Growth Des.*, 2004, **4**(5), 1045–1052.
150. C. M. Chew and R. I. Ristic, Crystallization by oscillatory and conventional mixing at constant power density, *AIChE J.*, 2005, **51**(5), 1576–1579.
151. R. I. Ristic, Oscillatory mixing for crystallization of high crystal perfection pharmaceuticals, *Chem. Eng. Res. Des.*, 2007, **85**(A7), 937–944.
152. C. J. Brown and X. Ni, Determination of metastable zone width, mean particle size and detectable number density using video imaging

in an oscillatory baffled crystallizer, *CrystEngComm*, 2012, **14**(8), 2944–2949.

153. X. Ni, A. Valentine, A. Liao, S. B. C. Sermage, G. B. Thomson and K. J. Roberts, On the crystal polymorphic forms of L-glutamic acid following temperature programmed crystallisation in a batch oscillatory baffled crystalliser, *Cryst. Growth Des.*, 2004, 1129–1135.
154. X. Ni and A. Liao, Effects of cooling rate and solution concentration on solution crystallisation of L-glutamic acid in an oscillatory baffled crystalliser, *Cryst. Growth Des.*, 2008, **8**, 2875–2881.
155. X. Ni and A. Liao, Effects of Mixing, Seeding, Material and Final Temperature on Solution Crystallisation of L-glutamic Acid in an Oscillatory Baffled Crystalliser, *Chem. Eng. J.*, 2010, **156**, 226–233.
156. R. Calderia and X. Ni, Evaluation and establishment of a cleaning protocol for the Production of Vanisal Sodium and Aspirin using a Continuous Oscillatory Baffled Reactor, *Org. Process Res. Dev.*, 2009, **13**, 1080–1087.
157. S. Lawton, G. Steele, P. Shering, L. Zhao, I. Laird and X. Ni, Continuous crystallisation of pharmaceuticals using a continuous oscillatory baffled crystalliser, *Org. Process Res. Dev.*, 2009, **13**(6), 1357–1363.
158. R. Macleod and X. Ni, Chiral symmetry by surface scraping in seeded cooling crystallisation of sodium chlorate, *CrystEngComm*, 2018, **20**, 3696–3701.
159. C. J. Brown, Y. C. Lee, Z. Nagy and X. Ni, Evaluation of crystallisation kinetcis of adipic acid in an oscillatory baffled crystalliser, *CrystEngComm*, 2014, **16**, 8008–8014.
160. H. R. McLachlan and X. Ni, An investigation into parameters affecting crystal purity of urea in a stirred tank and an oscillatory baffled crystallizer, *Chem. Eng. Commun.*, 2016, **203**, 1189–1197.
161. H. R. McLachlan and X. Ni, On the effect of added impurity on crystal purity of urea in the oscillatory baffled crystallizer and stirred tank crystallizer, *J. Cryst. Growth*, 2016, **442**, 81–88.
162. L. Zhao, V. Raval, N. E. B. Briggs, R. M. Bhardwaj, T. McGlone, I. D. H. Oswald and A. J. Florence, From discovery to scale-up: alpha-lipoic acid: nicotinamide co-crystals in a continuous oscillatory baffled crystalliser, *CrystEngComm*, 2014, **16**(26), 5769–5780.
163. J. A. Adelakun and X. Ni, On the kinetics of palm oil crystallisation, *Int. J. Eng. Res. Sci. Technol.*, 2016, **2**(10), 1–12.
164. C. J. Brown and X. Ni, Evaluating the rate of cyclopentane hydrate formation in an oscillatory baffled column using laser induced fluorescence and energy balance, *Chem. Eng. J.*, 2010, **157**, 131–139.
165. L. R. Agnew, T. McGlone, H. P. Wheatcroft, A. Robertson, A. R. Parsons and C. C. Wilson, Continuous crystallisation of paracetamol (acetaminophen) form II: selective access to a metastable solid form, *Cryst. Growth Des.*, 2017, **17**, 2418–2427.
166. C. J. Brown and X. Ni, *Scientific Understanding of Seeding in Continuous Oscillatory Baffled Crystallizer*, 2012.

167. C. Tachtatzis, R. Sheridan, C. Michie, R. C. Atkinson, A. Cleary, J. Dziewierz, I. Andonovic, N. E. B. Briggs, A. J. Florence and J. Sefcik, Image-based monitoring of encrustation in the crystallisation process, *Chem. Eng. Sci.*, 2015, **133**, 82–90.
168. A. Majumder and Z. K. Nagy, Dynamic modeling of encrust formation and mitigation strategy in a continuous plug flow crystallizer,, *Cryst. Growth Des.*, 2015, **15**(3), 1129–1140.
169. C. J. Brown and X. Ni, Evaluation of growth kinetcis of antisolvent crystallisation of paracetamol in an oscillatory baffled crystalliser utilizing video imaging, *Cryst. Growth Des.*, 2011, **11**, 3994–4000.
170. C. J. Brown, J. A. Adelakun and X. Ni, Characterisation and modelling of antisolvent crystallisation of salicylic acid in a continuous oscillatory baffled crystallizer, *Chem. Eng. Process.*, 2015, **97**, 180–186.
171. Q. Su, B. Benyahia, Z. K. Nagy and C. D. Rielly, Mathematical modeling, design, and optimization of a multi-segment multi-addition plug-flow crystallizer for anti-solvent crystallizations, *Org. Process Res. Dev.*, 2015, **19**(12), 1859–1870.
172. P. MacFhionnghaile, V. Svoboda, J. McGinty, A. Nordon and J. Sefcik, Crystallisation diagram for antisolvent crystallisation of lactose: using design of experiments to investigate continuous mixing-induced supersaturation, *Cryst. Growth Des.*, 2017, **17**, 2611–2621.
173. K. A. Powell, A. N. Saleemi, C. D. Rielly and Z. K. Nagy, Periodic steady-state flow crystallization of a pharmaceutical drug using MSMPR operation, *Chem. Eng. Process.*, 2015, **97**, 195–212.
174. CMAC Bulletin Internet, 2017.
175. J. Engstrom, C. Wang, C. Lai and J. Sweeney, Introduction of a new scaling approach for particle size reduction in toothed rotor-stator wet mills, *Int. J. Pharm.*, 2013, **456**(2), 261–268.
176. S. C. Abrahams and J. L. Bernstein, Remeasurement of Optically Active NaClO3 and NaBrO3, *Acta Crystallogr., Sect. B: Struct. Sci.*, 1977, **33**, 3601–3604.
177. F. S. Kipping and W. J. Pope, Stereochemistry and Vitalism, *Nature*, 1898, **59**, 53.
178. D. K. Kondepudi, R. J. Kaufman and N. Singh, Chiral Symmetry Breaking in Sodium Chlorate Crystallization, *Science*, 1990, **250**(4983), 975–976.
179. J. M. McBride and R. L. Carter, Spontaneous resolution by stirred crystallization, *Angew. Chem., Int. Ed.*, 1991, **30**(3), 293–295.
180. D. K. Kondepudi, K. L. Bullock, J. A. Digits, J. K. Hall and J. M. Miller, Kinetics of chiral symmetry breaking in crystallization, *J. Am. Chem. Soc.*, 1993, **115**(22), 10211–10216.
181. C. Viedma, Experimental evidence of chiral symmetry breaking in crystallization from primary nucleation, *J. Cryst. Growth*, 2004, **261**, 118–121.
182. D. K. Kondepudi, K. L. Bullock, J. A. Digits and P. D. Yarborough, Stirring Rate as a Critical Parameter in Chiral Symmetry Breaking Crystallization, *J. Am. Chem. Soc.*, 1995, **117**(1), 401–404.

183. C. Viedma, Chiral symmetry breaking during crystallization: complete chiral purity induced by nonlinear autocatalysis and recycling, *Phys. Rev. Lett.*, 2005, **94**(065504), 1–4.
184. W. L. Noorduin, T. Izumi, A. Millemaggi, M. Leeman, H. Meekes, W. J. P. Van Enckevort, R. M. Kellogg, B. Kaptein, E. Vlieg and D. G. Blackmond, Emergence of a single solid chiral state from a nearly racemic amino acid derivative, *J. Am. Chem. Soc.*, 2008, **130**, 1158–1159.
185. Y. Song, W. Chen and X. Chen, Ultrasonic field induced chiral symmetry breaking of NaClO3 crystallisation, *Cryst. Growth Des.*, 2008, **8**(5), 1448–1450.
186. Y. Song, W. Chen and X. Chen, Crystal chiral symmetry breaking: a self-seed inducing effect controlled by kinetics, *Cryst. Growth Des.*, 2012, **12**, 8–11.
187. P. Cintas, On cavitation and chirality: a further assessment, *Cryst. Growth Des.*, 2008, **8**(8), 2626–2627.
188. C. Viedma and P. Cintas, Homochirality beyond grinding: deracemizing chiral crystals by temperature gradient under boiling, *Chem. Commun.*, 2011, **47**, 12786–12788.
189. Z. Wu, S. Yang and W. Wu, Application of temperature cycling for crystal quality control during crystallization, *CrystEngComm*, 2016, **18**, 2222–2238.
190. K. Suwannasang, A. E. Flood, C. Rougeot and G. Coquerel, Using Programmed Heating–Cooling Cycles with Racemization in Solution for Complete Symmetry Breaking of a Conglomerate Forming System, *Cryst. Growth Des.*, 2013, **13**(8), 3498–3504.
191. S. Veintemillas-Verdaguer, S. Osuna Esteban and M. A. Herrero, The effect of stirring on sodium chlorate crystallization under symmetry breaking conditions, *J. Cryst. Growth*, 2007, **303**, 562–567.
192. B. Martin, A. Tharrington and X.-l. Wu, Chiral Symmetry Breaking in Crystal Growth: Is Hydrodynamic Convection Relevant?, *Phys. Rev. Lett.*, 1996, 77(13), 2826–2829.
193. G. D. Botsaris, R.-Y. Qian and A. Barrett, New insights into nucleation through chiral crystallization, *AIChE J.*, 1999, **45**(1), 201–203.
194. T. Buhse, D. Durand, D. Kondepudi, J. Laudadio and S. Spilker, Chiral Symmetry Breaking in Crystallization: The Role of Convection, *Phys. Rev. Lett.*, 2000, **84**(19), 4405–4408.
195. S. Nagata, *Mixing: Principles and Applications*, Kodansha Ltd, Tokyo, 1975.
196. E. G. Denk and G. D. Botsaris, Fundamental studies in secondary nucleation from solution, *J. Cryst. Growth*, 1972, **13/14**, 493–499.
197. E. G. Denk and G. D. Botsaris, Mechanism of contact nucleation, *J. Cryst. Growth*, 1972, **15**, 57–60.
198. K. J. Sypek, I. Burns, A. Florence and J. Sefcik, In situ monitoring of stirring effects on polymorphic transformations during cooling

crystallization of carbamazepine, *Cryst. Growth Des.*, 2012, **12**(10), 4821–4828.
199. E. Kaisersberger, S. Knappe, J. Opfermann and H. Möhler, The influence of recycled material on the crystallization kinetics of thermoplastic polymers, *J. Therm. Anal.*, 1993, **40**(2), 821–829.
200. K. Liang, G. White, D. Wilkinson, L. J. Ford, K. J. Roberts and W. M. L. Wood, An Examination into the Effect of Stirrer Material and Agitation Rate on the Nucleation of L-Glutamic Acid Batch Crystallized from Supersaturated Aqueous Solutions, *Cryst. Growth Des.*, 2003, **4**(5), 1039–1044.
201. X. Ni, R. Shepherd, J. Whitehead and T. Liu, Chiral symmetry breaking due to impellor size in cooling crystallisation of sodium chlorate, *CrystEngComm*, 2018, **20**, 6894–6899.
202. J. Ahn, D. H. Kim, G. Coquerel and W.-S. Kim, Chiral symmetry breaking and deracemization of sodium chlorate in turbulent flow, *Cryst. Growth Des.*, 2018, **18**, 297–306.
203. A. E. Nielsen, *Kinetics of Precipitation*, ed. R. Belcher and L. Gordon, Permanon Press, Oxford, London, Edinburgh, New York, Paris, Frankfurt, 1965.
204. C. Forsyth, I. S. Burns, P. A. Mulheran and J. Sefcik, Scaling of glycine nucleation kinetics with shear rate and glass–liquid interfacial area, *Cryst. Growth Des.*, 2016, **16**, 16–144.
205. J. Xiao, J. Han, F. Zhang and X. D. Chen, Numerical simulation of crystallisation fouling: taking into account fouling layer structures, *Heat Transfer Eng.*, 2017, **38**(7–8), 775–785.
206. Influence of different surface materials on nucleation and crystal growth in heat exchangers, in *Proceedings of 7th International Conference on Heat Exchanger Fouling and Cleaning – Challenges and Opportunities*, ed. T. Geddert, S. Kipp, W. Augustin and S. Scholl, ECI Symposium Series, Tomar, Portugal, 2007.
207. W. Beckmann, Seeding the desired polymorph: background, possibilities, limitations, and case studies, *Org. Process Res. Dev.*, 2000, **4**(5), 372–383.
208. N. Doki, N. Kubota, A. Sato, M. Yokota, O. Hamada and F. Masumi, Scaleup experiments on seeded batch cooling crystallization of potassium alum, *AIChE J.*, 1999, **45**(12), 2527–2533.
209. D. S. Tsekova, D. R. Williams and J. Y. Y. Heng, Effect of surface chemistry of novel templates on crystallization of proteins, *Chem. Eng. Sci.*, 2012, **77**, 201–206.
210. Y. Diao, A. S. Myerson, T. A. Hatton and B. L. Trout, Surface design for controlled crystallization: the role of surface chemistry and nanoscale pores in heterogeneous nucleation, *Langmuir*, 2011, **27**, 5324–5334.
211. K. Liang, G. White, D. Wikinson, L. J. Ford, K. J. Roberts and W. M. L. Wood, An Examination into the Effect of Stirrer Material and Agitation

Rate on the Nucleation of L-Glutamic Acid Batch Crystallized from Supersaturated Aqueous Solutions, *Cryst. Growth Des.*, 2004, **4**(5), 1039–1044.
212. M. Takasuga and H. Ooshima, Control of Crystal Size during Oiling Out Crystallization of an API, *Cryst. Growth Des.*, 2014, **14**, 6006–6011.
213. M. Muschol and F. Rosenberger, Liquid–liquid phase separation in supersaturated lysozyme solutions and associated precipitate formation/crystallization, *J. Chem. Phys.*, 1997, **107**, 1953–1962.
214. Binary liquid phase separation and critical phenomena in a protein/water solution, in *Proceedings of the National Academy of Sciences of the United States of America*, ed. J. A. Thomson, P. Schurtenberger, G. M. Thurston and G. B. Benedek, 1987.
215. Control of protein crystal nucleation around the metastable liquid–liquid phase boundary, in *Proceedings of the National Academy of Sciences of the United States of America*, ed. O. Galkin and P. G. Vekilov, 2000.
216. D. Vivares and F. Bonnete, Liquid–liquid phase separations in urate oxidase/PEG mixtures: Characterization and implications for protein crystallization, *J. Phys. Chem. B*, 2004, **108**, 6498–6507.
217. J. Lu, Y.-P. Li, J. Wang, Z. Li, S. Rohani and C.-B. Ching, Study on the Oiling-out and Crystallization for the Purification of Idebenone, *Org. Process Res. Dev.*, 2012, **16**(3), 442–446.
218. K. Kiesow, F. Tumakaka and G. Sadowski, Experimental investigation and prediction of oiling out during crystallization process, *J. Cryst. Growth*, 2008, **310**(18), 4163–4168.
219. I. de Albuquerque and M. Mazzotti, Crystallization Process Design Using Thermodynamics To Avoid Oiling Out in a Mixture of Vanillin and Water, *Cryst. Growth Des.*, 2014, **14**(11), 5617–5625.
220. S. Daver, N. Rodeville, F. Pineau, J.-M. Arlabosse, C. Moureou and F. Muller, *et al.*, Process Development and Crystallization in Oiling-Out System of a Novel Topical Antiandrogen, *Org. Process Res. Dev.*, 2017, **21**, 231–240.
221. K. Li, S. Wu, S. Xu, S. Du, K. Zhao and L. Lin, *et al.*, Oiling out and Polymorphism Control of Pyraclostrobin in Cooling Crystallization, *Ind. Eng. Chem. Res.*, 2016, **55**(44), 11631–11637.
222. X. Li, Q. Yin, M. Zhang, B. Hou, Y. Bao and J. Gong, *et al.*, Process Design for Antisolvent Crystallization of Erythromycin Ethylsuccinate in Oiling-out System, *Ind. Eng. Chem. Res.*, 2016, **55**(27), 7484–7492.
223. L. Derdour, A method to crystallize substances that oil out, *Chem. Eng. Res. Des.*, 2010, **88**, 1174–1181.
224. J. A. Adelakun, *Characterisation of Steady State of Crystallisation from Melt in a Continuous Oscillatory Baffled Crystalliser*, Heriot-Watt University, Edinburgh, Scotland, 2018.
225. M. Jiang and X. Ni, Reactive crystallisation of paracetamol in a continuous oscillatory baffled reactor, *Org. Process Res. Dev.*, 2019, **23**(5), 882–890.

226. K. A. Powell, A. N. Saleemi, C. D. Rielly and Z. K. Nagy, Monitoring continuous crystallization of paracetamol in the presence of an additive using an integrated PAT array and multivariate, *Org. Process Res. Dev.*, 2016, **20**(3), 626–636.
227. K. Powell, C. Rielly, Z. Nagy and D. Croker, PAT-based design of agrochemical co-crystallization processes: A case-study for the selective crystallization of 1:1 and 3:2 co-crystals of p-Toluenesulfonamide/triphenylphosphine oxide, *Chem. Eng. Sci.*, 2016, **152**, 95–108.
228. O. S. Agimelen, A. Jawor-Baczynska, J. McGinty, J. Dziewierz, C. Tachtatzis, A. Cleary, I. Haley, C. Michie, I. Andonovic, J. Sefcik and A. J. Mulholland, Integration of *in situ* imaging and chord length distribution measurements for estimation of particle size and shape, *Chem. Eng. Sci.*, 2016, **144**, 87–100.
229. A. Borsos, Z. K. Nagy, B. Szilagyi and P. S. Agachi, Real-Time Image Processing Based Online Feedback Control System for Cooling Batch Crystallization, *Org. Process Res. Dev.*, 2017, **21**(4), 511–519.
230. E. Simone, A. R. Klapwijk, C. C. Wilson and Z. K. Nagy, Investigation of the Evolution of Crystal Size and Shape during Temperature Cycling and in the Presence of a Polymeric Additive Using Combined Process Analytical Technologies, *Cryst. Growth Des.*, 2017, **17**(4), 1695–1706.
231. Q. Su, C. D. Rielly, K. Powell and Z. K. Nagy, Mathematical modelling and experimental validation of a novel periodic flow crystallization using MSMPR crystallizers, *AIChE J.*, 2016, **63**(4), 1313–1327.
232. B. J. Ridder, A. Majumder and Z. K. Nagy, Population balance model-based multiobjective optimization of a multisegment multiaddition (MSMA) continuous plug-flow antisolvent crystallizer, *Ind. Eng. Chem. Res.*, 2014, **53**(11), 4387–4397.
233. Q. Su, B. Benyahia, Z. K. Nagy and C. D. Rielly, Mathematical modeling, design, and optimization of a multi-segment multi-addition plug-flow crystallizer for anti-solvent crystallizations, *Org. Process Res. Dev.*, 2015, **19**(12), 1859–1870.
234. B. J. Ridder, A. Majumder and Z. K. Nagy, Parametric, Optimization-Based Study on the Feasibility of a Multisegment Antisolvent Crystallizer for *in Situ* Fines Removal and Matching of Target Size Distribution, *Ind. Eng. Chem. Res.*, 2016, **55**(8), 2371–2380.
235. O. S. Agimelen, A. J. Mulholland and J. Sefcik, Modelling of artefacts in estimations of particle size of needle-like particles from laser diffraction measurements, *Chem. Eng. Sci.*, 2017, **158**, 445–452.
236. A. Majumder and Z. K. Nagy, Fines Removal in a Continuous Plug Flow Crystallizer by Optimal Spatial Temperature Profiles with Controlled Dissolution, *AIChE J.*, 2013, **59**(12), 4582–4594.
237. H. Yang, X. Yu, V. Raval, Y. Makkawi and A. Florence, Effect of oscillatory flow on nucleation kinetics of butyl paraben, *Cryst. Growth Des.*, 2016, **16**(2), 875–886.

238. H. Siddique, C. J. Brown, I. Houson and A. J. Florence, Establishment of a continuous sonocrystallization process for lactose in an oscillatory baffled crystallizer, *Org. Process Res. Dev.*, 2015, **19**(12), 1871–1881.
239. S. Melendi, S. Bonyadi, P. Castell, M. T. Martinez and M. R. Mackley, The functionalisation of carbon nanotubes using a batch oscillatory flow reactor, *Chem. Eng. Sci.*, 2012, **84**, 544–551.
240. M. Jiang and X. Ni, Effects of Water and Temperature on Reaction Mechanism and Crystal Properties in a Reactive Crystallization of Paracetamol, *Chem. Eng. Process.*, 2018, **131**, 20–26.
241. M. Jiang and X. Ni, Effect of solvents and impurity on crystallization kinetics and crystal properties in a reactive crystallization of paracetamol, *J. Cryst. Growth*, 2019, **523**, 125150.
242. Z. Q. Yu, P. S. Chow and R. B. H. Tan, Operating Regions in Cooling Cocrystallization of Caffeine and Glutaric Acid in Acetonitrile, *Cryst. Growth Des.*, 2010, **10**(5), 2382–2387.
243. X. Shi, S. Y. Wong, X. Yang and A. S. Myerson, Concomitant Cocrystallization on Engineered Surfaces, *CrystEngComm*, 2013, **15**, 7450–7455.
244. V. Svoboda, P. MacFhionnghaile, J. McGinty, L. E. Connor, I. D. H. Oswald and J. Sefcik, Continuous cocrystallization of benzoic acid and isonicotinamide by mixing-induced supersaturation: exploring opportunities between reactive and antisolvent crystallization concepts, *Cryst. Growth Des.*, 2017, **17**(4), 1902–1909.
245. U. Teipel, *Energetic Materials – Particle Processing and Characterization*, ed. U. Teipel, Wiley-Vch, Weinheim, Germany, 2005.
246. A. E. D. M. van der Heijden, Y. L. M. Creyghton, E. Marino, R. H. B. Bouma, G. J. H. G. Scholtes and W. Duvalois, Energetic Materials: Crystallization, Characterization and Insensitive Plastic Bonded Explosives, *Propellants, Explos., Pyrotech.*, 2008, **33**(1), 25–32.
247. J. C. Bennion, *Multicomponent Crystallisation of Energetic Materials*, Michigan, Michigan, 2017.
248. J. Beckvermit, T. Harman, A. Bezdjian and C. Wight, Modeling Deflagration in Energetic Materials using the Uintah Computational Framework, *Procedia Comput. Sci.*, 2015, **51**, 552–561.
249. Y. Wang, J. Zhang, H. Su, S.-H. Li, S. Zhang and S.-P. Pang, A Simple Method for the Prediction of the Detonation Performances of Metal-Containing Explosives, *J. Phys. Chem. A*, 2014, **118**(25), 4575–4581.
250. J. J. Sabatini and K. D. Oyler, Recent advances in the synthesis of high explosive materials, *Crystals*, 2016, **6**(1), 5–26.
251. D. I. A. Millar, H. E. Maynard-casely, D. R. Allan, A. S. Cumming, A. R. Lennie and A. J. Mackay, *et al.*, Crystal engineering of energetic materials: Co-crystals of CL-20, *CrystEngComm*, 2012, **14**(10), 3742–3749.
252. A. E. D. M. van der Heijden and R. H. B. Bouma, Crystallization and Characterization of RDX, HMX, and CL-20, *Cryst. Growth Des.*, 2004, **4**(5), 999–1007.

253. T. M. Tillotson, L. H. Hrubesh, G. L. Fox, R. L. Simpson, R. W. Lee, R. W. Swansiger, *et al.*, Sol-gel processing of energetic materials, *5th International Symposium on Aerogels*, Montpellier, France, 1997.
254. P. A. Cattani, T. J. Fleck, J. F. Rhoads, S. F. Son and I. Emre Gunduz, Applications of Additive Manufacturing Techniques in Making Energetic Materials, *The Summer Undergraduate Research Fellowship Symposium*, Purdue University, 2016.
255. E. Papadakis, A. K. Tula and R. Gani, Solvent selection methodology for pharmaceutical processes: Solvent swap, *Chem. Eng. Res. Des.*, 2016, **115**(B), 443–461.
256. S. Elgue, L. Prat, M. Cabassud and J. Cezerac, Optimisation of solvent replacement procedures according to economic and environmental criteria, *Chem. Eng. J.*, 2006, **117**, 169–177.

CHAPTER 4

Process Control

R. LAKERVELD*[a] AND B. BENYAHIA[b]

[a]The Hong Kong University of Science and Technology, Department of Chemical and Biological Engineering, Clear Water Bay, Hong Kong, China; [b]Loughborough University, Department of Chemical Engineering, Leicestershire, UK
*E-mail: r.lakerveld@ust.hk

4.1 Introduction

Crystallization processes can deliver a highly pure and often stable material within a single processing step at mild operating conditions. However, in contrast to liquid-phase products, the quality attributes of crystalline materials not only involve bulk properties such as purity and chemical identity, but also intrinsic properties such as crystal size distribution (CSD), shape, and solid-state form, which complicates product quality control. The quality attributes of crystalline materials can affect process and product performance in two main ways. First, downstream processes such as filtration, drying, mixing, granulation, and possibly tablet compression in case of pharmaceuticals are affected by crystal quality attributes. Therefore, improved economic performance of the process and potentially a higher product quality can be achieved indirectly by having a fundamental understanding and tight control over the crystallization process. Second, a crystallized compound may be one of the elements of a blended final product. For example, in case of a pharmaceutical, quality attributes of a tablet with an active pharmaceutical ingredient (API) such as bioavailability and shelf life depend at least to some extent on the crystal properties such as

The Handbook of Continuous Crystallization
Edited by Nima Yazdanpanah and Zoltan K. Nagy

Published by the Royal Society of Chemistry, www.rsc.org

CSD and solid-state form of the primary particles, which are shaped during the crystallization process. Other examples include pigments of which the colour intensity depends on the CSD or fertilizers of which the impact on safety, health, and environment is determined by the CSD. Therefore, improved product performance can be obtained for those cases in a direct way by having tight control over the crystallization process while avoiding the need for corrective processing steps. Finally, especially for bulk products produced with evaporative crystallization, energy consumption may also have a significant impact on the economic performance of a crystallization process. Although energy consumption is usually related to throughput (*e.g.*, in case of evaporative crystallization), which is determined by the thermodynamic properties of the crystallization system, avoiding the production of off-spec material will obviously also improve the performance of an energy-intensive crystallization process and multiple-effect evaporative crystallization systems can also reduce energy consumption.

Considering the wide variety of chemical products and processes that involve crystallization, it is clear that process control of crystallization processes is crucial for the product and process performance of many chemicals. Furthermore, there is a clear overall trend towards continuous crystallization instead of batch crystallization, particularly in the pharmaceutical industry,[1–5] which has important implications for process control as well. Process control in the batch mode of operation is substantially different from process control in the continuous mode of operation. In particular, product quality control in batch crystallization is typically achieved by following a fixed recipe, which focuses on controlling process parameters with a known impact on crystal product quality such as the dynamic generation of supersaturation and batch time. The former is determined by the crystallization method, *e.g.*, solvent evaporation rate in the case of evaporative crystallization, anti-solvent addition rate in case of anti-solvent crystallization, reactant addition rate in case of reactive crystallization, or the dynamic temperature profile in case of cooling crystallization. The latter is determined by the time required to approach solid–liquid equilibrium sufficiently close such that most of the attainable yield can be recovered.

Control of continuous crystallization processes is challenging due to a number of reasons. First, the number and type of process variables that are available in conventional continuous crystallizers (*i.e.*, mixed vessels) to manipulate the process is limited. In fact, one could argue that the number of manipulated variables to control crystal quality attributes in the most basic continuous crystallization configuration, *i.e.*, the mixed-suspension mixed-product removal (MSMPR) crystallizer, is practically zero for a fixed throughput and feed conditions. Therefore, extensions to the MSMPR concept *via* classification devices, milling devices, and fines dissolution loops are commonly employed. Furthermore, there is a high interest in the development of completely novel actuators (*e.g.*, based on external fields), dynamic operating conditions, and different types of crystallizers that fundamentally offer more opportunities for process actuation comparable to batch crystallization (*e.g.*, plug-flow crystallizers). Second, crystal quality attributes are shaped by a number of physical phenomena such as crystal growth, nucleation, agglomeration, and breakage, which cannot be well manipulated independently in conventional continuous crystallizers, but only at best collectively *via* bulk variables.[6] Therefore, trade-offs often have to be made, which limit the range of attainable product qualities.[7] The development of novel actuators that selectively target specific crystallization phenomena is a promising area to improve the flexibility and predictability of crystallization process control as well as the development of dynamic process models. Third, due to the nature of continuous processing, material has to be further processed immediately to benefit from integrated continuous manufacturing, which calls for a need to measure crystal quality attributes in an on-line fashion. The spectacular advancement of process analytical technologies (PAT) is highly encouraging in that regard.[8] Furthermore, since continuous crystallization is usually embedded within a larger continuous process, control objectives at the level of a unit operation have to be aligned with so-called plant-wide control objectives, which for example have been studied in detail for continuous pharmaceutical processes involving crystallization steps.[3,9,10] Fourth, scale-up of crystallization processes is complicated due to scale-dependency of mixing, which is particularly a challenge for open-loop control strategies. Finally, crystallization processes typically exhibit highly non-linear behaviour, which complicates tuning of conventional PID controllers and limits standard model-based control strategies using model linearization. The great advancements in nonlinear dynamic modelling of crystallization processes and associated numerical methods nowadays offer many opportunities for the development and feasible implementation of model-based control strategies in practice.

When developing a control strategy for continuous crystallization, the definition of the control objectives is one of the first questions that arise. As discussed above, typical control objectives for continuous crystallization may be related to various types of product quality control, process stabilization (*e.g.*, in case of oscillatory or unstable open-loop behaviour), the economic performance of the process (*e.g.*, related to the use of utilities or transient behaviour) or environmental criteria such as an E-factor.[11] Just as for any unit operation of a chemical process, the basic questions to be answered to design a control strategy for crystallization processes for a given control objective are:

- What variables do we wish to control?
- What variables can (or should) be measured?
- What manipulated variables are available?
- How do we pair them in automated control loops (structurally and mathematically)?

This chapter aims to address each of these basic control questions. An overview of typical variables to be controlled in continuous crystallization processes is provided followed by an overview on how to measure process variables for control applications. Furthermore, a discussion on available manipulated variables will be presented, which will be structured according to the common types of crystallizers available in industry, as available manipulated variables are strongly related to the basic crystallizer design. Finally, common control strategies will be discussed that bring together controlled, measured, and manipulated variables in automated control loops. A distinction is made between three main control strategies that have been developed and implemented for crystallization processes: model-free, model-based and hybrid techniques.[12] The key features of model-free and model-based control strategies are presented. The focus is on solution crystallization, which is the most common and versatile type of industrial crystallization. The chapter ends with an outlook for future developments.

4.2 Controlled Variables

Variables that are of interest to be controlled for continuous crystallization processes include simple process variables such as temperature, pressure, volume, and liquid-phase composition and variables related to the intrinsic crystal quality attributes. With the exception of liquid-phase composition, automated control loops can be developed readily for simple process variables, as they are not unique to crystallization processes. However, crystal quality attributes are important to be controlled to improve downstream processing and product functionality, as discussed, and are much more challenging to control effectively. Common crystal quality attributes that are of interest to be selected as controlled variables include:

1. *Purity*; solvent inclusion, presence of foreign molecules
2. *Solid-state form*; various polymorphic forms, solvates, hydrates, salts, co-crystals
3. *Crystal habit*; aspect ratio, shape distribution, specific surface area, *etc.*
4. *Crystal size distribution*; various size definitions exist depending on the measurement instrument.

The purity of a crystalline product can be affected by mother liquor adhering to the crystal surface, inclusion of foreign molecules in the crystal lattice, or entrapment of mother liquor in a crystal particle. Adhesion of mother liquor is related to the crystal habit and CSD and can be mitigated *via* washing. The physical principle of crystallization does not easily allow for foreign molecules to be incorporated in a crystal lattice, but might be of significance if the foreign molecule has a similar molecular structure or when the crystal growth rate is fast favouring uptake of impurities in the crystal lattice. Similarly, solvent inclusion can be avoided by avoiding fast crystal growth rates, which occur at high supersaturation.

If the unit cell of a single-component crystal lattice can be structured in different ways, various polymorphic forms of the crystallizing compound can emerge. Consequently, although chemically identical, such materials exhibit different physicochemical properties and stability depending on the polymorphic form. For multi-component systems, crystal lattices may include solvent molecules (solvates) or other compounds (classified as salts if charge distribution is present or as co-crystals if the molecules have a neutral charge), which all may appear in different polymorphic forms. The classification of solid-state forms is not straightforward and subject to debate.[13,14] Nevertheless, since different solid-state forms have different physicochemical properties, control over the solid-state is important, as often the case for pharmaceutical applications. For example, Yu *et al.*[15] reported six solvent-free polymorphs for an API with much different shapes and sizes. Furthermore, the emergence of unexpected polymorphic forms or inadequate control over the formation of a desired polymorphic form during crystallization has been the source of significant supply shortages of pharmaceutical products and intensive research programs (*e.g.*, as in the well-known case of the protease inhibitor for human immunodeficiency virus Ritonavir[16]). Different solid-state forms have a different solubility and different nucleation kinetics. Therefore, manipulating the solubility of possible solid-state forms *via* control of the temperature and solvent composition and manipulating the nucleation kinetics and possible solid-state transitions provides routes for control over polymorphic forms. Similarly as for different polymorphic forms, solvates, salts and co-crystals have different physicochemical properties related to dissolution rates, density, shape, and brittleness, which calls for tight control over the solid-state form of multi-component crystals as well.

The shape of crystals can have a significant impact on downstream processing or final product quality. For example, the specific surface area of a crystal can differ significantly depending on the shape, which affects bioavailability in the case of pharmaceuticals. In general, a crystal habit can change due to different growth rates of the various facets of a crystal (*i.e.*, due to different surface properties), mechanical impacts, or chemical adsorption of foreign molecules. For example, the relative growth rate of facets of cyclotrimethylene trinitramine (RDX), and thus its crystal shape, can be controlled *via* manipulating the solvent composition.[17] Alternatively, temperature cycling can be used to induce a sequence of growth and dissolution cycles to control a crystal shape, which has been applied for batch crystallization[18–22] and is also feasible for continuous crystallization.[23,24] In general, manipulation of the supersaturation allows for some degree of control over the crystal shape. Finally, additives can potentially be used to control the crystal shape.[25]

The CSD has a large impact on downstream processing and product efficacy for many crystallization processes, and is, therefore, often chosen as a controlled variable. There are many metrics for crystal size (*e.g.*, length, sieve diameter, volume diameter, surface diameter, projected diameter), which are often associated to a certain principle of measurement. The CSD is shaped by an interplay of various physical phenomena, which are driven by supersaturation and shear forces. Therefore, some form of supersaturation control at homogeneous hydrodynamic conditions is usually employed to obtain control over the CSD.

Process variables do not change anymore at steady-state conditions in the absence of disturbances, which simplifies control. Therefore, the majority of control approaches aims to keep controlled variables close to a desired set point at steady state. However, it has also been noted that steady-state operation for continuous crystallization may not necessarily be a goal by itself. For example, for pharmaceutical applications, it has been argued that not steady state but rather a "state-of-control" is needed.[26] Furthermore, in view of the possible high added value of raw materials and products and short campaigns (*e.g.*, for pharmaceutical applications), transient operational phases such as start-up and shut-down may have to be considered. For example, Patrascu and Barton[27,28] demonstrated *via* dynamic optimization that optimizing a complete production campaign instead of only the steady-state operating point of an end-to-end continuous pharmaceutical manufacturing process including various crystallization steps can lead to a significant increase in productivity compared to when only considering steady state.

Powell *et al.*[24] introduced the periodic mixed-suspension mixed-product-removal (PMSMPR) crystallizer. In contrast to the conventional MSMPR crystallizer, product is transferred periodically from the crystallizer at high flow rates. A practical advantage of the PMSMPR crystallizer is that blockage in transfer lines can be minimized and classification in the product removal can be reduced due to the fast withdrawal of slurry. Selective removal of fines in the product outlet would not only violate the MSMPR concept, but could also lead to sustained oscillations.[29] A conceptual advantage of a seeded PMSMPR crystallizer is that the dynamic properties of the withdrawal cycle can be used as a manipulated variable to control CSD when PAT are available, which can be modeled.[30] They demonstrated that a multi-stage PMSMPR crystallizer system gave a better product quality in terms of size and shape compared to a single stage. Since the PMSMPR has dynamic properties that are comparable to batch operation, seeding is an important manipulated variable. Furthermore, the PMSMPR crystallizer reduces fouling, which can otherwise be an important practical problem for continuous crystallization. Dynamic operation for continuous crystallization can also be used to prevent fouling of a tubular crystallizer *via* model-based anti-fouling control by manipulating the properties of heating and cooling cycles.[31] Finally, for a different type of application, such heating and cooling cycles can also be used to control crystal shape.[32]

4.3 Measured Variables

Automated process control loops for continuous crystallization rely on the availability of real-time measurements of either controlled variables or process variables from which controlled variables can be inferred. In general, the recent development of more reliable, affordable and high performance PAT along with increased computing power have enabled a shift towards continuous manufacturing even in heavily regulated industries such as the pharmaceutical industry.[1,2,4] A significant number of new and advanced control strategies have been implemented in crystallization processes (and beyond) over the last decade, which have been enabled by PAT.[3,12,33–35]

Many control strategies for solution crystallization processes aim to control the crystal quality attributes indirectly by controlling supersaturation. Therefore, a significant improvement in crystallization process control has been enabled by *in situ* real-time PAT to measure solute concentration, such as attenuated total reflectance Fourier transform infrared (ATR-FTIR) spectroscopy[35–41] and attenuated total reflection ultraviolet/visible (ATR-UV/vis) spectroscopy.[42–46] The performance of these spectroscopic methods has further improved over the years *via* rigorous chemometric techniques, the use of multiple wavenumbers and the combination of temperature and absorbance. Various PAT are available for direct measurement of the crystal size. Arguably the most popular *in situ* and real-time measurement of crystal size is the focused beam reflectance measurement (FBRM), which measures the number-based chord length distribution (CLD). The advantages of FBRM are that the method is free of calibration, non-invasive, robust, and relatively affordable. However, the CLD is only a fingerprint of the CSD and mathematical transformations have to be developed if a CLD is to be used within a control method for CSD, which is challenging, but effective demonstrations of such conversion have recently been reported.[47] Laser diffraction is a common bench-top instrument for CSD measurements, which can be integrated with a continuous crystallizer in an on-line fashion using an automated sampling device, a flow cell, and a dilution loop to obtain a slurry density in the required range for laser diffraction.[48] Acoustic methods have also been used for on-line CSD measurements.[35] Neumann and Kramer[49] compared two on-line laser diffraction instruments (Sympatec HELOS and Malvern 2600c) and an acoustic instrument (Sympatec OPUS), which were all implemented on-line for continuous crystallization of ammonium sulphate in an 1100-L draft-tube baffle (DTB) crystallizer. The on-line measurements were compared with off-line measurements and they concluded that the shape of the CSD was measured consistently with all three instruments, but differences in absolute values existed. The latter was attributed to the different measurement principles of the instruments. The on-line implementation of some of these instruments were later used in model-predictive control methods for batch crystallization.[50–52] Such control frameworks could equally well be applied to continuous crystallization. Recently, particle vision measurement (PVM) probes have become popular for *in situ* measurement of the CSD and crystal morphology with ample examples presented in literature.[19,53,54] The obvious advantage of PVM probes is that full information of the crystal shape distribution can be obtained. However, dense crystal suspensions pose challenges for effective use of PVM probes, especially when automated software is used to extract size and shape information, due to overlapping crystals. Furthermore, the method is based on optical measurements with visible light and limited magnification, which reduces the ability to identify fines.

Raman spectroscopy is comparable to FTIR but uses different (and complementary) vibrational and rotational modes to generate scattering of electromagnetic radiation. Therefore, probes for Raman spectroscopy and ATR-FTIR can be combined to obtain complementary data from a crystallization process.[41,55] In particular, in addition to the measurement of solute concentration, Raman is capable of measuring *in situ* the polymorphic form of crystals in suspension.[56–59] Févotte[60] provided a review on the use of *in situ* Raman spectroscopy with a focus on pharmaceutical crystallization and industrial applications. The purity of crystals is difficult to measure in an on-line fashion as spectroscopic methods usually do not have sufficient resolution to measure small quantities of impurities. However, for purity, some form of inferential control can be used by limiting the crystal growth rate *via* supersaturation control.

A more comprehensive and multi-disciplinary review with current trends in PAT, including imaging, spectroscopy, acoustic and electronic sensors has recently been provided by Simon *et al.* (see also Chapter 9).[8] Finally, generally within the pharmaceutical industry, it has been observed that the role of PAT is different during development compared to commercial manufacturing.[34] In particular, during development the use of PAT is widespread and focuses on understanding by using multi-variate measurements from different types of instruments, whereas the use of PAT during commercial manufacturing is more limited and has a more targeted and simplified role with a focus on robustness and reliability. Reasons for the observed limited role during commercial manufacturing include more stringent quality requirements during manufacturing (*e.g.*, GMP compliance, data storage requirements, maintenance, *etc.*), lack of human resources, and the fact that the use of PAT during development may have led to an understanding of how simpler process variables that are critical for quality control can be used as controlled variables during commercial manufacturing.[34]

4.4 Model-free Control Strategies

All model-free strategies are in essence feedback control methods commonly based on real-time measurements of the concentration (ATR-FTIR, ATR-UV/Vis), temperature, or particle counts (FBRM) or their combination. The overall performance of these methods depends significantly on the accuracy and reliability of the PAT. The aim is to either control some property of the continuous or dispersed phase. For example, the aim can be to maintain operation within a certain supersaturation range (*i.e.*, between the solubility curve and the metastable limit) by manipulating the cooling duty, heating rate or anti-solvent addition rate. Such supersaturation control (SSC), also known as concentration control (C-control) or concentration feedback control (CFC) is simple and robust if suitable PAT is available. The SSC requires the measurement of concentration and a manipulated variable affecting the solubility of the system (*e.g.*, temperature, solvent/anti-solvent ratio) or solute concentration (*e.g.*, heat addition in case of evaporative crystallization). An optimal operating point requires an in-depth understanding of the crystallization mechanisms to mitigate the effects of nucleation or solid-state transitions. Another aim can be to control some property of the CSD directly. A traditional example is the use of a fines destruction loop, whereas a more recent approach is the so-called direct nucleation control (DNC) method that manipulates the kinetics of nucleation, growth, and dissolution through adaptive cooling and heating cycles.[61] The temperature cycles promote dissolution of the fines and growth of larger crystals, while reducing the width of the CSD, which is comparable to a fines dissolution strategy. The DNC has been extensively used for lab-scale crystallizers and demonstrated enhanced robustness and effective control over particle size, uniformity and polymorphic purity by eliminating fine particles and agglomeration.[62] Extensions to continuous crystallization are feasible in the case of plug-flow crystallizers, cascades of MSMPR crystallizers, or if additional actuation for heating is available in a single MSMPR.

The remainder of this section will discuss manipulated variables and illustrate typical feedback control loops that can operate without a process model. The available manipulated variables for continuous crystallization depend on the basic equipment configuration. Therefore, the discussion is organized by classifying the basic types of crystallizer hardware and their extensions.

4.4.1 MSMPR Crystallizer

Conventional continuous crystallizers such as the forced circulation crystallizer, DTB crystallizer, or the Oslo (fluidized-bed) crystallizer all aim to resemble the MSMPR concept. The advantages of MSMPR crystallizers are their robust operation and ability to provide the large residence times needed for crystallizing compounds with a slow growth rate or in case there is a need to produce large crystals. However, an MSMPR crystallizer operates at outlet conditions, which may limit attainable crystal growth rates, and inevitably leads to a high variability in the product quality due to the broad residence time distribution. Furthermore, the number of manipulated variables in a conventional MSMPR is limited. For example, consider a simple well-mixed jacketed draft-tube crystallizer operated in continuous flow mode for cooling crystallization as illustrated in Figure 4.1. There are two control valves present and possibly one variable speed drive. Therefore, we can construct three

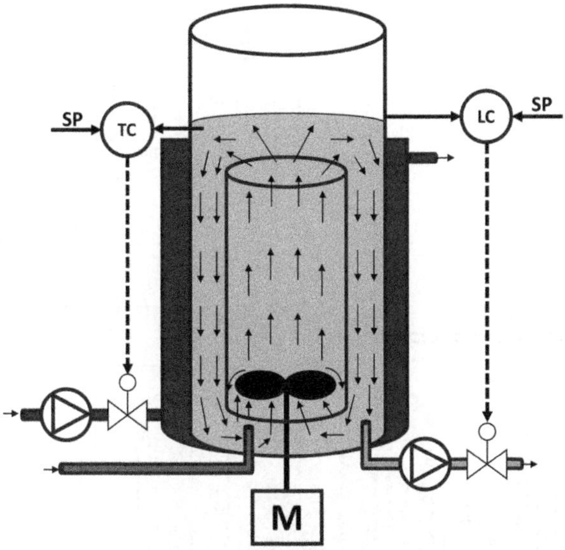

Figure 4.1 Schematic illustration of an MSMPR cooling crystallizer and control loops for cooling crystallization. SP – set point, TC – temperature controller, LC – level controller, M – motor.

single-input single-output control loops to control the crystallizer temperature, volume, and stirrer speed (see Figure 4.1). For an MSMPR crystallizer, it is clear that the set points of the temperature controller and level controller can be used to manipulate the CSD as an inner control loop within a cascade control structure. However, the flexibility to manipulate those set points for a basic MSMPR crystallizer is limited, as the level height is often restricted by draft-tube height (or other hydrodynamic considerations for other types of well-mixed crystallizer) and the operating temperature is restricted by practical considerations for the cooling medium and is primarily chosen for a desired attainable product yield. Finally, the impeller can be used as a potential actuator to induce secondary nucleation, but the effect on the CSD may be negligible depending on the system characteristics.[63] Furthermore, impeller frequency is in practice restricted by a lower bound related to the suspension criterion and possibly by an upper bound at which attrition fragments and highly strained crystals, which may easily dissolve, are produced.

To improve on the limited number of manipulated variables to control the crystal product quality in MSMPR crystallizers, a fines dissolution loop can be added such as in the DTB crystallizer illustrated in Figure 4.2. A classification zone created *via* vertical baffles operates at laminar flow conditions can be used to separate small crystals from larger crystals targeting a specific cut-off size, which can be manipulated to some extent *via* the flow rate in the fines dissolution loop. A DTB crystallizer with fines dissolution is typically used if the objective is to mitigate the impact of secondary nucleation with fines dissolution to create a coarse product. Therefore, the number of

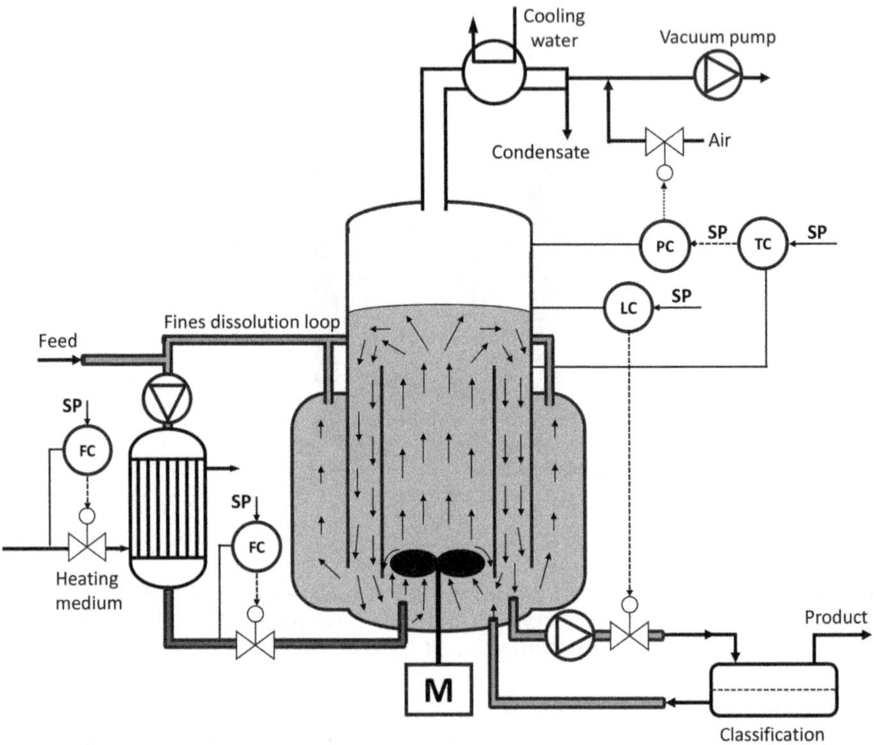

Figure 4.2 Schematic illustration of an MSMPR evaporative crystallizer with a fines dissolution loop, product classification and control loops. FC – flow controller, PC – pressure controller, SP – set point, TC – temperature controller, LC – level controller, M – motor.

fines (or a related CSD characteristic) can be used as controlled variable with the flow rate in the fines dissolution loop as manipulated variable in a feedback control loop. The fines dissolution loop may also be used to provide the latent heat of vaporization, which puts a constraint on the minimum flow rate for a fixed throughput to prevent a boiling solution at the outlet of the heat exchanger.

The dynamics of the process operation become more complicated when adding fines dissolution. In particular, the downside of the addition of a fines dissolution loop is the possibility of oscillatory behaviour of the CSD, which complicates steady-state operation and can cause loss of product.[64–67] The oscillatory behaviour is a well-reported observation and has been reproduced with dynamic simulations.[66,68] A complex interplay between nucleation and growth, which have a different dependency on supersaturation, and fines dissolution and product removal is the basis for this manifestation of strong nonlinear behaviour. Fines have a large specific surface area and the abundance of fines can consume the supersaturation effectively, which can lead to a low overall supersaturation in the MSMPR crystallizer below a nucleation threshold. Consequently, the nucleation rate will be low, which will suppress the formation of new crystals. Furthermore, the fines dissolution loop removes small crystals preferentially, which will significantly lower the available crystal surface area in the crystallizer. Consequently, a shortage of crystal surface area may occur, which causes the supersaturation in the MSMPR crystallizer to rise. Once the supersaturation passes the nucleation threshold, a burst of new nuclei can be created provided that the kinetics of nucleation rate are of a sufficiently high order, which completes the cycle. In order to match experimental data with predicted oscillations from dynamic simulations of a continuous DTB crystallizer with fines dissolution quantitatively, additional physical phenomena need to be considered such as strain in secondary nuclei[66] or dissolution near the inlet of the undersaturated hot stream in the crystallizer where predominantly large crystals accumulate.[68] The oscillations are stronger when the fines dissolution flow rate is higher.[67] Model-free control strategies can be employed to stabilize such a crystallizer. Rawlings *et al.*[29] reviewed the earlier literature on this topic and concluded that many studies have shown that oscillations can be dampened and disturbances can be rejected by using feedback control with the fines dissolution flow rate as manipulated variable and some characteristic process variable as controlled variable (*e.g.*, supersaturation, CSD characteristics or slurry density in the fines dissolution loop) using a proportional-integral (PI) controller law of the following well-known form:

$$\frac{\mathrm{d}\psi}{\mathrm{d}t} = e, u = K_{\mathrm{C}}\left(e + \frac{\psi}{\tau_{\mathrm{I}}}\right), \tag{4.1}$$

where e is the difference between the (measured) controlled variable and its set point, u is the calculated input to the process (*i.e.*, manipulated variable), and K_C and τ_I are tuning parameters. Studies have shown that a simple single-input single-output (SISO) PI controller can have a comparable behaviour to a SISO model-predictive controller (MPC), which suggests that the PI controller should be used for industrial applications due to its simplicity.[67] Finally, in a different application, fines dissolution in combination with an MSMPR crystallizer has also been demonstrated to be effective to control crystal shape for the case of lysozyme crystallization.[69]

Eek *et al.*[67,70] studied the extension of the DTB crystallizer with a product classification screen for crystallization of ammonium sulphate from water. Experimental data and dynamic simulations demonstrated that the addition of a product classification device leads to additional interactions between CSD dynamics and the production rate. The flow rate to the classification device and its design specifications have a significant impact on the product CSD. However, the product flow rate may not be suitable as manipulated variable, as this flow rate also affects throughput and may have to be used in a level control loop if the feed stream from upstream units cannot be manipulated. In conclusion, for an evaporative continuous crystallizer with a fines dissolution loop and product classification device and under the assumption of a fixed solute concentration in the feed stream, effective control of key process variables can be achieved by using four SISO PI controllers (Figure 4.2):

1. A temperature control loop to enable isothermal operation, which is often implemented as a cascade control loop with an inner pressure control loop and outer temperature control loop. The purpose of operating at low pressure is to minimize energy consumption and possibly to prevent product degradation.
2. A level control loop using the feed or product flow rate as manipulated variable. The objective is to obtain stable operation by rejecting disturbances of the system throughput and optimizing hydrodynamic conditions in the draft tube.
3. A production rate control loop of which the manipulated variable is the flow rate of a heating medium (*e.g.*, hot water, steam) and the controlled variable is the heat input to the system. The heat input may be split over different heat exchangers (*e.g.*, one heat exchanger in the fines dissolution loop and another one in the center of the draft tube or jacket) to decouple the fines destruction loop from the energy control loop. The objective is to obtain a certain slurry density (*i.e.*, third moment of the CSD) that matches a desired throughput and practical considerations related to slurry transport.
4. A fines destruction control loop of which the manipulated variable is the flow rate in the fines dissolution loop and the controlled variable is some characteristic of the CSD that depends on the amount of fines present in the system or supersaturation (*e.g.*, the number of crystals per unit volume of suspension). The objective is to obtain stable operation by avoiding oscillatory or unstable behaviour and obtaining a product with desired specification.

Similar control loops can be synthesized for different crystallization methods such as cooling crystallization or anti-solvent crystallization.

4.4.2 Plug-flow Crystallizer

To improve flexibility to achieve control objectives, a tubular crystallizer with multiple cooling zones can be used (see Figure 4.3), which can also be achieved by creating a cascade of MSMPR crystallizers in series.[71–73] Typical physical realizations of tubular crystallizers are based on static mixers,[74] segmented flows,[75–77] or oscillatory flow around orifices[78] and are designed to create plug-flow conditions under overall low flow velocities, which would otherwise lead to laminar flow violating the desired plug-flow behaviour needed for uniform crystal quality. For an ideal plug-flow system, which can be obtained by promoting radial mixing and eliminating axial mixing in a tubular crystallizer or by placing a large number of MSMPR crystallizers in series, the length coordinate acts as a time coordinate when comparing to batch crystallization. Therefore, different temperature profiles can be implemented, which is comparable to implementing a dynamic temperature profile for a batch crystallization with fixed batch time. The latter is a well-studied topic in literature for several decades, which can be

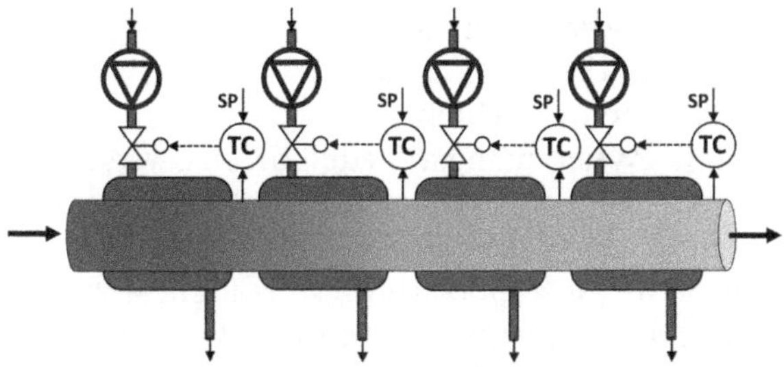

Figure 4.3 Schematic illustrations of a multi-segment plug-flow crystallizer and control loops for cooling crystallization.

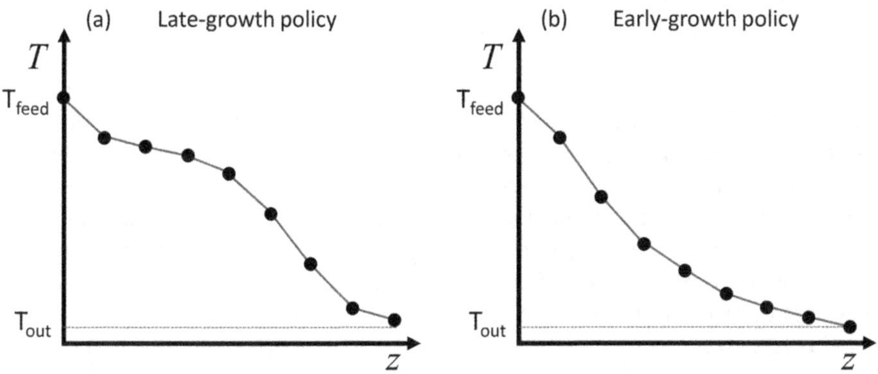

Figure 4.4 Schematic illustration of so-called early-growth (a) and late-growth (b) temperature profiles.

transferred to continuous crystallization if plug-flow behaviour is approximated. For unseeded operation, sufficient cooling should be provided in the first zone to trigger nucleation if alternative actuators for nucleation or seeds are not available at the inlet of the crystallizer. In general, the set points of the remaining temperature control loops can be designed in two opposite ways (see Figure 4.4 for schematic illustrations). One option is to provide slow cooling at the beginning of the cascade and rapid cooling at the end (so-called late-growth policy), which aims to align the generation of supersaturation with the growth of the total crystal surface area in the system (see Figure 4.4a) such that a constant supersaturation over the length of the system is achieved. Alternatively, a so-called early growth policy can be promoted, which is enabled by fast cooling in the beginning (see Figure 4.4b). Such a natural cooling curve is easier to implement, as it needs a large driving force for cooling when temperature is high. Therefore, a single temperature control loop could be used and temperature control for different zones is strictly speaking not needed. A disadvantage of using a single temperature control loop is the increased possibility for fouling and encrustation, which is a practical problem for tubular crystallizers for cooling crystallization with a relatively large temperature difference between cooling medium and crystallizer content.[79] Especially when secondary nucleation is important, the specific control objective will determine what policy should be implemented, which has been reviewed in detail for seeded batch crystallization.[80] When using a seeded batch crystallizer as reference for a continuous plug-flow crystallizer with a well-defined production of an initial crystal population or seeding, a late-growth policy will maximize secondary nucleation as the supersaturation is highest when most crystal mass is present and, thus, maximize the number of crystals. However, initial nucleation will be low and the largest volume of the final product is provided by the grown initial crystal population. In contrast, for an early growth policy, secondary nucleation will be minimized as the crystal mass is small when supersaturation is high. However, if any secondary nuclei are formed, they can grow to a large size as they form in the beginning of the tubular crystallizer. It is clear that the type of instrument for a CSD measurement also plays role here. In particular, using an instrument to measure a number-based CSD may lead to a different sequence of set points for the temperature control loops compared to when using an instrument to measure a volume-based CSD. For example, minimizing the number-based mean size of the crystals would lead to a sequence of set points for the temperature controllers in line with a late-growth policy and minimizing the volume-based mean crystal size would lead to a sequence of temperature control set points that approach an early-growth policy. Note that a volume-based CSD can be converted into a number-based CSD using shape factors, but a volume-based measurement device usually does not have high accuracy to measure a large number of crystals with very small volume.[63]

Similar growth profiles can be obtained when using different crystallization methods. For example, Alvarez and Myerson[74] distributed the addition of an anti-solvent over the length of a tubular crystallizer constructed from static mixers to control the supersaturation generation for various organic systems. Both an increase and decrease in particle size could be achieved depending on the kinetics of the system. For evaporative crystallization, the motivation to use a cascade of crystallization is traditionally energy savings by using vapour recompression. However, in principle, a supersaturation profile can also be controlled at a desired level using a multi-stage evaporative crystallizer. A desired supersaturation profile along the length of a tubular crystallizer does not only affect the CSD, but also crystal shape. Kwon *et al.*[81] used a dynamic model of a plug-flow crystallizer to design a feedforward control strategy to reject measured disturbances such that crystals with desired size and shape could be produced reliably.

Finally, the use of seeds is an important manipulated variable for batch crystallization,[63,82–84] which equally well applies to continuous plug-flow crystallization. For example, Eder *et al.*[85] investigated the effect of seed loading on the CSD of acetylsalicylic acid in a tubular crystallizer and concluded that a narrow product CSD could be achieved with control over the mean size. In follow-up work, Besenhard *et al.*[86] demonstrated that the flow rate of a seed suspension could be used effectively as manipulated in an automated feedback control loop with the mean crystal size of the product as controlled variable. In general, seeding is an important manipulated variable for continuous plug-flow crystallization due to various reasons. First, seeding provides control over the solid-state form of the final product. Second, seeding can minimize primary nucleation, which is often unpredictable and exhibits high variation, which makes control and reproducibility difficult. Third, seeding allows for a certain degree of control over the final CSD. In particular, if the seed surface area is sufficient to consume supersaturation at a sufficiently fast rate such that secondary nucleation is minimized, the final mean product size can be predicted from the so-called ideal growth expression,[87] which is derived from a mass balance over the seed population assuming a narrow CSD of the seeds, negligible nucleation, agglomeration, and no residual supersaturation at the end of the tubular crystallizer:

$$\frac{L_P}{L_S} = \left(\frac{C_S + 1}{C_S}\right)^{1/3} \tag{4.2}$$

where L_P is the product mean size, L_S is the seed mean size, and C_S is the seed load defined as the ratio of the seed mass over the product yield. Since the seed size and seed mass can be controlled and product yield can be estimated from solubility data, the final product size can be predicted and controlled without any feedback control. In the case where nucleation is not negligible, a product with a smaller mean can be expected, as the number of crystals increases. Since tubular crystallizers do not have moving mechanical parts, attrition may be small, which would improve the predictability of a seeded continuous plug-flow crystallizer.

4.4.3 Quality-by-design

Modern model-free process control strategies for pharmaceutical crystallization are often aligned with the concepts of so-called quality-by-design (QbD), which is a robust control strategy that ensures a built-in quality (*e.g.* safety and efficacy) for a drug product. QbD may involve control of the crystallization process within a so-called design space, which is defined as the range of critical process parameters (CPPs) that has been demonstrated to provide assurance of quality, described by a targeted range of critical quality attributes (CQAs).[88] Most of the published work has focussed on monitoring the CQAs using an array of PAT.[89] Since pharmaceutical manufacturing is currently almost exclusively conducted in batch mode of operation, most literature examples of QbD involve batch crystallization. For example, the QbD approach proved effective when operating a batch crystallization process within the design space to achieve the desired polymorphic form of tolbutamide using a combination of ATR-FTIR and Raman spectroscopy.[90] However, the QbD approach is also attractive for continuous crystallization by identifying those process parameters that have a significant impact on the steady-state behaviour of the continuous crystallization.

4.5 Model-based Control Strategies

With the rapid advance of predictive modelling of continuous crystallization during the last decades, model-based control strategies have become very attractive. Linear process models have the advantage of simplicity and well-defined characteristics. A crystallization process model can be linearized either for synthesis of feedback controllers or to be used in MPC. For example, Motz *et al.*[68] proposed a design of a controller based on a simplified process model, which is linearized around a desired steady state such that a transfer function can be derived. The controller uses the third moment of the CSD as controlled variable and the fines dissolution flow rate as manipulated variable. The design of the feedback controller is based on H_∞-theory for infinite-dimensional systems, as a simplified process model is used and plant-model mismatch needs to be mitigated. Dynamic simulations show that the controller can stabilize the oscillations in about 5 cycling periods. Continuous seeding can also be a suitable manipulated variable to stabilize the system.[91]

Predictive crystallization process models have a strongly nonlinear character. Therefore, despite the complications, controllers based on nonlinear process models inherently have better opportunity to provide superior closed-loop performance. Christofides and co-workers have developed a general framework for the design of bounded nonlinear predictive controllers for continuous crystallization processes. A key idea of their approach is to describe the dynamics of a continuous crystallization process by a low-order approximation of the full process model. The full crystallization model involves a population balance for the dispersed phase combined with material and energy balances for the continuous phase. The population balance is difficult to use directly for synthesis of practical model-based feedback controllers.[92] Furthermore, typical numerical solutions of the population balance based on a finite-dimensional approximation are of high-order, which are not suitable for the design of simple model-based feedback controllers. Therefore, they proposed to apply a model reduction based on the method of weighted residuals combined with the concept of approximate inertial manifold to construct a low-order approximation of the process model, which can be used for design of a nonlinear model-based feedback controller that can be implemented in a practical way.[93] The framework provides a basis for the design of robust nonlinear feedback controllers, which address the strong nonlinearity of continuous crystallization processes,[93,94] model uncertainty,[95] constraints on actuators,[96] sensor malfunctions,[97] asynchronous measurement sampling due to technical failures or different sampling rates of various instruments,[98] and fault detection, isolation, and compensation.[99,100] The key features of those controllers are summarized in this section. The reader is referred to the corresponding series of papers for more details.[93–100]

The motivating example for the development of the comprehensive framework from Christofides and co-workers for nonlinear feedback control of continuous crystallizers is an MSMPR crystallizer for cooling crystallization, possibly with fines dissolution. Such crystallizers can have a similar configuration as illustrated in Figure 4.2 if evaporation is used for cooling (*i.e.*, so-called flash cooling) or can be augmented with a cooling jacket. The former configuration is preferred for large-scale applications. When assuming constant temperature and nucleation at infinitesimal size only, the following mathematical model based on a population balance and a solute material balance describes the dynamic evolution of the dispersed and continuous phase in the MSMPR crystallizer:

$$\begin{aligned}
&\frac{\partial n(L,t)}{\partial t}+\frac{\partial\left[G(L,t)n(L,t)\right]}{\partial L}=-\frac{n(L,t)}{\tau}-h(L)\frac{n(L,t)}{\overline{\tau}}+\delta(L-0)B(t),\\
&\frac{\mathrm{d}c(t)}{\mathrm{d}t}=\frac{(c_0-\rho)}{\varepsilon(t)\tau}+\frac{(\rho-c(t))}{\tau}+\frac{(\rho-c(t))}{\varepsilon(t)}\frac{\mathrm{d}\varepsilon(t)}{\mathrm{d}t},
\end{aligned} \tag{4.3}$$

where $n(L,t)$ is the number of crystals with length L per unit of suspension volume as a function of time t, G is the crystal growth rate, τ is the residence time of the crystallizer, $\overline{\tau}$ is the residence time in the fines dissolution loop defined by V/F_F with F_F representing the flow rate through the fines dissolution loop and V representing the crystallizer volume, $h(L)$ is the fines classification function, $\delta(L-0)$ is the Dirac function, and B is the nucleation rate. The solute concentration as a function of time is represented by c and c_0 is the feed concentration. Furthermore, ρ is the density of the liquid phase. Finally, the connection between the population balance and the solute material balance is provided by ε, which is the volume of the liquid per unit volume of suspension expressed as a function of the third moment of the CSD as follows:

$$\varepsilon(t)=1-k_v\int_0^{\infty}n(L,t)L^3\,\mathrm{d}L \tag{4.4}$$

where k_v is a shape factor. Kinetic relations for the crystal growth rate typically are empirical relations involving supersaturation, temperature, and parameters fitted from experimental data, for example for the case of size-independent growth:

$$G(t)=k_G\left(c(t)-c_s\right)^{n_G} \tag{4.5}$$

where k_G and n_G are constants and c_s is the solubility, which is constant for an isothermal MSMPR. Kinetic expressions for nucleation may have a similar form as eqn (4.5), but usually also involve slurry density and the power input of an impeller to better describe secondary nucleation as a function of operating conditions and a higher-order dependency on the supersaturation. The reader is referred to Chapter 2 of this book for more details on kinetic modelling of continuous crystallization processes. The difference in dependency on supersaturation between crystal growth and nucleation inherently leads to strong nonlinear process behaviour (*e.g.*, oscillatory behaviour) and a steady state corresponding to a desired CSD may not be stable,[93] which motivates the use of feedback control to stabilize the process operation and obtain the desired properties of the CSD. As discussed earlier, a typical control loop would use some property of the CSD as measured and controlled variable and would use the flow rate through the fines dissolution loop as manipulated variable. An alternative control structure could use the inlet solute concentration as manipulated variable.

The first step in the framework from Christofides and co-workers is to apply a model reduction step. The aim is to capture the dominant dynamics of the process described by a limited number of ordinary differential equations (ODEs) so that practical nonlinear model-based feedback controllers can be designed. A general model reduction approach based on the method of the weighted residuals and the concept of inertial manifold is provided by Chiu and Christofides.[93] The method coincides with the well-known method of moments when the basis functions to expand the CSD are Laguerre polynomials and the weighting functions have a specific form. The *i*th moment of the CSD is defined as follows:

$$m_i = \int_0^\infty L^i n(L,t)\mathrm{d}L,\ i = 0,1,2,\ldots, \tag{4.6}$$

multiplying the population balance in eqn (4.3) with L^i, substituting (4.4), assuming spherical particles (*i.e.*, $k_\mathrm{v} = 4/3\pi$), assuming growth with a first-order dependency on supersaturation, and using the following kinetic relation for crystal nucleation,

$$B = \varepsilon k_\mathrm{A} \mathrm{e}^{\frac{-k_\mathrm{B}}{[(C/C_\mathrm{s})-1]^2}}, \tag{4.7}$$

and the following (sharp) classification function:

$$h(L) = \begin{cases} 1, \text{ if } L \le L_\mathrm{m}, \\ 0, \text{ if } L > L_\mathrm{m}, \end{cases} \tag{4.8}$$

and, finally, integrating over all crystal sizes gives the following set of ODEs:[94]

$$\begin{aligned} \frac{\mathrm{d}m_0}{\mathrm{d}t} &= -\frac{m_0}{\tau} + \left(1 - \frac{4}{3}\pi m_3\right) k_\mathrm{A} \mathrm{e}^{\frac{-k_\mathrm{B}}{[(C/C_\mathrm{S})-1]^2}} - \int_0^{L_\mathrm{m}} \frac{n(L,t)}{\overline{\tau}} \mathrm{d}L, \\ \frac{\mathrm{d}m_i}{\mathrm{d}t} &= -\frac{m_i}{\tau} + \mathrm{i}k_\mathrm{G}(c - c_\mathrm{S})m_{i-1} - \int_0^{L_\mathrm{m}} L^i \frac{n(L,t)}{\overline{\tau}} \mathrm{d}L,\ i = 1,2,\ldots, \\ \frac{\mathrm{d}c}{\mathrm{d}t} &= \frac{c_0 - c - 4\pi\tau(c - c_\mathrm{S})m_2(\rho - c)}{\tau\left(1 - \frac{4}{3}\pi m_3\right)}. \end{aligned} \tag{4.9}$$

Note that the moment transformation is exact and leads to an infinitely large set of ODEs. In the absence of fines dissolution, a closed and finite system can be obtained when considering the moments of up to the third order together with the material balance. However, in the presence of fines dissolution, no closure of the moment equations can be obtained due to the integral over the CSD. One solution when considering fines dissolution is to use an approximate analytical function for the CSD using a Laguerre polynomial expression, which can be written as follows by considering only the leading terms of the polynomial:[94]

$$n(L,t) = \frac{\lambda}{a} p^{(\lambda)}\left(\frac{\lambda L}{a}\right) m_0, \tag{4.10}$$

where λ and a are functions of the CSD as follows:

$$\lambda = \frac{a^2}{m_2/m_0 - a^2},\ a = \frac{m_1}{m_0}. \tag{4.11}$$

When substituting eqn (4.10) into eqn (4.9) and introducing the following dimensionless variable

$$\begin{aligned}
&\tilde{t} = \frac{t}{\tau},\ \tilde{x}_0 = 8\pi\sigma^3 m_0,\ \tilde{x}_1 = 8\pi\sigma^2 m_1,\ \tilde{x}_2 = 4\pi\sigma m_2 \\
&\tilde{x}_3 = \frac{4}{3}\pi m_3,\ \sigma = k_G \tau (c_0 - c_S),\ \mathrm{Da} = 8\pi\sigma^3 k_A \tau, \\
&F = \frac{k_B c_S{}^2}{(c_0 - c_S)^2},\ \alpha = \frac{(\rho - c_S)}{(c_0 - c_S)},\ \tilde{y} = \frac{(c - c_S)}{(c_0 - c_S)},
\end{aligned} \tag{4.12}$$

a low-order dynamic model of a continuous crystallizer with fines dissolution is obtained as follows:[94]

$$\begin{aligned}
\frac{d\tilde{x}_0}{d\tilde{t}} &= -\tilde{x}_0 + (1 - \tilde{x}_3)\mathrm{Da}\, e^{\frac{-F}{\tilde{y}^2}} - 8\pi\sigma^3 m_0 \frac{\tau}{\tilde{\tau}} \frac{1}{(\lambda - 1)!} \left(\frac{\lambda}{a}\right)^{\lambda} \int_0^{L_m} L^{\lambda-1} e^{-\lambda L/a}\, dL, \\
\frac{d\tilde{x}_1}{d\tilde{t}} &= -\tilde{x}_1 + \tilde{y}\tilde{x}_0 - 8\pi\sigma^2 m_0 \frac{\tau}{\tilde{\tau}} \frac{1}{(\lambda - 1)!} \left(\frac{\lambda}{a}\right)^{\lambda} \int_0^{L_m} L^{\lambda} e^{-\lambda L/a}\, dL, \\
\frac{d\tilde{x}_2}{d\tilde{t}} &= -\tilde{x}_2 + \tilde{y}\tilde{x}_1 - 4\pi\sigma m_0 \frac{\tau}{\tilde{\tau}} \frac{1}{(\lambda - 1)!} \left(\frac{\lambda}{a}\right)^{\lambda} \int_0^{L_m} L^{\lambda+1} e^{-\lambda L/a}\, dL, \\
\frac{d\tilde{x}_3}{d\tilde{t}} &= -\tilde{x}_3 + \tilde{y}\tilde{x}_2 - \frac{4}{3}\pi m_0 \frac{\tau}{\tilde{\tau}} \frac{1}{(\lambda - 1)!} \left(\frac{\lambda}{a}\right)^{\lambda} \int_0^{L_m} L^{\lambda+2} e^{-\lambda L/a}\, dL, \\
\frac{d\tilde{y}}{d\tilde{t}} &= \frac{1 - \tilde{y} - (\alpha - \tilde{y})\tilde{y}\tilde{x}_2}{1 - \tilde{x}_3},
\end{aligned} \tag{4.13}$$

which is suitable for controller design.

A typical MIMO controller for a continuous MSMPR crystallizer with fines dissolution would aim to control a critical property of the CSD that relates to the intended application or downstream processing of the crystals and would aim to control a property of the continuous phase (*e.g.*, solute concentration to obtain a sufficient yield) using two controlled variables. Christofides and co-workers consider the case in which the flow rate in the fines dissolution loop and the inlet concentration to the crystallizer are used as manipulated variables, *i.e.*,[94]

$$\tilde{u}_1(t) = \frac{1}{\overline{\tau}} - \frac{1}{\overline{\tau}_S}, \quad \tilde{u}_2 = \frac{c_0 - c_{0,S}}{c_0 - c_S}, \tag{4.14}$$

where $1/\overline{\tau}_S$ and $c_{0,S}$ represent the circulation rate in the fines dissolution loop and the inlet concentration at the intended steady state, respectively. The controlled variables are the number of particles per unit of volume suspension (*i.e.*, zeroth moment) and the solute concentration, *i.e.*,[94]

$$\begin{aligned} y_1(t) &= 8\pi\sigma^3 \int_0^\infty n(L,t)\,\mathrm{d}L = \tilde{x}_0, \\ y_2(t) &= \frac{(c(t) - c_S)}{(c_0 - c_S)} = \tilde{y}(t). \end{aligned} \tag{4.15}$$

When substituting eqn (4.14) and (4.15) into the dimensionless dynamic model (4.13), a system of equations with the following form is obtained:

$$\begin{aligned} \frac{\mathrm{d}\tilde{x}}{\mathrm{d}\tilde{t}} &= \tilde{f}(\tilde{x}) + \tilde{g}_1(\tilde{x})\tilde{u}_1 + \tilde{g}_2(\tilde{x})\tilde{u}_2, \\ y_1 &= \tilde{h}_1(\tilde{x}), \\ y_2 &= \tilde{h}_2(\tilde{x}), \end{aligned} \tag{4.16}$$

where $\tilde{x}$ is the state vector defined as $[\tilde{x}_0, \tilde{x}_1, \tilde{x}_2, \tilde{x}_3, \tilde{y}]^T$. The vector functions $\tilde{f}(\tilde{x})$, $\tilde{g}_1(\tilde{x})$ and $\tilde{g}_2(\tilde{x})$ and scalar functions $\tilde{h}_1(\tilde{x})$ and $\tilde{h}_2(\tilde{x})$ can be derived by inspection. A nonlinear feedback controller for systems of the form (4.16) can be defined as follows:[93]

$$\begin{aligned} \tilde{u}_1(\tilde{t}) &= \left[\beta_{12} L_{\tilde{g}_1} \tilde{h}_1(\omega)\right]^{-1} \left[v_1 - \beta_{10} y_1 - \beta_{11} L_{\tilde{f}} \tilde{h}_1(\omega)\right], \\ \tilde{u}_2(\tilde{t}) &= \left[\beta_{22} L_{\tilde{g}_2} \tilde{h}_2(\omega)\right]^{-1} \left[v_2 - \beta_{20} y_1 - \beta_{21} L_{\tilde{f}} \tilde{h}_2(\omega)\right], \end{aligned} \tag{4.17}$$

where β_{10}, β_{11}, β_{12}, β_{20}, β_{21}, β_{22} are tuning parameters and v_1 and v_2 are the set points for the two controlled variables. Furthermore, $L_{\tilde{g}i}\tilde{h}_i(\omega)$ denotes the standard Lie derivative of scalar function $\tilde{h}_i(\omega)$ with respect to vector function $\tilde{g}_i$ and $L_{\tilde{f}}\tilde{h}_i(\omega)$ is the standard Lie derivative of scalar function $\tilde{h}_i$ with respect to vector function $\tilde{f}$. An implementation of the controller (4.17) requires state estimation for the unmeasured states, which can be obtained *via* an extended Luenberger-type of nonlinear observer as follows:[94]

$$\begin{aligned} \frac{\mathrm{d}\omega}{\mathrm{d}\tilde{t}} &= \tilde{f}(\omega) + \tilde{g}_1(\omega)u_1 + \tilde{g}_2(\omega)u_2 \\ &+ L_1\left(y_1 - \tilde{h}_1(\omega)\right) + L_2\left(y_2 - \tilde{h}_2(\omega)\right) \end{aligned} \tag{4.18}$$

where ω represents the estimated state vector, which is to be used for computation of the controller output from eqn (4.17) and has the same dimension as the state vector $\tilde{x}$. Note that the state observer is constructed from the original model plus two linear terms with vector gains L_1 and L_2 to account for the discrepancy between the two measured and estimated outputs. Furthermore, state observers are often referred to as soft sensors, which emphasizes their ability to estimate process variables without physical measurement instruments. The nonlinear controller (4.17) implemented *via* the nonlinear state observer (4.18) has been demonstrated to provide superior control performance compared to a conventional PI controller for representative cases in which a desired property of the product CSD is obtained at an open-loop unstable operating point and actuator dynamics were included in the detailed process model used for controller performance assessment, but not in the state estimator.[92] Furthermore, the control framework can be extended to account for aspects such as model uncertainty and constraints on the inputs as discussed below.

Significant uncertainties in the kinetic parameters of a crystallization model involving a population balance often exist for real-life applications due to complications such as spatial inhomogeneity and the presence of impurities and foreign particles and likely due to limited resources for the development of reliable first-principle models. Therefore, the design of nonlinear model-based controllers should ideally take model uncertainty into account for robust performance. Chiu and Christofides[94,95] extended their control framework[93] to account for such uncertainties. In particular, for the motivating example of a single MSMPR with two controlled variables discussed earlier (particle concentration and solute concentration) and two manipulated variables (fines dissolution flow rate and inlet concentration), they considered the following parameter uncertainty:[94,95]

$$\begin{aligned} k_A &= k_{A,\mathrm{norm}} + 0.5k_{A,\mathrm{norm}}\sin(0.5t), \\ \rho &= \rho_{\mathrm{norm}} + 0.1\rho_{\mathrm{norm}}, \end{aligned} \tag{4.19}$$

in the nucleation kinetics and crystal density. The expressions of eqn (4.19) can be substituted into the definitions of eqn (4.12) to obtain:

$$\begin{aligned} \mathrm{Da} &= \mathrm{Da}_{\mathrm{norm}} + 0.5\mathrm{Da}_{\mathrm{norm}}\sin(0.5t), \\ \alpha &= \alpha_{\mathrm{norm}} + \frac{0.1\rho_{\mathrm{norm}}}{c_0 - c_s}. \end{aligned} \tag{4.20}$$

When considering parametric uncertainties, the general form of the process model (4.16) can be written as follows:

$$\begin{aligned} \frac{d\tilde{x}}{d\tilde{t}} &= \tilde{f}(\tilde{x}) + \tilde{g}_1(\tilde{x})\tilde{u}_1 + \tilde{g}_2(\tilde{x})\tilde{u}_2 + \tilde{w}_1(\tilde{x})\theta_1 + \tilde{w}_2(\tilde{x})\theta_2, \\ y_1 &= \tilde{h}_1(\tilde{x}), \\ y_2 &= \tilde{h}_2(\tilde{x}), \end{aligned} \tag{4.21}$$

where $\tilde{w}_1(\tilde{x})$ and $\tilde{w}_2(\tilde{x})$ are vector functions and θ_1 and θ_2 are the nominal values of the uncertain process parameters. Subsequently, the controller expression of eqn (4.17) can be augmented using a method developed by Christofides and Daoutidis[101] to account for the parametric uncertainty as follows:

$$\begin{aligned}
\tilde{u}_1(\tilde{t}) &= \left[\beta_{12} L_{\tilde{g}_1} \tilde{h}_1(\omega)\right]^{-1} \begin{bmatrix} \nu_1 - \beta_{10} y_1 - \beta_{11} L_{\tilde{f}} \tilde{h}_1(\omega) \\ -\chi_1 \dfrac{y_1 - \nu_1}{|y_1 - \nu_1| + \phi_1} 0.5 Da_{\text{norm}} \left|(1-\omega_3) e^{-F/\omega_4^2}\right| \end{bmatrix}, \\
\tilde{u}_2(\tilde{t}) &= \left[\beta_{22} L_{\tilde{g}_2} \tilde{h}_2(\omega)\right]^{-1} \begin{bmatrix} \nu_2 - \beta_{20} y_1 - \beta_{21} L_{\tilde{f}} \tilde{h}_2(\omega) \\ -\chi_2 \dfrac{y_2 - \nu_2}{|y_2 - \nu_2| + \phi_2} \dfrac{0.1 \rho_{\text{norm}}}{(c_0 - c_S)} \left|\dfrac{\omega_4 \omega_2}{1-\omega_3}\right| \end{bmatrix},
\end{aligned} \tag{4.22}$$

where $\chi_1, \chi_2, \phi_1, \phi_2$ are additional parameters. Implementation of the nonlinear model-based controller (4.22) requires the use of state observer (4.18) and measurement of the two controlled outputs. Simulations have demonstrated that the controller can maintain the controlled variables near set points while mitigating the impact of the time-varying uncertainty. Furthermore, the controller has demonstrated robustness with respect to actuator dynamics that are only present in a process simulator and not used for controller synthesis. Finally, comparison with a conventional PI controller showed superior behaviour of controller (4.22), as PI controllers are not designed for handling time-varying uncertainty.[95]

The manipulated variables of typical control structures for continuous crystallization are often limited by constraints. For example, a fines dissolution flow rate may be limited by the required heat transfer to evaporate a certain amount of solvent in case of evaporative crystallization or to prevent a boiling solution at the outlet of a heat exchanger in a fines dissolution loop. Other examples include constraints on the attainable process temperature, resulting from the available cooling utilities or the practical need to avoid scaling on cooled surfaces, or inlet concentration or flow rates of reactants or anti-solvent. When such input constraints are present, two important questions for control emerge:

1. What is the range of the attainable set points for the controlled variables for given input constraints? The answer to this question is independent of the implemented controller.
2. What is the range of initial conditions that provide closed-loop stability for the attainable set points?

El-Farra *et al.*[96] provided systematic answers to those two questions and extended the general control framework proposed by Chiu and Christofides[93] to provide a robust model-based feedback controller in the presence of constraints on the process inputs. The hard constraints on the input variables can be modelled using a saturation function as follows:[96]

$$\mathrm{sat}(u_i)=\begin{cases} u_{i,\min} & \text{if } u_i < u_{i,\min} \\ u_i & \text{if } u_{i,\min} \le u_i \le u_{i,\max} \\ u_{i,\max} & \text{if } u_i > u_{i,\max} \end{cases} \tag{4.23}$$

The saturation of input variables limits the attainable steady states of the crystallizer, which is independent of the controller used. Furthermore, the process dynamics may become sluggish or unstable when dynamic trajectories of the input variables are restricted by the saturation function when guiding the system from an initial point to a desired set point. An additional problem during saturation of the input variable is so-called integral "windup". El-Farra *et al.*[96] proposed to use the low-order dynamic model (4.13) of a continuous crystallizer as a basis for a systematic analysis of the attainable set points and robust controller design, which in the presence of input constraints has the following general form:[96]

$$\begin{aligned} \frac{\mathrm{d}\tilde{x}}{\mathrm{d}\tilde{t}} &= \tilde{f}(\tilde{x})+\sum_{i=1}^{m}\tilde{g}_i(\tilde{x})\,\mathrm{sat}(u_i), \\ \tilde{y}_i &= \tilde{h}_i(\tilde{x}),\ i=1,2,\ldots,m \end{aligned} \tag{4.24}$$

Finding the attainable set points can be done in a straightforward manner by identifying all steady states of the system (4.24) for the set of admissible input variables, *i.e.*,

$$0=\tilde{f}(\tilde{x}_S)+\sum_{i=1}^{m}\tilde{g}_i(\tilde{x}_S)u_i^*,\ \text{with: } u_i^*\in\left[u_{i,\min},u_{i,\max}\right], \tag{4.25}$$

which can be obtained, for example, by a nonlinear algebraic solver in combination with a continuation algorithm. The set of attainable set points for each controlled variable follows directly from the computed state vectors at steady state:

$$v_i=\tilde{h}_i(\tilde{x}_S),\ i=1,2,\ldots,m. \tag{4.26}$$

A practical benefit of identifying the attainable set points in the presence of input constraints is that an operator can know in advance the range of set points that can be achieved irrespective of the type of controller to be used. If a set point is attainable, the next question is how to design a robust controller that can track a set point in the presence of the input constraints. El-Farra *et al.*[96] proposed the use of Lyapunov-based control methods to synthesize such a feedback controller. In particular, they proposed a nonlinear feedback controller that has certain optimality properties in its ability to accomplish the control objectives using small control actions of the following form:

$$u=-\frac{1}{2}R^{-1}(\tilde{x})\left(L_{\tilde{g}}V\right)^{T}, \tag{4.27}$$

where R^{-1} is a strictly positive nonlinear scalar function given as follows:

$$\frac{1}{2}R^{-1}(\tilde{x})=\frac{L_f^*V+\sqrt{\left(L_f^*V\right)^2+\left(u_{\max}^2\left(L_{\tilde{g}}V\right)\left(L_{\tilde{g}}V\right)^T\right)^2}}{\left(L_{\tilde{g}}V\right)\left(L_{\tilde{g}}V\right)^T\left[1+\sqrt{1+u_{\max}^2\left(L_{\tilde{g}}V\right)\left(L_{\tilde{g}}V\right)^T}\right]}, \quad (4.28)$$

where $L_{\tilde{g}}V$ is the row vector $[L_{\tilde{g}1}V\cdots L_{\tilde{g}m}V]$ in which V is a quadratic Lyapunov function defined as:

$$V=\mathrm{e}^T P\mathrm{e},\ P=\begin{bmatrix}1 & c'\\ c' & 1\end{bmatrix},\ c'\in(0,1) \quad (4.29)$$

where P is a positive definite matrix that satisfies the Riccati inequality.[96] Furthermore, $L_f^*V=L_{\tilde{f}}V+\rho|\mathrm{e}|^2$ with $\rho>0$. For practical implementation of the controller given by eqn (4.27), once more a state observer is needed to estimate the state variables that are not measured. For example, an observer of the form (4.18) can be used, which is constructed from the process model including input constraints with the addition of linear gains to account for any differences between the measured and estimated output variables and has the following form:

$$\frac{\mathrm{d}\omega}{\mathrm{d}\tilde{t}}=\tilde{f}(\omega)+\tilde{g}(\omega)\mathrm{sat}(u)+L\left(y-\tilde{h}(\omega)\right). \quad (4.30)$$

In a case where the process model described by eqn (4.24) obeys a number of basic assumptions (see El-Farra *et al.*[96] for precise definitions), then given a certain initial condition and applicable input constraints, the following inequality should be verified

$$L_f^*V\le u_{\max}\left|\left(L_{\tilde{g}}V\right)^T\right| \quad (4.31)$$

and if the initial condition is within the region of closed-loop stability, desired closed-loop behaviour can be obtained by using the explicit form of the nonlinear model-based controller as follows:[96]

$$\begin{aligned}\frac{\mathrm{d}\omega}{\mathrm{d}\tilde{t}}&=\tilde{f}(\omega)-\frac{1}{2}\tilde{g}(\omega)R^{-1}(\omega)\left[L_{\tilde{g}}V(\omega)\right]^T+L\left(y-\tilde{h}(\omega)\right),\\ u&=-\frac{1}{2}R^{-1}(\omega)\left[L_{\tilde{g}}V(\omega)\right]^T.\end{aligned} \quad (4.32)$$

Dynamic simulations of a continuous crystallizer, which included actuator and sensor dynamics, with either a conventional PI controller or the nonlinear model-based controller described by eqn (4.32) have demonstrated that the latter controller has excellent stability properties and is able to drive the output of the system much faster to a desired set point, which is open-loop unstable, compared to when using a conventional PI controller.[96]

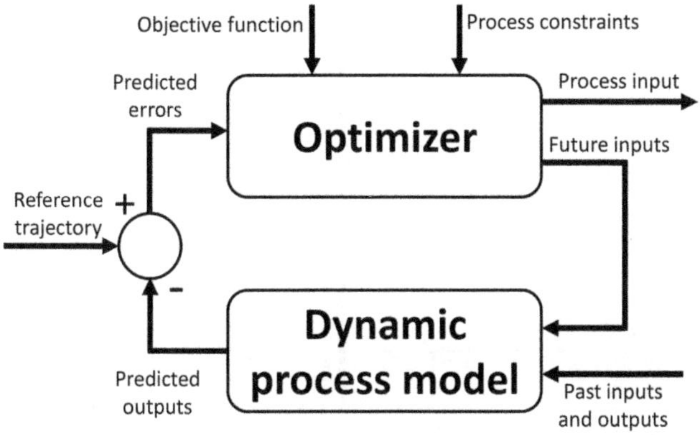

Figure 4.5 Schematic illustration of MPC algorithm.

The control strategies discussed so far all have the objective to drive controlled variables to set points using selected manipulated variables either based on model-free or model-based control laws. The model-based feedback controllers discussed earlier are able to handle input constraints. However, general process constraints cannot be handled by such controllers. For example, in addition to input constraints, constraints regarding process safety, equipment limitations, or bounds on operating conditions may have to be imposed. The latter may be important for pharmaceutical applications as a practical way of enforcing operation in a pre-defined design space.[10] Furthermore, in cases where several manipulated variables are available, decoupling of control loops may become difficult when simple control structures are used and a high interaction between the various manipulated and controlled variables exist. MPC is an alternative model-based control strategy, which involves an objective function that is to be optimized in an on-line fashion with explicit definition of process constraints. MPC control strategies have become popular over the last two decades, and have demonstrated enhanced robustness and flexibility for multivariate crystallization systems.[102,103] The objective function is subject to a mathematical model of the crystallization process and a set of operating constraints, such as maximum and minimum cooling rates, temperature, flow rates *etc.* (see Figure 4.5). The overall performance of MPC obviously relies on the quality of the mathematical model and the optimization approach (*e.g.* local or global dynamic optimizer) to deliver theoretically optimal trajectories. The control actions aim to guide the process variables along a pre-specified trajectory from the present point to a target point in the future (see Figure 4.6). The method relies on current measurements and future predictions. In particular, MPC can be formulated as follows:

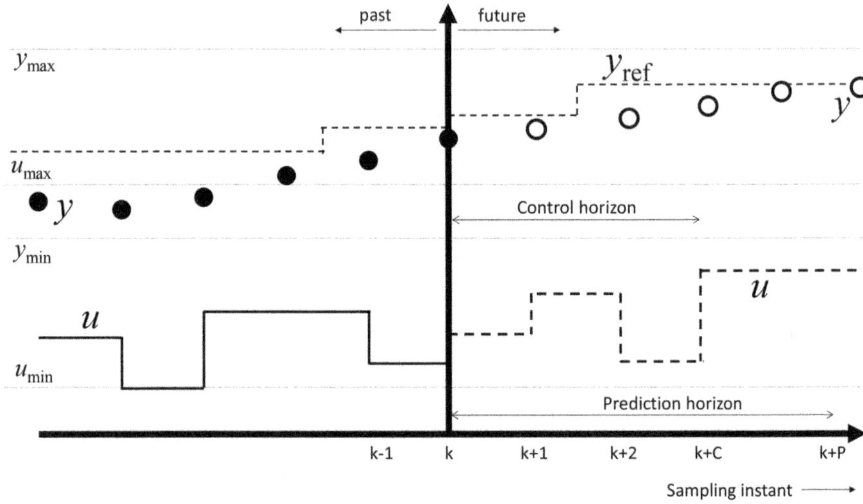

Figure 4.6 Basic concept of model-predictive control.

$$
\begin{aligned}
&\min_{u} J := \sum_{j=1}^{P}\left[y(k+j)-y_{\text{ref}}(k+j)\right]^T Q\left[y(k+j)-y_{\text{ref}}(k+j)\right] \\
&+\sum_{j=1}^{C}\Delta u(k+j-1)^T R\Delta u(k+j-1) \\
&s.t. \\
&\text{Process model} \\
&y_{\min} \le y(k) \le y_{\max} \\
&u_{\min} \le u(k) \le u_{\max}
\end{aligned}
\tag{4.33}
$$

where y and Δu are the output and changes in manipulated variables, respectively, and Q and R are weighting matrices. The first term of the objective function aims to guide the future outputs close to a desired reference trajectory and the second term penalizes any big changes in inputs to minimize control efforts. The balance between both objectives can be adjusted using the weighting matrices. The time horizon over which control actions are determined (*i.e.*, the control horizon) is usually shorter than the time horizon over which predictions are made (*i.e.*, the prediction horizon). Increasing the control horizon is likely to smoothen the control action at the expense of increased computational burden. The output of the optimizer will be the input to the process, which can either be direct manipulation of an actuator or a change in the set point of a PID controller. The future and past inputs and outputs are used by the dynamic process model for prediction of the outputs, which will be compared with the desired reference trajectory. Only the first control action will be implemented in the process. At the subsequent sampling step, the whole exercise will be repeated with the updated information from the process to update the sequence of control steps, which is referred to as the receding horizon principle. For implementation of an MPC approach, a state observer is necessary to estimate the unmeasurable states of the system. Note that the objective function of MPC is not an economic function. However, an outer optimization may be conducted (often based on steady-state models) to inform the MPC on desired trajectories of process variables to maximize profit, which is referred to as real-time optimization.

Rohani *et al.*[104,105] developed an MPC strategy for an MSMPR crystallizer with fines dissolution and classification, which involved a similar process configuration as discussed before (see Figure 4.2) and was based on cooling crystallization of KCl. First, they developed a nonlinear process model based on time series and neural network approaches, which was successfully used for a multivariable MPC. They concluded that linearization of a model is a simple approach but suffers from inaccuracies due to the nonlinearities in a continuous crystallization process model. In contrast, a nonlinear model gave a better description both for steady-state prediction and dynamic conditions. In addition, they showed the advantages of using MIMO control compared to SISO control.

Although non-linear models can provide much more accurate predictions compared to linear models, the use of highly nonlinear process models in MPC for continuous crystallization may also lead to infeasible solutions and complicates the analysis of operating points that lead to stable closed-loop behaviour. Motivated by these complications, Shi *et al.*[106] developed a hybrid control framework for continuous crystallization, based on earlier work from El-Farra *et al.*,[107] which uses an MPC as the main control approach, but exhibits a supervisory system that can activate a nonlinear model-based controller of the form given by eqn (4.32) as backup option (Figure 4.7). The aim is to combine the superior control performance of an MPC with the robustness of certain nonlinear model-based feedback controllers. The first step is to reduce a full population balance model (eqn (4.3)) to lower-order model, *e.g.*, moment equations (eqn (4.9)) to capture the dominant dynamics of the key characteristics of the CSD and continuous phase, which can be used to design an MPC and a bounded controller. The system is initialized with the MPC active and a supervisory system enforces a switch to the bounded controller when either a certain stability criterion is satisfied (see El-Farra *et al.*[107]) or when the MPC fails to compute future control actions.[106] Model-based simulations have demonstrated the effectiveness of such hybrid model-predictive control structures for the case of an MSMPR crystallizer with inlet concentration as manipulated variable in stabilizing an open-loop unstable steady state with a desired characteristic of the CSD.[106]

Liu *et al.*[98] applied an MPC framework to a continuous crystallizer with asynchronous measurements, which may either be the result of transmission failures or different sampling rates of analytical instruments. Their simulation data demonstrated that a Lyapunov-based MPC provides robust closed-loop performance in the presence of asynchronous measurements.

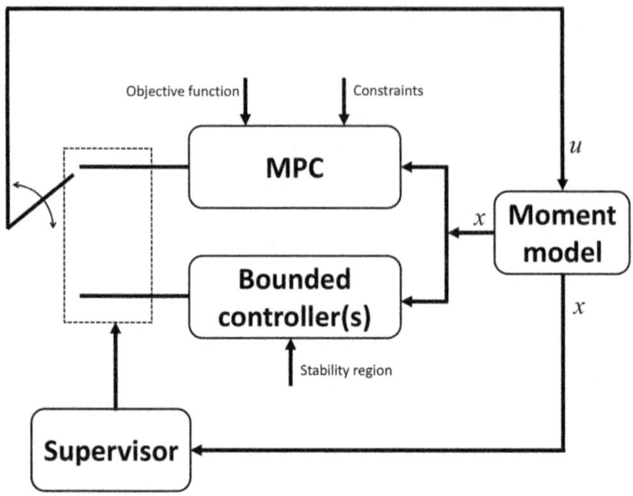

Figure 4.7 Hybrid model-predictive control framework for continuous crystallization.[106]

A linear process model avoids problems typically encountered during non-convex optimization. Silva *et al.*[108] developed a linear MPC framework for a multiple-effect evaporative crystallizer. For such cases, in addition to constraints on the CSD, energy costs are a main driver for the economic performance. They used the slurry mass in each stage as controlled variable and temperature and brine flow rates were selected as manipulated variables. Step tests were used to construct a linear input–output model from a full nonlinear process model based on the moment equations and material and energy balances. A real-time optimizer (RTO) was used to send desired manipulated variables to the MPC, which needed to be achieved within constraints on the controlled variables. An economic objective function based on energy costs was used for the RTO. A Kalman filter was used as state observer. The performance of the MPC was compared to a scheme with conventional PID controllers. The MPC was able to achieve steady state much faster than the PID controllers and prevented oscillations in slurry mass, which could be seen when PID controllers were applied. Furthermore, the conventional scheme with PID controllers could not follow the optimal values informed by the RTO. Finally, this case also demonstrated the inherent benefit of MPC to include process constraints, for example, by limiting the maximum supersaturation to avoid excessive nucleation.

A current challenge for MPC is to take into account the impact of parameter uncertainties, model-plant mismatch and process perturbations. The reliability and prediction capabilities of the mathematical model depend, besides the model structure, on the quality and information content of the experimental data used for the parameter identification, as well as the parameter estimation and validation strategies. Different parameter estimability or practical identifiability techniques were developed in many disciplines and research areas to enhance the prediction capability of the mathematical models, mitigate the effect of parameter correlation and inform the design of more optimal experiments that maximize the information content of the collected data.[109,110] A combination of a sequential orthogonalization techniques and Sobol proved particularly effective to enhance 1D and 2D population balance models and demonstrated more reliable predictions of the aspect ratio and effect of crystal shape modifiers.[111] As mentioned, the mathematical model of the crystallization process can be linearized to reduce the computational burden inherent to real time solution of the nonlinear dynamic optimization problem provided that the model remains sufficiently accurate. For instance, the Jacobian linearization can be performed over the prediction horizon at discrete operating points, in a successive linearization sequence. More recently, a more effective global state feedback linearization (SFL) technique was implemented as a multi-input multi-output MPC for batch and continuous crystallization.[112] Furthermore, the computation costs inherent to NMPC can be addressed by implementing a parallel solver based on for instance the parallel Schur complement method.[113]

4.6 Fault Detection and Isolation

Continuous crystallizers often exhibit a high level of automation, which gives a distinct advantage compared to batch crystallizers. However, automation also introduces risks due to instrumentation malfunctioning. Therefore, when defects occur in process equipment, sensors, actuators or somewhere else in the control loop, appropriate action should be taken to prevent significant loss. An effective fault control strategy should include at least reliable identification of fault occurrence (*i.e.*, detection) and some form of fault handling procedure (*i.e.*, controller reconfiguration after possibly fault isolation in case of multiple actuators), which all has to be supported by robust controllers. For example, in the case of an MSMPR crystallizer without fines dissolution and a single feed stream used as manipulated variable, a second redundant feed stream could be present as backup in case of an actuator fault in the default feed stream. Alternatively, switching between different types of control loops can be considered. For example, using either an inlet flow rate or a rate of fines dissolution as manipulated variable to control a characteristic of the CSD.

El-Farra and Giridhar proposed a systematic framework for actuator fault detection and fault-tolerant control for continuous crystallization.[99] The framework is based on a process model involving a population balance, control constraints, and a limited number of process measurements. The framework builds upon nonlinear model-based controllers and state estimation as discussed in Section 4.5, by adding model-based fault detection and a supervisory system for fault handling. Once a low-order dynamic model has been derived from the full population balance model, nonlinear model-based feedback controllers have to be designed, for example, using the methods discussed earlier and possibly considering input constraints. The stability region for each controller has to be defined. Subsequently, a fault detection filter has to be designed.

The fault detection can be based on a simulation of the reduced-order dynamic model without faults and a comparison of the model output with the actual process outputs. In case a difference larger than a certain threshold is observed, a fault is detected and the control structure is reconfigured *via* the supervisory system to preserve closed-loop stability. The fault-detection threshold has to be chosen carefully to discriminate between artificial modelling errors and the true occurrence of an actuator defect. In particular, El-Farra and Giridhar propose the following test based on the following form of the reduced-order process model:[99]

$$\frac{\mathrm{d}x}{\mathrm{d}t} = f(x) + g(x)\left[u^k + f_a^k\right], \\ y = h(x), \tag{4.34}$$

where the state vector x is defined as before involving characteristics of the dispersed phase (*e.g.*, moments) and state variables that define the continuous phase (*e.g.*, solute concentration). Furthermore, u^k is the constrained manipulated variable of control structure k and f_a^k is any potential contribution of an actuator fault of control structure k. The fault detection can be accomplished by comparing the behaviour of process model (4.34) with a simulation of the closed-loop performance in absence of any actuator faults, *i.e.*, by comparison with a simulation based on the following process model:[99]

$$\frac{\mathrm{d}w}{\mathrm{d}t} = f(w) + g(w)\,p\left(w, u_{\min}^k, u_{\max}^k\right), \tag{4.35}$$

where w is the filter state, which has the same dimension as x, and $p\left(w, u_{\min}^k, u_{\max}^k\right)$ is the feedback control law defining the manipulated variable of control configuration k. Subsequently, a residual is defined by comparing the fault-free closed-loop behaviour with the actual behaviour of the process:[99]

$$r(t) = \left\|w(t) - x(t)\right\|. \tag{4.36}$$

Note that an actuator fault will affect the dynamic development of $x(t)$ but not of $w(t)$; therefore, a residual larger than zero will be observed. Furthermore, the detection method has general applicability in the sense that partial defects will also be observed and more broadly speaking any defect in the instrumentation of the control system can be detected as long as the dynamic development of the state variables of the system are affected. The measurement of all state variables is unlikely to be available for a practical application of continuous crystallization. Therefore, a state observer, for example, an observer of the extended Luenberger type discussed before (*i.e.*, eqn (4.30)), is needed to estimate the full state vector based on a process model and measured output variables. Therefore, for most practical applications, calculation of the residual would be based on the estimated state vector ω instead of the true state vector x. Criteria have been developed for the design of such observers that guarantee a drop in state estimation error below a certain threshold after a short period of time,[99] which is important for fault detection purposes. Consequently, threshold values proposed in literature take into account a threshold on the state estimation error such that robust fault detection can be achieved.[99] Inherently, there is always a trade-off between choosing a small threshold avoiding faults to go undetected and choosing a large threshold avoiding false alarms. Since state estimation errors are larger initially, a time-varying threshold can also be used with a larger threshold shortly after initialization of the residual filter and a smaller threshold when time progresses justified by the smaller estimation error. Switching has to be initiated once a fault has been confirmed such that closed-loop stability is preserved, which has been successfully demonstrated for cases involving a single MSMPR crystallizer with redundant feed streams for controller switching when faults were detected by a supervisory system based on eqn (4.36).[99] Giridhar and El-Farra[100] extended the fault-detection framework to cases where several manipulated variables are present, which requires the supervisory system to identify the defect actuator, to avoid shutting functional actuators, and to decide on which backup actuator to activate. Furthermore, they introduced model uncertainty such as the time-varying uncertainty described by eqn (4.19) into the fault-detection framework, which addresses the challenge of avoiding false alarms due to time-varying model uncertainty. They demonstrated such unified frameworks for the case of an isothermal MSMPR crystallizer with fines dissolution loop.[100]

The loss of sensor data is another example of typical faults that may occur in automated feedback control loops for continuous crystallization. Gani *et al.*[97] conducted rigorous simulation to understand how a closed-loop feedback control system around an MSMPR crystallizer behaves when data loss from sensor failure occurs. They assumed that the measurement remained at an old value during sensor loss, which essentially rendered the system in open-loop mode. They considered the case of a single isothermal MSMPR crystallizer, which can be modelled by the first five moment equations in closed form similar to eqn (4.13) to capture the dominant dynamics of the crystallization process, but without the terms related to fines dissolution and considering the inlet concentration as manipulated variable as follows:[93]

$$\begin{aligned}
\frac{\mathrm{d}\tilde{x}_0}{\mathrm{d}\tilde{t}} &= -\tilde{x}_0 + (1-\tilde{x}_3)\mathrm{Da}\,\mathrm{e}^{\frac{-F}{\tilde{y}^2}},\\
\frac{\mathrm{d}\tilde{x}_1}{\mathrm{d}\tilde{t}} &= -\tilde{x}_1 + \tilde{y}\tilde{x}_0,\\
\frac{\mathrm{d}\tilde{x}_2}{\mathrm{d}\tilde{t}} &= -\tilde{x}_2 + \tilde{y}\tilde{x}_1,\\
\frac{\mathrm{d}\tilde{x}_3}{\mathrm{d}\tilde{t}} &= -\tilde{x}_3 + \tilde{y}\tilde{x}_2,\\
\frac{\mathrm{d}\tilde{y}}{\mathrm{d}\tilde{t}} &= \frac{1-\tilde{y}-(\alpha-\tilde{y})\tilde{y}\tilde{x}_2}{1-\tilde{x}_3} + \frac{u}{1-\tilde{x}_3},
\end{aligned} \tag{4.37}$$

where $u = (c_0 - c_{0,\mathrm{S}})/(c_{0,\mathrm{S}} - c_\mathrm{S})$. The dynamic model (4.37) can be combined with a state observer and nonlinear model-based control according to eqn (4.32) and assuming a measurement of the zeroth moment of the CSD to obtain:[96]

$$\begin{aligned}
\frac{\mathrm{d}\omega_0}{\mathrm{d}\tilde{t}} &= -\omega_0 + (1-\omega_3)\mathrm{Da}\,\mathrm{e}^{\frac{-F}{\omega_4^2}} + L_0(\tilde{x}_0 - \omega_0),\\
\frac{\mathrm{d}\omega_1}{\mathrm{d}\tilde{t}} &= -\omega_1 + \omega_4\omega_0 + L_1(\tilde{x}_0 - \omega_0),\\
\frac{\mathrm{d}\omega_2}{\mathrm{d}\tilde{t}} &= -\omega_2 + \omega_4\omega_1 + L_2(\tilde{x}_0 - \omega_0),\\
\frac{\mathrm{d}\omega_3}{\mathrm{d}\tilde{t}} &= -\omega_3 + \omega_4\omega_2 + L_3(\tilde{x}_0 - \omega_0),\\
\frac{\mathrm{d}\omega_4}{\mathrm{d}\tilde{t}} &= \frac{1-\omega_4-(\alpha-\omega_4)\omega_4\omega_2}{1-\omega_3} - \frac{1}{2}\frac{R^{-1}(\omega)L_{\bar{g}}V(\omega)}{1-\omega_3} + L_4(\tilde{x}_0 - \omega_0),\\
u &= -\frac{1}{2}R^{-1}(\omega)L_{\bar{g}}V(\omega)
\end{aligned} \tag{4.38}$$

Dynamic simulations of a process model and closed-loop system (4.38) demonstrated that process stabilization could be achieved up to a 90% probability of sensor losses.[97] It was also observed that for larger constraints on the input (*i.e.*, a higher $u_{\max}$), the system became more sensitive for data loss (*e.g.*, the process destabilized already at a 75% probability of sensor losses when $u_{\max}$ was doubled in their simulations). These simulation results suggest that $u_{\max}$ can be reduced artificially to make the system more robust in the presence of significant data losses due to measurement failures. Finally, in general, the relatively high values for sensor loss probability indicate that the used model-based controllers are rather robust with respect to sensor failures.

4.7 Actuators

Basic crystallizer configurations may lack sufficient and flexible process actuators. Therefore, a great amount of research on the potential of novel process actuators has been conducted, which often follows the fundamental principles of process intensification.[114,115] The use of external fields is particularly interesting, as it potentially provides a direct means of actuation on the CSD without strong coupling with other control loops. The use of ultrasound is the most widely studied technology to manipulate the CSD with *in situ* actuation. The ability of ultrasound to discharge a metastable fluid, such as a supersaturated solution, has been demonstrated almost a century ago[116] and remains a subject of much research. Ultrasound induces cavitation which creates a high temperature and pressure locally. The small bubbles also provide a surface for heterogeneous primary nucleation. Furthermore, temperature increases due to the energy input, which may dissolve crystals. Indeed it has been observed that ultrasound may inhibit nucleation[117–119] (*e.g.*, due to dissolution of fines) or promote nucleation (*e.g.*, due to heterogeneous primary nucleation[120–124]). One of the key challenges for effective use of ultrasound as a process actuator is scale-up. The penetration depth of ultrasound is typically only a few centimetres,[125] which creates inherent difficulties to predict the scale-up process behaviour. The scale dependency of ultrasound is particularly challenging for batch crystallization, as all material has to be processed at the same time, and seeding might be a more effective process actuator for the initial population.[117] Continuous crystallization provides more opportunities for the use of ultrasound, or external fields in general, as for a continuous process not all material has to be processed at the same time. Ultrasound has indeed been employed successfully for continuous crystallization,[75,126–131] but demonstrated applications within automated feedback control loops are still limited. Electric fields have also been used as a novel actuator for crystallization processes. Electric fields allow for classification of crystals[132] and can be used to manipulate nucleation of large molecules such as proteins.[133–136] Recently, the use of electric fields for protein crystallization has been extended for use in continuous flow, which provides a novel actuator to manipulate nucleation and achieve a higher yield.[137] Similarly as for the application of ultrasound, the use of electric fields is particularly interesting for continuous crystallization, as the large surface-to-volume of a channel can be used for effective exposure of the solution to the electric field. Microwaves have also been used as a novel actuator for solution crystallization either to provide rapid and localized heating or fast evaporation of the solution.[138–142] The former has been used in automated control loops for rapid dissolution of fines.[142] The benefit of using microwaves compared to conventional heating is twofold. First, microwaves provide instantaneous heating, which is important as the crystal growth of fines should be limited for rapid dissolution when controlling the fraction of fines of the CSD. Second, microwaves provide an energy source that does not compete with conventional heating or cooling, which provides opportunities for flexible operation and improved decoupling of automated control loops. The latter can be used to create supersaturation faster compared to conventional evaporative crystallization, which can be used to make small crystals.[141]

Instead of promoting primary nucleation with an external field, secondary nucleation can also be induced by mechanical means. Milling has been demonstrated to be effective as an additional actuator for continuous crystallization.[143] Yang *et al.*[144] used wet milling as an actuator to manipulate the CSD in automated control loops. They used an FBRM probe to measure the CLD *in situ* and to control the particle count at a desired set point *via* a feedback control loop (*i.e.*, according to the so-called direct-nucleation control approach[145]). They investigated two schemes (see Figure 4.8). In the first scheme, the wet mill

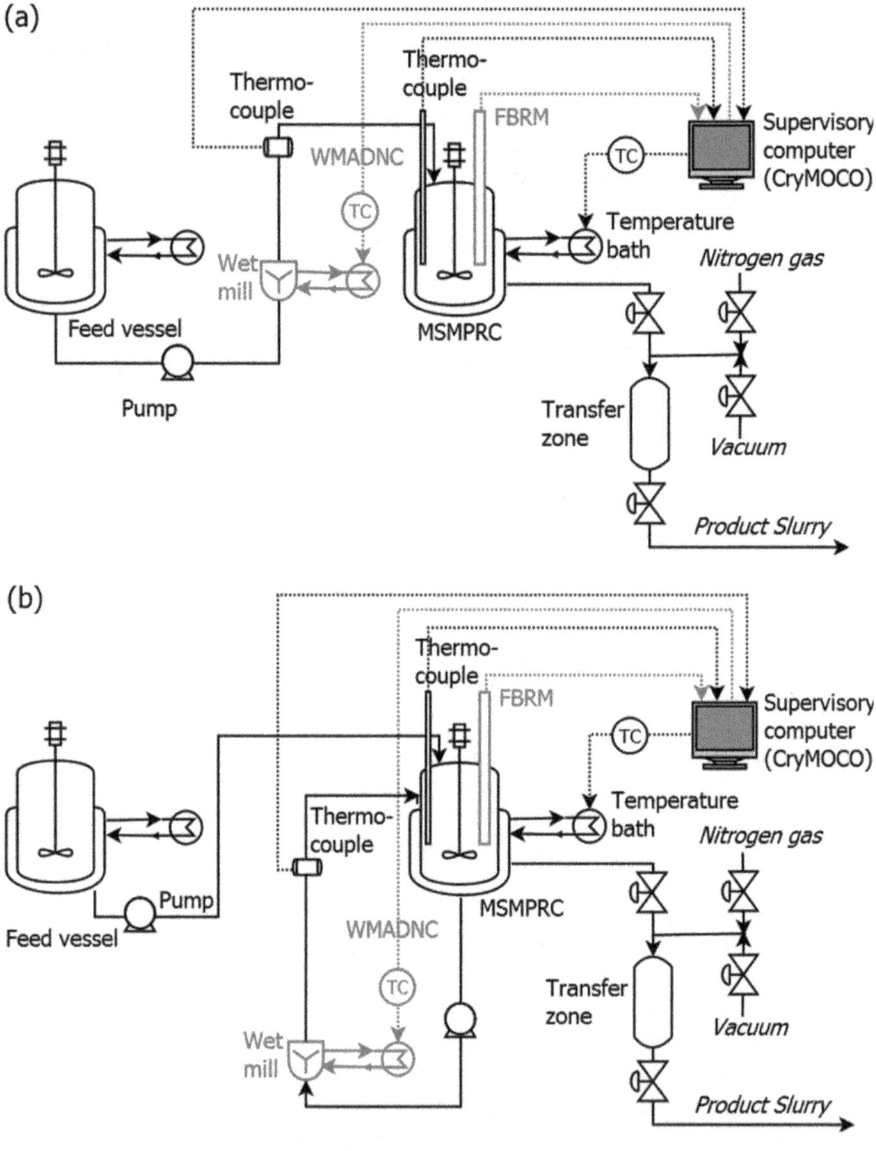

Figure 4.8 Experimental setups for wet milling-based DNC (a) process 1 and (b) process 2. Reproduced from ref. 144 with permission from American Chemical Society, Copyright 2016.

was used upstream to generate seed particles. In the second scheme, the wet mill was integrated in a recycle loop. Although the wet mill could not improve start-up dynamics, they demonstrated that a high-quality product and effective disturbance rejection could be achieved in closed-loop mode.

4.8 Conclusions and Perspective

Process control of continuous crystallization has been studied for several decades and remains an active area of research. In the earlier work, existing types of crystallizers were studied, sometimes even on unique pilot–plant scale, mainly for the production of bulk chemicals. Those earlier studies already demonstrated the importance of suitable and flexible process actuators for continuous crystallization as well as the need for robust and reliable process analytical technologies and dynamic process models. More recently, a number of fast-developing branches of research provide the necessary tools to address those fundamental challenges in different ways when all put together. Therefore, it is not surprising that the trend in important industries such as the pharmaceutical industry is towards continuous processing. In particular, new types of crystallizers have become available. For example, a new generation of plug-flow crystallizers allow for more flexible process control and fast learning from the broad literature on advanced process control of batch crystallization due to their conceptual similarity. The rapid advancement of PAT has provided new opportunities to measure process variables *in situ* and in real-time such that they can be used in automated control loops. Regulatory support and knowledge transfer is important to move those PAT tools from the current research and development environment to commercial manufacturing. Furthermore, new types of process actuators have also been developed, which are often based on the principles of process intensification. For example, a whole new generation of methods based on external fields (*e.g.*, ultrasound, electric, microwave, laser) are becoming available to provide targeted and non-invasive actuation. Finally, there has been much progress on the development of reliable dynamic models and numerical methods to optimize dynamic models rapidly such that they can be used for controller design and for real-time model-based control methods. The various combinations of all these recent developments provide ample opportunities to develop new and improved process control strategies for continuous crystallization.

References

1. K. P. Cole, J. M. Groh, M. D. Johnson, C. L. Burcham, B. M. Campbell and W. D. Diseroad, *et al.*, Kilogram-scale prexasertib monolactate monohydrate synthesis under continuous-flow CGMP conditions, *Science*, 2017, **356**(6343), 1144–1150.
2. A. Adamo, R. L. Beingessner, M. Behnam, J. Chen, T. F. Jamison and K. F. Jensen, *et al.*, On-demand continuous-flow production of pharmaceuticals in a compact, reconfigurable system, *Science*, 2016, **352**(6281), 61–67.

3. R. Lakerveld, B. Benyahia, P. L. Heider, H. Zhang, A. Wolfe and C. J. Testa, *et al.*, The Application of an Automated Control Strategy for an Integrated Continuous Pharmaceutical Pilot Plant, *Org. Process Res. Dev.*, 2015, **19**(9), 1088–1100.
4. S. Mascia, P. L. Heider, H. T. Zhang, R. Lakerveld, B. Benyahia and P. I. Barton, *et al.*, End-to-End Continuous Manufacturing of Pharmaceuticals: Integrated Synthesis, Purification, and Final Dosage Formation, *Angew. Chem., Int. Ed.*, 2013, **52**(47), 12359–12363.
5. K. Plumb, Continuous processing in the pharmaceutical industry – Changing the mind set, *Chem. Eng. Res. Des.*, 2005, **83**(A6), 730–738.
6. R. Lakerveld, H. J. M. Kramer, P. J. Jansens and J. Grievink, The application of a task-based concept for the design of innovative industrial crystallizers, *Comput. Chem. Eng.*, 2009, **33**(10), 1692–1700.
7. T. Vetter, C. L. Burcham and M. F. Doherty, Regions of attainable particle sizes in continuous and batch crystallization processes, *Chem. Eng. Sci.*, 2014, **106**, 167–180.
8. L. L. Simon, H. Pataki, G. Marosi, F. Meemken, K. Hungerbuhler and A. Baiker, *et al.*, Assessment of Recent Process Analytical Technology (PAT) Trends: A Multiauthor Review, *Org. Process Res. Dev.*, 2015, **19**(1), 3–62.
9. R. Lakerveld, B. Benyahia, R. D. Braatz and P. I. Barton, Model-Based Design of a Plant-Wide Control Strategy for a Continuous Pharmaceutical Plant, *AIChE J.*, 2013, **59**(10), 3671–3685.
10. Plant-wide model predictive control for a continuous pharmaceutical process, in *American Control Conference, 2015*, ed. A. Mesbah, J. A. Paulson, R. Lakerveld and R. D. Braatz, IEEE, Chicago, IL, USA, 2015.
11. R. A. Sheldon, The E factor 25 years on: the rise of green chemistry and sustainability, *Green Chem.*, 2017, **19**(1), 18–43.
12. Z. K. Nagy, G. Fevotte, H. Kramer and L. L. Simon, Recent advances in the monitoring, modelling and control of crystallization systems, *Chem. Eng. Res. Des.*, 2013, **91**(10), 1903–1922.
13. S. Aitipamula, R. Banerjee, A. K. Bansal, K. Biradha, M. L. Cheney and A. R. Choudhury, *et al.*, Polymorphs, Salts, and Cocrystals: What's in a Name?, *Cryst. Growth Des.*, 2012, **12**(5), 2147–2152.
14. E. Grothe, H. Meekes, E. Vlieg, J. H. ter Horst and R. de Gelder, Solvates, Salts, and Cocrystals: A Proposal for a Feasible Classification System, *Cryst. Growth Des.*, 2016, **16**(6), 3237–3243.
15. L. Yu, G. A. Stephenson, C. A. Mitchell, C. A. Bunnell, S. V. Snorek and J. J. Bowyer, *et al.*, Thermochemistry and Conformational Polymorphism of a Hexamorphic Crystal System, *J. Am. Chem. Soc.*, 2000, **122**(4), 585–591.
16. S. R. Chemburkar, J. Bauer, K. Deming, H. Spiwek, K. Patel and J. Morris, *et al.*, Dealing with the Impact of Ritonavir Polymorphs on the Late Stages of Bulk Drug Process Development, *Org. Process Res. Dev.*, 2000, **4**(5), 413–417.

17. J. H. ter Horst, R. M. Geertman, A. E. van der Heijden and G. M. van Rosmalen, The influence of a solvent on the crystal morphology of RDX, *J. Cryst. Growth*, 1999, **198–199**, 773–779.
18. H. Eisenschmidt, N. Bajcinca and K. Sundmacher, Optimal Control of Crystal Shapes in Batch Crystallization Experiments by Growth-Dissolution Cycles, *Cryst. Growth Des.*, 2016, **16**(6), 3297–3306.
19. M. Jiang, X. Zhu, M. C. Molaro, M. L. Rasche, H. Zhang and K. Chadwick, *et al.*, Modification of Crystal Shape through Deep Temperature Cycling, *Ind. Eng. Chem. Res.*, 2014, **53**(13), 5325–5336.
20. E. Simone, A. R. Klapwijk, C. C. Wilson and Z. K. Nagy, Investigation of the Evolution of Crystal Size and Shape during Temperature Cycling and in the Presence of a Polymeric Additive Using Combined Process Analytical Technologies, *Cryst. Growth Des.*, 2017, **17**(4), 1695–1706.
21. S. Bötschi, D. R. Ochsenbein, M. Morari and M. Mazzotti, Multi-Objective Path Planning for Single Crystal Size and Shape Modification, *Cryst. Growth Des.*, 2017, **17**(9), 4873–4886.
22. G. Yang, N. Kubota, Z. Sha, M. Louhi-Kultanen and J. Wang, Crystal Shape Control by Manipulating Supersaturation in Batch Cooling Crystallization, *Cryst. Growth Des.*, 2006, **6**(12), 2799–2803.
23. Y. Yang, L. Song and Z. K. Nagy, Automated Direct Nucleation Control in Continuous Mixed Suspension Mixed Product Removal Cooling Crystallization, *Cryst. Growth Des.*, 2015, **15**(12), 5839–5848.
24. K. A. Powell, A. N. Saleemi, C. D. Rielly and Z. K. Nagy, Periodic steady-state flow crystallization of a pharmaceutical drug using MSMPR operation, *Chem. Eng. Process.*, 2015, **97**, 195–212.
25. K. A. Powell, A. N. Saleemi, C. D. Rielly and Z. K. Nagy, Monitoring Continuous Crystallization of Paracetamol in the Presence of an Additive Using an Integrated PAT Array and Multivariate Methods, *Org. Process Res. Dev.*, 2016, **20**(3), 626–636.
26. A. S. Myerson, M. Krumme, M. Nasr, H. Thomas and R. D. Braatz, Control Systems Engineering in Continuous Pharmaceutical Manufacturing. May 20-21, 2014 Continuous Symposium, *J. Pharm. Sci.*, 2015, **104**(3), 832–839.
27. M. Patrascu and P. I. Barton, Optimal campaigns in end-to-end continuous pharmaceuticals manufacturing. Part 1: Nonsmooth dynamic modeling, *Chem. Eng. Process.*, 2018, **125**, 298–310.
28. M. Patrascu and P. I. Barton, Optimal campaigns in end-to-end continuous pharmaceuticals manufacturing. Part 2: Dynamic optimization, *Chem. Eng. Process.*, 2018, **125**, 124–132.
29. J. B. Rawlings, S. M. Miller and W. R. Witkowski, Model identification and control of solution crystallization processes: a review, *Ind. Eng. Chem. Res.*, 1993, **32**(7), 1275–1296.
30. Q. Su, C. D. Rielly, K. A. Powell and Z. K. Nagy, Mathematical modelling and experimental validation of a novel periodic flow crystallization using MSMPR crystallizers, *AlChE J.*, 2017, **63**(4), 1313–1327.

31. A. Koswara and Z. K. Nagy, Anti-Fouling Control of Plug-Flow Crystallization via Heating and Cooling Cycle, *IFAC-PapersOnLine*, 2015, **48**(8), 193–198.
32. M. A. Lovette, M. Muratore and M. F. Doherty, Crystal shape modification through cycles of dissolution and growth: Attainable regions and experimental validation, *AIChE J.*, 2012, **58**(5), 1465–1474.
33. Z. Gao, S. Rohani, J. Gong and J. Wang, Recent Developments in the Crystallization Process: Toward the Pharmaceutical Industry, *Engineering*, 2017, **3**(3), 343–353.
34. A. Chanda, A. M. Daly, D. A. Foley, M. A. LaPack, S. Mukherjee and J. D. Orr, *et al.*, Industry Perspectives on Process Analytical Technology: Tools and Applications in API Development, *Org. Process Res. Dev.*, 2015, **19**(1), 63–83.
35. S. S. Kadam, E. van der Windt, P. J. Daudey and H. J. M. Kramer, A Comparative Study of ATR-FTIR and FT-NIR Spectroscopy for In-Situ Concentration Monitoring during Batch Cooling Crystallization Processes, *Cryst. Growth Des.*, 2010, **10**(6), 2629–2640.
36. M. Fujiwara, P. S. Chow, D. L. Ma and R. D. Braatz, Paracetamol Crystallization Using Laser Backscattering and ATR-FTIR Spectroscopy: Metastability, Agglomeration, and Control, *Cryst. Growth Des.*, 2002, **2**(5), 363–370.
37. N. Doki, H. Seki, K. Takano, H. Asatani, M. Yokota and N. Kubota, Process Control of Seeded Batch Cooling Crystallization of the Metastable α-Form Glycine Using an In-Situ ATR-FTIR Spectrometer and an In-Situ FBRM Particle Counter, *Cryst. Growth Des.*, 2004, **4**(5), 949–953.
38. L. Feng and K. A. Berglund, ATR-FTIR for Determining Optimal Cooling Curves for Batch Crystallization of Succinic Acid, *Cryst. Growth Des.*, 2002, **2**(5), 449–452.
39. T. Togkalidou, H.-H. Tung, Y. Sun, A. Andrews and R. D. Braatz, Solution Concentration Prediction for Pharmaceutical Crystallization Processes Using Robust Chemometrics and ATR FTIR Spectroscopy, *Org. Process Res. Dev.*, 2002, **6**(3), 317–322.
40. C. Lindenberg, M. Krättli, J. Cornel, M. Mazzoti and J. Brozio, Design and optimization of a combined cooling/antisolvent crystallization process, *Cryst. Growth Des.*, 2009, **9**(2), 1124–1136.
41. J. Cornel, C. Lindenberg and M. Mazzotti, Quantitative Application of in Situ ATR-FTIR and Raman Spectroscopy in Crystallization Processes, *Ind. Eng. Chem. Res.*, 2008, **47**(14), 4870–4882.
42. D. R. Thompson, E. Kougoulos, A. G. Jones and M. W. Wood-Kaczmar, Solute concentration measurement of an important organic compound using ATR-UV spectroscopy, *J. Cryst. Growth*, 2005, **276**(1), 230–236.
43. L. L. Simon, Z. K. Nagy and K. Hungerbuhler, Comparison of external bulk video imaging with focused beam reflectance measurement and ultra-violet visible spectroscopy for metastable zone identification in

food and pharmaceutical crystallization processes, *Chem. Eng. Sci.*, 2009, **64**(14), 3344–3351.
44. E. Simone, A. N. Saleemi, N. Tonnon and Z. K. Nagy, Active Polymorphic Feedback Control of Crystallization Processes Using a Combined Raman and ATR-UV/Vis Spectroscopy Approach, *Cryst. Growth Des.*, 2014, **14**(4), 1839–1850.
45. A. N. Saleemi, C. D. Rielly and Z. K. Nagy, Comparative Investigation of Supersaturation and Automated Direct Nucleation Control of Crystal Size Distributions using ATR-UV/vis Spectroscopy and FBRM, *Cryst. Growth Des.*, 2012, **12**(4), 1792–1807.
46. A. N. Saleemi, C. D. Rielly and Z. K. Nagy, Monitoring of the combined cooling and antisolvent crystallisation of mixtures of aminobenzoic acid isomers using ATR-UV/vis spectroscopy and FBRM, *Chem. Eng. Sci.*, 2012, 77, 122–129.
47. B. Szilágyi, P. Ş. Agachi and Z. K. Nagy, Chord Length Distribution Based Modeling and Adaptive Model Predictive Control of Batch Crystallization Processes Using High Fidelity Full Population Balance Models, *Ind. Eng. Chem. Res.*, 2018, **57**(9), 3320–3332.
48. J. Jager, H. J. M. Kramer, B. Scarlett, E. J. D. Jong and S. D. Wolf, Effect of scale of operation on CSD dynamics in evaporative crystallizers, *AlChE J.*, 1991, **37**(2), 182–192.
49. A. M. Neumann and H. J. M. Kramer, A Comparative Study of Various Size Distribution, *Part. Part. Syst. Charact.*, 2002, **19**(1), 17–27.
50. A. Mesbah, J. Landlust, A. E. M. Huesman, H. J. M. Kramer, P. J. Jansens and P. M. J. Van den Hof, A model-based control framework for industrial batch crystallization processes, *Chem. Eng. Res. Des.*, 2010, **88**(9), 1223–1233.
51. A. Mesbah, A. E. M. Huesman, H. J. M. Kramer and P. M. J. Van den Hof, A comparison of nonlinear observers for output feedback model-based control of seeded batch crystallization processes, *J. Process Control*, 2011, **21**(4), 652–666.
52. A. Mesbah, A. E. M. Huesman, H. J. M. Kramer and Z. K. Nagy, Hof PMJVd. Real-time control of a semi-industrial fed-batch evaporative crystallizer using different direct optimization strategies, *AlChE J.*, 2011, **57**(6), 1557–1569.
53. S. S. Kadam, J. A. W. Vissers, M. Forgione, R. M. Geertman, P. J. Daudey and A. I. Stankiewicz, *et al.*, Rapid Crystallization Process Development Strategy from Lab to Industrial Scale with PAT Tools in Skid Configuration, *Org. Process Res. Dev.*, 2012, **16**(5), 769–780.
54. G. Hou, G. Power, M. Barrett, B. Glennon, G. Morris and Y. Zhao, Development and Characterization of a Single Stage Mixed-Suspension, Mixed-Product-Removal Crystallization Process with a Novel Transfer Unit, *Cryst. Growth Des.*, 2014, **14**(4), 1782–1793.
55. C. Herman, B. Haut, S. Douieb, A. Larcy, V. Vermylen and T. Leyssens, Use of in Situ Raman, FBRM, and ATR-FTIR Probes for the Understanding of the Solvent-Mediated Polymorphic Transformation

of II–I Etiracetam in Methanol, *Org. Process Res. Dev.*, 2012, **16**(1), 49–56.
56. Y. Hu, J. K. Liang, A. S. Myerson and L. S. Taylor, Crystallization Monitoring by Raman Spectroscopy: Simultaneous Measurement of Desupersaturation Profile and Polymorphic Form in Flufenamic Acid Systems, *Ind. Eng. Chem. Res.*, 2005, **44**(5), 1233–1240.
57. T. Ono, J. H. ter Horst and P. J. Jansens, Quantitative Measurement of the Polymorphic Transformation of l-Glutamic Acid Using In-Situ Raman Spectroscopy, *Cryst. Growth Des.*, 2004, **4**(3), 465–469.
58. A. Dandeu, B. Humbert, C. Carteret, H. Muhr, E. Plasari and J. M. Bossoutrot, Raman Spectroscopy – A Powerful Tool for the Quantitative Determination of the Composition of Polymorph Mixtures: Application to CaCO3 Polymorph Mixtures, *Chem. Eng. Technol*, 2006, **29**(2), 221–225.
59. H. Qu, H. Alatalo, H. Hatakka, J. Kohonen, M. Louhi-Kultanen and S.-P. Reinikainen, *et al.*, Raman and ATR FTIR spectroscopy in reactive crystallization: Simultaneous monitoring of solute concentration and polymorphic state of the crystals, *J. Cryst. Growth*, 2009, **311**(13), 3466–3475.
60. G. Févotte, In Situ Raman Spectroscopy for In-Line Control of Pharmaceutical Crystallization and Solids Elaboration Processes: A Review, *Chem. Eng. Res. Des.*, 2007, **85**(7), 906–920.
61. Z. K. Nagy, E. Aamir and C. D. Rielly, Internal Fines Removal Using Population Balance Model Based Control of Crystal Size Distribution under Dissolution, Growth and Nucleation Mechanisms, *Cryst. Growth Des.*, 2011, **11**(6), 2205–2219.
62. Z. K. Nagy and R. D. Braatz, Advances and New Directions in Crystallization Control, *Annu. Rev. Chem. Biomol. Eng.*, 2012, **3**(1), 55–75.
63. A. N. Kalbasenka, Model-Based Control of Industrial Batch Crystallizers: Experiments on Enhanced Controllability by Seeding Actuation, PhD thesis, TU Delft, Delft, 2009.
64. P. K. Pathath and A. Kienle, A numerical bifurcation analysis of nonlinear oscillations in crystallization processes, *Chem. Eng. Sci.*, 2002, **57**(20), 4391–4399.
65. P. K. Pathath and A. Kienle, Nonlinear Oscillations in Ammonium Sulfate Crystallization: A Comparison of Different Model Predictions, *Ind. Eng. Chem. Res.*, 2003, **42**(26), 6949–6955.
66. A. R. Menon, H. J. M. Kramer, J. Grievink and P. J. Jansens, Modelling the cyclic behaviour in a DTB crystallizer—a two-population balance model approach, *J. Cryst. Growth*, 2005, **275**(1), e1373–e1381.
67. R. A. Eek, A. J. Hoogenboezem and O. H. Bosgra, Dcsign issues related to the control of continuous crystallizers, *Comput. Chem. Eng.*, 1996, **20**(4), 427–435.
68. S. Motz, A. Mitrović, E. D. Gilles, U. Vollmer and J. Raisch, Modeling, simulation and stabilizing $H\infty$-control of an oscillating continuous crystallizer with fines dissolution, *Chem. Eng. Sci.*, 2003, **58**(15), 3473–3488.

69. J. S.-I. Kwon, M. Nayhouse, P. D. Christofides and G. Orkoulas, Modeling and control of crystal shape in continuous protein crystallization, *Chem. Eng. Sci.*, 2014, **107**(0), 47–57.
70. R. A. Eek and J. A. Both, Hof PMJVd. Closed-loop identification of a continuous crystallization process, *AIChE J.*, 1996, **42**(3), 767–776.
71. H. Zhang, J. Quon, A. J. Alvarez, J. Evans, A. S. Myerson and B. Trout, Development of Continuous Anti-Solvent/Cooling Crystallization Process using Cascaded Mixed Suspension, Mixed Product Removal Crystallizers, *Org. Process Res. Dev.*, 2012, **16**(5), 915–924.
72. A. J. Alvarez, A. Singh and A. S. Myerson, Crystallization of Cyclosporine in a Multistage Continuous MSMPR Crystallizer, *Cryst. Growth Des.*, 2011, **11**(10), 4392–4400.
73. T.-T. C. Lai, J. Cornevin, S. Ferguson, N. Li, B. L. Trout and A. S. Myerson, Control of Polymorphism in Continuous Crystallization via Mixed Suspension Mixed Product Removal Systems Cascade Design, *Cryst. Growth Des.*, 2015, **15**(7), 3374–3382.
74. A. J. Alvarez and A. S. Myerson, Continuous Plug Flow Crystallization of Pharmaceutical Compounds, *Cryst. Growth Des.*, 2010, **10**(5), 2219–2228.
75. R. J. P. Eder, S. Schrank, M. O. Besenhard, E. Roblegg, H. Gruber-Woelfler and J. G. Khinast, Continuous Sonocrystallization of Acetylsalicylic Acid (ASA): Control of Crystal Size, *Cryst. Growth Des.*, 2012, **12**(10), 4733–4738.
76. M. Jiang, Z. Zhu, E. Jimenez, C. D. Papageorgiou, J. Waetzig and A. Hardy, *et al.*, Continuous-Flow Tubular Crystallization in Slugs Spontaneously Induced by Hydrodynamics, *Cryst. Growth Des.*, 2014, **14**(2), 851–860.
77. P. Neugebauer and J. G. Khinast, Continuous Crystallization of Proteins in a Tubular Plug-Flow Crystallizer, *Cryst. Growth Des.*, 2015, **15**(3), 1089–1095.
78. S. Lawton, G. Steele, P. Shering, L. H. Zhao, I. Laird and X. W. Ni, Continuous Crystallization of Pharmaceuticals Using a Continuous Oscillatory Baffled Crystallizer, *Org. Process Res. Dev.*, 2009, **13**(6), 1357–1363.
79. T. McGlone, N. E. B. Briggs, C. A. Clark, C. J. Brown, J. Sefcik and A. J. Florence, Oscillatory Flow Reactors (OFRs) for Continuous Manufacturing and Crystallization, *Org. Process Res. Dev.*, 2015, 1186–1202.
80. J. D. Ward, D. A. Mellichamp and M. F. Doherty, Choosing an operating policy for seeded batch crystallization, *AIChE J.*, 2006, **52**(6), 2046–2054.
81. J. S. I. Kwon, M. Nayhouse, G. Orkoulas and P. D. Christofides, Crystal shape and size control using a plug flow crystallization configuration, *Chem. Eng. Sci.*, 2014, **119**, 30–39.
82. D. Jagadesh, N. Kubota, M. Yokota, N. Doki and A. Sato, Seeding Effect on Batch Crystallization of Potassium Sulfate under Natural Cooling Mode and a Simple Design Method of Crystallizer, *J. Chem. Eng. Jpn.*, 1999, **32**(4), 514–520.

83. N. Doki, M. Yokota, K. Kido, S. Sasaki and N. Kubota, Reliable and Selective Crystallization of the Metastable α-Form Glycine by Seeding, *Cryst. Growth Des.*, 2004, **4**(1), 103–107.
84. A. N. Kalbasenka, L. C. P. Spierings, A. E. M. Huesman and H. J. M. Kramer, Application of Seeding as a Process Actuator in a Model Predictive Control Framework for Fed-Batch Crystallization of Ammonium Sulphate, *Part. Part. Syst. Charact.*, 2007, **24**(1), 40–48.
85. R. J. P. Eder, E. K. Schmitt, J. Grill, S. Radl, H. Gruber-Woelfler and J. G. Khinast, Seed loading effects on the mean crystal size of acetylsalicylic acid in a continuous-flow crystallization device, *Cryst. Res. Technol.*, 2011, **46**(3), 227–237.
86. M. O. Besenhard, P. Neugebauer, C.-D. Ho and J. G. Khinast, Crystal size control in a continuous tubular crystallizer, *Cryst. Growth Des.*, 2015, **15**, 1683–1691.
87. N. Kubota, N. Doki, M. Yokota and A. Sato, Seeding policy in batch cooling crystallization, *Powder Technol.*, 2001, **121**(1), 31–38.
88. R. A. Lionberger, S. L. Lee, L. Lee, A. Raw and L. X. Yu, Quality by design: Concepts for ANDAs, *AAPS J.*, 2008, **10**(2), 268–276.
89. H. Wu, M. White and M. A. Khan, Quality-by-Design (QbD): An integrated process analytical technology (PAT) approach for a dynamic pharmaceutical co-precipitation process characterization and process design space development, *Int. J. Pharm.*, 2011, **405**(1), 63–78.
90. S. Thirunahari, P. S. Chow and R. B. H. Tan, Quality by Design (QbD)-Based Crystallization Process Development for the Polymorphic Drug Tolbutamide, *Cryst. Growth Des.*, 2011, **11**(7), 3027–3038.
91. J. Nývlt and J. W. Mullin, The periodic behaviour of continuous crystallizers, *Chem. Eng. Sci.*, 1970, **25**(1), 131–147.
92. P. D. Christofides, N. El-Farra, M. Li and P. Mhaskar, Model-based control of particulate processes, *Chem. Eng. Sci.*, 2008, **63**(5), 1156–1172.
93. T. Chiu and P. D. Christofides, Nonlinear control of particulate processes, *AlChE J.*, 1999, **45**(6), 1279–1297.
94. T. Chiu and P. D. Christofides, Robust nonlinear control of a continuous crystallizer, *Comput. Chem. Eng.*, 1999, **23**, S257–S260.
95. T. Y. Chiu and P. D. Christofides, Robust control of particulate processes using uncertain population balances, *AlChE J.*, 2000, **46**(2), 266–280.
96. N. H. El-Farra, T. Y. Chiu and P. D. Christofides, Analysis and control of particulate processes with input constraints, *AlChE J.*, 2001, **47**(8), 1849–1865.
97. A. Gani, P. Mhaskar and P. D. Christofides, Handling sensor malfunctions in control of particulate processes, *Chem. Eng. Sci.*, 2008, **63**(5), 1217–1229.
98. J. Liu, D. Muñoz de la Peña, P. D. Christofides and J. F. Davis, Lyapunov-based Model Predictive Control of Particulate Processes Subject to Asynchronous Measurements, *Part. Part. Syst. Charact.*, 2008, **25**(4), 360–375.

99. N. H. El-Farra and A. Giridhar, Detection and management of actuator faults in controlled particulate processes using population balance models, *Chem. Eng. Sci.*, 2008, **63**(5), 1185–1204.
100. A. Giridhar and N. H. El-Farra, A unified framework for detection, isolation and compensation of actuator faults in uncertain particulate processes, *Chem. Eng. Sci.*, 2009, **64**(12), 2963–2977.
101. P. D. Christofides and P. Daoutidis, Robust control of multivariable two-time-scale nonlinear systems, *J. Process Control*, 1997, **7**(5), 313–328.
102. C. Damour, M. Benne, L. Boillereaux, B. Grondin-Perez and J.-P. Chabriat, NMPC of an industrial crystallization process using model-based observers, *J. Ind. Eng. Chem.*, 2010, **16**(5), 708–716.
103. J. S.-I. Kwon, M. Nayhouse, G. Orkoulas, D. Ni and P. D. Christofides, Run-to-Run-Based Model Predictive Control of Protein Crystal Shape in Batch Crystallization, *Ind. Eng. Chem. Res.*, 2015, **54**(16), 4293–4302.
104. S. Rohani, M. Haeri and H. C. Wood, Modeling and control of a continuous crystallization process Part 2. Model predictive control, *Comput. Chem. Eng.*, 1999, **23**(3), 279–286.
105. S. Rohani, M. Haeri and H. C. Wood, Modeling and control of a continuous crystallization process Part 1. Linear and non-linear modeling, *Comput. Chem. Eng.*, 1999, **23**(3), 263–277.
106. D. Shi, N. H. El-Farra, M. Li, P. Mhaskar and P. D. Christofides, Predictive control of particle size distribution in particulate processes, *Chem. Eng. Sci.*, 2006, **61**(1), 268–281.
107. N. H. El-Farra, P. Mhaskar and P. D. Christofides, Hybrid predictive control of nonlinear systems: method and applications to chemical processes, *Int. J. Robust Nonlin. Control*, 2004, **14**(2), 199–225.
108. A. L. Nunis da Silva, M. M. Seckler and O. Darci, Application of Model Predictive Control to a Continuous Multiple-Effect Crystallizer, *Chem. Eng. Technol.*, 2018, **41**(7), 1406–1416.
109. B. Benyahia, M. A. Latifi, C. Fonteix and F. Pla, Emulsion copolymerization of styrene and butyl acrylate in the presence of a chain transfer agent. Part 2: Parameters estimability and confidence regions, *Chem. Eng. Sci.*, 2013, **90**, 110–118.
110. I. I. Onyemelukwe, B. Benyahia, N. M. Reis, Z. K. Nagy and C. D. Rielly, The heat transfer characteristics of a mesoscale continuous oscillatory flow crystalliser with smooth periodic constrictions, *Int. J. Heat Mass Transfer*, 2018, **123**, 1109–1119.
111. D. Fysikopoulos, A. Borsos, W. Li, I. Onyemelukwe, B. Benyahia and Z. K. Nagy, *et al.*, Local vs Global Estimability Analysis of Population Balance Models for Crystallization Processes, in *Computer Aided Chemical Engineering*, ed. A. Espuña, M. Graells and L. Puigjaner, Elsevier, 2017, vol. 40, pp. 55–60; D. Fysikopoulos, B. Benyahia, A. Borsos, Z. K. Nagy, and C. D. Rielly, A Framework for Model Reliability and Estimability Analysis of Crystallization Processes with Multi-Impurity Multi-Dimensional Population Balance Models, *Comput. Chem. Eng.*, 2019, **1222**, 275–292.

112. P. Parekh, B. Benyahia and C. D. Rielly, A Global State Feedback Linearization and Decoupling MPC of a MIMO Continuous MSMPR Cooling Crystallization Process, *Comput.-Aided Chem. Eng.*, 2018, **43**, 1607–1612.
113. Y. Cao, J. Kang, Z. Nagy and C. Laird, Parallel Solution of Robust Nonlinear Model Predictive Control Problems in Batch Crystallization, *Processes*, 2016, **4**(3), 20.
114. R. Lakerveld, H. J. M. Kramer, A. I. Stankiewicz and J. Grievink, Application of generic principles of process intensification to solution crystallization enabled by a task-based design approach, *Chem. Eng. Process.*, 2010, **49**(9), 979–991.
115. J. Wang, F. Li and R. Lakerveld, Process intensification for pharmaceutical crystallization, *Chem. Eng. Process.*, 2018, **127**, 111–126.
116. W. T. Richards and A. L. Loomis, The chemical effects of high frequency sound waves i. a preliminary survey, *J. Am. Chem. Soc.*, 1927, **49**(12), 3086–3100.
117. R. Lakerveld, N. G. Verzijden, H. Kramer, P. Jansens and J. Grievink, Application of Ultrasound for Start-Up of Evaporative Batch Crystallization of Ammonium Sulfate in a 75-L Crystallizer, *AlChE J.*, 2011, **57**(12), 3367–3377.
118. K. Kakinouchi, H. Adachi, H. Matsumura, T. Inoue, S. Murakami and Y. Mori, *et al.*, Effect of ultrasonic irradiation on protein crystallization, *J. Cryst. Growth*, 2006, **292**(2), 437–440.
119. R. K. Bund and A. B. Pandit, Sonocrystallization: effect on lactose recovery and crystal habit, *Ultrason. Sonochem.*, 2007, **14**(2), 143–152.
120. S. L. Hem, The effect of ultrasonic vibrations on crystallization processes, *Ultrasonics*, 1967, 202–207.
121. A. Kordylla, S. Koch, F. Tumakaka and G. Schembecker, Towards an optimized crystallization with ultrasound: Effect of solvent properties and ultrasonic process parameters, *J. Cryst. Growth*, 2008, **310**(18), 4177–4184.
122. K. Wohlgemuth, A. Kordylla, F. Ruether and G. Schembecker, Experimental study of the effect of bubbles on nucleation during batch cooling crystallization, *Chem. Eng. Sci.*, 2009, **64**(19), 4155–4163.
123. K. Wohlgemuth, F. Ruether and G. Schembecker, Sonocrystallization and crystallization with gassing of adipic acid, *Chem. Eng. Sci.*, 2010, **65**(2), 1016–1027.
124. Z. Guo, A. G. Jones, H. Hao, B. Patel and N. Li, Effect of ultrasound on the heterogeneous nucleation of BaSO4 during reactive crystallization, *J. Appl. Phys.*, 2007, **101**(5), 054907.
125. G. Ruecroft, D. Hipkiss, T. Ly, N. Maxted and P. W. Cains, Sonocrystallization: The use of ultrasound for improved industrial crystallization, *Org. Process Res. Dev.*, 2005, **9**(6), 923–932.
126. M. Jiang, C. Gu and R. D. Braatz, Analysis of focused indirect ultrasound via high-speed spatially localized pressure sensing and its consequences on nucleation, *Chem. Eng. Process.*, 2017, **117**, 186–194.

127. O. Narducci, A. G. Jones and E. Kougoulos, Continuous crystallization of adipic acid with ultrasound, *Chem. Eng. Sci.*, 2011, **66**(6), 1069–1076.
128. O. Narducci, A. G. Jones and E. Kougoulos, An Assessment of the Use of Ultrasound in the Particle Engineering of Micrometer-Scale Adipic Acid Crystals, *Cryst. Growth Des.*, 2011, **11**(5), 1742–1749.
129. H. Siddique, C. J. Brown, I. Houson and A. J. Florence, Establishment of a Continuous Sonocrystallization Process for Lactose in an Oscillatory Baffled Crystallizer, *Org. Process Res. Dev.*, 2015, **19**(12), 1871–1881.
130. M. Furuta, K. Mukai, D. Cork and K. Mae, Continuous crystallization using a sonicated tubular system for controlling particle size in an API manufacturing process, *Chem. Eng. Process.*, 2016, **102**, 210–218.
131. J. Jordens, E. Canini, B. Gielen, T. Van Gerven and L. Braeken, Ultrasound Assisted Particle Size Control by Continuous Seed Generation and Batch Growth, *Crystals*, 2017, 7(7), 195.
132. W. W. Li, N. Radacsi and H. J. M. Kramer, Heijden AEDMvd, Horst JHt. Solid Separation from a Mixed Suspension through Electric-Field-Enhanced Crystallization, *Angew. Chem., Int. Ed.*, 2016, **55**(52), 16088–16091.
133. Z. Hammadi, J. P. Astier, R. Morin and S. Veesler, Protein crystallization induced by a localized voltage, *Cryst. Growth Des.*, 2007, **7**(8), 1472–1475.
134. D. Hou and H. C. Chang, ac field enhanced protein crystallization, *Appl. Phys. Lett.*, 2008, **92**(22), 223902.
135. T. Wakamatsu, Method and apparatus for characterization of electric field-induced aggregation in pre-crystalline protein solutions, *Rev. Sci. Instrum.*, 2015, **86**(1), 015112.
136. F. Li and R. Lakerveld, Influence of Alternating Electric Fields on Protein Crystallization in Microfluidic Devices with Patterned Electrodes in a Parallel-Plate Configuration, *Cryst. Growth Des.*, 2017, **17**(6), 3062–3070.
137. F. Li and R. Lakerveld, Electric-field-assisted protein crystallization in continuous flow, *Cryst. Growth Des.*, 2018, **18**(5), 2964–2971.
138. K. Mauge-Lewis, A. Mojibola, E. A. Toth, M. Mohammed, D. Seifu and K. Aslan, Metal-Assisted and Microwave-Accelerated Evaporative Crystallization: Proof-of-Principle Application to Proteins, *Cryst. Growth Des.*, 2015, **15**(7), 3212–3219.
139. M. A. Pinard and K. Aslan, Metal-Assisted and Microwave-Accelerated Evaporative Crystallization, *Cryst. Growth Des.*, 2010, **10**(11), 4706–4709.
140. N. Radacsi, G. D. Stefanidis, P. Szabó-Révész and R. Ambrus, Analysis of niflumic acid prepared by rapid microwave-assisted evaporation, *J. Pharm. Biomed. Anal.*, 2014, **98**, 16–21.
141. N. Radacsi, J. H. ter Horst and G. D. Stefanidis, Microwave-Assisted Evaporative Crystallization of Niflumic Acid for Particle Size Reduction, *Cryst. Growth Des.*, 2013, **13**(10), 4186–4189.

142. R. Kacker, P. M. Salvador, G. S. J. Sturm, G. D. Stefanidis, R. Lakerveld and Z. K. Nagy, *et al.*, Microwave Assisted Direct Nucleation Control for Batch Crystallization: Crystal Size Control with Reduced Batch Time, *Cryst. Growth Des.*, 2016, **16**(1), 440–446.
143. Y. Yang, L. Song, T. Gao and Z. K. Nagy, Integrated Upstream and Downstream Application of Wet Milling with Continuous Mixed Suspension Mixed Product Removal Crystallization, *Cryst. Growth Des.*, 2015, **15**(12), 5879–5885.
144. Y. Yang, L. Song, Y. Zhang and Z. K. Nagy, Application of Wet Milling-Based Automated Direct Nucleation Control in Continuous Cooling Crystallization Processes, *Ind. Eng. Chem. Res.*, 2016, **55**(17), 4987–4996.
145. M. R. Abu Bakar, Z. K. Nagy, A. N. Saleemi and C. D. Rielly, The Impact of Direct Nucleation Control on Crystal Size Distribution in Pharmaceutical Crystallization Processes, *Cryst. Growth Des.*, 2009, **9**(3), 1378–1384.

CHAPTER 5

Slug-flow Continuous Crystallization: Fundamentals and Process Intensification

J. CARL PIRKLE JR[a], MICHAEL L. RASCHE[a], RICHARD D. BRAATZ*[a] AND MO JIANG*[a,b]

[a]Massachusetts Institute of Technology, Department of Chemical Engineering, 77 Massachusetts Avenue, Cambridge, Massachusetts 02139, USA; [b]Virginia Commonwealth University, Department of Chemical and Life Science Engineering, 601 West Main Street, Box 843068, Richmond, Virginia 23284-3068, USA
*E-mail: jcpirkle@mit.edu, rascheml@gmail.com, braatz@mit.edu, mjiang3@vcu.edu

5.1 Introduction to Slug Flow Crystallization

5.1.1 State-of-the-art

Many continuous-flow crystallizer designs have been developed in recent years for increasing process efficiency and reproducibility, as reviewed in other chapters of this book. "Plug-flow" properties are usually desired in these crystallizers to narrow the residence time distribution and to provide easier scale-up and control of crystal size and shape distribution. Segmented/slug flow has close-to-plug-flow properties even in the laminar flow regime, with nearly uniform velocity profile on the length scale of a slug,[1–7]

The Handbook of Continuous Crystallization
Edited by Nima Yazdanpanah and Zoltan K. Nagy

Published by the Royal Society of Chemistry, www.rsc.org

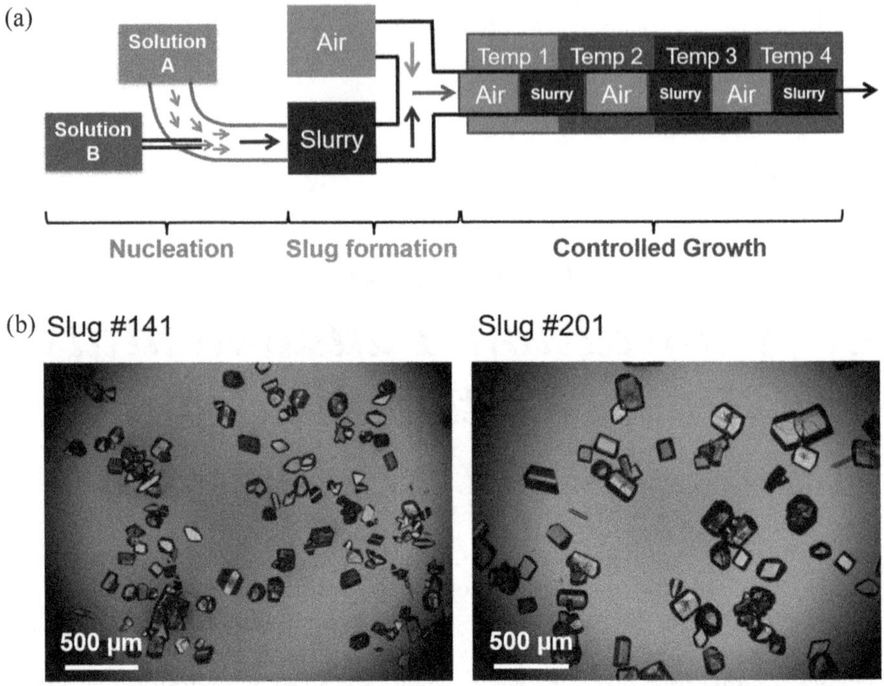

Figure 5.1 (a) Schematic of a slug-flow cooling crystallization process, with nucleation from a cooling coaxial micromixer.[2] Adapted from ref. 2 with permission from American Chemical Society, Copyright 2014. (b) Microscope images of product crystals in representative slugs (*e.g.*, #141 and 201) from the slug-flow crystallization design in (a).

as discussed in reviews of multiphase flow in microfluidic channels.[8,9] For millifluidic systems (the tubing inner diameter on the scale of millimeters), cooling crystallization,[1–3] anti-solvent crystallization,[10] and reaction precipitation[4–7] have been demonstrated with residence time in the order of minutes (and above) and with crystal productivity sufficiently high for economic manufacturing of pharmaceuticals and other products. These segmented-flow crystallizers also have most advantages of tubular crystallizers, including (1) improved heat/mass transfer[11] due to having small volume and large surface-area-to-volume ratio, (2) low shear, and (3) a high degree of spatial control of temperature along the tube.

Recently designed slug-flow crystallizers[1,2,12] (*e.g.*, Figures 5.1a and 5.2a) complement past crystallizers and provide many degrees of freedom for the control of crystal shape and size distribution (*e.g.*, Figures 5.1b and 5.2c,d). Based on properly selected co-current flow rates of gas and liquid, stable alternating slugs of liquid and gas are spontaneously generated at aspect ratios close to 1 (*e.g.*, Figure 5.3a and b), facilitating good mixing for

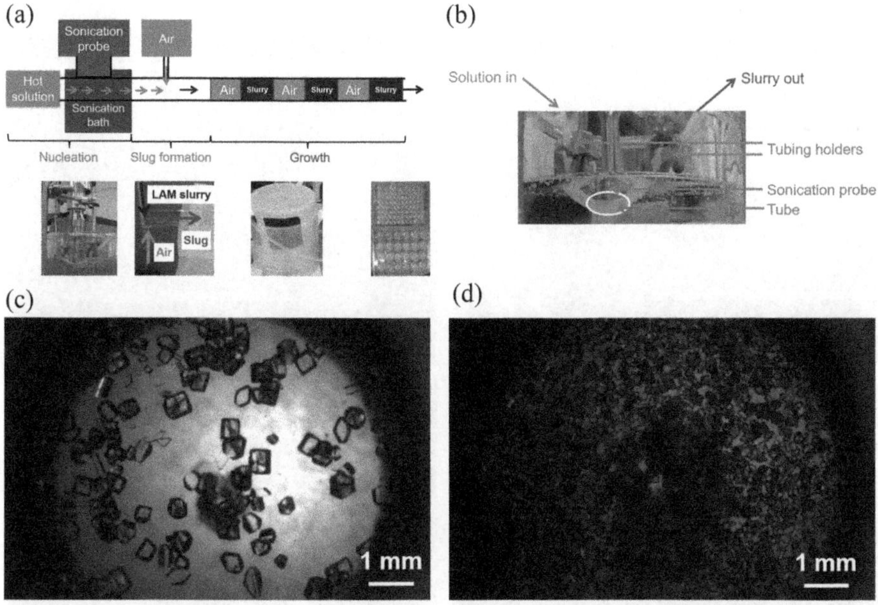

Figure 5.2 (a) Schematic of a slug-flow cooling crystallization process, with nucleation from indirect focused ultrasonication.[1,2] Photos are shown under corresponding schematic for the nucleation setup, micromixer for slug formation, parallel tubing for growth, and well plates for sample collection. Adapted from ref. 2 with permission from American Chemical Society, Copyright 2014. (b) Details of the indirect ultrasonication nucleation setup (side view). The focused sonication zone is circled, right under the ultrasonic probe. Adapted from ref. 44 with permission from Elsevier, Copyright 2017. (c–d): Microscope images of product crystals from the design in (a) with sonication (c) on or (d) off for nucleation. Images of both (c) and (d) are adapted from ref. 1 with permission from American Chemical Society, Copyright 2015.

enhanced heat and mass transport without mixing blades. These liquid slugs can serve as a series of independent crystallizers of very similar experimental conditions. In this way, past knowledge of batch crystallization is applied to continuously generate high-quality drug crystals, which enhance the control of product CSD with properly tuned parameters within the design space. The scale-up can be achieved by simply running the experiment for longer time (generating more slugs using the same slug and tube dimensions),[1,2] and/or lining up parallel tubular crystallizers at the same conditions.[6] Instead of specific complex segmenters[4,6,13] ("flow cutters"), common setup (tubing with fittings[1–3,14] such as in Figures 5.1a and 5.2a) can be used to generate alternative segments or slugs of compound slurry and a non-miscible phase (usually air). An additional improvement is that the slug-flow crystallizer does not require a solvent/solvent separation operation afterwards, thus further improving crystal product purity and process energy efficiency.[1,2]

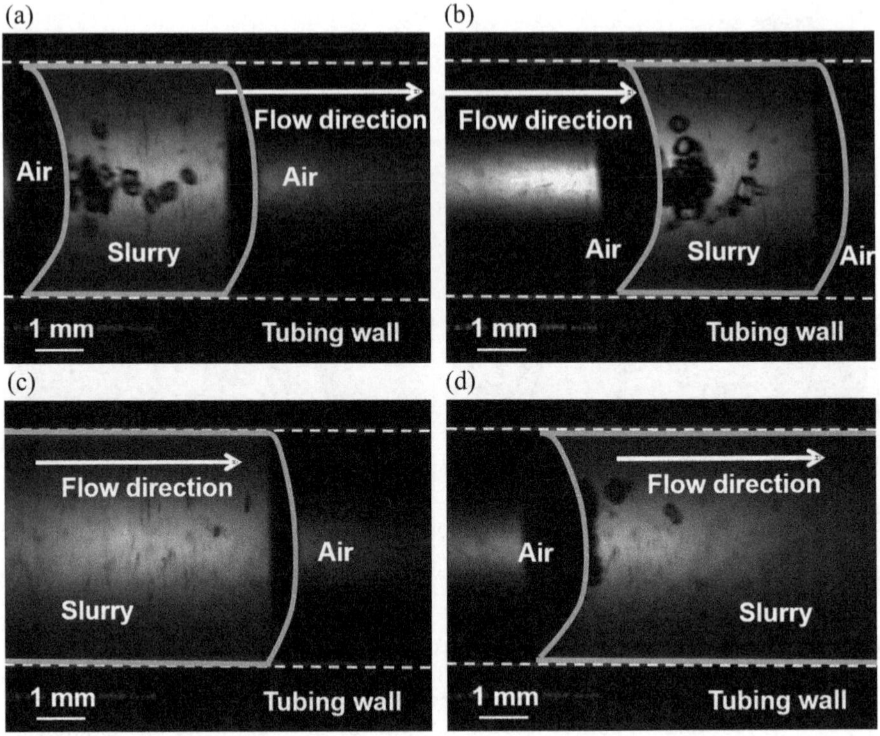

Figure 5.3 In-line stereomicroscope video snapshots of (a) an earlier time and (b) later time of a slurry slug with an aspect ratio of about 1, and (c) the front and (d) the end of another slurry slug with an aspect ratio close to 4. Slurry slugs are separated by air slugs inside a silicone tube (two inner walls, pale lines). The black shades at two air/liquid interfaces are due to curvature. Adapted from ref. 51 with permission from John Wiley and Sons, © 2017 WILEY-VCH Verlag GmbH & Co. KGaA, Weinheim.

5.1.2 Chapter Outline

The structure of this chapter is based on a process intensification and design strategy of an advanced continuous crystallizer, namely, decoupling nucleation and growth for individual control while suppressing undesired phenomena. For example, the suppression of secondary nucleation and aggregation in the growth stage has been demonstrated during slug-flow crystallization, when the equipment and operations are appropriately designed. For the sections in this chapter, experimental discovery is supported by theoretical review/analysis and design/operational advice. The specific sections are

- The slug formation process is detailed in Sections 5.2 and 5.3. Specifically, the stability and sizes of slugs can be tuned to improve the recirculation flow pattern, and the size distribution of product crystals inside.[2] Detailed review and analysis from fluid mechanics and transport phenomena perspectives are also included.

- The crystal growth (and/or dissolution) process inside individual slugs is discussed in Section 5.4, with control by both robust design and temperature zones.
- The pre-slug nucleation process is discussed in Section 5.5. Successful nucleation control can make the growth control much easier. As continuous nucleation methods are well reviewed in other chapters of the book, only recently developed methods are discussed, such as cooling micromixers and focused indirect ultrasonication.

5.2 Control Slug Stability

The slug-flow stability is analyzed for the enhanced control of the continuous crystallization process, with design and operational considerations, in Subsection 5.2.1. Detailed analysis for hydrodynamically stable flows is in Subsection 5.2.2. Transitions of different forms of slug flow are in Subsection 5.2.3.

5.2.1 Stable Slug Flow for Crystallization Purposes

Slug stability refers to the situation in which slugs do not combine or separate, nor switch to other flow patterns (*e.g.*, stratified flow, see Figure 5.4).[2] In this way, all slugs have the same residence time, without significant material transferred between slugs. Important factors that affect slug stability include: (1) tubing diameter, (2) velocity of air and liquid streams, and (3) tubing inner surface properties. The larger the tubing inner diameter, the more limited the operating conditions that generate and maintain stable slugs, that is, the smaller the operating regime of stable slug flow. For tubing with fixed diameter, the stable slug-flow regime is mainly affected by velocity, and the tubing inner surface properties are discussed in Subsection 5.3.2 for slug geometry. The tubing should be wrapped in parallel without kinks to reduce possible slug size change or slug combination.[2]

A common method for generating stable slugs is to search by trial-and-error for combinations of air and liquid flow rates. The slug stability (*e.g.*, whether any adjacent slugs combine, or slugs separate) can be easily checked (*e.g.*, visual observation) throughout the transparent crystallizer tubing.[2] If a "flow map" is available for systems of similar geometries, the air and liquid flow rates can be chosen within the stable operation regime. It is advisable to operate far away from regime boundaries to avoid unexpected flow transition (Subsection 5.2.3) or slug size change (Subsection 5.3.1). Note that liquid flow rate is also directly limited by the final production rate.

In practical continuous operation, slugs accelerated during the shutdown of experiments. The higher velocity likely comes from low resistance, with fewer liquid slugs in the system. If too fast, the slug flow rate can cause slug combination, or liquid left inside the tubing. To maintain nearly constant slug velocities throughout the entire experiment, slugs of water were used before and after the crystal-containing slugs so that the pressure drops from the inlets to the tube outlet are nearly constant.[1]

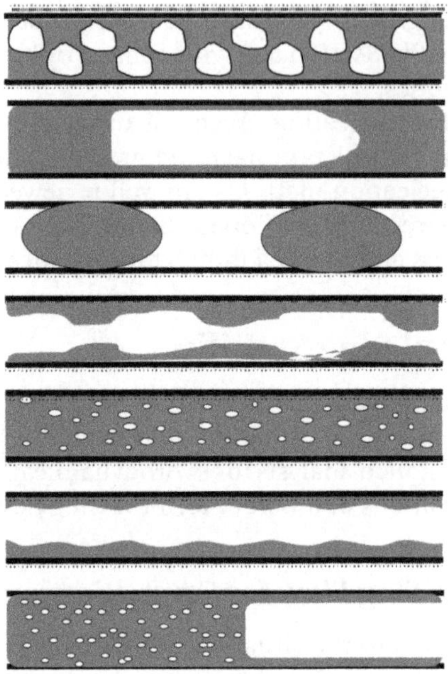

Figure 5.4 Schematics of hydrodynamically stable flow patterns of a gas (white color) and liquid (darker color) mixture in a horizontal round tubing. The notations and images of the flow were based on Ullmann and Brauner[15] and Jiang *et al.*[2] The air and liquid in slug flow can swap places depending on the affinity of the liquid for the material on the inner surface of the tube, *e.g.*, a water slug with a strongly hydrophobic surface would appear as shown, with convex/roundish water slugs moving through the tube. For a strongly hydrophilic surface with water as the liquid, the air slugs would be convex. Regardless of surface affinities, the slugs become asymmetric at sufficiently high velocities. Stratified flow regimes are not shown due to their absence for $U_{LS} > 0.01$ when $D \leq 0.004$ m.

5.2.2 Hydrodynamically Stable Regime Analysis for Slug Flow

Among the stable hydrodynamic regimes identified for gas–liquid flow (Figure 5.4), two suitable for crystal growth are the short-bubble slug flow (SB-slug) regime and the elongated-bubble slug flow (EB-slug) regime, and it is possible to operate in either. In mapping out the hydro-dynamically stable regions, experimentalists have made numerous observations at various values of U_{GS} and U_{LS}, the superficial velocities of the gas and liquid phases, respectively (the superficial velocity is the volumetric flow rate divided by the total cross-sectional area in the tube). Figure 5.5 gives a flow region map for an air–water mixture flowing through a 0.0031 m i.d. tube. The transition boundaries dividing up the map were generated by using relations advocated by Ullmann and Brauner.[15] The work by Ullmann and Brauner was a follow-up to that started by Taitel and Dukler,[16]

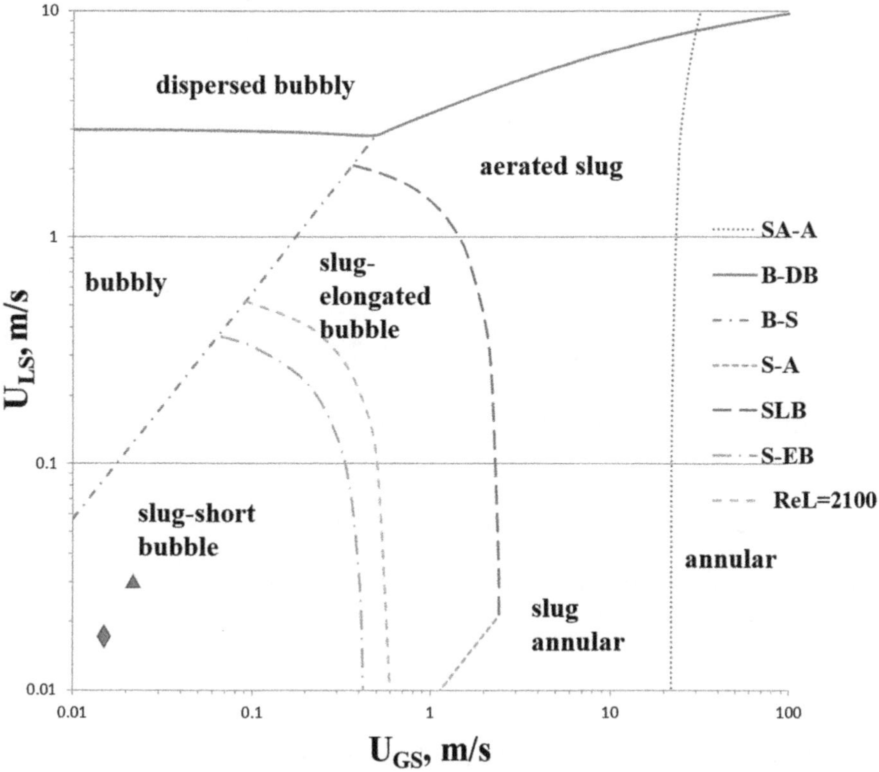

Figure 5.5 Flow regime map in horizontal 3.1 mm tube: air–water mixture with transition boundaries. Experimental Points (Jiang *et al.*[2]): Radial mixing ◆; Coaxial mixing ▲; Transition boundaries: SA-A, aerated slug (or slug annular) to annular; B-DB, bubbly, aerated slug, or annular to dispersed bubbly; B-S, bubbly to slug; S-A, slug to slug annular; TBW, slug to aerated slug *via* Taylor bubble wake model; SLB, slug to aerated slug *via* slug bulk model; S-EB, short bubble slug to elongated bubble-slug.

and carried forth by Barnea *et al.*[17,18] The work was based on more recent experimental work by Damianides and Westwater,[19] and Triplett *et al.*[20] The triangle and diamond marks in the short-bubble slug-flow region of Figure 5.5 indicate the location of two crystallization experiments carried out by Jiang *et al.*[2] Equipment dimensions and physical properties for the air-water-solute mixture are detailed in the nomenclature at the end of the chapter.

The remainder of this subsection reviews the correlations summarized and suggested by Ullmann and Brauner,[15] representing the transition curves in Figure 5.5 that surround the stable slug-flow region. Other transition curves are summarized by Ullmann and Brauner.[15] First, some key dimensionless parameters are defined that characterize the flow system. For most of the flow regimes, we designate liquid as the continuous phase and air as the dispersed phase. In this case, the mixture velocity is defined as

$$U_m = U_{cS} + U_{dS} = U_{LS} + U_{GS} \tag{5.1}$$

Reynolds, Weber, and Eötvös numbers are defined by

$$\mathrm{Re}_L = \frac{DU_m}{\nu_L}, \tag{5.2}$$

$$\mathrm{We}_L = \frac{\rho_L D U_m{}^2}{\sigma}, \tag{5.3}$$

$$\mathrm{Eo}_D = \frac{D^2(\rho_L - \rho_G) g \cos\beta'}{8\sigma}, \tag{5.4}$$

respectively. In the latter equation, β' is defined by

$$\beta' = \begin{Bmatrix} |\beta|; |\beta| < \pi/4 \\ \pi/2 - |\beta|; |\beta| > \pi/4 \end{Bmatrix} \tag{5.5}$$

where β is the inclination of the tube from the horizontal.

The gas-phase holdup is

$$\varepsilon_G = \frac{Q_G}{Q_G + Q_L}, \tag{5.6}$$

where Q_G and Q_L denote the volumetric flow rates of gas and liquid, respectively, through the tube.

5.2.3 Flow Transition of Slug Flow

The flow regime map shown in Figure 5.5 shows several transition boundaries. To maintain stable slug-flow conditions, the most important boundaries are those surrounding the slug-flow region. Thus only the envelope around the stable slug-flow region is discussed here. Information on the other transitions in Figure 5.5 is available in Ullmann and Brauner.[15]

5.2.3.1 Transition from Bubbly to Slug-flow Regime

Starting from the left side of the slug-flow region, the transition from bubbly to slug flow should involve surface tension through the Eötvös number, which is small for the corresponding crystallization in micro- and minitubes. Under this condition, Ullmann and Brauner suggest the relationship,

$$U_{LS} = \frac{1 - (\varepsilon_G)_{crit}}{(\varepsilon_G)_{crit}} U_{GS}, \tag{5.7}$$

where $(\varepsilon_G)_{crit}$ is 0.15. Eqn (5.7) is plotted in Figure 5.5 as the bubbly flow to slug-flow transition line B-S.

5.2.3.2 Transition from Short-bubble Slug Flow to Elongated-bubble Slug Flow

Figure 5.4 shows pictures of two types of slug flow: (1) one type with short symmetric bubbles (SB-slug flow) with rounded tails, which corresponds to low flow rates, and (2) the other type with elongated bullet-shaped bubbles (EB-slug flow) with flat tails, which corresponds to higher flow rates. In small-diameter tubes, surface tension force overcomes the inertial force of the trailing liquid slug and favors short bubbles. Short bubbles are preferable for continuous slug-flow crystallization, keeping the aspect ratios (l_s/D) of the slugs around 1 to maximize mixing. To maintain the curved tail bubbles, Ullmann and Brauner[15] propose the condition

$$(\mathrm{We_m})_{\mathrm{crit}} = \frac{\rho_L U_m{}^2 D}{\sigma} = 8. \tag{5.8}$$

Criterion (5.8) is the transition curve for short to elongated bubble shown in Figure 5.5. The laminar flow requirement $\mathrm{Re_L} < 2100$ is also plotted, and experience suggests that it is advisable to operate the crystallizer the left of the $\mathrm{Re_L} = 2100$. Conveniently, this condition is met by eqn (5.7). Note that the values of $\mathrm{Re_L}$ corresponding to the data points[2] in Figure 5.5 are 301 and 175 for the coaxial and radial mixer, respectively.

5.2.3.3 Transition from Slug Flow to Aerated Slug Flow

As the gas flow rate increases, the liquid slugs become aerated (see Figure 5.4). There are two models for transition from slug to aerated slug flow shown in Figure 5.5: (1) the Slug Bulk (SLB) model of Barnea and Brauner[18] and (2) the Taylor bubble wake (TBW) model of Ullmann and Brauner.[21,22]

The SLB model attributes the slug aeration to bubble breakage by turbulence in the liquid slug bulk flow. Details for the derivation of this model are supplied by Ullman and Brauner.[15] For conditions in Figure 5.5, the transition relationship involving superficial velocities is

$$U_{GS} + U_{LS} = 9.72 \frac{\sigma^{0.55}}{\rho_L{}^{0.55} \nu_L{}^{0.11} D^{0.44}}, \tag{5.9}$$

which is plotted as the line designated by SLB in Figure 5.6.

The TBW model attributes liquid slug aeration to fragmentation of the elongated Taylor bubbles by turbulence in the bubble wake. For low $\mathrm{Eo_D}$ systems, the TBW model simplifies[30] to

$$U_{GS} + U_{LS} = 16 \left[\frac{\sigma}{\rho_L D} \right]^{0.5}. \tag{5.10}$$

This relation is plotted as curve TBW in Figure 5.6. Operationally, it is advisable to use the SLB boundary, eqn (5.9), as the criterion to maintain slug flow due to its location to the left and below the TBW curve.

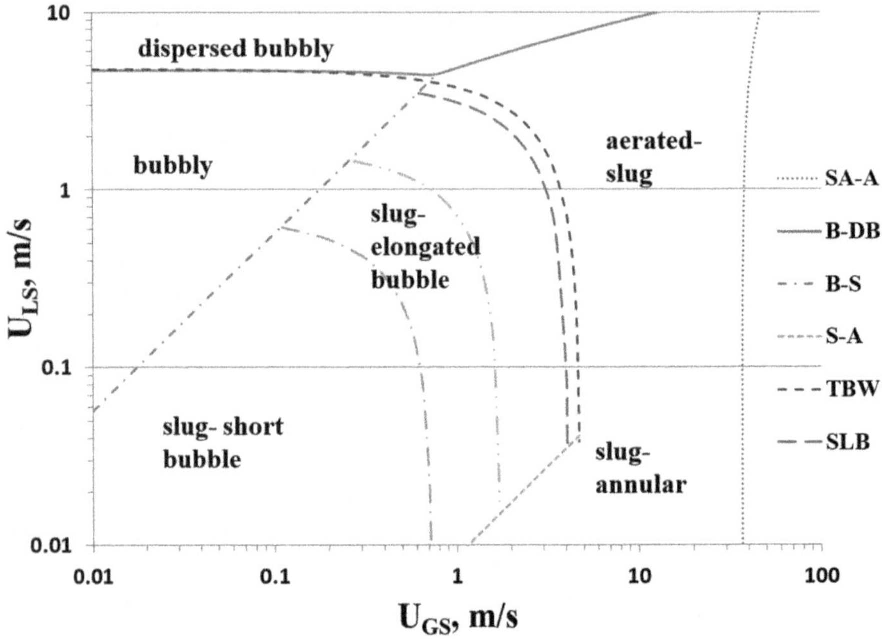

Figure 5.6 Flow regime map in horizontal 1.097 mm tube: air–water mixture with transition boundaries; Transition boundaries: SA-A, aerated slug (or slug annular) to annular; B-DB, bubbly, aerated slug, or annular to dispersed bubbly; B-S, bubbly to slug; S-A, slug to slug annular; TBW, slug to aerated slug *via* Taylor bubble wake model; SLB, slug to aerated slug *via* slug bulk model; S-EB, short bubble slug to elongated bubble-slug.

The Reynolds number $Re_L = DU_m/\nu_L$ must exceed 2100 for the turbulence assumptions of the SLB or TBW models to apply.

5.2.3.4 *Effect of Inner Surface Property of Tubing*

Whether the inner surface of the crystal growth tube is hydrophilic or hydrophobic can be determined by measuring the contact angle with the surface material. None of the above relations include the effect of contact angle. Barajas and Panton[23] experimentally studied the effect of contact angle on two phase flow in 0.0016 mm capillary tubes. Four materials of this diameter were tested: pyrex, with H_2O contact angle of 34°; polyethylene, with contact angle 61°; polyurethane, with contact angle 74°; and a fluoropolymer resin FEP, with contact angle 106°. The latter is similar to the silicone tubing, which has a contact angle of 108°, used for crystallization.

The experiments showed that contact angle influences the location of some transition boundaries in the two-phase flow regime map. Barajas and Panton[23] had a different nomenclature for their flow regimes. Reinterpreting their findings in terms of the Ullmann and Brauner[15] regimes, the affected boundaries are as follows. Barajas and Panton involve the boundary B-DB in Figure 5.6, which includes: the transition from the bubbly to the dispersed

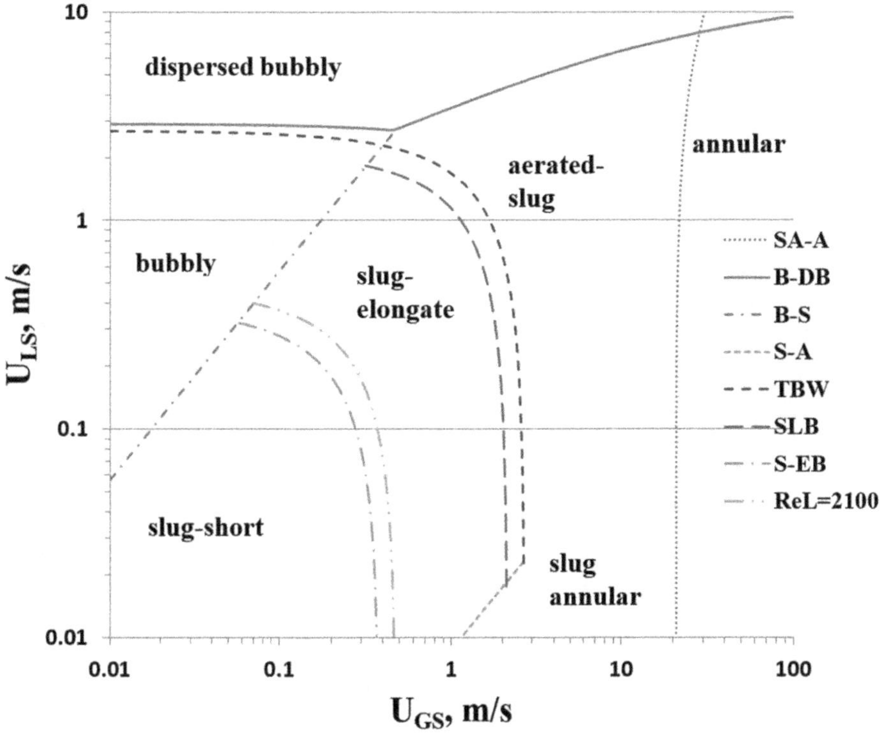

Figure 5.7 Flow regime map in horizontal 4.0 mm tube: air–water mixture with transition boundaries; Transition boundaries: SA-A, aerated slug (or slug annular) to annular; B-DB, bubbly, aerated slug, or annular to dispersed bubbly; B-S, bubbly to slug; S-A, slug to slug annular; TBW, slug to aerated slug *via* Taylor bubble wake model; SLB, slug to aerated slug *via* slug bulk model; S-EB, short bubble slug to elongated bubble-slug.

bubble regime that occurs at high values of superficial liquid velocity U_{LS}; the slug to dispersed bubble transition at mid-range U_{GS}; the aerated slug to dispersed bubble transition; and the annular to dispersed bubble transition. Other affected boundaries were the slug to slug annular or aerated slug regimes (SLB or TBW boundaries). The boundary between bubbly flow and slug flow (B-S) did not change with contact angle.

The left boundary (B-S) of the slug-flow regime does not change with contact angle, whereas the boundary (SLB) above and to the right of the slug-flow regime shifts with the contact angle. The lower branch of SLB increases slightly, allowing values of U_{GS} to reach 6 m s^{-1}. These trends can be observed in Figures 7 and 8 of Barajas and Panton,[23] although some of the flow regimes are labeled differently from this book chapter.

In the crystallization paper by Jiang *et al.*,[2] slug flow was easily maintained with a silicone tube possessing a contact angle of 108°. As such, even the most hydrophobic surfaces are conducive to slug flow and may be preferable. Also, the hydrophobic surface leads to a more convex liquid slug, which reduces the tendency of crystals to stick to the wall.

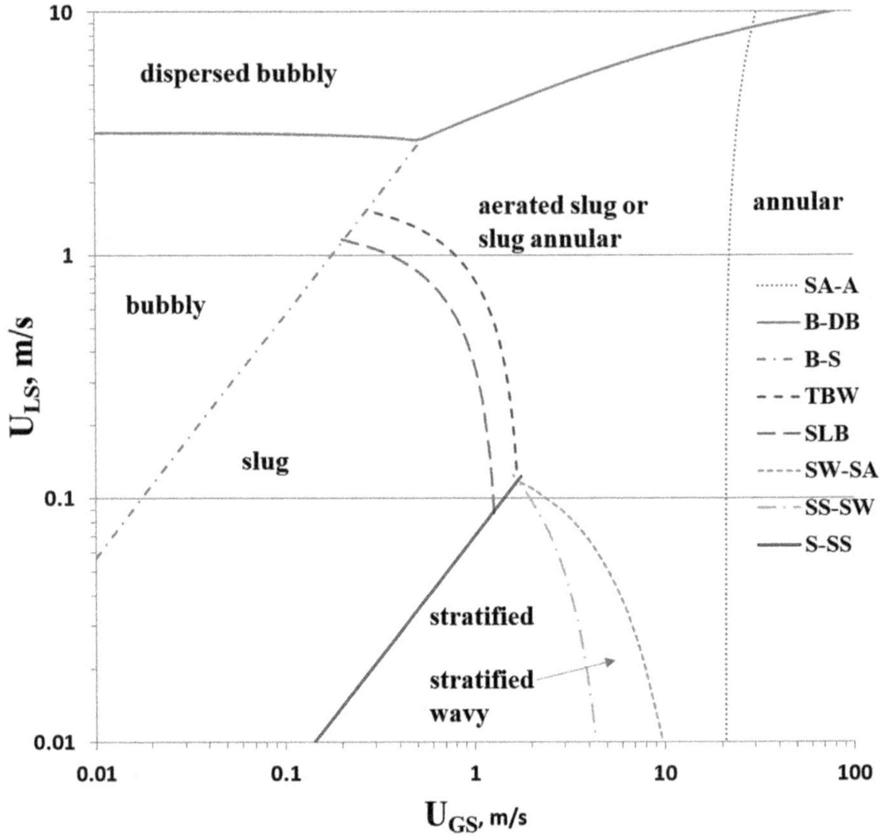

Figure 5.8 Flow regime map in horizontal 10.0 mm tube: air–water mixture with transition boundaries; Transition boundaries: SA-A, aerated slug (or slug annular) to annular; B-DB, bubbly, aerated slug, or annular to dispersed bubbly; B-S, bubbly to slug; TBW, slug to aerated slug *via* Taylor bubble wake model; SLB, slug to aerated slug *via* slug bulk model; SW-SA stratified wavy to aerated slug (or slug annular); SS-SW stratified to stratified wavy; S-SS slug to stratified.

5.2.3.5 Effect of Tubing Diameter

Flow regime diagrams, Figures 5.6, 5.7 and 5.8, were generated for three other tube diameters: 0.001097 m, 0.004 m, and 0.010 m, respectively. The smaller the tube diameter, the larger is the operational area of the slug-flow regime. According to eqn (5.7), the B-S boundary does not change with diameter *D*. In eqn (5.9), it is easy to see why the SLB curve shifts upward and to the right as diameter *D* decreases. If there were no lower limit on *D*, the use of micro-sized tubes would assure a very large operating regime with stable slug flow. For crystallization, however, there is the potential of plugging the tube if the diameter *D* is too small. If the maximum crystal diameter is d_{crys}, a rule of thumb is that the minimum value of the diameter *D* should be about $4d_{crys}$ to avoid clogging. For example, a 1 mm diameter tube would restrict maximal crystal size to 0.25 mm.

In Figure 5.8, when $D = 10$ mm, the slug-flow regime is further diminished in size by the stratified/stratified wavy flow regimes lying below the slug-flow regime. Thus, if the tube diameter is too large, the slug-flow region becomes too small to allow for flexible and robust operation. Also, the higher flow rate necessary for slug flow would require longer tubes to assure adequate residence times. The calculation of the stratified flow boundaries in the figure were made with procedures reported in Taitel and Dukler[16] and Barnea *et al.*[17] For tube diameters $D \leq 0.004$ m, the stratified flow regimes lie below $U_{LS} = 0.01$ m s^{-1} and are absent from Figures 5.5 to 5.7.

5.3 Control Slug Geometry for Recirculation

Internal recirculation is one of the key reasons for using slug flow (minimizing clogging). Internal recirculation depends on the slug size and shape (Subsection 5.3.1), whose determining factors also affect the slug stability. Flow analysis for recirculation inside slugs is detailed in Subsection 5.3.2.

5.3.1 Control Slug Size and Shape for Crystallization Purpose

The shape of slugs refers to both the size and the contact angle of slug. The slug geometry can be approximated as a cylinder (Figure 5.3), with slug length (along the tubing) the cylinder height. The slug size refers to the aspect ratio of the slug, which is the ratio between the length of slugs with respect to the inner diameter of tubing. Once the tubing size is chosen, the slug cylinder's bottom area is fixed, and the slug size is largely determined by the flow rate ratio between the air and liquid streams within the stable operation regime in Section 5.2. Generally speaking, the larger the air/liquid ratio, the smaller the slug size.

In practice, slug size has some variability. Potential causes of variability include the oscillating flow of both air and liquid from peristaltic pumps. Flow pulsations can be reduced by replacing the peristaltic pump for cold liquid solution with a syringe pump, and by implementing offsetting dual heads of the peristaltic pump for the gas flow.[2] Minimization of the slug size variability assures similar crystallization conditions among different slugs.

The slug shape affects the recirculation flow pattern, which not only provides the mixing (distribution of temperature and concentration), but also pushes particles near the tubing wall towards the tubing center, without using a mixing blade (thus minimizing attrition).[2] The closer the slug shape is to a sphere (with an aspect ratio close to 1), the smaller the volume of any dead zone and the better the mixing, favoring uniform distribution of crystals inside slugs (Figures 5.3a and b).[2]

The slug contact angle depends on the surface affinity of liquid for the inner tubing wall. When the aqueous slugs are not moving, the surface properties dominate. For example, the front and back of the slugs are flat surfaces for the Pharma-80 silicone tubing (that is, each slug is very close to a cylinder), compared to a concave shape in hydrophilic tubing materials (*e.g.*, quartz tubing) with high affinity of solvent water to the material of the inner tubing wall. In flow systems with wall dragging the liquid, the front of a moving slug shows different curvature from the concave back of the slurry slug with respect to flow direction (Figures 5.3a, b and d).

5.3.2 Flow Analysis for Recirculation within Slugs

5.3.2.1 Dimensionless Recirculation Time

The rapid internal circulation of liquid slugs in capillary flow benefits crystallization by enhanced mixing as well as limiting adherence of crystals to the wall.[2] This internal circulation has been investigated for more than 50 years.[24–27] Taylor[24] carried out some simple experiments and suggested some possible flow patterns. Later Thulasidas *et al.*[25] used particle imaging velocimetry (PIV) to visualize recirculation patterns with a high degree of mixing. Depending on the capillary number N_{Ca}, where

$$N_{Ca} = \frac{\rho_L \nu_L U_b}{\sigma}, \tag{5.11}$$

either counter-rotating flow inside the liquid slug or a complete bypass of liquid around the bubble were observed, in agreement with Taylor. For counter-rotating flow, they developed a theory for average recirculation time within the slug vortices. They defined a dimensionless recirculation time τ_{cir}, as the time for the liquid to move from one end of the slug to the other divided by a reference time. The reference time was the slug length l_s divided by the bubble velocity U_b, or l_s/U_b.

Imposing laminar conditions and using analysis from an earlier paper,[28] Thulasidas *et al.*[25] derived an expression for the dimensionless recirculation time,

$$\tau_{cir} = \frac{\psi}{1-\psi/2}, \tag{5.12}$$

where $\psi = U_b/U_m$. With the help of previous work,[28,29] Liu *et al.*[30] proposed an empirical correlation for N_{Ca},

$$N_{Ca} = 4.472\frac{(\psi-1)^3}{\psi^2}. \tag{5.13}$$

This correlation allows $\psi = U_b/U_m$ to be calculated from the capillary number, and the dimensionless recirculation time to be calculated from eqn (5.12).

For the dimensions and physical properties of the slug-flow crystallization unit of Jiang *et al.*,[2] the calculated dimensionless recirculation time *versus* N_{Ca} behavior is shown in the top curve in Figure 5.9, and $\psi = U_b/U_m$ is also plotted as the bottom curve. The plots in Figure 5.9 are confined to Reynolds numbers less than $Re_L = 2100$, as laminar conditions must hold not only for eqn (5.12) to apply, but also for the desired range of slug flow to exist.

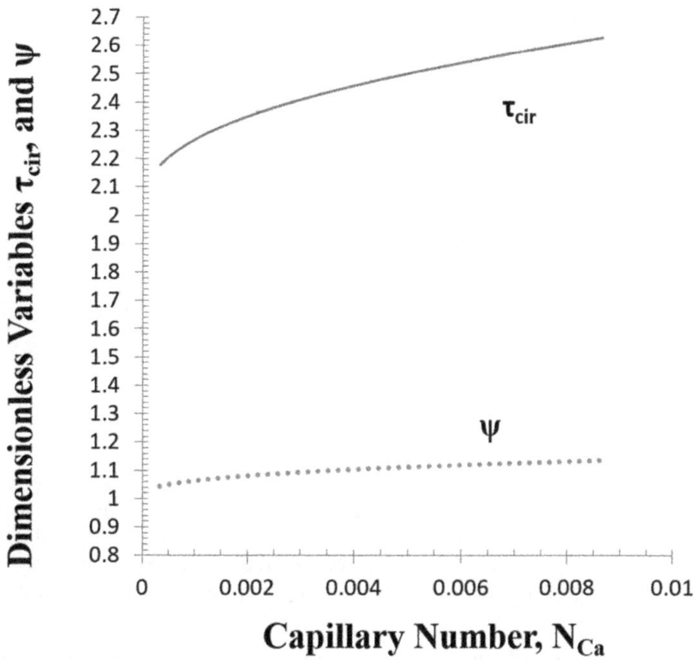

Figure 5.9 Dimensionless recirculation time τ_{cir} and dimensionless bubble velocity $\psi = U_b/U_m$ *versus* Capillary number: Upper limit of Capillary number corresponds to a total superficial velocity U_m yielding a Reynolds number $Re_L = 2100$.

5.3.2.2 *Absolute Recirculation Times*

Now that dimensionless liquid slug recirculation times and bubble velocities are available for a set of system conditions and properties, only slug lengths are needed to compute the reference time l_s/U_b and then the absolute recirculation time from $\tau_{cir}l_s/U_b$. Liu *et al.*[30] proposed a dimensional correlation for slug length *versus* gas- and liquid-phase Reynolds numbers, and Su *et al.*[31] reformulated this correlation to a dimensionless form as

$$\frac{l_s}{D} = 0.3694\left(Re_L^{'}\right)^{0.585} Re_G^{-0.483}. \tag{5.14}$$

where Re'_L and Re_G are Reynolds numbers of the gas and liquid phases based on tube diameter and the respective superficial velocities:

$$Re_L^{'} = \frac{U_{LS}D}{\nu_L}, \; Re_G = \frac{U_{GS}D}{\nu_G}. \tag{5.15}$$

For superficial liquid velocities corresponding to slug-flow crystallizers seeded by upstream coaxial and radial mixers by Jiang *et al.*,[2] the absolute slug length is plotted *versus* superficial gas velocity in Figure 5.10. For the actual values of the superficial gas velocity U_{GS} used in the two experiments, the measured and calculated liquid slug lengths are given in Table 5.1. Eqn

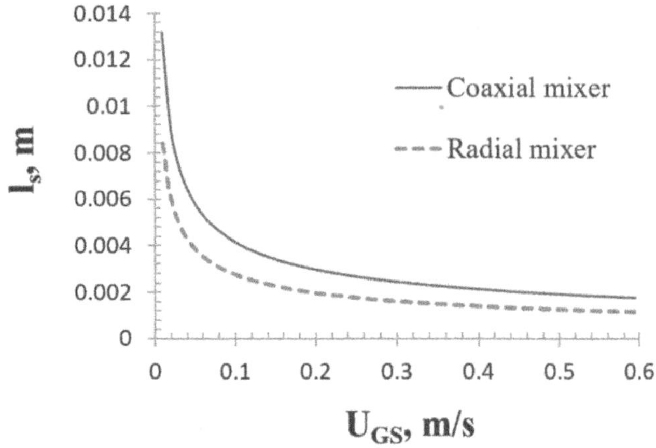

Figure 5.10 Absolute liquid slug length *versus* superficial gas velocity: U_{LS} = 0.030915 m s^{-1}, the superficial liquid velocity used in the crystallizer with coaxial mixing of hot solution with cold liquid; U_{LS} = 0.015457 m s^{-1}, the superficial liquid velocity used in the crystallizer with radial mixing of hot solution with cold liquid; Reynolds number $Re_L < 2100$.

Table 5.1 Comparison of calculated and experimental liquid slug lengths in crystallizer.

Type of mixer	U_{LS}, m s^{-1}	U_{GS}, m s^{-1}	Measured l_s, m	Predicted l_s, m
Coaxial	0.030915	0.056442	0.0060	0.00545
Radial	0.015457	0.035209	0.0040	0.00456

(5.14) gives a reasonable estimate of liquid slug length, with some underprediction for the configuration that had an upstream coaxial mixer and some overprediction for the configuration that had an upstream radial mixer.

Absolute recirculation times τ_{cir} *versus* superficial gas velocity U_{GS} are plotted in Figure 5.11 for the superficial liquid flow rates used for the coaxial and radial mixing experiments.[2] The values of τ_{cir} decline exponentially with U_{GS}, reflecting the similar decline in slug length.

5.3.2.3 *Mixing Efficiency*

To estimate the completeness of micro-mixing from the previous calculations, we introduce the micro-mixedness ratio α,[31] which is a measure of the perfectly mixed volume to the totally segregated volume in the mixing process. This ratio is defined in terms of a segregation variable X_s, which is 0 for perfect mixing and 1 for complete segregation:

$$\alpha = \frac{1 - X_s}{X_s}. \tag{5.16}$$

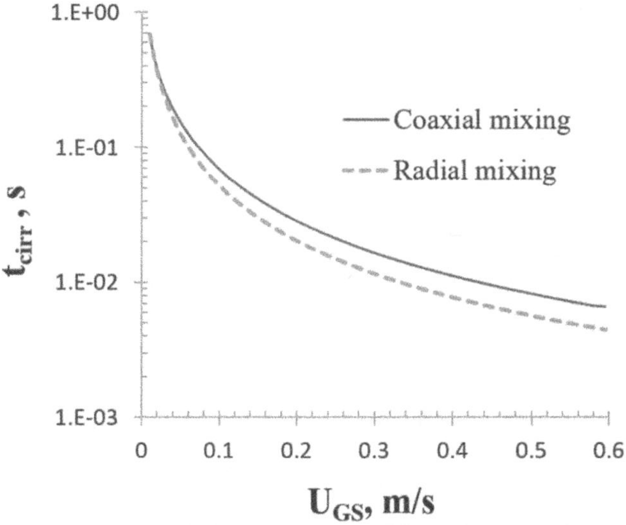

Figure 5.11 Absolute recirculation time *versus* superficial gas velocity: U_{LS} = 0.030915 m s^{-1}, the superficial liquid velocity used in the crystallizer with coaxial mixing of hot solution with cold liquid; U_{LS} = 0.015457 m s^{-1}, the superficial liquid velocity used in the crystallizer with radial mixing of hot solution with cold liquid; Reynolds number $Re_L < 2100$.

The value of α equals zero for no mixing and infinite for perfect mixing.

A correlation that has been proposed[31] for α in terms of a modified Peclet number Pe* is

$$\alpha = 0.00115(\mathrm{Pe}^*)^{1.3},$$
$$\mathrm{Pe}^* = \frac{U_\mathrm{m} D^2}{l_\mathrm{s} D_\mathrm{diff}} \frac{l_\mathrm{b} + l_\mathrm{s}}{2l_\mathrm{b} + l_\mathrm{s}}, \tag{5.17}$$

where D_{diff} is the molecular diffusivity of the solute and l_b is the average bubble length. Within the experimental slug-flow range for crystallization of LAM, Jiang *et al.*[2] observed that values of l_b were about three to four times those for l_s. Taking the 4 to 1 value, and substituting eqn (5.14) into eqn (5.17) yields

$$\alpha = 0.00115(\mathrm{Pe}^*)^{1.3} = 0.001955\left(\frac{U_\mathrm{m} D}{D_\mathrm{diff}}\right)^{1.3} (\mathrm{Re}_\mathrm{G})^{0.6279} (\mathrm{Re}'_\mathrm{L})^{-0.7605}. \tag{5.18}$$

For the superficial liquid velocity U_{LS} corresponding to the co-axial and radial mixing experiments, α is plotted *versus* U_{GS} in Figure 5.12. To assure laminar flow in the viable slug-flow region, the superficial gas velocity U_{GS} is

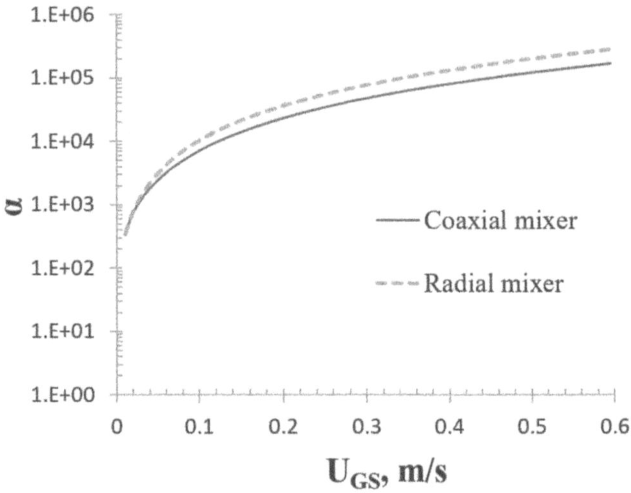

Figure 5.12 Micro-mixedness ratio α *versus* superficial gas velocity: $U_{LS} = 0.030915$ m s^{-1}, the superficial liquid velocity used in the crystallizer with coaxial mixing of hot solution with cold liquid; $U_{LS} = 0.015457$ m s^{-1}, the superficial liquid velocity used in the crystallizer with radial mixing of hot solution with cold liquid; Reynolds number $Re_L < 2100$.

restricted to satisfy $Re_L < 2100$. Note that the value of α is high even for small values of U_{GS}. The corresponding values of the segregation X_s indicate almost perfect mixing, as X_s is below 0.01 for the range of conditions reported in the continuous crystallizer.

5.4 Controlled Crystal Growth in Slugs with Temperature Zones

When the size distribution of product crystals is not satisfactory, there are ways to adjust the distribution (to some extent) during the growth stage. For a compound with temperature-dependent solubility, spatially varying temperature profiles can be applied. The temperature zones along the length of the tubular crystallizer were set with tubing submerged in heat baths[2] or heat exchangers.[32]

5.4.1 Heat Baths for T Zones

In order to narrow the size distribution and increase average size (*e.g.*, for faster filtration), temperature cycles were applied, with fast heating followed by slower cooling, to dissolve smaller crystals and grow on larger crystals (Figure 5.13).[2] Even with uncontrolled nucleation (*e.g.*, cooling in laminar flow), temperature cycles based on two heat baths (*e.g.*, using peristaltic pumps and Proportional-Integral controllers[2]) effectively narrowed the final product crystal size distribution and reduced aggregation.

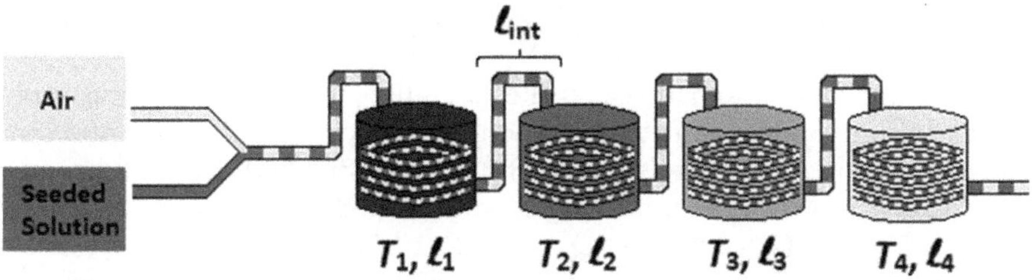

Figure 5.13 Schematic for a multiple bath SFC fed with liquid solution with seed crystals, which can be continuously generated by micromixers or an ultrasonic probe.[1,2] A typical tube is made of silicone of Teflon with an inner diameter of 3.1 mm. During cooling, a typical tank temperature ranges from 60 to 20 °C.

A well-mixed batch crystallizer can be modeled using the population balance equation[33,34]

$$\frac{\partial f}{\partial \tau}+\frac{\partial (Gf)}{\partial L}=B\delta(L-L_0), \tag{5.19}$$

$$f(\tau=0, L)=f_0(L), \tag{5.20}$$

where G and B are growth and nucleation rates, respectively, f is the distribution of particle sizes at residence time, τ, L is the particle size, L_0 is the size of nuclei, and δ is the Dirac delta function.

The tube enters the first bath of temperature T_1, where the bath is agitated to provide spatially uniform temperature and to promote heat transfer between the liquid in the bath and the outer surface of the tube. The length of tubing in the first bath is denoted by ℓ_1. The tube then passes into a second bath at a different temperature, T_2. The length of tubing in the second bath is denoted by ℓ_2, and the length of tubing in the interval between (and outside of) adjacent baths is denoted by ℓ_{int}. An arbitrary number of baths can be included in the experimental configuration.

Experimental evidence indicates that each slug is well-mixed,[2,35] so each slug operates as an individual batch crystallizer that is physically transported down the tube. For batch systems under low supersaturation, where nucleation can be considered negligible, the term on the right-hand side of eqn (5.19) can be neglected for the batch step. With common assumptions of size-independent growth and no growth rate dispersion, the population balance model describing the evolution of the crystal size distribution for each slug is reduced to

$$\frac{\partial f}{\partial \tau}+G\frac{\partial f}{\partial L}=0 \tag{5.21}$$

where τ is the time from when a slug enters the first bath. The growth rate can be defined for this system as[34]

$$G = k_g[C - C_{sat}(T)]^g \tag{5.22}$$

where k_g and g are fit to data, C_{sat} is the solubility (aka saturation concentration) as a function of temperature, and C and T are the bulk concentration and temperature, respectively. A typical slug in the first bath starts at the concentration C_0 and temperature T_0 at supersaturated or saturated conditions ($C_0 \geq C_{sat}(T_0)$), inlet seed mass m_{seed}, and inlet CSD, f_0.

Attrition, aggregation, agglomeration, breakage, and nucleation within each slug are considered to be negligible, as has been observed in experiments.[2] The low levels of these phenomena are associated with the lack of any mixing blade, static mixers, or other internals to induce such phenomena for the levels of supersaturation that occur in the experiments. The system of ordinary differential equations

$$\begin{aligned} \frac{d\mu_0}{d\tau} &= 0 \\ \frac{d\mu_1}{d\tau} &= G\mu_0 \\ \frac{d\mu_2}{d\tau} &= 2G\mu_1 \\ &\vdots \end{aligned} \tag{5.23}$$

is obtained by applying the method of moments[33,34] to the population balance model, eqn (5.21), where the kth-order moment, μ_k, is defined by

$$\mu_k(\tau) = \int_0^\infty f(\tau, L) L^k \, dL \tag{5.24}$$

and the inlet conditions are given by the moments calculated from the inlet CSD, f_0. The zeroth-order, first-order, second-order, and third-order moments are proportional to the number, total length, total surface area, and total volume of crystals in a slug. A solute balance for the system is given by

$$\frac{dC}{d\tau} = -3G\rho_c k_v \mu_2 \tag{5.25}$$

where ρ_c is the crystal density and k_v is the volumetric shape factor. Assuming that the slug is thermally well-mixed and its heat capacity $\hat{C}_p$ is spatially uniform, and that the temperature in the bath is spatially uniform, the total mass and energy balances for a slug are

$$\frac{dm}{d\tau} = 0, \tag{5.26}$$

$$m\hat{C}_p \frac{dT}{d\tau} = UA(T_i - T) + \frac{m\Delta\tilde{H}_c}{M_c}\frac{dC}{d\tau}, \tag{5.27}$$

$$T(\tau = 0) = T_0 \tag{5.28}$$

where m is the mass of a slug, T is the steady-state temperature as a function of residence time τ, U is the overall heat transfer coefficient, A is the surface area for heat transfer, T_i is the temperature of bath i, and $\Delta\tilde{H}c$ and M_c are the heat of crystallization and molecular weight of the solute molecule.

The analytical solution for the energy balance, eqn (5.27) and (5.28), in the first bath is

$$T = T_1 + \exp\left(\frac{-UA\tau}{m\hat{C}_p}\right)\left[\int_0^\tau \exp\left(\frac{UA\tau'}{m\hat{C}_p}\right)\frac{\Delta\tilde{H}_c}{\hat{C}_p M_c}\frac{dC}{d\tau}d\tau' + T_0 - T_1\right]. \tag{5.29}$$

Similarly, for downstream baths, assuming temperature does not drop during intervals between adjacent baths (the connecting tube is insulated and ℓ_{int} is minimized), the temperature can be calculated as

$$T = T_i + \exp\left(\frac{-UA(\tau-\tau_i)}{m\hat{C}_p}\right)\left[\int_{\tau_i}^\tau \exp\left(\frac{UA(\tau'-\tau_i)}{m\hat{C}_p}\right)\frac{\Delta\tilde{H}_c}{\hat{C}_p M_c}\frac{dC}{d\tau}d\tau' + T_{i-1} - T_i\right]. \tag{5.30}$$

The time for which the slug enters bath i,

$$\tau_i = \frac{1}{v}\left[\left(\sum_{j=1}^{i}\ell_{i-1}\right) + \ell_{int}(i-1)\right], \tag{5.31}$$

is calculated by assuming a constant slug velocity, v. The overall heat transfer coefficient U is composed of three parts: convective heat transfer within the slug surface, conduction through the tube wall, and convective transfer in the bulk cooling water (bath), that is,

$$\frac{1}{U} = \frac{1}{h_{slug}} + \frac{d_1 \ln\frac{d_2}{d_1}}{2k_w} + \frac{1}{h_{bath}}, \tag{5.32}$$

where d_1 and d_2 are the inside and outside diameters of the tubing, k_w is the thermal conductivity of the tube wall, and h_{slug} and h_{bath} are respective convective heat transfer coefficients. The Sieder-Tate correlation for the Nusselt number,[36]

$$\mathrm{Nu} = \frac{h_{slug}d_1}{k_s} = 0.023\mathrm{Re}^{0.8}\mathrm{Pr}^{1/3}\left(\frac{\mu_{c,s}}{\mu_{w,s}}\right), \tag{5.33}$$

can be used to determine h_{slug}, where Re and Pr are the Reynolds and Prandtl numbers, respectively, k_s is the thermal conductivity of the slug solution, and the ratio of dynamic viscosities in the last term for the slug solution at the center and wall is assumed to be unity. The value for h_{bath} can be calculated from ref. 37.

$$\mathrm{Nu} = \frac{h_{\mathrm{bath}} D}{k_{\mathrm{b}}} = 0.87 \mathrm{Re}_{\mathrm{N}}^{2/3} \mathrm{Pr}^{1/3} \left(\frac{\mu_{\mathrm{c,b}}}{\mu_{\mathrm{w,b}}} \right)^{0.14} \tag{5.34}$$

where D is the inside diameter of the agitated vessel, k_{b} is the thermal conductivity of the bath fluid (typically water), the last term is the ratio of dynamic viscosities for the bath fluid, and the Reynolds number Re_{N} for an agitated vessel is given by

$$\mathrm{Re}_{\mathrm{N}} = \frac{N \rho_{\mathrm{b}} L_{\mathrm{a}}{}^{2}}{\mu_{\mathrm{c,b}}} \tag{5.35}$$

where N is the agitator speed, L_{a} is the agitator diameter, and ρ_{b} and $\mu_{\mathrm{c,b}}$ are the density and dynamic viscosity of the bath fluid.

For an example of multi-bath system design, an optimal choice for the bath temperatures and tube lengths could be based on the objective of low maximum supersaturation level within the metastable zone, to maintain purity and avoid secondary nucleation within the liquid slugs passing through the tube. In the case of a four-bath system, the optimization is given by

$$\min_{\substack{T_1, T_2, T_3 \\ \ell_1, \ell_2, \ell_3, \ell_4}} \; w_1 \max\{0, [C_{\mathrm{f}} - C_{\mathrm{sat}}(T_4)]\} + w_2 S_{\max} + w_3 \sum_i \ell_i \tag{5.36}$$

where the first term compares the final concentration C_{f} in the system to $C_{\mathrm{sat}}(T_4)$, the saturation concentration at the final temperature, to force high yield. In the second term, $S_{\max}$ is the maximum supersaturation within the system. The third term is the total length of tubing. The values of weights, w_i, are dependent upon the experimental system.

5.4.2 Heat Exchangers for T Zones

An alternative system replaces the constant-temperature baths with counterflow single-pass double-pipe heat exchangers (Figure 5.14). With the shell-side temperature for each heat exchanger held constant, $T_{c,i}$, the length of tubing, ℓ_i, and cooling water flowrate, $\dot{m}_{\mathrm{c},i}$, can differ, offering degrees of freedom for design and control.[32]

The outlet temperatures of a double pipe heat exchanger can be calculated using the effectiveness defined as[38]

$$\eta = \frac{1 - \mathrm{e}^{-\alpha}}{1 - \dfrac{W_{\mathrm{c}}}{W_{\mathrm{s}}} \mathrm{e}^{-\alpha}}, \tag{5.37}$$

for $W_{\mathrm{c}} \neq W_{\mathrm{s}}$, where W_{s} and W_{c} are the heat capacity rates of slugs and cooling water ($W = \dot{m}\hat{C}_{\mathrm{p}}$), and the exponent α is defined by

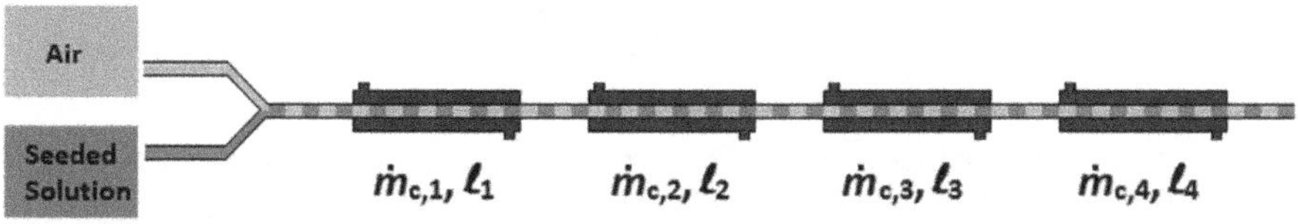

Figure 5.14 Schematic for a multiple double pipe heat exchanger SFC. The temperature of the inlet cooling water in the shell is constant, typically 25 °C or lower.

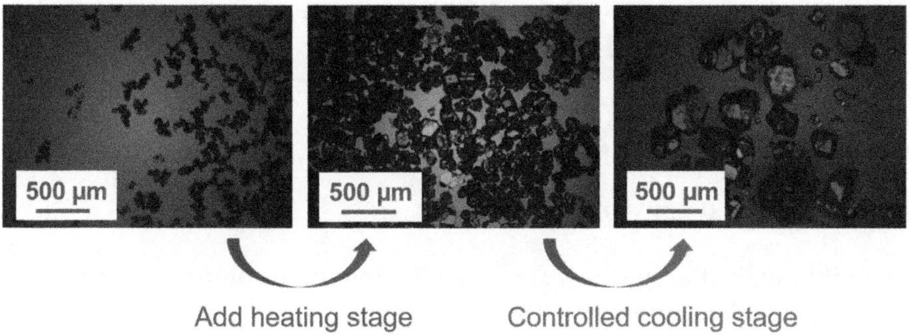

Figure 5.15 Microscope images of product crystals from slug-flow crystallization without nucleation control (nucleation in laminar flow) (left) no temperature-controlled stages; (middle) adding heating stage, followed by uncontrolled cooling in air; and (right) controlled heating and cooling stage with water baths. Adapted from ref. 2 with permission from American Chemical Society, Copyright 2014.

$$\alpha = U\pi d_1 \ell \beta \left(\frac{1}{W_c} - \frac{1}{W_s} \right) \tag{5.38}$$

where U is the overall heat transfer coefficient, d_1 is the tube diameter and ℓ is the length of the tube. Then the outlet temperature of the slug stream is

$$T_{s,out} = T_{s,in} + \frac{W_c}{W_s} \eta \left(T_{c,in} - T_{s,in} \right) \tag{5.39}$$

where the inlet temperatures $T_{c,in}$ and $T_{s,in}$ of the cooling water and slug stream, respectively, are specified. For $W_c = W_s$, the application of L'Hopital's rule to eqn (5.37) gives the expression

$$\lim_{W_c \to W_s} \eta = \frac{U\pi d_1 \ell \beta}{U\pi d_1 \ell \beta + W_s}. \tag{5.40}$$

If curvature caused by the non-constant ΔT_{LM} is neglected, a linear approximation to the temperature in the tube as a function of distance x from the heat exchanger entrance can be derived as[32]

$$T_s(x) = T_{s,in} - \frac{T_{s,out} - T_{s,in}}{\ell}x. \tag{5.41}$$

For a given inlet temperature of cooling water and total length of tubing, the lengths of tubing in each heat exchanger can be chosen to minimize the supersaturation over the total length of tube while also maximizing the total yield. In an analogous example to the design of the multi-bath system above (Eq. (5.36)), design of a slug-flow system with four double pipe heat exchangers in series would use the optimization:

$$\begin{aligned} &\min_{\substack{\ell_1,\ell_2,\ell_3 \\ \dot{m}_{c,1},\dot{m}_{c,2},\dot{m}_{c,3},\dot{m}_{c,4}}} \left[S_{max} + \varepsilon_1 \max\{C_f - C_{sat}(T_{c,in}),0\}\right] \\ &0 < \ell_i \le \ell_{total},\ i = 1,\ldots,n \\ &\ell_n = \ell_{total} - \sum_i^{n-1} \ell_i \\ &0 < \dot{m}_{c,i} \le \dot{m}_{c,max},\ i = 1,\ldots,n. \end{aligned} \tag{5.42}$$

where the value of ε_1 specifies the tradeoff between maximum supersaturation and yield in a system with fixed tube length.

5.5 Controlled Nucleation before Slug Formation

Two of the most effective methods are discussed for continuous nucleation of uniform non-aggregating seeds, with detailed design criteria and operational suggestions (Figure 5.15).

5.5.1 Micromixers

Micromixers were well studied and commonly used to generate uniform particles/crystals from combining solution streams for antisolvent or reaction crystallizations.[39,40] In continuous crystallization, these slurry-borne crystals will then go through further growth towards target size and shape.

Recently, laminar-regime micromixers were also developed for cooling crystallization, which combines two solutions of hot and cold temperature for generating crystals.[41–43] High local supersaturation (much higher than the average supersaturation) can be obtained near the interface between hot and cold solutions when the thermal diffusivity is much larger than the molecular diffusivity, as experimentally demonstrated and theoretically explained.[41,42] Cooling micromixers of various configurations (*e.g.*, Figure 5.2a) were tested, resulting in crystals of improved size uniformity and reduced aggregation, compared to direct cooling of hot solution in tubing (poor mixing in laminar flow).[2]

Design criteria based on physicochemical properties for cooling micromixers suitable for slug-flow crystallization are summarized below, with suggestions included for controlled nucleation:[42]

- The Lewis number $\alpha/D \gg 1$, so that the heat transfer is much faster than mass transfer, resulting in a supersaturation sufficiently high to nucleate crystals even at low spatially averaged supersaturation;
- The solute concentration on the hot side of the interface should be as high as allowed for chemical stability of the solute molecule, in order to increase the probability of nucleation;
- The residence time of solution can be increased by decreasing the inlet jet velocity, and/or increasing the distance between the two jets.

5.5.2 Sonication

Ultrasonication has been widely used to facilitate crystallization processes, including continuous crystallization. The sonication intensity is intrinsically non-uniform in spatial distribution (*e.g.*, with exponential decay along distance[44]), making it very difficult to narrow product crystal size distribution while avoiding contact contamination. A recent sonication-aided nucleation design[1,44] narrowed the sonication time and intensity distribution, and was used in a continuous slug flow crystallization process. In the design, the nucleation zone is localized to only a small portion of the tubing ("sonication zone", right under the probe tip, Figures 5.2a and b) which receives the highest sonication intensity.[44] Additional control can be achieved by increasing the degree of freedom such as the sonication amplitude.[1]

Below are main justifications and operational considerations for the focused indirect sonication design for the slug-flow crystallization process:[1,44]

- The inlet air and liquid flow rates are selected so that stable slugs form spontaneously and there is enough time for the liquid solution to experience cavitation (this time is the *ultrasonication residence time*).[1] For example, the mass flow rate of the inlet liquid solution in a designed experiment was 4.03 g min^{-1} with a linear velocity of 0.89 cm s^{-1}, and ultrasonication residence time of 3 seconds.[1] A reasonable sonication amplitude was chosen, so that the energy was enough for nucleation induction, but not so high that the system temperature would increase and dissolve some nuclei.[1]
- The probe tip was pressed tightly against the tubing wall to minimize the distance between the fluid in the tubing and the probe tip to maximize acoustic energy intensity. A tube outer diameter (~6 mm) was chosen to be smaller than the tip diameter of the sonication probe (~10 mm), to ensure that all of the fluid within the tube right under the probe is under high-intensity sonication.[44]

- The fluid outside the tube, water, is selected to have a high thermal diffusivity so that heat generated during sonication is conducted away from the region of high-intensity sonication, to limit the increase in local temperature. Any fluid with high thermal diffusivity could be used in place of water.[44]

5.6 Conclusions and Future Perspectives

This book chapter reviews the current status of segmented/slug-flow continuous crystallization, with detailed elaboration and demonstration for an advanced slug-flow cooling crystallization process and its associated process intensification strategies. While slug-flow crystallization is a continuous process, each slug is like an individual batch crystallizer, with past knowledge and experience readily applicable. The slug formation, nucleation, and growth processes are decoupled for individual control of each phenomenon. For the slug formation process, the effect of slug stability and geometry on crystallization outcome was analyzed, together with corresponding design and operational parameters and flow fields. The growth process in slug flow has been controlled from both experiments and modeling perspectives, with temperature zone designs set by heat baths or heat exchangers. The nucleation process in laminar flow (before slug formation) focuses on two recent designs: cooling micromixers and focused indirect ultrasonication. These advanced growth control and continuous nucleation designs work for other continuous flow systems as well.

The slug flow and continuous crystallization process is expected to continue to receive high interest.[1] One improvement would be to employ fully automated startup and shutdown, in addition to quasi-steady operations. Also of interest is providing simultaneous control of multiple crystal properties (*e.g.*, size distribution, shape, and polymorphic identity), especially in the presence of process disturbances and variations in crystallization kinetics (*e.g.*, due to changes in the contaminant profile in the feed streams). The design would benefit from advanced process monitoring and controls[45] (with suitable in-line process analytical technology[46,47]) and predictive models[47–49] based on deeper mechanistic understanding of the crystallization phenomena.[50]

Roman Symbols

	Property	Value
D	diameter of tube (m)	0.0031
D_{diff}	diffusivity of LAM ($m^2\ s^{-1}$)	10^{-9}
g	gravitational acceleration ($m\ s^{-2}$)	9.80665
L_{tube}	tube length (m)	15.2

Greek Symbols

	Property	Value
ν_L	kinematic viscosity of liquid ($m^2\ s^{-1}$)	9.00×10^{-7}
ν_G	kinematic viscosity of gas ($m^2\ s^{-1}$)	1.55×10^{-5}
ρ_L	density of liquid ($kg\ m^{-3}$)	1000
ρ_G	density of gas ($kg\ m^{-3}$)	1.1684
σ	surface tension ($N\ m^{-1}$)	7.20×10^{-2}

References

1. M. Jiang, C. D. Papageorgiou, J. Waetzig, A. Hardy, M. Langston and R. D. Braatz, *Cryst. Growth Des.*, 2015, **15**, 2486–2492.
2. M. Jiang, Z. Zhu, E. Jimenez, C. D. Papageorgiou, J. Waetzig, A. Hardy, M. Langston and R. D. Braatz, *Cryst. Growth Des.*, 2014, **14**, 851–860.
3. R. J. P. Eder, S. Schrank, M. O. Besenhard, E. Roblegg, H. Gruber-Woelfler and J. G. Khinast, *Cryst. Growth Des.*, 2012, **12**, 4733–4738.
4. R. Vacassy, J. Lemaître, H. Hofmann and J. H. Gerlings, *AIChE J.*, 2000, **46**, 1241–1252.
5. P. Neugebauer and J. G. Khinast, *Cryst. Growth Des.*, 2015, **15**, 1089–1095.
6. S. Guillemet-Fritsch, M. Aoun-Habbache, J. Sarrias, A. Rousset, N. Jongen, M. Donnet, P. Bowen and J. Lemaître, *Solid State Ionics*, 2004, **171**, 135–140.
7. T. Yonemoto, M. Kubo, T. Doi and T. Tadaki, *Chem. Eng. Res. Des.*, 1997, **75**, 413–419.
8. C. X. Zhao, L. He, S. Z. Qiao and A. P. J. Middelberg, *Chem. Eng. Sci.*, 2011, **66**, 1463–1479.
9. P. Moschou, M. H. J. M. De Croon, J. Van Der Schaaf and J. C. Schouten, *Rev. Chem. Eng.*, 2014, **30**, 127–138.
10. S. Kudo and H. Takiyama, *J. Chem. Eng. Jpn.*, 2012, **45**, 305–309.
11. T. McGlone, N. E. B. Briggs, C. A. Clark, C. J. Brown, J. Sefcik and A. J. Florence, *Org. Process Res. Dev.*, 2015, **19**, 1186–1202.
12. M. O. Besenhard, P. Neugebauer, C. Da-Ho and J. G. Khinast, *Cryst. Growth Des.*, 2015, **15**, 1683–1691.
13. J. Schiewe and B. Zierenberg, Process and apparatus for producing inhalable medicaments, *US Pat.*, 20030015194 A1, 2003.
14. R. J. P. Eder, S. Radl, E. Schmitt, S. Innerhofer, M. Maier, H. Gruber-Woelfler and J. G. Khinast, *Cryst. Growth Des.*, 2010, **10**, 2247–2257.
15. A. Ullmann and N. Brauner, *Multiphase Sci. Technol.*, 2007, **19**, 49–73.
16. Y. Taitel and A. E. Dukler, *AIChE J.*, 1976, **22**, 47–55.
17. D. Barnea, Y. Luninski and Y. Taitel, *Can. J. Chem. Eng.*, 1983, **61**, 617–620.
18. D. Barnea and N. Brauner, *Int. J. Multiphase Flow*, 1985, **11**(1), 43–49.
19. C. A. Damianides and J. M. Westwater, *Proceedings 2nd UK National Conference On Heat Transfer*, Glasgow, Scotland, 1988, vol. 2, pp. 1257–1268.
20. K. A. Triplett, S. M. Ghiaasiaan, S. I. Abdel-Khalik and D. L. Sadowski, *Int. J. Multiphase Flow*, 1999, **25**(3), 377–394.

21. A. Ullmann and N. Brauner, *Multiphase Sci. Technol.*, 2004, **16**, 355–387.
22. N. Brauner and A. Ullmann, *Int. J. Multiphase Flow*, 2004, **30**, 273–290.
23. A. M. Barajas and R. L. Panton, *Int. J. Multiphase Flow*, 1993, **19**, 337–346.
24. G. I. Taylor, *J. Fluid Mech.*, 1961, **10**, 161–165.
25. T. C. Thulasidas, M. A. Abraham and R. L. Cerro, *Chem. Eng. Sci.*, 1997, **52**, 2947–2962.
26. M. H. Kashid and L. Kiwi-Minsker, *Ind. Eng. Chem. Res.*, 2009, **48**, 6465–6485.
27. S. Kececi, M. Worner, A. Onea and H. S. Soyhan, *Catal. Today*, 2009, **147S**, S125–S131.
28. T. C. Thulasidas, M. A. Abraham and C. L. Cerro, *Chem. Eng. Sci.*, 1995, **50**(2), 183–199.
29. D. A. Reinelt, *J. Fluid Mech.*, 1987, **175**, 557–565.
30. H. Liu, C. O. Vandu and R. Krishna, *Ind. Eng. Chem. Res.*, 2005, **44**, 4884–4897.
31. Y. Su, G. Chen and Q. Yuan, *AIChE J.*, 2012, **58**(6), 1660–1670.
32. M. L. Rasche, M. Jiang and R. D. Braatz, *Comput. Chem. Eng.*, 2016, **95**, 240–248.
33. H. M. Hulbert and S. Katz, *Chem. Eng. Sci.*, 1964, **19**(8), 555–574.
34. A. D. Randolph and M. A. Larson, *Theory of Particulate Processes: Analysis and Techniques of Continuous Crystallization*, Academic Press, New York, 2nd edn, 1974.
35. M. N. Kashid, L. Gerlach, S. Goetz, J. Franzke, J. F. Acker, F. Platte, D. W. Agar and S. Turek, *Ind. Eng. Chem. Res.*, 2005, **44**(14), 5003–5010.
36. E. N. Sieder and G. E. Tate, *Ind. Eng. Chem. Res.*, 1936, **28**(12), 1429–1435.
37. T. H. Chilton, T. B. Drew and R. H. Jenkins, *Ind. Eng. Chem. Res.*, 1944, **36**(6), 510–516.
38. W. M. Kays and A. L. London, *Compact Heat Exchangers*, McGraw-Hill, New York, 3rd edn, 1984.
39. A. S. Myerson, *Handbook of Industrial Crystallization*, Butterworth-Heinemann, Woburn, MA, 2nd edn, 2002.
40. H.-H. Tung, E. L. Paul, M. Midler and J. A. McCauley, *Crystallization of Pharmaceuticals: An Industrial Perspective*, Wiley, Hoboken, NJ, 2009.
41. M. Jiang, M. H. Wong, Z. Zhu, J. Zhang, L. Zhou, K. Wang, A. N. Ford Versypt, T. Si, L. M. Hasenberg, Y. E. Li and R. D. Braatz, *Chem. Eng. Sci.*, 2012, **77**, 2–9.
42. M. Jiang, C. Gu and R. D. Braatz, *Chem. Eng. Process.*, 2015, **97**, 187–194.
43. M. Jiang, Y.-E. Li, H.-H. Tung and R. D. Braatz, *Chem. Eng. Process.*, 2015, **97**, 242–247.
44. M. Jiang, C. Gu and R. D. Braatz, *Chem. Eng. Process.*, 2017, **117**, 186–194.
45. Z. K. Nagy and R. D. Braatz, *Annu. Rev. Chem. Biomol. Eng.*, 2012, **3**, 55–75.
46. L. L. Simon, H. Pataki, G. Marosi, F. Meemken, K. Hungerbuhler, A. Baiker, S. Tummala, B. Glennon, M. Kuentz, G. Steele, H. J. M. Kramer, J. W. Rydzak, Z. Chen, J. Morris, F. Kjell, R. Singh, R. Gani, K. V. Gernaey, M. Louhi-Kultanen, J. Oreilly, N. Sandler, O. Antikainen, J. Yliruusi, P. Frohberg, J. Ulrich, R. D. Braatz, T. Leyssens, M. Von Stosch, R. Oliveira,

R. B. H. Tan, H. Wu, M. Khan, D. Ogrady, A. Pandey, R. Westra, E. Delle-Case, D. Pape, D. Angelosante, Y. Maret, O. Steiger, M. Lenner, K. Abbou-Oucherif, Z. K. Nagy, J. D. Litster, V. K. Kamaraju and M. Sen Chiu, *Org. Process Res. Dev.*, 2015, **19**, 3–62.

47. K. A. Powell, A. N. Saleemi, C. D. Rielly and Z. K. Nagy, *Chem. Eng. Process.*, 2015, **97**, 195–212.
48. G. Power, G. Hou, V. K. Kamaraju, G. Morris, Y. Zhao and B. Glennon, *Chem. Eng. Sci.*, 2015, **133**, 125–139.
49. H. Zhang, J. Quon, A. J. Alvarez, J. Evans, A. S. Myerson and B. Trout, *Org. Process Res. Dev.*, 2012, **16**, 915–924.
50. M. Jiang and R. D. Braatz, *CrystEngComm*, 2019, **21**, 3534–3551.
51. M. Jiang and R. D. Braatz, *Chem. Eng. Technol.*, 2018, **41**, 143–148.

CHAPTER 6

Continuous Crystallization of Bulk and Fine Chemicals

MATTHIAS KIND*

Karlsruhe Institute of Technology, Department of Chemical and Process Engineering, Kaiserstr. 12, Karlsruhe, D-76131, Germany
*E-mail: matthias.kind@kit.edu

6.1 Introduction

Although, this chapter is about continuous crystallization and precipitation of bulk and fine chemicals, it must be admitted that bulk chemicals and crystallization thereof are no well-defined terms. Here, we understand the production of "bulk chemicals" as the large-scale production of several hundred kilograms per hour to several tens of tons per hour. In this context, crystallization of bulk chemicals is the production of particulate crystalline solid as bulk material at production rates of the said order of magnitude. Compared to batch-operation such productions are closer to optimal when designed to operate continuously,[1] and when all of the integrated unit-operations are continuously operated as well. This is particularly true for the crystallization step.

In contrast, the term "fine chemicals" is rather well defined: Fine chemicals are complex, single, pure chemical substances. They are produced mainly by traditional organic synthesis in multipurpose plants in limited tonnage (<1000 t/a) and at relative high prices according to exacting specifications. Fine chemicals are used as starting materials for specialty chemicals, particularly pharmaceuticals and agrochemicals.[2] In the context of this chapter

The Handbook of Continuous Crystallization
Edited by Nima Yazdanpanah and Zoltan K. Nagy

Published by the Royal Society of Chemistry, www.rsc.org

crystallization of fine chemicals shall address the particular situation of multipurpose plants. For such limited tonnage of fine chemicals multipurpose plants are used, because it is rather difficult to justify the construction of a continuous plant comprised of small vessels and pipes.[1] Therefore, batch and semi-batch or semi-continuous multi-purpose processes are utilized often when production rates are small. Furthermore, batch and semi-batch or semi-continuous multi-purpose processes are employed when the product demand is intermittent, especially when the demand for a chemical is interspersed with the demand for one or more other products and the quantities needed and the timing of the orders are uncertain. Furthermore, design teams would often opt for batch or semi-continuous processes when the chemicals are hazardous or toxic or when safety aspects were of great concern.

Within a production process of bulk chemicals, crystallization is only one step of an entire train of unit operations. Due to several reasons discussed later, this crystallisation step may be carried out in a single crystallizer or in a group of crystallizers. Often, the feedstock to the crystallizer(s) is provided by a number of up-stream operations like chemical reactions, mechanical or thermal separation and pre-concentration. Down-stream from the crystallizer(s) the next unit-operations are solid–liquid separation, drying, classification, storage and shipment. A number of rather considerable recycle streams, like mother liquor from the solid–liquid separation, dust from the dryer or reject from the classifier (fines and crushed coarse particles) may have to be considered.

Typical examples for such large-scale bulk chemicals are table salt (NaCl from brine) or fertilizers (*e.g.* KCl from brine or $(NH_4)_2SO_4$ from caprolactam synthesis). Also some organics are crystallized continuously, like adipic acid (DuPont, BASF) in evaporative cooling crystallizers,[3] melamine (DSM/Stamicarbon) in a vacuum crystallizer,[4] or p-xylene (Krupp-Koppers, Maruzen Oil, Sohio/BP, Arco Technology, BP/Lummus) in at least one scraped surface crystallizer.[5] Further examples include the crystallization of ferrous sulfate, $FeSO_4$, from spent pickle fluid in the steel industry. For an exemplary long-list, see ref. 6. Furthermore, there are numerous "hidden" crystallization processes in the chemical and process industries, which are related to side-products or waste streams. Take for example the solids formation in the reboiler of a distillation column or in the sump of an absorber.

Compared to this multitude of continuous crystallization processes, such processes can only be found in the fine chemical industry if the production is carried out in a dedicated plant, or if certain product properties cannot be obtained by batch, semi-batch or semi-continuous processes. Such properties may be a particular structure of the crystalline bulk material (porosity, purity, crystallinity, polymorphic structure or particle size distribution). In this case a steady state continuous crystallization process may be advantageous because a certain window of operation may be reached and kept, which could not be obeyed in a batch process with permanently changing process conditions.

Continuous precipitation is found as well in the bulk as in the fine chemicals industry. One bulk chemicals example is the continuous hydrolysis in the production of titanium dioxide, TiO_2, in the sulfate process. During this process a hierarchy of primary to quaternary nanoparticles forms.[7] These hydrolysates are further processed to the final product, the white pigment. In fine chemicals production, the continuous precipitation of beta-carotene can be mentioned.[8] This process is fast and mixing controlled. Therefore, a mixing nozzle is employed to produce bioavailable nano-particles of beta-carotene.

6.2 Recommended General Literature

In the following, some selected literature is recommended which is of general interest to the field of continuous crystallization.

Until the beginning of the 20th century most large scale crystallization processes were done batch- or semi-batch-wise in pans. Industrialization found technical solutions for efficiently evaporating large amounts of water (steam engine) in tube-and-shell boilers. This concept was adapted to evaporators and last but not least to evaporative crystallizers in the process industry. One of the first books systematically presenting the various designs of built industrial crystallizers was *Kristallisation* by Matz.[9] It contains a number of rather detailed technical sketches of crystallizers. Many of these apparatuses are best operated at constant liquid level and are an integral part of the process steam net of a production plant. Presumably, these advantages were the driving reasons for developing continuous crystallization processes.

Nevertheless, no good understanding about the formation of the particle size distribution was common until the seminal book *Theory of Particulate Processes* by Randolph and Larson.[10,11] In this book not only a thorough treatment of the foundation of the theory of population balances is given, but also are presented a number of examples of high interest and value for continuous crystallization processes. The concept of *Mixed Suspension Mixed Product Removal (MSMPR)* is explained and extended to classified product removal, to fines dissolution and to non-steady state operation, see also Chapter 2 by Z. Nagy. Later on, of large influence for the scientific and engineering community was (and is) the book *Crystallization* by Mullin.[12] It covers many relevant aspects from crystallography, thermodynamics of solid–liquid equilibria and kinetics to technical and design considerations. Of similar scope, value and impact was the book *Precipitation* by Söhnel and Garside.[13] These books triggered tremendous advances in science, understanding, technology and know-how in the field. This advance is documented in *Crystallization Technology Handbook* by Mersmann.[14] Other handbooks with particular focus each are provided by Myerson[15] and Lewis *et al.*[16]

Exemplary summaries of the state of the art are given as well by Mullin in *Ullmann's Encyclopedia of Industrial Chemistry*,[17] as in *Perry's Chemical*

Engineer's Handbook[18] and for instance in *Thermal Separation Technology* by Mersmann *et al.*[19] Probably of high value, although in German language, is *Kristallisation in der industriellen Praxis* by Hofmann (ed.).[20]

Over the years, the proceedings of the triannual *International Symposium on Industrial Crystallization (ISIC)* are a good documentation of the advances made.[21–36]

6.3 Challenges

Important challenges in industrial bulk processes are energy efficiency, space time yield, yield, purity and product properties and last but not least the compatibility to the process as a whole. To all of them continuous crystallization offers established solutions.

If it comes to energy efficiency, then continuous crystallization is the stronger variant compared to batch and semi-batch, because heat transfer processes can be designed for low temperature differences (minimum exergy loss), and because efficient measures for saving heat are applicable (multi-effect evaporation, vapour recompression, integration into a heat transfer network which may be optimized according to pinch-point analysis[37]).

Space-time-yield (production rate per installed volume) is essential for minimizing the investment costs. Unproductive pause time and costly storage volume with perhaps even more costly product inventory can be avoided in continuous processes. Furthermore, the solids fraction (suspension density = mass of crystals per volume of suspension) may be set up to its technological limit, and hence, the crystallizer volume may be minimized further.

Yield and purity often go hand in hand. In continuous crystallization processes, a proven measure for increasing the yield is to install recycle streams which in turn increase the impurity level in the system. The maximum impurity level in the system is reached, when the resulting impurity level in the product is no longer acceptable. For continuous crystallization, the distribution coefficient between the impurity level in the crystal related to the impurity level in the mother liquor is favourable due to the well-controlled steady state conditions.

The particle size distribution is a decisive product property. Continuously operated well mixed crystallizers are in principle not good in this respect. Nevertheless, particular measures are well known for tackling this drawback: One such measure is sieving to fulfill the customer's demand for a narrow size distribution. Other measures are fines dissolution, classified product removal or cascading.

These solutions to the various challenges are not only applicable for continuous crystallization of bulk chemicals, but for continuous crystallization of fine chemicals, too.

6.4 Fundamentals

6.4.1 Solubility, Supersaturation and Particle Size

The leading property of bulk crystalline material is its particle size (distribution). Obviously, a high nucleation rate will lead to fine material and *vice versa* a low nucleation rate is required for coarse product. Experience teaches that sparingly soluble substances lead to fine material and only rather well soluble substances can be crystallized to coarse bulk material. The reason behind this observation lies in the definition of supersaturation: The driving force for nucleation is the difference in chemical potential between the material in the dissolved state and the crystalline state. By some rearrangement and simplification this difference may be expressed as supersaturation ratio S_A of the crystallizing substance A

$$S_A = \frac{x_A}{x_A^*}, \tag{6.1}$$

with x_A being the mass fraction of A in the solution and x_A^* being the mass fraction of A in equilibrium with its crystalline phase, *i.e.* the solubility of A. From eqn (6.1) it is obvious that rather high supersaturation is obtained at low solubility and low supersaturation is obtained at high solubility. Hence, well soluble substances are crystallized at low nucleation rates compared to sparingly soluble substances and thus become coarse.[38] This is the explanation for the observation that crystallization of well soluble substances leads to product in the size range above 10 μm while precipitation of sparingly soluble substances leads to fine material even in the nano-range.[14]

6.4.2 Growth Rate, Particle Size, Residence Time and Crystallizer Volume

Furthermore, it is observed that continuous crystallization of well soluble substances often leads to linear growth rates G in the order of 5–10 × 10^{-8} m s^{-1}, see ref. 38. This observation allows for a crude first estimation of the required mean residence time (also called: draw-down time) for crystals of a targeted mean size $L_{4,3}$

$$\tau \equiv \frac{M_{\text{A,cryst in crystallizer}}}{\dot{M}_\text{A}} \approx \frac{L_{4,3}}{G}. \tag{6.2}$$

The mean residence time, τ, of the crystals in the crystallizer is the mean time which is available for growth of the crystals. It is defined by the mass of crystalline substance A suspended in the crystallizer, $M_{\text{A,cryst in crystallizer}}$, divided by the mass flow of A produced by the crystallizer, $\dot{M}_\text{A}$. In reality each crystal has its own residence time and, hence, there is a residence time distribution. Typical technical values for the mean residence time for a 100–500 μm sized product is in the order of 1–5 h. More exact figures have to be obtained from adequate laboratory experiments.

6.4.3 Reaction Crystallization, Precipitation and Drowning-out Crystallization

There are numerous processes, which comprise of a reaction step in direct combination with a crystallization step, see Table 6.1. In such processes the reaction time may be the determining step.

Nevertheless, there are many cases, where the reaction time is not pivotal, because of a rather high reaction rate. This is in particular true for ionic reactions. In this case, and if further on the product of this reaction in sparingly soluble, then the crystallizing substance is said to precipitate. Reaction crystallization and precipitation imply a chemical reaction between two well soluble educts producing the crystallizing substance *in situ*. If its solubility is low, then it will be highly supersaturated and "precipitate" rapidly at high nucleation rate and low particle size. An example is precipitation of $BaSO_4$ from aqueous Na_2SO_4- and $BaCl_2$-solution, where the $BaSO_4$-particles are obtained in the nanometer range.[39]

$$Na_2SO_{4,aq} + BaCl_{2,aq} \rightarrow BaSO_4 \downarrow + 2NaCl_{aq}$$

A rather similar process for producing a crystalline product at high yield is achieved when adding an anti-solvent, for instance an alcohol, to the (aqueous) solution containing the dissolved substance. The anti-solvent reduces the solubility of this dissolved substance and "drowning-out" of this substance takes place. An example for drowning-out crystallization is the crystallization of sodium sulphate using methanol as the drowning-out agent.[40]

Table 6.1 Examples of reaction crystallization processes.[19]

Homogeneous reaction
$Ba(OH)_2 + H_2SO_4 \rightleftharpoons BaSO_4 \downarrow + 2H_2O$
$BaCl_2 + Na_2SO_4 \rightleftharpoons BaSO_4 \downarrow + 2NaCl$
$AgNO_3 + KCl \rightleftharpoons AgCl \downarrow + KNO_3$
$NaClO_4 + KCl \rightleftharpoons KClO_4 \downarrow + NaCl$
$Ti(OC_2H_5)_4 + 4H_2O \xrightarrow{C_2H_5OH} TiO_2 \downarrow + 2H_2O + 4C_2H_5OH$
$MgCl_2 + Na_2C_2O_4 \rightleftharpoons MgC_2O_4 \downarrow + 2NaCl$
$Ba(NO_3)_2 + 2NH_4F \rightleftharpoons BaF_2 \downarrow + 2NH_4NO_3$
$NiSO_4 + (NH_4)_2SO_4 + 6H_2O \rightleftharpoons NiSO_4(NH_4)_2SO_4 . 6H_2O \downarrow$
Heterogeneous reaction
$Ca(OH)_2 + CO_2(g) \rightleftharpoons CaCO_3 \downarrow + H_2O$
$Ca(OH)_2 + 2HF(g) \rightleftharpoons CaF_2 \downarrow + 2H_2O$
$K_2CO_3 + CO_2(g) + H_2O \rightleftharpoons 2KHCO_3 \downarrow$
$Ca(OH)_2 + SO_2 \rightleftharpoons CaSO_3 + H_2O$

All these processes, *i.e.* reaction-, precipitation- and drowning-out-crystallization, require the mixing of two educt streams. As well in batch, semi-batch and continuous operation, this mixing step is often crucial for the quality of the product, see below.

6.4.4 Importance of Mixing and Classification

Two aspects of mixing have to be considered. One is that of mixing two educts to react with each other on the molecular level (meso- and micro-mixing). The other aspect is that of macro-mixing of the crystallizing solids in the crystallizer volume. This aspect is well known from chemical reaction technology.[41]

Solids in a continuous crystallizer may be ideally well mixed. Then the crystallizer does not exhibit spatial variation of the solids content or of the crystal size distribution. Furthermore, the exit stream may be representative. This means that the product in the crystallizer is identical in composition and particle size distribution to the product in the exiting product removal line. Typical types of such well mixed apparatuses are the ideally mixed stirred vessel and the ideally mixed forced circulation crystallizer. In principle, they are the same. Note: It is not at all an easy task to design a crystallizer (laboratory or production) for representative product removal.

In terms of macro-mixing, the other extreme of a crystallizer is an idealized plug-flow crystallizer, where all volume elements in the crystallizer exhibit the same distinct residence time. The residence time distribution is (infinitively) narrow. Typical apparatuses would be a pipe in full turbulent flow regime, a pipe with slugs of suspension segmented by an immiscible fluid (air, oil), or a cascade of an infinite number of well mixed crystallizers. All of them help in narrowing the time distribution available for the growth of the crystals. One principal problem of continuously operated plug-flow crystallizers is nucleation. If not seeded, then the locus of nucleation in such a crystallizer is not well defined. At the entrance of such an apparatus a high supersaturation builds up or already exists and gives rise to uncontrolled nucleation and in many cases to incrustation of the crystallizer internals.[42] Note: It is not at all an easy task to design a plug-flow crystallizer for no-slip between the solid and the solution. Only at no-slip are the residence time of the crystals and the solution the same.

Currently, an intense discussion takes place about the prerequisites which have to be fulfilled for making a continuously operated plug-flow crystallizer a reasonable alternative to a continuously operated well-mixed crystallizer. Besides the nucleation issue addressed above, the required residence time for crystal growth is also an issue. Eqn (6.2) holds for the plug-flow configuration, too. The required crystallizer volume, which for a given production rate and residence time is estimated from the required mass of crystals in the crystallizer (see Eqn. (6.2)) and the volumetric hold-up of crystals in the suspension, must be accommodated for as well in a linear arrangement (pipe or cascade of stirred vessels) as in the well-mixed crystallizer. This may lead to very long linear apparatuses which then are designed as coils.[43] Besides the cost issue, sedimentation of solids in these long pipes and slip between the solid and the liquid phase are problems of ongoing research and development endeavour,[44] see several of the other Chapters of this book.

Obviously, the particle size distribution of these two extreme types of crystallizers is different. In the latter case, all crystals are of the same age when they leave the crystallizer, and hence, in theory represent a perfectly narrow particle size distribution. The well mixed crystallizer exhibits a wide residence time distribution, and hence, a wide particle size distribution.

Population balance modelling teaches,[10] that even though the inventory of a crystallizer may be well mixed, either the product exit stream may be classifying, or one or more additional classifying exit stream may be installed for fines removal, mother liquor withdrawal or other purposes like removal of a second liquid phase (VLLS-system). The latter is the case in the cyclohexanone oximation of the caprolactam process.[45]

Rather common is the case that a classifying exit stream takes out from the crystallizer only crystals that are larger than a certain minimum size. This measure narrows the size distribution of the product crystals by cutting off and rejecting the fines back into the crystallizer. Also common is that a second exit stream is used to wash-out crystals below a certain size from the inventory and dissolve them (fines dissolution). This measure reduces the number of crystals in a crystallizer and hence, leads to a considerable increase of the particle size distribution, see Figure 6.1.

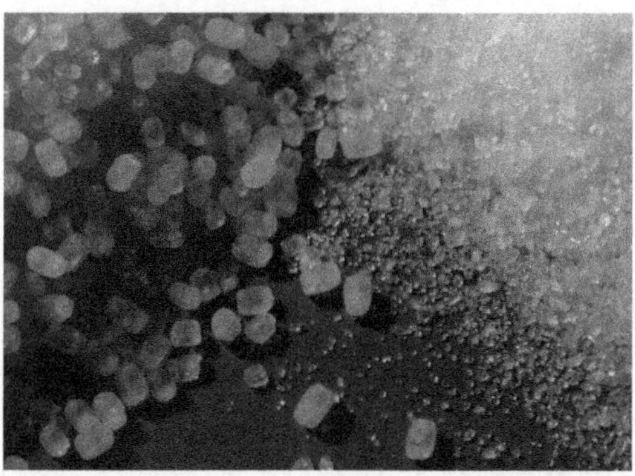

Figure 6.1 Different samples of bulk $(NH_4)_2SO_4$-crystals from continuous crystallization with fines dissolution (left) and without fines dissolution (right).

6.5 The Idealized Continuous Crystallizer– MSMPR

The simplest case of continuous crystallisation is that of a continuously operated stirred vessel at steady state. This type of crystallizer is named a MSMPR-crystallizer (mixed suspension mixed product removal) if the following conditions are fulfilled:

- crystal-free feed
- representative product removal
- uniform growth rate G, *i.e.* no size-dependent growth rate, no growth rate dispersion.

In technical reality, MSMPR-crystallizers often are designed as one or the other of the two apparatuses depicted in Figure 6.2. Both apparatuses are suited for cooling or evaporative crystallization. Only the stirred vessel is limited in its volume-specific heat transfer surface and therefore cannot be used above a crystallizer-volume of a few cubic meters. At larger volumes/production rates the forced circulation crystallizer must be used. Its heat transfer area can easily be adapted to the aimed at thermal duty.

By population balance modelling of the crystallizer, see Chapter 2, it can be shown that MSMPR-crystallizers deliver a particle size distribution of the form

$$n(L) \equiv \frac{\mathrm{d}N}{V \cdot \mathrm{d}L} = n_0 \exp\left(-\frac{L}{G \cdot \tau}\right) \tag{6.3}$$

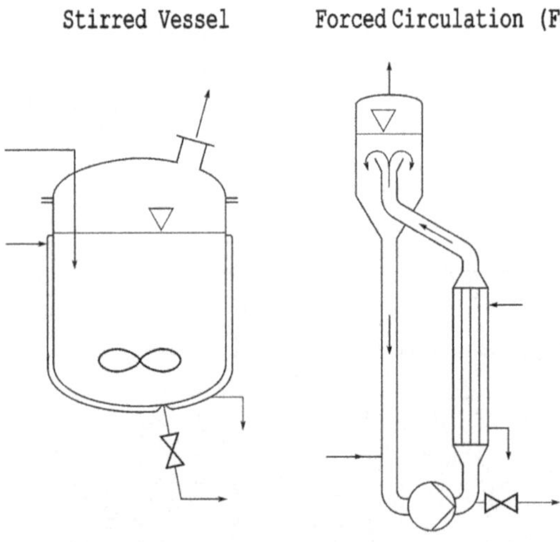

Figure 6.2 Two technical representations of well mixed crystallizers (left: stirred-vessel-crystallizer; right: forced-circulation-crystallizer (FC-)). In cases where they are operated in steady state, with crystal-free feed and with representative product removal, then both of them may be considered as evaporative MSMPR-crystallizers. Reproduced from ref. 19 with permission from Springer Nature, Copyright 2011.

with n being the number density (number of crystals dN per size class dL and volume V), which is an exponential function of crystal size L. n_0 is the population density of the crystals at $L = 0$, and G is the uniform linear growth rate of the crystals and τ their mean residence time in the crystallizer.

The absolute value of the growth rate G depends on the prevailing supersaturation, which itself is the result of the balance between creation of supersaturation by cooling or evaporation and its depletion by growth of the existing crystal surface. From the viewpoint of modelling and simulation, bulk crystallization processes can only be fully described by coupling of the population balance to the mass and component balances of the crystallizer, see Chapter 2. In turn, the mass and component balances may be coupled to the heat balance of the crystallizer.

In practice, a size distribution of a crystalline product is characterized rather by the mass density $q_3(L)$ than by the number density $n(L)$. Both characteristics of a size distribution can be converted into each other. Figure 6.3 shows the result of this conversion. For the MSMPR-crystallizer the maximum number of crystals is found at particle size $L/(G{\cdot}\tau) = 0$. The mass density distribution has its mean at much higher particle sizes[10,14] $L/(G{\cdot}\tau) = 3.67$, and its maximum at $L/(G{\cdot}\tau) = 3$. Hence, the particle distribution of any MSMPR-crystallizer is rather broad!

Although a MSMPR-crystallizer does not deliver a uniform and narrow particle size distribution it is a most valuable reference crystallization process, because it is well-defined. In the laboratory it is used for the determination of kinetic data (growth and nucleation rate) as a function of process parameters.[14,17]

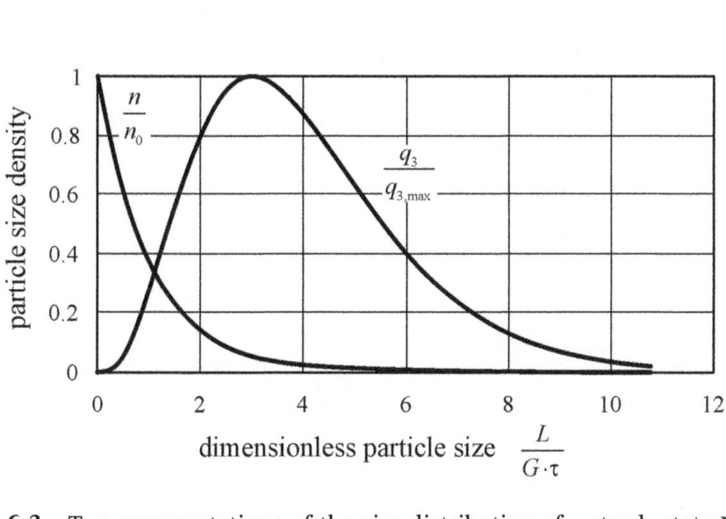

Figure 6.3 Two representations of the size distribution of a steady-state MSMPR crystallizer. The dimensionless number density n/n_0 according to eqn (6.3) has its maximum at $L/(G{\cdot}\tau) = 0$. The dimensionless mass density $q_3/q_{3,\max}$ has its maximum at $L/(G{\cdot}\tau) = 3$.

6.6 Variants of Crystallizers for Satisfying Special Product Requirements

Starting from the idealized MSMPR-case the following variants of crystallizers are possible and are widely used in practice:

6.6.1 Classified Product Removal

An "elutriation leg" may be used in the exit line for classifying the product to be taken out from the crystallizer.[46,47] Such a leg is an add-on to the crystallizer, preferably a vertical cylinder. In this cylinder an upward fluid flow is induced by saturated mother liquor entering at the lower part of the cylinder. Crystals have to sediment by gravitation against this fluid flow. Only crystals above a certain particle size successfully reach the product withdrawal-line at the bottom of the leg, see Figure 6.4. The design of this "elutriation leg" follows that of a classifying thickener.[48,49] The effect on the particle size distribution in the crystallizer must be discussed in view of the population balance.[10]

6.6.2 Fines Dissolution

Crystallizers may possess more than one product removal line. A typical technical design is that of a draft-tube-baffled-crystallizer (DTB-), see Figure 6.5. This crystallizer is equipped with an integrated axial-symmetric classifying zone (baffle). Only fine crystals follow the upward flow in the

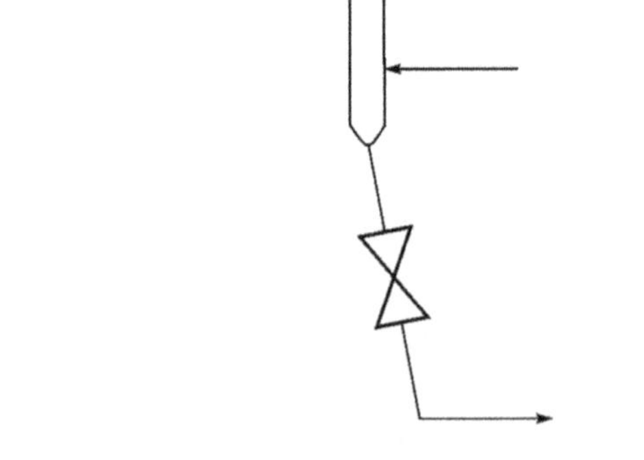

Figure 6.4 "Elutriation leg" for classified product removal.

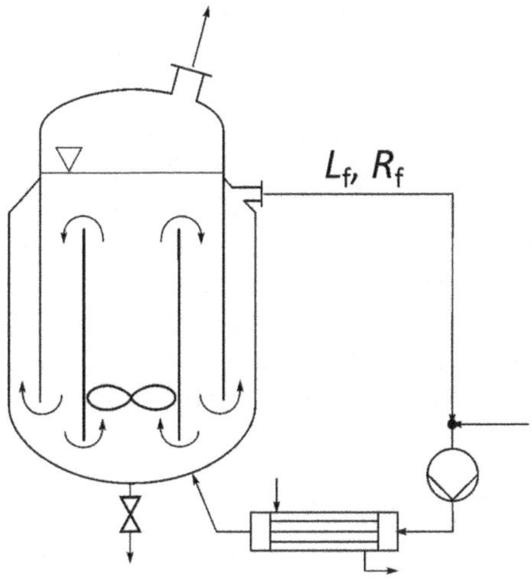

Figure 6.5 Draft-tube-baffled-(DTB-) crystallizer with fines dissolution. The cut size L_f and the flow ratio R_f are the decisive parameters for the effect of fines dissolution on the particle size distribution. Reproduced from ref. 19 with permission from Springer Nature, Copyright 2011.

baffle and are removed from the crystallizer. These fines are dissolved in the heater of the crystallizer. Besides fines dissolution this heater has to bear the thermal duty of the apparatus. From the heater, clear superheated solution is fed back into the crystallizer. Product withdrawal may be representative or classified. Essentially, the centre volume of the apparatus is a stirred vessel, which is equipped with a draft tube. The draft tube facilitates the suspending of the crystals. The two main parameters governing the effect of fines dissolution on the particle size distribution are the cut size L_f of the classifying zone and the ratio R_f of the flow rate of the fines containing stream $\dot{V}_f$ to the sum of the flow rates of the fines containing stream $\dot{V}_f$ and the product stream $\dot{V}_p$.[10]

$$R_f = \dot{V}_f/(\dot{V}_f + \dot{V}_p) \tag{6.4}$$

6.6.3 Minimisation of the Nucleation Rate

An immanent characteristic of continuous crystallizers is the steady withdrawal of particles with the product stream. For steady state operation, this steady loss in particle number has to be balanced by the nucleation rate. This rate is pivotal for the particle size distribution. A high nucleation rate leads to a fine particle size distribution and, *vice versa*, a low rate leads to coarse product. Hence, minimisation of nucleation rate is a measure for increasing the product particle size. In technical crystallizers which produce particles in the 100 μm range and above the source of nuclei are the particles themselves, because these particles experience mechanical abrasion in the apparatus, preferentially at the stirrer and in the pumps (secondary nucleation). A measure to reduce this mechanical stress is to suspend the crystals in a fluidized bed without coming into contact with agitators or pumps. A classic fluidized-bed-crystallizer design (Type OSLO) is depicted in Figure 6.6. The pump drives the crystal-free solution trough the heater in the evaporation zone. From there the supersaturated solution flows downward in the central pipe and is deflected at the bottom of the apparatus. The resulting upward flow fluidizes the crystal bed. After leaving the bed, the crystal-free solution is again sucked into the external loop by the pump.

6.6.4 Mother Liquor Advance

Eqn (6.2) provides an argument for increasing the residence time of the solids (*i.e.* draw-down time), if an increased particle size is desired. From the definition of the residence time it follows that at given production rate $\dot{M}_{\mathrm{A}}$ of crystalline compound A the mass of A in the crystallizer $M_{\mathrm{A,cryst\ in\ crystallizer}}$ must be increased. This can be done by extracting mother liquor from the crystallizer independently from the production rate $\dot{M}_{\mathrm{A}}$. In order to do so, a similar internal classification device has to be used as was shown in Figure 6.5 for the DTB-crystallizer. The only difference is that the flow of advanced mother liquor is much smaller than the circulating flow in the DTB-case.

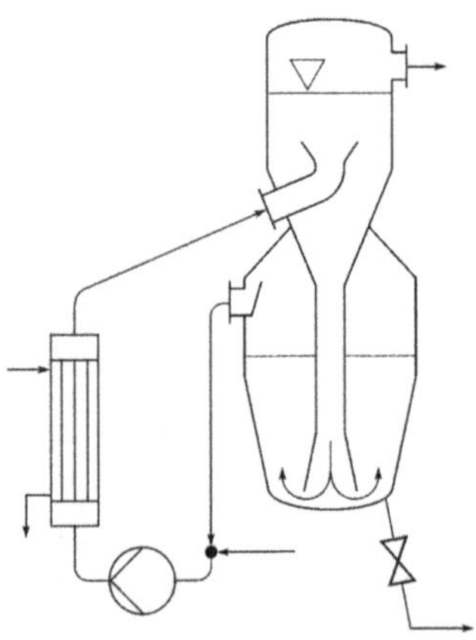

Figure 6.6 Fluidized-bed-crystallizer (Type OSLO). No crystals pass by the pump. Reproduced from ref. 19 with permission from Springer Nature, Copyright 2011.

6.7 Energy Consumption

Well known measures for saving energy in evaporation processes are multi-effect-evaporation (co-current or counter-current) or vapour recompression (thermal or mechanical).[14,49,50] These concepts are applied to crystallization processes, too. In addition, crystallization processes may be part of a larger installation, for which a global optimum related to energy consumption is sought. This optimum of the entire production system can be found by pinch-point-analysis.[37] Minimization of exergy losses by, for instance, unnecessary high temperature differences for heat transfer are part of the optimization considerations. The objective function is the economy of the entire plant. This implies weighing against each other of the cost of energy and the cost of depreciation of the investment.

6.8 Process Integration

A continuously operated crystallizer is part of a larger process with a number of recycle streams. Figure 6.7 depicts a rather simple academic example of such a plant. The recycle streams shown are necessary for maximizing the yield and keeping all process parameters within their specified range. For instance, the solid content in the crystallizer and in the solid–liquid separation section has to be chosen such that the suspension is not too "thin" and not too "thick". A typical value would be for instance 20 to 30 volume-% of solids. Furthermore, the crystallizer ought to be fed with crystal free solution. Therefore, the dissolver is required. As a rule, the feed solution contains impurities which would level-up in the process, if no solution was purged

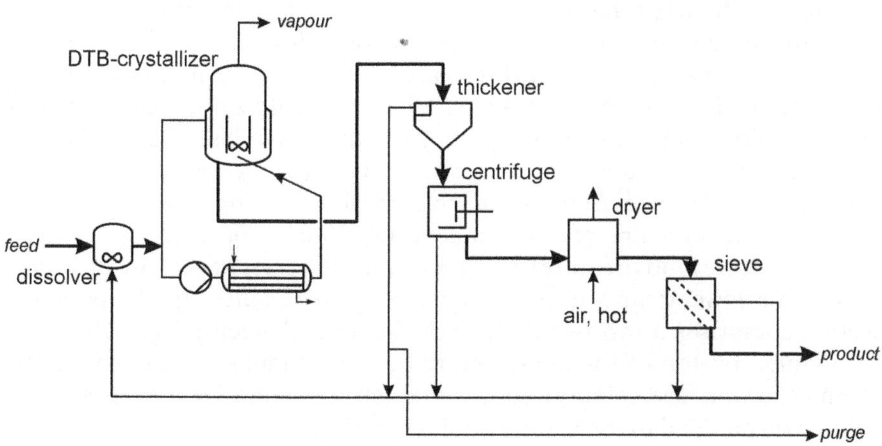

Figure 6.7 Simplified continuous bulk crystallization process comprising a DTB-crystallizer, solid–liquid separation (thickener and centrifuge), dryer (*i.e.* fluidized bed), and sieve. The solids in the recycle stream are dissolved before being fed to the crystallizer. The purge stream is required to limit the impurity level in the process.

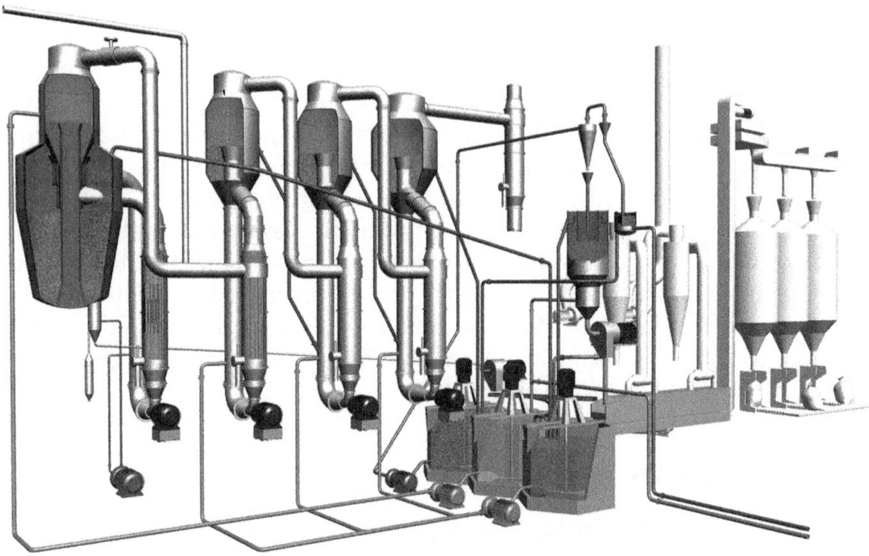

Figure 6.8 Salt recovery from solar brine (3D-sketch reproduced with permission from GEA-Messo, Duisburg (D), taken from https://www.gea.com/en/products/solution-crystallizer.jsp (accessed March 2018).

from the process. Finally, it must be considered that such a process is characterized by process elements which are creating a lag-time between some cause and the resulting effect. Such a characteristic is the underlying reason for unintended "cycling of the process". This unwanted phenomenon must be tamed by thoughtful process control strategies.

Figure 6.8 depicts a 3D-sketch of a realistic industrial plant for salt recovery from solar brine. This plant comprises one fluidized bed crystallizer for producing coarse product and three cascaded forced circulation crystallizers for producing a more fine product. All four crystallizers are connected to build a multi-effect evaporation system. The first crystallizer (left) is heated with live steam. The vapour produced is used to heat the second crystallizer and so forth. The steam from the fourth crystallizer is condensed in a jet condenser. Whereas the first crystallizer is the one which is operated at the highest temperature (and pressure), the three following crystallizers are operated at gradually lower temperatures (and pressures). The solid–liquid separation section comprises a hydro-cyclone, a thickener and a centrifuge. The sieve is mounted on top of the silos. The three stirred tanks are the collecting volumes for the feed solution, for the mother liquor and for the suspension before it is pumped to the solid–liquid separation section.

6.9 Summary

Continuous crystallization is a well-established technology for bulk chemicals. If properly designed and operated, these crystallization plants can be operated for weeks without having to be shut down due to incrustation or other mishaps. The situation is slightly different for crystallization of fine chemicals. Here, small production rates and the necessity for flexible response to market demands have led to multi-purpose, multi-product plants which are operated batch- or semi-batch-wise. One mayor disadvantage of such operation is their transient nature. During a batch- or semi-batch-crystallization the process conditions are not constant, and the first crystal is formed under rather different conditions than the last one. This gives rise to unclear product properties. Hence, for the future it can be expected that the necessity of improved product quality will be the main driver for more continuous crystallization processes to come, even in the production of fine chemicals.

References

1. W. D. Seider, J. D. Seader and D. R. Lewin, *Product and Process Design Principles: Synthesis, Analysis, and Evaluation*, John Wiley & Sons, Inc., 1999.
2. P. Pollak and R. Vouillamoz, Fine Chemicals, in *Ullmann's Encyclopedia of Industrial Chemistry*, Wiley-VCH, 2002, DOI: 10.1002/14356007.q11_q01.
3. M. T. Musser, Adipic Acid, in *Ullmann's Encyclopedia of Industrial Chemistry*, Wiley-VCH, 2005, DOI: 10.1002/14356007.a01.
4. G. M. Crews *et al.*, Melamine and Guanamines, in *Ullmann's Encyclopedia of Industrial Chemistry*, Wiley-VCH Verlag GmbH & Co. KGaA, 2012, p. 378392, DOI: 10.1002/14356007.a16_171.pub2.
5. J. Fabri, U. Graeser and T. A. Simo, Xylenes, in *Ullmann's Encyclopedia of Industrial Chemistry*, Wiley-VCH Verlag GmbH & Co. KGaA, 2012, p. 643, DOI: 10.1002/14356007.a28_433.
6. https://www.gea.com/en/products/solution-crystallizer.jsp, last accessed March 2018.
7. G. Auer, P. Woditsch, A. Westerhaus, J. Kischkewitz, W.-D. Griebler, M. Rohe and M. Liedekerke, Pigments, Inorganic, 2. White Pigments, in *Ullmann's Encyclopedia of Industrial Chemistry*, Wiley-VCH, 2002.
8. L. End, D. Horn, E. Lueddecke, J. U. Schneider, P. P. Hoppe and F.-W. Rensmann, Preparation of stable injectable beta-carotene solubilizates, *US Pat.*, US005,453,447A, 1995.
9. G. Matz, *Kristallisation: Grundlagen und Technik*, Springer, Berlin, 1969.
10. A. D. Randolph and M. A. Larson, *Theory of Particulate Processes: Analysis and Techniques of Continuous Crystallization*, Academic Press, San Diego, 1988.
11. A. D. Randolph and M. A. Larson, *Theory of Particulate Processes*, Academic Press, 1971.
12. J. W. Mullin, *Crystallization*, Butterworth-Heinemann, 1993.
13. O. Söhnel and J. Garside, *Precipitation: Basic Principles and Industrial Applications*, Butterworth-Heinemann, Oxford, 1992.
14. A. Mersmann, *Crystallization Technology Handbook*, Marcel Dekker Inc, Hoboken, 2001.
15. *Handbook of Industrial Crystallization*, ed. A. S. Myerson, Butterworth-Heinemann, Boston, 2002.

16. A. E. Lewis, M. M. Seckler, H. J. Kramer and G. van Rosmalen, *Industrial Crystallization: Fundamentals and Applications*, Cambridge University Press, Cambridge, 2015.
17. J. W. Mullin, Crystallization and Precipitation, in *Ullmann's Encyclopedia of Industrial Chemistry*, Wiley-VCH, 2002, DOI: 10.1002/14356007.b02_03.
18. W. J. Genck *et al.*, Liquid-Solid Operations and Equipment – Crystallization from Solution, in *Perry's Chemical Engineers' Handbook*, ed. D. W. Green and R. H. Perry, McGraw-Hill Professional Publishing, 8th edn, 2007.
19. A. Mersmann, M. Kind and J. Stichlmair, *Thermal Separation Technology: Principles, Methods, Process Design*, Springer-Verlag, Berlin, Heidelberg, 2011.
20. *Kristallisation in der industriellen Praxis*, ed. G. Hofmann, Wiley-VCH, Weinheim, 2005.
21. *19th International Symposium on Industrial Crystallization: ISIC 19, September 16–19, 2014 in Toulouse (F)*, ed. B. Biscans, Elsevier, Amsterdam, 2015.
22. *18th International Symposium on Industrial Crystallization: ISIC 18, September 13–16, 2011 in Zurich (CH)*, ed. B. Biscans, Elsevier, Amsterdam, 2012.
23. *17th International Symposium on Industrial Crystallization: ISIC17, September 14–17, 2008, in Maastricht (NL)*, ed. P. Jansens, EFCE, Maastricht, 2008.
24. *16th International Symposium on Industrial Crystallization,: ISIC16, September, 11–14, 2005 in Dresden (D)*, ed. J. Ulrich, VDI-Verl., Dresden, 2005.
25. *15th International Symposium on Industrial Crystallization: ISIC15, September, 15–18, 2002 in Sorrento (I)*, ed. A. Chianese, AIDIC Associazione Italiana di Ingegneria Chimica, Milano, 2002.
26. M. Hounslow, R. Davey and J. Garside, 14th International Symposium on Industrial Crystallization: ISIC14, September 12–16, 1999 in Cambridge (GB), *Chem. Eng. Sci.*, 2001, **56**(7), 2243–2588.
27. *13th International Symposium on Industrial Crystallization: ISIC13, 16–19 September 1996, Toulouse (F)*, ed. B. Biscans, ChERD, vol. 74(A7), 1996, pp. 713–876.
28. *12th International Conference on Industrial Crystallization: ISIC12, September 1993 in Warsaw (PL)*, ed. Z. H. Rojkowski, 1993.
29. *11th Symposium on Industrial Crystallization: ISIC11, September 18–20, 1990, in Garmisch-Partenkirchen (D)*, ed. A. Mersmann, VDI-Verlag, Düsseldorf, 1990.
30. *10th International Symposium on Industrial Crystallization: ISIC10, September 21–25, 1987 in Bechyně (CZ)*, ed. J. Nývlt and S. Žáček, Academia, Praha, 1989.
31. *9th International Symposium on Industrial Crystallization: ISIC9, September 25–28, 1984 in the Hague (NL)*, ed. S. J. Jančić and E. J. de Jong, Elsevier, Amsterdam, 1984.

32. *8th International Symposium on Industrial Crystallization: ISIC8, September 28–30, 1981 in Budapest (H)*, ed. S. J. Jančić, North-Holland Publishing Company, Amsterdam, 1982.
33. *7th International Symposium on Industrial Crystallization: ISIC7, September 25–27, 1978 in Warsaw (PL)*, ed. E. J. d. Jong, North Holland Publishing Company, Amsterdam, 1979.
34. *5th International Symposium on Industrial Crystallization: ISIC5, 1972*, North Holland Publishing Company, Amsterdam, 1972.
35. *6th International Symposium on Industrial Crystallization: ISIC6, September 1–3, 1975 in Ústí nad Labem (CZ)*, ed. J. W. Mullin, Plenum Press, New York, 1976.
36. *4th International Syposium on Industrial Crystallization: ISIC4, April 15–16, 1969 in London (GB)*, Institution of Chemical Engineers, London, 1969.
37. B. Linnhoff and E. Hindmarsh, *Chem. Eng. Sci.*, 1983, **38**(5), 745.
38. A. Mersmann and M. Kind, *Chem. Eng. Technol.*, 1988, **11**(1), 264.
39. R. T. Kügler, S. Doyle and M. Kind, *Chem. Eng. Sci.*, 2015, **133**, 140.
40. W. Fleischmann and A. Mersmann, Drowning-out Crystallization of Sodium Sulphate Using Methanol, in *9th International Symposium on Industrial Crystallization: ISIC9, September 25–28, 1984 in the Hague (NL)*, ed. S. J. Jančić and E. J. de Jong, Elsevier, 1984, p. 165.
41. O. Levenspiel, *The Chemical Reactor Omnibook*, lulu.com, Raleigh, NC, 2013.
42. N. E. B. Briggs, U. Schacht, V. Raval, T. McGlone, J. Sefcik and A. J. Florence, *Org. Process Res. Dev.*, 2015, **19**(12), 1903.
43. L. Hohmann, R. Gorny, O. Klaas, J. Ahlert, K. Wohlgemuth and N. Kockmann, *Chem. Eng. Technol.*, 2016, **39**(7), 1268.
44. T. McGlone, N. E. B. Briggs, C. A. Clark, C. J. Brown, J. Sefcik and A. J. Florence, *Org. Process Res. Dev.*, 2015, **19**(9), 1186.
45. J. Tinge, *et al.*, Caprolactam, in *Ullmann's Encyclopedia of Industrial Chemistry*, Wiley-VCH Verlag GmbH & Co. KGaA, 2018, DOI: 10.1002/14356007.a05_031.pub3.
46. F. Bella and R. C. Bennett, Elutriation leg and method, *US Pat.*, US4113552A, 1976.
47. T. G. Bernhardi, Hydraulic classifier, *US Pat.*, US3071447A, 1958.
48. Y. Zimmels, Sedimentation, in *Ullmann's Encyclopedia of Industrial Chemistry*, Wiley-VCH, 2002, p. 277, DOI: 10.1002/14356007.b02_12.
49. *Perry's Chemical Engineers' Handbook*, ed. D. W. Green and R. H. Perry, 8th edn, McGraw-Hill Professional Publishing, Blacklick, USA, 2007.
50. R. Billet, *Evaporation in Ullmann's Encyclopedia of Industrial Chemistry*, Wiley-VCH, 2002, DOI: 10.1002/14356007.b03_03.

CHAPTER 7

Process Intensification in Continuous Crystallization

T. STELZER*[a,b], R. LAKERVELD[c] AND A. S. MYERSON[d]

[a]University of Puerto Rico, Department of Pharmaceutical Sciences, PO Box 365067, San Juan, PR 00936, USA; [b]Crystallization Design Institute, Molecular Sciences Research Center, 1390 Ponce De León Avenue, San Juan, PR 00926, USA; [c]The Hong Kong University of Science and Technology, Department of Chemical and Biological Engineering, Clear Water Bay, Kowloon, Hong Kong; [d]Massachusetts Institute of Technology, Department of Chemical Engineering, 77 Massachusetts Ave., MA 02139, Cambridge, USA
*E-mail: torsten.stelzer@upr.edu

7.1 Introduction

Process intensification (PI) is gaining much attention in academia and industry (*e.g.*, chemical,[1] food,[2] and pharmaceutical[3]) as one of the key objectives for retrofitting existing equipment and as a progress area for new process and processing designs with the promise to significantly improve the economic and ecological efficiency of these processes by increased heat and mass transfer as well as space-time yield. Over the past 20 years, several PI definitions of this interdisciplinary topic have been proposed.[4–10] Generally, PI might be referred to as an integrated approach that replaces large, expensive, energy-intensive equipment or processes with one that is smaller, less costly, more efficient, with minimized environmental footprint,

The Handbook of Continuous Crystallization
Edited by Nima Yazdanpanah and Zoltan K. Nagy

Published by the Royal Society of Chemistry, www.rsc.org

increased safety and improved control, and/or combines multiple operations into a single or fewer devices.[5] Moreover, Hessel[9] emphasized that PI needs to be distinguished from (i) process optimization, defined as performance improvement of existing concepts, and (ii) process system engineering, defined as multi-scale integration of existing and new concepts using PI as a "provider of building blocks" from which process systems are further developed (Table 7.1).[7,9]

Beyond a mere PI definition, van Gerven and Stankiewicz[7] have defined the fundamentals of PI according to four basic principles, so called domains: 'time', 'space', 'function', and 'energy'. Wang *et al.*[11] further defined these principles with respect to the unit operation crystallization (Table 7.2) but did not distinguish particularly between batch and continuous processing.

Continuous crystallization, as outlined in previous chapters of this book, is already one approach to PI within the time domain (Table 7.2). This chapter, however, focuses on the application of PI from all four domains to the concept of continuous crystallization because PI can also be obtained using batch crystallization processing.[3,11] Although it becomes obvious that each concept in Table 7.2 contains elements of multiple PI domains,

Table 7.1 Definitions and descriptions of process optimization, process systems engineering, and process intensification.[9]

	Process optimization	Process systems engineering	Process intensification (PI)
Aim	Performance improvement of existing concepts	Multiscale integration of existing & new concepts	Development of new concepts of process steps & equipment
Focus	Model & numerical methods	Model, software	Experiment, phenomenon, interphase
Interdisciplinarity	**Weak** *Applied mathematics*	**Modest** *Chemistry, applied mathematics, and informatics*	**Strong** *Chemistry, applied physics, materials design, chemical & mechanical engineering, automation, etc.*

Table 7.2 Fundamental principles of process intensification (PI) for continuous crystallization.[11]

Time domain	Space domain	Function domain	Energy domain
Continuous operation • MSMPR • PFC Periodic operation • Temperature cycling • Periodic flow	Structure • Static mixers • Segmented flow • OBC Miniaturization • Porous materials • Patterned surfaces • Emulsions • Microfluidic crystallizers	Hybrid processes • Chromatography • Membrane • Distillation Process integration • Spherical crystallization • Wet milling • Multifunctional equipment	External fields • Ultrasound • Electric • Microwave

these domains are distinguished based on the most dominant PI principle for crystallization processes.[11] The chapter is not intended to provide an exhaustive list of PI examples that have been documented, but rather will draw attention to the recent developments in PI concepts applied in continuous crystallization. Finally, some perspectives on these PI approaches in terms of their readiness for implementation in industrial settings are discussed.

7.2 Time Domain

The currently ongoing paradigm shift from batch to continuous processing for the manufacture of pharmaceuticals to improve the manufacturing efficiency is a PI approach within the time domain.[3,11–18] Continuous processing has the potential to improve product quality, process efficiency and safety as well as aid in minimizing waste and maintenance. In addition, it can reduce the number of unit operations, size of equipment, and plant footprint allowing for the process development and manufacturing at the laboratory scale ranging from grams to kilograms per day, which aids to reduce the time to market.[14,19–21] Generally, the PI principle time follows two common strategies (i) the manipulation of process time scales at which different process steps proceed and (ii) the introduction of dynamic states within a process, typically in the form of periodicity.[7] To comply with these two strategies different crystallizer designs have been developed, periodic operations are applied, and mixers for faster mixing (mass transfer) of process streams have been invented.

7.2.1 Crystallizer Designs

A continuous crystallizer is an apparatus that enables the creation of a suitable environment for the nucleation and growth of the crystalline material with the desired CSD and operates with continuous feed and continuous withdrawal of crystallized product (contained in mother liquor). It requires the ability (i) to generate supersaturation at a desired temperature, (ii) to provide sufficient residence time for the crystals to nucleate and grow, and (iii) to provide some extent of mixing that allows ideally a uniform environment throughout the entire crystallizer volume (temperature, concentration, slurry density), which ensures that the composition in the crystallizer is the same as that of the effluent (slurry). Such a continuous crystallizer is called a mixed suspension, mixed product removal crystallizer (MSMPR).[22]

A wide variety of continuous crystallizer designs exist and even more of specialized crystallizers. The two main designs currently studied for PI, especially for fine chemicals and pharmaceuticals, are the MSMPR[23–27] and the tubular crystallizer.[28–34] The choice of whether to use a MSMPR or a tubular crystallizer is primarily driven by the kinetics and slurry density of the process. Generally, the MSMPR can be used for low conversions, long residence times, and high slurry density, while the tubular crystallizer is favored for higher conversions with short residence times and low slurry density at any residence time due to settling of crystals. The MSMPR provides a robust operation with intensive mechanical mixing at a small surface-to-volume ratio, which reduces fouling and settling. Moreover, the MSMPR permits the relatively simple conversion of existing batch capacity to continuous capacity to obtain PI.[35] However, the residence time distribution (RTD) is broader in a single MSMPR compared to a tubular crystallizer, which leads to a broader crystal size distribution (CSD) and thus non-uniform product quality. To produce narrower CSDs, MSMPRs can be connected in series, also known as cascade, which offers the ability to gain degrees of freedom for process design and operation. Although leading to increased equipment costs, a MSMPR cascade allows for flexible process temperatures, targeted supersaturation profiles over the sequence of MSMPRs in the cascade, increased yield, and economies of energy consumption.[36–38] MSMPRs, single or cascade, represent the classic workhorse for continuous crystallization (cooling,[39] antisolvent,[14,21,40,41] reactive[20,42]) in academia and industry. Tubular crystallizer strategies have also been employed but their examples remain mostly in the academic literature.[28,30–32,43,44] A tubular crystallizer has the advantage of a narrower CSD at a typically higher efficiency compared to a MSMPR of the same volume due to the narrower RTD.[31,45–47] This is important as it may aid in avoiding additional downstream operations such as milling to reduce the crystal size, which contributes to PI.[48,49] In addition, the tubular crystallizer enables (i) supersaturation profile manipulation along the tube through temperature control[33,43] or multiple antisolvent additions[32] in different segments of the tube and (ii) a more rapid approach to steady state simplifying process dynamics. The tubular crystallizer concept can be distinguished in (i) plug flow crystallizers (PFC),[28–34] (ii) segmented flow crystallizer,[28,46,50,51] and (iii) oscillatory baffled crystallizers (OBC),[49,52,53] illustrated in Figure 7.1a–c, respectively. Further discussions are provided within the PI principle space domain (Section 7.3).

The crystallizer designs introduced can include advanced control strategies such as mother liquid recycling,[27,54] solids recycling,[36] and fines destruction (fines dissolution)[33,34] to reduce waste and increase crystal yields[22,55–57] or be combined with other technologies, such as membranes[23,58,59] for further PI (see function domain in Section 7.4).

7.2.2 Periodic Operation

Periodic operation (*e.g.*, temperature cycling, solvent cycling, flow cycling) is a common operation strategy for PI of continuous crystallization processes. This periodic operation is aimed at controlling critical quality attributes (CQA) of the crystalline products (*e.g.*, CSD, morphology, polymorphism, and chirality) or process performance (prevention of

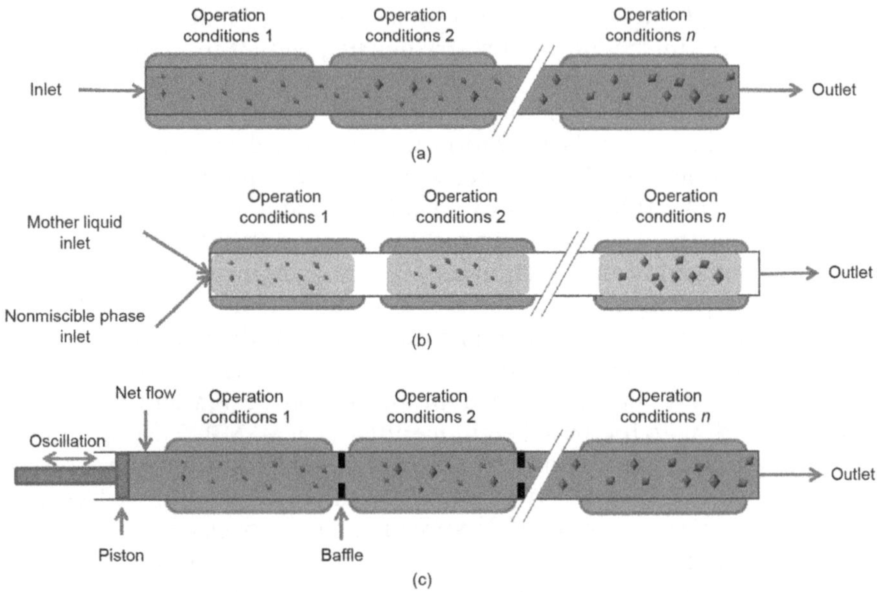

Figure 7.1 Schematic diagrams of (a) plug flow crystallizer (PFC), (b) segment flow crystallizer, and (c) oscillatory baffled crystallizer (OBC). Reproduced from ref. 54, https://doi.org/10.1021/cg301221q, under the terms of the CC BY 4.0 licence, https://creativecommons.org/licenses/by/4.0/.

incrustation and clogging). The temperature cycling approach exposes the crystal suspension to alternating cooling and heating cycles.[60] During the heating cycle the solution becomes undersaturated, which results in (partial) dissolution of existing crystals (preferably only small crystals) leading to an increase in the solute concentration. This is supported by the fact that the molecule packing on the surface is energetically less stable than the ordered packing within the crystal lattice. Since small crystals possess a larger surface-to-volume ratio compared to big crystals, they will dissolve faster leading to the removal of fines. The remaining big crystals grow during the subsequent cooling period by consumption of the newly generated supersaturation in the solution phase accompanied with nucleation of new crystals (if well controlled within the metastable region through secondary nucleation). The cycles of sequential dissolution–recrystallization can be considered as an intensified form of Ostwald ripening,[61] which aids to accelerate the conversion. However, while temperature cycling has been broadly applied and reported as a PI strategy to control CSD,[62] morphology,[63] polymorphism,[64] and chirality[65] in batch crystallization processes, literature on temperature cycling in continuous crystallization is limited. The available studies focus on the control of CSD[28,33] and the prevention of incrustation (fouling)[66] in PFCs.

Another strategy to mitigate fouling in PFCs is based on dissolution of the incrusted layer using pure solvent (solvent cycles). Majumder and Nagy[47] have shown in a modeling study the benefits of this approach, which basically allows decisions on when to stop the crystallization process in a given PFC segment to conduct dissolution of the crust formed in order to maintain the crystal properties. Thus, this approach allows maximizing the crystallization cycles with consistent CQA before necessary cleaning cycles need to be performed.

Continuous crystallization processes conducted in MSMPRs can also be intensified utilizing the concept of periodic flow. Instead of adding and withdrawing materials (solution, slurry) at a constant flow rate, periodic material transfer can be applied (Figure 7.2).

While MSMPRs have been in operation in many large-scale, high volume settings for many decades in the chemical industry,[56] recent continuous crystallizer developments, aimed at the pharmaceutical industry, operate at working volumes as little as 30 mL,[26,40,42,67] leading to significant technical difficulties for the slurry withdrawal at the outlet (*e.g.* blockage, crystal classification).[35,42,68,69] Unlike the conventional continuous MSMPR operation, described by Randolph and Larson,[22] in which product slurry is continuously withdrawn, the small scale of the aforementioned MSMPRs results in very low flow rates that leads to (i) the sedimentation of crystals in the transfer lines and over the course of time to blocking, as well as (ii) the preferential removal of small crystals from the MSMPRs due to the slower settling velocities and faster response times of small crystals compared to larger crystals. For instance, Quon *et al.*[42] reported a theoretical continuous flow rate of 0.17 mL min^{-1} at the outlet of the 30 mL MSMPR for the reactive crystallization of aliskiren hemifumerate. Consequently, low withdrawal velocities failed to achieve isokinetic (representative) withdrawal.[68] To mitigate this issue, intermittent (periodic) withdrawal has been reported.[23,26,41,70] Intermittent withdrawal permits slurry withdrawal at very high flow rates of up to 10% of the suspension volume build-up within the MSMPR at repeating time intervals every one tenth of a residence time using a reasonably wide outlet tube. The position (height) of

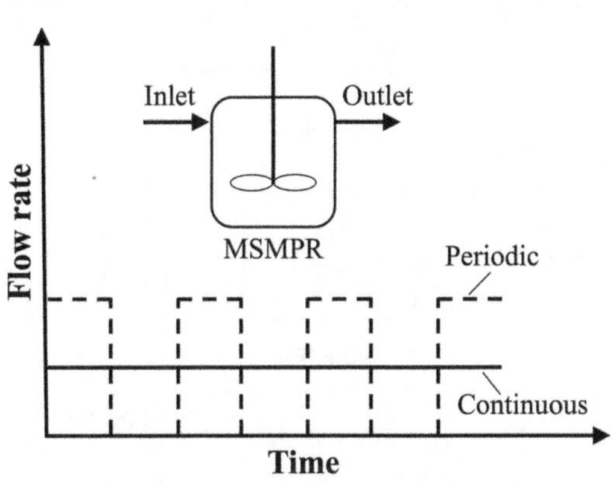

Figure 7.2 Schematic flow rate profiles at the inlet and outlet of an mixed suspension, mixed product removal crystallizer (MSMPR).

the dip tube governs the operating volume of the MSMPR (Figure 7.3). The internal diameter of the dip tube should at least be 10 to 20 times larger than the largest significant crystal size in the CSD to prevent blockage of the transfer lines.[68] Ferguson *et al.*[68] reported an intermittent pneumatic MSMPR withdrawal method for an antisolvent crystallization (Figure 7.3). This concept sets up a differential pressure across the crystallization system with a positive pressure applied to the MSMPR headspace (34.5 kPa) and a partial vacuum (−50 kPa) applied in a product hold tank, which drives the rapid withdrawal of 10% of the MSMPR volume within about 2 s.[68] By maintaining the pressure difference when the solution level drops below that of the dip tube, air from the headspace is sucked through the transfer line, which removes any remaining slurry. To mitigate solvent evaporation during pressurization and subsequent transfers, air can be substituted with nitrogen or other inert gases saturated with the solvents. A similar approach has been used in industrial research & development (R&D) for the development of continuous crystallization processes by Elli Lilly and Company.[18,20,21,41] Intermittent slurry withdrawal from MSMPRs has also been reported employing periodically operating peristaltic pumps.[23,40,42,69]

Most recent research efforts aimed at maintaining a small environmental footprint for pharmaceutical manufacturing (~1/40) compared to the current industrial size[14,71] required innovations in compact crystallization systems including slurry withdrawal.[26,72] While MSMPRs have been demonstrated to operate with a working volume as little as 20 mL,[67] the size of pumps or combined vacuum/pressure transfer systems[20,21,41,70] responsible for the slurry transfer between MSMPRs or subsequent unit operations (*e.g.*, filtration) can in general be much greater than the size of the MSMPRs used. To tackle this issue, Cui *et al.*[26] developed a pressure-driven flow crystallizer

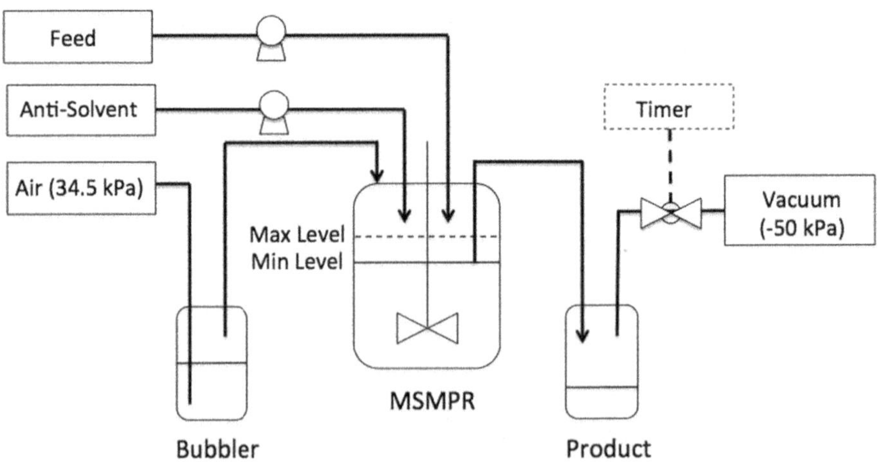

Figure 7.3 Schematic of a mixed suspension, mixed product removal crystallizer (MSMPR) with intermittent pneumatic slurry transport mechanism. Reprinted from ref. 68, Copyright 2013, with permission from Elsevier.

(V = 40 mL) with a customized suspension transfer pumping system, utilizing a dip tube for the slurry handling, controlled only by the generation of head pressure within the MSMPR using a small scale diaphragm metering pump. The total footprint of a two-stage MSMPR (each 40 mL) with introduced pressure-driven transfer scheme is only 4800 cm^3, including valves and pumps. In comparison, the volume of a commonly used single Masterflex peristaltic pump (Cole-Parmer, Masterflex L/S Digital Drive 600 rpm) for intermittent withdrawal is 7360 cm^3.[26] In addition, this compact system has been demonstrated to operate on the bench scale for at least 24 h[26] and was integrated into a continuous flow platform for on-demand pharmaceutical manufacturing.[14]

While all previous examples were driven to enable representative and isokinetic withdrawal of product crystals from a MSMPR without blocking the transfer lines, Nagy and coworkers[73,74] studied the effect of periodic flow operation applied also to the feed stream. The periodic addition and withdrawal of solution and slurry (synchronous or asynchronous)[74] is followed by an equilibration or pause period without any material transfer, which allows the system to stay undisturbed until the next cycle begins. Consequently, this concept of MSMPR operation is a hybrid of continuous and batch operations leading to periodically oscillating supersaturation between an upper and lower limit compared to a fixed supersaturation in true continuous MSMPR. Moreover, the mean residence time of crystals in the periodic operating MSMPR is extended by the period of batch operation, which can be manipulated by the holding time in between the cycles. Thus, higher yield and larger product crystals can be achieved compared to truly continuous operating MSMPR at steady-state given that the oscillations of the supersaturation are kept small by a "controlled state of operation" (periodic steady-state).[73,74] In addition, periodic operation of the feed stream aids in the mitigation of fouling issues in the transfer lines due to much higher flow rates.[73] The concept of an OBC (Chapter 3) intrinsically applies periodic flow to provide enhanced mixing at low flow rates resulting in enhanced heat and mass transfer and particle mixing with reduced particle settling by providing structures (see Section 7.3 for further details on the structure).[52]

7.3 Space Domain

In general, spatial randomness (inhomogeneity) in crystallizers needs to be avoided for the sake of process control (*e.g.*, supersaturation, nucleation, growth) and robust operation (*e.g.*, no process fluctuation beyond the process design under steady state operation condition[75]). For instance, homogenization through mixing during the addition of reactants and antisolvents is crucial to obtain desired CQAs of the resulting crystalline material (*e.g.* CSD, size, morphology, purity, polymorph, solvate). Although mixing is the manipulation of time, PI methods applied to intensify the mixing process for crystallization are typically based on the introduction of structures in the continuous crystallizers (*e.g.*, draft tube baffle (DTB) crystallizer,[56,76] OBC,[52] static mixers[32,77,78]). Another PI approach within the space domain controls the crystallization in confined spaces using the principle of miniaturizing the crystallization environment.

7.3.1 Structure

It is a well-known fact that the crystallization kinetics (*e.g.*, level of supersaturation) is greatly influenced by the hydrodynamic conditions caused through rather vigorous, turbulent mixing because it controls heat and mass transfer. Thus, spatial randomness needs to be avoided by providing structure.[7] For instance, mixing plays a decisive role in reactive crystallization/precipitation or often also in antisolvent crystallizations where locally very high supersaturation levels can be generated rapidly with temporal and spatial gradients. Consequently, the micromixing needs to be faster than the crystallization kinetics to prevent non-uniform and unpredictable CQAs (CSD, morphology, and purity).[55,56,79–81] The intrinsic multiphase nature of crystallization processes (keeping crystals well suspended while mixing liquid streams) complicates this topic. Large density differences between solid and liquid phases may cause suspension density gradients in the axial direction of a stirred tank (*e.g.*, MSMPR)[82–85] or different RTD of the solid phase compared to the liquid phase (*e.g.*, PFC).[44,86,87] Thus, strong mixing is required to overcome the gravitational settling of the crystals for ideal homogeneous suspensions within continuous crystallizers aimed at preventing classifying product removal and ensuring that all crystals have the same processing history (RTD) to achieve uniform product CQAs.[22,88] On this account, the complex process mixing needs to be distinguished based on different scales: (i) macromixing, which refers to the overall mixing performance of the crystallizer described by the RTD that sets a general environment for micromixing and indirectly influences the distribution of supersaturation, (ii) mesomixing, which refers to the disintegration of large eddies that forms a direct environment for micromixing, and (iii) micromixing, which refers to the mixing on the molecular level and is the fastest of the three length scales of mixing (Figure 7.4).[56,57,79,81,88,89] For extended discussion of these and other mixing topics the interested reader is referred to specialist literature.[55–57,79,90,91]

In general, the concepts of intensifying mixing encompasses various designs of structures that can be applied either at the entrance or within the continuous crystallizer to generate rapid mixing as well as to eliminate local regions of uncontrolled supersaturation or back mixing (*e.g.*, in PFC). In addition, the focus of this section is on PI using passive structures and not for instance impellers.

Confined impinging jet (CIJ) mixers,[78,92] T-mixers,[93] Y-mixers,[30,94–96] mixing elbows,[20,97] and vortex-type mixers (*e.g.*, Roughton[80,95,98,99] and Multi-Inlet Vortex Mixer[100]) are non-helical shaped static mixers applied for PI in continuous crystallization processes. These type of static mixers are typically

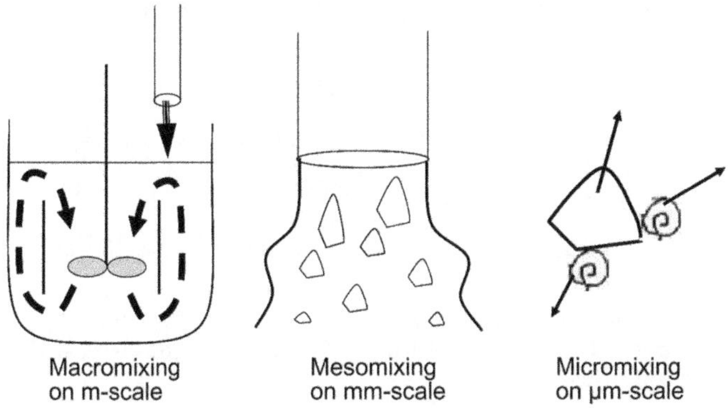

Figure 7.4 Different length scales for mixing of two fluid streams. Reproduced from ref. 58 with permission from John Wiley and Sons, © 2013 Wiley-VCH Verlag GmbH & Co. KGaA.

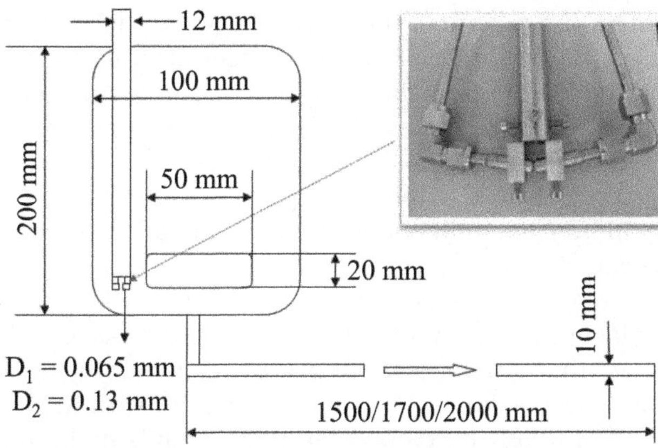

Figure 7.5 Sketch of continuous reactive crystallization process (1 L) with submerged impinging jet for production of sodium cefuroxime. Reproduced from ref. 79 with permission from John Wiley and Sons, © 2016 American Institute of Chemical Engineers.

employed at the entrance of PFCs[30,31,99] or MSMPRs.[20,78] The main concept underlying these mixers is the impingement or tangential contact of two streams (multiple possible for vortex-type) in a confined space where mixing proceeds without external energy (*e.g.*, mechanical, pneumatic, or ultrasonic forces).[92,95,96,101,102] Liu *et al.*[78] reported a continuous reactive crystallization for the antibiotic sodium cefuroxime using an impinging jet mixer submerged as close as possible to the impeller in the solution of a 1 L continuous stirred tank reactor (CSTR), analogous to an MSMPR (Figure 7.5). This approach combines intensified micromixing with the favorable mixing conditions near the impeller.[103] A crucial design decision of this approach relates to the angles of the impinging jets.[78]

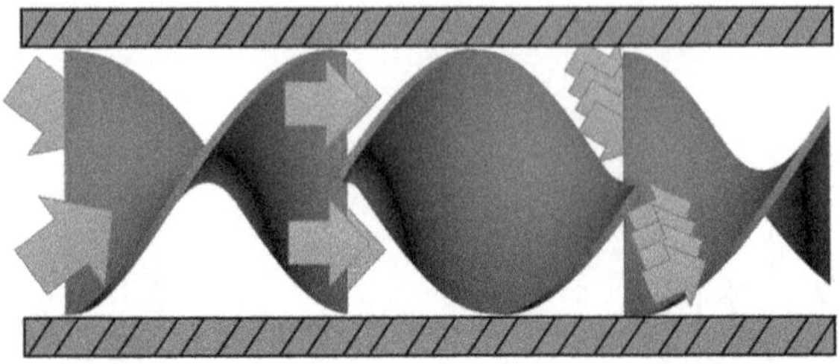

Figure 7.6 Schematics of mixing of two streams flowing through a Kenics type static mixer. Reprinted from ref. 11, Copyright 2018, with permission from Elsevier.

An industrial application employed a mixing elbow for PI of a continuous reactive crystallization (5 L MSMPR crystallizer) within an integrated pilot scale production (synthesis and purification) of an API in phase 2 clinical trials with a production rate of 3 kg day^{-1} over a period of 72 h continuous operation.[20] This study demonstrates the importance of the mixing intensity of the reactant for the control of the crystal CQAs (particle size, CSD, aspect ratio, tap density, flowability) to generate powder blends for direct compression of the product while providing acceptable purity and yield. Moreover, this industrial study emphasizes that despite the increased mechanical complexity of this combination of PI approaches (continuous, static mixer, process integration) the operational complexity is reduced due to the lower number of equipment and unit operations (*e.g.*, co-addition, wet milling, and pH cycles) needed.[20]

However, the majority of recent studies that apply the concept of structure for PI are focused on tubular crystallizers. For instance, helical shaped static mixers, another type of static mixers, are commonly applied as a PI method to improve the mixing in tubular reactors without moving parts (Figure 7.6).[104–106] Initially developed for mixing fluids in laminar flow and later applied for heat transfer, turbulence enhancement, and multiphase systems, static mixers are based on well-defined structures that redistribute the fluid through splitting and twisting the flow of streams mainly in the radial direction.

Consequently, axial mixing can be reduced leading to a RTD that approaches plug-flow conditions, which favours the production of crystals with uniform CSD. In this context, Hobbs and Muzzio[107,108] demonstrated that the number of mixing elements in static mixers is decisive to approach plug flow (Figure 7.7).

Alvarez and Myerson[32] have demonstrated that a combination of enhanced mixing of antisolvent with solutions of ketoconazole, flufenamic acid, and L-glutamic acid using static mixers and the addition of antisolvent at multiple points along the tubular crystallizer is a vital PI strategy to maintain low supersaturation at all points along the PFC and thus to control the CSD (smaller and narrower). For a comprehensive overview of the key parameters

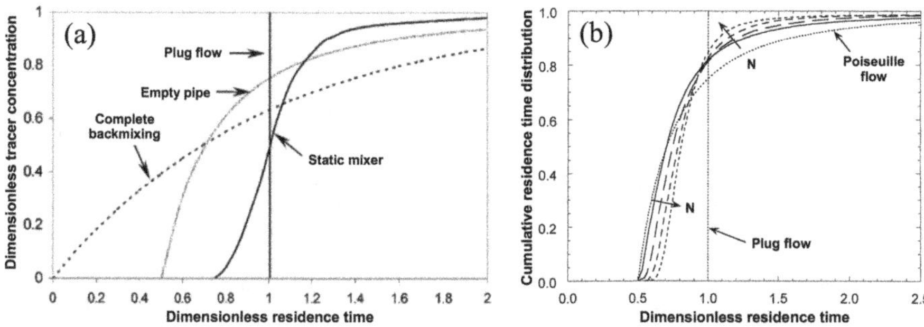

Figure 7.7 (a) Cumulative residence time distributions for various tubular flow systems. Reprinted from ref. 106, Copyright 2003, with permission from Elsevier and (b) Cumulative residence time distribution *versus* the number of static mixer elements (N = 4, 8, 16, 32). Reprinted from ref. 107, Copyright 1997, with permission from Elsevier.

(*e.g.*, pressure drop, mixing homogeneity, axial mixing) needed to select a suitable static mixer, the interested reader is referred to the review by Thakur *et al.*[106]

Another strategy to provide structure for PI is the concept of segmented flow (Figure 7.8a) in PFCs. Segmented flow aids to enable a favorable plug flow condition in a tubular crystallizer with low Reynolds numbers (low volumetric flow rates), which are often required for emerging small scale flow applications in the pharmaceutical industry. Simply increasing the velocity to approach more turbulent flow is not favorable because it would require increasing the tube length proportional to the velocity increase in order to maintain the relative long residence times often needed for crystallization of organic substances (*e.g.*, pharmaceuticals). Thus, PFCs are prone to clogging of the tube through settling of crystals, which together with incrustation, often represents a challenge for PFCs. On this account, alternative PFC strategies have been studied that divide the flow into small subsequent segments of suspension/slurry and fluid (air or immiscible liquid).[28,46,50,109,110] Each segment, also known as a slug, can be assumed being perfectly mixed due to the small size of each slug and the re-circulating flow field while the segments are moving along the tube. In addition, the segmented structures aid to push the settling crystals through the PFC, thus avoid back mixing because each segment acts as a small well-mixed crystallizer independent from other segments (Figure 7.8a).[29] Various examples have been reported in the academic literature.[28,46,50,51,109,110] In general, care has to be taken to control the flow rate ratio of suspension/slurry and fluid phase as it can lead to different flow patterns (*e.g.*, segmented, bubbly, and annular flow with the latter two being undesired).[28,111] The interested reader is referred to Chapter 5 of this book for an extended discussion on segmented-flow (slug-flow) continuous crystallizers.

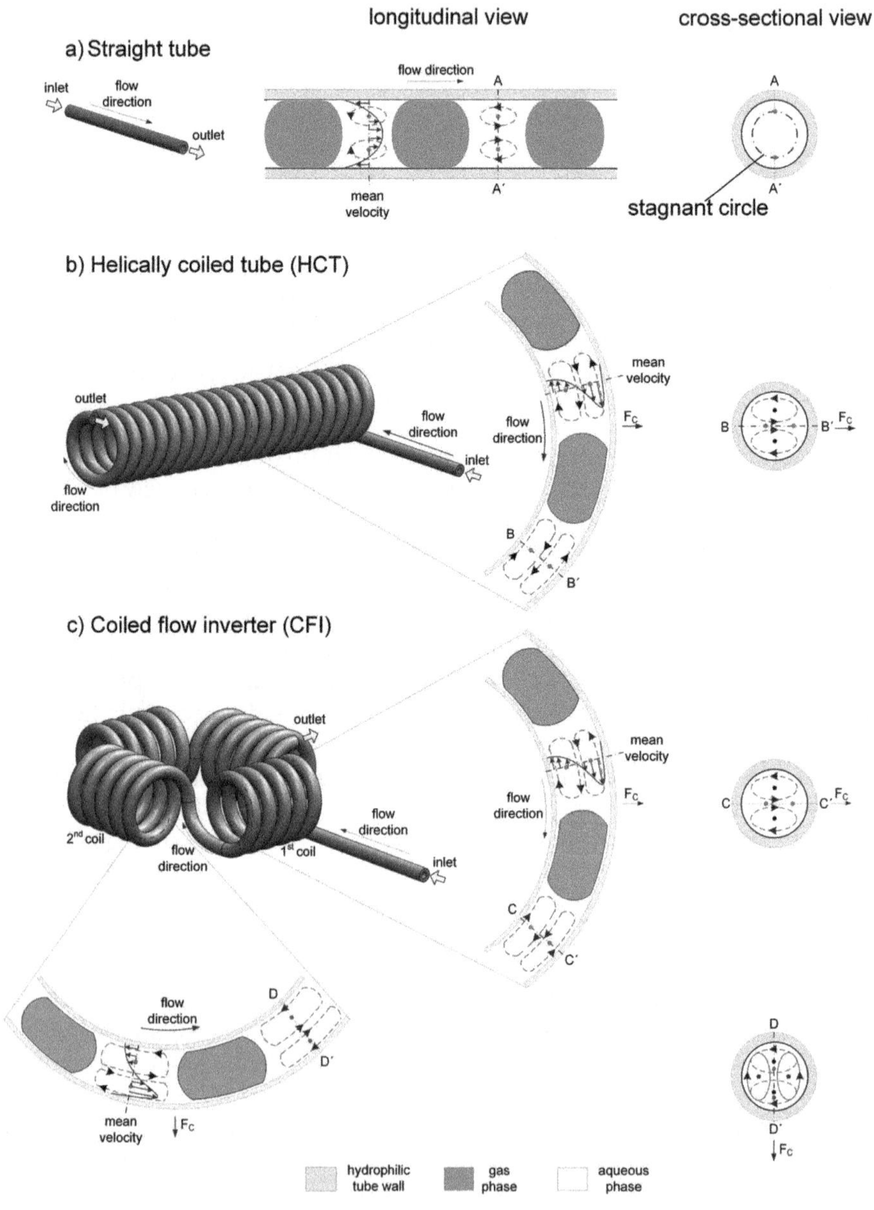

Figure 7.8 Schematic of slug flow mixing behavior in (a) straight tube (b) helically coiled tube (HCT) (c) coiled flow inverter (CFI). Reprinted from S. K. Kurt, F. Warnebold, K. D. P. Nigam, and N. Kockmann, Gas–Liquid Reaction and Mass Transfer in Microstructured Coiled Flow Inverter, *Chem. Eng. Sci.*, 2017, **169**, 164–178, Copyright 2017, with permission from Elsevier.

Another approach of structure for PFCs is adopted from fluid (gas, liquid) reaction systems, which induces centrifugal forces through coiling the tubes (Figure 7.8b) leading to secondary flow profiles, so-called 'Dean vortices'.[44,112] Theoretically, these vortices intensify particle fluidization[113,114] and decrease wall scaling[115] through enhanced radial mixing, which leads also to narrower RTDs compared to straight tubes (Figure 7.8a).[116] The positive effect of coiling can be enhanced further by frequently bending the coils with an angle of 90° in a so-called 'coiled flow inverter' (CFI) crystallizer, demonstrated by Kockmann and co-workers (Figure 7.8c).[44,117,118] Due to the direction changes of the centrifugal forces, the Dean vortices reorganize sequentially, which breaks-up stagnant regions in the center of the vortices leading to RTDs close to ideal plug flow.[119] However, despite these promising studies of structures applied in tubular crystallizers, it needs to be emphasized that relatively high flow rates are still necessary to maintain particulate suspension.[28,44,117]

Similarly to the segmented flow mechanism, which provides mild and uniform mixing conditions that can help to reduce secondary nucleation, air-lift crystallizers apply also an inert fluid phase to induce mild mixing by sparging air bubbles in a column (*e.g.*, riser). Specifically, the air-lift mechanism creates a density difference with any column in an open connection to the riser that leads to liquid circulation without moving parts.[120,121] Although not realized in a continuous operation yet, this concept could be operated in a continuous flow setting simply by connecting a number of air-mixed compartments in series to approach plug-flow behaviour while maintaining the favourable properties leading to reduced secondary nucleation.[120]

Oscillatory baffled crystallizers (OBCs), a particular type of tubular crystallizer with plug flow capability, provide more sophisticated auxiliary structures compared to the aforementioned concepts, in order to intensify radial and axial mixing. OBCs are equipped with periodically spaced restrictions, generally spatial orifice baffles, with oscillatory motion of the fluid (Figure 7.9). The fluid motion can be generated either by periodic movement of the intrinsic baffles[122] or periodic motion of the fluid where

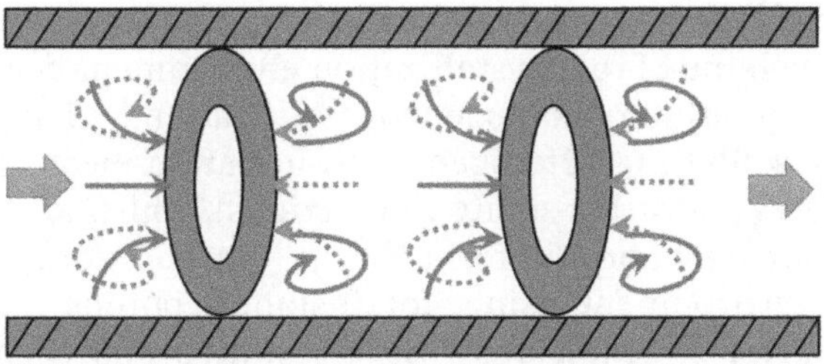

Figure 7.9 Schematic of oscillatory flow through periodically spaced spatial orifice baffles in oscillatory baffled crystallizers (OBCs). Reprinted from ref. 11, Copyright 2018, with permission from Elsevier.

the intrinsic baffles are fixed.[123–125] The oscillatory fluid flow leads to the generation and decay of eddies around the baffles that provide enhanced mixing at low flow rates resulting in enhanced heat and mass transfer, particle mixing with reduced particle settling compared to other tubular crystallizer strategies and at lower slurry densities compared to MSMPRs at any residence time.[52,87] However, this improvement comes at the cost of significantly greater mechanical complexity.[126] In addition, the oscillation amplitude and frequency should be selected carefully as the oscillation can introduce axial back mixing, broaden the residence time distribution,[86] and can increase the secondary nucleation due to the high shearing of the oscillation flow.[49] Recent studies have demonstrated the capabilities of OBC in continuous operation to produce salicylic acid by anti-solvent/cooling,[124] as well as α-lactose monohydrate[127] and β-L-glutamic acid[123] by cooling crystallization. For technical details and the design consideration of OBCs, the interested reader is referred to Chapter 3 of this book and to specialist literature.[52,87]

In conclusion to the concept of spatial structures, this approach can be employed to realize any of the four PI principles (Table 7.2)[7] and will be included in further discussions within this chapter.

7.3.2 Miniaturization

The PI concept of miniaturization considers the crystallization in confined spaces to create well-defined environments that allow controlling the solution properties as a function of spatial coordinates. For instance, as the volume of solution containing droplets decreases, the evaporation rate increases and the droplets become highly supersaturated favouring the nucleation of high energy metastable polymorphic forms, while low levels of supersaturation lead typically to the thermodynamically stable forms.[128,129] Moreover, crystallization in confined spaces (physical barrier, *e.g.*, nanopores) is a concept where polymorph selectivity can be directed by suppressing the nucleation of polymorphic forms whose critical nucleus size is larger than the pore volume dimensions. As a consequence, nucleation of polymorphic forms with critical sizes less than the pore volume can be achieved.[130–132] It is also well known that nucleation becomes more difficult as the volume of the crystallization environment decreases due to the suppression of primary homogeneous nucleation.[133] This provides the opportunity to exploit the heterogeneous nucleation mechanism through templating effects (substrate-solute interactions), which allows control of the formation of polymorphic form.[128,134–138] In addition, confining the crystallization environment using, for instance, porous materials,[139–143] patterned surfaces, or emulsions[144] is a common approach to produce nano-sized particles to enhance the physicochemical properties of substances with poor aqueous solubility (*e.g.*, APIs).[145–147] Although nano-sized crystals can also be obtained using top-down methods such as milling and high pressure homogenization to break large crystals into smaller ones, these methods have associated problems with rather broad, non-uniform CSD, contamination, high-energy consumption, and possible polymorphic phase transformations.[144,148,149]

A simple approach to confine the crystallization environment is the application of porous materials such as pore glass, fumed silica, metal organic frameworks (MOFs), and biodegradable polymers (*e.g.*, alginate, chitosan) to load APIs into these matrices. However, the literature reporting on the application of continuous flow procedures is very limited. Dwyer *et al.*[142] reported a two-stage MSMPR strategy for the loading of fenofibrate and griseofulvin (APIs with poor aqueous solubility), in porous silica. The first stage is employed to continuously wet the porous silica leading to nucleation of the APIs within the confined pores. The second stage intensifies the loading process of the API by submerging the already drug-loaded silica in slightly supersaturated API solutions, which allows the nanocrystals to grow further within the pores. The *in vitro* dissolution studies demonstrated significantly enhanced performance when compared to the bulk crystals. A similar concept, but with the aim to eliminate unneeded unit operations (*e.g.*, milling, granulation, blending) in integrated end-to-end continuous manufacturing settings, uses surfaces of a crystalline excipient (D-mannitol) to directly crystallize acetaminophen on the heterosurface.[150] The nucleation on interfaces through epitaxy is energetically more favorable than spontaneous nucleation and allows suppression of bulk nucleation.[150,151] One of the key criteria of this PI approach is the solvent selection.[150] The solvent needs to be able to dissolve the API to achieve a crystallization solution, but should have negligible solubility for the substrate to provide the necessary confined surface. In addition, spontaneous nucleation (bulk nucleation) at high supersaturations, secondary nucleation, mixing, as well as attrition and fragmentation through pumping and particle–impeller interactions are further critical process parameters (CPPs) that have to be carefully designed to control such a continuous crystallization process.[150] Both PI approaches using surfaces and porous material allow reduction of the number of unit operations needed for downstream processing of solid dosage forms, which also adds to the miniaturization of the process's footprint and thus to lower capital and operational costs.[150,152]

Use of patterned surfaces is another approach to miniaturization, which combines confined environments through physical barriers (array of isolated wells that act as crystallizers) with heterosurfaces that allow tuning of molecular interactions at the substrate-solute interface for nucleation and polymorphic form control.[128,129,134–137,146,153,154] In general, the choice of the heterosurface material,[128,129,134,137,146,153,154] surface roughness[135] as well as the angle[136] or curvature[135] of the imprinted patterns have a significant impact on the crystallization outcome. Despite the potential of confining the crystallization process using patterned surfaces documented in the scientific literature, to the best of our knowledge, none are currently pursuing continuous crystallization approaches. However, patterned surfaces are used for high-throughput screening of crystallization conditions in wells as little as 20 nL, which minimizes the consumption of precious material.[155–157]

The last enabling technique for confining the crystallization environment uses emulsions. An emulsion is a mixture of at least two immiscible liquids in which one is the liquid phase dispersed in a continuous liquid phase. The emulsion droplets (each acting as an independent crystallizer) can be prepared with a uniform and controllable micro- to nano-scale size using high energy (mechanical) and low energy (chemical) methods.[158–161] Emulsions can also be produced using microfluidic devices.[159,162–165] In general, emulsion crystallization is employed for the production of micro- and nano-sized crystals within confined spaces,[144,166,167] encapsulation and delivery of substances,[158,168–170] as well as screening of crystallization conditions.[171–173] However, the concept of continuous emulsion crystallization, to the best of our knowledge, is in its beginnings[165,174] and deserves more attention as a PI strategy.

7.3.2.1 Microfluidic Devices

Microfluidic flow devices are ubiquitous in the scientific literature because they provide a unique environment for the study of crystallization processes at the meso or nano scale with precise control over process parameters.[175] However, due to the propensity of solid phases and precipitating systems to block small channels associated with microfluidic devices, the type of applications are limited. In general, the microfluidic technique as a PI approach is not sufficiently scalable for separation and purification of reaction streams in commercial activities that require isolation of target products. On the other hand, microfluidic crystallization provides the capacity to explore hundreds of experimental conditions through a single experiment while minimizing the consumption of precious substances, which cannot be achieved with conventional crystallization processes. Therefore, microfluidic crystallization is primarily associated with screening of crystalline forms[176,177] and extraction of kinetic information for crystallization including proteins[178–181] and not for continuous isolation and purification from reaction streams. Droplet-based microfluidic applications using the concept of emulsion crystallization may provide a route to access a wider microfluidic crystallization aimed at producing tailored spherical particles for small-scale production of formulated drugs.[165] Recently, Shi *et al.*[175] provided a review of the state of the art in the field of crystallization under microfluidic flow.

7.4 Function Domain

The implementation of multiple processing functions into one single physical space is another PI principle in crystallization.[11] Any unit operation for separation and purification has its thermodynamic, kinetic, and economic boundaries within which they can operate optimally. However, combining different unit operations (functions) offers the potential to exceed the boundaries and achieve enhanced performance compared to separately operating functions. A simple example in the area of crystallization is reactive crystallization, also known as precipitation, which combines reaction (liquid) and separation through the generation of a new solid phase within a crystallizer.[3,182] Combining crystallization and other separation and purification technologies (*e.g.*, chromatography, membranes, distillation) into hybrid processes is a strategy that generates synergistic effects, often attributed with cost efficiency by reducing process time, equipment size, and energy costs. Generally, hybrid processes are characterized by the combination of at least two different unit operations that contribute to the separation task by different physical separation principles and the design and optimization of these processes are highly complex due to the large number of structural and operational degrees of freedom.[183–187] While the concept of hybrid processes is generally aimed to overcome the limitations of the individual unit operations, the process integration approach within the function domain is aimed to reduce processing time, energy consumption, and equipment cost by integrating multiple unit operations (*e.g.* crystallization and agglomeration) into one.

7.4.1 Hybrid Processes

7.4.1.1 Chromatography-crystallization Process

Chromatography is a unit operation commonly used to separate mixtures of chemical substances into their individual components based on their differential partitioning, which leads to a difference in the residence time of the individual component within the chromatography column. However, due to the relatively high cost of the stationary phases (adsorbents), the capital cost of chromatography remains high compared to crystallization.[188] On the other hand, crystallization is strongly affected by the presence of impurities, especially those possessing similar/identical solid–liquid equilibria (*e.g.*, structurally similar impurities and enantiomers). Therefore, in recent years, there have been an increasing number of studies on the combination of continuous chromatography and crystallization as a hybrid separation process aimed at generating a synergistic effect, in particular for the separation of enantiomers.[188–190] The basic principle of this approach will be briefly discussed in this section. For further fundamental details on enantiomer separation by chiral crystallization in hybrid settings, the interested reader is referred to Chapter 12 and to the review by Lorenz and Seidel-Morgenstern.[190]

Generally, chiral separation can be realized by countercurrent moving bed chromatography, where the mobile phase (fluid) and stationary phase (solid) move in the opposite directions. The more retained enantiomer adsorbs to the stationary phase and the less retained enantiomer moves with the mobile phase. Consequently, collecting the concentrated enantiomers at both ends of the chromatographic column leads to the separation

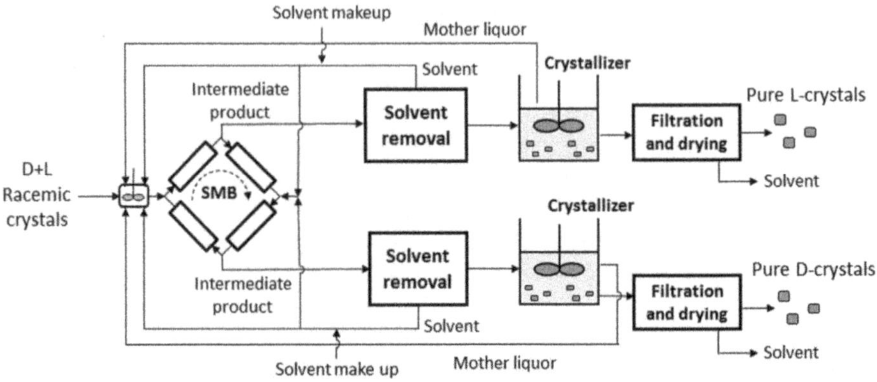

Figure 7.10 Schematic setup of simulated moving bed (SMB)-crystallization hybrid process with solvent removal. Adapted with permission from ref. 188, Copyright 2017, American Chemical Society.[190]

of the racemic mixture. For continuous processing, instead of moving the stationary phase, the simulated moving bed (SMB) simulates the movement of the stationary phase by rotating the ports for injection and withdrawal of materials with the flow direction of the mobile phase (Figure 7.10).[191] Thus, SMB consists of a series of chromatography columns separated by valves and ports for inlets and outlets followed by crystallization (preferential) of the enantiomerically enriched mixtures to obtain the pure enantiomers.[188,189,192,193] The racemic mixture is supplied to the SMB, which purifies the enantiomers to a purity that can be notably lower than the final product purity resulting from the crystallization.[188] Instead of feeding the SMB streams directly into a continuous crystallizer, the streams can further be processed prior to the crystallization by partial removal of the solvent to increase the concentration of the enantiomers without altering their purities (Figure 7.10).[188,189,193]

Application of the hybrid SMB-crystallization process for the separation of enantiomers has been discussed including for praziquantel,[193] bicalutamide,[194] Tröger's base,[192] 2′,6′-pipecoloxylidide[195] as well as the epimers of a diastereomeric compound.[196] However, the available literature is relatively limited on examples where both SMB and crystallization were operated continuously in hybrid settings, *e.g.*, phenylalanine[188] and mandelic acid.[197]

7.4.1.2 Membrane-crystallization Process

Membrane technology, as a stand-alone unit operation, has a wide range of applications including the production of potable water, pharmaceuticals, chemicals, and electronics.[10] The membrane technology is based on a thin barrier of nano-structured/functionalized material (membrane), which controls the exchange between two phases not only by external forces and under the effect of fluid properties, but also through the intrinsic characteristics of the membrane material itself.[10] In general, membrane separations offer attractive options for ease of performance as well as clean, energy efficient, and continuous operation. In addition, membranes can be used to promote crystallization through heterogeneous nucleation, control crystallization kinetics and as aids in the crystallization of specific polymorphs[198] and proteins[199] directly inside the membrane module where the supersaturation is generated.[200,201] Thus, this strategy, known as integrated or *in situ* membrane crystallization, is aimed at developing novel crystallizers (Figure 7.11a).[202] For further details on this aspect the interested reader is referred to Chapter 8. The focus of this section is on the application of the membrane technology as assisted operation in a hybrid process setting together with crystallization (Figure 7.11b)[202] for (i) the evaporative mass transfer of volatile solvents at low temperature to concentrate feed solutions above their saturation limit,[23] (ii) the retention of accumulating process impurities,[23,59,203] and (iii) the enantiomer separation of chiral molecules.[204,205] Thus, the membrane is used to generate supersaturation and/or retain molecules but the nucleation and crystal growth takes place in the crystallizer (Figure 7.11b).[202]

Traditionally, mother liquor concentration in crystallization processes uses solvent evaporation. This method has been demonstrated, for instance, for the chiral separation of phenylalanine in a continuous SMB-crystallization hybrid process with solvent removal (Figure 7.11)[188] and for the purification of cyclosporine (an immunosuppressant API) in a single stage MSMPR with mother liquor recycle.[54] However, concentration *via* evaporation has two major weaknesses: (i) the evaporation rate may be insufficient due to temperature limits of heat sensitive APIs, especially with high boiling point solvents and (ii) the mother liquor concentration by evaporation is not applicable for mixed solvent or antisolvent systems without changing significantly the makeup of the solvent or solvent/antisolvent ratios. To overcome these problems a pressure-driven membrane technique can be used (*e.g.*, reverse osmosis (RO)[206,207] and organic solvent

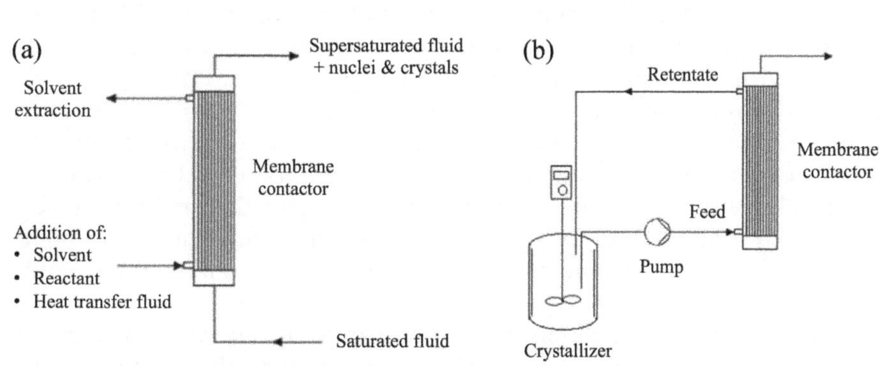

Figure 7.11 Schematic illustration of (a) integrated membrane crystallization process and (b) hybrid membrane crystallization process. Reprinted from ref. 202, Copyright 2016, with permission from Elsevier.

nanofiltration (OSN).[23,58,208]) While RO is a process where only solvent can permeate, OSN has separation capabilities in the molecular weight range between 200 and 2000 Da[208] that allows preferential purging of impurities out of the system while simultaneously concentrating the mother liquor when the molecular weight differences between the API and the main impurity of interest is large. In addition, the separation by RO and OSN is not temperature-driven and thus there is no risk of temperature-related API degradation. For many membranes there should also be no or limited effect on the solvent/antisolvent concentrations.[23] For a comprehensive overview of OSN in molecular separations, types of membranes, characterization methods, industrial membrane configurations, operating strategies, membrane preparation, multi-scale modeling and simulation the interested reader is referred to the review by Marchetti *et al.*[209]

In an experimental study Ferguson *et al.*[23] combined a single MSMPR ($T = 20$ °C) with OSN to recycle the mother liquor containing Deferasirox and separate the impurity 4-hydrazinobenzoic acid (4-HBA) from the API to meet regulatory and process specification set for the commercial product. To achieve this goal the slurry of the continuous antisolvent crystallization was withdrawn intermittently and passed through an inline micro-filter unit to extract crystals from the mother liquor. The mother liquor, then enriched in impurities relative to the API, is fed into a holding vessel and from there into an OSN for concentration. The membrane in the OSN is preferentially retaining the API (rejection factor of 0.96) compared to the primary impurity (rejection factor of 0.57) thus reducing the amount of API that is lost *via* the membrane separation. Following the OSN, the mother liquor is returned to the MSMPR. The application of membrane technique as a stand-alone unit operation would result in a yield of only 25% to meet the target process purity of the main impurity 4-HBA. Batch crystallization achieved yields of 89% (at equilibrium) and purity of 0.32 ppm while continuous crystallization (1 h residence time) resulted in yields of 70% and purity of 0.13 ppm (at steady state). However, when integrating membrane recycle of concentrated mother liquor during the continuous crystallization in a hybrid process setting, yields of 99% with 0.15 ppm of the main impurity (max. 3 ppm allowed) were achieved. Since the only route out of the system is through the membrane for liquids and the micro-filter for solids, the purity of the solid product is in effect controlled solely by the membrane (although this can be supplemented with mother liquor purging), thus the preferential concentration of impurities by crystallization enables purification *via* nanofiltration with a much wider set of product-impurity systems than would otherwise be possible.

A model-based optimization study from Wang and Lakerveld,[58] using acetaminophen (paracetamol) as model system, supports these findings with respect to the capability of hybrid membrane–crystallization processes to enlarge the range of attainable product qualities (*e.g.*, crystal size) and design spaces (*e.g.*, yield, shorter residence times, temperature) compared to conventional continuous MSMPR (Figure 7.12).

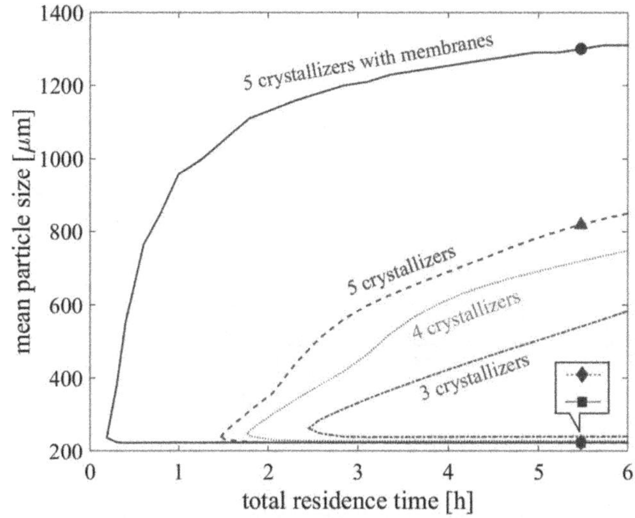

Figure 7.12 Comparison of attainable regions of mean crystal size for different configurations of mixed suspension, mixed product removal crystallizer (MSMPR) cascades and a five MSMPR cascade with membranes.[59] Reprinted with permission from ref. 11, Copyright 2017 American Chemical Society.

The removal of impurities *via* OSN can further be enhanced using tailored complexation, especially when the difference of the molecular diameter between the compound of interest and the impurity is similar. Those compound-impurity systems with similar structures and molecular weights are characterized by the strong tendency of the impurities to incorporate into the crystal lattice, making it difficult to obtain a high purity with a single crystallization or even multiple crystallization steps. Alternatively, complexation of the impurity with an additive can be used to sterically prevent impurity incorporation in the host lattice where OSN is utilized to preferentially reject the higher molecular weight impurity complex in solution, while allowing the lower molecular weight compound to permeate through. The larger dimension of the complex translates into a higher rejection ratio and better separation of the OSN where the permeate stream exiting the membrane contains now less impurity and can be concentrated and recycled in a continuous mode to enhance the yield (Figure 7.13).

This PI approach for enhanced purification forms the basis of the application of complexation-assisted OSN with continuous crystallization in a hybrid setting while taking advantage of two synergistic effects.[59] Firstly complexation of impurity reduces the ability of a lattice-substituting impurity to partition into the solid phase product, which increases the power of the crystallization to concentrate impurities relative to the product in the stream sent to OSN for recycle. Secondly, large differences in membrane rejection factor can be generated between the compound and impurity, potentially broadening the scope for separation of any system *via* OSN.

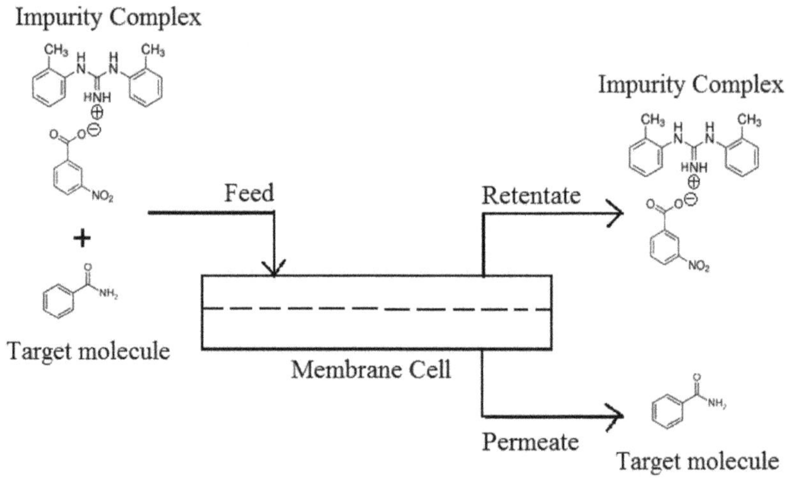

Figure 7.13 Schematic of membrane separation of target molecule (benzamide) from the impurity complex (3-nitrobenzoic acid (impurity)–1,3-di-*o*-tolylguanidine (additive)). Reprinted with permission from ref. 59, Copyright 2017 American Chemical Society.

This builds upon a large synergistic performance gain by combining the two operations without complexation, making it a very powerful purification strategy that can combine the broad purging power of crystallization with targeted enhancement of persistent impurities.[59,203] Vartak and Myerson,[59,203] have demonstrated the capabilities of the complexation-assisted OSN with continuous crystallization to improve the crystal purity for the four model systems 4-nitrophenol, acetaminophen, benzamide, and ketoprofen.

In addition, it has been demonstrated that OSN-assisted chiral resolution is possible. One enantiomer permeates through the membrane while a supramolecular complex is formed between the other enantiomer and a catalyst, which is retained by the membrane.[210] Racemic phenethyl alcohol was also resolved using OSN separation *via* complexation.[211] However, to the best of our knowledge, separation of supramolecular complexes[212,213] using membranes[210,211] has not been coupled yet in a hybrid setting with continuous crystallization.[190,202,209]

Nevertheless, membranes have been combined with preferential crystallization for chiral separations of conglomerate forming systems[214–216] and racemic compound forming systems.[204,205] While membranes have been used for conglomerates to prevent the mere transport of enantiopure crystals between two compartments of a crystallizer[214,215] or two crystallizers,[216] membranes have been used for racemic compound forming system as an enantiomeric enrichment step coupled with preferential crystallization.[204,205] Unlike conglomerates, racemic compound forming systems cannot be separated directly by preferential crystallization as a stand-alone technique.[217] Consequently, a hybrid setting of the membrane technique for enantiomeric enrichment and preferential crystallization of the enantiopure crystals generates beneficial synergistic effects. Gou *et al.*[204] and Robl *et al.*[205] demonstrated this approach for the chiral separation of mandelic acid from racemic solutions. Although, the results showed that enantiomeric excesses of 10–20% could be achieved, to date, such a hybrid membrane-crystallization setting has not been established for continuous operation to the best of our knowledge. General drawbacks of the membrane technology that currently limits its application potential are the relatively low transport rates through the membranes (transmembrane flux), the risk of membrane fouling, the selectivities offered by the available chiral selectors contained in liquid supported membranes, and the instability over long periods of time when liquid membranes are used.[190,205]

7.4.1.3 Distillation-crystallization Process

Distillation is the workhorse of the chemical process industry for the separation of liquid mixtures.[218] However, the separation by distillation is hampered for systems with close boiling points, azeotropes and heat-sensitivity, while crystallization is limited for systems with eutectic compositions, which prevents purification into the pure components.[219] On this account, hybrid process settings of distillation coupled with melt crystallization, first proposed by Rittner and Steiner[220] in 1985, have a significant potential for PI because they provide an efficient pathway to bypass thermodynamic boundaries that neither distillation nor crystallization can overcome alone when used separately (*e.g.*, high energy-consumption at small separation factors and limitation of yield by eutectics, respectively).[184,186,187,219,221] The synergistic effects lie in the high capacity of distillation and the high selectivity of melt crystallization. For fundamentals of melt crystallization the interested readers are referred to Chapter 11 and specialist literature.[222–225] Such hybrid separations have found industrial application for the separation of isomers of xylenes [2], dichlorobenzenes [3], carbon acids [4,5] and diphenylmethane diisocyanates [6].

Berry and Ng[219] identified two classes of hybrid distillation–crystallization processes for binary mixtures. While class I is aimed to bypass barriers imposed by the vapor–liquid equilibrium (*e.g.*, azeotropes), class II is used to bypass barriers imposed by solid–liquid equilibrium (*e.g.*, eutectic points). Figure 7.14 shows the principle of distillation-crystallization hybrid configurations used to tackle these two classes. The numbers in the flow sheets correspond to the process points in the phase diagrams.

Generally, the feed composition of the mixture and the physical properties determine which flow-sheet structure should be considered for the separation task. In the following, the two settings will be briefly explained and the interested reader is referred to specialist literature for further details.[219] The Class I configuration as shown in Figure 7.14a is designed to recover

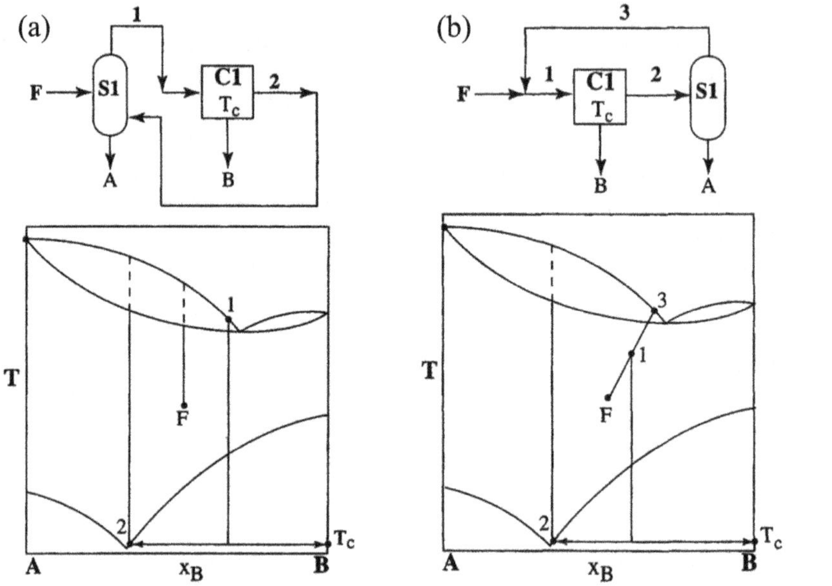

Figure 7.14 Distillation-crystallization configuration to separate an (a) eutectic and (b) azeotrope. Reproduced from ref. 221 with permission from John Wiley and Sons, Copyright © 1997 American Institute of Chemical Engineers.

the heavy component in the bottom of the distillation column (S1) when the composition of component B in the feed ($x_{B,F}$) and at the eutectic point ($x_{B,e}$) is below the composition of B in the azeotrope ($x_{B,a}$). In the case where $x_{B,F}$ and $x_{B,e}$ are both greater than $x_{B,a}$, the light component is recovered as the distillate in the distillation column. The distillate (stream 1) in Figure 7.14a is a resulting mixture enriched with B in respect to the eutectic point. The cooled stream 1 is then sent to the crystallizer C1 that operates at temperature T_c slightly above the eutectic temperature to prevent cocrystallization of A while maximizing the yield. Crystals of B are discharged from the system by filtration while the mother liquor is preheated before being recycled back to the distillation column. The systematic study by Berry and Ng revealed that such a hybrid process for a base-case system requires 43% and 9% of the number of stages and the vapor boilup compared to distillation, respectively.

The class II configuration requires that both $x_{B,F}$ and $x_{B,a}$ are either smaller or greater than $x_{B,e}$. The example shown in Figure 7.14b recovers B by crystallization and A by distillation and the process proceeds as follows. The feed is combined with the distillate of the distillation column S1 (stream 3) that results in stream 1, which is cooled before entering the crystallizer C1 at T_c. Component B crystallizes and is discharged from the system by filtration. The mother liquor (stream 2) is preheated and fed into the distillation column where component A is recovered as the bottom product and the distillate (cooled stream 3) is recycled.[219] Such a hybrid process setting requires 34% and 45% of the number of stages and vapor boilup, respectively, given that the composition of B in stream 1 should be approximately 10 mol% above the eutectic composition to be crystallized.[219] An abundance of work exists on hybrid distillation–crystallization processes.[183,184,186,187,219,221] However, although melt crystallization can be operated continuously[222,223,225] (see also Chapter 11) or semi-continuously, the majority of examples in literature use a static layer melt crystallization process.

7.4.2 Process Integration

Originally conceived as a purification and separation process, crystallization has also become an important technique in particle design from macro to nano-sized particles. Generally, crystallization processes are tailored (i) to obtain the product purity needed with maximized yield and (ii) to improve the downstream operations such as filtration, drying, handling, milling, powder mixing, and tableting. Besides processing, physicochemical properties of the final drug product, such as dissolution rate and bioavailability, also depend on the size and the morphology of the crystals.[143,226,227] Up to date, solid dosage forms (*e.g.*, tablets, capsules) are the most convenient and industrially relevant drug delivery form for peroral administration due to their low cost, good stability, and patient-friendly administration.[228,229] On this account, appropriate flowability and handling properties of pharmaceutical powders in downstream processing such as direct tablet-compression or capsule-filling are paramount. Conventionally, milling and granulation/agglomeration are often employed as unit operations subsequent to crystallization either for better bioavailability or for improving flowability, compressibility, and content-uniformity by avoiding demixing of powders when transported. Despite the fact that milling and granulation improve the overall crystal size, it comes often at the expense of a wider CSD[24] and can only be performed on mixtures that are not moisture sensitive and may potentially cause polymorphic transformations of the API.[230–232] However, instead of operating crystallization, milling, and granulation separately, process integration by spherical crystallization (crystallization combined with agglomeration)[233] as well as wet milling combined with crystallization[234] generate synergistic effects that permit some of these challenges to be overcome and minimize equipment cost, process time, and energy compared to when they are operated as stand-alone unit operations.[235]

7.4.2.1 Spherical Crystallization

Spherical crystallization, developed by Kawashima *et al.*[233] using salicylic acid, is a method that increases the size of crystals by transforming the crystals directly into compact spherical particles through agglomeration during the crystallization process leading to improved crystal handling properties. Thus, spherical crystallization is inherently a PI technique due to its ability to reduce cost and labor associated with pharmaceutical production by eliminating the downstream particle property enhancing unit operations like granulation.[24,27] Various review articles provide an overview of the different spherical crystallization methods commonly employed.[236–238] Initially, the spherical agglomeration and quasi-emulsion solvent diffusion methods were employed. Later, the crystallo-co-agglomeration and the ammonia diffusion system have been introduced to overcome the limitations of the previous methods.

The *spherical agglomeration method* directly agglomerates crystalline material into spherical agglomerates by the addition of a bridging liquid to a precrystallized broth under agitation. The crystals are commonly obtained by mixing a solution (solvent containing an API) with an antisolvent, provided that the solvent and antisolvent are freely miscible and the affinity between the solvent and antisolvent is stronger than the affinity between the API and the solvent. The bridging liquid preferentially wets the crystals[239,240] and causes the wetted crystals to adhere and create agglomerates upon collision.[233]

The *quasi-emulsion solvent diffusion method*, first introduced by Kawashima *et al.* in 1989,[241] requires that the interactions between the API and the solvent are stronger than those between the solvent and antisolvent. When the solution is dispersed into the antisolvent, transient-emulsion (quasi-emulsion) droplets are produced, even though they are miscible. The transient emulsion is unstable due to increasing interfacial tensions between solvent and antisolvent. The solvent gradually diffuses to the outside of the emulsion droplets while the antisolvent diffuses into the droplets, which reduces the solubility of API inside the droplets leading to supersaturation and eventually causes the API to crystallize. The remaining solvent in the droplet acts as the bridging liquid that assists in the formation of the spherical agglomerates.

The *crystallo-co-agglomeration method* is a modified form of the spherical agglomeration method, first described by Kadam *et al.*[242] This method is aimed at overcoming the limitations of spherical agglomeration to obtain directly compressible agglomerates of low-dose and poorly compressible APIs and the design of API combinations.[243–246] In particular, an API is crystallized and agglomerated with an excipient[243–245] or another API, which may or may not be crystallized in the system.[246]

The *ammonia diffusion method* is employed for amphoteric substances, when spherical agglomeration and quasi-emulsion solvent diffusion methods cannot be used because amphoteric substances are insoluble in neutral aqueous solutions or organic solvents but soluble in acidic or alkaline solutions.[236,237] Ammonia water, which is alkaline, serves as the solvent and bridging liquid.[247] In particular, similar to the quasi-emulsion solvent diffusion method, the ammonia water diffuses out of a droplet and the antisolvent diffuses into the droplet causing crystallization. The remaining ammonia water aids the formation of the spherical agglomerates.

The literature on continuous spherical crystallization has been very limited until recently.[248,249] However, the currently ongoing paradigm shift from batch to continuous processing for pharmaceutical manufacturing has led to new studies applying the spherical agglomeration method using a two stage MSMPR cascade[24] and an OBC[53] as well as the quasi-emulsion solvent diffusion method using an MSMPR.[27] Nevertheless, to control the quality of products manufactured *via* the use of simultaneous crystallization and agglomeration processes leading to particle granulation requires typically extensive research and development,[237] since for each API and specific property desired the optimization of solvents, excipients, and process parameters must be performed.[27]

7.4.2.2 Integrated Wet Mill Crystallization

While large crystals are preferred for the efficiency of downstream operations, small and uniform crystals are also of great importance, especially for the pharmaceutical industry, due to their enhanced solubility and dissolution rate that ultimately improves the bioavailability of an API. Low bioavailability of poorly water-soluble APIs is a challenge for API formulation and even for the selection in the drug discovery phase.[250,251] About 40% of currently marketed drugs are poorly soluble and nearly 90% of drugs in the discovery pipeline have low aqueous solubility.[252] Typical crystallization approaches to manipulate the particle size encompass seeding combined with controlled temperature or/and antisolvent addition profiles, supersaturation control employing various types of mixers or large crystal dissolution in a recycle loop.[253] In addition, it is common to use milling (*e.g.*, jet milling, wet milling) for further size reduction when crystals cannot be produced in the desired size range through crystallization.[251,254–257] However, instead of employing the unit operations crystallization and milling one after the other, recent work by Yang *et al.*[25,258] and Acevedo *et al.*[234] have demonstrated that the process integration of wet milling (rotor stator wet mill) and continuous crystallization allows synergistic effects for PI. In particular, the process integration (i) helps to reduce the number of unit operations, (ii) improves the chemical and physical stability of the product and (iii) enhances the safety by reducing the risk of API dust exposure and API dust explosions.[259] In addition, both primary and secondary nucleation rates are dramatically increased under the high shear environment,[260–263] which enables the application of wet milling either as an *in situ* seed generator or as a size reducing tool for existing crystals within continuous crystallization processes (Figure 7.15).[25,258]

When a wet mill is integrated in the feed stream of an MSMPR (upstream), a higher yield, narrower CSD (either small or large depending on the tip speed of the rotor stator wet mill), and quicker startup of an MSMPR can be achieved due to the improved nucleation kinetics (Figure 7.15a).[25] Thus, this concept offers an attractive alternative to the commonly used dry seeding technique, which requires the introduction of foreign material that has to

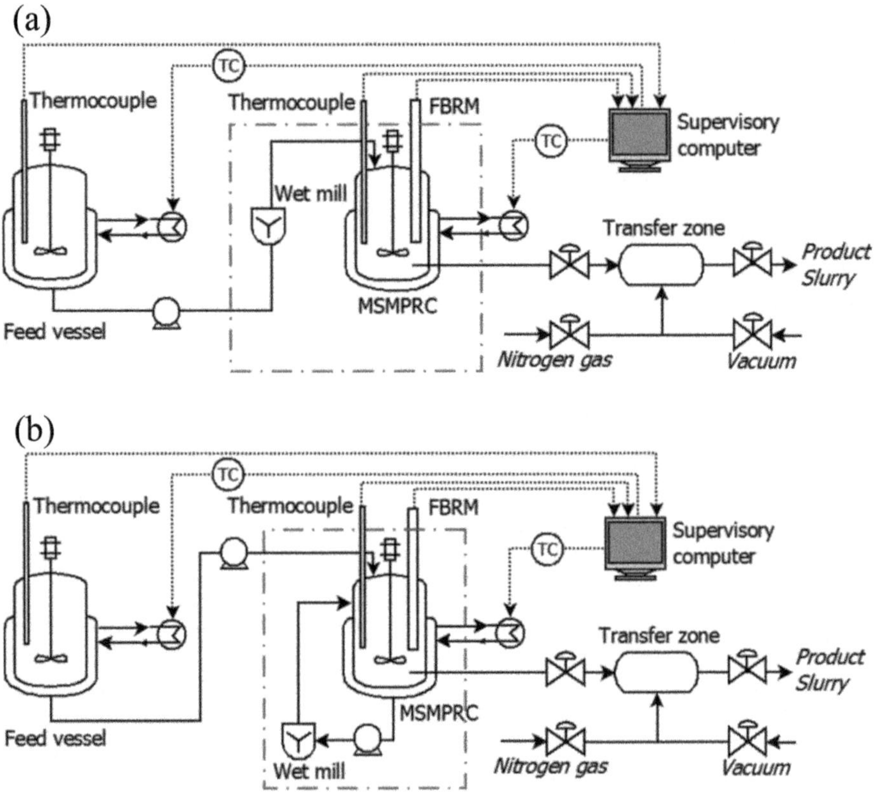

Figure 7.15 Process integration of wet mill and continuous crystallization (MSMPRC). (a) Wet mill operated upstream and (b) wet mill operated downstream. Reprinted and adapted with permission from ref. 25, Copyright 2015 American Chemical Society.

be tightly regulated.[11] Theoretically this concept should also be applicable to PFC but to the best of our knowledge this aspect has not been reported in the literature yet. On the other hand, when a wet mill is applied either in the recycling loop (downstream) of an MSMPR (Figure 7.15b)[25,258] or directly *in situ*[234] smaller average crystal sizes and narrower CSDs compared to MSMPRs operated without wet milling can be achieved due to the enhanced secondary nucleation and breakage mechanisms.

7.4.2.3 Multifunctional Equipment

A novel approach of process integration is the use of multifunctional equipment that combines multiple unit operations into a single unit operation with the purpose of eliminating challenging slurry and solid handling (*e.g.*, filtration, drying, powder mixing, tableting) for more robust continuous operation.[264] Coupling the purification and solid–liquid separation into a single unit operation without filtration may be beneficial when specific powder-free formulation techniques are utilized that require impurity-free solutions to make composite or specialized solid dosage formulations. In recent years, several powder-free formulation techniques have been developed that could enable powder-/slurry-free production of orally dosed tablets, such as crystallization (i) within cast polymer thin films,[265] (ii) within co-dissolved polymers using spray drying, electro-spraying, or electro-spinning,[266] (iii) on the surface of polymers,[136,138] and (iv) within hydrogels.[165,174,267,268]

Yazdanpanah *et al.*[264] demonstrated the capability of the so-called filtration avoidance approach that combines crystallization, filtration, drying, and dissolution in a single unit to avoid the transfer and isolation of solids. The basic concept is adapted from falling film melt layer crystallization.[222,223,225] The falling film solution layer crystallization involves the crystallization on a surface from a superheated thin liquid film of crude mother liquor flowing over the cooled surface (Figure 7.16). The solution quickly cools as it flows down the cold surface and becomes supersaturated leading to firstly nucleation followed by growth onto the nucleated layer. The residual solution is collected in a superheated section beneath where it can be recirculated to improve the yield. Once the crystal layer has been grown to the desired thickness or the mother liquor has reached equilibrium, the crude solution is removed from the

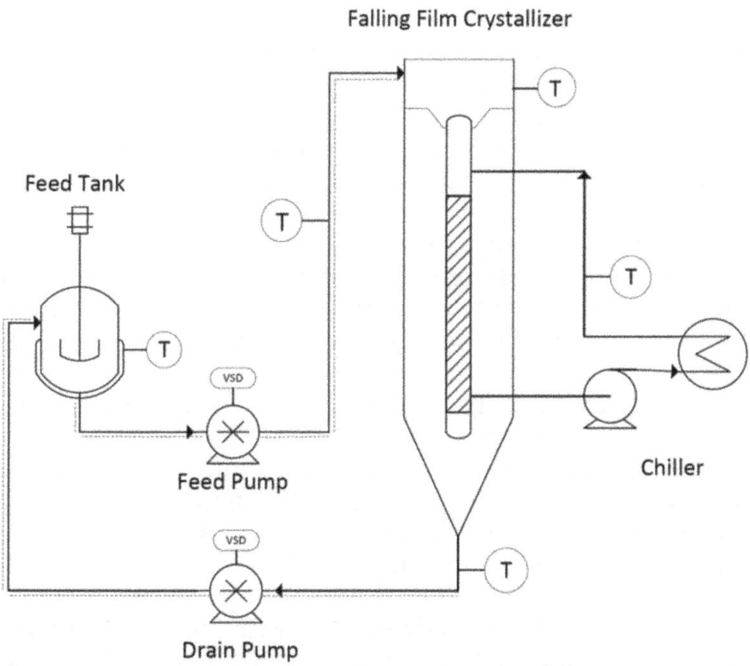

Figure 7.16 Process flow diagram of falling film layer crystallization. Reprinted with permission from ref. 264, Copyright 2016 American Chemical Society.

column. The remaining solid crystal layer can then undergo a secondary purification operation (washing), if desired, to remove entrained solvents and impurities. Subsequently, clean solvent can be added to dissolve or partially dissolve the layer, to provide a purified solution to be sent to the next unit operation without the generation of a slurry. The next unit operation might be another reaction, a powder-free formulation, another falling film stage, or further purification.

The column containing the tube is sealed, and the second outside interface to the crystal layer is composed of air but could also be a gas phase of some type in which the vapor pressure can be controlled. Yazdanpanah *et al.*[264] discussed this feature as a possible further PI approach to foster evaporation from the surface (vacuum) or for the stripping of solvent (dry gas flow) to boost the process yield or to drive the crystallization process. Here, the stripping of solvent with dry gas as the primary driving force for crystallization would enable the generation of thicker crystal layers because the heat transfer limitations are eliminated. Moreover, an anti-solvent or a reagent for a reactive crystallization could be added theoretically through the gas phase to broaden the application of the filtration avoidance approach to other forms of crystallization. The purification capability of the filtration avoidance concept was tested with a solution composed of 98% fenofibrate and 2% fenofibric acid (impurity) in pure ethanol, achieving an average final purity of 99.4% and a relatively high yield of 72%. A drawback of this process integration approach is the limited surface area, which is governed by the surface area of the cooling tube, compared to suspension crystallization (surface area provided by the crystals). Although, more surface area could be supplied by adding more tubes in a single unit to boost the productivity of the device, the surface area will still be about two orders of magnitude lower than in suspension crystallization.[222] Moreover, the combination of all unit operations in a single unit leads intrinsically to a consecutive (batch) operation. To compensate, multiple single units might run in parallel at different stages (semi-batch) to maintain overall continuous operation from crystallization to dissolution. While one might operate in the crystallization mode, others would already be in the washing, drying, or dissolution phase. In an effort to identify optimized process conditions a numerical study was conducted. The mathematical model can be used to predict and model the solute concentration, temperature profile, crystal layer thickness, yield, and purity over time for further optimization by determining the results for a combination of process parameters.[269]

In addition, recent efforts in compact (~1/40 of the current industrial size), on-demand continuous flow manufacturing platforms for the end-to-end manufacturing of pharmaceuticals have led to further innovations in crystallization equipment through process integration.[12,14] For instance, Wong *et al.*,[72] reported a crystallization process combined with a filtration-drying-dissolution unit that could be utilized to maintain a small environmental footprint of compact manufacturing platforms.[12,14] An advantage of this approach is that the post-crystallization operations (filtration, drying, dissolution) can be operated within the same customized setup once the crystallization is finalized. The results using diphenhydramine hydrochloride, fluoxetine hydrochloride, and ibuprofen as model systems demonstrated satisfying operation (yields between 78 and 99%) with reduced loss of material, due to the fact that slurry and/or solid transport to the next unit operation is not needed. However, similar to the filtration avoidance approach, it comes with the expenses of consecutive batch operation for each unit operation within the setup. To maintain an overall continuous operation and to boost the production capacity of the innovative crystallization system, multiple units might operate in parallel in a semi-continuous mode.

7.5 Energy Domain

The focus of PI approaches in the energy domain is on the development of novel concepts that deliver energy from the source to the intended place of action in the required form and amount.[7] Transfer of energy for crystallization processes may be needed (i) to generate supersaturation (*e.g.*, in case of cooling or evaporative crystallization), (ii) to dissolve crystals, (iii) to overcome the energy threshold for primary nucleation, or (iv) to induce secondary nucleation *via* crystal breakage or attrition. A significant number of studies exist in which alternative forms of energy such as ultrasound, electric, and magnetic fields are applied to crystallization processes. In general, by varying the properties of such fields, additional control over the crystallization process can be obtained.

Although the majority of studies on the effect of alternative energy forms involve batch crystallization, PI methods in the energy domain can typically benefit from the inherent scale-up advantages of continuous crystallization when conducted in flow channels. In particular, batch crystallization requires by definition all material to be processed at the same time, which may pose inherent limitations for transfer of alternative forms of energy to larger volumes due to, for example, rapid energy dissipation (*e.g.*, ultrasound) or impracticality of establishing potential energy fields of high strength (*e.g.*, in case of electric fields or microwaves). In contrast, energy can be transferred during the complete operational time to a smaller volume when operating in continuous mode. Furthermore, when batch crystallization is conducted in a well-mixed tank, the area-to-volume ratio will decrease with increasing scale, which may also pose limitations in terms of effective transfer of alternative energy forms in contrast to continuous crystallization (*e.g.*, in a flow channel), where a larger surface-to-volume ratio is available for the energy transfer. This difference in the surface-to-volume ratio can be exploited to intensify the energy transfer, for instance, by using multiple transducers for ultrasound or electrodes for electric fields.

7.5.1 Ultrasound

Ultrasound is the most studied alternative form of energy to be used for solution crystallization processes with dedicated literature spanning several decades. The ability of ultrasound to discharge metastable systems was already reported in 1927.[270] Since then many more studies have been conducted on so-called sonocrystallization and there is broad consensus that ultrasound can have a significant impact on crystallization kinetics for a wide variety of systems. However, the precise mechanism is still a topic of debate.[271] The general understanding is that ultrasound can induce the formation of cavitational bubbles, which can implode when instable and thereby locally create high temperature and pressure. There are several mechanisms through which the crystallization kinetics can be affected as a result of such cavitation: (i) acoustic streaming, (ii) microjets,[272] (iii) heterogeneous primary nucleation,[273–275] (iv) heat, and (v) pressure.[276] The observed effects of ultrasound on crystallization systems vary and include (i) enhanced or (apparent) inhibited primary nucleation, (ii) enhanced secondary nucleation (*e.g.*, attrition), (iii) breakage, (iv) reduced fouling, and (v) de-agglomeration.[277,278]

Different crystallization configurations and hardware to induce ultrasound into continuous crystallization have been reported (Table 7.3). A typical objective is to supply sufficient energy to overcome the threshold for cavitation but avoid excessive heating of the solution. In addition, the equipment should facilitate a uniform cavitation field to obtain a predictable influence when scaling up. For instance, Eder *et al.*[34] combined sonocrystallization for seed generation with a segmented flow crystallizer as dedicated growth compartment for continuous cooling crystallization of acetylsalicylic acid from solution. Specifically, a segmented flow crystallizer was placed inside an ultrasonic bath to induce nucleation for seed generation, which allowed control over the final CSD.[34,43] Jiang and coworkers[279,280] made a variation to that concept by replacing the ultrasonic bath with an ultrasonic probe that was pressed against the outside of the tube, which enabled a more localized generation of the ultrasound field. Furuta *et al.*[281] placed an ultrasonic generator inside a temperature-controlled tank that contained a tubular crystallizer (20 m to 40 m long) for the crystallization of a commercial API. This PI approach helped to mitigate fouling issues while achieving a desired particle size (1–7 μm), which was not possible when using a conventional fed-batch system. Siddique *et al.*[127] describe how the crystallization of α-lactose monohydrate in an OBC can be intensified by applying sonocrystallization for seed generation. Although, the CSD of the final product was bimodal in the OBC, similar to a batch crystallization process, the CSD was narrower with an adjustable mean crystal size and higher yield.

Sonocrystallization has also been applied in continuous milli- and microfluidic settings. Hattoria *et al.*[282] demonstrated sonocrystallization in a continuous microfluidic channel leading to precipitation of

Table 7.3 Overview of crystallization configuration and ultrasound equipment for sonocrystallization in continuous flow.

Crystallization Configuration	Estimated scale	Acoustic field properties	Ultrasound hardware	Reference
PFC				
1	500 mL (seed batch) 50–100 mL (tubular crystallizers)	35 kHz	Ultrasonic bath (seed generation)	34 and 43
2	60–120 mL	40 kHz 500 W	Cylindrical immersible sonicator	281
3	100 mL	20 kHz	Probe in contact with outside of tubing for seed generation	279 and 280
Milli- and Microfluidic Channels				
1	8 μm^3	40 kHz	Digital ultrasonic generator to vibrate a microreactor	282
2	1.0 mL	42–660 kHz	Piezoelectric element	284
3		20 kHz	Probe with tip placed at 2 mm from the capillary	283
4	0.25 mL	20 kHz	Sonotrode mounted to the side of the micro heat exchanger	308
Oscillatory Baffled Crystallizer				
1	10 L	20 kHz (10–70W)	Probe system	127
MSMPR				
1	200 mL (one MSMPR)	24–41 kHz	Lab-scale probe system (30 kHz or 36 kHz)	290
	870 mL (two MSMPRs in series with recycle)	50 W/L (lab scale)	Lab-scale transducer glued to the bottom of a glass plate (41 kHz)	
	50 L (semi-pilot MSMPR)	16 W/L (semi-pilot scale)	Radial oscillating probe at semi-pilot scale (24 kHz, 2 kW, and/or 30 kHz, 1 kW)	
2	300 mL	20 kHz (90 W/100 g or 9 W/100 g)	Probe system	287 and 288
3	500 mL	20 kHz (8–60W)	Probe system	289

indomethacin along a liquid–liquid interface, while no precipitation was observed in the absence of ultrasound. While Hattoria *et al.*[282] applied the ultrasound by vibrating the entire microreactor with an ultrasonic oscillator, Rossi *et al.*[283] have studied a targeted sonocrystallization by placing the probe at short distance from the microfluidic channel after a cooling section for adipic acid crystallization (Figure 7.17). This PI approach resulted in narrower CSD, smaller mean crystal size (*ca.* 15 μm), and higher crystal production rate.

On the other hand, applying ultrasound over the complete length of the channel has the advantage that fouling can be reduced. This remains a major problem of tubular crystallizers compared to an MSMPR crystallizer (Figure 7.18). In an effort to further understand the influence of ultrasound on the crystallization kinetics of this millifluidic tubular crystallizer, Jamshidi *et al.*[284] conducted experimental and computational studies to prove that ultrasound mainly affects the nucleation but does not significantly affect crystal growth and/or induce crystal breakage.[284]

Adding seeds to crystallization processes is a classical PI approach to mitigate fouling problems, control the supersaturation level, and to gain control over CSD. Here, Devarakonda *et al.*[285] used a flow cell coupled with ultrasound to generate seeds followed by their growth in a batch stirred crystallizer. Although this study used a batch crystallizer setting, it shows a practical way of conducting sonocrystallization in continuous flow, which could be easily fed into a continuous crystallizer. Moreover, this study demonstrated that the residence time in the flow cell had a significant impact on the mean size of the seed crystals and that the ultrasound effectively disrupts agglomerated seed crystals. Gielen *et al.*[271] integrated a similar flow cell concept in a recirculation loop of a stirred batch crystallizer, which could also be replaced with an MSMPR. In addition, this PI approach revealed that secondary nucleation had a larger impact compared to primary nucleation on the final CSD when using ultrasound in the presence of crystals. However, the results presented are system dependent because exposure of crystals to ultrasound in absence of significant growth may reduce the apparent secondary nucleation rate due to ripening from heat effects of the ultrasound, as it has been demonstrated for other systems.[286]

While many studies have reported the application of sonocrystallization in batch operated well-mixed vessels employing a probe system, this PI approach can be extended easily to MSMPRs. For instance, Narducci *et al.*[287] investigated a MSMPR crystallizer with a probe system for continuous cooling crystallization of adipic acid from water to produce small spherical crystals consistently, which can have important implication for applications in which otherwise milling for size reduction would be needed. It was also found that ultrasound could shorten the time to steady state, produce a more uniform product quality, and reduce fouling on the crystallizer hardware.[288] Sayan *et al.*[289] investigated the sonocrystallization of potassium dihydrogen phosphate in an MSMPR crystallizer with a probe system and concluded that

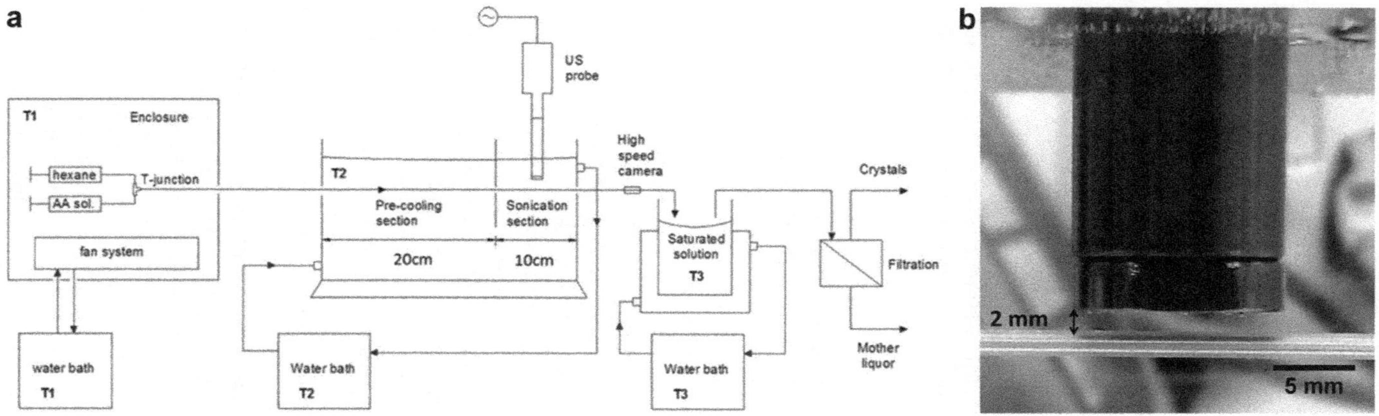

Figure 7.17 (a) Experimental setup of a continuous droplet-based capillary sonocrystallizer and (b) tip of an ultrasound probe placed above a microfluidic channel. Reprinted with permission from ref. 283, Copyright 2015 American Chemical Society.

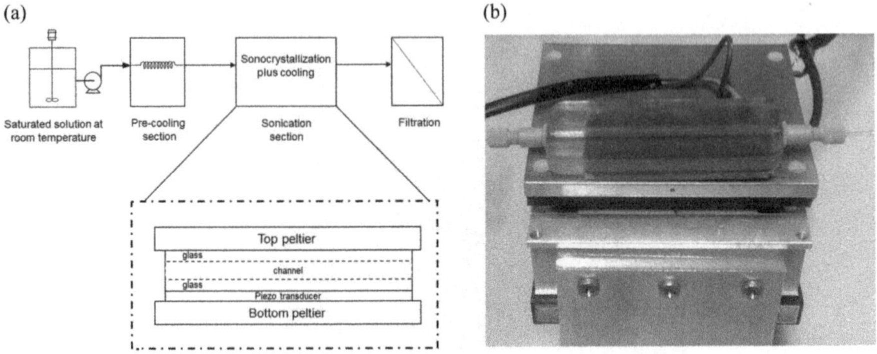

Figure 7.18 Illustration of continuous sonocrystallization in a millifluidic device (a) integrating a piezoelectric element for transfer of ultrasound and Peltier elements for removing heat (b). Reprinted with permission from ref. 284. Copyright 2016 American Chemical Society.

ultrasound has a significant effect both on nucleation as well as on crystal growth kinetics. The mean crystal size decreased by a factor of four when ultrasound above a minimum energy threshold of 16 W/L was applied. In addition, the crystal surfaces contained defects and the crystal shape was affected significantly by the ultrasound.

The scale-up of sonocrystallization of manganese carbonate in an MSMPR crystallizer was studied by Gielen *et al.*[290] It was found that the use of a single transducer at larger scale is likely to be ineffective due to the fast dissipation of ultrasound and limited penetration depth.[278] Therefore, it is recommended to design the equipment for sonocrystallization based on multiple transducers. First, the authors tested both a probe and a transducer at lab scale to determine the best performing design. Second, a 50 L MSMPR crystallizer was constructed in which two probes could be placed (2 kW at 24 kHz and 1 kW at 40 kHz). The two probes emitted the ultrasonic field predominantly in different directions (*i.e.*, one in axial direction and one in radial direction) to improve uniformity of the ultrasonic field. They found that a comparable size was obtained under silent and sonicated conditions in contrast to experiments on a smaller scale, which was attributed to attrition in the pump of a recirculation system and the lower power input at larger scale (*i.e.*, 16 W/L at semi-pilot scale compared to 50 W/L on lab-scale). At a larger scale, multiple transducers and radial oscillating probes are better capable of producing an effective ultrasound field. For example, modular flow cells with a height of 1 m each and 40 transducers operating at a maximum power of 50 W have been presented in literature.[278] Despite the growing number of academic studies in the application of ultrasound in continuous settings, reports on the use of ultrasound on an industrial scale are highly limited in the open literature.

7.5.2 Electric Fields

Electric fields have mostly been applied to crystallization of biomacromolecules such as proteins in order to alter the nucleation and growth kinetics of proteins. Large and high quality protein crystals are needed for structural studies based on X-ray diffraction. However, spontaneous nucleation of large biomolecules is often accompanied with a long induction time even in the presence of high supersaturation. This is supported by the fact that the nucleation mechanism of proteins in solution is itself not fully understood yet. Consequently, electric-field-assisted protein nucleation is at its beginning in terms of understanding. However, it has been widely observed that electric fields of various forms (*e.g.*, external, internal, alternating current, direct current, *etc.*) can have a positive effect on protein nucleation. A recent review of electric-field-assisted protein crystallization has been provided by Nanev.[291] While the majority of studies employing this PI concept focused on generating single crystals for X-ray studies in small batch system with electrodes, protein crystallization as a separation and purification process (*e.g.*, for the manufacture of biopharmaceuticals) suffers scale-up issues (ineffective and unpredictable at best) of the electric-field-assisted batch protein crystallization. Thus, continuous crystallization offers PI opportunities to employ electric-field-assisted protein crystallization as a separation and purification method of proteins. Similarly to ultrasound, continuous crystallization allows for the application of the external field over a smaller volume. Furthermore, when continuous crystallization is executed in a flow channel, a large surface-to-volume ratio provides flexibility to design electrodes with a large contact area with a supersaturated solution. Li and Lakerveld[292] extended the electric-field-assisted protein crystallization from a conventional batch configuration to a configuration based on continuous flow by constructing a microfluidic device with two co-planar electrodes that were separated by a gap in the center of the flow channel (Figure 7.19).

It was demonstrated that the use of electric fields can deliver a higher and more consistent yield compared to control experiments in the flow channel without any electric field due to increased nucleation. In general, such intensified protein crystallization processes in flow hold great potential to replace or supplement conventional chromatography-based separation and purification processes in industry. In addition to applications related to protein crystallization, electric fields have also been used for electrochemically induced crystallization[293,294] and crystal separation[295] of smaller molecules. While the former is often based on a pH shift to change the solubility of an organic molecule with an acidic or basic functional group to create supersaturation, the latter is centred on the notion that particle movements in strong electric fields are based on specific properties such as size and shape, which can be exploited for *in situ* particle separation during crystallization. However, the translation of such concepts towards continuous crystallization is still in its infancy.

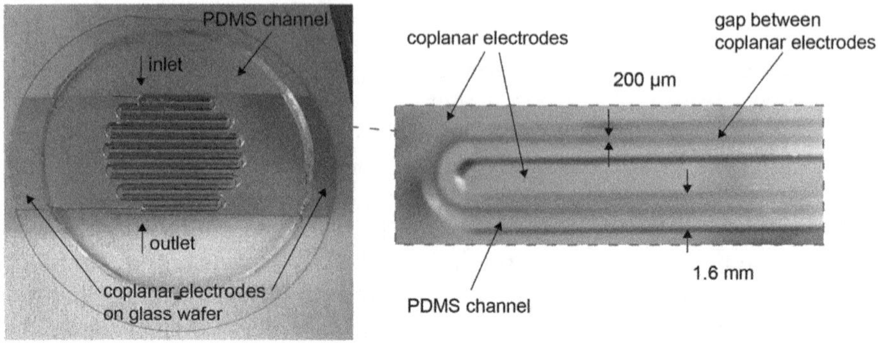

Figure 7.19 Microfluidic device enabling electric-field-assisted protein crystallization in continuous flow. A low-voltage AC electric field is placed over the co-planar electrodes to enhance the crystallization process. Reprinted with permission from ref. 292, Copyright 2016 American Chemical Society.

7.5.3 Microwave Fields

The use of microwave irradiation allows for the rapid heating of a process medium and the manipulation of crystal nucleation and growth kinetics by increasing mass transfer phenomena, which has been widely applied for chemical reactions and in equipment for microwave-assisted processing in continuous flow.[296,297] Microwaves enhanced mass transfer in an MSMPR crystallizer for crystallization of magnesium carbonate from water, which increased the nucleation rate and crystal growth rate.[298] Rodríguez-Clemente and Gómez-Morales[299] studied the reactive crystallization of calcium carbonate from aqueous solution in an MSMPR crystallizer. The crystallizer was made out of microwave transparent PTFE and placed inside a microwave furnace in such a way that the feed and product could be transferred easily. They found that microwaves did not have a significant effect on crystal growth, but that the nucleation rate was several orders of magnitude higher if microwave irradiation was used. Furthermore, steady state was reached faster with microwave irradiation (within 2–3 residence times instead of within 6–10 residence times for a conventional MSMPR crystallizer). This effect was explained with the enhanced nucleation rate by a faster dehydration of the hydration shell of the cations, which is known to be a rate-limiting step for the precipitation of calcium carbonate. A similar mechanism was considered responsible for the observation that microwaves provided increased control over the polymorphic form of calcium carbonate.[300]

Similarly as for the application of ultrasound, the penetration depth of microwaves is limited to a few centimeters,[297] therefore, flow cells may be needed to obtain a uniform irradiation and to avoid hot spot formation. Kacker *et al.*[301] used a custom-made circulation loop through a microwave cavity (Figure 7.20) for fines dissolution during a batch cooling crystallization of paracetamol from isopropanol.

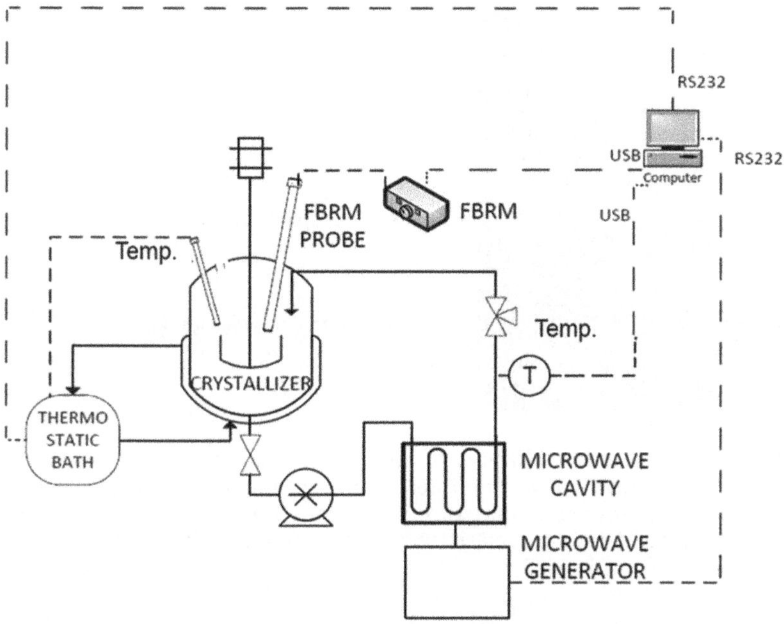

Figure 7.20 Circulation loop through a microwave cavity used for rapid fines dissolution in direct-nucleation control framework of a batch crystallizer. Reprinted with permission from ref. 301, Copyright 2016 American Chemical Society.

For such a system, heat can be added rapidly once fines are detected *via* some *in situ* measurement preventing their growth and, thus, facilitating rapid dissolution. In addition, any conventional heating medium that is used for cooling is not affected by this operation. Therefore, the circulation loop can be designed flexibly and ultimately microwave heating can even be fully integrated within commercially available crystallizer hardware (Figure 7.21).[302]

Although the reporting literature on microwave-assisted crystallization is predominantly focused on batch operation, the commercial availability of instrumentation and the vast experience in microwave-assisted chemical synthesis can support the development of future microwave-assisted continuous crystallization processes to achieve PI for industrial crystallization processes.

7.6 New Challenges for Process Intensification in Continuous Crystallization

Pharmaceutical manufacturing is historically dominated by batch processing, which typically takes place in multiple locations for synthesis, purification, and final drug product formulation resulting in prolonged production times and potential supply chain disruptions. On this account, the interest

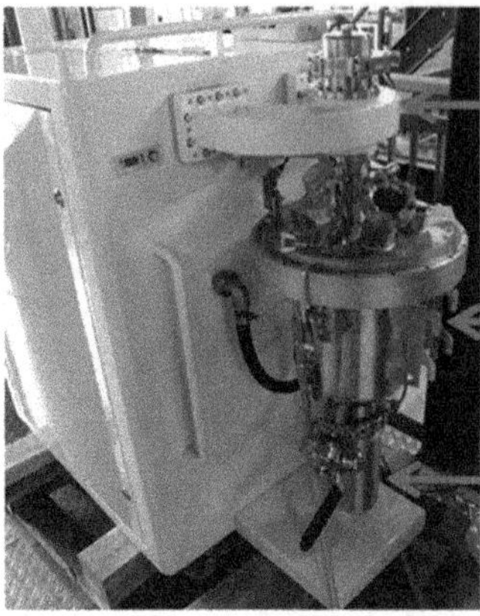

Figure 7.21 Jacketed crystallizer with integrated microwave heating (Siarem SAS Labotron 4000). Reprinted with permission from ref. 302, https://pubs.acs.org/doi/abs/10.1021/acs.cgd.7b00368. Further permissions related to the material excerpted should be directed to American Chemical Society.

in applying PI principles in this industry has grown in the past decade. These efforts have cumulated, thus far, in few proof-of-principle systems capable of end-to-end continuous flow production that integrate synthesis, purification, and formulation in uninterrupted reactor networks to improve efficiency and product-quality control.[12–14,18] An important aspect of this PI approach is the tight integration of the upstream and downstream process streams. Specifically, it could be demonstrated that an integrated crystallization design requires simple yet effective organic flow synthesis routes controlled at the very upstream end (synthesis and work-up) of the process to limit the formation of impurities in the first place, which resulted in reduced complexity of the downstream purification by crystallization and separation.[75,303] While current PI efforts in continuous crystallization are predominantly focused on the integration into downstream processing schemes aiming to replace or reduce the number of downstream unit operations, the integration with upstream flow synthesis unit operations and its impact on continuous crystallization is highly limited.[12,14,303] However, the lack of connectivity between flow synthesis and continuous crystallization leading to overall batch operation hampers advantage being taken of the full benefits of the concept of continuous operation in the PI time.

Moreover, in an effort to address the challenges associated with limited response to sudden changes in demand *e.g.*, epidemic or crisis outbreaks, a compact refrigerator-sized [1.0 m (width) × 0.7 m (length) × 1.8 m (height)], on-demand end-to-end continuous flow manufacturing platform is reported, capable of producing hundreds to thousands of oral or topical liquid doses of four APIs (diphenhydramine hydrochloride, lidocaine hydrochloride, diazepam, and fluoxetine hydrochloride).[12] The second-generation platform is also reconfigurable and 25% smaller [0.5 m (width) × 1.0 m (length) × 1.3 m (height)],[14] which was achieved among other things by applying PI principles for continuous crystallization (*e.g.*, a compact two-stage MSMPR cascade with pressure-driven slurry transfer).[26]

Ultimately, the increasing trend towards continuous manufacturing in the pharmaceutical industry and the accompanying academic research for compact manufacturing platforms[12,14] requires the attention of the vendors of process analytical technology (PAT) to provide solutions for the customers' needs.[304,305] For instance, Mettler Toledo has developed their in-line ATR FT-IR system, known as FlowIR, to meet the demands of flow chemistry.[306] Similar necessary innovative solutions will be needed to support the PI efforts in small-scale continuous crystallization integrated in compact platforms to monitor and maintain the process within its design space (see Chapter 4 for further details on control strategies). For instance, key challenge often overlooked is the physical size of probes in combination with control boxes. Thus, PATs suitable in size (miniaturization) but with the same accuracy as laboratory bench devices are needed.[305] In this context, handheld PATs, already implemented in the pharmaceutical industry for raw material validation and cleanliness checking,[307] could be a potential solution besides innovative compact sensor arrays. However, commercially available handheld devices are currently lacking in adaptors to allow them to be used as plug-and-play PAT in everyday crystallizer setups and thus require further development efforts in the future.

References

1. ACS, Sustainable US Manufacturing: Chemical and Allied Industries Technology, Technology Area 4: Next Generation Chemical Manufacturing, https://www.acs.org/content/dam/acsorg/sustainability/acsandsustainability/sustainablemanufacturing/roadmaps/smrt-technology-area-4-next-generation-chemical-manufacturing-pdf.pdf, accessed September 28, 2017.
2. K. Knoerzer, P. Juliano and G. W. Smithers, *Woodhead Publishing Series in Food Science, Technology and Nutrition*, Woodhead Publishing, 2016.
3. C. Jiménez-González, P. Poechlauer, Q. B. Broxterman, B.-S. Yang, D. am Ende, J. Baird, C. Bertsch, R. E. Hannah, P. Dell'Orco, H. Noorman, S. Yee, R. Reintjens, A. Wells, V. Massonneau and J. Manley, *Org. Process Res. Dev.*, 2011, **15**(4), 900–911.
4. A. Stankiewicz, *Chem. Eng. Prog.*, 2000, **96**(January), 22–34.
5. J.-C. Charpentier, *Chem. Eng. J.*, 2007, **134**(1–3), 84–92.
6. S. Becht, R. Franke, A. Geißelmann and H. Hahn, *Chem. Eng. Process.*, 2009, **48**(1), 329–332.
7. T. van Gerven and A. Stankiewicz, *Ind. Eng. Chem. Res.*, 2009, **48**(5), 2465–2474.

8. J. M. Ponce-Ortega, M. M. Al-Thubaiti and M. M. El-Halwagi, *Chem. Eng. Process.*, 2012, **53**, 63–75.
9. V. Hessel, *Chem. Eng. Technol.*, 2009, **32**(11), 1655–1681.
10. E. Drioli, A. I. Stankiewicz and F. Macedonio, *J. Membr. Sci.*, 2011, **380**(1), 1–8.
11. J. Wang, F. Li and R. Lakerveld, *Chem. Eng. Process.*, 2018, **127**, 111–126.
12. A. Adamo, R. L. Beingessner, M. Behnam, J. Chen, T. F. Jamison, K. F. Jensen, J.-C. M. Monbaliu, A. S. Myerson, E. M. Revalor, D. R. Snead, T. Stelzer, N. Weeranoppanant, S. Y. Wong and P. Zhang, *Science*, 2016, **352**(6281), 61–67.
13. S. Mascia, P. L. Heider, H. Zhang, R. Lakerveld, B. Benyahia, P. I. Barton, R. D. Braatz, C. L. Cooney, J. M. B. Evans, T. F. Jamison, K. F. Jensen, A. S. Myerson and B. L. Trout, *Angew. Chem., Int. Ed. Engl.*, 2013, **52**(47), 12359–12363.
14. P. Zhang, N. Weeranoppanant, D. A. Thomas, K. Tahara, T. Stelzer, M. G. Russell, M. O'Mahony, A. S. Myerson, H. Lin, L. P. Kelly, K. F. Jensen, T. F. Jamison, C. Dai, Y. Cui, N. Briggs, R. L. Beingessner and A. Adamo, *Chem. - Eur. J.*, 2018, **24**(11), 2776–2784.
15. S. D. Schaber, D. Gerogiorgis, R. Ramachandran, J. M. B. Evans, P. I. Barton and B. L. Trout, *Ind. Eng. Chem. Res.*, 2011, **50**(17), 10083–10092.
16. K. Plumb, *Chem. Eng. Res. Des.*, 2005, **83**(6), 730–738.
17. P. McKenzie, S. Kiang, J. Tom, A. E. Rubin and M. Futran, *AIChE J.*, 2006, **52**(12), 3990–3994.
18. K. P. Cole, J. M. Groh, M. D. Johnson, C. L. Burcham, B. M. Campbell, W. D. Diseroad, M. R. Heller, J. R. Howell, N. J. Kallman, T. M. Koenig, S. A. May, R. D. Miller, D. Mitchell, D. P. Myers, S. S. Myers, J. L. Phillips, C. S. Polster, T. D. White, J. Cashman, D. Hurley, R. Moylan, P. Sheehan, R. D. Spencer, K. Desmond, P. Desmond and O. Gowran, *Science*, 2017, **356**(6343), 1144–1150.
19. T. D. White, C. A. Alt, K. P. Cole, J. M. Groh, M. D. Johnson and R. D. Miller, *Org. Process Res. Dev.*, 2014, **18**(11), 1482–1491.
20. C. S. Polster, K. P. Cole, C. L. Burcham, B. M. Campbell, A. L. Frederick, M. M. Hansen, M. Harding, M. R. Heller, M. T. Miller, J. L. Phillips, P. M. Pollock and N. Zaborenko, *Org. Process Res. Dev.*, 2014, **18**(11), 1295–1309.
21. K. P. Cole, B. M. Campbell, M. B. Forst, J. McClary Groh, M. Hess, M. D. Johnson, R. D. Miller, D. Mitchell, C. S. Polster, B. J. Reizman and M. Rosemeyer, *Org. Process Res. Dev.*, 2016, **20**(4), 820–830.
22. A. D. Randolph and M. A. Larson, *Theory of Particulate Processes*, Academic Press, Inc., New York, NY, 2nd edn, 1988.
23. S. Ferguson, F. Ortner, J. Quon, L. Peeva, A. Livingston, B. L. Trout and A. S. Myerson, *Cryst. Growth Des.*, 2013, **14**(2), 617–627.
24. R. Peña and Z. K. Nagy, *Cryst. Growth Des.*, 2015, **15**(9), 4225–4236.
25. Y. Yang, L. Song, T. Gao and Z. K. Nagy, *Cryst. Growth Des.*, 2015, **15**(12), 5879–5885.

26. Y. Cui, M. O'Mahony, J. J. Jaramillo, T. Stelzer and A. S. Myerson, *Org. Process Res. Dev.*, 2016, **20**(7), 1276–1282.
27. K. Tahara, M. O'Mahony and A. S. Myerson, *Cryst. Growth Des.*, 2015, **15**, 5149–5156.
28. M. O. Besenhard, P. Neugebauer, O. Scheibelhofer and J. G. Khinast, *Cryst. Growth Des.*, 2017, **17**(12), 6432–6444.
29. M. O. Besenhard, R. Hohl, A. Hodzic, R. J. P. Eder and J. G. Khinast, *Cryst. Res. Technol.*, 2014, **49**(2–3pdf), 92–108.
30. R. J. P. Eder, S. Radl, E. Schmitt, S. Innerhofer, M. Maier, H. Gruber-Woelfler and J. G. Khinast, *Cryst. Growth Des.*, 2010, **10**(5), 2247–2257.
31. R. J. P. Eder, E. K. Schmitt, J. Grill, S. Radl, H. Gruber-Woelfler and J. G. Khinast, *Cryst. Res. Technol.*, 2011, **46**(3), 227–237.
32. A. J. Alvarez and A. S. Myerson, *Cryst. Growth Des.*, 2010, **10**(5), 2219–2228.
33. A. Majumder and Z. K. Nagy, *AIChE J.*, 2013, **59**(12), 4582–4594.
34. R. J. P. Eder, S. Schrank, M. O. Besenhard, E. Roblegg, H. Gruber-Woelfler and J. G. Khinast, *Cryst. Growth Des.*, 2012, **12**(10), 4733–4738.
35. J. Chen, B. Sarma, J. M. B. Evans and A. S. Myerson, *Cryst. Growth Des.*, 2011, **11**(4), 887–895.
36. J. Li, B. L. Trout and A. S. Myerson, *Cryst. Growth Des.*, 2016, **20**(2), 510–516.
37. T. Vetter, C. L. Burcham and M. F. Doherty, *Chem. Eng. Sci.*, 2014, **106**, 167–180.
38. T. Vetter, C. L. Burcham and M. F. Doherty, *Ind. Eng. Chem. Res.*, 2015, **54**(42), 10350–10363.
39. T.-T. C. Lai, J. Cornevin, S. Ferguson, N. Li, B. L. Trout and A. S. Myerson, *Cryst. Growth Des.*, 2015, **15**(7), 3374–3382.
40. H. Zhang, J. Quon, A. J. Alvarez, J. Evans, A. S. Myerson and B. Trout, *Org. Process Res. Dev.*, 2012, **16**(5), 915–924.
41. M. D. Johnson, S. A. May, J. R. Calvin, J. Remacle, J. R. Stout, W. D. Diseroad, N. Zaborenko, B. D. Haeberle, W.-M. Sun, M. T. Miller and J. Brennan, *Org. Process Res. Dev.*, 2012, **16**(5), 1017–1038.
42. J. L. Quon, H. Zhang, A. Alvarez, J. Evans, A. S. Myerson and B. L. Trout, *Cryst. Growth Des.*, 2012, **12**(6), 3036–3044.
43. M. O. Besenhard, P. Neugebauer, C.-D. Ho and J. G. Khinast, *Cryst. Growth Des.*, 2015, **15**(4), 1683–1691.
44. L. Hohmann, T. Greinert, O. Mierka, S. Turek, G. Schembecker, E. Bayraktar, K. Wohlgemuth and N. Kockmann, *Cryst. Growth Des.*, 2018, **18**(3), 1459–1473.
45. A. J. Alvarez and A. S. Myerson, *Cryst. Growth Des.*, 2010, **10**(5), 2219–2228.
46. M. Jiang, Z. Zhu, E. Jimenez, C. D. Papageorgiou, J. Waetzig, A. Hardy, M. Langston and R. D. Braatz, *Cryst. Growth Des.*, 2014, **14**(2), 851–860.
47. A. Majumder and Z. K. Nagy, *Cryst. Growth Des.*, 2015, **15**(3), 1129–1140.
48. S. Ferguson, G. Morris, H. Hao, M. Barrett and B. Glennon, *Chem. Eng. Sci.*, 2012, **77**, 105–111.

49. S. Lawton, G. Steele, P. Shering, L. Zhao, I. Laird and X.-W. Ni, *Org. Process Res. Dev.*, 2009, **13**(6), 1357–1363.
50. R. Vacassy, J. Lemaıtre and H. Hofmann, *AIChE J.*, 2000, **46**(6), 1241–1252.
51. N. Jongen, M. Donnet, P. Bowen, J. Lemaître, H. Hofmann, R. Schenk, C. Hofmann, M. Aoun-Habbache, S. Guillemet-Fritsch, J. Sarrias, A. Rousset, M. Viviani, M. T. Buscaglia, V. Buscaglia, P. Nanni, A. Testino and J. R. Herguijuela, *Chem. Eng. Technol.*, 2003, **26**(3), 303–305.
52. T. McGlone, N. E. B. Briggs, C. A. Clark, C. J. Brown, J. Sefcik and A. J. Florence, *Org. Process Res. Dev.*, 2015, **19**(9), 1186–1202.
53. R. Peña, J. A. Oliva, C. L. Burcham, D. J. Jarmer and Z. K. Nagy, *Cryst. Growth Des.*, 2017, **17**(9), 4776–4784.
54. S. Y. Wong, A. P. Tatusko, B. L. Trout and A. S. Myerson, *Cryst. Growth Des.*, 2012, **12**(11), 5701–5707.
55. A. Mersmann, *Crystallization Technology Handbook*, Marcel Dekker, Inc., New York, NY, 2nd edn, 2001.
56. A. S. Myerson, *Handbook of Industrial Crystallization*, Butterworth Heinemann, 2001.
57. W. Beckmann, *Crystallization: Basic Concepts and Industrial Applications*, Wiley-VCH Verlag GmbH & Co. KGaA, Weinheim, Germany, 2013.
58. J. Wang and R. Lakerveld, *Ind. Eng. Chem. Res.*, 2017, **56**(19), 5705–5714.
59. S. Vartak and A. S. Myerson, *Org. Process Res. Dev.*, 2017, **21**(2), 253–261.
60. Z. Wu, S. Yang and W. Wu, *CrystEngComm*, 2016, **18**(13), 2222–2238.
61. P. W. Voorhees, *J. Stat. Phys.*, 1985, **38**(1), 231–252.
62. A. Saleemi, C. Rielly and Z. K. Nagy, *CrystEngComm*, 2012, **14**(6), 2196–2203.
63. R. C. Snyder, S. Studener and M. F. Doherty, *AIChE J.*, 2007, **53**(6), 1510–1517.
64. E. Simone, A. N. Saleemi, N. Tonnon and Z. K. Nagy, *Cryst. Growth Des.*, 2014, **14**(4), 1839–1850.
65. K. Suwannasang, A. E. Flood, C. Rougeot and G. Coquerel, *Cryst. Growth Des.*, 2013, **13**(8), 3498–3504.
66. A. Koswara and Z. K. Nagy, *IFAC-PapersOnLine*, 2015, **48**(8), 193–198.
67. S. Nishio, J. P. Kavanagh and J. Garside, *Chem. Eng. Sci.*, 1991, **46**(2), 709–711.
68. S. Ferguson, G. Morris, H. Hao, M. Barrett and B. Glennon, *Chem. Eng. Sci.*, 2013, **104**, 44–54.
69. T.-T. C. Lai, S. Ferguson, L. Palmer, B. L. Trout and A. S. Myerson, *Org. Process Res. Dev.*, 2014, **18**, 1382–1390.
70. G. Hou, G. Power, M. Barrett, B. Glennon, G. Morris and Y. Zhao, *Cryst. Growth Des.*, 2014, **14**(4), 1782–1793.
71. K. F. Jensen, T. F. Jamison, A. S. Myerson, J.-C. Monbaliu, M. Behnam, S. Y. Wong, N. Weeranoppanant, E. Revalor, T. Stelzer, J. Chen, A. Adamo, D. R. Snead and P. Zhang, Systems and Methods for Synthesizing Chemical Products, Including Active Pharmaceutical Ingredients, *WO Pat.*, WO 2016/025803 A1, 2016.

72. S. Y. Wong, J. Chen, L. E. Forte and A. S. Myerson, *Org. Process Res. Dev.*, 2013, **17**(4), 684–692.
73. K. A. Powell, A. N. Saleemi, C. D. Rielly and Z. K. Nagy, *Chem. Eng. Process.*, 2015, **97**, 195–212.
74. Q. Su, C. D. Rielly and K. A. Powell, *AIChE J.*, 2017, **63**(4), 1313–1327.
75. A. S. Myerson, M. Krumme, M. Nasr, H. Thomas and R. D. Braatz, *J. Pharm. Sci.*, 2015, **104**(3), 832–839.
76. H. J. Kramer, S. K. Bermingham and G. M. van Rosmalen, *J. Cryst. Growth*, 1999, **198/199**, 729–737.
77. M. Raphael, *Can. J. Chem. Eng.*, 1999, **77**(3), 540–554.
78. W. J. Liu, C. Y. Ma, J. J. Liu, Y. Zhang and X. Z. Wang, *AIChE J.*, 2017, **63**(3), 967–974.
79. J. Baldyga and J. R. Bourne, *Turbulent Mixing and Chemical Reactions*, John Wiley and Sons, Inc., Chichester, United Kingdom, 1999.
80. O. Söhnel and J. Garside, *Precipitation: Basic Principles and Industrial Application*, Butterworth-Heinemann, Oxford, 1992.
81. A. G. Jones, *Crystallization Process Systems*, Butterworth-Heinemann, Oxford, 2002.
82. A. Tamburini, A. Cipollina, G. Micale, A. Brucato and M. Ciofalo, *Chem. Eng. J.*, 2013, **223**, 875–890.
83. D. Wadnerkar, R. P. Utikar, M. O. Tade and V. K. Pareek, *Adv. Powder Technol.*, 2012, **23**(4), 445–453.
84. P. Ayazi Shamlou and E. Koutsakos, *Chem. Eng. Sci.*, 1989, **44**(3), 529–542.
85. A. Barresi and G. Baldi, *Chem. Eng. Sci.*, 1987, **42**(12), 2949–2956.
86. R. Kacker, S. I. Regensburg and H. J. M. Kramer, *Chem. Eng. J.*, 2017, **317**, 413–423.
87. L. N. Ejim, S. Yerdelen, T. McGlone, I. Onyemelukwe, B. Johnston, A. J. Florence and N. M. Reis, *Chem. Eng. J.*, 2017, **308**, 669–682.
88. N. S. Tavare, *AIChE J.*, 1986, **32**(5), 705–732.
89. P. V. Danckwerts, *Chem. Eng. Sci.*, 1958, **8**(1), 93–102.
90. E. L. Paul, V. A. Atiemo-Obeng and S. M. Kresta, *Handbook of Industrial Mixing: Science and Practice*, John Wiley & Sons Inc., New York, NY, 2004.
91. S. M. Kresta, A. W. Etchells III, D. S. Dickey and V. A. Atiemo-Obeng, *Advances in Industrial Mixing: A Companion to the Handbook of Industrial Mixing*, John Wiley & Sons Inc., New York, NY, 2015.
92. B. K. Johnson and R. K. Prud'homme, *AIChE J.*, 2003, **49**(9), 2264–2282.
93. M. Roelands, J. ter Horst, H. Kramer and P. J. Jansens, *Cryst. Growth Des.*, 2006, **6**(6), 1380–1392.
94. M. Roelands, J. Derksen, J. ter Horst, H. Kramer and P. Jansens, *Chem. Eng. Technol.*, 2003, **26**(3), 296–302.
95. C. Lindenberg, M. Mazzotti, J. Schöll, L. Vicum, M. Mazzotti and J. Brozio, *Chem. Eng. Sci.*, 2008, **63**(16), 4135–4149.
96. C. Lindenberg and M. Mazzotti, *AIChE J.*, 2011, **57**(4), 942–950.
97. U. K. Singh, G. Spencer, R. Osifchin, J. Tabora, O. A. Davidson and C. J. Orella, *Ind. Eng. Chem. Res.*, 2005, **44**(11), 4068–4074.

98. C. Lindenberg, *Chem. Eng. Sci.*, 2009, **64**, 4286–4294.
99. S. Ferguson, *Chem. Eng. Sci.*, 2012, **77pdf**, 105–111.
100. Y. Liu, C. Cheng, Y. Liu, R. K. Prud'homme and R. O. Fox, *Chem. Eng. Sci.*, 2008, **63**(11), 2829–2842.
101. M. Engler, N. Kockmann, T. Kiefer and P. Woias, *Chem. Eng. J.*, 2004, **101**(1), 315–322.
102. K. Krupa, M. A. Sultan, C. P. Fonte, M. I. Nunes, M. M. Dias, J. C. B. Lopes and R. J. Santos, *Chem. Eng. J.*, 2012, **207–208**, 931–937.
103. D. O'Grady, M. Barrett, E. Casey and B. Glennon, *Chem. Eng. Res. Des.*, 2007, **85**(7), 945–952.
104. J. R. Bourne, J. Lenzner and S. Petrozzi, *Ind. Eng. Chem. Res.*, 1992, **31**(4), 1216–1222.
105. J. Bałdyga, J. R. Bourne and S. J. Hearn, *Chem. Eng. Sci.*, 1997, **52**(4), 457–466.
106. R. K. Thakur, C. Vial, K. D. P. Nigam, E. B. Nauman and G. Djelveh, *Chem. Eng. Res. Des.*, 2003, **81**(7), 787–826.
107. D. M. Hobbs and F. J. Muzzio, *Chem. Eng. J.*, 1997, **67**(3), 153–166.
108. D. M. Hobbs and F. J. Muzzio, *Chem. Eng. Sci.*, 1998, **53**(18), 3199–3213.
109. S. Kudo and H. Takiyama, *J. Chem. Eng. Jpn.*, 2012, **45**(4), 305–309.
110. P. Neugebauer and J. G. Khinast, *Cryst. Growth Des.*, 2015, **15**(3), 1089–1095.
111. H. Liu, C. O. Vandu and R. Krishna, *Ind. Eng. Chem. Res.*, 2005, **44**(14), 4884–4897.
112. A. K. Saxena and K. D. P. Nigam, *AIChE J.*, 1984, **30**(3), 363–368.
113. K. P. Sandeep, C. A. Zuritz and V. M. Puri, *J. Food Sci.*, 1997, **62**(4), 647–652.
114. T. K. Palazoglu and K. P. Sandeep, *LWT--Food Sci. Technol.*, 2004, **37**(4), 387–393.
115. H. Mallubhotla and G. Belfort, *J. Membr. Sci.*, 1997, **125**(1), 75–91.
116. A. K. Saxena and K. D. P. Nigam, *Chem. Eng. Sci.*, 1979, **34**, 425–426.
117. L. Hohmann, R. Gorny, O. Klaas, J. Ahlert, K. Wohlgemuth and N. Kockmann, *Chem. Eng. Technol.*, 2016, **39**(7), 1268–1280.
118. S. K. Kurt, M. Akhtar, K. D. P. Nigam and N. Kockmann, *Ind. Eng. Chem. Res.*, 2017, **56**(39), 11320–11335.
119. S. Klutz, S. K. Kurt, M. Lobedann and N. Kockmann, *Chem. Eng. Res. Des.*, 2015, **95**, 22–33.
120. R. Lakerveld, J. J. H. van Krochten and H. J. M. Kramer, *Cryst. Growth Des.*, 2014, **14**(7), 3264–3275.
121. A. Soare, R. Lakerveld, J. van Royen, G. Zocchi, A. I. Stankiewicz and H. J. M. Kramer, *Ind. Eng. Chem. Res.*, 2012, **51**(33), 10895–10909.
122. C. J. Callahan and X.-W. Ni, *CrystEngComm*, 2014, **16**(4), 690.
123. N. E. B. Briggs, U. Schacht, V. Raval, T. McGlone, J. Sefcik and A. J. Florence, *Org. Process Res. Dev.*, 2015, **19**(12), 1903–1911.
124. C. J. Brown, J. A. Adelakun and X. W. Ni, *Chem. Eng. Process.*, 2015, **97**, 180–186.
125. M. J. Hounslow and X. Ni, *Chem. Eng. Sci.*, 2004, **59**(4), 819–828.

126. I. R. Baxendale, R. D. Braatz, B. K. Hodnett, K. F. Jensen, M. D. Johnson, P. Sharratt, J.-P. Sherlock and A. J. Florence, *J. Pharm. Sci.*, 2015, **104**(3), 781–791.
127. H. Siddique, C. J. Brown, I. Houson and A. J. Florence, *Org. Process Res. Dev.*, 2015, **19**(12), 1871–1881.
128. A. Y. Lee, I. S. Lee, S. S. Dette, J. Boerner and A. S. Myerson, *J. Am. Chem. Soc.*, 2005, **127**(43), 14982–14983.
129. I. S. Lee, A. Y. Lee and A. S. Myerson, *Pharm. Res.*, 2008, **25**(4), 960–968.
130. J.-M. Ha, J. H. Wolf, M. A. Hillmyer and M. D. Ward, *J. Am. Chem. Soc.*, 2004, **126**(11), 3382–3383.
131. J. M. Ha, B. D. Hamilton, M. A. Hillmyer and M. D. Ward, *Cryst. Growth Des.*, 2012, **12**(9), 4494–4504.
132. Q. Jiang and M. D. Ward, *Chem. Soc. Rev.*, 2014, **43**(7), 2066–2079.
133. S. S. Kadam, S. A. Kulkarni, R. Coloma Ribera, A. I. Stankiewicz, J. H. ter Horst and H. J. M. Kramer, *Chem. Eng. Sci.*, 2012, **72**, 10–19.
134. Y. Diao, A. S. Myerson, T. A. Hatton and B. L. Trout, *Langmuir*, 2011, **27**(9), 5324–5334.
135. Y. Diao, T. Harada, A. S. Myerson, T. Alan Hatton and B. L. Trout, *Nat. Mater.*, 2011, **10**, 867.
136. V. López-Mejías, A. S. Myerson and B. L. Trout, *Cryst. Growth Des.*, 2013, **13**(8), 3835–3841.
137. V. López-Mejías, J. W. Kampf and A. J. Matzger, *J. Am. Chem. Soc.*, 2012, **134**(24), 9872–9875.
138. J. Stojaković, F. Baftizadeh, M. A. Bellucci, A. S. Myerson and B. L. Trout, *Cryst. Growth Des.*, 2017, **17**(6), 2955–2963.
139. M. O. Mahony, A. K. Leung, S. Ferguson, B. L. Trout and A. S. Myerson, *Org. Process Res. Dev.*, 2015, **19**(9), 1109–1118.
140. J.-M. Ha, B. D. Hamilton, M. A. Hillmyer and M. D. Ward, *Cryst. Growth Des.*, 2009, **9**(11), 4766–4777.
141. L. M. Dwyer, V. K. Michaelis, M. O'Mahony, R. G. Griffin and A. S. Myerson, *CrystEngComm*, 2015, **17**(41), 7922–7929.
142. L. Dwyer, S. Kulkarni, L. Ruelas and A. Myerson, *Crystals*, 2017, **7**(5), 131.
143. X. Yang, T.-C. Ong, V. K. Michaelis, S. Heng, R. G. Griffin and A. S. Myerson, *CrystEngComm*, 2015, **17**, 6044–6052.
144. B. Sjöström, B. Bergenståhl, M. Lindberg and Å. C. Rasmuson, *J. Dispersion Sci. Technol.*, 1994, **15**(1), 89–117.
145. D. P. Otto and M. M. de Villiers, *Nanotechnology in Drug Delivery*, ed. P. Aramwit and G. S. Kwon, Springer New York, New York, NY, 2009, pp. 3–33.
146. K. Kim, I. S. Lee, A. Centrone, T. A. Hatton and A. S. Myerson, *J. Am. Chem. Soc.*, 2009, **131**, 18212–18213.
147. R. Shegokar and R. H. Müller, *Int. J. Pharm.*, 2010, **399**(1), 129–139.
148. J. P. Möschwitzer, *Int. J. Pharm.*, 2013, **453**(1), 142–156.
149. W. Kaialy and M. Al Shafiee, *Adv. Colloid Interface Sci.*, 2016, **228**, 71–91.
150. N. Yazdanpanah, C. J. Testa, S. R. K. Perala, K. D. Jensen, R. D. Braatz, A. S. Myerson and B. L. Trout, *Cryst. Growth Des.*, 2017, **17**(6), 3321–3330.

151. J. L. Quon, K. Chadwick, G. P. F. Wood, I. Sheu, B. K. Brettmann, A. S. Myerson and B. L. Trout, *Langmuir*, 2013, **29**(10), 3292–3300.
152. D. A. Acevedo, J. Ling, K. Chadwick and Z. K. Nagy, *Cryst. Growth Des.*, 2016, **16**(8), 4263–4271.
153. V. López-Mejías, J. W. Kampf and A. J. Matzger, *J. Am. Chem. Soc.*, 2009, **131**(13), 4554–4555.
154. E. Curcio, V. López-Mejías, G. Di Profio, E. Fontananova, E. Drioli, B. L. Trout and A. S. Myerson, *Cryst. Growth Des.*, 2014, **14**(2), 678–686.
155. Y. Yang, L. Song and Z. K. Nagy, *Cryst. Growth Des.*, 2015, **15**(12), 5839–5848.
156. X. Zhou, L. Lau, W. W. L. Lam, S. W. N. Au and B. Zheng, *Anal. Chem.*, 2007, **79**(13), 4924–4930.
157. F. Chen, G. Du, D. Yin, R. Yin, H. Zhang, W. Zhang and S.-M. Yang, *AIP Conf. Proc.*, 2017, **1820**(1), 30003.
158. D. J. McClements, *Adv. Colloid Interface Sci.*, 2012, **174**, 1–30.
159. A. Gupta, H. B. Eral, T. A. Hatton and P. S. Doyle, *R. Soc. Chem.*, 2016, **12**(11), 2826–2841.
160. Y. Yang, C. Marshall-Breton, M. E. Leser, A. A. Sher and D. J. McClements, *Food Hydrocolloids*, 2012, **29**(2), 398–406.
161. J. Komaiko and D. J. McClements, *J. Food Eng.*, 2015, **146**, 122–128.
162. R. K. Shah, H. C. Shum, A. C. Rowat, D. Lee, J. J. Agresti, A. S. Utada, L.-Y. Chu, J.-W. Kim, A. Fernandez-Nieves, C. J. Martinez and D. A. Weitz, *Mater. Today*, 2008, **11**(4), 18–27.
163. J. H. Xu, S. W. Li, J. Tan, Y. J. Wang and G. S. Luo, *Langmuir*, 2006, **22**(19), 7943–7946.
164. C.-X. Zhao and A. P. J. Middelberg, *Chem. Eng. Sci.*, 2011, **66**(7), 1394–1411.
165. R. A. L. Leon, W. Y. Wan, A. Z. M. Badruddoza, T. A. Hatton and S. A. Khan, *Cryst. Growth Des.*, 2014, **14**(1), 140–146.
166. B. Y. Shekunov, P. Chattopadhyay, J. Seitzinger and R. Huff, *Pharm. Res.*, 2006, **23**(1), 196–204.
167. M. Trotta, M. Gallarate, M. E. Carlotti and S. Morel, *Int. J. Pharm.*, 2003, **254**(2), 235–242.
168. J. S. Negi, P. Chattopadhyay, A. K. Sharma and V. Ram, *Eur. J. Pharm. Sci.*, 2013, **48**(1), 231–239.
169. A. Gupta, A. Z. M. Badruddoza and P. S. Doyle, *Langmuir*, 2017, **33**(28), 7118–7123.
170. A. Z. M. Badruddoza, A. Gupta, A. S. Myerson, B. L. Trout and P. S. Doyle, *Adv. Ther.*, 2018, **1**(1), 1700020.
171. P. S. Dittrich and A. Manz, *Nat. Rev. Drug Discovery*, 2006, **5**, 210.
172. B. T. C. Lau, C. A. Baitz, X. P. Dong and C. L. Hansen, *J. Am. Chem. Soc.*, 2007, **129**(3), 454–455.
173. J. Leng and J.-B. Salmon, *Lab Chip*, 2009, **9**(1), 24–34.
174. H. B. Eral, M. O. Mahony, R. Shaw, B. L. Trout, A. S. Myerson, P. S. Doyle, M. O'Mahony, R. Shaw, B. L. Trout, A. S. Myerson and P. S. Doyle, *Chem. Mater.*, 2014, **26**(21), 6213–6220.

175. H. Shi, Y. Xiao, S. Ferguson, X. Huang, N. Wang and H. Hao, *Lab Chip*, 2017, **17**, 2167–2185.
176. S. Goyal, M. R. Thorson, G. G. Z. Zhang, Y. Gong and P. J. A. Kenis, *Cryst. Growth Des.*, 2012, **12**(12), 6023–6034.
177. V. Bhamidi, S. H. Lee, G. He, P. S. Chow, R. B. H. Tan, C. F. Zukoski and P. J. A. Kenis, *Cryst. Growth Des.*, 2015, **15**(7), 3299–3306.
178. M. Sultana and K. F. Jensen, *Cryst. Growth Des.*, 2012, **12**(12), 6260–6266.
179. B. G. Abdallah, S. Roy-Chowdhury, R. Fromme, P. Fromme and A. Ros, *Cryst. Growth Des.*, 2016, **16**(4), 2074–2082.
180. A. Feuerborn, A. Prastowo, P. R. Cook and E. Walsh, *Lab Chip*, 2015, **15**(18), 3766–3775.
181. C. W. Carruthers, C. Gerdts, M. D. Johnson and P. Webb, *PLoS One*, 2013, **8**(11), 1–12.
182. D. A. Berry and K. M. Ng, *AIChE J.*, 1997, **43**(7), 1737–1750.
183. W. Marquardt, S. Kossack and K. Kraemer, *Chin. J. Chem. Eng.*, 2008, **16**(3), 333–342.
184. M. B. Franke and N. Nowotny, *AIChE J.*, 2008, **54**(11), 2925–2942.
185. M. Kaspereit, S. Swernath and A. Kienle, *Org. Process Res. Dev.*, 2012, **16**(2), 353–363.
186. C. Bravo-Bravo, J. G. Segovia-Hernández, S. Hernández, F. I. Gómez-Castro, C. Gutiérrez-Antonio and A. Briones-Ramírez, *Chem. Eng. Process.*, 2013, **67**, 25–38.
187. J. Micovic, T. Beierling, P. Lutze, G. Sadowski and A. Górak, *Chem. Eng. Process.*, 2013, **67**, 16–24.
188. B. Sreedhar, B. Shen, H. Li, R. Rousseau and Y. Kawajiri, *Org. Process Res. Dev.*, 2017, **21**(1), 31–43.
189. S. Swernath, M. Kaspereit and A. Kienle, *Chem. Eng. Technol.*, 2013, **36**(8), 1417–1429.
190. H. Lorenz and A. Seidel-Morgenstern, *Angew. Chem., Int. Ed.*, 2014, **53**(5), 1218–1250.
191. G. Ströhlein, M. Schulte and J. Strube, *Sep. Sci. Technol.*, 2003, **38**(14), 3353–3383.
192. M. Amanullah and M. Mazzotti, *J. Chromatogr. A*, 2006, **1107**(1–2), 36–45.
193. B.-G. Lim, C.-B. Ching, R. B. H. Tan and S.-C. Ng, *Chem. Eng. Sci.*, 1995, **50**(14), 2289–2298.
194. H. Kaemmerer, Z. Horvath, J. W. Lee, M. Kaspereit, R. Arnell, M. Hedberg, B. Herschend, M. J. Jones, K. Larson, H. Lorenz and A. Seidel-Morgensten, *Org. Process Res. Dev.*, 2012, **16**(2), 331–342.
195. J. Von Langermann, M. Kaspereit, M. Shakeri, H. Lorenz, M. Hedberg, M. J. Jones, K. Larson, B. Herschend, R. Arnell, E. Temmel, J. E. Bäckvall, A. Kienle and A. Seidel-Morgenstern, *Org. Process Res. Dev.*, 2012, **16**(2), 343–352.
196. K. Gedicke, M. Kaspereit, W. Beckmann, U. Budde, H. Lorenz and A. Seidel-Morgenstern, *Chem. Eng. Res. Des.*, 2007, **85**(7), 928–936.

197. H. Lorenz, P. Sheehan and A. Seidel-Morgenstern, *J. Chromatogr. A*, 2001, **908**(1), 201–214.
198. G. Di Profio, E. Curcio, S. Ferraro, C. Stabile and E. Drioli, *Cryst. Growth Des.*, 2009, **9**(5), 2179–2186.
199. X. Zhang, P. Zhang, K. Wei, Y. Wang and R. Ma, *Desalination*, 2008, **219**, 101–117.
200. C. Charcosset, R. Kieffer, D. Mangin and F. Puel, *Ind. Eng. Chem. Res.*, 2010, **49**(12), 5489–5495.
201. G. Di Profio, E. Curcio and E. Drioli, *Comprehensive Membrane Science and Engineering*, ed. E. Drioli and L. Giorno, Elsevier, Oxford, 2017, pp. 21–46.
202. E. Chabanon, D. Mangin and C. Charcosset, *J. Membr. Sci.*, 2016, **509**, 57–67.
203. S. Vartak and A. S. Myerson, *Cryst. Growth Des.*, 2017, **17**(10), 5506–5516.
204. L. Gou, S. Robl, K. A. I. Leonhard, H. Lorenz, M. Sordo, A. Butka, S. Kesselheim, M. Wolff, A. Seidel-morgenstern and K. Schaber, *Chirality*, 2011, **23**(2), 118–127.
205. S. Robl, L. Gou, A. Gere, M. Sordo, H. Lorenz, A. Mayer, C. Pauls, K. Leonhard, A. Bardow, A. Seidel-Morgenstern and K. Schaber, *Chem. Eng. Process.*, 2013, **67**, 80–88.
206. R. Lakerveld, J. Kuhn, H. J. M. Kramer, P. J. Jansens and J. Grievink, *Chem. Eng. Sci.*, 2010, **65**(9), 2689–2699.
207. H.-Y. Wang, K.-L. Tung and J. D. Ward, *J. Taiwan Inst. Chem. Eng.*, 2017, **81**, 159–169.
208. E. M. Rundquist, C. J. Pink and A. G. Livingston, *Green Chem.*, 2012, **14**(8), 2197–2205.
209. P. Marchetti, M. F. Jimenez Solomon, G. Szekely and A. G. Livingston, *Chem. Rev.*, 2014, **114**(21), 10735–10806.
210. C. Roengpithya, D. A. Patterson, P. C. Taylor and A. G. Livingston, *Desalination*, 2006, **199**(1), 195–197.
211. N. F. Ghazali, F. C. Ferreira, A. J. P. White and A. G. Livingston, *Tetrahedron: Asymmetry*, 2006, **17**(12), 1846–1852.
212. A. Grandeury, S. Petit, G. Gouhier, V. Agasse and G. Coquerel, *Tetrahedron: Asymmetry*, 2003, **14**, 2143–2152.
213. A. Grandeury, L. Renou, F. Dufour, S. Petit, G. Gouhier and G. Coquerel, *J. Therm. Anal. Calorim.*, 2004, **77**, 377–390.
214. M. P. Elsner, G. Ziomek and A. Seidel-morgenstern, *AIChE J.*, 2009, **55**(3), 640–649.
215. A. Svang-Ariyaskul, W. J. Koros and R. W. Rousseau, *Chem. Eng. Sci.*, 2009, **64**(9), 1980–1984.
216. A. Svang-Ariyaskul, W. J. Koros and R. W. Rousseau, *Chem. Eng. Sci.*, 2012, **77**, 35–41.
217. H. Lorenz, D. Polenske and A. Seidel-morgenstern, *Chirality*, 2006, **18**(10), 828–840.

218. N. Kockmann, *History of Distillation*, Elsevier Inc., 2014.
219. D. A. Berry and K. M. Ng, *AIChE J.*, 1997, **43**(7), 1751–1762.
220. S. Rittner and R. Steiner, *Chem. Ing. Tech.*, 1985, **57**(2), 91–102.
221. B. Lunelli, *Inorg. Chim. Acta*, 2007, **360**(3), 1217–1220.
222. J. Ulrich and T. Stelzer, in *Crystallization: Basic Concepts and Industrial Applications*, ed. W. Beckmann, Wiley-VCH Verlag GmbH & Co. KGaA, Weinheim, Germany, 2013, pp. 289–304.
223. J. Ulrich and T. Stelzer, in *Crystallization: Basic Concepts and Industrial Applications*, ed. W. Beckmann, Wiley-VCH Verlag GmbH & Co. KGaA, Weinheim, Germany, 2013, pp. 325–335.
224. G. F. Arkenbout, *Melt Crystallization Technology*, Technomic Publishing Company, Inc., Lancaster, Pennsylvania, USA, 1995.
225. J. Ulrich and H. Glade, *Melt Crystallization - Fundamentals, Equipment and Applications*, Shaker Verlag, Aachen, 2003.
226. A. S. Paulino, G. Rauber, C. E. Campos, M. H. Mauracio, R. R. de Avillez, G. Capobianco, S. G. Cardoso, S. L. Cuffini, M. H. Maurício, R. R. de Avillez, G. Capobianco, S. G. Cardoso and S. L. Cuffini, *Eur. J. Pharm. Sci.*, 2013, **49**(2), 294–301.
227. S. R. Modi, A. K. R. Dantuluri, V. Puri, Y. B. Pawar, P. Nandekar, A. T. Sangamwar, S. R. Perumalla, C. C. Sun and A. K. Bansal, *Cryst. Growth Des.*, 2013, **13**(7), 2824–2832.
228. M. G. Ierapetritou and R. Ramachandran, *Process Simulation and Data Modeling in Solid Oral Drug Development and Manufacture*, Humana Press, 2016.
229. K. Wening and J. Breitkreutz, *Int. J. Pharm.*, 2011, **404**(1), 1–9.
230. M. W. Y. Wong and A. G. Mitchell, *Int. J. Pharm.*, 1992, **88**(1), 261–273.
231. R. Hilfiker, *Polymorphism*, Wiley-VCH Verlag GmbH & Co. KGaA, Weinheim, Germany, 2006.
232. *Polymorphism in Pharmaceutical Solids*, ed. H. G. Brittain, Informa Healthcare USA, Inc., New York, 2nd edn, 2009.
233. Y. Kawashima, M. Okumura and H. Takenaka, *Science*, 1982, **216**(4550), 1127–1128.
234. D. Acevedo, V. K. Kamaraju, B. Glennon and Z. K. Nagy, *Org. Process Res. Dev.*, 2017, **21**(7), 1069–1079.
235. Y. Kawashima, *Arch. Pharmacal Res.*, 1984, **7**(2), 145–151.
236. A. Chatterjee, M. Gupta and B. Srivastava, *Int. J. Pharm. Invest.*, 2017, 7(1), 4–9.
237. B. Kovacic, F. Vrecer and O. Planinsek, *Acta Pharm.*, 2012, **62**(1), 1–14.
238. S. V. Patil and S. K. Sahoo, *Res. J. Pharm. Technol.*, 2009, **2**(2), 234–237.
239. Y. Kawashima and C. E. Capes, *Powder Technol.*, 1974, **10**(1), 85–92.
240. Y. Kawashima, F. Cui, H. Takeuchi, T. Niwa, T. Hino and K. Kiuchi, *Int. J. Pharm.*, 1995, **119**(2), 139–147.
241. Y. Kawashima, T. Niwa, T. Handa, H. Takeuchi, T. Iwamoto and K. Itoh, *J. Pharm. Sci.*, 1989, **78**(1), 68–72.

242. S. S. Kadam, K. R. Mahadik and A. R. Paradkar, A process for making agglomerates for use as or in a drug delivery system, *Indian Pat.*, 183,036, 1997.
243. M. Maghsoodi, O. Taghizadeh, G. P. Martin and A. Nokhodchi, *Int. J. Pharm.*, 2008, **351**(1), 45–54.
244. N. Jadhav, A. Pawar and A. Paradkar, *AAPS PharmSciTech*, 2007, **8**(3), E61–E67.
245. A. Pawar, A. Paradkar, S. Kadam and K. Mahadik, *AAPS PharmSciTech*, 2004, **5**(4), 30–35.
246. A. P. Pawar, A. R. Paradkar, S. S. Kadam and K. R. Mahadik, *AAPS PharmSciTech*, 2004, **5**(3), 57–64.
247. M. Ueda, Y. Nakamura, H. Makita, Y. Imasato and Y. Kawashima, *Chem. Pharm. Bull.*, 1990, **38**(9), 2537–2541.
248. Y. Kawashima, Y. Kurachi and H. Takenaka, *Powder Technol.*, 1982, **32**(2), 155–161.
249. M. Ueda, Y. Nakamura, H. Makita and Y. Kawashima, *J. Microencapsulation*, 1993, **10**(1), 25–34.
250. C. A. Lipinski, *J. Pharmacol. Toxicol. Methods*, 2000, **44**(1), 235–249.
251. S. Stegemann, F. Leveiller, D. Franchi, H. de Jong and H. Lindén, *Eur. J. Pharm. Sci.*, 2007, **31**(5), 249–261.
252. T. Loftsson and M. E. Brewster, *J. Pharm. Pharmacol.*, 2010, **62**(11), 1607–1621.
253. D. W. Griffin, D. A. Mellichamp and M. F. Doherty, *Chem. Eng. Sci.*, 2010, **65**(21), 5770–5780.
254. S. Kalepu and V. Nekkanti, *Acta Pharm. Sin. B*, 2015, **5**(5), 442–453.
255. P. Khadka, J. Ro, H. Kim, I. Kim, J. T. Kim, H. Kim, J. M. Cho, G. Yun and J. Lee, *Asian J. Pharm. Sci.*, 2014, **9**(6), 304–316.
256. Z. H. Loh, A. K. Samanta and P. W. Sia Heng, *Asian J. Pharm. Sci.*, 2015, **10**(4), 255–274.
257. P. Liu, X. Rong, J. Laru, B. van Veen, J. Kiesvaara, J. Hirvonen, T. Laaksonen and L. Peltonen, *Int. J. Pharm.*, 2011, **411**(1), 215–222.
258. Y. Yang, L. Song, Y. Zhang and Z. K. Nagy, *Ind. Eng. Chem. Res.*, 2016, **55**(17), 4987–4996.
259. C. L. Burcham, P. C. Collins, D. J. Jarmer and K. D. Seibert, in *Formulation and Analytical Development for Low-Dose Oral Drug Products*, ed. J. Zheng, John Wiley & Sons Inc., Hoboken, NJ, 2008, pp. 205–222.
260. R. J. Allen, C. Valeriani, S. Tănase-Nicola, P. R. ten Wolde and D. Frenkel, *J. Chem. Phys.*, 2008, **129**(13), 134704.
261. T. Kamahara, M. Takasuga, H. H. Tung, K. Hanaki, T. Fukunaka, B. Izzo, J. Nakada, Y. Yabuki and Y. Kato, *Org. Process Res. Dev.*, 2007, **11**(4), 699–703.
262. D. Reguera and J. M. Rubí, *J. Chem. Phys.*, 2003, **119**(18), 9888–9893.
263. R. A. Gray, P. B. Warren, S. Chynoweth, Y. Michopoulos and G. S. Pawley, *Proc. R. Soc. London, Ser. A*, 1932, **1995**(448), 113–120.
264. N. Yazdanpanah, S. T. Ferguson, A. S. Myerson and B. L. Trout, *Cryst. Growth Des.*, 2016, **16**(1), 285–296.

265. A. Mesbah, A. N. Ford Versypt, X. Zhu and R. D. Braatz, *Ind. Eng. Chem. Res.*, 2014, **53**(18), 7447–7460.
266. B. K. Brettmann, K. Cheng, A. S. Myerson and B. L. Trout, *Pharm. Res.*, 2013, **30**(1), 238–246.
267. H. B. Eral, V. López-Mejías, M. O'Mahony, B. L. Trout, A. S. Myerson and P. S. Doyle, *Cryst. Growth Des.*, 2014, **14**(4), 2073–2082.
268. T. Gu, E. W. Q. Yeap, Z. Cao, D. Z. L. Ng, Y. Ren, R. Chen, S. A. Khan and T. A. Hatton, *Adv. Healthcare Mater.*, 2017, **1700797**, 1–9.
269. N. Yazdanpanah, A. Myerson and B. Trout, *Ind. Eng. Chem. Res.*, 2016, **55**(17), 5019–5029.
270. W. T. Richards and A. L. Loomis, *J. Am. Chem. Soc.*, 1927, **49**(12), 3086–3100.
271. B. Gielen, T. Claes, J. Janssens, J. Jordens, L. C. J. Thomassen, T. van Gerven and L. Braeken, *Chem. Eng. Technol.*, 2017, **40**(7), 1300–1308.
272. T. Y. Wu, N. Guo, C. Y. Teh and J. X. W. Hay, *Advances in Ultrasound Technology for Environmental Remediation*, Springer Netherlands, Dordrecht, 2013, pp. 5–12.
273. D. Kashchiev and G. M. van Rosmalen, *Cryst. Res. Technol.*, 2003, **38**(7–8), 555–574.
274. S. L. Hem, *Ultrasonics*, 1967, **5**(4), 202–207.
275. K. Wohlgemuth, F. Ruether and G. Schembecker, *Chem. Eng. Sci.*, 2010, **65**(2), 1016–1027.
276. C. Virone, H. J. M. Kramer, G. M. van Rosmalen, A. H. Stoop and T. W. Bakker, *J. Cryst. Growth*, 2006, **294**(1), 9–15.
277. L. de los Santos Castillo-Peinado and M. D. Luque de Castro, *J. Pharm. Pharmacol.*, 2016, **68**(10), 1249–1267.
278. G. Ruecroft, D. Hipkiss, T. Ly, N. Maxted and P. W. Cains, *Org. Process Res. Dev.*, 2005, **9**(6), 923–932.
279. M. Jiang, C. Gu and R. D. Braatz, *Chem. Eng. Process.*, 2017, **117**, 186–194.
280. M. Jiang, C. D. Papageorgiou, J. Waetzig, A. Hardy, M. Langston and R. D. Braatz, *Cryst. Growth Des.*, 2015, **15**(5), 2486–2492.
281. M. Furuta, K. Mukai, D. Cork and K. Mae, *Chem. Eng. Process.*, 2016, **102**, 210–218.
282. Y. Hattoria, K. Kadotab, T. Yanoa, A. Shimosakaa, H. Ichikawac, Y. Fukumoric, Y. Shirakawaa and J. Hidaka, *Chem. Eng. Process.*, 2015, **97**, 233–241.
283. D. Rossi, R. Jamshidi, N. Saffari, S. Kuhn, A. Gavriilidis and L. Mazzei, *Cryst. Growth Des.*, 2015, **15**(11), 5519–5529.
284. R. Jamshidi, D. Rossi, N. Saffari, A. Gavriilidis and L. Mazzei, *Cryst. Growth Des.*, 2016, **16**(8), 4607–4619.
285. S. Devarakonda, J. M. B. Evans and A. S. Myerson, *Cryst. Growth Des.*, 2004, **4**(4), 687–690.
286. R. Lakerveld, N. G. Verzijden, H. Kramer, P. Jansens and J. Grievink, *AIChE J.*, 2011, **57**(12 LB-54), 3367–3377.
287. O. Narducci, A. G. Jones and E. Kougoulos, *Cryst. Growth Des.*, 2011, **11**(5), 1742–1749.

288. O. Narducci, A. G. Jones and E. Kougoulos, *Chem. Eng. Sci.*, 2011, **66**(6), 1069–1076.
289. P. Sayan, S. T. Sargut and B. Kiran, *Ultrason. Sonochem.*, 2011, **18**(3), 795–800.
290. B. Gielen, Y. Thimmesch, J. Jordens, G. Janssen, L. C. J. Thomassen, T. Van Gerven and L. Braeken, *Chem. Eng. Res. Des.*, 2016, **115**, 131–144.
291. C. Nanev, *Crystals*, 2017, **7**(10), 310.
292. F. Li and R. Lakerveld, *Cryst. Growth Des.*, 2018, **18**(5), 2964–2971.
293. J. Urbanus, R. J. M. Bisselink, K. Nijkamp, J. H. ter Horst, D. Verdoes and C. P. M. Roelands, *J. Membr. Sci.*, 2010, **363**(1), 36–47.
294. J. Urbanus, C. P. M. Roelands, J. Mazurek, D. Verdoes and J. H. ter Horst, *CrystEngComm*, 2011, **13**(8), 2817–2819.
295. W. W. Li, N. Radacsi, H. J. M. Kramer, A. E. D. M. van der Heijden and J. H. ter Horst, *Angew. Chem., Int. Ed.*, 2016, **55**(52), 16088–16091.
296. T. N. Glasnov and C. O. Kappe, *Macromol. Rapid Commun.*, 2007, **28**(4), 395–410.
297. L. Estel, M. Poux, N. Benamara and I. Polaert, *Chem. Eng. Process.*, 2017, **113**, 56–64.
298. Z. Guo, W. Han, W. Zhao, L. Li, B. Wang, Y. Xiao and V. Alopaeus, *Powder Technol.*, 2018, **328**, 358–366.
299. R. Rodríguez-Clemente and J. Gómez-Morales, *J. Cryst. Growth*, 1996, **169**(2), 339–346.
300. A. Rizzuti and C. Leonelli, *Powder Technol.*, 2008, **186**(3), 255–262.
301. R. Kacker, P. M. Salvador, G. S. J. Sturm, G. D. Stefanidis, R. Lakerveld, Z. K. Nagy and H. J. M. Kramer, *Cryst. Growth Des.*, 2016, **16**(1), 440–446.
302. R. Kacker, M. Radoiu and H. J. M. Kramer, *Cryst. Growth Des.*, 2017, **17**(7), 3766–3774.
303. J.-C. M. Monbaliu, T. Stelzer, E. Revalor, N. Weeranoppanant, K. F. Jensen and A. S. Myerson, *Org. Process Res. Dev.*, 2016, **20**(7), 1347–1353.
304. T. Page, H. Dubina, G. Fillipi, R. Guidat, S. Patnaik, P. Poechlauer, P. Shering, M. Guinn, P. McDonnell and C. Johnston, *J. Pharm. Sci.*, 2015, **104**(3), 821–831.
305. T. Stelzer, S. Y. Wong, J. Chen and A. S. Myerson, *Org. Process Res. Dev.*, 2016, **20**(8), 1431–1438.
306. C. F. Carter, H. Lange, S. V. Ley, I. R. Baxendale, B. Wittkamp, J. G. Goode and N. L. Gaunt, *Org. Process Res. Dev.*, 2010, **14**(2), 393–404.
307. B. Üstün, *Eur. Pharm. Rev.*, 2013, **13**(3), 39–41.
308. M. Schoenitz, S. Joseph, H. Bunjes and S. Scholl, *Chem. Eng. Technol.*, 2013, **36**(6), 1075–1079.

CHAPTER 8

Continuous Membrane Crystallization

EFREM CURCIO*[a,b] AND GIANLUCA DI PROFIO*[b]

[a]Department of Environmental and Chemical Engineering DIATIC, University of Calabria, Via P. Bucci Cubo 45A, 87036 Rende, CS, Italy; [b]Institute on Membrane Technology, National Research Council of Italy ITM-CNR, Via P. Bucci Cubo 17C, 87036 Rende, CS, Italy
*E-mail: e.curcio@unical.it, g.diprofio@itm.cnr.it

8.1 Introduction

It is today well recognized that membrane processes have the potential to replace many conventional unit operations in the chemical industry. The ability to promote an efficient and selective mass transport of specific components under low energy input, operational simplicity for an easy and safe process control, modularity for high flexibility and simple scale-up, compatibility between different membrane units in integrated systems, and low environmental impacts represent the premise for the rationalization of industrial plants in the logic of process intensification.[1]

In a general sense, membrane is a nano-structured – eventually functionalized – interface, whose physico-chemical properties are able to promote a controlled mass and energy exchange between two phases contacting the membrane itself.

The use of membranes to assist crystallization operations has been proposed and systematically investigated in recent years. The potential of polymeric membranes: (i) to concentrate a solution by solvent evaporation up to

The Handbook of Continuous Crystallization
Edited by Nima Yazdanpanah and Zoltan K. Nagy

Published by the Royal Society of Chemistry, www.rsc.org

supersaturation under a partial pressure gradient, and (ii) to promote heterogeneous nucleation on the membrane surface, so decreasing the energy barrier to the formation of critical nuclei and, ultimately, accelerating the crystallization process, has been clearly envisaged for the first time by Curcio *et al.* (2001).[2]

In principle, several membrane operations can be used to drive a solution to supersaturation by selective removal of solvent. Hybrid processes based on pressure-driven membrane operations, such as nanofiltration (NF) or reverse osmosis (RO) are used to concentrate an aqueous solution by removing the solvent in liquid phase while retaining the solute; crystals are then recovered in a separate tank often operated at lower temperature and with seeding.[3,4] However, due to the progressive increase of osmotic pressure at high concentration, exacerbated by concentration polarization phenomenon, these membrane-assisted operations are limited to the recovery of sparingly soluble salts.[5]

In addition, a solid product can be formed by reactive crystallization as a result of active or passive diffusion of reactants through a membrane. Takiguchi *et al.* (2006) investigated the reactive crystallization of $CaCO_3$ promoted by the use of ion exchange membranes for a selective transport of $CaCl_2$ and $NaCO_3$.[6] It was observed that calcium carbonate crystals grew in the form of tube-structured agglomerates (self-assembled calcite microcrystals) on the surface of a Nafion N-117 cation-exchange membrane within a sodium carbonate solution.

Ye *et al.* (2013) investigated the crystallization of Na_2CO_3 by reactive absorption of CO_2 into alkaline NaOH solution by using a gas–liquid membrane contactor.[7] The membrane is microporous and hydrophobic, not wetted by the polar phase; carbon dioxide, diffusing through the membrane, reacts with hydroxyl ions leading to the formation of HCO_3^- and CO_3^{2-} species.

In this chapter the attention is focused on three membrane crystallization configurations which promise applicability to a wide variety of crystallization processes from solutions:

1. Temperature-driven membrane crystallization, in which supersaturation is generated by the transmembrane evaporation of solvent from the warm feed side to the opposite side (distillate), taking place under a vapour pressure gradient generated by a temperature difference;
2. Osmotic-driven membrane crystallization, in which supersaturation is generated by the transmembrane evaporation of solvent from the feed solution taking place under a partial pressure gradient promoted by a concentration difference, often generated by a hypersaline (draw) solution kept at the same temperature as the feed solution;
3. Antisolvent membrane crystallization, in which an antisolvent is transferred in vapor phase to the crystallizing solution across a membrane, according to the same principle described in 1).

8.2 Principles of Membrane Crystallization Technology

In a membrane crystallizer, the level of supersaturation and its generation rate are controlled by acting on the transmembrane flux of solvent (or antisolvent) through the membrane. As a foreign interface, the membrane also induces heterogenous nucleation whose extent depends on its physicochemical properties. Moreover, the possibility to promote specific interactions between solute molecules and the membrane surface (by selecting the appropriate membrane material, or by properly functionalizing the membrane surface) might guide the nucleation mechanism towards polymorph selection or crystal habit modification.[8–10]

Temperature-driven membrane crystallization is a thermal process in which a microporous hydrophobic membrane separates two solutions kept at different temperatures.[11] The hydrophobic nature of the membrane prevents the filtration of liquid but allows the transport of the volatile solvent in vapor phase (Figure 8.1a) under a vapor pressure gradient. The mass transfer mechanism consists of three steps: solvent evaporation at membrane interface exposed to the warm feed side, diffusion of the vaporized solvent through the membrane, and condensation on the opposite side (distillate).

Osmotic-driven membrane crystallization is a technological configuration in which the removal of the volatile solvent from the feed side is obtained by using a hypertonic draw solution contacting the opposite side of the membrane. This configuration is used for the crystallization of thermo-labile species, such as biological macromolecules, since the system is operated under isothermal bulk condition. The diffusion of vapour through the membrane pores is driven by the partial pressure difference across the membrane generated by the different activities of feed and draw solutions (Figure 8.1b). Since the mass transport involves the evaporation of the solvent, the latent heat flux slightly decreases the temperature from bulk to the membrane interfaces (the temperature difference is usually lower than 1 °C for operations at room temperature), thus leading to a negligible reduction of the transmembrane flux.[12]

Solvent/antisolvent demixing mediated by a membrane can take place under different configurations. In Figure 8.1c, due to the higher volatility of antisolvent with respect to solvent, a net transmembrane flux of antisolvent moves towards the feed side under a temperature gradient. Here, the fraction of antisolvent in the feed solution increases progressively and, above a certain threshold, the lower solubility of the solute generates supersaturation.

Additional requirements for this configuration are that: (i) the antisolvent and the solvent are miscible; (ii) the solute is under its solubility limit in the initial solvent/antisolvent mixture; and (iii) the solvent and the antisolvent do not form azeotropes or, if they do, supersaturation has to be reached below the azeotropic composition.

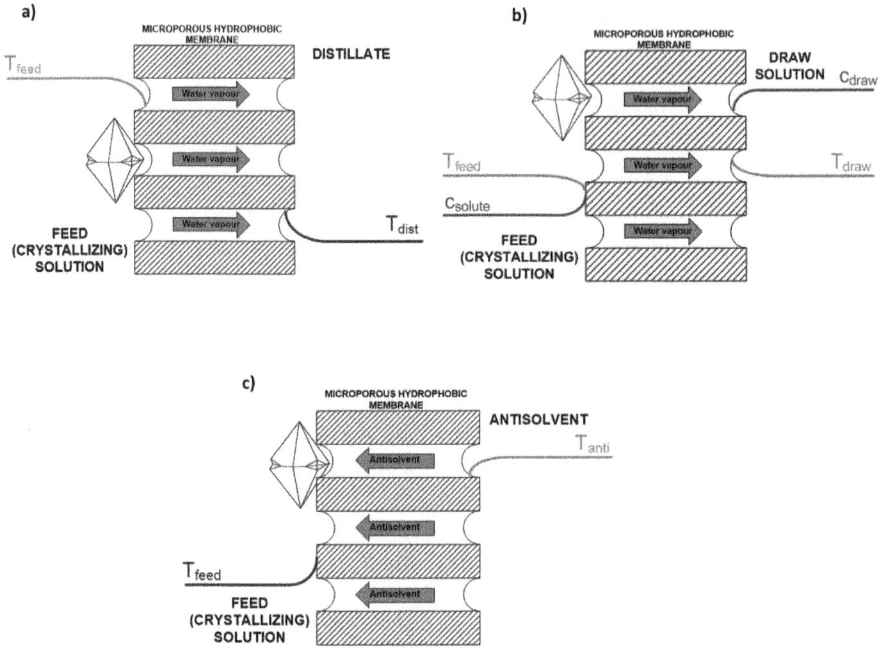

Figure 8.1 Operational principles of a membrane crystallizer. (a) Temperature-driven membrane crystallizer (direct contact configuration), in which a microporous hydrophobic membrane is contacted by the warm crystallizing solution at one side, and by cold pure water on the opposite side; (b) osmotic-driven membrane crystallizer, in which a microporous hydrophobic membrane is contacted by the crystallizing solution at one side, and by a hypersaline solution on the opposite side; (c) antisolvent membrane crystallization: antisolvent is added to the feed solution by diffusion in vapor phase from the opposite side, and the consequent decrease of solubility induces the crystallization.

8.3 Membrane Materials and Transport Phenomena

Although, in principle, a large variety of membranes (having different morphology, surface topology, physico-chemical properties, selectivity *etc.*) can be used in membrane-assisted crystallization applications, microporous hydrophobic membranes are the most appropriate when treating aqueous solutions and, so far, are the most utilized.

Among hydrophobic polymers, fluoropolymers represent a unique class of materials with an interesting combination of properties – such as high thermal stability and chemical resistance, and low surface tension as determined by the low polarizability and the strong electronegativity of fluorine atoms – that have attracted significant attention over the past few decades.[13,14]

Fluorinated polymers and copolymers (the most common are listed in Table 8.1) are widely employed to fabricate membranes at industrial level, often manufactured by non-solvent induced phase separation (NIPS) and thermally induced phase separation (TIPS) methods.

Table 8.1 Polymeric materials typically used for the fabrication of microporous hydrophobic membranes.

Polymer	Chemical structure	Main characteristics
Polypropylene (PP)	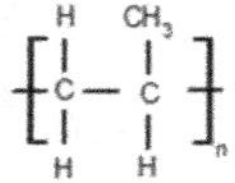	Chemically resistant, inherently hydrophobic.
Polyvinylidenefluoride (PVDF)	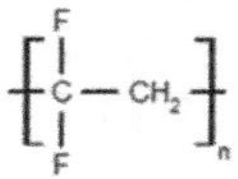	Homopolymers of PVDF are semi crystalline with long chain macromolecules containing 59.4 wt% fluorine and 3 wt% hydrogen. Typical crystallinity of 35–70%. High temperature resistant, inherently hydrophobic. All common extrusion and molding techniques can be used to process it.
Polytetrafluoroethylene (PTFE)	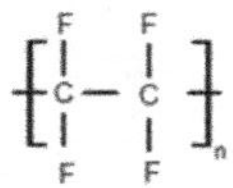	Low surface energy for high hydrophobicity. Outstanding thermal resistance and good chemical stability. It does not flow easily even above its melting point 327 °C; insoluble at room temperature. PTFE membranes are usually produced through extrusion, rolling and stretching or sintering procedures.
Poly(ethylene chlorotrifluoroethylene) (PECTFE)	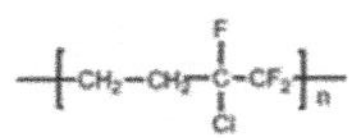	PECTFE is a 1 : 1 alternating copolymer of ethylene and chlorotrifluoroethylene. Excellent chemical stability at pH values from 2 to 13, limited solubility in organic solvents. PECTFE is a semi-crystalline, thermoplastic copolymer with a brittleness temperature lower than −76 °C and a melting point between 200 and 260 °C.
HYFLON	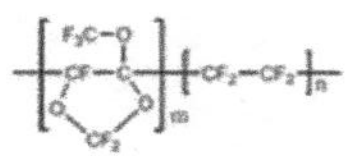	Produced by microemulsion free radical polymerization of tetrafluoroethylene (TFE) and 2,2,4-trifluoro-5-trifluoromethoxy-1,3-dioxole (TTD). Highly hydrophobic, temperature resistant, good processability.
TEFLONAF	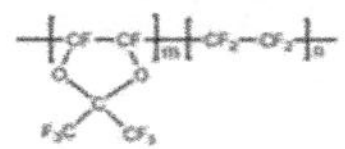	Copolymer of perfluoro-2,2 dimethyldioxole (PDD) with TFE. Highly hydrophobic, temperature resistant, good processability.
CYTOP	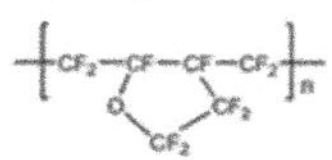	Homopolymer obtained by cyclopolymerization of perfluorobutenylvinylether (BVE), characterized by a controlled alternating structure with T_g of 108 °C. Highly hydrophobic, temperature resistant, good processability

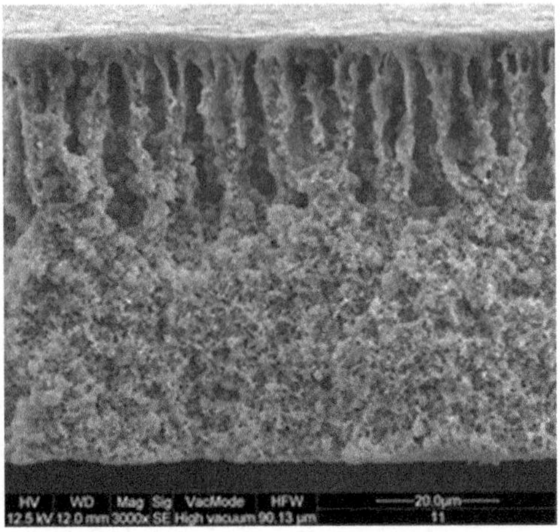

Figure 8.2 SEM micrographs of the cross-section of a PVDF membrane prepared by NIPS.

Most common solvents for NIPS process are *N*,*N*-dimethyl acetamide (DMAc), *N*,*N*-dimethyl formamide (DMF), dimethyl sulfoxide (DMSO), hexamethyl phosphoramide (HMPA), *N*-methyl-2-pyrrolidone (NMP) and tetrahydrofuran (THF),[15] while the applicable diluents for the TIPS process are dimethyl phthalate (DMP), diethyl phthalate (DEP), dibutyl phthalate (DBP), dihexyl phthalate (DHP), glycerin triacetate (GTA), ethyl acetoacetate (EAA), propylene glycol carbonate (PGC), diphenyl ketone (DPK), diphenyl carbonate (DPC), cyclohexanone, camphor, and butyrolactone.[16]

In a NIPS process, the membrane is formed by the precipitation of the polymer due to an increase of its concentration in the solvent as a result of the diffusive interchange between the solvent and the non-solvent. The final structure of the membrane is determined by the precipitation rate of the polymer:[17] a low precipitation rate leads to the formation of a symmetric structure, whereas a high precipitation rate produces an asymmetric membrane with large spongy sublayer and/or finger-like cavities below a microporous or dense upper layer (Figure 8.2). A NIPS process is articulated in three simple steps:

1. dissolution of the polymer in an appropriate solvent to form a solution typically containing 10–30 wt% of polymer;
2. casting the solution on a suitable support into a film of 100–300 μm thickness;
3. quenching of the film in the non-solvent (typically water or an aqueous solution).

TIPS methodology leads to solid–liquid phase separation by removing thermal energy from a polymer–solvent binary system. A TIPS process basically consists of four simple steps:[18]

1. formation of a homogeneous solution by melt-bending the polymer with a high-boiling, low-molecular weight solvent;
2. casting of the solution;
3. cooling of the cast solution to induce phase separation and solidification of the polymer;
4. removal of the solvent (typically by extraction) to produce the membrane structure.

TIPS allows isotropic microporous structures to be obtained.

According to the operational principle of a membrane crystallizer, the mass transfer through a microporous hydrophobic membrane must take place in vapor phase. Dusty gas model (DGM) theory is conventionally used to describe the transport of gases in a porous medium:[19] under the assumption of negligibility of surface diffusion[1] and in absence of viscous resistance (no filtration of liquid), the transmembrane flux is determined by the Knudsen diffusion resistance (due to collisions between molecules and membrane walls) and by the ordinary diffusion resistance (due to collisions among diffusing molecules).[20] Predominance or coexistence of these different mechanisms is determined by the Knudsen number (Kn) that compares the mean free path (λ) of diffusing molecules to the mean pore size (d_p) of the membrane.

Specifically, if the free mean path of the diffusing gas is small with respect to the average membrane pore diameter, Kn < 1 and molecule–molecule collisions predominate over molecule-wall collisions. *Vice versa*, in the so-called "Knudsen diffusion regime" (Kn > 1), the mean free path of the gas is large with respect to the average membrane pore diameter and molecule-wall collisions predominate over molecule–molecule collisions.

Kinetic theory of ideal gases calculates λ (m) as:

$$\lambda = \frac{k_B T}{P\sqrt{2}\pi\sigma^2} \tag{8.1}$$

where k_B is the Boltzmann constant (1.380×10^{-23} J K^{-1}), T (K) is the absolute temperature, P (J m^{-3}) the pressure, and σ is the collision diameter of the molecule (*i.e.* 2.7 Å for water, 3.7 Å for air).

Membrane modules are usually not deaerated; therefore, water vapour diffuses in air and, in this circumstance, the free mean path $\lambda_{w/a}$ can be evaluated at the average membrane temperature $\bar{T}$ as:[21]

$$\lambda_{w/a} = \frac{k_B \bar{T}}{\pi\left[\left(\sigma_w + \sigma_a\right)/2\right]^2 P} \frac{1}{\sqrt{1+\left(M_w / M_a\right)}} \tag{8.2}$$

where σ_a and σ_w are the collision diameters, and M_a and M_w (g mol^{-1}) the molecular weight for air (average) and water, respectively. At room temperature, the mean free path of diffusing water vapour is in the order of 10^{-1} μm, that is comparable to the typical pore size of microporous membranes; as a consequence, both diffusional mechanisms have to be considered.

The transmembrane flux is expressed as:[19]

$$\frac{J_i^{\mathrm{D}}}{D_{ie}^{\mathrm{K}}}+\sum_{j=1\neq i}^{n}\frac{p_j J_i^{\mathrm{D}}-p_i J_j^{\mathrm{D}}}{D_{ije}^{0}}=-\frac{1}{RT}\nabla p_i \tag{8.3a}$$

$$J_i^{\mathrm{V}}=-\frac{\varepsilon r^2 p_i}{8RT\tau\mu}\nabla P \tag{8.3b}$$

$$D_{ie}^{\mathrm{K}}=\frac{2\varepsilon r}{3\tau}\sqrt{\frac{8RT}{\pi M_i}} \tag{8.3c}$$

$$D_{ije}^{0}=\frac{\varepsilon}{\tau}PD_{ij}^{0} \tag{8.3d}$$

where J^{D} (mol m^{-2}s) is the diffusive flux, J^{V} the viscous flux, D^{K} the Knudsen diffusion coefficient (m^2 s^{-1}), D^0 the ordinary diffusion coefficient, M_i (kg mol^{-1}) is the molecular weight, p (J m^{-3}) the partial pressure, R the gas constant (8.314 J mol^{-1} K^{-1}), T the temperature, P the total pressure, μ (m^2 s^{-1}) the gas viscosity, r (m) the membrane radius, ε the membrane porosity and τ the membrane tortuosity. Subscript "*e*" indicates the effective diffusion coefficient.

Although DGM is strictly derived for an isothermal system, it is successfully applied whenever a membrane crystallizer is operated under relatively small thermal gradients (10–50 °C); in this case, the average temperature across the membrane is assumed as T value.

8.4 Heterogeneous Nucleation on Membranes

Membrane crystallization combines the principles of mass and heat transport through microporous hydrophobic membranes with the theory of heterogeneous nucleation promoted by polymeric films.

Transport phenomena related to the removal of solvent or to the controlled addition of anti-solvent have been elucidated in Section 8.2. In this section, the classical nucleation theory is adapted to the case of porous or rough membranes in order to clarify – from a mathematical point of view – the relationships existing between the physico-chemical properties of the polymeric surface and the nucleation kinetics.

The formation of clusters (or nuclei) from solution can be imagined as resulting from collision and aggregation of a certain number of solute molecules moving among the molecules of solvent. At the early stage these clusters have a large probability to dissolve; however, under specific conditions, nuclei are able to reach a critical size characterized by the same probability (50%) to grow or to dissolve (critical nuclei). According to the classical nucleation theory (CNT),[22] crystallization is an activated process and requires overcoming an energy barrier (whose peak corresponds to the Gibbs free energy for a critical nucleus).

The total Gibbs free energy ΔG of a spherical cap nucleating from solution on a solid surface is composed by two contributions: volumetric (ΔG_{vol}) and superficial (ΔG_{surf}):

$$\Delta G = \Delta G_{\text{vol}} + \Delta G_{\text{surf}} \tag{8.4a}$$

While the term ΔG_{vol} is negative, since the formation of a solid phase decreases the overall energy of the system, the term ΔG_{surf} has a dual character: positive for what concerns the interface energy that sustains the solid phase from the surrounding liquid medium, negative for the interface energy related to the portion of the cap surface laid on the membrane:

$$\Delta G_{\text{vol}} = -\frac{\Delta\mu}{\Omega}V \tag{8.4b}$$

$$\Delta G_{\text{surf}} = \gamma_{\text{L}}A_{\text{L}} - (\gamma_{\text{s}} - \gamma_{\text{i}})A_{\text{SL}} \tag{8.4c}$$

In eqn (8.4b) $\Delta\mu$ (J mol^{-1}) is the chemical potential gradient, Ω (m^3 mol^{-1}) the molar volume and V (m^3) is the volume of the sphere cap. In eqn (8.4c) γ (N m^{-1}) is the surface tension, subscripts "L", "i" and "s" refers to the nucleus-liquid, nucleus-membrane and liquid–membrane interfaces, respectively, A_{L} (m^2) is the lateral area of the nucleus-solution surface, and A_{SL} (m^2) is the contact area of the nucleus-membrane surface.

In eqn (8.4b), the gradient of chemical potential $\Delta\mu$ – the driving force to crystallization – can be expressed as:

$$\Delta\mu \approx k_{\text{B}}T\ln\frac{c}{c^*} \tag{8.5}$$

where k is the Boltzmann constant (1.380×10^{-23} J K^{-1}), T the absolute temperature, c (mol m^{-3}) the actual concentration of the solute in the mother liquor, and c^* (mol m^{-3}) the saturation. The ratio c/c^* is a quantitative measurement of the supersaturation.

Figure 8.3 illustrates the three case-studies here considered:

CASE A (Figure 8.3a): sphere cap nucleating on an ideally flat solid substrate;
CASE B (Figure 8.3b): sphere cap nucleating on a porous solid substrate;
CASE C (Figure 8.3c): sphere cap nucleating on a rough solid substrate.

Referring to the specific geometry of the systems under investigation:

for case A:

$$V = \frac{\pi R^3}{3}(1-\cos\theta)^2(2+\cos\theta) \tag{8.6a}$$

$$A_{\text{L}} = 2\pi R^2(1-\cos\theta) \tag{8.6b}$$

$$A_{\text{SL}} = \pi R^2\sin^2\theta \tag{8.6c}$$

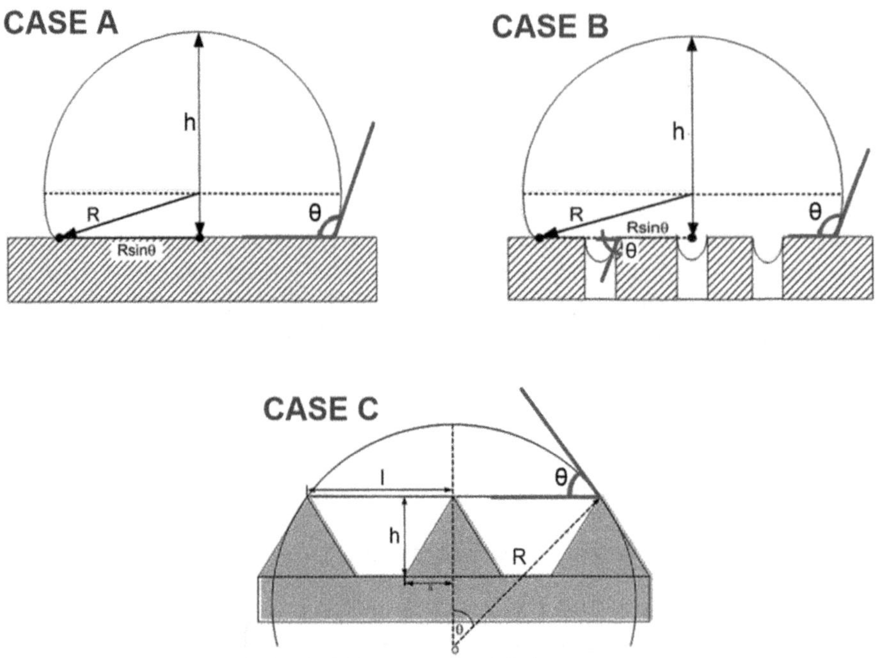

Figure 8.3 Geometry of a sphere cap nucleating on: (a) solid flat surface; (b) porous surface; (c) rough surface.

After eqn (8.6a–c) and eqn (8.4a–c):

$$\Delta G = -\frac{\pi R^3}{3}\frac{\Delta\mu}{\Omega}(1-\cos\theta)^2(2+\cos\theta) \\ +2\pi R^2\gamma_L(1-\cos\theta)-\pi R^2(\gamma_s-\gamma_i)\sin^2\theta \quad (8.7)$$

for case B:

$$V = \frac{\pi R^3}{3}(1-\cos\theta)^2(2+\cos\theta) \quad (8.8a)$$

$$A_L = 2\pi R^2[(1-\cos\theta)+\epsilon(1+\cos\theta)] \quad (8.8b)$$

$$A_{SL} = \pi R^2(1-\varepsilon)\sin^2\theta \quad (8.8c)$$

where R (m) is the radius of the sphere cap, ε is the membrane porosity, and θ is the contact angle (°).

After eqn (8.8a–c) and eqn (8.4a–c):

$$\Delta G = -\frac{\pi R^3}{3}\frac{\Delta\mu}{\Omega}(1-\cos\theta)^2(2+\cos\theta) \\ +2\pi R^2\gamma_L\left[(1-\cos\theta)+\epsilon(1+\cos\theta)\right]-\pi R^2(\gamma_s-\gamma_i)(1-\epsilon)\sin^2\theta \quad (8.9)$$

for case C:

$$V = \frac{\pi R^3}{3}(1-\cos\theta)^2(2+\cos\theta) + \pi R^2 h \sin^2\theta - \frac{\pi}{3} m s^2 h \tag{8.10a}$$

$$A_L = 2\pi R^2(1-\cos\theta) \tag{8.10b}$$

$$A_{SL} = \pi R^2 \sin^2\theta - m\pi s^2 + ms\sqrt{s^2+h^2} \tag{8.10c}$$

where m is the number of asperities on the surface, h (m) is the profile height, l (m) is the distance between two consecutive peeks, s (m) is the half-base length of an asperity.

After eqn (8.4a–c) and eqn (8.10a–c):

$$\begin{aligned}\Delta G = &-\frac{\pi R^3}{3}\frac{\Delta\mu}{\Omega}(1-\cos\theta)^2(2+\cos\theta)\\ &+2\pi R^2\gamma_L(1-\cos\theta) - \pi R^2(\gamma_s-\gamma_i)\sin^2\theta\end{aligned} \tag{8.11}$$

To progress further, a mathematical correlation is required to link γ_L and the term $(\gamma_s - \gamma_i)$. When nucleation occurs on a non-porous (dense) surface, the Young equation can be used to express the mechanical equilibrium between interfaces:

$$(\gamma_s - \gamma_i) = \gamma_L \cos\theta \tag{8.12}$$

For a porous membrane, a modified form of the Young equation correlating the surface porosity to the measured and equilibrium contact angles can be adopted:[23]

$$(\gamma_s - \gamma_i) = \gamma_L\left[\cos\theta + \frac{4\varepsilon(1+\cos\theta)}{(1-\varepsilon)(1-\cos\theta)}\right] \tag{8.13}$$

For a rough membrane surface, the wetting behavior can be adequately described by the Wenzel equation:

$$(\gamma_s - \gamma_i) = \frac{1}{r_W}\gamma_L \cos\theta \tag{8.14}$$

where r_W is the Wenzel roughness factor ($r_W > 1$). For the case illustrated in Figure 8.3c, when assuming that asperities on the surface have a conical shape, the roughness factor r can be evaluated as:

$$r_W = \frac{(\pi l^2 - m\pi s^2) + m\pi s\sqrt{h^2+s^2}}{\pi l^2} \tag{8.15}$$

In order to determine the Gibbs free energy barrier ΔG^* reached for a critical nucleus having radius R^*, the following maximization condition has to be applied:

$$\frac{d(\Delta G)}{dR} = 0 \tag{8.16}$$

When an agglomerate reaches the critical size, its further growth leads to a reduction of Gibbs free energy: therefore, all clusters having a radius higher than R^* will be likely to grow spontaneously.

Homogenous nucleation is mathematically described when combining eqn (8.7) and (8.12) and applying eqn (8.16) under the condition that θ = 180°. In this case:

$$R^* = \frac{2\gamma_L \Omega}{\Delta\mu} \tag{8.17a}$$

and

$$\Delta G^*_{hom} = \frac{16}{3}\pi\gamma_L{}^3\left(\frac{\Omega}{\Delta\mu}\right)^2 \tag{8.17b}$$

where ΔG^*_{hom} is the value of the Gibbs energy threshold for nucleation occurring in solution and absence of a solid surface.

The presence of a foreign interface in the crystallizing system decreases the work required to generate a critical nucleus and increases locally the probability of nucleation with respect to other locations in the bulk of the system. This phenomenon is known as heterogeneous nucleation; for an ideally smooth surface (case A):

$$\Delta G^*_{het} = \frac{16}{3}\pi\gamma_L{}^3\left(\frac{\Omega}{\Delta\mu}\right)^2\left[\frac{1}{2} - \frac{3}{4}\cos\theta + \frac{1}{4}\cos^3\theta\right] \tag{8.18}$$

From eqn (8.18), it can be observed that the work of nucleation occurring on a non-porous membrane having a contact angle θ = 90° is half with respect to that required to form a critical cluster in homogeneous phase. As expected, the values of Gibbs free energy barrier for homogeneous and heterogeneous nucleation become equal when θ = 180°.

According to eqn (8.18), the maximum of the Gibbs free energy associated to formation of a critical cluster on porous membranes is lower than on solid and dense surfaces (Figure 8.4a). Monte Carlo simulations based on the Ising model confirm that the nucleation in a single pore is faster than on a smooth surface, and the logarithm of the nucleation rate shows an almost linear dependence with the pore size.[24] Theoretical predictions about the favorable effect of surface porosity on nucleation rate are experimentally confirmed in protein crystallization.[25] Figure 8.4b illustrates the ratio between the energy barrier for heterogeneous nucleation occurring on rough surfaces and the energy barrier for homogeneous nucleation. The effect of roughness is predicted to have a different influence on nucleation kinetics depending on the contact angle: for $\theta < 90°$ (hydrophilic membrane surface), a value of Wenzel coefficient $r > 1$ increases the energy barrier and – consequently – decreases the nucleation rate; for $\theta > 90°$ (hydrophobic membrane surface) the behavior is opposite. For the limiting value of θ = 180°, the curves converge to

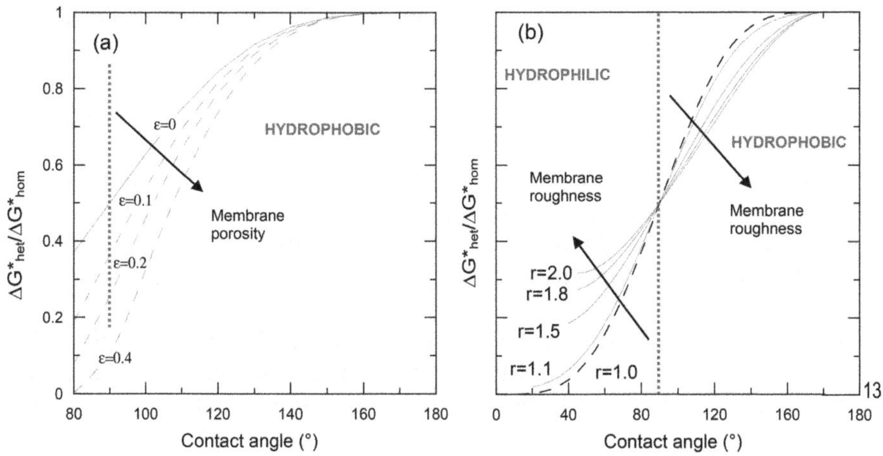

Figure 8.4 Trend of the energy barrier to the formation of a critical nucleus on a membrane as a function of contact angle (θ), for different: (a) porosity (ε); (b) Wenzel roughness coefficient (r).

1 (nucleation in homogeneous phase). For a perfectly smooth surface (r = 1) the energy ratio is expressed by eqn (8.18). Experimental investigations confirmed the enhanced nucleation rate of calcium sulphate dihydrate on heated rough stainless-steel surfaces.[26]

According to CNT, nucleation is an activated process requiring that the Gibbs free energy barrier (ΔG^*) has to be overcome in order to form a stable nucleus. The correlation between ΔG^* and nucleation rate N is mathematically expressed by the Arrhenius-like correlation:

$$N = \Gamma e^{-\frac{\Delta G^*}{k_B T}} \tag{8.19}$$

where Γ is a pre-exponential kinetic factor. Therefore, the possibility of designing membranes with specific chemical, topological and structural properties offers a powerful and unique tool for controlling the nucleation stage in a crystallization process.

8.5 Membrane Crystallization of Proteins

Biological macromolecules and proteins crystallize from solutions according to the same principles and by the same mechanisms as small molecules; however, their peculiarities make their crystallization process very challenging. Due to their structural complexity, the surface of macromolecules displays a rather asymmetric and weak bonding configuration, thus significantly reducing the probability of an ordered attachment of growing units. Moreover, proteins show a significant attitude to aggregate in n-mers that diversify shape and size of lattice units and exhibit very low diffusivities so that their crystallization kinetics are extremely slow.

Overall, protein crystallization is a complicated physical-chemical process: a specific macromolecule generally crystallizes only in a narrow range of numerous parameters, including concentration, temperature, pH, type and concentration of precipitating agent or other chemical additives, method and rate of supersaturation generation.[27]

In order to avoid the risk of protein denaturation at high temperature, osmotic membrane crystallization is adopted as operational configuration. Figure 8.5 shows the influence of the draw solution ($MgCl_2$) on water flux through the membrane in the crystallization tests of hen egg white lysozyme.[28]

Real-time tuning of the transmembrane flux can be easily obtained by adjusting the concentration of the salt in the stripping solution, and higher evaporation rate at constant temperature is obtained when increasing the salt content.

With respect to the evaporation of the solvent from the protein solution, the vapour–liquid equilibrium is mathematically expressed through the Raoult law in terms of partial pressure (p_i), vapour pressure of the pure component (p^0_i) and activity coefficient ζ_i that considers the non-ideal behavior of the system:

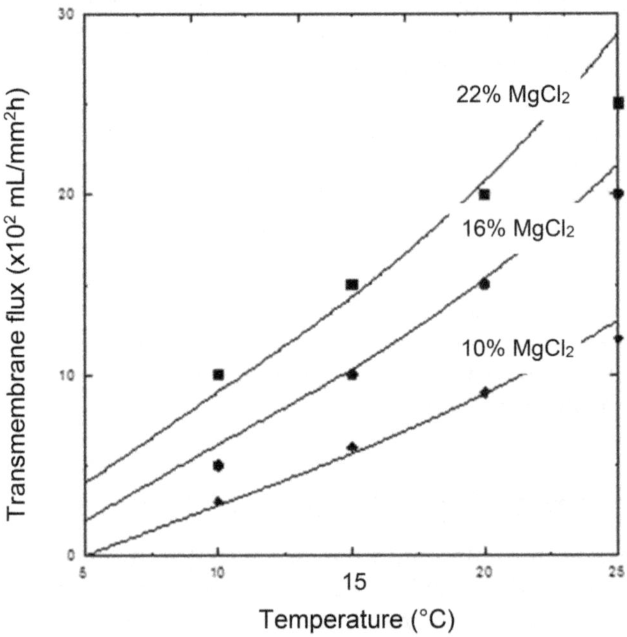

Figure 8.5 Transmembrane flux *versus* temperature (5–25 °C) at different stripping concentrations (10–22% $MgCl_2$). Theoretical (lines) and experimental (symbols) results obtained at constant hen egg white lysozyme concentration (20 mg mL^{-1}), 4.5% NaCl and pH = 4.6. Reprinted from *J. Cryst. Growth*, **247/1–2**, E. Curcio, G. Di Profio and E. Drioli, A new membrane-based crystallization technique: tests on lysozyme, 166–176, Copyright 2003, with permission from Elsevier.[28]

$$p_i = p^0{}_i x_i \zeta_i \tag{8.20}$$

The activity coefficient ζ_i for proteins is commonly estimated from measurements of the virial coefficients B_{22} for a given concentration (c) under the assumption that, in relatively diluted solutions, binary interactions between pairs of particles prevail over ternary and other more complex interactions:[29]

$$\ln \zeta \approx 2B_{22}c \tag{8.21}$$

The determination of B_{22} coefficient (for example, by Dynamic Light Scattering) can provide useful guidelines to drive crystallization attempts towards success; in general, the progression of the second virial coefficient systematically towards more negative values indicates an increase of protein–protein attractions. Selection of solution composition requires a delicate balance and, in this respect, membrane crystallization offers the possibility to adjust solution concentration by acting on the transmembrane flux.

Figure 8.6 illustrates the influence of membrane porosity on the nucleation of lysozyme (see Section 8.4): the ratio between the maximum Gibbs free energy associated with the formation of a critical cluster and the

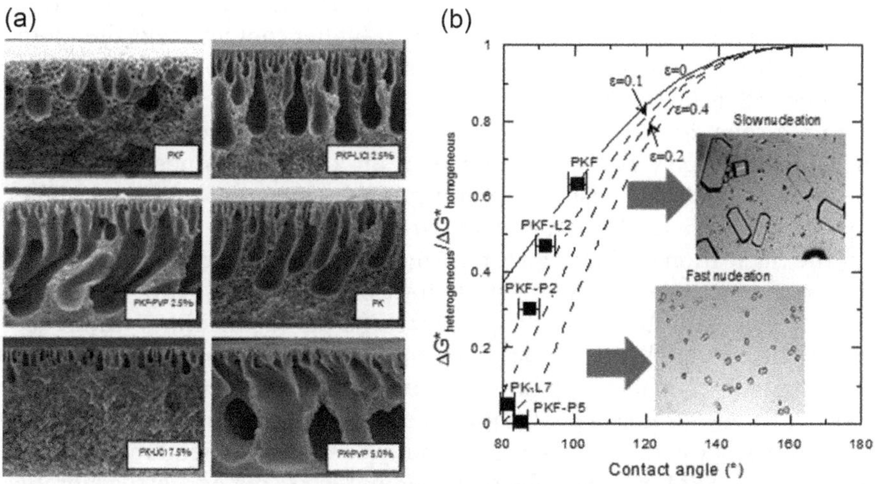

Figure 8.6 (a) Cross section of poly(vinylidenefluoride-*co*-hexafluoropropylene) (PKF) and polyvinylidenefluoride homopolymer (PK) prepared with different wt% of LiCl and PVP in the casting solution; (b) $\Delta G_{het}/\Delta G_{hom}$ ratio as a function of the contact angle of HEWL solution (4 mg mL^{-1}, 2% w/v NaCl, pH 4.6) on PKF and PK membranes at different porosity (P2: 2.5 wt %PVP, P5: 5.0 wt %PVP, L2: 2.5wt %LiCl, L7: 7.5wt %LiCl). Reprinted with permission from E. Curcio, E. Fontananova, G. Di Profio and E. Drioli, Influence of the Structural Properties of Poly(vinylidene fluoride) Membranes on the Heterogeneous Nucleation Rate of Protein Crystals, *J. Phys. Chem. B*, 2006, **110**, 12438, Copyright 2006 American Chemical Society.[23]

Figure 8.7 SEM image of crystals nucleated and grown on a microporous polypropylene membrane surface. (A) tetragonal lysozyme. Reprinted from ref. 30, Copyright 2003, with permission from Elsevier. (B) trypsin crystals. Reprinted with permission from ref. 10, Copyright 2005 American Chemical Society.

corresponding energy barrier in homogeneous phase is plotted against the contact angle. Points are located on the diagram according to the contact angle measured for a lysozyme solution (40 mg mL^{-1}, 2% w/v NaCl, pH 4.6) on different lab-made PVDF membranes. In general, the energetic barrier decreases at higher porosity and increases at higher contact angle; therefore, in a membrane crystallizer, nucleation rate is accelerated by highly porous and moderately hydrophobic substrates.

From a theoretical point of view, Monte Carlo simulations confirm that nucleation in a pore is always faster than on a perfectly smooth surface.[24] The scientific literature also provides empirical evidence of a favorable effect of porosity on the nucleation rate; in particular, Chayen *et al.* (2006) proved that the nucleation of thaumatin, trypsin, lobster α-crustacyanin, lysozyme, c-phycocyanin, myosin-binding protein-C, and α-actinin actin binding is enhanced when occurring in the presence of a porous medium, while non-porous surfaces are less successful in promoting nucleation.[25]

Accelerated crystallization kinetics and reduced induction time with respect to conventional crystallization techniques, while preserving diffracting quality of protein crystals, are generally observed in membrane crystallization.[28,30,31] SEM micrographs of tetragonal lysozyme and orthorhombic trypsin crystals embedded on microporous polypropylene membrane surface are shown in Figure 8.7.

8.6 Crystal Morphology and Polymorphism

The evolution of nuclei to macroscopic crystals can follow several pathways finally resulting in different product characteristics such as particle size distribution, shapes and structures (polymorphism) that strongly depend on the nucleation and the growth rates. In addition, the morphology of a crystal can be influenced by the supersaturation, the nature of the solvent, the presence of impurities, the fluid dynamic regime *etc.*; at industrial level, the crystalline habit plays a crucial role in downstream processes such as filtration, washing, drying, flow, compaction, dissolution, and packaging.[32]

Polymorphism is the ability of a solid substance to exist in at least two different crystalline arrangements of the molecules having the same chemical composition; the term pseudo-polymorphism is used to describe the ability of crystals to incorporate variable quantities of solvent with a defined stoichiometry. Polymorphism often plays a significant role in the preparation of active pharmaceutical ingredients (APIs) and other products of the pharmaceutical industry, since different polymorphs can exhibit important differences in solubility, dissolution rate, stability, melting point, density and many other properties that significantly affect the efficacy, bioavailability and safety of APIs.[33]

According to the Ostwald step rule,[34] a crystalline system evolves from the least stable (more soluble, highest Gibbs free energy barrier) to the most thermodynamically stable form (less soluble, lowest Gibbs free energy barrier). In some cases, if the energy of the system is not sufficient to overcome the nucleation barrier of unstable polymorphs, only the stable form is obtained.

A practical implication of Ostwald's rule is that – when manipulating in an appropriate way the energy barrier to nucleation – different polymorphs can be isolated. Referring to the energy/reaction coordinate diagram for a dimorphic system in Figure 8.8, the reduction of the activation energy induced by heterogeneous nucleation on a membrane can enable the formation of both polymorphs A and B.

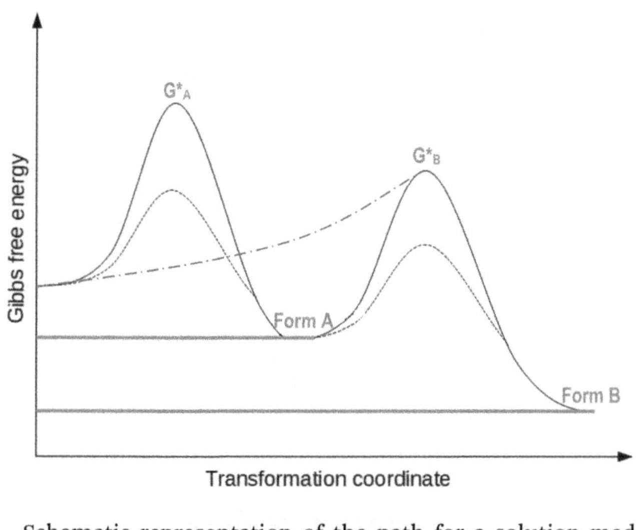

Figure 8.8 Schematic representation of the path for a solution mediated phase transformation from an unstable form A to the thermodynamically stable form B. Two pathways are possible: directly to the stable form (dot-dash line) or – if the energy of the system is sufficient to overcome the nucleation barrier for A – to form B *via* metastable form A. Membrane crystallization promotes heterogeneous nucleation, and the decrease of energy barrier can facilitate the appearance of polymorphic forms.

8.6.1 Influence of the Transmembrane Flux

The first evidence for polymorph selection promoted by membranes was obtained by Di Profio *et al.* (2007) during glycine crystallization tests.[9] Glycine exhibits three polymorphic forms: α, β and γ.[35] Two additional forms, δ and ε, were obtained under high-pressure conditions.[36]

The relative stability at ambient temperature of the three forms was found to be: $\gamma > \alpha > \beta$.[37]

Figure 8.9 illustrates the selection of glycine polymorphs made by membrane crystallization operated at pH 6.2: experimental results show a preferential crystallization of the γ form for an evaporation rate lower than 1.4×10^{-2} mL h^{-1}, while for an evaporation rate higher than 1.8×10^{-2} mL h^{-1} the kinetic product, α-glycine, is obtained.

Paracetamol (acetaminophen) is a common antipyretic and analgesic drug that can be crystallized in three polymorphic forms: monoclinic form I (space group $P2_1/n$, thermodynamically stable at room temperature), orthorhombic form II (space group *Pcab*, metastable at ambient conditions), and a least stable form III obtained by crystallization from the melt.[38–40]

Since the crystal structure of the commercial monoclinic form shows poor compression properties and requires binding agents for tableting, attention has been focused on the production of the orthorhombic form II,

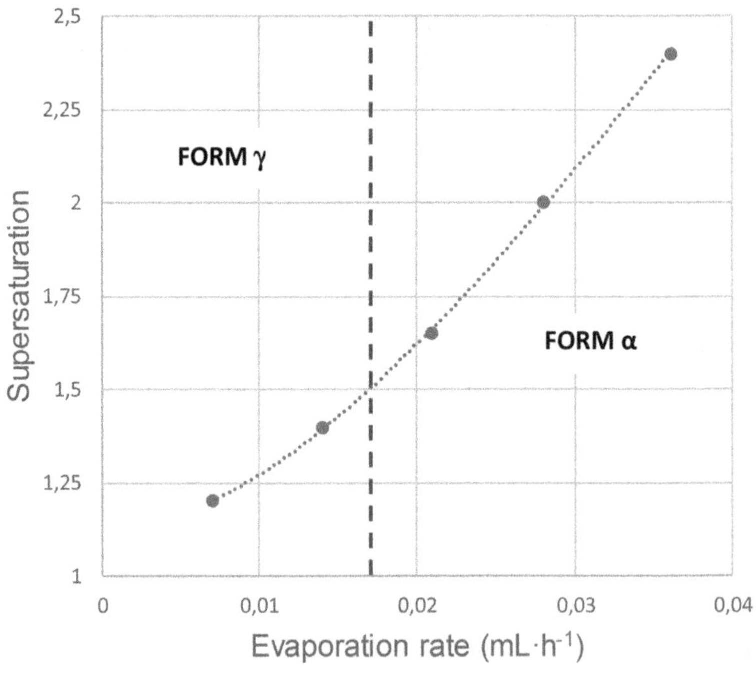

Figure 8.9 Glycine polymorphic selection based on the control of the transmembrane flux in a membrane crystallizer. Reprinted with permission from ref. 9, Copyright 2007 American Chemical Society.

characterized by enhanced plastic deformation due to parallel hydrogen-bonded sheets along the longitudinal axis. However, while form I is easily grown from solution in various solvents, form II is conventionally obtained only by complicated or multistep processes.[41,42]

The shift of some specific peaks and infrared absorptions patterns in the range of 1700–650 cm^{-1} allows discrimination between the monoclinic and orthorhombic forms:[43] the form I spectrum is characterized by a peak around to 1515 cm^{-1} and a strong absorption at 807 cm^{-1}, whilst the form II spectrum is characterized by absorptions at 1666 cm^{-1}, 1622 cm^{-1} and 1423 cm^{-1}.

Table 8.2 reports the occurrence of polymorphs I and II when Membrane Crystallization is carried out at different transmembrane evaporation rates (J).[8] For $J \geq 7.90 \times 10^{-2}$ mL h^{-1} the metastable form II is obtained, while for $4.33 \times 10^{-2} \leq J \leq 6.60 \times 10^{-2}$ mL h^{-1}, the thermodynamic monoclinic product appears. At low supersaturation, where heterogeneous nucleation on the membrane prevails, crystals exhibit the polymorphic structure of form II: therefore, the presence of polypropylene membrane surface in the crystallizing system provides – for very low values of evaporation rate ($J = 1.37 \times 10^{-2}$) – low energy nucleation sites the increase the chance of nucleation for the less stable orthorhombic form of paracetamol. A concomitant presence of both polymorphs is found at $J = 1.94 \times 10^{-2}$ mL h^{-1}. This experimental investigation demonstrates that polymorphic selection can be achieved by a careful control of the solvent flux through the microporous membrane.

Experimental evidence of polymorph selection by Membrane Crystallization has been also obtained for Carbamazepine (CBZ), a water-insoluble drug used in the treatment of epilepsy and neuropathic pain, and that requires high dosage (>100 mg day^{-1}) for therapeutic effect.[44] CBZ exists in a dihydrate form (CBZ DH) and in at least five anhydrous forms: primitive monoclinic (CBZ III), triclinic (CBZ I), C-centered monoclinic (CBZ

Table 8.2 Paracetamol polymorphs obtained at different transmembrane evaporation rate (J); corresponding values of induction time and supersaturation are reported. Reprinted with permission from ref. 8, Copyright 2007 American Chemical Society.

Transmembrane flow (mL h^{-1})	Induction time (h)	Supersaturation ([a])	Polymorphic form
1.37×10^{-2}	64.3	1.34	II
1.94×10^{-2}	53.6	1.52	II (+I)
4.33×10^{-2}	23.5	1.91	I
4.37×10^{-2}	31.0	1.83	I
5.89×10^{-2}	24.3	1.93	I
6.60×10^{-2}	23.0	1.96	I
7.90×10^{-2}	19.5	2.00	II
8.98×10^{-2}	17.5	2.05	II

[a]Evaluated at the induction time.

IV) and trigonal (CBZ II), mentioned by decreasing thermodynamic stability, and orthorhombic form V.[45,46] Interestingly, in membrane crystallization of CBZ at increasing transmembrane flux, the amount of CBZ I in the solid precipitate decreases while the amount of less stable form CBZ IV increases.

8.6.2 Influence of the Chemistry of the Surface

Curcio *et al.* (2014) investigated the effect of polymer–solute interactions on the kinetics of heterogeneous nucleation on membranes; in particular, experimental efforts were focussed on the acetaminophen (ACM) crystallization as a result of specific solute–surface interactions occurring on polydimethylsiloxane (PDMS) with siloxane functional groups Si–O–Si forming the backbone of silicones, partially fluorinated elastomer polyvinylidenedifluoride (PVDF), polystyrene (PS) having weak electron-donor phenyl rings, poly(*n*-butyl methacrylate) (PnBMA) showing ester functionality, polyimide (PI) that includes both hydrogen-bond acceptor imide functionality and carbonyl groups and ethylene/acrylic acid (EAA) copolymer with a carboxyl moiety.[47]

Powder X-ray diffraction (PXRD) analysis demonstrated the occurrence of preferential orientation (PO) as a function of the strength of intermolecular interactions at the polymer–crystal interface taking place at the early stage of nucleation, ranging from non-specific adsorption (PDMS, PnBMA, and PS) to oriented arrangement of molecules in the crystalline lattice (EAA, PI, and PVDF).

All investigated polymeric membranes promoted the formation of the thermodynamically stable monoclinic ACM form I, with the interesting exception of polyimide.

PXRD analysis of ACM crystals nucleating on hydrophobic PVDF shows a main peak at 13.9° (001), and two predominant reflections at 14.1° (001) and 28.1° (002) on the EAA surface (Figure 8.10).

These experimental data indicate a strong orientation of crystals along {001}: hydroxyl groups of the ACM molecules aligned along the plane {001} are perpendicularly oriented with respect to the polymeric surface (interactions through hydrogen-bonding).

Referring to Figure 8.11, the PXRD pattern of ACM crystals nucleated on hydrophobic PDMS surface does not give evidence for a prevalent crystal orientation along a specific crystallographic plane, and four major reflections at 12.1° (110), 15.4° (20$\bar{1}$), 18.9° (020), and 24.4° (220) are observed. Reflections whose linear combinations correspond to vector direction {110} indicate that some ACM molecules are parallel to the membrane surface, while others have hydroxyl and methyl groups alternatively redirected towards the polymeric substrate. The main peak at 15.4°, corresponding to orientation along {20$\bar{1}$}, reveals that methyl groups and the amide portion in some ACM molecules are perpendicularly aligned to the PDMS surface. The PXRD diffractogram of PnBMA shows a similar behavior, although the presence of

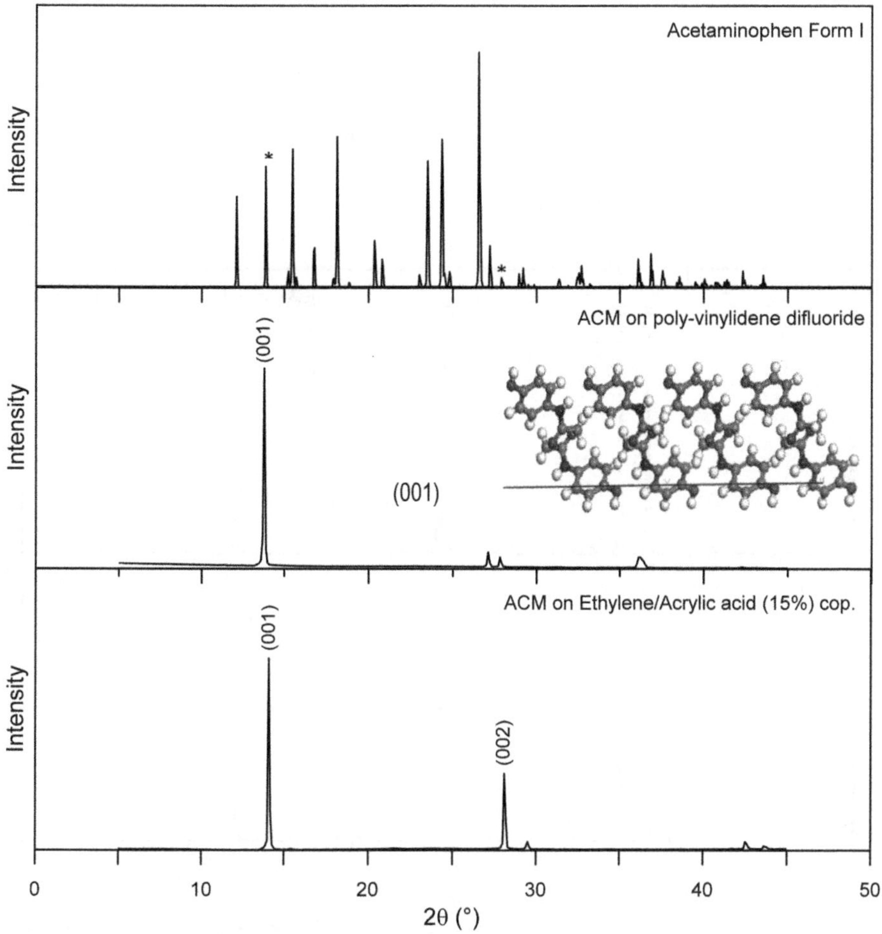

Figure 8.10 Preferential orientation of acetaminophen (ACM) crystals nucleated on PVDF and EAA copolymer membranes. Calculated PXRD pattern for ACM Form I shown on the top. In ACM molecular structures: [o: hydrogen, •: oxygen, •: nitrogen, •: carbon]. Reprinted with permission from ref. 47, Copyright 2014 American Chemical Society.

three peaks at 12.1° (110), 24.4° (220) and 37.1° (330) confirms that mixed hydrophobic–hydrophilic interactions occur along the direction {110}: the absence of specific hydrogen-bonding interactions is consistent with the observation that surface–molecular interactions are dominated by terminal methyl groups of the ester side chain when in contact with water. There is no evidence of PO for ACM crystals grown on PS.

PXRD patterns of ACM crystals nucleated on a PI surface revealed the concomitant presence of two polymorphs: the metastable orthorhombic form II and the stable monoclinic form I (Figure 8.12). In particular, the major reflection at 13.9° gives evidence for a preferential orientation of

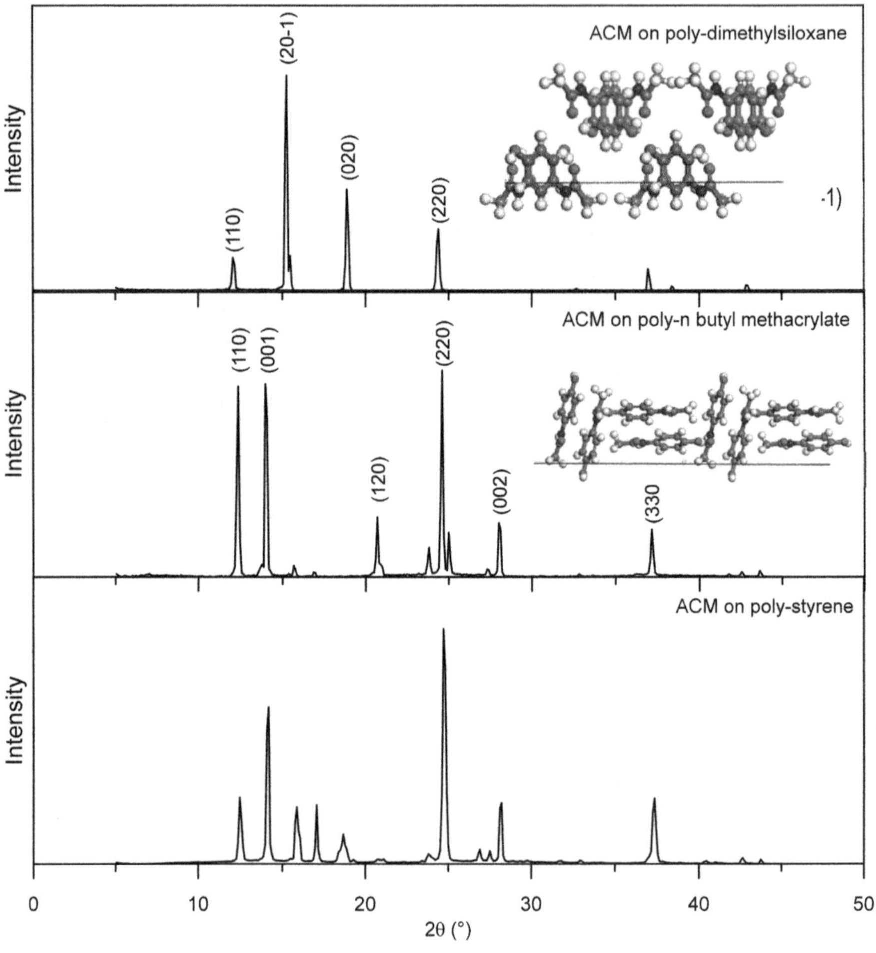

Figure 8.11 Preferential orientation of acetaminophen (ACM) crystals nucleated on PDMS, PnBMA and PS membranes. In ACM molecular structures: [o: hydrogen, •: oxygen, •: nitrogen, •: carbon]. Reprinted with permission from ref. 47, Copyright 2014 American Chemical Society.

form I crystals along {001} with hydroxyl groups of ACM oriented towards the polymeric surface. The major reflection at 15.0° is related to the preferential growth of form II crystals along the plane (111) with hydroxyl groups of ACM obliquely oriented towards the substrate. Interfacial interactions are likely to involve both imide functionality and carbonyl groups in polyimide.

8.7 Continuous Membrane Crystallization Processes

A typical configuration of a Membrane Crystallizer is illustrated in Figure 8.13.

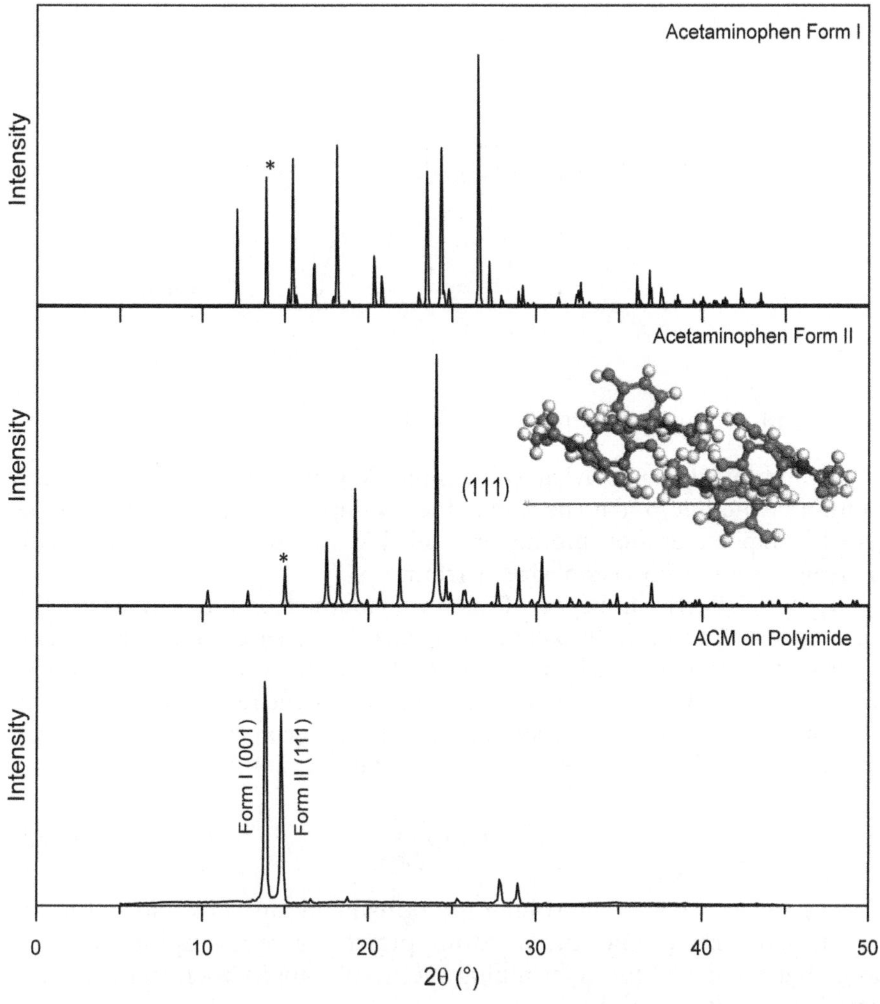

Figure 8.12 PXRD patterns revealing the concomitant occurrence of polymorphism in ACM crystals grown on polyimide (PI) membrane. Calculated PXRD pattern for Form I on the top and for Form II in the center. In ACM molecular structures: [o: hydrogen, •: oxygen, •: nitrogen,• : carbon]. Reprinted with permission from ref. 47, Copyright 2014 American Chemical Society.

The membrane module, hosting microporous hydrophobic membranes, is the core of the Membrane Crystallization plant. Typical polymeric membranes have a nominal pore size between 0.1 and 0.55 μm, porosity > 70%, and are manufactured in the form of hollow fibers or flat sheets.

Crystallization kinetics are characterized in terms of growth and nucleation rates of crystals formed from solution. From and industrial point of view, the main advantages of a membrane crystallizer with respect to conventional equipment are: (i) fine control of supersaturation related to the

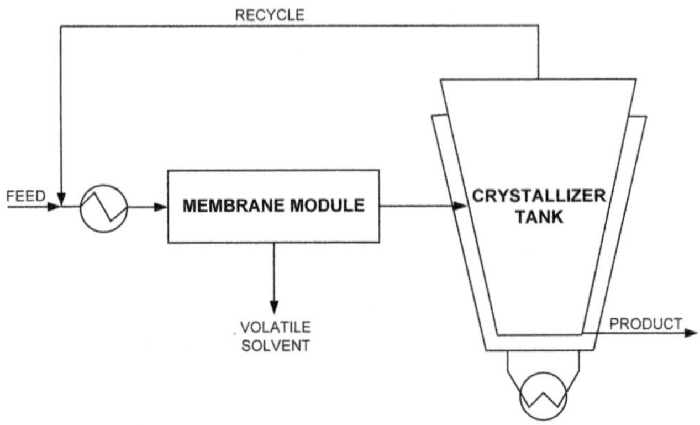

Figure 8.13 Scheme of a Membrane Crystallizer.

transmembrane flux of solvent; (ii) laminar flow of the solution through the membrane module, and no need of baffles or impellers to guarantee a homogeneous supersaturation profile; and (iii) low risk of attrition and crystal breakage for a narrow crystal size distribution.

Due to the selective transport of the solvent at the membrane interface, the concentration profile of the solute is expected to increase from bulk to membrane surface through the boundary layer. The term "concentration polarization" indicates the increase of the concentration of the retained components at the membrane surface (c^{m}) compared to the bulk concentration (c^{b}). It can be demonstrated that, in the ideal case of total rejection:[48]

$$c^{\mathrm{m}} = c^{\mathrm{b}} \exp\left(\frac{J}{k_x}\right) \tag{8.22}$$

where J (m^3 m^{-2} s^{-1}) is the volumetric transmembrane flux and k_x (m s^{-1}) the mass transfer coefficient. Literature provides several empirical correlations that are useful for determining the mass transfer coefficient, usually expressed in the form:[49]

$$\mathrm{Sh} = \alpha \mathrm{Re}^{\beta} \mathrm{Sc}^{\gamma} \tag{8.23}$$

where:

Sh, Sherwood number $\mathrm{Sh} = \frac{k_x d_{\mathrm{h}}}{D}$ (d_{h}: hydraulic diameter, D: diffusion coefficient);

Re, Reynolds number $\mathrm{Re} = \frac{\rho v d_{\mathrm{h}}}{\mu}$ (ρ: fluid density, v: fluid velocity, μ: fluid viscosity);

Sc, Schmidt number $\mathrm{Sc} = \frac{\mu}{\rho D}$

Some values of coefficients α, β and γ are reported in ref. 50–52.

Because the solvent flux across membranes takes place in the vapour phase, the design of a continuous membrane crystallizer involves the simultaneous transport of mass and heat.

The heat fluxes Q (J m^{-2} s^{-1}) for the feed (subscript "F") and – in direct contact configuration – distillate (subscript "D") streams are determined by changes in enthalpy and estimated from temperatures (T) at the module inlet (subscript "in") and outlet (subscript "out"):

$$Q_F = m_F c_{P,F}(T_{F,in} - T_{F,out}) \tag{8.24a}$$

$$Q_D = m_D c_{P,D}(T_{D,in} - T_{D,out}) \tag{8.24b}$$

where m (kg m^{-2} s^{-1}) is the mass flow rate of the liquid stream, and c_p (J kg^{-1} K^{-1}) is the specific heat capacity. Under steady-state conditions, and assuming that no heat is lost to the environment:

$$Q_F = Q_D = Q \tag{8.25}$$

Analogously, heat fluxes through boundary layers are:

$$Q_F = h_F A_F(T_F - T_F^*) \tag{8.26a}$$

$$Q_D = h_D A_D(T_D - T_D^*) \tag{8.26b}$$

where h (W m^{-2} K^{-1}) is the heat transfer coefficient, A is the membrane surface area, and T^* identifies the temperature at the membrane interface.

Combining eqn (8.25) and (8.26a–b):

$$T_F^* = T_F - \frac{Q}{h_F A_F} \tag{8.27a}$$

$$T_D^* = T_D + \frac{Q}{h_D A_D} \tag{8.27b}$$

According to eqn (8.27a), the temperature profile of the feed stream progressively decreases within the boundary layer from bulk to membrane interface, eventually influencing the supersaturation value in proximity of the membrane where nucleation occurs. This phenomenon is known as "temperature polarization".

Several empirical correlations can be used to estimate the heat transfer coefficient based on the following dimensionless numbers:[53–55]

Nusselt number $\mathrm{Nu} = \dfrac{hD}{k}$ (h: heat transfer coefficient, k: thermal conductivity)

Grashof number $\mathrm{Gr} = \frac{D^3 \rho^2 g \beta \Delta T}{\mu^2}$ (g: gravity acceleration, β: thermal expansion coefficient)

Prandtl number $\mathrm{Pr} = \frac{c_p \mu}{k}$ (c_p: specific heat)

Graetz number (for hollow fiber membrane modules with length L and mass flowrate $\dot{m}$): $\mathrm{Gr} = \frac{\dot{m} c_p}{kL}$

Since the membrane acts as hetero-nucleant interface with low surface, and due to concentration polarization that increases the solute concentration in proximity of the membrane, the nucleation process in a membrane crystallizer preferentially occurs on the membrane module; crystals are then flushed and recirculated into a separate crystallizer tank by the retentate flow. The combination of membrane modules and temperature-controlled crystallization should be considered as one system consisting of both nucleation and growth steps. Moreover, the retentate flowrate plays a crucial role in determining the detachment of crystals from the membrane surface and, in general, on the overall crystallization process.

For a continuous process, the growth rate (G) is estimated from the variation of the overall crystal length ($\bar{\mathrm{L}}$) as a function of the residence time (τ):

$$G = \frac{\mathrm{d}\bar{L}(\tau)}{\mathrm{d}\tau} \approx \frac{\bar{L}(\tau_{n+1}) - \bar{L}(\tau_n)}{(\tau_{n+1} - \tau_n)} \tag{8.28}$$

where:

$$\bar{L}(\tau) \approx \sum_i \alpha_i(\tau) L_i \tag{8.29a}$$

$$\alpha_i(\tau) = \frac{N_i(\tau)}{N_{\mathrm{total}}(\tau)} \tag{8.29b}$$

The crystal growth rate is empirically related to the supersaturation (S):

$$G = k_g S^g \tag{8.30}$$

where k_g is the growth rate constant (function of temperature, fluid-dynamics, presence of impurity *etc.*) and exponent g represents the kinetic order of the growth process. For NaCl crystallization on microporous polypropylene membranes carried out at low supersaturation ratio ($S = 0.065$–0.097), least square multiple linear regression analysis of data resulted in $g = 1.17$ and $k_g = 1.92 \times 10^{-5}$ (G expressed in m s^{-1}).[2]

The nucleation rate N (# m^{-3} s^{-1}) is determined by the change of the total number of crystals (N_{total}) as a function of the residence time per unit of mother liquor volume V (m^3) in the crystallizer tank:

$$B = \frac{1}{V(\tau)} \frac{N_{total}(\tau_{n+1}) - N_{total}(\tau_n)}{(\tau_{n+1} - \tau_n)} \tag{8.31}$$

The crystal size distribution (CSD) is usually expressed on a (counted) number basis, thus in terms of relative percentage frequency curve $f(L_i)$ *versus* the size classes L_i. The function $f(L_i)$ is then defined as the ratio between the number of crystals belonging to a certain size class (N_i) to the total number of crystals measured ($\Sigma_i N_i$) divided by the width of the size class (ΔL_i):

$$f(L_i) = \frac{N_i}{\sum_i N_i} \cdot \frac{1}{\Delta L_i} \tag{8.32}$$

An example of CSD is provided by Figure 8.14 for NaCl crystals obtained in a membrane crystallization experiment carried out on seawater concentrate by Reverse Osmosis.

A coefficient of variation (CV) is usually employed to evaluate the dispersion of CSD. CV is defined as:

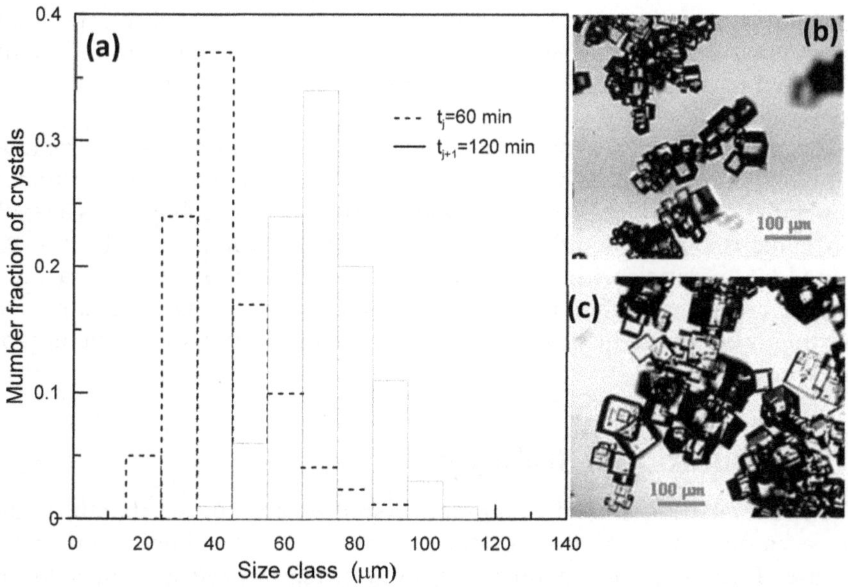

Figure 8.14 (a) CSD histogram of NaCl crystals from seawater desalination brine. Images of crystals collected (b) 60 min/(c) 120 min after reaching supersaturation. Reprinted from ref. 56, Copyright 2010, with permission from Elsevier.

$$CV = \frac{\overline{L}_{80\%} - \overline{L}_{20\%}}{2\overline{L}_{50\%}} \times 100 \quad (8.33)$$

Experimental CV values measured in membrane crystallization tests carried out on NaCl range within 15–35%;[57,58] as a comparison, the coefficient of variation for ideal MSMPR is 50% for size-independent growth.

From an industrial point of view, advantages of membrane processes are generally related to their unique properties of modularity for an easy scale-up and a large operational flexibility, compatibility between different membrane operations in pre-treatment and downstream processes, robustness and stability under operating conditions, low environmental impact and low energy consumption. In membrane crystallization technology, the interfacial membrane area for solvent removal or antisolvent addition per unit volume is very high if compared to conventional forced circulation or draft tube baffled crystallizers, especially in the case of hollow fibers membrane modules (packing density > 20.000 m^2 m^{-3}). Moreover, the gentle flow of the mother solution through membrane modules in laminar regime guarantees a high mixing degree and reduces the risk of attrition and breakage of crystals with moving parts of the equipment.[59]

Flux decline with increased feed concentration, in conjunction with concentration and temperature polarization phenomena, is an inherent characteristic of membrane crystallization. Due to occurrence of heterogeneous nucleation, attention should be paid to the hydrodynamics of the system in order to assure that the shear stress generated by flow recirculation is sufficient to remove crystals deposited on the membrane surface, so preserving the stability of the process.

The main concerns about a membrane relates to the long-term maintenance of its hydrophobic behavior. In case of membrane wetting determined by the loss of hydrophobicity, the crystallization process collapses as a result of the dispersion of liquid phases. This potential problem can be circumvented by operating with moderate transmembrane fluxes (1–5 L m^{-2} h^{-1}) and by using hydrophobic polymeric materials, whose non-wettability is eventually enhanced by physical or chemical modification of the membrane surface.

8.8 Operational Stability

The operational stability of membrane crystallization imposes that the liquid phase cannot penetrate through the microporous hydrophobic membrane; in case of wetting, since membrane is not intrinsically selective, all solution components would be transferred from feed to the permeate side, thus leading to process failure. From an operative point of view, a fluid will not pass through a porous medium if its pressure is kept below a critical threshold, known as "breakthrough pressure" (ΔP_{entry}) and estimated for cylindrical pores by the Young–Laplace equation:

$$\Delta P_{\text{entry}} = -\frac{2\gamma_{\text{L}}\cos\theta}{r_{\text{p,max}}} \tag{8.34}$$

In eqn (8.34), γ_L is the surface tension of the liquid, θ is the contact angle, and $r_{p,max}$ is the radius of the largest pores of the membrane. Literature provides several examples of breakthrough pressure; for most commercially available microporous polymeric membranes, ΔP_{entry} varies between 100 and 400 kPa.[60] For a membrane having a contact angle of 130°, the penetration pressure of water through an ideally cylindrical pore with diameter of 1 μm is 185 kPa.[61]

Breakthrough pressure can be drastically reduced in the presence of organic molecules or surfactants, even at trace level; therefore, in membrane crystallization operations, the impact of fouling is not evaluated in terms of pressure drop or transmembrane flux decrease, but with respect to occurrence of wettability. Experimental practice demonstrated that, if the membrane is wetted by the liquid phase, unwetted conditions cannot be simply restored by decreasing the hydrostatic pressure.

As discussed in Section 8.7, as long as the solvent molecules of the crystallizing solution evaporate at the feed side and diffuse across the membrane, retained solutes accumulate at the membrane surface. The resulting concentration polarization phenomenon described by eqn (8.22), although not determining a significant decrease in transmembrane flux (contrary to what happens in Reverse Osmosis, where osmotic phenomena play a critical role), influences the supersaturation profile and, ultimately, the nucleation rate. As a consequence, crystallization occurring at the membrane surface requires accurate kinetic control in order to avoid excessive nucleation and consequent pore encrustation.

Abbreviations

ACM	acetaminophen
APIs	active pharmaceutical ingredients
CBZ	carbamazepine
CNT	classical nucleation theory
CSD	crystal size distribution
DBP	dibutyl phthalate
DEP	diethyl phthalate
DHP	dihexyl phthalate
DMAc	*N*,*N*-dimethyl acetamide
DMF	*N*,*N*-dimethyl formamide
DMP	dimethyl phthalate
DMSO	dimethyl sulfoxide
DPC	diphenyl carbonate
DPK	diphenyl ketone
DGM	dusty gas model
EAA	ethyl acetoacetate

EAA	ethylene/acrylic acid
GTA	glycerin triacetate
HMPA	hexamethyl phosphoramide
NMP	*N*-methyl-2-pyrrolidone
NF	nanofiltration
NIPS	non-solvent induced phase separation
BVE	perfluorobutenylvinylether
PDD	perfluoro-2,2 dimethyldioxole
PnBMA	poly(*n*-butyl methacrylate)
PDMS	polydimethylsiloxane
PECTFE	poly(ethylene chlorotrifluoroethylene)
PI	polyimide
PP	polypropylene
PS	polystyrene
PTFE	polytetrafluoroethylene
PVDF	polyvinylidenedifluoride
PKF	poly(vinylidenefluoride-co-hexafluoropropylene)
PK	polyvinylidenefluoride homopolymer
PXRD	powder X-ray diffraction
PO	preferential orientation
PGC	propylene glycol carbonate
RO	reverse osmosis
TFE	tetrafluoroethylene
THF	tetrahydrofuran
TIPS	thermally induced phase separation
TTD	2,2,4-trifluoro-5-trifluoromethoxy-1,3-dioxole

References

1. E. Drioli and E. Curcio, *J. Chem. Technol. Biotechnol.*, 2007, **82**, 223.
2. E. Curcio, A. Criscuoli and E. Drioli, *Ind. Eng. Chem. Res.*, 2001, **40/12**, 2679.
3. R. Azoury, J. Garside and W. G. Robertson, *J. Cryst. Growth*, 1986, **79**, 654.
4. R. Azoury, W. G. Robertson and J. Garside, *Chem. Eng. Res. Des.*, 1987, **65**, 342.
5. C. A. C. Van de Lisdonk, B. M. Rietman, S. G. J. Heijman, G. R. Sterk and J. C. Schippers, *Desalination*, 2001, **138**, 259.
6. M. Takiguchi, K. Igarashi, M. Azuma and H. Ooshima, *Cryst. Growth Des.*, 2006, **6/7**, 1611.
7. W. Ye, J. Lin, J. Shen, P. Luis and B. Van der Bruggen, *Cryst. Growth Des.*, 2013, **13**, 2362.
8. G. Di Profio, S. Tucci, E. Curcio and E. Drioli, *Chem. Mater.*, 2007, **19**, 2386.
9. G. Di Profio, S. Tucci, E. Curcio and E. Drioli, *Cryst. Growth Des.*, 2007, **7**, 526.
10. G. Di Profio, G. Perrone, E. Curcio, A. Cassetta, D. Lamba and E. Drioli, *Ind. Eng. Chem. Res.*, 2005, **44**, 10005.

11. E. Curcio and E. Drioli, *Sep. Purif. Rev.*, 2005, **34/1**, 35.
12. E. Curcio, G. Di Profio and E. Drioli, *J. Cryst. Growth*, 2003, **247**, 166.
13. J. G. Drobny, *Technology of Fluoropolymers*, CRCPress, New York, 2nd edn, 2009.
14. B. Ameduri, *Chem. Rev.*, 2009, **109**, 6632.
15. F. Liu, N. A. Hashim, Y. T. Liu, M. R. M. Abed and K. Li, *J. Membr. Sci.*, 2011, **375**, 1.
16. J. A. Yang, X. L. Wang, Y. Tian, Y. K. Lin and F. Tian, *J. Polym. Sci., Part B: Polym. Phys.*, 2010, **48**, 2468.
17. H. Strathmann and K. Kock, *Desalination*, 1977, **21**, 241.
18. D. R. Lloyd, K. E. Kinzer and H. S. Tseng, *J. Membr. Sci.*, 1990, **52**, 239.
19. E. A. Mason and A. P. Malinauskas, *Gas Transport in Porous Media: The Dusty-Gas Model*, Elsevier, New York, 1983.
20. W. Kast and C. R. Hohenthanner, *Int. J. Heat Mass Transfer*, 2000, **43**, 807.
21. H. Kuhn and H. D. Fostering, *Principles of Physical Chemistry*, Wiley, New York, 2000.
22. M. Volmer and A. Weber, *Z. Phys. Chem.*, 1926, **119**, 277.
23. E. Curcio, E. Fontananova, G. Di Profio and E. Drioli, *J. Phys. Chem. B*, 2006, **110**, 12438.
24. A. J. Page and R. P. Sear, *Phys. Rev. Lett.*, 2006, **97**, 065701.
25. N. E. Chayen, E. Saridakis and R. P. Sear, *Proc. Natl. Acad. Sci. U. S. A.*, 2006, **103**, 597.
26. D. J. Gunn, *J. Cryst. Growth*, 1980, **50**, 533.
27. A. McPherson, *Crystallization of Biological Macromolecules*, Cold Spring Harbor Laboratory Press, Cold Spring Harbor, 1999.
28. E. Curcio, G. Di Profio and E. Drioli, *J. Cryst. Growth*, 2003, **247**, 166.
29. M. L. Grant, *J. Cryst. Growth*, 2000, **209**, 130.
30. G. Di Profio, E. Curcio, A. Cassetta, D. Lamba and E. Drioli, *J. Cryst. Growth*, 2003, **257**, 359.
31. E. Curcio, S. Simone, G. Di Profio, E. Drioli, A. Cassetta and D. Lamba, *J. Membr. Sci.*, 2005, **257**, 134.
32. A. Mersmann, *Thermische Verfahrenstechnik*, Springer, Heidelberg, 1980.
33. A. Llinàs and J. M. Goodman, *Drug Discovery Today*, 2008, **13**, 198.
34. W. Z. Ostwald, *J. Phys. Chem.*, 1897, **22**, 289.
35. R. E. Marsh, *Acta Crystallogr.*, 1958, **11**, 654.
36. A. Dawson, D. R. Allan, S. A. Belmonte, S. J. Clark, W. I. F. David, P. A. McGregor, S. Parsons, C. R. Pulham and L. Sawyer, *Cryst. Growth Des.*, 2005, **5**, 1415.
37. E. V. Boldyreva, V. A. Drebushchak, T. N. Drebushchak, I. E. Paukov, Y. A. Kovalevskaya and E. S. Shutova, *J. Therm. Anal. Calorim.*, 2003, **73**, 409.
38. M. Haisa, S. Kashino, R. Kawai and H. Maeda, *Acta Crystallogr., Sect. B: Struct. Crystallogr. Cryst. Chem.*, 1976, **32**, 1283.
39. M. Haisa, S. Kashino and H. Maeda, *Acta Crystallogr., Sect. B: Struct. Crystallogr. Cryst. Chem.*, 1974, **30**, 2510.
40. J. C. Burley, M. J. Duer, R. S. Stein and R. M. Vrcelj, *Eur. J. Pharm. Sci.*, 2007, **31**, 271.

41. G. Nichols and C. S. Frampton, *J. Pharm. Sci.*, 1998, **87**, 684.
42. M. A. Mikhailenko, *J. Cryst. Growth*, 2004, **265**, 616.
43. B. B. Ivanova, *J. Mol. Struct.*, 2005, **738**, 233.
44. A. Caridi, G. Di Profio, R. Caliandro, A. Guagliardi, E. Curcio and E. Drioli, *Cryst. Growth Des.*, 2012, **12/7**, 4349.
45. A. L. Grzesiak, M. Lang, K. Kim and A. J. Matzger, *J. Pharm. Sci.*, 2003, **92**, 2260.
46. J. P. Arlin, L. S. Price, S. L. Price and A. Florence, *J. Chem. Soc., Chem. Commun.*, 2011, **47**, 7074.
47. E. Curcio, V. López-Mejías, G. Di Profio, E. Fontananova, E. Drioli, B. L. Trout and A. S. Myerson, *Cryst. Growth Des.*, 2014, **14/2**, 678.
48. J. Mulder, *Basic Principles of Membrane Technology*, Kluver Academic Publishers, Dordrecht, 1996.
49. C. Bennet and J. Myers, *Momentum Heat and Mass Transfer*, McGraw-Hill, New York, 1982.
50. S. R. Wickramasinghe, M. J. Semmens and E. L. Cussler, *J. Membr. Sci.*, 1992, **69**, 235.
51. K. L. Wang and E. L. Cussler, *J. Membr. Sci.*, 1993, **85**, 265.
52. E. Drioli, A. Criscuoli and E. Curcio. *Membrane Contactors. Fundamentals, Potentialities and Applications*, Elsevier, The Netherlands, 2005.
53. E. Curcio and E. Drioli, *Sep. Purif. Rev.*, 2005, **34**, 35.
54. M. Gryta and M. Tomaszewska, *J. Membr. Sci.*, 1998, **144**, 211.
55. J. I. Mengual, M. Khayet and M. P. Godino, *Int. J. Heat Mass Transfer*, 2004, **47**, 865.
56. X. Ji, E. Curcio, A. Al Obaidani, G. Di Profio, E. Fontananova and E. Drioli, *Sep. Purif. Technol.*, 2010, **71/1**, 76.
57. E. Drioli, E. Curcio, A. Criscuoli and G. Di Profio, *J. Membr. Sci.*, 2004, **239**, 27.
58. A. König and D. Weckesser, *VDI-Ber.*, 2005, **1901.2**, 1171.
59. G. Di Profio, E. Curcio and E. Drioli, *Ind. Eng. Chem. Res.*, 2010, **49/23**, 11878.
60. K. Schneider, W. Hol and R. Wollbeck, *J. Membr. Sci.*, 1988, **39**, 25.
61. K. W. Lawson and D. R. Lloyd, *J. Membr. Sci.*, 1997, **124**, 1.

CHAPTER 9

Process Analytical Technology in Continuous Crystallization†

L. L. SIMON*[a] AND E. SIMONE*[b]

[a]New Active Ingredients Process Technology, Syngenta Crop Protection AG, Breitenloh 5, CH-4333 Münchwilen, Switzerland; [b]University of Leeds, School of Food Science and Nutrition, Woodhouse Ln, LS29JT, Leeds, UK
*E-mail: levente.simon@syngenta.com, e.simone@leeds.ac.uk

9.1 Introduction

A first definition of process analytical technology (PAT) can be found in the "Guidance for Industry PAT—A Framework for Innovative Pharmaceutical Development, Manufacturing, and Quality Assurance" published in 2004 by the Food and Drug Administration. In this document the process analytical technology term describes "a system for designing, analysing, and controlling manufacturing through timely measurements (*i.e.*, during processing) of critical quality and performance attributes of raw and in-process materials and processes, with the goal of ensuring final product quality".[1] PAT instruments are a fundamental element for the implementation of the quality by design concept, which is a scientific and systematic approach and should replace trial-and-error based process development activities. According to QbD the desired product quality is achieved through the correct design

†Electronic supplementary information (ESI) available. Colour version of Figure 9.7. See DOI: 10.1039/9781788013581

The Handbook of Continuous Crystallization
Edited by Nima Yazdanpanah and Zoltan K. Nagy

Published by the Royal Society of Chemistry, www.rsc.org

of both the product and the manufacturing process. Therefore, process control, product and process understanding are key elements of QbD. The main QbD steps are summarized below:

1. Identification of the critical quality attributes of the target drug product;
2. Design and development of drug formulation and manufacturing process;
3. Determination of critical process parameters and sources of variability;
4. Control of the manufacturing process to produce consistent quality over time.

The QbD implementation in the pharmaceutical and biopharmaceutical industries is based on four elements as shown in Figure 9.1:

1. Multivariate tools for design, data acquisition and analysis;
2. Process analysers;
3. Process control tools;
4. Continuous improvement and knowledge management tools.

The first element deals with the use of multivariate mathematical approaches – statistical design of experiments, response surface methodologies, process simulation, and pattern recognition algorithms – to understand multi-factorial relationships between process, formulation and quality. Process analysers include on- and in-line equipment that allows obtaining

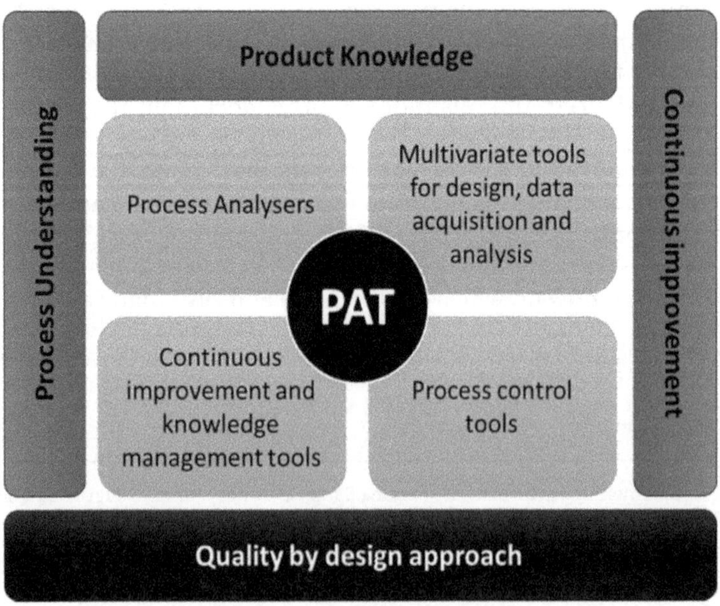

Figure 9.1 The PAT concepts based QbD framework.

instant measurements and avoiding the time delay typical of off-line analysis. The process control strategies take into account the attributes of input materials, the characteristics of process analysers and the achievement of process end points in order to ensure the desired quality of the final product. Measurement and science-based continuous improvement and knowledge management will continue to play important roles during the process and product life cycle.

PAT has become a "must have" tool during the design and operation of batch pharmaceutical and fine chemical processes.[2,3] PAT instruments are also relevant in the context of continuous manufacturing since they: (i) provide on-line measurements that are required to determine whether the process parameters are within the design space, (ii) can be used to shorten start up and shut down times and (iii) speed up the system response to disturbances and re-establish steady-state conditions faster than in uncontrolled operation.[2,4,5] The main PAT instruments used for monitoring and control of continuous crystallization processes are the focused beam reflectance measurement (FBRM), imaging, attenuated total reflectance Fourier-transform infrared spectroscopy (ATR-FTIR), ultraviolet-visible (ATR-UV/VIS) and Raman spectroscopy. A detailed description of these PAT instruments is presented in the following section.

9.2 Process Analytical Technology Instruments

9.2.1 Focused Beam Reflectance Measurement

FBRM is the most common *in situ* PAT tool for monitoring nucleation and crystal growth during crystallization processes.[2] The working principle of FBRM is laser backscattering: a rotating, high speed scanning infrared laser beam emanates through the probe window inserted in a slurry, which is reflected to the instrument whenever it hits a particle. By multiplying the length of time during which a continuous signal is reflected back to the probe by the velocity of the laser beam it is possible to calculate the chord length distribution of the measured slurry. Such chord length depends on both the size and shape of the particles. FBRM is an important sensor for the implementation of automated direct nucleation control (ADNC), a feedback control strategy that allows crystal growth and fines removal for continuous crystallization processes in mixed-suspension, mixed-product removal crystallizers.[4,6]

9.2.2 Ultraviolet-visible and Attenuated Total Reflectance Fourier-transform Infrared Spectroscopy

Ultraviolet-visible light (190 to 750 nm) can be absorbed by molecules containing valence electrons. These are promoted from their ground states to higher energy states and the magnitude of such absorption can be related to the concentration of the molecule itself within a sample through the Beer–Lambert law. Infrared (IR) instead is a vibrational spectroscopy technique based on the absorption of light in the region between 780 nm and 1 mm. IR spectroscopy is more common since most compounds can absorb radiation in the IR while not all molecules contain chromophores that can be excited by UV/Vis light. For crystallization processes attenuated total reflectance probes (ATR) are used in order to analyse only the liquid phase, with negligible interference from the suspended particles. The presence of an ATR crystal positioned at the end of the *in situ* probe determines a depth of penetration of the light beam between 0.5 and 2 μm. Both ATR-FTIR and ATR-UV/Vis have been used for monitoring the solute concentration during batch and continuous crystallization processes.[7–10]

9.2.3 Raman Spectroscopy

Raman spectroscopy is a form of vibrational spectroscopy based on the inelastic scattering of monochromatic laser light. It is complementary to infrared spectroscopy but, while IR bands are generated from a change in the dipole moment of molecules, Raman bands arise from a change in their polarizability. This form of spectroscopy does not require sample preparation and can be used to analyse gaseous, solid and liquid samples. Raman spectra can be used to determine the crystal structure (*e.g.*, polymorphism) of a solid sample or to analyse the solvent composition and measure solute concentration.[11]

9.2.4 Imaging and Particle Vision Measurement (PVM)

Imaging instruments provide real-time information of the crystals' size and shape. Additionally, agglomeration, breakage and polymorphic transformation can be detected using imaging techniques. The low cost of high speed cameras and their reduced size make these sensors the preferred choice for on-line monitoring of continuous plug-flow and oscillatory baffled crystallizers.[12,13] *In situ* particle vision and measurement (PVM) probes can be used to collect and analyse images of the slurry in mixed suspension mixed product removal (MSMPR) crystallizers. Quantitative information, such as size and shape distributions, can be extracted from the recorded images and used for on-line feedback control.

9.3 Data Analysis and Management

PAT instruments are designed to provide on-line measurements of quality attributes to ensure the final product quality. The simultaneous use of several measurement technologies generates a large amount of complex data that need to be managed and analysed in order to extract relevant information.

Multivariate data analysis, also called chemometrics, is a method widely used to deal with large datasets. The purpose of multivariate data analysis is to find a relation between two or more input variables *e.g.* principal component analysis, or to provide a multivariate input–output space mapping *e.g.* partial least squares. In the context of PAT these methods are typically applied to spectroscopic data. One of the biggest challenges when dealing with a large amount of raw data is the efficient integration of several sensors, for which real-time data acquisition and supervisory control platforms need to be implemented. Such control systems should integrate the signals from all different PAT instruments and provide real-time process monitoring and control capabilities.[14] An example of a data acquisition, synchronization and control platform for batch and crystallization processes is the Crystallization Process Informatics System (CryPRINS) software developed at Loughborough University, UK.[15] The CryPRINS software provides a communication interface *via* file transfer, object linking and embedding for process control (OPC), RS232 serial communication or dynamic data exchange with a variety of PAT instruments. This platform allows the application of feedback control strategies *e.g.* supersaturation control, direct nucleation control, during both batch and continuous crystallization processes.

9.4 Systematic Steady-state Detection Using Econometrics

Steady-state (SS) is a fundamental property of continuously operated processes. The rigorous statistical definition of stationarity implies that several statistical descriptors (mean, variance, and covariance) have defined and constant values for a series of measurements. The automated and on-line SS detection is motivated by process modeling, monitoring and control applications. Since continuous processes are modeled using SS models, SS data should be used for model identification. In this case the automated adjustment of SS models is performed when the process is at SS. Once SS is identified, further data processing methods that can be triggered include fault detection and data reconciliation. Another on-line SS detection application is related to unsupervised process optimization: according to this strategy, after the perturbation of input variables the process is run until SS is automatically detected. The method triggers then the implementation of new inputs. This type of experimental plan automation fits within the PAT-based process monitoring and control strategies. Since the time to achieve steady-state changes with operating conditions, this flexible process control method has the advantage of being able to cope with different process dynamics.

An inherent draw-back of continuous operation is that during start-up the product characteristics are not uniform and the amount of off-spec material that needs to be reworked is proportional to the time needed to reach steady-state.[16] Therefore, it is important to systematically detect SS conditions and to separate off-spec and in-spec material.

From a process qualification point of view, it is important that the samples and measurements that are used to qualify the production process are representative and taken during SS operation. Ideally, each sample or measurement, should have a "certificate", which indicates that at the time when the sample was taken the process was at SS. The importance of SS detection is reflected by the Food and Drug Administration (FDA) guideline on the design of multiple-dose *in vivo* bioavailability studies for determining the bioavailability or bioequivalence of drug products (21 CFR 320.27). More exactly, section *d1* of this guideline states that "... appropriate dosage administration and sampling should be carried out to document attainment of steady state". Therefore, in the context of continuous process monitoring the following questions arise: how can it be systematically concluded that the process is at SS and what is the appropriate statistics-based "certificate", which indicates that the process was at SS during sampling?

The work of Simon introduced a systematic SS detection framework based on time series analysis and econometrics concepts to address the questions formulated above.[17] The model system is a plug-flow crystallizer and the measurement method is FBRM with a 2 s data acquisition time. Since in the case of crystallization processes the product properties are defined by the solid phase, the SS detection should be made using a solid phase measurement. The solid phase measurements contain all particle related phenomena: nucleation, growth, attrition and agglomeration so the SS detection is more challenging compared to liquid phase monitoring, which looks at a homogeneous phase.

The hypothesis tests under discussion are the augmented Dickey–Fuller (aDF) and Philips-Perron (PP) unit-root tests, while statistical stationarity is tested using the Leybourne–McCabe and Kwiatkowski–Phillips–Schmidt–Shin tests. The unit-root tests consider a non-stationary null hypothesis, while the stationarity tests assume a stationary process as null hypothesis. Both types of tests are often used in conjunction, in particular to identify fractional integration in time series. Indication for fractional integration or long-range dependence is observed in the case when both the unit-root and stationarity tests reject the null hypothesis. The tests discussed above are univariate.

The systematic steady-state detection method proposed by Simon[17] is presented schematically in Figure 9.2. After a univariate data set that reflects SS operation has been acquired the next step is to determine the number of autocorrelation lags. After this the unit-root and stationarity testing are performed. Assuming that the unit-root tests reject the null and the stationarity tests accept it the process can be classified as mean-stationary. The raw time series trend of the plug-flow crystallizer under discussion, as well as the test probability p values are presented in Figure 9.3. The on-line steady-state detection capabilities of the tests under discussion have been tested using a moving window of 29 data points, thus the

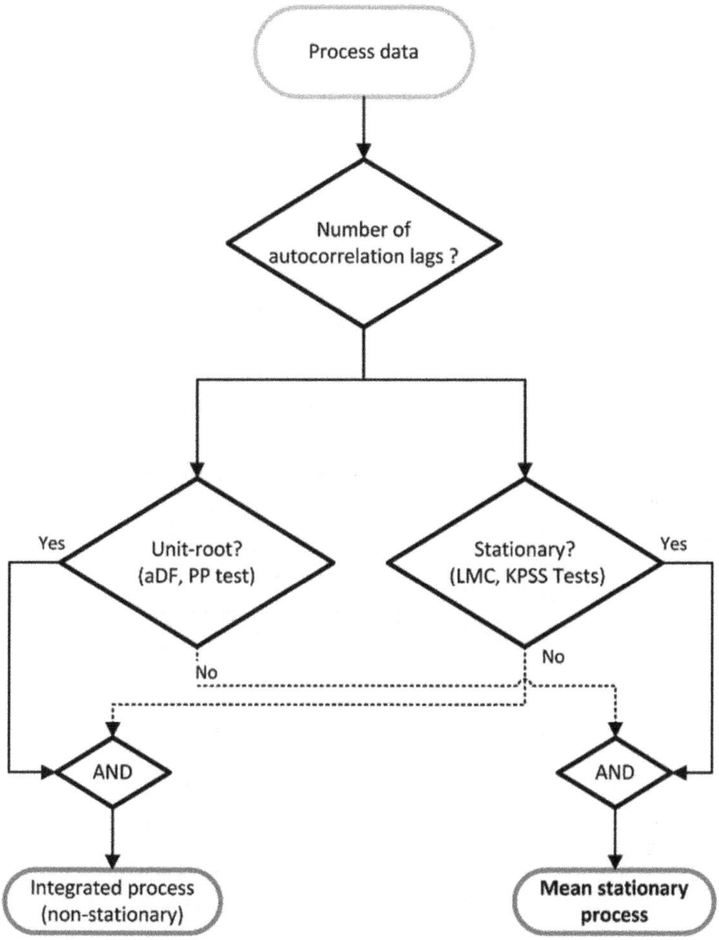

Figure 9.2 Block diagram of the systematic mean stationarity detection framework. Reproduced from ref. 17 with permission from John Wiley and Sons, © 2018 American Institute of Chemical Engineers.

first test results are calculated after 58 s. The time series models used for hypothesis testing include three lags for the unit-root tests and four lags for the stationarity tests. It is concluded that the unit-root tests reject the null after 130 s and that the aDF test unit root coefficient is not significantly different than zero after 150 s and the unit-root null hypothesis is rejected with high confidence. Another observation is that the estimated unit-root coefficients of the aDF and PP are different. The PP unit-root coefficient remains significantly different than zero, however, to low extent.

The KPSS test indicates stationarity between 70 and 90 s, which is false, and then correctly after 130 s. The LMC test shows large sensitivity and it

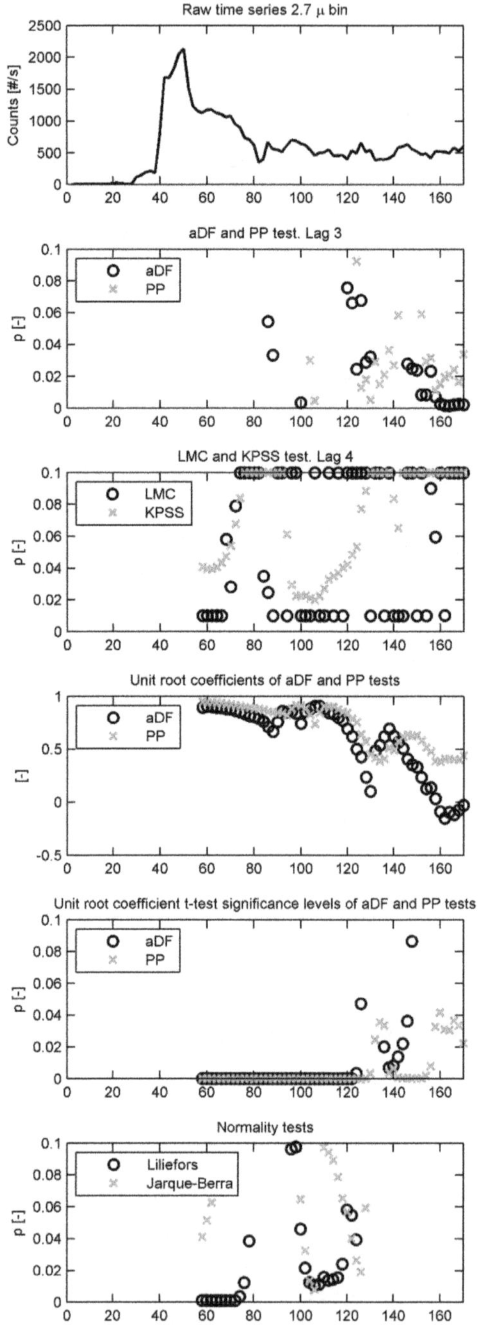

Figure 9.3 On-line unit-root, stationarity and normality monitoring results for the 2.7 micron bin time series. Reproduced from ref. 17 with permission from John Wiley and Sons, © 2018 American Institute of Chemical Engineers.

oscillates between rejecting and accepting the stationarity null hypothesis. By jointly analysing the unit-root and stationarity tests it can be stated that the process has reached stationarity after *ca.* 130 s. The Lilliefors and Jarque–Berra tests show that the data is normally distributed after 130 s, which is an indication of strict stationarity.

These results have shown that the hypothesis tests can be used to detect steady-state conditions on-line in a plug-flow crystallizer. The statistical framework described above can be applied to any continuously operated unit operation or sensor measurement. Besides on-line stationarity detection, the method described above can also be used as a steady-state "certificate". In order to indicate that samples (solid or liquid) have been taken while the process was in stationary conditions, the probability of null-hypothesis acceptance p values returned by the unit-root, stationarity and normality tests should be documented and reported.

9.5 Model-free PAT-based Control Strategies

Model-free control strategies rely solely on the signals from PAT instruments and do not require any information about the kinetics of crystallization (*e.g.*, nucleation, growth, agglomeration) of the studied system. These feedback controllers can be used to achieve consistent product quality during batch and continuous crystallization. For continuous processes, PAT-based model-free control strategies have been used to achieve optimal crystal size distribution as well as crystal shape and polymorphism.[18] FBRM, ATR-FTIR and UV/Vis are the most common instruments for model-free approaches, although Raman spectroscopy has also recently been used.[11] Automated direct nucleation control is a common strategy based on the use of an FBRM probe and the CryPRINS platform. The ADNC control approach is designed to keep the number of solid particles in a vessel constant during a crystallization process, using the FBRM signal and CryPRINS. Usually the total counts/s statistic is sent to CryPRINS and the temperature in the crystallizer is decreased if the measured total counts/s is lower than the setpoint range or increased if it is higher. In this way fine crystals in excess are dissolved and growth of the larger particles is promoted. The control method uses an algorithm with the following variables to be set: total counts setpoint, upper and lower total counts limits, proportional gains for the heating and cooling phase, maximum cooling and heating rates, and temperature at the end of the process. This strategy has recently been used for the operation optimization of MSMPR crystallizers with and without a wet milling unit.[4,6]

Model-free strategies have the advantage that they require little information about the controlled system. In fact, a full knowledge of the kinetics of every crystallization mechanism is not needed to develop a model-free approach. However, these techniques rely on the quality of the measured signal, which can be affected by fouling of the probes.

9.6 MSMPR Crystallizer Monitoring

PAT instruments are used in MSMPR crystallizers to determine steady-state conditions as well as to monitor and control, *in situ*, crystal size, shape and polymorphism.[18] Powell *et al.* (2016) used an integrated PAT array to monitor a continuous single stage MSMPR crystallizer.[8] This system of PAT instruments combined with the use of the CryPRINS software constitutes the automated intelligent decision support (IDS) system, Figure 9.4. The IDS framework controls the temperature in the MSMPR vessels, monitors solute concentration in the liquid phase and tracks changes in the solid phase *e.g.* particle number, size and shape. The IDS system was applied to study the effect of a polymeric additive on the crystallization of paracetamol.[8] FBRM, ATR-UV/Vis, PVM and Raman spectroscopy were used for process monitoring. The use of a PAT array was found to provide more robust monitoring capabilities for the characterization of steady-state conditions compared to the use of standalone PAT instruments.

Recently, a modified version of the ADNC model-free control strategy was used to provide quick start-up and robust control of the crystal size distribution as well as automated and effective suppression of disturbances for MSMPR crystallizers.[6] In the work of Yang *et al.* (2015) a modified version of ADNC was applied to single and multi-stage MSMPR crystallizers as shown in Figure 9.5. An FBRM probe is inserted in each stage and the temperatures

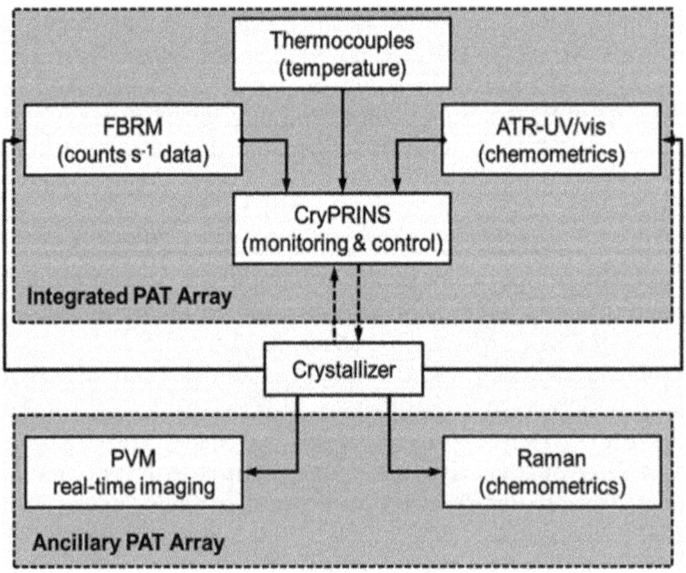

Figure 9.4 Automated intelligent decision support system with integrated and ancillary PAT arrays using the CryPRINS software.[8] Reproduced from ref. 8 with permission from American Chemical Society, Copyright 2016.

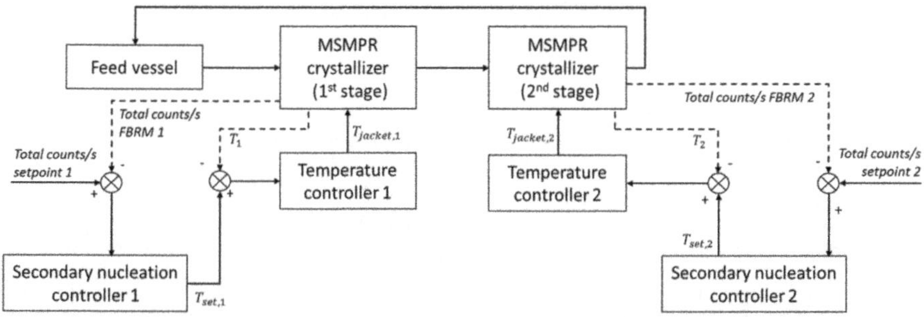

Figure 9.5 Schematic of a two stage MSMPR crystallizer using ADNC.[6]

of the crystallizers are controlled independently by the ADNC after selecting a setpoint for the total counts. The temperature in each vessel is raised to dissolve the excess crystals in case the total counts is higher than the setpoint. On the contrary, if the measured total counts is lower than the desired setpoint, the temperature in the vessel is lowered to promote secondary nucleation. A feed vessel at constant temperature is positioned before the first stage crystallizer to provide a continuous flow of seed crystals.

Another example of PAT-based model-free control of an MSMPR crystallizer is the wet-milling based automated direct nucleation control (WMADNC).[5] In this approach FBRM is used to measure and control the total counts/s in the crystallizer and the manipulated variable is the temperature in the jacket of the wet milling unit, where primary or secondary nucleation takes place. Two different WMADNC setups were proposed and tested, as shown in Figure 9.6, ADNC over an upstream wet milling unit and over a downstream unit. In the upstream version the temperature of the MSMPR vessel is kept constant and a supersaturated solution is fed to the milling unit that behaves as a seed generator for the MSMPR crystallizer. In this configuration the mill induces and controls primary nucleation: if the total counts in the MSMPR vessel is higher than the selected setpoint, the temperature in the jacket of the wet mill is increased in order to dissolve part of the produced seeds; in the opposite case the wet mill jacket is cooled down promoting further nucleation.

The control strategy of the downstream WMADNC is similar to the upstream approach: if the total counts/s in the MSMPR crystallizer is higher than the desired setpoint the wet mill is heated up, while it is cooled when the total counts is lower. In this configuration, the wet mill is positioned on an external loop originating from the vessel and it acts as a crystal size reducer, where high shear promotes secondary nucleation and breakage, reducing the size of the crystals in the MSMPR crystallizer and increasing their number. Both approaches were found to provide robust control of the chord length distribution as well as effective disturbance rejection *e.g.* a change in the wet mill angular speed. However, both these strategies did not improve the start-up dynamics.

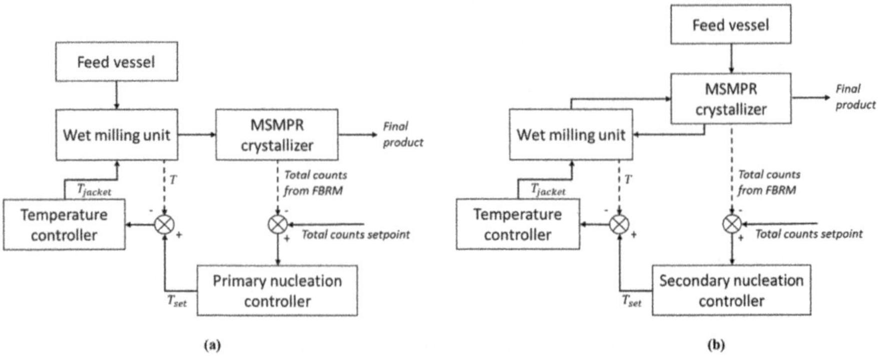

Figure 9.6 Schematic of a one stage MSMPR crystallizer with WMADNC implemented (a) upstream, for the control of primary nucleation and (b) downstream to control secondary nucleation.[4]

An array of PAT tools and CryPRINS software was used by Powell and co-workers[7] to develop and monitor a periodic MSMPR crystallization process. This operation method involves the periodic high flowrate suspension addition and withdrawal. The rapid suspension transfers prevent the sedimentation of crystals and the blockage of transfer lines. This process achieves a periodic steady-state operation or "state of controlled operation" (SCO). A system of PAT instruments involving FBRM, PVM and *in situ* Raman spectroscopy was used to determine the time necessary to achieve the SCO as well as to characterize the operating conditions and the crystal properties within such state.[7] The signals from one of the PMSMPR crystallization processes is shown in Figure 9.7.

PAT instruments have been used to monitor polymorphism in MSMPR crystallizers and to determine the best operating conditions for polymorph selection and control. Lai and co-authors[19] conducted a study to control steady-state polymorphism of *para*-aminobenzoic acid (PABA) in single and two-stage MSMPR crystallizers. PABA has two known polymorphic forms, α and β, which are enantiotropically related with a transition temperature of around 15 °C (the β form is the most stable below this temperature while α is more stable above it). Lai *et al.*[19] were able to control polymorphism of PABA and they could obtain both the pure α form and the β form at the end of single stage MSMPR experiments by varying the temperature, residence time and feed composition, as shown in Table 9.1. Two stage MSMPR experiments were also carried out by the same authors. In this setup pure crystals of the α form were produced in the first stage at steady state conditions. These crystals were continuously fed to the second stage in order to increase the global yield of PABA recovery from solution. The temperature of the first stage was set to 30 °C while the second stage had a temperature of 5 °C. Therefore, the desired α form was thermodynamically stable in the first stage, but not in

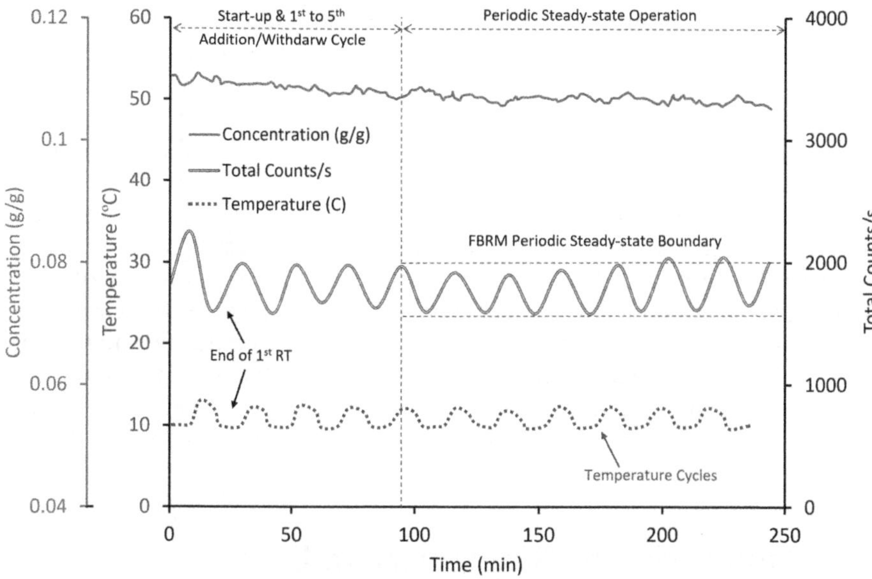

Figure 9.7 Signals from FBRM, and Raman spectroscopy during a PMSMPR crystallization process. Adapted from ref. 7, https://doi.org/10.1016/j.cep.2015.01.002, under the terms of the CC BY 4.0 licence, https://creativecommons.org/licenses/by/4.0/.

Table 9.1 Summary of the experimental conditions and steady-state results for the experiments conducted by Lai and co-workers.[19]

	Two-stage MSMPR		Single stage MSMPR (exp. 1)	Single stage MSMPR (exp. 5)	Single stage MSMPR (exp. 6)	Single stage MSMPR (exp. 7)
Stage number	1	2	1	1	1	1
Stage temperature (°C)	30	5	5	15	20	30
Residence time (min)	60	60	60	60	60	60
Feed concentration (g kg^{-1} solvent)	10	10	10	10	10	10
Polymorphism at steady-state	Pure α	Pure α	Pure β	Pure β	Pure α	Pure α
Polymorph stability at steady-state temperature	Stable	Metastable	Stable	Stable	Stable	Stable

the second one. However, the continuous feeding of α crystals to the second stage provided enough surface area to dominate the growth or nucleation of the undesired β polymorph and ensure the purity of the final product. The concentration of mother liquor in the MSMPR crystallizers was measured *via* Fourier transform infrared spectroscopy while the chord length distribution was monitored using FBRM. The produced crystals were characterized offline using X-ray powder diffraction (XRPD) in order to determine their polymorphic form. Data from FTIR and XRPD were also used in the second part of the same work to estimate the kinetic parameters necessary to conduct dynamic simulations on the studied system and predict the process yield as a function of the operating conditions.

Preferential crystallization in MSMPR crystallizers has also been a subject of investigation. Galan and co-worker used on-line polarimetry to measure the optical rotation of the liquid phase and estimate the amount of each enantiomer in solution.[20] The clear solution from each vessel was circulated in an external loop where a polarimeter in series with a density meter was positioned. More recently FBRM and PVM have been used by Steendam and ter Horst to monitor the size and shape of chirally pure sodium bromate crystals.[21] The chirality of the solid crystals produced was evaluated off-line using polarized microscopy.

9.7 Monitoring of Tubular Crystallizers

An alternative to MSMPR crystallizers for continuous processes are tubular crystallizers. In order to minimize attrition and improve mixing, slug flow[22] and oscillatory operation *e.g.* the continuous oscillatory baffled crystallizers (COBC) have been investigated.[23–25] Other alternatives are tubular crystallizers equipped with static mixers.[26]

On-line chord length distribution monitoring of a tubular crystallizer equipped with a Kenics static mixer has been presented by Simon and Myerson using an FBRM probe,[27] which then was used for systematic on-line steady-state detection.[17] Other monitoring methods are based on stereomicroscopes[28] and high speed cameras to monitor the size and number of crystals,[12,13,25] aggregation,[29] mixing quality,[30] fouling[31] as well as flow type.[23] Besenhard and co-workers[23] have used high speed imaging and on-line laser diffraction (a HELOS system equipped with a LIXELL flow cell installed on-line) to monitor the number of fines, the size of crystals produced and their polymorphic form in a segmented flow tubular crystallizer. The recorded images gave information about the quality of the mixing and the type of flow, gas bubbles and continuous liquid phase. Ferguson *et al.*[10] used FBRM, PVM and ATR-FTIR to monitor crystals produced in a plug-flow anti-solvent crystallizer. In this case the instruments were positioned in a specifically designed flow cell situated outside the end section of the plug-flow crystallizer. Furthermore, Brown and Li used video imaging to study the crystallization of paracetamol in a COBC system. Image analysis algorithms have been developed to determine the metastable zone width, the mean particle size and the number density of crystals during the process.[12] The growth

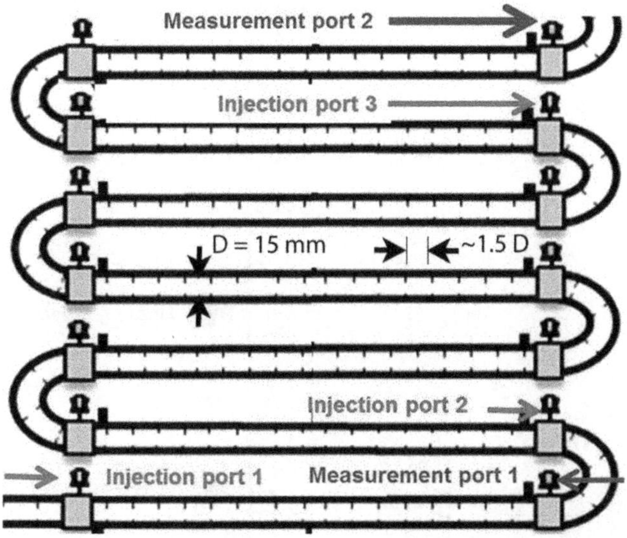

Figure 9.8 Schematic representation of the COBC used by Kacker *et al.*[32] Reproduced from ref. 32 with permission from Elsevier, Copyright 2017.

rate of the same compound was determined at different operating conditions.[13,25] In another imaging based application, the effect of the geometry of the inner tube baffles on the quality of solid dispersion within the COBC was investigated.[30] Another PAT application included the use of on-line near-infrared probes for the determination of the residence time distribution at different oscillatory conditions in a COBC, as shown in Figure 9.8.[32]

Fouling or encrustation is another phenomenon which was the subject of investigation. Fouling can lead to increasing thermal resistance, impeding fluid flow, and frequent shut-downs. The encrustation process is characterized by an initiation stage followed by growth of the deposits, crust removal or detachment due to fluid shear or erosion, and aging which can affect the thermal and mechanical properties of the deposited layer.[33,34]

Tachtatzis *et al.* have recently used a commodity web camera to monitor fouling on the walls of a tubular crystallizer.[31] A statistical analysis of the pixel intensity of images captured during the formation of an encrustation layer was used to distinguish, in each image, the pixels of the bulk solution from those of the fouled surface. The number of pixels of the fouled surface was used as an indicator of the degree of fouling during crystallization. A schematic of the rig used for this study is shown in Figure 9.9.

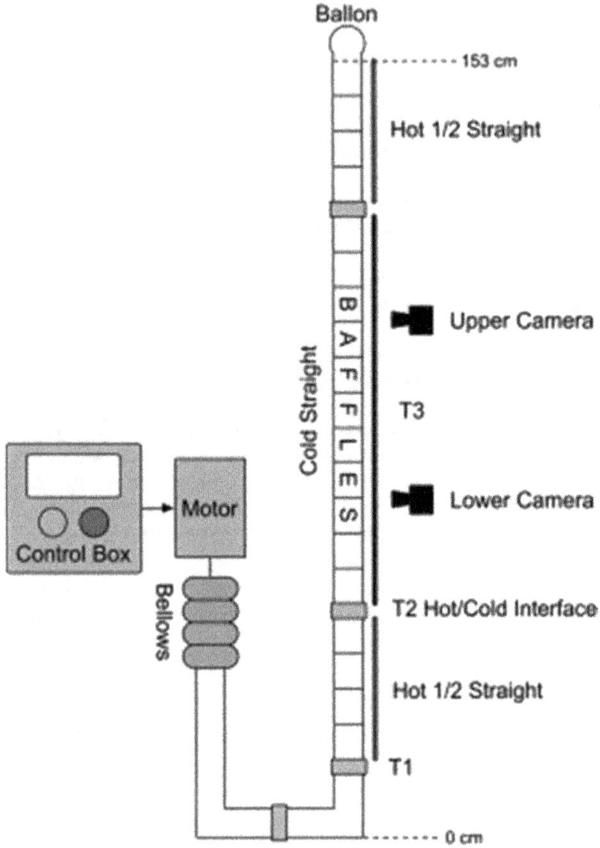

Figure 9.9 Experimental setup showing positions of temperature measurement, cameras, glass sections and collars which connect the glass components together.[31] Reproduced from ref. 31 with permission from Elsevier, Copyright 2015.

References

1. FDA, *Guidance for Industry PAT — A Framework for Innovative Pharmaceutical Development, Manufacturing, and Quality Assurance*, FDA Off Doc, 2004, p.16.
2. L. L. Simon, H. Pataki, G. Marosi, F. Meemken, K. Hungerbühler and A. Baiker, *et al.*, Assessment of Recent Process Analytical Technology (PAT) Trends: A Multiauthor Review, *Org. Process Res. Dev.*, 2015, **19**(1), 3–62.
3. A. R. Klapwijk, E. Simone, Z. K. Nagy and C. C. Wilson, Tuning Crystal Morphology of Succinic Acid Using a Polymer Additive, *Cryst. Growth Des.*, 2016, **16**(8), 4349–4359.

4. Y. Yang, L. Song, T. Gao and Z. K. Nagy, Integrated Upstream and Downstream Application of Wet Milling with Continuous Mixed Suspension Mixed Product Removal Crystallization, *Cryst. Growth Des.*, 2015, **15**(12), 5879–5885.
5. Y. Yang, L. Song, Y. Zhang and Z. K. Nagy, Application of wet milling-based automated direct nucleation control in continuous cooling crystallization processes, *Ind. Eng. Chem. Res.*, 2016, **55**(17), 4987–4996.
6. Y. Yang, L. Song and Z. K. Nagy, Automated Direct Nucleation Control in Continuous Mixed Suspension Mixed Product Removal Cooling Crystallization, *Cryst. Growth Des.*, 2015, **15**(12), 5839–5848.
7. K. A. Powell, A. N. Saleemi, C. D. Rielly and Z. K. Nagy, Periodic steady-state flow crystallization of a pharmaceutical drug using MSMPR operation, *Chem. Eng. Process.*, 2015, **97**, 195–212.
8. K. A. Powell, A. N. Saleemi, C. D. Rielly and Z. K. Nagy, Monitoring Continuous Crystallization of Paracetamol in the Presence of an Additive Using an Integrated PAT Array and Multivariate Methods, *Org. Process Res. Dev.*, 2016, **20**(3), 626–636.
9. S. Ferguson, G. Morris, H. Hao, M. Barrett and B. Glennon, Characterization of the anti-solvent batch, plug flow and MSMPR crystallization of benzoic acid, *Chem. Eng. Sci.*, 2013, **104**, 44–54.
10. S. Ferguson, G. Morris, H. Hao, M. Barrett and B. Glennon, *In situ* monitoring and characterization of plug flow crystallizers, *Chem. Eng. Sci.*, 2012, **77**, 105–111.
11. K. A. Powell, D. M. Croker, C. D. Rielly and Z. K. Nagy, PAT-based design of agrochemical co-crystallization processes: A case-study for the selective crystallization of 1:1 and 3:2 co-crystals of p-toluenesulfonamide/triphenylphosphine oxide, *Chem. Eng. Sci.*, 2016, **152**, 95–108.
12. C. J. Brown and X.-W. Ni, Determination of metastable zone width, mean particle size and detectable number density using video imaging in an oscillatory baffled crystallizer, *CrystEngComm*, 2012, **14**(8), 2944.
13. C. J. Brown and X. Ni, Online evaluation of paracetamol antisolvent crystallization growth rate with video imaging in an oscillatory baffled crystallizer, *Cryst. Growth Des.*, 2011, **11**(3), 719–725.
14. S. Chatterjee, *FDA Perspective on Continuous Manufacturing*, IFPAC Annual Meeting, 2012.
15. Z. K. Nagy, G. Fevotte, H. Kramer and L. L. Simon, Recent advances in the monitoring, modelling and control of crystallization systems, *Chem. Eng. Res. Des.*, 2013, **91**(10), 1903–1922.
16. A. S. Myerson, M. Krumme, M. Nasr, H. Thomas and R. D. Braatz, Control Systems Engineering in Continuous Pharmaceutical Manufacturing May 20-21, 2014 Continuous Manufacturing Symposium, *J. Pharm. Sci.*, 2015, **104**(3), 832–839.

17. L. L. Simon, Continuous Manufacturing: Is the Process Mean Stationary?, *AIChE J.*, 2018, **64**(7), 2426–2437.
18. T. Wang, H. Lu, J. Wang, Y. Xiao, Y. Zhou and Y. Bao, *et al.*, Recent progress of continuous crystallization, *J. Ind. Eng. Chem.*, 2017, **54**, 14–29.
19. T. C. Lai, J. Cornevin, S. Ferguson, N. Li, B. L. Trout and A. S. Myerson, Control of polymorphism in continuous crystallization *via* mixed suspension mixed product removal systems cascade design, *Cryst. Growth Des.*, 2015, **15**(7), 3374–3382.
20. K. Galan, M. J. Eicke, M. P. Elsner, H. Lorenz and A. Seidel-Morgenstern, Continuous preferential crystallization of chiral molecules in single and coupled mixed-suspension mixed-product-removal crystallizers, *Cryst. Growth Des.*, 2015, **15**(4), 1808–1818.
21. R. R. E. Steendam and J. H. ter Horst, Continuous Total Spontaneous Resolution, *Cryst. Growth Des.*, 2017, **17**(8), 4428–4436.
22. M. Jiang, C. D. Papageorgiou, J. Waetzig, A. Hardy, M. Langston and R. D. Braatz, Indirect Ultrasonication in Continuous Slug-Flow Crystallization, *Cryst. Growth Des.*, 2015, **15**(5), 2486–2492.
23. M. O. Besenhard, P. Neugebauer, O. Scheibelhofer and J. G. Khinast, Crystal Engineering in Continuous Plug-Flow Crystallizers, *Cryst. Growth Des.*, 2017, 6432–6444.
24. T. McGlone, N. E. B. Briggs, C. A. Clark, C. J. Brown, J. Sefcik and A. J. Florence, Oscillatory Flow Reactors (OFRs) for Continuous Manufacturing and Crystallization, *Org. Process Res. Dev.*, 2015, **19**, 1186–1202.
25. C. J. Brown and X. W. Ni, Evaluation of growth kinetics of antisolvent crystallization of paracetamol in an oscillatory baffled crystallizer utilizing video imaging, *Cryst. Growth Des.*, 2011, **11**(9), 3994–4000.
26. A. J. Alvarez and A. S. Myerson, Continuous plug flow crystallization of pharmaceutical compounds, *Cryst. Growth Des.*, 2010, **10**(5), 2219–2228.
27. L. L. Simon and A. S. Myerson, in *Continuous Antisolvent Plug-flow Crystallization of a Fast Growing API*, ed. M. Mazzotti and B. Biscans, International Symposium on Industrial Crystallization (ISIC), Zurich, 2011.
28. M. Jiang, Z. Zhu, E. Jimenez, C. D. Papageorgiou, J. Waetzig and A. Hardy, *et al.*, Continuous-Flow Tubular Crystallization in Slugs Spontaneously Induced by Hydrodynamics, *Cryst. Growth Des.*, 2014, **14**(2), 851–860.
29. C. Borchert and K. Sundmacher, Crystal Aggregation in a Flow Tube: Image-Based Observation, *Chem. Eng. Technol.*, 2011, **34**(4), 545–556.
30. L. N. Ejim, S. Yerdelen, T. McGlone, I. Onyemelukwe, B. Johnston and A. J. Florence, *et al.*, A factorial approach to understanding the effect of inner geometry of baffled meso-scale tubes on solids suspension and axial dispersion in continuous, oscillatory liquid–solid plug flows, *Chem. Eng. J.*, 2017, **308**, 669–682.

31. C. Tachtatzis, R. Sheridan, C. Michie, R. C. Atkinson, A. Cleary and J. Dziewierz, *et al.*, Image-based monitoring for early detection of fouling in crystallisation processes, *Chem. Eng. Sci.*, 2015, **133**, 82–90.
32. R. Kacker, S. I. Regensburg and H. J. M. Kramer, Residence time distribution of dispersed liquid and solid phase in a continuous oscillatory flow baffled crystallizer, *Chem. Eng. J.*, 2017, **317**, 413–423.
33. A. Majumder and Z. K. Nagy, Dynamic modeling of encrust formation and mitigation strategy in a continuous plug flow crystallizer, *Cryst. Growth Des.*, 2015, **15**(3), 1129–1140.
34. M. Vendel and A. C. Rasmuson, Initiation of incrustation by crystal collision, *Trans. Inst. Chem. Eng.*, 2000, **78**, 749–755.

CHAPTER 10

Continuous Protein Crystallization

WENQIAN CHEN, HUAIYU YANG AND JERRY YONG YEW HENG*

Imperial College London, Department of Chemical Engineering, South Kensington Campus, London SW7 2AZ, UK
*E-mail: jerry.heng@imperial.ac.uk

10.1 Downstream Processing of Proteins

Proteins are polymers of amino acids with more than 50 residues according to the IUPAC nomenclature.[1] Numerous proteins have been used for medicinal purposes due to their biological activities. The most famous example is insulin, whose human version consists of one acidic polypeptide (21 amino acids) and one basic polypeptide (30 amino acids) joined by disulfide bonds.[2] The discovery of canine insulin in 1921–1922 and the subsequent industrial production of insulin were important breakthroughs in the history of medicine, providing the treatment for diabetes which had remained incurable for centuries.[3–11]

Over the past three decades, protein-based biological products (*i.e.* biologics) have been an important channel of innovation in the pharmaceutical industry.[12] Between 1993 and 2015, a total of 107 biologics have been approved,[13] accounting for a global market of $175 billion.[14] With the growing significance of protein-based biologics, the large-scale production of these compounds has gained more industrial interest.

The Handbook of Continuous Crystallization
Edited by Nima Yazdanpanah and Zoltan K. Nagy

Published by the Royal Society of Chemistry, www.rsc.org

The downstream processing of proteins consists of purification and formulation, which has a significant contribution to the overall production cost (*e.g.* nearly 70% in the case of monoclonal antibody (mAb)).[15] The purification stage starts with the removal of particulates such as cell debris by centrifugation or filtration, and proceeds to the further purification of the protein by chromatography and filtration.[16–18] Although chromatography can achieve high protein purity, scaling up chromatography is significantly more costly than other purification methods such as filtration and crystallisation due to the high cost of packing material and buffer. For instance, mAb is industrially purified by protein-A chromatography, whose resin costs €5000 to €14000 per litre.[19] As a result, protein crystallisation has become a topic of interest for the large-scale bioseparation of proteins, in order to selectively isolate certain proteins from mixtures.[20,21]

Building on the crystallisation theories discussed in the previous chapters of this book as well as existing reviews and books on protein crystallisation,[20,22–30] this chapter aims to offer a process development perspective on continuous protein crystallisation. The general protein crystallisation process is briefly discussed to establish the background knowledge for a more detailed review of the current developments in this field.

10.2 Protein Crystals

Since the reported crystallisation of haemoglobin in 1840,[31] the general principles for protein crystallisation are still not fully understood after nearly two centuries of intensive research.[20,26] The comprehensive characterisation of protein crystals can provide useful insight into the crystallisation mechanisms, but it is often difficult to grow crystals beyond micrometre size for accurate structure determination.[27]

Similar to small-molecule pharmaceutical compounds, proteins are more stable in the crystalline form.[32–36] However, the crystallisation process is much more complicated for proteins due to their complex three-dimensional configurations.[37–42] In order for a protein to crystallise, the protein needs to remain in the active form, whose three-dimensional structure allows the protein to perform the intended biological functions. The hydrogen bonds and hydrophobic interactions among protein crystals are much weaker in comparison to crystalline small molecules due to the large amount of incorporated water molecules, which can account for 51 ± 14% of the total crystal volume.[25,43–48] The resulting flexibility of protein molecules leads to the variance of the final crystalline state, making the nucleation during the protein crystallisation process much more difficult to predict than for small molecules.

10.3 Development of Continuous Protein Crystallisation

The general strategy for developing protein crystallisation is shown in Figure 10.1. It starts with the high-throughput screening of crystallisation conditions and then proceeds to the scale-up of the crystallisation process up from the nanolitre/microlitre scale of screening experiments to a typical scale of 1 mL in batch shaking mode. Once successful, the crystallisation process can be further scaled up in batch mode in stirred tank or tubular crystallisers before converting the process into the continuous mode. The following sections will discuss these steps in depth, especially the scale-up of batch crystallisation and the conversion from batch to continuous mode.

10.3.1 Screening and Phase Diagram

Protein crystallisation starts with the screening of suitable conditions, which is often achieved with high-throughput vapour diffusion experiments in either the hanging drop or sitting drop mode.[49–52] The results are used to construct a phase diagram (Figure 10.2), in which the solubility line indicates

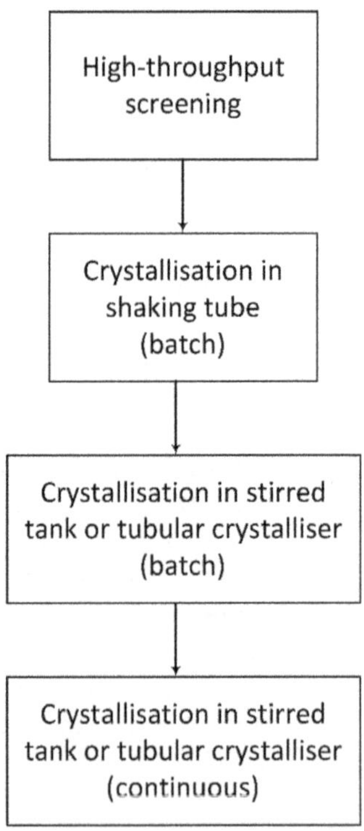

Figure 10.1 General workflow for developing continuous protein crystallisation.

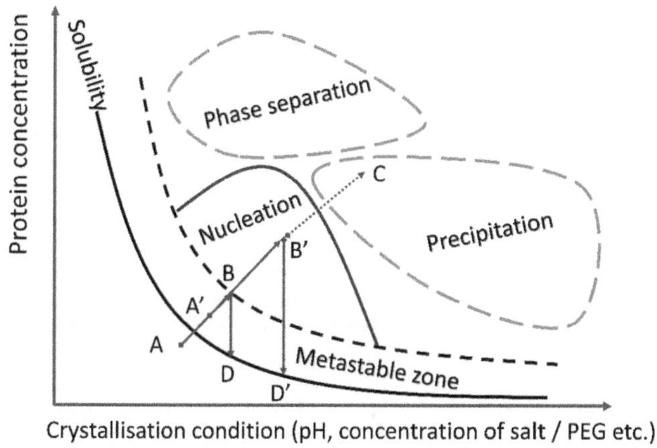

Figure 10.2 A typical phase diagram for proteins.

the maximum quantity of dissolved protein at a particular combination of conditions such as temperature, pH, concentrations of precipitants (*e.g.* salt and polyethylene glycol (PEG)).[20]

In a typical experiment of vapour diffusion, a protein solution is prepared by dissolving the protein in a buffer solution, which is then mixed with a precipitation solution that has higher concentrations of additives such as salt and PEG. The protein in the resulting droplet can be located in the under-saturation or meta-stable zones as represented by points A and A′ respectively in the phase diagram (Figure 10.2). The droplet is then exposed to a sealed atmosphere that is shared with an excess quantity of the same precipitation solution. Due to the lower osmotic pressure inside the droplet, water gradually diffuses from the droplet into the shared atmosphere, reaching higher concentrations of protein and additives. This transition is represented by the line connecting points A and A′ with points B and B′ in the phase diagram.

With slow diffusion of water out of the droplet, the protein in the droplet reaches point B and nucleation occurs, decreasing the protein concentration to point D. When the diffusion is fast, point B′ can be reached when nucleation starts and point D′ is the resulting end point of the crystallisation process. A higher rate of water diffusion from the droplet can push the protein in the droplet to point C in the precipitation zone, where amorphous protein aggregates are formed.[20]

As shown in the phase diagram, the suitable crystallisation conditions are determined by temperature, pH and the concentrations of salt and PEG. These conditions should avoid extreme values that can cause the denaturation of protein. Table 10.1 summarises the range of crystallisation conditions according to the existing literature. It is advisable to start the screening experiments within the recommended range for a particular condition before extending its lower and upper limits to the full operating range.

Table 10.1 Summary of key variables for protein crystallisation.[54,59]

Process variable	Operating range	Recommended range	Note
Temperature	0–40 °C	4–25 °C	Protein melting temperature is the upper limit.
pH	4–10	4–10	Protein pI is the reference pH.
Salt concentration	0.1–10 M	0.1–1 M	Homogeneous precipitant.
PEG concentration	1–40 w/v %	1–20 w/v %	

Although some exceptions such as horse serum albumin have been reported, the solubility of most proteins decreases with temperature.[53] The operating temperature during the crystallisation is often reduced at a controlled rate or remains constant at a target temperature between 4 °C and 25 °C. A study of over 2500 reported protein crystallisations showed that majority of successful protein crystallisations occur at 20 °C and 4 °C.[54] The maximum range for the target temperature is suggested to be between 0 °C and 40 °C, in order to avoid denaturing the protein as the higher melting temperature of most proteins tends to be between 40 °C and 80 °C.[55]

The isoelectric point (pI) of a protein is an important intrinsic property that influences the suitable pH range for the crystallisation.[56,57] In general, the solubility of a protein is the lowest at its pI as the protein molecules become neutral in charge and the protein–protein interactions become stronger than protein-water interactions which account for the solubility of protein. This trend can be illustrated with the human monoclonal antibody (mAb01), which has a pI at 6.8.[58] The most commonly used pH range for protein crystallisation is between 4 and 10 according to a study of over 35 000 proteins in the protein data bank (PDB).[59] Another study of nearly 24 000 proteins in PDB showed that the pH for crystallising acidic proteins (*i.e.* pI < 7) is usually one unit above their pI, whereas the pH for crystallising basic proteins (*i.e.* pI > 7) is 1.5 to 3 units below their pI.[59]

Proteins are commonly precipitated with additives such as ammonium sulfate, polyethyleneimine and polyethylene glycols (PEG) in order to preserve protein folding and the related bio-activeness.[60–63] A study of over 44 000 data sets in the PDB showed that PEG 3350, 4000 and 8000 are the most frequently used PEGs, while ammonium sulfate, sodium chloride, sodium acetate and magnesium chloride are the most commonly used in protein crystallisation studies as precipitants.[54] The effects of cations and anions on negatively charged proteins follow the Hofmeister series,[64,65] but positively charged proteins follow the reversed series.[66] Such effects in general can be modelled by the osmotic second virial coefficient and preferential binding parameter.[67] It is noticed that the solubility is not linearly correlated with salt concentration.[68,69] Solubility increases with increasing salt concentration at low salt level (salting-in), however, decreases with further increase in salt concentration (salting-out).[70]

Seeding in the context of crystallisation refers to the introduction of heterogeneous additives into a crystallising solution. A recent development of heterogeneous protein crystallisation is the design of nucleant which provides some degree of control over the protein nucleation step with its porosity and surface chemistry.[71–74] In particular, mesoporous silicates (MPS) with pore size between 2 and 50 nm have shown promising nucleation performance for a wide range of proteins.[71,75–81] In general, the pore size of MPS should be larger than the hydrodynamic or gyration diameter of the target protein.[82–84] It is reported that MPS with narrow pore size distributions between 2 and 20 nm can crystallise seven proteins which have gyration/hydrodynamic diameter smaller than the pore size.[79]

10.3.2 Scale-up and Mixing

Although high throughput screening enables the crystallisation of thousands of proteins for structural determination purposes,[85–89] only a few have been scaled up beyond the microliter scale due to the limited availability of protein as well as the difficulty of scaling up from the screening conditions.[90]

Tables 10.2 and 10.3 summarise the literature on the bulk crystallisation of 11 proteins, which include antibodies, aprotinin, insulin, lipase, L-methionine γ-lyase, lysozyme, ovalbumin, protease, rubisco, subtilisin and urease. The majority of literature report on the crystallisation of these proteins in batch stirred tank crystallisers (Table 10.2), which are simple to operate and control as compared to the tubular crystallisers that involve direct and oscillatory flow (Table 10.3). These studies demonstrate the feasibility of large-scale protein crystallisation, whose scale can reach up to 100 L as in the case of L-methionine γ-lyase.[91] In general, the crystallisations were conducted within the temperature range of 0–40 °C and the pH range of 4–10. These mild conditions mean the temperature control is relatively simple and no special construction material is required for corrosion resistance.

In large scale protein crystallisation, mixing is necessary for enhancing the mass and heat transfer within the crystalliser. The size and shape of crystals are affected by the mixing conditions, as the growth rates of different crystal faces are strongly dependent on the flow conditions.[92–96] The nucleation process is also strongly influenced by the mixing conditions. Mild agitation of protein droplets during vapour diffusion experiments delays the induction of crystallisation due to higher homogeneity of protein concentration within the droplets.[97–99] In contrast, more vigorous mixing by the impeller in a stirred tank crystalliser was found to shorten the induction time of crystallisation and lead to smaller protein crystals.[100]

More importantly, the mixing conditions have significant influence on the stability of the protein, as the denaturation of protein in high flow rate was observed in some cases.[101] The cause might be the presence of air bubbles and the resulting formation of protein foam, since some studies showed that proteins are stable under high shear rates.[102] For example, immunoglobulin-G1 monoclonal antibodies were found to be stable at 250 000 s^{-1}, whereas

Table 10.2 Summary of literature on scaled up protein crystallisation in batch stirred tank crystallisers.

Protein[a]									
Type	Initial concentration	Buffer	Precipitant	Temp.	pH	Crystalliser	Scale	Yield	Reference
Antibody (mAb01)	25 g L^{-1}	• Histidine/acetate • Trehalose	—	10 °C	7	Stirred tank	1 L	98%	58
Aprotinin	5 g L^{-1}	• Sodium citrate	• Sodium chloride	20–25 °C	5	Stirred tank	1 L	86%[b]	133
Insulin	—	—	• Alkali metal hydroxide/ammonium hydroxide	RT	7–10	Stirred tank	180 mL–2 L	—	134
Lipase	100 g L^{-1}	• Sodium acetate	• Sodium chloride • Monosodium phosphate • PEG-10000	20 °C	4	Stirred tank	1 L	95%	135
L-Methionine γ-lyase	—	• Sodium phosphate • EDTA • PLP • DTT • 2-ME	• Ammonium sulfate • Sodium chloride • PEG-6000	<15 °C	7	Stirred tank	100 L	74–85%	91
Lysozyme	25 g L^{-1}	• Sodium acetate	• Sodium chloride • 2-HEAF	0–20 °C	4	Stirred tank	1 L	97%	135
Ovalbumin	25 g L^{-1}[b]	—	• Ammonium sulfate	30 °C	5	Stirred tank	600 mL[b]	80%[b]	136
Protease	40–60 g L^{-1}	—	• Sodium sulfate	20–40 °C	5–9	Stirred tank	10–800 mL	70–80%	137
Rubisco	—	• Sodium metabisulfite	—	4 °C	6	Unstirred tank	200 mL	—	138
Subtilisin	65–99 g L^{-1}	—	• Halide salts	3–5 °C	5	Unstirred tank	1 mL	34–84%	139
Urease	—	• Potassium dihydrogen phosphate • Dipotassium hydrogen phosphate	• PEG-4000 • PEG-6000	4–22 °C	5–8	Unstirred tank	50 mL	—	140

[a]Including the variants of the proteins.
[b]Estimated value based on the literature.

Table 10.3 Summary of literature on scaled up protein crystallisation in batch tubular crystallisers.

Protein									
Type	Initial concentration	Buffer	Precipitant	Temp.	pH	Crystalliser	Scale[a]	Yield	Reference
Lysozyme	10–50 g L^{-1}	• Sodium acetate	• Sodium chloride	20 °C	5	Oscillatory flow	4 mL	8–76%	114
	15 g L^{-1}			0 °C		Direct flow	4 mL	25–80%	96
				10 °C			20 mL	—	141
	50 g L^{-1}			16–35 °C			1 mL		142
	40–80 g L^{-1}			0–25 °C			6 mL		143
	5–70 g L^{-1}[b]			24 °C			3 mL		144
Insulin	3	• Sodium citrate	• Zinc sulfate • Acetone	10 °C	6	Oscillatory flow		40–50%	101

[a]Either the flow cell volume or system volume that includes both flow cell and reservoir.
[b]Estimated value based on the literature.

recombinant human growth hormone and recombinant human deoxyribonuclease were found to remain stable beyond 100 000 s^{-1}.[103,104] Since the typical shear rate in a production process is only 20 000 s^{-1}, shear rate itself should not be a concern for protein stability in general.[104]

Despite its importance, mixing is not discussed in most of the literature for protein crystallisations in stirred tank crystallisers. The type, dimensions and stirring speed of impeller are important design and operating variables for achieving sufficient mixing.[97–100,105] The pitched-blade and marine-blade impellers are suitable for stirring protein solutions, because they are designed for stirring shear-sensitive substances such as cells.[106–108]

Once the impeller type is selected, the dimensions and stirring speed of the impeller can be optimised to achieve the crystal size distribution within the crystalliser. The mixing should ideally produce a homogeneous suspension, where all crystals are lifted to expose their entire surface for crystal growth. The corresponding stirring speed is referred to as the minimum stirrer speed,[109] which can be relatively high.[110] With the suspension of crystals and the homogeneity of protein concentration, sufficient mixing narrows the size distribution of the protein crystals.[111]

For the crystallisation of mAb where high concentrations of PEG and salt are often used, the protein solution often phase separates into an aqueous phase and a PEG phase. Vigorous mixing is needed in these scenarios to disperse the PEG phase homogeneously within the aqueous phase for the crystallisation to proceed.[112]

For tubular crystallisers, mixing can be enhanced by increasing the flow rate, introducing bubbles and oscillating the flow. Although inducing turbulence can achieve thorough mixing, most of the existing studies were conducted under laminar flow since proteins such as insulin are sensitive to high flow rate.[101] Increasing the linear flow velocity in the laminar region is beneficial for ensuring homogeneity in protein concentration and hence the narrowing of size distribution. In a study of lysozyme crystallisation under forced flow, a linear flow velocity of 12.8 mm min^{-1} could sharpen the crystal size distribution of lysozyme in comparison to the quiescent condition.[96]

Another way to improve the mixing along a tubular crystalliser is to operate the system in slug flow where air bubbles are introduced intermittently to push the crystal slurry along the tube. A study of lysozyme crystallisation in this configuration showed that a narrow size distribution of protein crystal was achieved with peak values of 20 and 25 μm.[113]

Due to the limited mixing control in direct flow tubular crystallisers, oscillatory flow crystallisers have been studied as alternatives,[101,114] where the tubular crystalliser has multiple internal constrictions that are equally far apart and the flow moves back and forth with a set distance at a particular frequency. Similar to the mixing in a stirred tank crystalliser, the oscillatory flow was found to shorten the induction time of crystallisation and even double the yield of crystal in comparison with the quiescent condition.[101]

10.3.3 Transition from Batch to Continuous Crystallisation

Although there are limited studies on continuous protein crystallisation (Table 10.4), the existing studies of batch protein crystallisation prove that it is technically feasible, since the batch operation of a crystalliser can be converted into the continuous mode with relative ease. Five experimental studies have been reported so far: four conducted the crystallisation of lysozyme in a tubular crystalliser,[113,115–117] whereas the remaining one performed the crystallisation of lysozyme and mAb01 in a mixed-suspension, mixed-product-removal (MSMPR) crystalliser with tubular bypass.[118] The four simulation studies constructed the models for the size and shape distribution of lysozyme protein in MSMPR and plug flow crystallisers, as well as the corresponding control schemes.[110,119–121]

The two early studies of continuous protein crystallisation focused on the crystal growth behaviour under flow conditions instead of the crystallisation process itself.[115,116] One of the more recent experimental studies investigated the crystallisation of lysozyme under slug flow in a tubular crystalliser and achieved a decent yield of 68%.[113] The supersaturation was manipulated through temperature to control the nucleation and growth of lysozyme crystals. The tubular crystalliser was separated into three segments for temperature control, where the first segment was designed for nucleation and the remaining two were for the growth of crystals. Although the liquid flow was laminar, the introduction of air bubbles ensured sufficient mixing along the crystalliser, preventing the deposition of crystals on the inner wall of the crystalliser and achieving a narrow size distribution of protein crystals. The continuous crystallisation of lysozyme was also demonstrated in an oscillatory flow crystalliser,[117] which will be discussed in detail in the next section.

In another recent experimental study, lysozyme and mAb01 were crystallised separately in an MSMPR crystalliser with the high yield of 93% in both cases.[118] Unlike the conventional configuration, the MSMPR crystalliser in this study was unjacketed and its cooling was provided by the cooled liquid flowing out from a tubular bypass. The advantage of such a configuration

Table 10.4 Summary of published studies on continuous protein crystallisation.

	Crystalliser			
Protein	Category	Type	Nature of study	Reference
Lysozyme	Stirred tank	MSMPR	Experimental	118
			Modelling	110, 119 and 120
	Tubular	Direct flow	Experimental	113, 115 and 116
			Modelling	121
		Oscillatory flow	Experimental	117
Antibody (mAb01)	Stirred tank	MSMPR	Experimental	118

was the decoupling of the stirring speed from the heat transfer within the MSMPR crystalliser. As a result, the stirring speed was manipulated to control the size of large crystals that settled down to the bottom of the crystalliser. In comparison to the study of lysozyme crystallisation in the tubular crystalliser, the yield in this study was significantly higher (93% *vs.* 68%) due to the higher residence time (30 hours *vs.* < 2 hours).

Together with the existing studies on batch protein crystallisation, these experimental studies demonstrate the technical feasibility of continuous protein crystallisation. The real challenges in this area are in fact related to process control and optimisation, in which the first step is to monitor the process continuously. This means the real-time measurement of protein concentration as well as crystal size and shape must be made with suitable equipment.

When the liquid system is one-phase and well-mixed, the protein concentration can be determined by attenuated total reflection-Fourier transform infrared (ATR-FTIR) spectrometry.[122–124] Since the stirring speed in a stirred tank crystalliser or the flow velocity in a tubular crystalliser is often kept low purposely to avoid the denaturation of protein,[101,125] the absence of turbulence leads to a protein concentration distribution over space, which is difficult to capture accurately by a point-based measurement such as ATR-FTIR. Furthermore, liquid–liquid phase separation can occur during protein crystallisation,[126,127] making it difficult to obtain accurate measurements of protein concentration. On the other hand, the real-time measurement of crystal size and shape is an area for development, as manual sampling is still required to determine the crystal size and shape by image analysis[128–131] or light scattering measurements.[132]

With real-time monitoring of the crystallisation process, the development and optimisation of a control strategy can be implemented. The four existing simulation studies serve as a good starting point for the further study of these aspects of continuous protein crystallisation.[110,119–121]

The continuous crystallisation of lysozyme in MSMPR and plug flow crystallisers were simulated with nucleation and growth rates that were dependent on the supersaturation.[110,120,121] The supersaturation was controlled by the jacket temperature in both configurations. Based on the knowledge of nucleation and growth rates, model predictive and feed-forward controllers were implemented to control the size and shape of lysozyme crystals.

The simulated studies of the MSMPR crystalliser found that implementing a product classification unit was useful for achieving monodispersity of crystal size.[110,120] The product classification unit allowed crystals larger than the cut-off size to enter the final product stream and could effectively mitigate the aggregation of protein crystals to sharpen the size distribution of protein crystals. On the other hand, a low flow rate to the fine trap that captured and dissolved small crystals was found to enhance aggregation.

The simulation of a plug flow crystalliser showed that the crystal size and shape could be controlled with both jacket temperature and flow velocity.[121] The plug flow crystalliser had segmented temperature control that was similar to the experimental study.[113] The feed-forward controller could effectively mitigate the fluctuation in the inlet protein concentration, improving the monodispersity of crystal size.

10.3.4 Case Study: Development of Oscillatory Flow Protein Crystallisation

Over the past few years, our research group has been working on the development of a tubular crystalliser for protein crystallisation.[96,101,117] In our early work, lysozyme was crystallised in a tubular crystalliser under direct flow.[96] The crystals were grown on the inner wall of glass capillaries, which were located within a water flow cell in a configuration similar to a shell and tube heat exchanger. The water flow cell was maintained at 0 °C to achieve supersaturation within the glass capillaries, while the other parts of the system were maintained at room temperature (20–22 °C) to avoid supersaturation and hence crystallisation. All the tubing and flow cell were transparent to allow the direct observation of crystals with a microscope. With this simple set-up, the impact of flow conditions on protein crystallisation was studied by the direct observation of crystal appearance on the inner wall of the crystallising section.

The experimental results showed that lysozyme crystals grew faster and larger under direct flow without the recirculation of protein solution than the quiescent condition and a narrower crystal size distribution was obtained. For example, lysozyme crystals grew up to 0.37 mm under direct flow in 27 hours as compared to only 0.16 mm under the quiescent condition. When the protein solution was circulated, the crystal growth rate increased with the flow velocity until an optimum value, after which a decrease was observed. The decrease in crystal growth rate at high flow velocity was attributed to the presence of impurities in the protein solution and the shear effects as well as the temperature lag in the crystallising capillaries. There were mainly two technical challenges of this system, namely the temperature lag at the entrance of the glass capillaries and the narrowing of the inner side of the glass capillaries. The temperature lag could be reduced by decreasing the inner diameter of the glass capillaries as well as the flow velocity. The first approach increases the surface area to volume ratio, whereas the second approach increases the contact time of the protein solution with the heat exchange surface. The narrowing of glass capillaries was due to the growth of crystals on their inner wall. It increased the effective flow velocity through the glass capillaries as the volumetric flow rate was kept constant, causing a control issue for the flow through the capillaries.

In a later study by our research group, the same glass capillaries/water flow cell was used to study the crystallisation of insulin under direct flow, but the protein was surprisingly denatured under the same flow velocities that had been used in the lysozyme study.[101] This situation prompted us to study oscillatory flow with an aim to achieve sufficient mixing with moderate flow velocities. In the new set-up, the original glass capillaries/water flow cell kit was operated in single passage mode (*i.e.* no recirculation). The glass capillaries were connected to syringes and the water flow cell was maintained at 10 °C. The protein/precipitant mixture was stored at room temperature in the syringes and pushed into the glass capillaries by the syringe pumps, which then generated the oscillatory flow by pushing and withdrawing the protein/precipitant solution at fixed speed and frequency. The appearance of crystals within the glass capillaries was observed directly under a microscope. The process yield was determined by measuring the protein concentration in the supernatant of the protein/precipitant mixture.

The experimental data of this study showed that the oscillatory flow enhanced the nucleation rate, as a large number of small crystals were obtained under various flow velocities in comparison to the smaller number of large crystals obtained under the quiescent condition. The fluid flow was modelled with the physics modelling and simulation software COMSOL in order to study the velocity and shear rate distribution within the glass capillaries. The simulation results showed that the shear rate generated by the oscillatory flow was within the range where the ordered arrangement of colloids was favoured, which was in line with the experimental observation that the oscillatory flow conditions favoured the nucleation step. In addition, the oscillatory flow conditions also doubled the yield of crystals in 48 hours in comparison to the quiescent condition, showing that the flow conditions not only enhanced the rate of nucleation, but also the growth of crystals. The technical challenge in this study was the complexity of the oscillatory flow, as the velocity and shear rate distribution was difficult to measure physically in the current set-up.

With the experience accumulated from the insulin study, we studied the development of continuous protein crystallisation under oscillatory flow using lysozyme as the model protein.[117] An unjacketed polyvinyl chloride (PVC) tube was connected to syringes that contained the protein and precipitant solutions. This unbaffled crystalliser configuration was similar to the glass capillaries/water flow cell set-up except for the temperature control component, allowing the easy manipulation of residence time at a fixed flow velocity by controlling the length of the PVC tube. The batch crystallisation of lysozyme was first studied before it was converted into the continuous mode, where the system consisted of two additional syringes that constantly supplied protein and precipitant solutions into the PVC tube. These syringes were located upstream of the one that was responsible for generating the oscillatory flow.

The batch experiments showed that the flow conditions enhanced the rate of nucleation, as higher oscillation frequency and amplitude both shortened the induction time of nucleation. This was in agreement with the findings of the insulin study.[101] In the continuous mode, this enhancement effect of high shear rate on the nucleation and growth rates helped to reduce the residence time of the protein solution in the tubular crystalliser required for achieving a particular yield. Higher shear rate produced better mixing and was supposed to give a narrower crystal size distribution, but the experimental data showed that the effect of better mixing was overshadowed by the enhanced nucleation and growth rates, resulting in poorer crystal quality and a wider crystal size distribution. Although the current oscillatory flow process still needs further optimisation, this study demonstrated the feasibility of continuous protein crystallisation under oscillatory flow in a tubular configuration. The effects of different operation variables such as residence time, oscillation frequency and amplitude on yield, crystal quality and size distribution have been investigated in our on-going studies, which will be published in the near future.

10.4 Outlooks and Perspectives

Unlike small molecules such as sodium chloride, proteins are much more difficult to crystallise. As protein-based biologics continue to gain importance in the pharmaceutical industry, protein crystallisation for purification and formulation purposes will become an important part in the pharmaceutical manufacturing process. The key process variables include protein concentration, temperature, additive concentrations, pH, pressure and flow conditions, which can be controlled to achieve the desired yield and crystal quality.

Although the development of continuous protein crystallisation is still at the early stage, the existing experimental studies demonstrated its feasibility at millilitre scale in MSMPR and tubular crystallisers, as relatively high yields (93% and 68% respectively) and narrow crystal size distributions were achieved in these studies. On the other hand, the existing simulation studies showed that different process control strategies could effectively obtain the desired crystal size and shapes.

Looking forward, new strategies for ensuring sufficient mixing within the crystalliser while accommodating the delicate nature of proteins should be further investigated. Such new strategies include slug flow and oscillatory flow in a tubular crystalliser, whose usefulness has been demonstrated in batch mode studies. The suitable scale-up criteria for crystallisations that apply these mixing methods should be investigated.

For a more robust size and shape control of crystals, it is advisable to perform seeding crystallisation where the protein solution lies in the metastable region to ensure only crystal growth occurs within the crystalliser, since crystal growth can be modelled more accurately than nucleation for protein crystallisation.

As proteins are expensive to produce and purify, the general limitation of availability will continue to hinder the study of continuous protein crystallisation on the wide variety of proteins that have been successfully crystallised at microliter scale. The development of microfluidic crystallisers can potentially alleviate this problem by performing continuous protein crystallisation at microliter or even nanoliter scale, but there are technical challenges to be overcome such as the delayed induction of crystallisation at small scale.

Finally, heterogeneous seeds such as MPS can be developed and optimised for the selective purification of proteins in the presence of impurities. This can be achieved by controlling the particle size, pore size and surface chemistry of the heterogeneous seeds to enhance the seed-protein interaction.

Abbreviations

2-HEAF	2-hydroxyethylammonium formate
2-ME	2-mercaptoethanol
ATR-FTIR	attenuated total reflection-Fourier transform infrared
mAb	monoclonal antibody
MES	2-morpholinoethanesulfonic acid
MPD	2-methyl-2,4-pentanediol
MSMPR	mixed-suspension, mixed-product-removal
PDB	Protein Data Bank
PEG	polyethylene glycol

Acknowledgements

We would like to acknowledge the financial support from the Engineering and Physical Sciences Research Council (EPSRC) for the Seeding and Continuous Biopharmaceutical Crystallisation (SCoBiC) programme (EP/N015916/1) as well as the European Union's Horizon 2020 research and innovation programme within the Revolutionising Downstream Processing of Monoclonal Antibodies by Continuous Template-Assisted Membrane Crystallization (AMECRYS) project (grant agreement number 712965).

References

1. A. Cornishbowden, *Int. J. Pept. Protein Res.*, 1984, **24**, 9–37.
2. C. N. Nanev, F. V. Hodzhaoglu and I. L. Dimitrov, *Cryst. Growth Des.*, 2011, **11**, 196–202.
3. L. Rosenfeld, *Clin. Chem.*, 2002, **48**, 2270–2288.
4. D. S. Alymer, *US Pat.*, 2,143,590, 1939.
5. S. Melville, *US Pat.* 2,174,862, 1939.
6. K. Petersen, *US Pat.*, 2,626,228, 1953.
7. H. J. D. Herman, *US Pat.*, 2,663,666, 1953.
8. J. Schlichtkrull and I. N. Nöring, *US Pat.*, 2,819,999, 1958.
9. K. Petersen and J. Schlichtkrull, *US Pat.*, 2,836,542, 1958.
10. K. Petersen and J. Schlichtkrull, *US Pat.*, 2,920,014, 1960.
11. H. M. Christensen, *US Pat.*, 3,102,077, 1963.
12. D. J. Newman and G. M. Cragg, *J. Nat. Prod.*, 2012, **75**, 311–335.
13. J. K. Jenkins, FDA/CMS Summit, https://www.fda.gov/media/101930/download, last accessed: 17th Nov 2019.

14. Protein therapeutics market – Technology advances spur market growth of protein therapies, http://www.drug-dev.com/Main/Back-Issues/PROTEIN-THERAPEUTICS-MARKET-Technology-Advances-Sp-1211.aspx, accessed 12 May 2018.
15. M. D. Costioli, C. Guillemot-Potelle, C. Mitchell-Logean and H. Broly, *BioPharm Int.*, 2010, **23**, 26–35.
16. M. Kalyanpur, *Mol. Biotechnol.*, 2002, **22**, 087–098.
17. A. Jungbauer, *Trends Biotechnol.*, 2013, **31**, 479–492.
18. A. A. Shukla, B. Hubbard, T. Tressel, S. Guhan and D. Low, *J. Chromatogr. B: Anal. Technol. Biomed. Life Sci.*, 2007, 28–39.
19. M. Franzreb, E. Muller and J. Vajda, *BioProcess Int.*, 2014, **12**(9), 44–52.
20. D. Hekmat, *Bioprocess Biosyst. Eng.*, 2015, **38**, 1209–1231.
21. R. dos Santos, A. L. Carvalho and A. C. A. Roque, *Biotechnol. Adv.*, 2017, **35**, 41–50.
22. A. McPherson and A. McPherson, *Crystallization of Biological Macromolecules*, Cold Spring Harbor Laboratory Press, New York, 1999, vol. 586.
23. T. M. Bergfors, *Protein Crystallization*, Internat'l University Line, 2009.
24. B. Rupp, *Biomolecular Crystallography: Principles, Practice, and Application to Structural Biology*, Garland Science, 2009.
25. A. McPherson and J. A. Gavira, *Acta Crystallogr., Sect. F: Struct. Biol. Commun.*, 2014, **70**, 2–20.
26. S. D. Durbin and G. Feher, *Annu. Rev. Phys. Chem.*, 1996, **47**, 171–204.
27. J. M. Wiencek, *Annu. Rev. Biomed. Eng.*, 1999, **1**, 505–534.
28. A. M. Kierzek and P. Zielenkiewicz, *Biophys. Chem.*, 2001, **91**, 1–20.
29. N. E. Chayen and E. Saridakis, *Nat. Methods*, 2008, **5**, 147–153.
30. I. R. Krauss, A. Merlino, A. Vergara and F. Sica, *Int. J. Mol. Sci.*, 2013, **14**, 11643–11691.
31. A. McPherson, *J. Cryst. Growth*, 1991, **110**, 1–10.
32. J. Drenth and C. Haas, *J. Cryst. Growth*, 1992, **122**, 107–109.
33. M. J. Pikal and D. R. Rigsbee, *Pharm. Res.*, 1997, **14**, 1379–1387.
34. B. Shenoy, Y. Wang, W. Shan and A. L. Margolin, *Biotechnol. Bioeng.*, 2001, **73**, 358–369.
35. A. A. Elkordy, R. T. Forbes and B. W. Barry, *Int. J. Pharm.*, 2002, **247**, 79–90.
36. A. A. Elkordy, R. T. Forbes and B. W. Barry, *Int. J. Pharm.*, 2004, **278**, 209–219.
37. K. A. Dill, S. B. Ozkan, M. S. Shell and T. R. Weikl, *Annu. Rev. Biophys.*, 2008, **37**, 289–316.
38. K. A. Dill and J. L. MacCallum, *Science*, 2012, **338**, 1042–1046.
39. A. M. Morris, M. A. Watzky and R. G. Finke, *Biochim. Biophys. Acta*, 2009, **1794**, 375–397.
40. H. C. Mahler, W. Friess, U. Grauschopf and S. Kiese, *J. Pharm. Sci.*, 2009, **98**, 2909–2934.
41. C. J. Roberts, *Biotechnol. Bioeng.*, 2007, **98**, 927–938.
42. C. Frieden, *Protein Sci.*, 2007, **16**, 2334–2344.

43. M. C. Bellissent-Funel, A. Hassanali, M. Havenith, R. Henchman, P. Pohl, F. Sterpone, D. van der Spoel, Y. Xu and A. E. Garcia, *Chem. Rev.*, 2016, **116**, 7673–7697.
44. L. R. Pratt and A. Pohorille, *Chem. Rev.*, 2002, **102**, 2671–2692.
45. M. B. Hillyer and B. C. Gibb, *Annu. Rev. Phys. Chem.*, 2016, **67**, 307–329.
46. T. Simonson, *Rep. Prog. Phys.*, 2003, **66**, 737–787.
47. C. X. Weichenberger and B. Rupp, *Acta Crystallogr., Sect. D: Biol. Crystallogr.*, 2014, **70**, 1579–1588.
48. B. W. Matthews, *J. Mol. Biol.*, 1968, **33**, 491–497.
49. M. Weselak, M. G. Patch, T. L. Selby, G. Knebel and R. C. Stevens, *Methods Enzymol.*, 2003, **368**, 45–76.
50. S. L. Morissette, Ö. Almarsson, M. L. Peterson, J. F. Remenar, M. J. Read, A. V. Lemmo, S. Ellis, M. J. Cima and C. R. Gardner, *Adv. Drug Delivery Rev.*, 2004, **56**, 275–300.
51. D. E. McRee, *Practical Protein Crystallography*, Academic press, 1999.
52. M. Jaskolski, Z. Dauter and A. Wlodawer, *FEBS J.*, 2014, **281**, 3985–4009.
53. F. Rosenberger, S. B. Howard, J. W. Sowers and T. A. Nyce, *J. Cryst. Growth*, 1993, **129**, 1–12.
54. M. Charles, S. Veesler and F. Bonneté, *Acta Crystallogr., Sect. D: Biol. Crystallogr.*, 2006, **62**, 1311–1318.
55. A. Razvi and J. M. Scholtz, *Protein Sci.*, 2006, **15**, 1569–1578.
56. K. A. Kantardjieff and B. Rupp, *Bioinformatics*, 2004, **20**, 2162–2168.
57. J. Kirkwood, D. Hargreaves, S. O'Keefe and J. Wilson, *Bioinformatics*, 2015, **31**, 1444–1451.
58. B. Smejkal, N. J. Agrawal, B. Helk, H. Schulz, M. Giffard, M. Mechelke, F. Ortner, P. Heckmeier, B. L. Trout and D. Hekmat, *Biotechnol. Bioeng.*, 2013, **110**, 2452–2461.
59. J. Kirkwood, D. Hargreaves, S. O'keefe and J. Wilson, *Acta Crystallogr., Sect. F: Struct. Biol. Commun.*, 2015, **71**, 1228–1234.
60. R. R. Burgess, *Protein Precipitation Techniques*, Academic Press, Cambridge, 2009, vol. 463, 331–342–.
61. S. Englard and S. Seifter, *Methods Enzymol.*, 1990, **182**, 285–300.
62. K. C. Ingham, *Methods Enzymol.*, 1990, **182**, 301–306.
63. R. K. Scopes, *Protein Purification: Principles and Practice*, Springer Science & Business Media, Berlin, 1994, vol. 2.
64. F. Hofmeister, *Naunyn-Schmiedeberg's Arch. Pharmacol.*, 1888, **25**, 1–30.
65. R. L. Baldwin, *Biophys. J.*, 1996, **71**, 2056–2063.
66. N. Schwierz, D. Horinek, U. Sivan and R. R. Netz, *Curr. Opin. Colloid Interface Sci.*, 2016, **23**, 10–18.
67. E. Ruckenstein and I. L. Shulgin, *Adv. Colloid Interface Sci.*, 2006, **123**, 97–103.
68. T. Arakawa and S. N. Timasheff, *Biochemistry*, 1982, **21**, 6545–6552.
69. T. Arakawa and S. N. Timasheff, *Biochemistry*, 1984, **23**, 5912–5923.
70. A. A. Green and W. L. Hughes, *Methods Enzymol.*, 1955, **1**, 67–90.
71. S. Khurshid, E. Saridakis, L. Govada and N. E. Chayen, *Nat. Protoc.*, 2014, **9**, 1621–1633.

72. L. Gránásy, F. Podmaniczky, G. I. Tóth, G. Tegze and T. Pusztai, *Chem. Soc. Rev.*, 2014, **43**, 2159.
73. U. V. Shah, C. Amberg, Y. Diao, Z. Yang and J. Y. Y. Heng, *Curr. Opin. Chem. Eng.*, 2015, **8**, 69–75.
74. H. Yang, C. L. Song, Y. X. S. Lim, W. Chen and J. Y. Y. Heng, *CrystEngComm*, 2017, **19**(44), 6573–6578.
75. N. E. Chayen, E. Saridakis, R. El-Bahar and Y. Nemirovsky, *J. Mol. Biol.*, 2001, **312**, 591–595.
76. N. E. Chayen, E. Saridakis and R. P. Sear, *Proc. Natl. Acad. Sci. U. S. A.*, 2006, **103**, 597–601.
77. C. N. Nanev, E. Saridakis and N. E. Chayen, *Sci. Rep.*, 2017, **7**, 35821.
78. U. V. Shah, M. C. Allenby, D. R. Williams and J. Y. Y. Heng, *Cryst. Growth Des.*, 2012, **12**, 1772–1777.
79. U. V. Shah, D. R. Williams and J. Y. Y. Heng, *Cryst. Growth Des.*, 2012, **12**, 1362–1369.
80. U. V. Shah, J. V. Parambil, D. R. Williams, S. J. Hinder and J. Y. Y. Heng, *Powder Technol.*, 2015, **282**, 10–18.
81. U. V. Shah, N. H. Jahn, S. Huang, Z. Yang, D. R. Williams and J. Y. Y. Heng, *J. Cryst. Growth*, 2017, **469**, 42–47.
82. J. Lei, J. Fan, C. Yu, L. Zhang, S. Jiang, B. Tu and D. Zhao, *Microporous Mesoporous Mater.*, 2004, **73**, 121–128.
83. S. Hudson, E. Magner, J. Cooney and B. Kieran, *J. Phys. Chem. B*, 2005, **109**, 19496–19506.
84. S. Hudson, J. Cooney and E. Magner, *Angew. Chem., Int. Ed.*, 2008, **47**, 8582–8594.
85. H. Berman, K. Henrick, H. Nakamura and J. L. Markley, *Nucleic Acids Res.*, 2007, **35**, D301.
86. K. Henrick, Z. Feng, W. F. Bluhm, D. Dimitropoulos, J. F. Doreleijers, S. Dutta, J. L. Flippen-Anderson, J. Ionides, C. Kamada, E. Krissinel, C. L. Lawson, J. L. Markley, H. Nakamura, R. Newman, Y. Shimizu, J. Swaminathan, S. Velankar, J. Ory, E. L. Ulrich, W. Vranken, J. Westbrook, R. Yamashita, H. Yang, J. Young, M. Yousufuddin and H. M. Berman, *Nucleic Acids Res.*, 2007, **36**, D426–D433.
87. E. L. Ulrich, H. Akutsu, J. F. Doreleijers, Y. Harano, Y. E. Ioannidis, J. Lin, M. Livny, S. Mading, D. Maziuk, Z. Miller, E. Nakatani, C. F. Schulte, D. E. Tolmie, R. Kent Wenger, H. Yao and J. L. Markley, *Nucleic Acids Res.*, 2008, **36**, D402–D408.
88. A. Gutmanas, Y. Alhroub, G. M. Battle, J. M. Berrisford, E. Bochet, M. J. Conroy, J. M. Dana, M. A. Fernandez Montecelo, G. Van Ginkel, S. P. Gore, P. Haslam, R. Hatherley, P. M. S. Hendrickx, M. Hirshberg, I. Lagerstedt, S. Mir, A. Mukhopadhyay, T. J. Oldfield, A. Patwardhan, L. Rinaldi, G. Sahni, E. Sanz-García, S. Sen, R. A. Slowley, S. Velankar, M. E. Wainwright and G. J. Kleywegt, *Nucleic Acids Res.*, 2013, **42**, D285–D291.
89. P. W. Rose, A. Prlić, C. Bi, W. F. Bluhm, C. H. Christie, S. Dutta, R. K. Green, D. S. Goodsell, J. D. Westbrook, J. Woo, J. Young, C. Zardecki, H.

M. Berman, P. E. Bourne and S. K. Burley, *Nucleic Acids Res.*, 2015, **43**, D345–D356.
90. S. K. Basu, C. P. Govardhan, C. W. Jung and A. L. Margolin, *Expert Opin. Biol. Ther.*, 2004, **4**, 301–317.
91. T. Takakura, T. Ito, S. Yagi, Y. Notsu, T. Itakura, T. Nakamura, K. Inagaki, N. Esaki, R. M. Hoffman and A. Takimoto, *Appl. Microbiol. Biotechnol.*, 2006, **70**, 183–192.
92. L. B. Wang, N. I. Wakayama and W. Q. Tao, *J. Cryst. Growth*, 2008, **310**, 5370–5374.
93. Y. Yu, Y. Liu, X. Wang, D. Oberthür, K. Dierks and C. Betzel, *Cryst. Res. Technol.*, 2010, **45**, 380–386.
94. M. Maruyama, H. Kawahara, G. Sazaki, S. Maki, Y. Takahashi, H. Y. Yoshikawa, S. Sugiyama, H. Adachi, K. Takano and H. Matsumura, *et al.*, *Cryst. Growth Des.*, 2012, **12**, 2856–2863.
95. Y. Hayashi, M. Maruyama, M. Yoshimura, S. Okada, H. Y. Yoshikawa, S. Sugiyama, H. Adachi, H. Matsumura, T. Inoue, K. Takano, S. Murakami and Y. Mori, *Cryst. Growth Des.*, 2015, **15**, 2137–2143.
96. M. M. Roberts, J. Y. Y. Heng and D. R. Williams, *Cryst. Growth Des.*, 2010, **10**, 1074–1083.
97. R. Murai, H. Y. Yoshikawa, H. Kawahara, S. Maki, S. Sugiyama, T. Kitatani, H. Adachi, K. Takano, H. Matsumura, S. Murakami, T. Inoue, T. Sasaki and Y. Mori, *J. Cryst. Growth*, 2008, **310**, 2168–2172.
98. Q.-Q. Lu, D.-C. Yin, Y.-M. Liu, X.-K. Wang, P.-F. Yang, Z.-T. Liu and P. Shang, *J. Appl. Crystallogr.*, 2010, **43**, 473–482.
99. Q.-Q. Lu, B. Zhang, L. Tao, L. Xu, D. Chen, J. Zhu and D.-C. Yin, *Cryst. Growth Des.*, 2016, **16**, 4869–4876.
100. B. Smejkal, B. Helk, J. M. Rondeau, S. Anton, A. Wilke, P. Scheyerer, J. Fries, D. Hekmat and D. Weuster-Botz, *Biotechnol. Bioeng.*, 2013, **110**, 1956–1963.
101. J. V. Parambil, M. Schaepertoens, D. R. Williams and J. Y. Y. Heng, *Cryst. Growth Des.*, 2011, **11**, 4353–4359.
102. A. Hawe, M. Wiggenhorn, M. van de Weert, J. H. O. Garbe, H. C. Mahler and W. Jiskoot, *J. Pharm. Sci.*, 2012, **101**, 895–913.
103. Y.-F. Maa and C. C. Hsu, *Biotechnol. Bioeng.*, 1996, **51**, 458–465.
104. J. S. Bee, J. L. Stevenson, B. Mehta, J. Svitel, J. Pollastrini, R. Platz, E. Freund, J. F. Carpenter and T. W. Randolph, *Biotechnol. Bioeng.*, 2009, **103**, 936–943.
105. M. Yaoi, H. Aadachi, K. Takano, H. Matsumura, T. Inoue, Y. Mori and T. Sasaki, *Jpn. J. Appl. Phys.*, 2004, **43**, L686.
106. R. Mirro and K. Voll, *BioProcess Int.*, 2009, **7**, 52–57.
107. S. Sandadi, H. Pedersen, J. S. Bowers and D. Rendeiro, *Bioprocess Biosyst. Eng.*, 2011, **34**, 819–832.
108. M. C. C. Bustamante, M. O. Cerri and A. C. Badino, *Chem. Eng. Sci.*, 2013, **90**, 92–100.
109. T. N. Zwietering, *Chem. Eng. Sci.*, 1958, **8**, 244–253.

110. J. S. I. Kwon, M. Nayhouse, G. Orkoulas and P. D. Christofides, *Ind. Eng. Chem. Res.*, 2014, **53**, 15538–15548.
111. D. Hekmat, D. Hebel, H. Schmid and D. Weuster-Botz, *Process Biochem.*, 2007, **42**, 1649–1654.
112. H. Huettmann, M. Berkemeyer, W. Buchinger and A. Jungbauer, *Biotechnol. Bioeng.*, 2014, **111**, 2192–2199.
113. P. Neugebauer and J. G. Khinast, *Cryst. Growth Des.*, 2015, **15**, 1089–1095.
114. F. Castro, A. Ferreira, J. A. Teixeira and F. Rocha, *Cryst. Growth Des.*, 2016, **16**, 3748–3755.
115. M. Pusey, W. Witherow and R. Naumann, *J. Cryst. Growth*, 1988, **90**, 105–111.
116. M. L. Grant and D. A. Saville, *J. Cryst. Growth*, 1995, **153**, 42–54.
117. H. Yang, P. Peczulis, P. Inguva, X. Li and J. Y. Y. Heng, *Chem. Eng. Res. Des.*, 2019, **136**, 529–535.
118. D. Hekmat, M. Huber, C. Lohse, N. Von Den Eichen and D. Weuster-Botz, *Cryst. Growth Des.*, 2017, **17**, 4162–4169.
119. C. Borchert, N. Nere, D. Ramkrishna, A. Voigt and K. Sundmacher, *Chem. Eng. Sci.*, 2009, **64**, 686–696.
120. J. S. Il Kwon, M. Nayhouse, P. D. Christofides and G. Orkoulas, *Chem. Eng. Sci.*, 2014, **107**, 47–57.
121. J. Sang-Il Kwon, M. Nayhouse, G. Orkoulas and P. D. Christofides, *Chem. Eng. Sci.*, 2014, **119**, 30–39.
122. D. D. Dunuwila, L. B. Carroll and K. A. Berglund, *J. Cryst. Growth*, 1994, **137**, 561–568.
123. D. D. Dunuwila and K. a. Berglund, *J. Cryst. Growth*, 1997, **179**, 185–193.
124. F. Lewiner, J. P. Klein, F. Puel and G. Févotte, *Chem. Eng. Sci.*, 2001, **56**, 2069–2084.
125. D. Hebel, S. Huber, B. Stanislawski and D. Hekmat, *J. Biotechnol.*, 2013, **166**, 206–211.
126. B. D. Mason, J. Zhang-Van Enk, L. Zhang, R. L. Remmele and J. Zhang, *Biophys. J.*, 2010, **99**, 3792–3800.
127. K. Reiche, J. Hartl, A. Blume and P. Garidel, *Biophys. Chem.*, 2017, **220**, 7–19.
128. P. A. Larsen, J. B. Rawlings and N. J. Ferrier, *Chem. Eng. Sci.*, 2006, **61**, 5236–5248.
129. S. Schorsch, D. R. Ochsenbein, T. Vetter, M. Morari and M. Mazzotti, *Chem. Eng. Sci.*, 2014, **105**, 155–168.
130. R. Zhang, C. Y. Ma, J. J. Liu and X. Z. Wang, *Chem. Eng. Sci.*, 2015, **137**, 9–21.
131. Y. Huo, T. Liu, H. Liu, C. Y. Ma and X. Z. Wang, *Chem. Eng. Sci.*, 2016, **148**, 126–139.
132. W. W. Wilson, *J. Struct. Biol.*, 2003, **142**, 56–65.
133. J. Peters, T. Minuth and W. Schröder, *Protein Expression Purif.*, 2005, **39**, 43–53.
134. R. Jackson, *US Pat.*, 3,719,655, 1973.

135. D. Hebel, M. Ürdingen, D. Hekmat and D. Weuster-Botz, *Cryst. Growth Des.*, 2013, **13**, 2499–2506.
136. R. A. Judge, M. R. Johns and E. T. White, *Biotechnol. Bioeng.*, 1995, **48**, 316–323.
137. E. H. Gros and J. L. Cunefare, *US Pat.*, 6,207,437, 2001.
138. P. Kwanyuen, S. M. Allina, A. K. Weissinger and R. F. Wilson, *J. Proteome Res.*, 2002, **1**, 471–473.
139. T. Becker and V. B. Lawlis Jr, *US Pat.*, 5,041,377, 1991.
140. M. Webei, M. J. Jones and J. Ulrich, *Cryst. Growth Des.*, 2008, **8**, 711–716.
141. A. Kadowaki, I. Yoshizaki, L. Rong, H. Komatsu, O. Odawara and S. Yoda, *J. Synchrotron Radiat.*, 2004, **11**, 38–40.
142. P. G. Vekilov and F. Rosenberger, *J. Cryst. Growth*, 1998, **186**, 251–261.
143. T. A. Nyce and F. Rosenberger, *J. Cryst. Growth*, 1991, **110**, 52–59.
144. S. D. Durbin and G. Feher, *J. Cryst. Growth*, 1986, **76**, 583–592.

CHAPTER 11

Continuous Melt Crystallization

HONGXUN HAO*[a,b] AND YAN XIAO[a,b]

[a]Tianjin University, National Engineering Research Center of Industrial Crystallization Technology, School of Chemical Engineering and Technology, Tianjin 300072, PR China; [b]Collaborative Innovation Center of Chemical Science and Chemical Engineering (Tianjin), Tianjin 300072, PR China
*E-mail: hongxunhao@tju.edu.cn

11.1 Introduction

11.1.1 Definitions for Melt Crystallization

The crystallization process, as one of the most important separation unit operations, could be divided into two subcategories: crystallization from solution and crystallization from melt.[1–4] Normally, a solution is a homogeneous mixture of different species, while a melt refers to a pure molten solid. As for melt crystallization, the melt is used in a more general way to describe mixtures of melt materials that are usually solid state at room temperature.

Thermodynamically, both solution crystallization and melt crystallization processes happen from the deviation of the system from a phase equilibrium state. Generally, for solution crystallization, the concentration of target products is relatively low. Also, it is driven by supersaturation and is usually operated at a temperature which is much lower than the melting

The Handbook of Continuous Crystallization
Edited by Nima Yazdanpanah and Zoltan K. Nagy

Published by the Royal Society of Chemistry, www.rsc.org

point of the product. Melt crystallization, growing crystals from melt, is often applied to systems with high percentage of target product and operated at temperature close to the melting temperature of the components. Melt crystallization is driven by supercooling (temperature is lower than the equilibrium one). It is suggested by Ulrich *et al.* that "solution" could be used for a process where mass transfer effects dominate, while "melt" could be used for a process where heat transfer effects dominate. Besides, the growth rate of crystals for melt crystallization could be 3 orders higher than that of solution crystallization while still producing highly pure products.[2]

Compared to solution crystallization which could be used for modifying crystal size distribution and crystal morphology and help to optimize the manufacturing downstream processes, melt crystallization is generally used for purification, separation and concentration, instead of changing particle properties, and in most circumstances, purified product is obtained in liquid form.[5–8]

11.1.2 Features of Melt Crystallization

Melt crystallization has advantages and features that make it a potential and competitive separation method, compared to distillation and solution crystallization.[1,2,9]

Firstly, phase change happens between liquid and solid state in melt crystallization, which means the energy consumption is much smaller (usually one third to one seventh) than that for liquid–vapor phase change in distillation or evaporation crystallization processes. Meanwhile, compared to vapor, liquid occupies much less space for the same amount of substance. Therefore, the equipment volume for melt crystallization could be smaller and need less construction work, meaning less capital cost and less safety issues concerning pressure and leakage, and so on. Besides, about 60% of organics have melting points between 0 and 210 °C, which falls in the low-level energy range and could be operated using waste heat from other processes to further save energy cost. And heat sensitive substances, such as pharmaceuticals, food and vitamins, could be protected from decomposition or deterioration due to the low process temperatures.

Secondly, solvents are no more a problem in melt crystallization processes. In addition to the problems of solvent residue, solvent toxicity, as well as solvent consumption and recovery, the absence of solvent also means a smaller crystallizer. More importantly, melt mixtures normally have a relatively higher concentration of the product component, which will give rise to a higher purification effect and larger production capacity.

Furthermore, in these days, isomer separation remains a difficult challenge for distillation which will need many theoretical plates and high reflux ratio as a result of similar boiling points. As a result, if distillation is applied to separate isomers, strict requirements about the equipment and high cost are inevitable. However, the large difference between the melting temperatures of isomers makes melting crystallization an applicable and highly efficient separation method, with 2 to 10 times smaller reflux ratio as well as smaller energy consumption.

Finally, high separation efficiency is also one of the most remarkable advantages of melt crystallization. Theoretically, absolute pure product could be obtained within one separation step for eutectic systems which account for 54.3% of the known two-component organic mixtures. Although 31.6% of the binary systems tend to form molecular compounds or to have peritectic and eutectic points, only 14.1% of them belong to a solid solution forming system, which could only be purified step by step and be operated like a distillation process to achieve high purity.

However, melt crystallization also has its disadvantages. The uncrystallized melt or impurities are easily included into the crystal layers because of the fast crystal growth rate. Also, impurities or uncrystallized melt can readily adhere to the crystal surface as a result of the high viscosity and high slurry density of the melt system. Meanwhile, high viscosity and high slurry density of the melt system also makes the transportation and downstream processing more difficult than in solution crystallization. It also results in a more complicated structure of melt crystallizer. Therefore, multistage crystallization or post crystallization treatments are necessary for a melt crystallization process.

11.1.3 Material Selection

Theoretically, melt crystallization could be used for all the systems which are stable in their molten state. However, as mentioned before, since the crystalline products are usually obtained in their liquid form for melt crystallization, the melt crystallization technique is mainly used for the purposes of purification or ultra-purification, separation or concentration. Nowadays, distillation has been developed as the standard and mature unit operation in most chemical separations. Therefore, melt crystallization has been typically used in purification applications where distillation becomes difficult,[10,11] such as:

1. Isomers with close boiling points
2. Azeotropic systems
3. Temperature sensitive substances
4. Components that tend to polymerize
5. Explosive substances

So far, some successful industrial applications of melt crystallization include acrylic acid, bisphenol A, caprolactam, cresol, dichlorobenzene, fatty acids, naphthalene, paraffin, xylene, juice, beer, dairy, and so on.[12]

11.2 Theoretical Basis

Similar to other crystallization processes, phase diagrams are the thermodynamic basis for the design and optimization of melt crystallization and will determine the separation limit. Crystallization kinetics as well as the model establishment are helpful for better understanding of the process and to provide a design basis for industrial melt crystallization.

11.2.1 Phase Diagram

The most basic two types of equilibrium phase diagrams for binary mixtures are the eutectic and solid solution systems, as shown in Figure 11.1.[1,13,14] And other complicated systems could always be divided into several parts, which are either eutectic ones or solid solution forming ones.

As for the eutectic system (Figure 11.1a), when the mixture (C) is cooled down from the liquid phase region (L) into the solid phase region (S), pure A will crystallize out from one crystallization step ideally and the composition of melt liquid will gradually move along to the eutectic point. The highest separation limit would be the eutectic composition for the melt residue, which means that mixture with eutectic concentration is not separable any more under the same operation conditions. However, in real processes, crystallization is not happening under the phase equilibrium state, and the separation efficiency would be decreased by kinetic effects. As a result, mother liquor could be trapped in or adhere to the fast-growing crystals. Therefore, multiple crystallization steps or post crystallization processes, such as washing or sweating, are necessary to get high purity crystal products.

As we can see from Figure 11.1b, for solid solution forming systems, it is not possible to get pure crystals from only one crystallization step. Theoretically, cooling of melt C_0 could only give crystals with composition of C_1. By melting the crystals and cooling it again, crystals with composition of

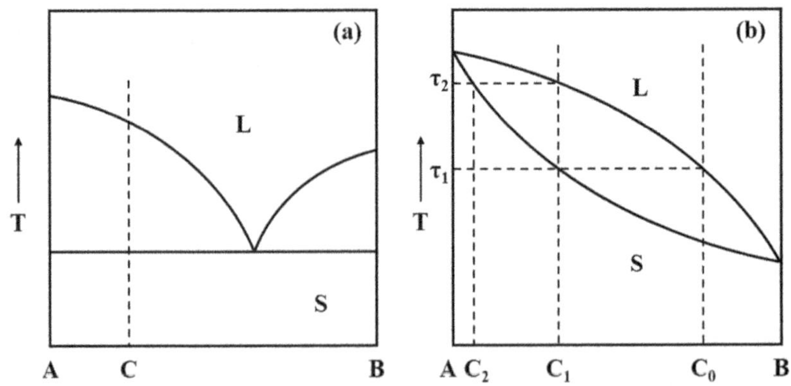

Figure 11.1 Phase diagrams of (a) eutectic and (b) solid solution systems.

C_2 could be obtained. By repeating the heating and cooling thermal cycles, purity of products could be increased step by step. Although multiple theoretical stages are needed to purify the materials, there is no separation limit for solid solution forming systems. Similarly, crystallization kinetic effects and the non-complete solid–liquid separation call for post crystallization processes to improve the purification efficiency.

11.2.2 Crystallization Kinetics

As mentioned above, crystallization kinetics could significantly influence the separation efficiency of melt crystallization. Therefore, a good understanding of crystal nucleation and growth will help to improve the process development and design.

11.2.2.1 Crystal Nucleation[15,16]

Nyvlt characterized nucleation kinetics by an empirical relationship between the metastable zone width Δc_{max} and the nucleation rate N_m, expressed as:

$$N_m = k_n \Delta c^m{}_{max} \tag{11.1}$$

where k_n is the nucleation rate constant and m is the apparent order of crystal nucleation.

With respect to the nucleation process with a constant cooling rate, the relationship between the biggest supercooling degree and cooling rate could be written as:

$$\log b = \log\left(k_n \left(dc^*/dT\right)^{m-1}\right) + m \log \Delta T_m \tag{11.2}$$

where b is the cooling rate, ΔT_m is the maximum allowable supercooling, taken as the difference between the equilibrium temperature and the temperature at which nucleation happens, and c^* is the equilibrium concentration.

By plotting log b against log ΔT_m, crystal nucleation kinetic parameters could be obtained for the specific crystallization system.

11.2.2.2 Crystal Growth[15–17]

Crystal growth rate could also be expressed as an exponential equation as:

$$G = k_G \Delta w^a \tag{11.3}$$

where G is the growth rate, Δw is the supercooling defined as the difference between the weight fraction of the melt at any time and the equilibrium value, k_G is the growth rate constant and a is the order of crystal growth.

For continuous melt crystallization process, the crystal size distribution (CSD) of product could be derived from the population balance equation

$$n = n^0 \exp(-L/G\tau) \tag{11.4}$$

where n is the number of crystals per unit volume and size differential, τ is the residence time. The population density of nuclei, n^0, is related to the numerical nucleation rate B^0 and crystal growth rate G, according to

$$B^0 = n^0 G \tag{11.5}$$

Therefore, with CSD data, n^0, G, B^0 could be obtained for continuous melt crystallization process.

11.2.3 Model Description of Melt Crystallization

For the design and optimization of melt crystallization processes, a complete understanding of the process is necessary. To this end, a detailed mathematical model needs to be established to describe the impurity concentration distribution, separation efficiency and product purity during the melt crystallization processes. Nevertheless, limited model theories could be found in the literature, which is mainly caused by the complexity of mass and heat transfer processes in the crystallizers. The non-linear differential equations could only be solved numerically and also require a considerable number of simplifying assumptions.[18–21] In this chapter, a typical model established by Yazdanpanah *et al.* for layer crystallization process is introduced.[22–24]

Figure 11.2 shows the assignment of state variables for the melt crystallization process on cooled surface. T_m is assumed to be constant at the cooled surface, while r_s, T_s and r_f, T_f would change due to the growth of crystal layer. The length of cooled surface in the axial is L.

11.2.3.1 Mass Transfer

The product component mass balance equation in the melt can be represented by Fick's diffusion law:

$$\frac{\partial C}{\partial t} + u(r)\frac{\partial C}{\partial x} = D\frac{1}{r}\frac{\partial}{\partial r}\left(r\frac{\partial C}{\partial r}\right), r_s \le r \le r_f, 0 \le x \le L \tag{11.6}$$

where C is the product component concentration, $u(r)$ is the fluid velocity, D is the molecular diffusion coefficient, x is the axial distance, r is the radial distance, and L is the total length of cooled surface.

At the solid–liquid surface, the mass flux to the surface is equal to the crystallization rate. Therefore, the initial and boundary conditions for eqn (11.6) are IC:

$$t = 0, C(r,x,0) = C_0$$

$$x = 0, \quad \frac{dC(r,0,t)}{dt} = \frac{F(L,t)\left(C_{\text{melt}}(L,t) - C_{\text{melt}}(0,t)\right)}{V_{\text{tank}}}$$

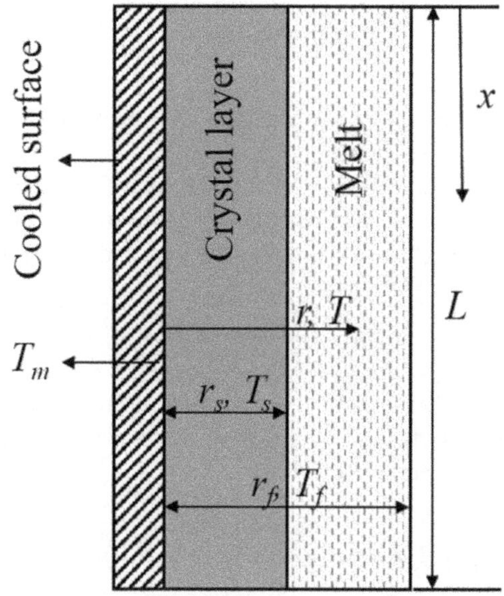

Figure 11.2 Schematic description of state variables in layer crystallization.

BC:

$$r = r_f, -D\frac{\partial C}{\partial r} = 0$$

$$r = r_s, -D\frac{\partial C}{\partial r} = -G = -k_i\left(C_r - C_e\right)^n$$

where F is the flow rate of the melt, C_0 is the initial concentration, C_{melt} is the average concentration of product component at distance x, G is the crystal growth rate, k_i and n are the crystal growth parameters, C_r is the concentration at the crystal layer surface, C_e is the equilibrium concentration, and V_{tank} is the volume of feeding tank.

Similarly, the mass transfer equation for impurity in the melt could be written as:

$$\frac{\partial C_{imp}}{\partial t} + u(r)\frac{\partial C_{imp}}{\partial x} = D_{imp}\frac{1}{r}\frac{\partial}{\partial r}\left(r\frac{\partial C_{imp}}{\partial r}\right), r_s \le r \le r_f, 0 \le x \le L \qquad (11.7)$$

IC:

$$t = 0, C_{imp}(r,x,0) = C_{imp,0}$$

$$x = 0, \frac{dC_{imp}(r,0,t)}{dt} = \frac{F(L,t)\left(C_{imp,melt}(L,t) - C_{imp,melt}(0,t)\right)}{V_{tank}}$$

BC:

$$r = r_f, -D_{imp}\frac{\partial C_{imp}}{\partial r} = 0$$

$$r = r_s, -D_{imp}\frac{\partial C_{imp}}{\partial r} = -G\left(C_{imp} - K\frac{C_{imp}}{C}\rho\right)$$

where $C_{imp,\,melt}$ is the average concentration of impurity, K is the distribution coefficient and ρ is the density of crystal.

The definition[1,8,23,25] for K, the distribution coefficient, is the ratio of the impurity concentration (C_{imp}) to product component concentration (C) in the crystal phase divided by that ratio in the melt phase, calculated as:

$$K = \frac{\left(C_{imp}/C\right)_{crystal}}{\left(C_{imp}/C\right)_{melt}} \tag{11.8}$$

A K close to 1 means almost no separation effect, while a K close to 0 means complete separation. Therefore, K is normally used to characterize the separation efficiency of melt crystallization.

11.2.3.2 Heat Transfer

In melt crystallization processes, supersaturation is created by supercooling and tiny temperature changes in the melt could lead to fast crystallization. Therefore, heat transfer processes could play a significant role in melt crystallization.[17,25] Similar to the mass transfer equation, the heat transfer equation with the initial and boundary conditions is defined as:

$$\frac{\partial T}{\partial t} + u(r)\frac{\partial T}{\partial x} = \alpha\frac{\partial}{\partial r}\left(r\frac{\partial T}{\partial r}\right), r_s \le r \le r_f, 0 \le x \le L \tag{11.9}$$

IC:

$t = 0$, $T(r,x,0) = T_m$

$x = 0$, $T(r,0,t) = T_{tank}$

BC:

$$r = r_f, -k_{melt}\frac{\partial T}{\partial r} = 0$$

$$r = r_s, -k_{melt}\frac{\partial T}{\partial r} = U\left(T_m - T_s\right)$$

where T is the temperature, α is the thermal diffusivity, k_{melt} is the thermal conductivity of the solution, T_m is the temperature of cooled surface and U is the overall heat transfer coefficient.

By solving the mass and heat transfer equations together, the product purity, distribution coefficient, yield and temperature profile of the melt crystallization process could be predicted and further insights could be gained to optimize the crystallization process.

11.3 Post-crystallization Processes

In practice, absolute solid–liquid separation is unrealistic. The separation efficiency of the melt crystallization process could be decreased dramatically as a result of the inevitable adhering and liquid inclusions of the residual melt in crystals forming by three mechanisms:[2]

1. Nucleation processes at high supercooling
2. High crystal growth rates
3. Adherence of residual melt at the end of process

Therefore, additional purification steps including sweating and washing (post-crystallization processes) are necessary to obtain the desired purity.

11.3.1 Sweating

Sweating is a temperature induced purification process.[2,18,26] After the crystallization process, the temperature of cooled surface is raised to close to the melting point of the pure component. Due to the lower melting point of impurity-rich part of the crystal layer, the impurities included during the nucleation step, as well as the liquid inclusions could remelt and flow out. All the liquid, including the adhering melt residue, will drain off under gravity, which leaves the crystal product with a higher purity. Besides, higher temperature decreases the viscosity of the liquid phase, which further helps the draining of contaminated liquid. According to Ulrich and Delannoy *et al.*, with the sweating process, the distribution coefficient could decrease from 0.4 to 0.2, or 0.6 to 0.3 under different crystal growth rates, which means that sweating could have the same purification effect as a recrystallization process in some cases.

Along with the purity increasing, more crystals will be remolten and drain off, causing a decrease of yield. Therefore, a balance needs to be achieved from both economic and technical points of view. Compared to recrystallization, the sweating step performs much faster (about 1/3 to 1/9 of a crystallization process) and it does not need another phase transformation (the product has to be remolten anyway), saving both time and energy. Hence, the sweating operation is widely used in almost every melt crystallization process.

11.3.2 Washing

When talking about washing, it can be divided into two different processes, rinsing and diffusion washing.[2,18,26] In a rinsing step, the contaminated liquid adhering to the crystal surfaces is substituted by a purer washing liquid, which is usually part of the remolten purified product or the feed liquid. This process happens because of the flowing of pure liquid and will be accomplished in several seconds. If the washing liquid keeps existing around the crystals, the liquid–liquid diffusion process from the pores of the crystals to the washing liquid would happen. Normally, the diffusion washing step takes more than 11 minutes. To prevent the washing liquid from crystallizing onto the crystals, it has to be superheated. Therefore, the diffusion washing is usually accompanied by a sweating process, which further promotes its purification effect. Considering the widely used washing columns in the melting crystallization industrial processes, its separation efficiency is obvious.

The parameters which need to be considered include the purity, temperature, amount, and the flow regime of the washing liquid. The purification of product is achieved through the contamination of the washing liquid which usually is part of the recycled purified product. In most cases, the contaminated liquid needs to be purified again. Compared to the crystallization process, a washing step needs much less operation time (11–20 minutes) and causes less energy consumption (just pumps and temperature control are needed).

As for the industrial melt crystallization, the selection of post-crystallization step (sweating or washing) depends on the properties of different systems and varies from case to case. Sweating is easy to operate and needs no additional equipment. However, we must make sure that the crystal layer will not slip off the cooled surface when partially molten. Washing is more efficient, but needs additional equipment and usually proceeds in a column, which means additional capital costs. Besides, accurate temperature control is necessary during washing processes to avoid too much remelting or crystallizing of the washing liquid.

11.4 Continuous Melt Crystallization

Compared to solution crystallization, melt crystallization is more difficult to realize in continuous operation mode due to the high concentration and high magma density of the system. Depending on the place where crystallization happens, melt crystallization processes can be divided into suspension crystallization and solid layer crystallization.[2,27–29] In suspension crystallization, crystals are suspended in the melt. For suspension crystallization, MSMPR (mixed suspension mixed product removal) crystallizers as well as their variations can be used in continuous mode. In solid layer crystallization, crystals nucleate and grow on a cooled surface. For commonly used solid layer crystallizers, such as MWB (Sulzer Metallwerk Buchs) and FFMC (falling film melt crystallization), pure products are harvested by melting the crystal layer. Therefore, it is difficult to operate them continuously. Nowadays, crystallizers with moving cooling surface have been developed to fulfill the continuous operating demand for solid layer melt crystallization. In this part, the continuous melt crystallizers are summarized for both suspension and solid layer crystallization.

11.4.1 Continuous Suspension Crystallization

This section presents several kinds of suspension melt crystallizers which can be operated in continuous mode. They are mainly different in their solid handling methods.

11.4.1.1 MSMPR Crystallizer

The MSMPR crystallizer, usually a tank or column, is one of the basic crystallizers for all crystallization processes.[7,16] Melt crystallization processes could be continuously operated with indirect contact cooling or direct contact cooling methods. The cooling medium circulates in the jacket to create sub cooling in the crystallizers by an indirect contact method. In this case, crystals tend to nucleate on the cooling surface and decrease the heat transfer coefficient, causing encrustations. Therefore, mechanical scrapers are used to remove the crystal layer on the crystallizer inner surface. As for direct contact cooling, immiscible coolants, like air, water or sodium chloride solution, remove the heat and create sub cooling by contacting the melt directly. In this case, encrustation is avoided and the heat transfer area is increased enormously. However, gas distributors (for gas coolants) or separation apparatus (for liquid coolants) increase the complexity of crystallization setups. Meanwhile, the crystallizer volume will be bigger to hold the coolants. For both cooling methods, the crystal-melt residue separation process is important to guarantee the quality of products. Vacuum filtration, pressure filtration, centrifugation, as well as washing could be applied to further increase the purity of crystal products.

11.4.1.2 Inclined Column Crystallizer

As shown in Figure 11.3, an inclined column crystallizer[30,31] consists of three sections: crystallization section, purification section and melting section. The crystallization section has a jacket for the coolant to create sub cooling and generate crude crystals, while the melting section includes a melter to produce the uprising washing liquid and molten product. To better control the crystallization process, the purification section also needs circulation utility to prevent heat loss. A scraper with blades is rotated in the crystallization section to prevent encrustation on the column wall and enhance the countercurrent contacts between the crystalline particles and washing liquid. Inclined columns work well for systems which have slow settling velocities, resulting from a small density difference between solid particles and mother liquid and high viscosity of mother liquid, when the crystals settle by gravity force.

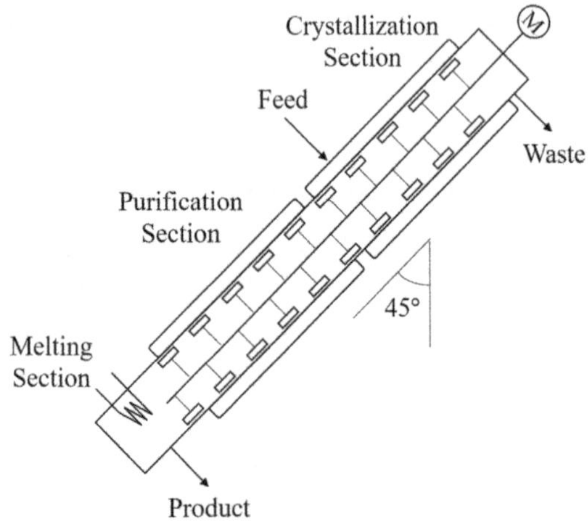

Figure 11.3 Inclined column crystallizer.

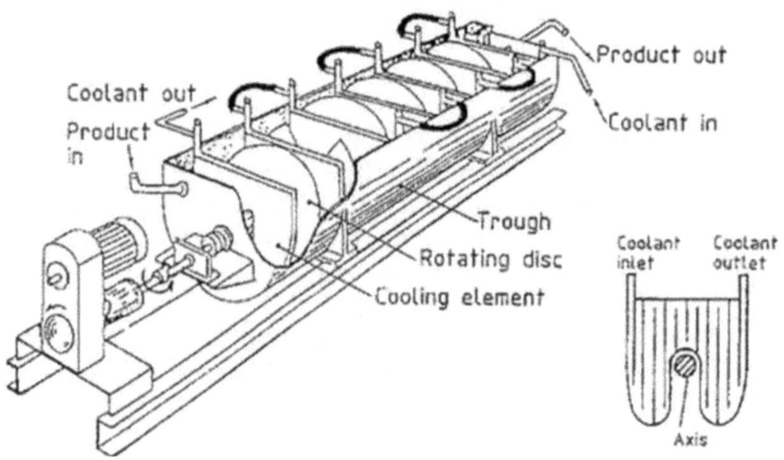

Figure 11.4 Cooling disk crystallizer. Reprinted from P. J. Diepen, O. S. L. Bruinsma and G. M. van Rosmalen, Crystal Size Engineering in Melt Suspension Crystallization, *Chemical Engineering Research and Design*, vol. 75, pp. 171–175, Copyright 1997, with permission from Elsevier.

11.4.1.3 Cooling Disk Crystallizer

A cooling disk crystallizer could be regarded as a cascade of several well mixed crystallizers with back mixing, as shown in Figure 11.4.[32] The horizontal crystallizer trough is divided into compartments by cooling elements and each of the compartments works like an MSMPR crystallizer. Slurries generated from the former compartment can continuously flow into the next one *via* the openings in the bottom of the cooling elements. By adjusting the

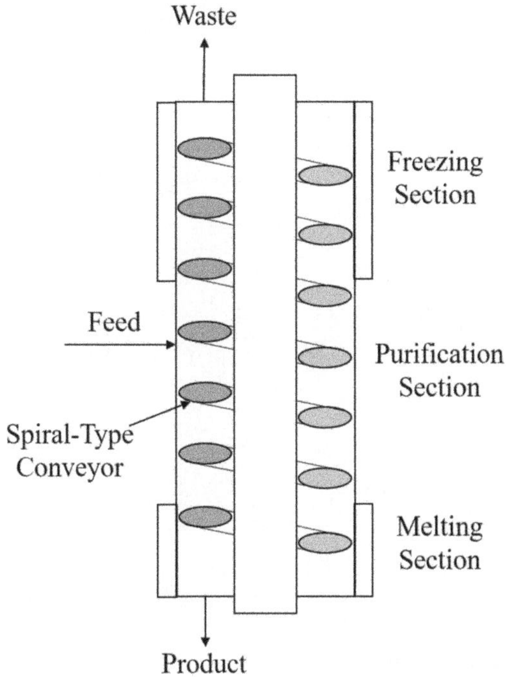

Figure 11.5 Schildknecht column.

size of openings and residence times, crystals with bigger size and higher permeability could be obtained and help to increase the efficiency of solid–liquid separation.

11.4.1.4 Schildknecht Column

Schildknecht developed a vertical column (Figure 11.5) for continuous melt suspension crystallization[1,33] and all the processes, including the crystal growth, countercurrent flow of solid and liquid and melting of the crystals, take place in the different parts of the same column. The column is annular in cross-section and the crystals are transported between the two tubes through a rotating helix, in the direction against the liquid. Openings on the helix could let the liquid flow through it while keep the crystals moving downwards. However, because of the difficulties concerning the complicated structures of helical coils, scaling-up of the Schildknecht column is difficult.

11.4.1.5 Philips Crystallizer

Philips crystallizers,[1,3,33,34] shown in Figure 11.6, were developed based on the continuous fractional crystallization concept, analogous to the distillation process, in the 1950s in America. In Philips crystallizers, crystals nucleate

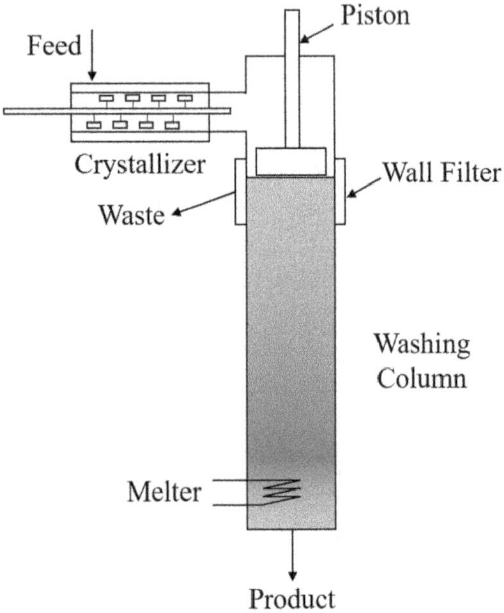

Figure 11.6 Philips crystallizer.

and grow in an external crystallizer with scrapers to prevent the crystals from encrustation. Then, the slurry is transported into the washing column and forced down by the reciprocating piston. The washing liquid reflux flows from the bottom (the melting zone) upwards counter currently through the crystal bed and out through the wall filter on the top of the column as the residue liquid. The product is collected at the bottom of the purifier. Pulse, instead of piston, could also be used to force the crystals to move downwards in the purifier.

11.4.1.6 Brodie Crystallizer

The Brodie crystallizer (Figure 11.7),[3,28,35] developed in 1971 in Australia, has a long, horizontal freezing section consisting of a set of cooling scraper crystallizers, which guarantees a long enough time to lower the crystal growth rate and get crystals as pure as possible. Besides, the feeding point is in the middle of crystallizing section and a countercurrent flow is formed between the crystals and the melt to avoid growing all crystals from the least pure melt residue. The slurry is transported by using a spiral conveyor to the vertical purification column and forms a countercurrent contact mass transfer process, with part of the molten product forming in the bottom melter. To avoid channeling and crystal caking

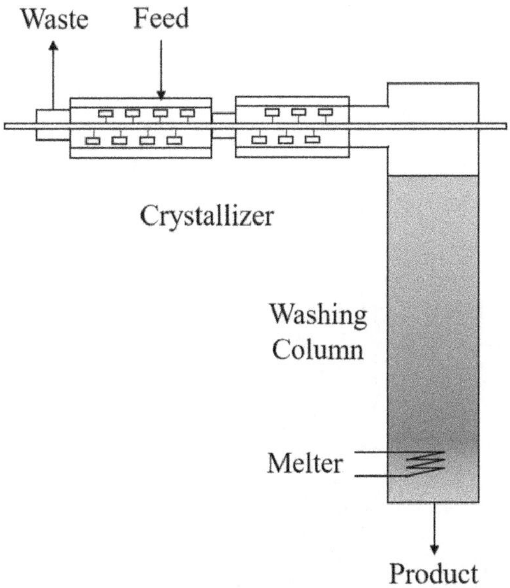

Figure 11.7 Brodie crystallizer.

phenomena, slow stirring is adopted in the purification column. As the first industrial melt crystallizer operated continuously in the world, the Brodie crystallizer has many advantages, including low operation cost, less safety problems, less capital and maintenance expenditure, higher purification effect, and so on. It can guarantee the purity of target materials if the raw materials concentration stays within 11%, which means a big operational flexibility.

11.4.1.7 TNO Purifier

The TNO purifier,[1,3] shown in Figure 11.8, was developed in the Netherlands in 1972. In addition to a countercurrent washing process, the mass transfer between liquid and solid is further promoted by the mechanical action on the crystals in the TNO purifier. Combinations of balls and sieve plates in the washing column could result in damaging, breaking, or grinding of the crystals. As a result, suspension of crystals with very different particle sizes will be obtained and Ostwald ripening processes happen, in which small crystals melt and big crystals grow. Therefore, impurities incorporated in the crystals could also be separated and the separation efficiency of per unit of length could be increased significantly. In the TNO purifier, slurry is fed in the middle of the column and the waste and product are taken out from the top and bottom of the column, respectively.

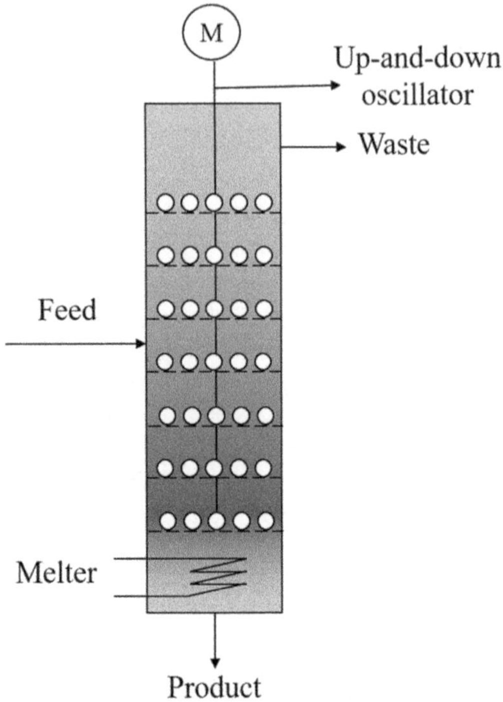

Figure 11.8 TNO purifier.

11.4.1.8 *Kureha Crystal Purifier (KCP)*

The Kureha crystal purifier[3,28,33,35] (KCP, shown in Figure 11.9) was developed in Japan in the 1960s. The crude crystals are formed in a subcooled crystallizer and filtered to be transported into the KCP Purifier from the bottom. There is a double screw with intermeshing blades in the column to force the crystals move upwards and prevent the crystal caking problem. In this purifier, the melter is on the top of the column and part of the molten crystals flow downwards counter currently through the rising crystals to finish the washing process. Crystal breakage and recrystallization processes could happen because of the double screw structure during crystal transportation, which could promote the purification efficiency. The waste melt comes out from the bottom while the purified product is harvested from the top of the column. Although a good separation effect is reported for organic systems, the complicated structure and difficult operation prevent its wide application.

11.4.1.9 *Brennan–Koppers Purifier*

The Brennan–Koppers purifier (Figure 11.10)[33] employs top melting like the KCP and wall filters like the Philips. The slurry is produced in a separate crystallizer and transported into the purifier from the bottom. Then, the

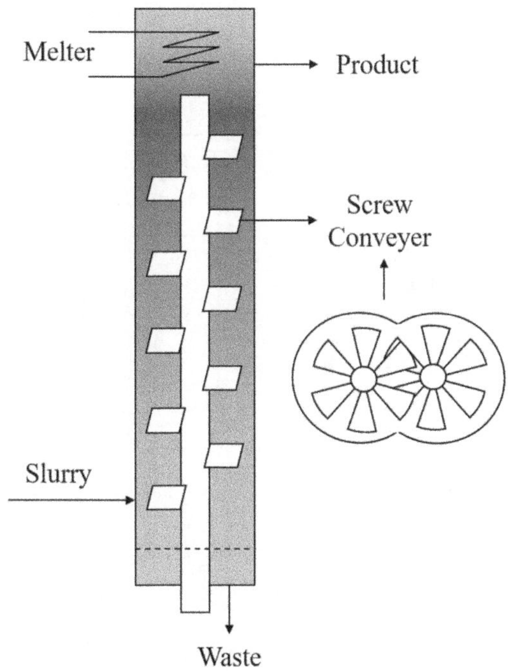

Figure 11.9 Kureha crystal purifier.

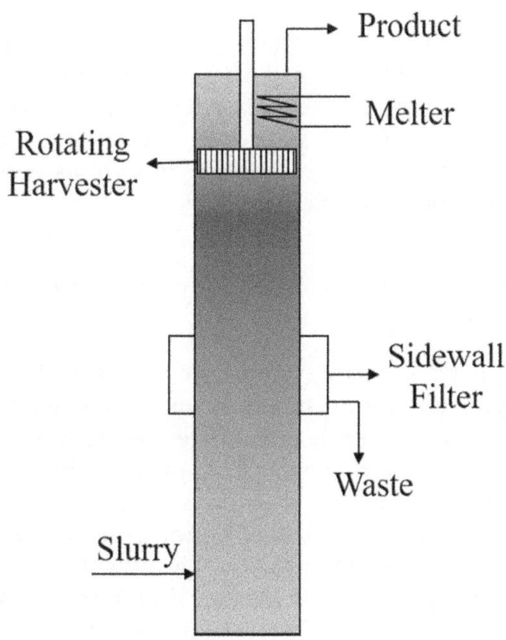

Figure 11.10 Brennan–Koppers purifier.

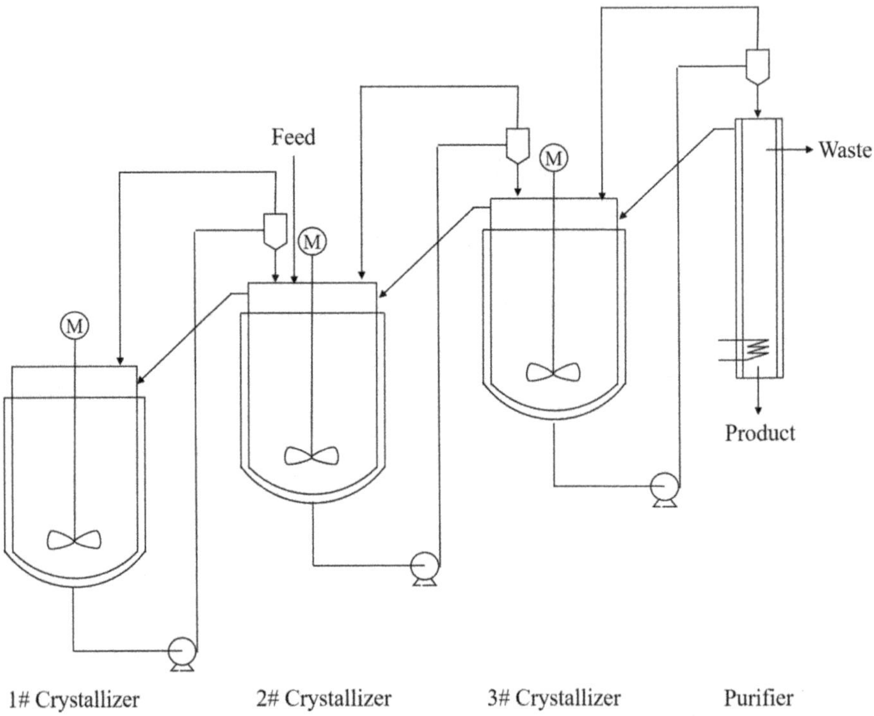

Figure 11.11 CCCC crystallizer.

crystals are dragged upwards by the flows to form the solid bed on the top. The top crystal layer could be scraped off by the harvester and conveyed to the melter. Similarly, part of the melt flows downwards to wash the upwards moving crystals and leaves from the sidewall filter. Results have shown that the movement of crystals has been positive and not as dependent on the crystal size as in some other purifiers.

11.4.1.10 Counter Current Cooling Crystallization (CCCC Crystallizer)

The CCCC crystallizer[3,28] was developed in Japan and commercialized in 1983. It is composed of 3 crystallizers and 1 purification column, as shown in Figure 11.11. In each crystallizer, stirrers with scrapers are employed to promote mass transfer and prevent encrustation on the inner supercooled surface. Feed flow of raw material is added into 2# crystallizer and mixed with the mother liquor from 3# crystallizer before undergoing the recrystallization process. The slurry is separated and the crystals go to 3# crystallizer, while mother liquor goes to 1# crystallizer. Crystals from 3# crystallizer are finally transported into the purification

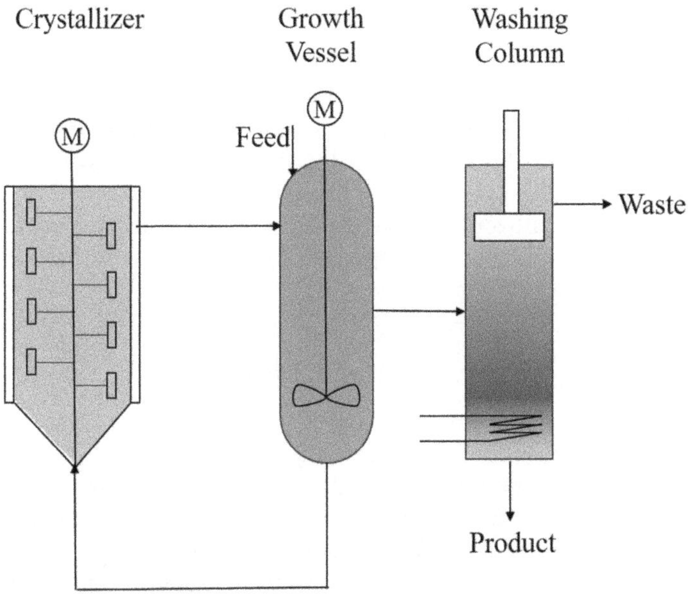

Figure 11.12 Sulzer Suspension Crystallizer.

column. Similarly, after the countercurrent washing process, product can be harvest at the bottom and the rising washing liquid could be added into 3# crystallizer from the top to increase productivity. Although CCCC crystallizers show a lower energy demand and can reach high purity of products, the control and operation system turn out to be difficult. So far, it has been used in the separation of organic isomers on an industrial scale.

11.4.1.11 Sulzer Suspension Crystallization Technology

The schematic Sulzer suspension crystallization process[12] is shown in Figure 11.12. The crystallization loop includes a crystallizer and a growth vessel. The feed flow is added into the growth vessel and circulated into the scraped crystallizer to generate crystals. Then, the mature crystals are transported into the piston type wash column. The suspension is compressed by the piston which has a filter screen to separate the melt residue and crystals. The residue leaves the system from the top of the wash column as waste, while the crystals are washed counter currently by the rising wash liquid coming from the melter. Purified crystals are scraped off from the bottom of wash column and melted in the melter as the final product. This process has been applied to a series of organics, such as acetic acid, carprolactam, ionic liquids, p-dichlorobenzene, phenol, and so on.

11.4.1.12 Sulzer Multiblok Suspension Melt Crystallizer

The Sulzer company developed the multiblok suspension melt crystallizer using a modular design concept with identical segments stacked on top of each other. The segments are either scraper ones, which prevent rapid crystal layer build-up and allow for higher loads, or mixer ones which effectively support radial and axial mixing. All the elements are mounted on one central shaft in the crystallizer. Modular design with standardized segments can increase the flexibility and plant capacity of the assembly with optimal engineering and fabrication costs.

11.4.1.13 Other Suspension Melt Crystallizers

Wang *et al.*[36] proposed a new multistage countercurrent melt crystallizer with sieve plates which combines the advantages of the TNO purifier and the inclined column crystallizer. Similar to normal column crystallizers, a stirrer with scrapers is used to prevent encrustation on the column walls and to enhance countercurrent contact between the crystals and reflux melts. Also, the crystallizer consists of three sections: crystallization, purification and melting. The modification includes several sieve plates fixed on the shaft to enhance heat and mass transfer, as well as crystal breakage in the purification section. However, unlike the TNO column, the sieves do not have balls on the trays. Besides, the sieves are inclined to facilitate the settlement of solid particles which includes the advantages of inclined column. Therefore, this new column crystallizer has better mass and heat transfer effects and can solve problems associated with solid particle sedimentation.

Zhong *et al.*[37] developed a melt crystallization process for the separation of *p*-xylene. After the slurry is formed using suspension crystallization, it is fed into a filtration and purification unit to obtain mother liquor, washing residue and pure *p*-xylene. The filtration and purification process could be accomplished by using a simulated moving bed or a combination of multiple moving beds to implement continuous feeding and discharging. The simulated moving bed implements functions of filtration, washing and melting by switching streams *via* a rotary valve, while the combination of multiple moving beds rotates the moving beds to switch streams. The application of melt crystallization and simulated moving beds or moving beds helps to lower the capital costs and make the operation easier for the separation of *p*-xylene.

11.4.2 Solid Layer Crystallization

An endless moving conveyer belt or rotating drum could provide the cooled surface for layer melt crystallization techniques continuously. Therefore, the normal falling film crystallization concept could be applied and modified to realize continuous operation. This section presents the continuous solid layer melt crystallizers based on moving belts and rotating drums. In addition, the zone melting method further increases the separation efficiency of both crystallization processes.

11.4.2.1 Crystallization on a Cooled Belt

A cooled conveyer belt could provide the sub cooling surface for continuous separation of melt mixtures.[38,39] As shown in Figure 11.13, the cooled belt has an angle with the ground and moves upwards, together with the crystal layer which could be scraped off after the countercurrent mass transfer process. The melt is added on the belt to flow downwards and is gradually crystallized on the belt, while uncrystallized residue leaves the system at the lowest part. By adjusting the belt speed and slope, optimal crystal productivity and purity combination could be obtained.

11.4.2.2 Crystallization on a Rotary Drum

In this apparatus, the rotating drum supplies the cooled surface for crystallization.[40,41] As shown in Figure 11.14, a small part of the drum is immersed in the melt mixture and the sub cooling degree is created by coolant inside the drum. As the cooled surface passes through the molten container, part

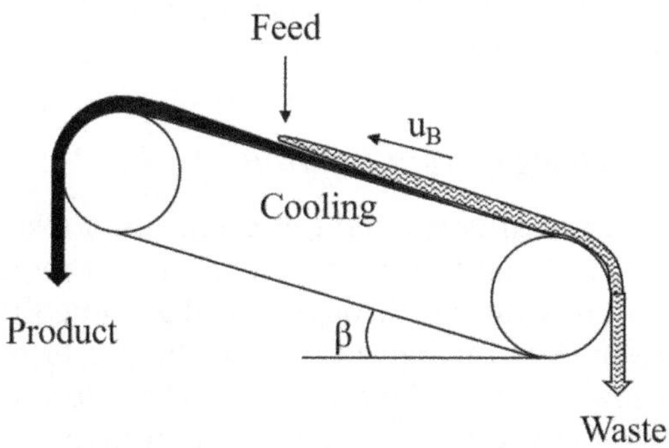

Figure 11.13 Crystallization on a cooled belt. Reprinted from ref. 38, Copyright 1998, with permission from Elsevier.

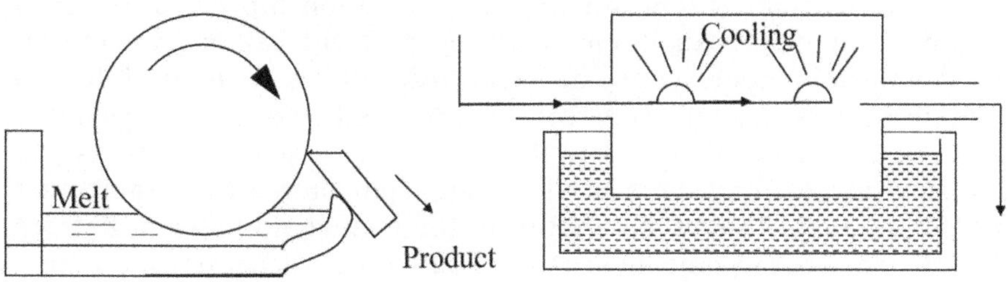

Figure 11.14 Crystallization on a rotary drum.

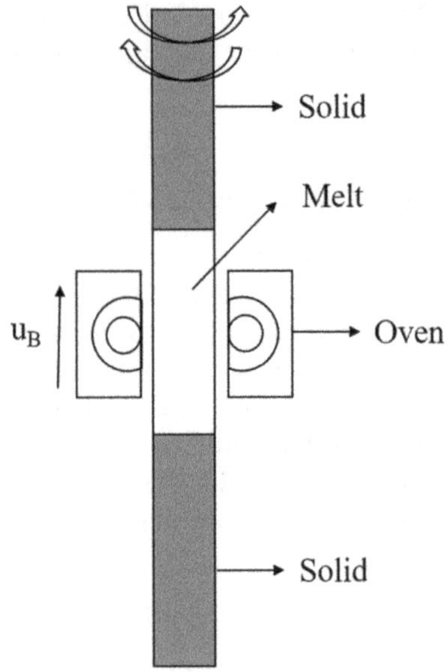

Figure 11.15 Zone melting crystallizer.

of the substance crystallizes on the drum and grows until the crystals leave the bath of melt. Products can be harvested by using a scraper and further purified by washing or sweating.

11.4.2.3 Zone Melting Crystallization

Zone melting[2,42,43] is a multistage melt crystallization technique based on a repeated formation of solid layers in a continuous mode. As shown in Figure 11.15, the oven is moved through the glass tube filled with sample mixtures. The substance in the part of the tube heated by the oven is in the molten state and could be mixed effectively by the intermittent rotation of the tube. Other parts of the tube are cooled down until the substances recrystallize. As the molten substance leaves the oven, purer crystal layer forms and the impurities are forced to move upwards. By repeating the process, purity of the samples gets higher and higher.

In real continuous zone melting crystallization processes, the zone melting technique is usually employed together with belt or drum crystallizers, shown in Figure 11.16 and 11.17, respectively. With multiple zone melting heaters, crystal layers formed from the belt or drum crystallizer could be further purified before they can be harvested using the scrapers. However, as for materials with high vapor pressures at their melting points, rapid loss of material from the melt could happen. Besides, the efficiency of the zone melting method strongly depends on the mass transfer between the liquid at the freezing interface and the liquid in the

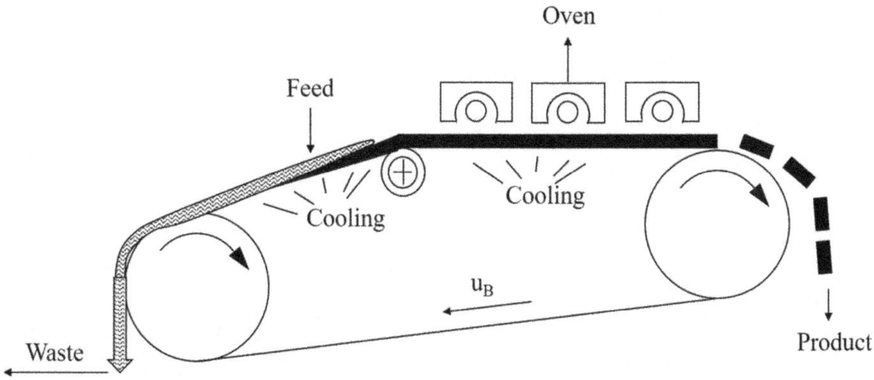

Figure 11.16 Belt crystallizer together with zone melting purifier. Reprinted by permission of Springer Nature from ref. 43, Copyright 2015.

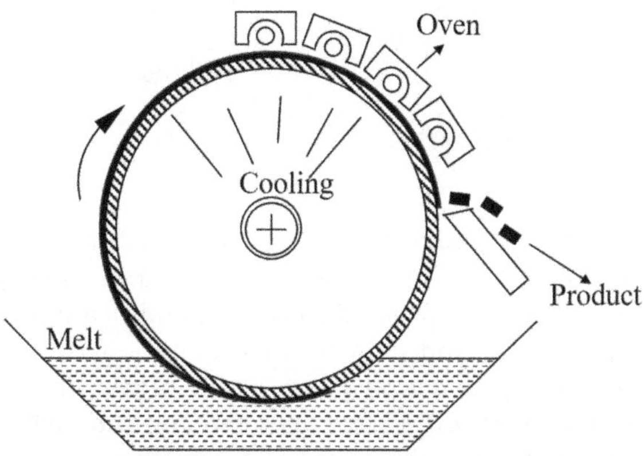

Figure 11.17 Drum crystallizer together with zone melting purifier. Reprinted by permission of Springer Nature from ref. 43, Copyright 2015.

bulk zone, which means that additional mixing techniques should be adopted for systems with small mass transfer coefficients, or slower moving speed of the oven or crystals should be used to guarantee the purity of product.

11.4.3 Other Crystallization Methods

11.4.3.1 Pastille Crystallization Method

The pastille crystallization method[34,44] combines the suspension and solid layer melt crystallization methods. It includes a granulation process to make the melt droplets fully solidified on the subcooled belt and a purification process in countercurrent washing columns, as shown in Figure 11.18. In

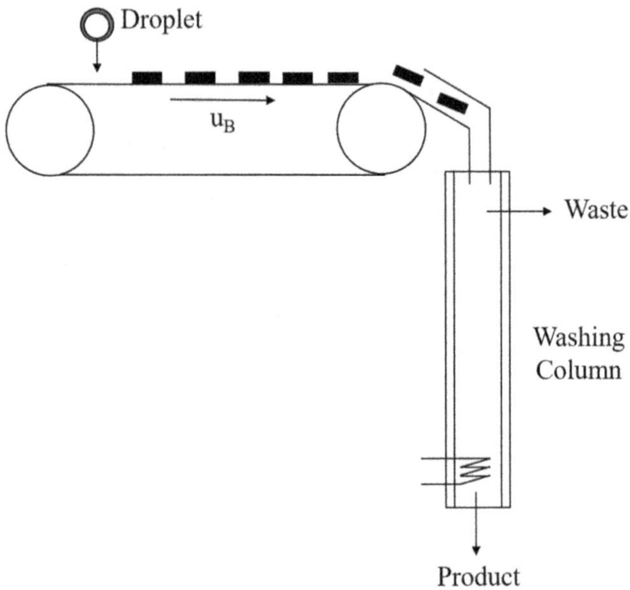

Figure 11.18 Pastille crystallization process.

this case, pastilles which have similar size could help to improve the flow conditions and promote washing efficiency in the purification process afterwards. Besides, crystals grow faster during the solidification process under higher sub cooling degree and help to increase production capacity. Also, with an easily realized granulation process and better washing effect, the pastille crystallization method can be operated continuously. However, the granulation device could be complicated and expensive, which will increase the capital cost.

11.4.3.2 Eutectic Freeze Crystallization (EFC)

The eutectic freeze crystallization method (EFC),[45–47] a double crystallization method (generates two kinds of crystals simultaneously), was firstly proposed for brine disposal in the 1970s. The basic principle is that when the aqueous solution is cooled slowly, both ice and salt will crystallize out due to the eutectic property of the system, and they can be easily separated because of the huge density difference between salt and ice. In this case, the eutectic composition is no longer the separation limit for freeze (melt) crystallization. This concept could also be used for a ternary or multi-component system. Hengstermann[9] found that the addition of ice to an acrylic acid bed could optimize the properties of crystal bed and improve the performance of the solid–liquid separation process significantly.

11.5 Applications of Continuous Melt Crystallization

With the specific advantages of melt crystallization mentioned in Section 2, as well as the merits of continuous processes (including stable product quality, higher productivity, higher automation, *etc.*), intensive investigations have been devoted to the applications of continuous melt crystallization. In this section, processes for separation of organic mixtures, production of ultra-pure products and concentration are summarized, separately.

11.5.1 Separation of Organic Mixtures

Isomers and organic compounds with similar structures and properties could be separated by using the continuous melt crystallization method advantageously. For example, Kim *et al.*[8] applied the direct contact cooling method to separate *n*-dodecanol and *n*-decanol. Results showed that the obtained crystals were of good quality and significantly larger than those from scraped crystallizers. The separation effect was higher than that of film crystallization in terms of purity and production. 99.97% (mass purity) p-dichlorobenzene could be obtained from 87% feed, with an annual production capacity of 7000 metric tons using a Kureha purifier.[33] By using a cooling disk crystallizer, Diepen *et al.*[32] investigated the influence of different operation conditions on the permeability of caprolactam products, obtained from caprolactam-water mixtures. Misztal[48] designed a continuous MSMPR crystallizer which could remove the fines from melt crystallization of caprolactam and developed an algorithm to calculate the crystal size and its distribution.

11.5.2 Production of Ultra-pure Inorganic Products

Phosphoric acid, one of the most essential chemical materials, is widely used in the manufacture of pharmaceuticals, foodstuffs, semiconductor elements and in other fields. Because of the high efficiency and low operation cost, crystallization has been one of the preferential options for the purification of phosphoric acid. Chen *et al.*[7] applied a continuous melt suspension crystallization process which includes a MSMPR crystallizer and a gravity washing column to purify phosphoric acid and found the operation window to produce food grade products. In addition, solid layer crystallization has been successfully used for the purification of inorganic materials, such as silicon and germanium. Boutin *et al.*[41] applied a continuous fractional crystallization on a rotary drum and a zone melting process to purify metals. By adjusting the relative position and distance between the drum and the bottom of the trough, the apparatus could be used to produce pure aluminum and protect it from oxidation during the melt crystallization.

11.5.3 Concentration

Continuous melt crystallization could also be used to concentrate juice, beer and desalinate sea water by freezing. Margolis *et al.*[6] studied the desalination process in a well stirred continuous crystallizer to produce ice by direct contact refrigeration. By investigating the nucleation and growth kinetics as well as the permeabilities of ice beds, they found that larger ice production rates per unit crystallizer volume could be achieved without sacrifice of crystal size. Huige *et al.*[49] investigated the nucleation and growth kinetics of ice from dextrose solutions in a continuous stirred tank crystallizer with supercooled feed. Results showed that supercooling is the key factor to influence the crystallization kinetics, while crystal surface area also played an important role in the crystal nucleation process, which helped to verify that nucleation occurs by breakage of dendrites from the surface of parent crystals.

11.6 Outlook

As a green separation technique, continuous melt crystallization has drawn intense attention for the purification of fine chemicals, food, pharmaceuticals, inorganic materials, and so on. However, current work has mainly focused on the feasibility or separation effectiveness of given mixtures or materials. Further theoretical research has not been fully conducted. To fully understand and better apply continuous melt crystallization, more work needs to be done on the following aspects.[1,5,34,50–52]

1. Further understanding of the mechanisms of the melt crystallization process, such as the combined mass and heat transfer processes. Establishment of common mathematical models of melt crystallization will help to predict the crystallization results and guide the industrialization process.
2. Development of new separation techniques, including better solid–liquid separation (application of magnetic fields, new slurry transportation methods, *etc.*) and hybrid separation methods which combine melt crystallization and distillation, extraction, *etc.* to deal with low concentration systems and take advantage of different separation methods.
3. Development of effective crystallizers which are easy to scale-up and have higher separation effects. Currently, most of the commercial continuous melt crystallizers include complicated internal components which make the scale up of laboratory or pilot apparatus difficult. Besides, the maintenance cost is also quite high for industrial processes.
4. Energy integration technology to save energy costs, especially for systems with low melting points. As for materials with low melting points, refrigeration costs could increase operation fee profoundly. Cascade

utilization of energy would widen the application of the melt crystallization separation technique.

Roman Symbols

a	order of crystal growth (/)
b	cooling rate (Ks^{-1})
B^0	numerical nucleation rate ($1/(m^3\ s)$)
C	product component concentration (kg kg^{-1})
C_0	initial concentration (kg kg^{-1})
C_e	equilibrium concentration (kg kg^{-1})
C_{imp}	impurity concentration (kg kg^{-1})
$C_{imp,\ melt}$	average concentration of impurity (kg kg^{-1})
C_{melt}	average concentration of product component at distance x (kg kg^{-1})
Δc_{max}	metastable zone width (mol m^{-3})
C_r	concentration at crystal layer surface (kg kg^{-1})
c*	equilibrium concentration (mol m^{-3})
D	molecular diffusion coefficient ($m^2\ s^{-1}$)
F	flow rate of melt ($m^3\ s^{-1}$)
G	growth rate (m s^{-1})
K	distribution coefficient (/)
k_G	growth rate constant (m s^{-1})
k_i	crystal growth parameter (m s^{-1})
k_{melt}	thermal conductivity of the solution (W/(m K))
k_n	nucleation rate constant (kg/(m s))
L	length of cooled surface (m)
m	apparent order of crystal nucleation (/)
n	number of crystals per unit volume and size differential (eqn (11.4)) (m^{-4})
n	crystal growth parameters (/)
n^0	population density of nuclei (m^{-4})
N_m	mass nucleation rate (kg/(kg s))
r	radial distance (m)
r_f	total thickness of solid and melt layers (m)
r_s	thickness of solid layer (m)
T	temperature (K)
T_f	temperature of melt surface (K)
T_m	cooled surface temperature (K)
T_s	temperature of solid layer surface (K)
ΔT_m	maximum allowable supercooling (K)
U	overall heat transfer coefficient ($W/(m^2\ K)$)
$u(r)$	fluid velocity (m s^{-1})
V_{tank}	volume of feeding tank (m^3)
Δw	supercooling (kg kg^{-1})
x	axial distance (m)

Greek Symbols

α	thermal diffusivity ($m^2\ s^{-1}$)
τ	residence time (s)
ρ	density of crystal ($kg\ m^{-3}$)

References

1. G. J. Arkenbout, *Sep. Purif. Methods*, 1978, **7**, 99–134.
2. J. Ulrich and T. Stelzer, *Melt Crystallization*, Wiley-VCH Verlag GmbH & Co. KGaA, Germany, 2013.
3. G. Fu, *Purifying of PNCB and MNCB by Melt Crystallization*, Tianjin University, 2005.
4. J. Wang, *Chemical Engineering Handbook*, Chemical industry press, Beijing, 1996.
5. S. He, *Study on Separation of Amine Compounds by Melt Crystallization*, Qingdao University of Science & Technology, 2012.
6. G. Margolis, T. K. Sherwood, P. L. T. Brian and A. F. Sarofim, *Ind. Eng. Chem. Fundam.*, 1971, **10**, 439–452.
7. A. Chen, J. Zhu, K. Chen, B. Wu, L. Ji and Y. Wu, *Asia-Pac. J. Chem. Eng.*, 2013, **8**, 354–361.
8. K.-J. Kim, J.-M. Lee and A. Mersmann, *Sep. Sci. Technol.*, 1996, **31**, 1859–1875.
9. A. Hengstermann, *A New Approach to Industrial Melt Crystallization of Acrylic Acid*, Delft University of Technology, 2010.
10. P. J. Jansens and M. Matsuoka, *Melt Crystallization*, Academic Press, 2000.
11. GEA, http://www.gea.com.
12. Sulzer Chemtech, http://www.sulzer.com.
13. X. Feng, Z. Li and Y. Wang, *Fine Spec. Chem.*, 2012, **20**, 42–45.
14. Q. Li and Y. Zhang, *J. Beijing Univ. Chem. Technol.*, 1997, **24**, 1–8.
15. K. J. Kim and A. Mersmann, *J. Ind. Eng. Chem.*, 1999, **5**, 204–211.
16. K. Bartosch and A. Mersmann, *Chem. Eng. Sci.*, 2001, **56**, 2347–2356.
17. K.-J. Kim and A. Mersmann, *Chem. Eng. Commun.*, 1999, **172**, 125–149.
18. J. Ulrich and H. C. Bülau, in *Handbook of Industrial Crystallization*, Butterworth-Heinemann, Woburn, 2nd edn, 2002, pp. 161–179.
19. S. W. Gilbert, *AIChE J.*, 1991, **37**, 1205–1218.
20. X. Jiang, B. Hou, G. He and J. Wang, *Chem. Eng. Sci.*, 2013, **91**, 111–121.
21. X. Jiang, W. Xiao and G. He, *Chem. Eng. Sci.*, 2014, **117**, 198–209.
22. N. Yazdanpanah, S. T. Ferguson, A. S. Myerson and B. L. Trout, *Cryst. Growth Des.*, 2016, **16**, 285–296.
23. N. Yazdanpanah, A. Myerson and B. Trout, *Ind. Eng. Chem. Res.*, 2016, **55**, 5019–5029.
24. L. Zhou, M. Su, B. Benyahia, A. Singh, P. I. Barton, B. L. Trout, A. S. Myerson and R. D. Braatz, *AIChE J.*, 2013, **59**, 1308–1321.

25. X. Jiang, B. Hou, G. He and J. Wang, *Chem. Eng. Sci.*, 2012, **84**, 120–133.
26. K. Wangnick and J. Ulrich, *Cryst. Res. Technol.*, 1994, **29**, 349–356.
27. J. Wang, *Chem. Eng. (China)*, 1992, 57–63.
28. S. Song, *The Crystallization Purification Process and Mathematical Model of para-Xylene,* Xiangtan University, 2016.
29. X. Ding, T. Li and J. Zhu, *Appl. Chem. Ind.*, 2012, **41**, 1430–1433.
30. M. Matsuoka, H. Takiyama and O. Soutome, *Chem. Eng. Res. Des.*, 1997, **75**, 206–212.
31. K. Funakoshi, H. Uchida, H. Takiyama and M. Matsuoka, *J. Cryst. Growth*, 2002, **237–239**, 2251–2256.
32. P. J. Diepen, O. S. L. Bruinsma and G.M. van Rosmalen, *Chem. Eng. Res. Des.*, 1997, **75**, 171–175.
33. J. R. Couper, W. R. Penney, J. R. Fair and S. M. Walas, in *Chemical Process Equipment*, Elsevier, 2012, pp. 561–589.
34. Q. Li and W. Zhang, *Chem. Prod. Technol.*, 2013, **20**, 29–31.
35. S. Li, W. Qin and J. Wang, *Chem. Equip. Technol.*, 1994, 1–4.
36. S. Wang, G. Zhao, Y. Du and Y. Qu, *Korean J. Chem. Eng.*, 2015, **32**, 1151–1157.
37. L. Zhong, J. Xiao, Y. Lu, Y. Guo and D. Kong, *US Pat.*, 8293963, 2012.
38. P. Ossipov, *Int. J. Heat Mass Transfer*, 1998, **41**, 691–697.
39. P. Ossipov, *Appl. Math. Modell.*, 1999, **23**, 419–436.
40. Y. Shao, *Research on the Separation and Purification of Phosphoric Acid by Continuous Layer Melt Crystallization,* East China University of Science and Technology, 2016.
41. F. R. Boutin, *US Pat.*, 4581062, 1986.
42. W. R. Wilcox, R. Friedenberg and N. Back, *Zone Melting of Organic Compounds*, Blackwell, New York, 1963.
43. Y. A. Taran, A. Y. Kholin, A. V. Taran and A. L. Taran, *Coke Chem.*, 2015, **58**, 353–361.
44. B. Wang, *Purification of* p-*Dichlorobenzene by Direct Contact Cooling Solidification,* Nanjing University of Technology, 2004.
45. P. M. Williams, M. Ahmad, B. S. Connolly and D. L. Oatley-Radcliffe, *Desalination*, 2015, **356**, 314–327.
46. F. van der Ham, G.J. Witkamp, J. de Graauw and G.M. van Rosmalen, *J. Cryst. Growth*, 1999, **198–199**, 744–748.
47. H. J. M. Kramer and P. J. Jansens, *Chem. Eng. Technol.*, 2003, **26**, 247–255.
48. S. Misztal, *Chem. Eng. Trans.*, 2013, **32**, 2101–2106.
49. N. J. J. Huige, M. M. G. Senden and H. A. C. Thijssen, *Cryst. Res. Technol.*, 1973, **8**, 785–801.
50. S. Sun, G. Yang, J. Ding and Y. Du, *Chem. Eng. (China)*, 2008, **36**, 18–20.
51. H. Liu and Q. Yin, *Chem. Ind. Eng.*, 2004, **21**, 367–371.
52. M. Gu, *Shanghai Chem. Ind.*, 2003, 26–29.

CHAPTER 12

Continuous Enantioselective Crystallization of Chiral Compounds

HEIKE LORENZ*[a], ERIK TEMMEL[a] AND ANDREAS SEIDEL-MORGENSTERN[a,b]

[a]Max Planck Institute for Dynamics of Complex Technical Systems, Sandtorstrasse 1, D-39106 Magdeburg, Germany; [b]Otto von Guericke University, Institute of Process Engineering, Universitätsplatz 1, D-39106 Magdeburg, Germany

*E-mail: lorenz@mpi-magdeburg.mpg.de

12.1 Introduction

Enantiomers are stereoisomers that are non-superimposable mirror images of each other. They are designated by classical notations as D or L, as (*R*) or (*S*) or as (+) or (−).[1] There is still debate as to why life on earth is constructed using essentially L-amino acids as building blocks. This fascinating and unsolved problem continues to stimulate a large spectrum of research activities.[2–5] Moreover, there is tremendous interest in producing pure enantiomers in the food, cosmetic and agrochemical industries, and in particular in manufacturing pharmaceuticals.[6–8] There is clear evidence regarding chiral drugs that often only one enantiomer provides the desired physiological effect. In many cases, the other enantiomer has

The Handbook of Continuous Crystallization
Edited by Nima Yazdanpanah and Zoltan K. Nagy

Published by the Royal Society of Chemistry, www.rsc.org

no effect or is even harmful. Regulators increasingly demand that chiral drugs are administered in optically pure form.[9–11] This requirement triggered intensive efforts from industrial and academic research devoted to developing techniques which are capable of producing pure enantiomers. The utilized methods can be divided into the "chiral approach", which is based on asymmetric synthesis of only the desired enantiomer and the "racemic approach", which is based on separation of mixtures of enantiomers.[12–14] The latter approach is considered below, while focusing on exploiting crystallization-based resolution methods, and, in particular, on applying the concept of preferential crystallization (PC), which is well suited to the development and implementation of continuous separation processes.

Direct crystallization of an enantiomer from a racemic solution is only feasible when the enantiomers in their mixtures form separate but pure crystals. This requires that the corresponding racemates are compositionally conglomerates, *i.e.* mixtures of the crystals of both enantiomers.[15] Such cases are the focus of this report. PC is an attractive approach to resolve mixtures of conglomerate forming chiral molecules. It allows for direct crystallization of the desired enantiomer without the need for any chiral auxiliary, which is typically required in case of exploiting other types of phase equilibria. PC is a kinetically driven separation that relies on different crystallization rates of the enantiomers in the presence of homochiral seeds. Such entrainment processes are usually applied for smaller scales exploiting the conventional batch mode. Examples of industrial use concern, in particular, the production of broad-spectrum antibiotics, such as chloramphenicol, thiamphenicol, and β-lactames.[15] Simultaneous crystallization is an option for larger scales. In this case, the enantiomers crystallize simultaneously from solutions, which are spatially separated and remain controlled close to the racemic composition. The process can be performed both in batch and continuous modes and has been industrially realized, *e.g.* for the production of (−)-menthol (Haarmann & Reimer) and an L-α-methyldopa intermediate (Merck). Ajinomoto produced L-glutamic acid at about 13 000 t/a over a period of ten years by simultaneous crystallization.[16] Various continuous resolution processes using different configurations of stirred tanks, tubes and fluidized bed systems were described in a monograph published in 2009, emphasizing the application of fluidized bed crystallizers and the implementation of ultrasonic cleavage of seed particles to achieve a steady state process regime.[17] Advantages of PC were illustrated successfully by resolving racemic calcium pantothenate (with the (*R*)-enantiomer being an important commercial precursor of vitamin B5) and an industrially required resolving agent, both for laboratory and technical scales (35 L and 3000 L).[18,19] More detailed information on PC can be found in several review articles (*e.g.* ref. 15, 20 and 21), book contributions[16,22] and the following sections.

12.2 Phase Equilibria of Chiral Systems

The specific solid–liquid phase equilibria (SLE) and the corresponding phase diagrams of a system form the fundamental thermodynamic basis to design crystallization processes. Detailed SLE data provide the information necessary to answer the question of which solid-state form of a target product can be crystallized under certain conditions (*i.e.* a certain polymorph or solvate) together with information regarding yields and purities achievable.

In binary systems one can distinguish between two groups of phase diagrams regarding solid–liquid equilibria: systems showing (1) immiscibility and (2) miscibility in the solid state. The first group can be further divided into systems exhibiting eutectic behaviour or the formation of intermediate molecular compounds. It is by far the most prominent group, representing about 86% of the organic systems, with the majority (*ca.* 54%) being simple eutectic systems.[23] The second group, systems with full miscibility in the solid state, is less frequently observed within organic systems (about 14%).[23] In addition, limited (partial) miscibility can occur in all types of group (1) systems. Concerning chiral systems, the frequency at which different phase diagrams occur is influenced by the unique similarity of the two chiral molecules and their mirror-image symmetry in particular. Eutectic systems are found in only 5 to 10% of the cases.[15] In 90 to 95% of the systems the two enantiomers form a 1 : 1 racemic compound, represented by an additional two-phase domain in the corresponding ternary solubility phase diagrams (Figure 12.1). Although full miscibility in the solid state is rare for chiral systems, limited miscibility can occur and is increasingly reported and analysed within the last years.[24–30] Figure 12.1 illustrates ternary solubility phase diagrams of a eutectic and a racemic compound-forming chiral system in a certain solvent as well as a eutectic system with limited miscibility in the solid state. An example of the latter case is described for the ethanolamine salt of 3-chloromandelic acid.[25] It should be noted that the phase equilibria depicted above can be much more diverse due to, for example, solid–solid or liquid–liquid equilibria that interfere with the SLE (*e.g.* polymorphy,[31] liquid–liquid demixing[32]) and the formation of non-equimolar discrete compounds of the enantiomers.[33]

The separation of enantiomers can be considered as a special case of purification by crystallization, because the counter-enantiomer can occur in the same amount as the target enantiomer and the two molecules to be separated show extraordinary similarity. In fact, thermodynamic and kinetic properties are identical, *e.g.* the melting points, the solubility and crystallization characteristics.

Due to these similarities, resolution of the racemates by crystallization under equilibrium conditions is not feasible. Resolution is possible, however, through the aforementioned preferential crystallization (PC) method: a kinetically controlled separation process relying on the

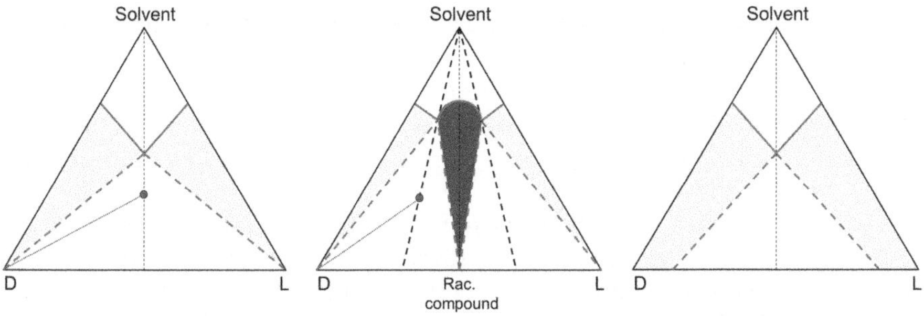

Figure 12.1 Schematic ternary solubility phase diagrams of a system of two enantiomers D and L in a solvent at constant temperature characterized by a simple eutectic (left), a racemic compound (middle) and eutectic behaviour with limited miscibility in the solid state (right). Light-grey colour marks two-phase domains where pure enantiomer crystals (left and middle) or solid solutions (right) are in equilibrium with saturated solution. White area(s) below the two-phase domains indicate three-phase domains where PC can be performed. There, the two enantiomers (left) or one enantiomer (D in the middle figure) and the racemic compound as solid phases and a liquid phase containing the enantiomers in racemic/eutectic composition are present in equilibrium. Points indicate potential exploitable starting compositions for PC.

different crystallization rates of the enantiomers in the presence of homochiral seeds. Fundamental knowledge of solid–liquid phase equilibria and crystallization kinetics as well as application of suitable inline/online process monitoring techniques facilitate robust, reproducible and successful application of PC. The essential limiting requirement for implementation of this method is eutectic behaviour of the compounds to be separated. As illustrated in Figure 12.1, PC can be applied as a direct crystallization method for conglomerates of chiral compounds (left figure) and for eutectic mixtures of one enantiomer (*e.g.* D indicated in Figure 12.1, middle) and the racemic compound. Thus, a racemic conglomerate may be directly resolved, providing pure enantiomers, while in a racemic compound-forming system, an enantiomerically enriched mixture may deliver a desired enantiomer by this method. In this and the following figures, the enantiomers are assigned with D and L as applied for amino acids, which are also the example systems we will consider below for process demonstration.

As mentioned above, there are in principal two basic techniques of PC: (1) the entrainment technique and (2) the technique of simultaneous crystallization, both applicable for separation from the melt and from solution. In entrainment, the enantiomers (or in general two components) crystallize consecutively from the supersaturated solution, significantly exceeding the eutectic composition, thus “entraining” the crystallizing enantiomer. In simultaneous crystallization, the enantiomers (or two components) crystallize simultaneously but are spatially separated while the solution permanently remains close to the initial eutectic composition. As shown in Section 12.5, this technique has large potential in the fine chemical and life-science industries for producing *e.g.* optically active pharmaceuticals, food additives and flavouring agents, in particular if used in a continuous mode.

As classical sources limit the occurrence of conglomerates to about 10%, several activities are currently directed toward developing methods for faster recognition of conglomerates as well as predicting potential conglomerates within larger families of derivatives of a target chiral system.[34] It is expected that these efforts will significantly increase the number of conglomerates available, which are then promising candidates for the application of continuous variants of direct crystallization.

The principle of racemate resolution by PC is shown in Figure 12.2 through illustration of the ternary solubility phase diagram of a conglomerate system. When starting from a racemic composition in the three-phase region of the ternary phase diagram (gray point) and adding homochiral seeds (*i.e.* seeds of only one of the enantiomers, *e.g.* the L-enantiomer), for a certain period of time, only the seeded enantiomer selectively crystallizes. The crystallization process follows a trajectory along the extension of the line connecting the crystallizing L-enantiomer and the start composition. The crystallization process has to be terminated and the crystalline product harvested by filtration before reaching the line of metastable solubility (black dashed line). Otherwise, as illustrated by the turning PC trajectory, the counter-enantiomer will nucleate and the system will approach equilibrium, where both the solid and

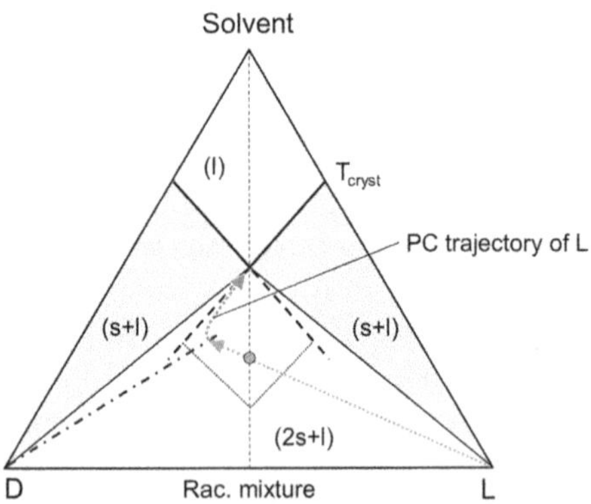

Figure 12.2 Solubility phase diagram of D and L enantiomers in a solvent and illustration of the PC principle. The trajectory of a PC process, performed at a crystallization temperature, T_{cryst}, providing a single enantiomer (here L) is shown as grey line with arrows. Black dashed lines are lines of metastable solubility of the enantiomers and mark the crystallization limits for successful PC.[16,22] In each characteristic region of the phase diagram, the state of aggregation of the equilibrated phases is depicted (s: solid, l: liquid).

liquid phases are of racemic composition. The PC process described can be performed also in a cyclic mode, *i.e.* successively, preferentially crystallizing the L- and D-enantiomers in an alternating manner with addition of racemic feed and the respective seed crystals in-between (see later Figure 12.8). Such a cyclic approach can also be applied to racemic compound-forming systems, in which one of the enantiomers and the racemic compound are alternately crystallized.[35,36]

More details and examples regarding various batchwise operated PC process variants are available in a review article.[21] Several process variants capable of performing PC in batch and continuous modes are introduced in Section 12.4. In Section 12.5 we will present case studies of continuous PC for the conglomerate systems of the two amino acids threonine and asparagine monohydrate.

12.3 Preferential Crystallization: Kinetics, Driving Forces and Metastable Zones

Crystallization kinetics are of essential importance for the resolution of chiral systems exploiting PC. They identify and mark possible regions for performing the kinetically controlled resolution process within a three-phase region of the phase diagram. The PC process illustrated in Figure 12.2 is redrawn in Figure 12.3 (left). After a certain process time, t, the seeded enantiomer has grown and in parallel the solution is enriched in its antipode. A current liquid phase composition, $x_{i,0}(t)$, is marked by a gray dot.

The course of the initially straight trajectory becomes curved after nucleation of the counter-enantiomer and subsequent growth of the nuclei (grey dotted line in Figure 12.3). In equilibrium a racemic composition is reached again for both the liquid ($x^{\mathrm{DL}}_{\mathrm{sat,L}} = x^{\mathrm{DL}}_{\mathrm{sat,D}}$, Figure 12.3) and the solid phases. Thus, after nucleation of the antipode, the crystallization kinetics of the unseeded component becomes faster to deplete the established enrichment. The differences in the kinetics are based on the two different and constantly changing driving forces, connected to the specific, actual chemical potentials of the particular enantiomer in the current state of the system (*e.g.* at $x_{i,0}(t)$) and the two corresponding saturated states (at $x_{\mathrm{sat,D}}$ and $x_{\mathrm{sat,L}}$, Figure 12.3). These saturated states are related to the intersection of the lines starting at the pure phase corners and going through the actual state ($\overline{\mathrm{L}x_{i,0}}$ and $\overline{\mathrm{D}x_{i,0}}$) and the extensions of the solubility isotherms for the applied temperature (dashed lines in Figure 12.3). It should be noted that the ratio between the mass fractions of the respective opposite enantiomer and the solvent is constant along these lines (isopleths, dotted grey and dotted black lines in Figure 12.3), which requires consideration in the calculation of the driving force. The dynamically changing specific supersaturations $S_i(t)$ can be expressed for polythermal conditions in a general way by eqn (12.1), where the mass (or molar) compositions of the states are used instead of the chemical potentials, which is typically sufficiently accurate for engineering purposes.

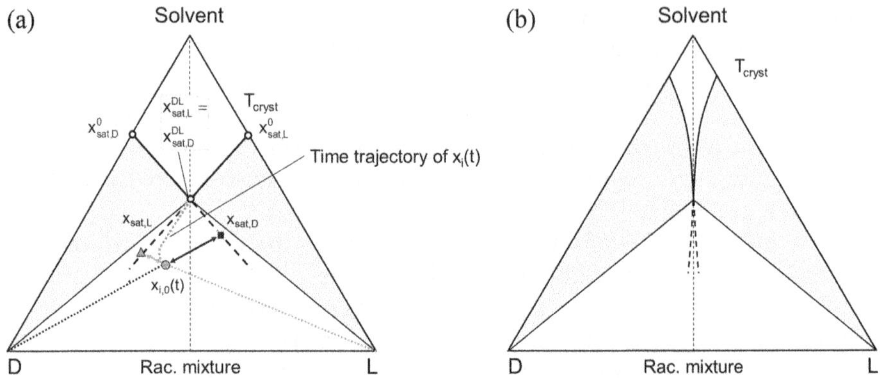

Figure 12.3 Ternary phase diagram of two enantiomers D and L in a solvent. (a): Illustration of PC after seeding with L. A specific actual composition on the indicated time trajectory $x_i(t)$ is marked as $x_{i,0}(t)$ (dot). The current driving force for phase transfer into the solid phase (solid arrows) is depicted for each enantiomer as the intersection of the lines $\overline{\text{L}x_{i,0}}$ and $\overline{\text{D}x_{i,0}}$ with the corresponding dashed prolongations of the solubility isotherms valid for the actual crystallization temperature. (b): Illustration of another possible shape of the phase diagram, where the solubility of one enantiomer is significantly increased in presence of the other, which causes steeper isotherms and reduces the region exploitable for PC. In addition, non-linearity of isotherm lines is assumed caused by potential non-idealities in the system.

$$S_i(T(t),t) = \frac{x_{i,0}(t)}{x_{\text{sat},i}\left(x^0_{\text{sat},i}(T(t)),\, x^{\text{DL}}_{\text{sat},i}(T(t))\right)} \qquad i = \text{D,L} \tag{12.1}$$

In eqn (12.1) S_i describes the relative supersaturation of enantiomer i, $T(t)$ the transient process temperature and $x_{i,0}(t)$ the current mass fractions in the system. As mentioned above, the saturation concentrations of the enantiomers i, $x_{\text{sat},i}$, at the specific temperature $T(t)$ and the composition $x_{i,0}(t)$ can be calculated by linear functions connecting the points of saturation of the pure enantiomer, $x^0_{\text{sat},i}$, and the racemic composition, $x^{\text{DL}}_{\text{sat},i}$, in the ternary phase diagram.

The maximal enrichment of a system and hence, the achievable yield of the depicted PC, is obviously strongly limited by the slopes of the solubility isotherms. This can be quantified by the so-called Meyerhoffer ratio defined as the ratio of solubilities of the racemate and an enantiomer for the same conditions.[16] The supersaturation calculation for curved saturation lines needs to apply a more general approach instead of the linear functions mentioned above.

A suitable way to improve the process output is to increase the driving force for growth of the preferred enantiomer by changing the initial or the final temperature. Nevertheless, the allowable sub-cooling remains bound by the specific borders of primary nucleation at which both components crystallize

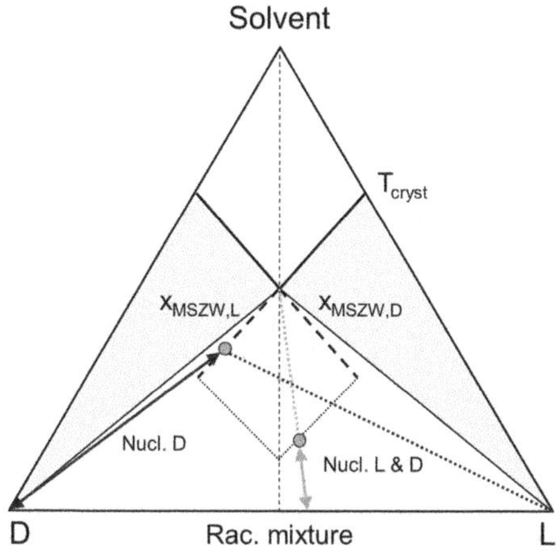

Figure 12.4 Ternary phase diagram of two enantiomers D and L in a solvent and illustration of the region of metastability. Points indicate states where either both enantiomers or only one enantiomer (here D) nucleate.

from a clear solution. Thus, two metastable zone widths (limits) exist. One is defined by the maximum sub-cooling at which spontaneous nucleation of D and L occurs (the two narrow dotted lines in Figure 12.4) and the other one by the tolerable enrichment of the entrainment process before the antipodes nucleate (dashed lines in Figure 12.4).

A successful resolution process starting from eutectic composition must be carried out respecting both borders of metastability to produce pure enantiomers. Since primary nucleation is of stochastic nature and affected by many variables (*e.g.* the cooling rate, the impurity content or the thermal history of the applied solution), a precise determination of the metastable zone with respect to temperature is difficult and requires large experimental efforts.[37] A statistically significant identification of the nucleation border as a function of enrichment requires numerous seeded batch crystallization runs. These runs need to be carried out under the same conditions and need measurements of both the occurrences of nuclei of the counter-enantiomer and the enantiomeric excess. Meeting these requirements is challenging and only possible if sufficient homochiral seed material is available. In conclusion, a considerable uncertainty regarding the knowledge of the widths of metastability will always remain, which necessitates conservative operation and further limits the field of operation with respect to the exploitable supersaturations, as illustrated in Figure 12.5.

Another prerequisite for a successful PC is an economically justifiable production rate of the seeded crystals. This requires a sufficient growth rate of the preferred enantiomer. It is well known that impurities or additives can retard or inhibit the growth of specific crystal faces.[38–40] The mechanisms behind these influences are complex and not precisely clarified.[39,41,42] A complete or partial structural similarity of the additives to the crystallizing substance, *i.e.* key-lock interaction options, are an

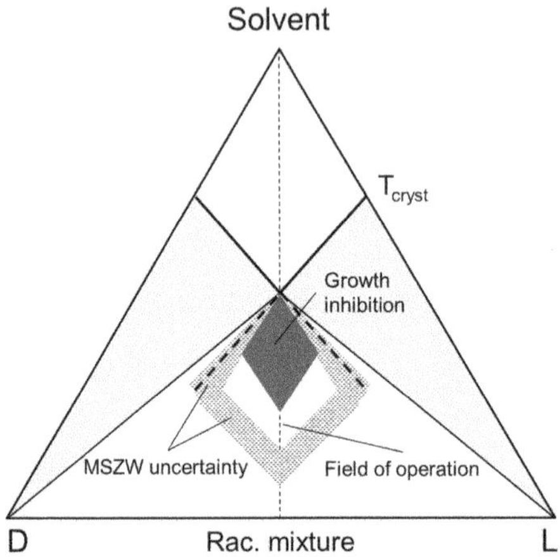

Figure 12.5 Region for successful operation of a PC-based resolution process framed by the inhibition of growth in presence of the counter-enantiomer and the uncertainty in knowing the metastable zone borders.

indicator that an impact of the foreign substance on the crystallization kinetics is probable. The concentration levels at which additives become effective vary from ppm ranges to several mass percent. The potential of influence also varies from a slight retardation of the kinetics of a specific crystal face to a complete dead zone,[43] or dead range of supersaturations at which no growth of all faces of the crystal is measurable.[44–46] A similar impact on the nucleation kinetics can occur as well, which alters the size of the metastable zone width or the rate of nuclei formation.[47–49]

A few substance systems have shown surprisingly no retardation when one enantiomer is crystallized from a racemic composition in comparison to the crystallization from the enantiopure solution, while others exhibit the expected interaction.[22] General conclusions can still not be drawn from the currently known data basis. However, three scenarios regarding the growth kinetics can be distinguished: If crystal growth is completely inhibited by the dissolved antipode in a supersaturation range within the metastable zone, the enantiomeric system cannot be separated by preferential crystallization.[50] In the opposite case, when no influence is noticeable, any PC based process should yield pure enantiomers if operated appropriately. In the intermediate range, where growth kinetics are significantly influenced, several options remain, which enable successful separation. A larger surface area of the seed crystals can lead, for example, to a productive entrainment before the counter-enantiomer nucleates or a change of the solvent can positively influence the substance transport or diffusion processes at the molecular level.[38] In any case the process variables should be chosen accordingly. Keeping the supersaturation at a high level slightly below the metastable limit (as possible *e.g.* in continuous processes, see below) can be beneficial.

12.4 Process Variants of PC

PC is currently performed typically in a batch operation. In the last years, a couple of variants capable of improving the standard batch process have been developed.[21,51–53] Progress has been achieved also in transferring the concept to continuous operation. Below, we will describe selected variants studied in our own laboratory. A common goal is that high purity of the produced enantiomers has to be guaranteed first. Indeed, for enantioseparation this requirement is of more relevance compared to designing particle size distributions and crystal shapes, which are often the target of other crystallization processes.

12.4.1 Batch Processes of PC

Different PC process options applying batch-wise operation have been applied successfully for the separation of enantiomers. These options are illustrated in Figure 12.6 and are described in the following.

12.4.1.1 Conventional and Cyclic Preferential Batch Crystallization (PC)

Conventional PC was briefly described in Section 12.2 (Figure 12.2). It is the basic concept of enantioseparation by crystallization, which is, however, limited in productivity and not easy to operate. Nevertheless, it is widely applied as the first approach toward producing enantiopure crystals on a small scale. Starting from a saturated solution at a certain temperature ($T_{cryst} + \Delta T$), the solution undergoes a cooling (ΔT) until the crystallization temperature (T_{cryst}) and hence, the desired supersaturation is reached (Figure 12.6a, PC 1st cycle). The process is usually subdivided by the achievement of the set temperature in an isothermal technique with a simple initial temperature step and a polythermal technique where cooling profiles like linear cooling or even optimized cooling profiles[22,54] are applied. The particular temperature differences and profiles are chosen based on the knowledge regarding crystallization kinetics in order to work within the field of operation, where nucleation of the counter-species is inhibited (Figure 12.5). Subsequently, the process is initiated by adding seed material of the desired enantiomer. Growth and secondary nucleation then yield selective crystallization of this component but simultaneously a relative enrichment of the counter-enantiomer occurs in the mother liquor (Figure 12.6a). The process is interrupted before the metastability limit of the antipode is exceeded by performing a solid–liquid separation to harvest the pure crystalline material. Conventional PC is an essentially delicate process with respect to purity, since the equilibrium is again the racemic composition (Figures 12.3 and 12.6a) and the supersaturation of the unseeded enantiomer remains at a high level during the separation. The choice of the optimal time to stop the process is therefore of key

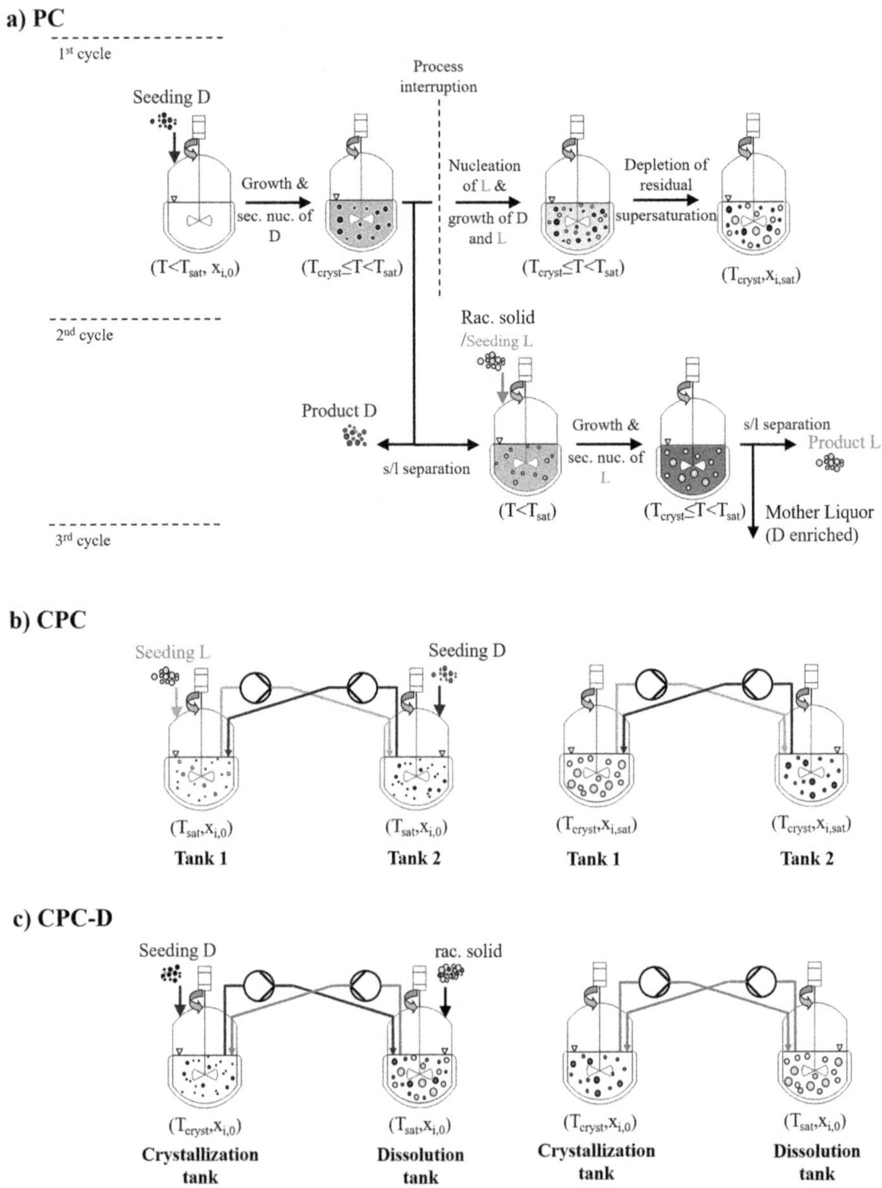

Figure 12.6 Schematic overview of different batch preferential crystallization processes. (a) PC: conventional preferential crystallization, (b) CPC: coupled preferential crystallization, (c) CPC-D: coupled preferential crystallization and selective dissolution. (● – D-enantiomer, ○ – L-enantiomer, □ – racemic solution, ▩ – L-enriched solution, ■ – D-enriched solution).

relevance. Its specification is challenging because of the stochastic nature of nucleation. Enhancement of the mass of pure enantiomer produced, which drives the process closer to the metastable limit, is consequently diametrically opposed to the robustness of the process.

There are various models available to simulate the course of PC and to find optimal process conditions. Most of the approaches exploit detailed population balance models.[55–59] The underlying basics and examples of these powerful models are given in other chapters of this book. Most crucial for the predictive power of all theoretical models is the correct incorporation of the two dynamically changing driving forces, S_i, of both enantiomers expressed by eqn (12.1).

Below, in order to illustrate the dynamics of a PC process, a simplified model discussed in more detail by Carneiro *et al.*,[60] is utilized. It contains at first two mass balances for the two enantiomers in the liquid phase. The driving forces are expressed *via* eqn (12.1) through implementing and permanently updating in a rigorous mathematical way the illustration given in Figure 12.3. Simple *n*th order growth and nucleation rate terms (G_i and B_i) are used to describe the overall process trend at the expense of including more specific mechanistic insights.

$$V_{\text{Liq}}\frac{\mathrm{d}c_i}{\mathrm{d}t} = -G_i - B_i = -k_{\text{G}}A_i\left(S_i - 1\right)^{n_{\text{G}}} - k_{\text{N}}A_{i,\text{Nuc}}\left(S_i - 1\right)^{n_{\text{N}}} \qquad i = \text{D,L} \qquad (12.2)$$

Due to the difficulties in determining reliable kinetic parameters for both growth and nucleation, and the fact that nucleation does not contribute significantly to the depletion of the concentrations c_i in the liquid phase in the essential initial period of PC, a further simplified kinetic sub-model, assuming lumped effective growth rates G_i^{eff} of the crystal surfaces, A_i, of the enantiomers with an effective growth rate coefficient $k_{\text{G}}^{\text{eff}}$ and an effective order n_{eff}, is applied.

$$V_{\text{Liq}}\frac{\mathrm{d}c_i}{\mathrm{d}t} = -k_{\text{G}}^{\text{eff}}A_i\left(S_i - 1\right)^{n_{\text{eff}}} \qquad i = \text{D,L} \qquad (12.3)$$

Furthermore, the initially introduced solid masses, $m^0{}_{\text{Seeds},i}$, are considered to consist of $N^0{}_i$ uniformly sized spherical particles with radii R_i:

$$N_i^0 = \frac{m_{\text{Seeds},i}^0}{\rho_{\text{Solid}}\frac{4}{3}\pi R_i^{0^3}} \qquad i = \text{D,L} \qquad (12.4)$$

To mimic unavoidable nucleation of the counter-enantiomer, very small particles are assumed to be present from the beginning. Their growth can be retarded by introducing in the corresponding mass balances an induction time, $t_{\text{induction}}$, as an additional free model parameter. It is assumed that later no new nuclei are formed. Thus, the numbers of both particle types are assumed to remain constant during the whole process $N^0{}_i$.

The numerous assumptions made lead to the following simple liquid phase mass balances for the two enantiomers:

$$\frac{dc_i}{dt} = -\frac{4\pi k_G^{\text{eff}}}{V_{\text{Liq}}} N_i R_i^2 (S_i - 1)^{n_{\text{eff}}} \qquad i = \text{D,L} \tag{12.5}$$

These equations are complemented by the following two mass balances of the enantiomer-specific solid particles. The mass gained by the solid particles due to effective growth is expressed by the increase of the two radii, R_i, of the spherical particles:

$$\frac{dR_i}{dt} = \frac{k_G^{\text{eff}}}{\rho_{\text{Sol}}} (S_i - 1)^{n_{\text{eff}}} \qquad i = \text{D,L} \tag{12.6}$$

After providing the relatively small amount of free parameters required by this simple model (namely, the effective growth rate constant, k_G^{eff}, the effective order, n_{eff}, the liquid volume, V_{Liq}, the saturation limits of the pure enantiomer and the racemic mixture required to calculate the two supersaturations, S_i, with eqn (12.1), *i.e.* $x_{\text{sat},i}^0, x_{\text{sat},i}^{\text{DL}}$, the solid density, ρ_{Sol}, and the solvent density, ρ_{Liq}, required to express the three mass fractions, x_i, by the liquid phase concentrations, c_i, (and *vice versa*), only the two initial liquid phase concentrations, c_i^0, the radii of the first particles, R_i^0, and the seed masses, $m_{\text{Seed},i}^0$ (or the corresponding particle numbers, N_i^0) have to be specified.

Characteristic features of a PC batch run as predicted by this short-cut model are illustrated in Figure 12.7. The transient behavior predicted for typical parameters reveals the aforementioned time window for successful operation. In this window the target enantiomer can be provided (here the D-enantiomer) with a very high purity (Pur). The extent of this operation window strongly depends on the induction time for the nucleation of the counter-enantiomer. In the specific case presented, the batch process must be stopped after the induction time (set in this case study to be one hour) in order to collect the target enantiomer. The transients shown in Figure 12.7 also reveal the complex changes in the two driving forces, S_i, during the process.

It should be noted that the simple model suggested above can be easily extended to include further batch-wise operated PC variants as discussed below and continuous PC operation performed in well-mixed crystallizers as described in Section 12.4.2.1.

An improvement of conventional PC can be achieved using a cyclic process mode[22] (Figures 12.6a and 12.8). After crystallization and recovery of the solid L-enantiomer (light grey arrow A → B, Figure 12.8), the mother liquor is mixed with racemic solid feed and heated to create a saturated but slightly enriched solution with respect to the counter D-enantiomer (point C). After adding the respective seeds, D is now preferentially crystallized (grey arrow C → D, Figure 12.8). The overall process performance is enhanced in this way by repeating the described cycling of both enantiomers sequentially, since more material can be harvested in one operation, and the initial driving force of the seeded enantiomer and therefore its growth rates, are larger. Thus, it is worthwhile to distribute the production

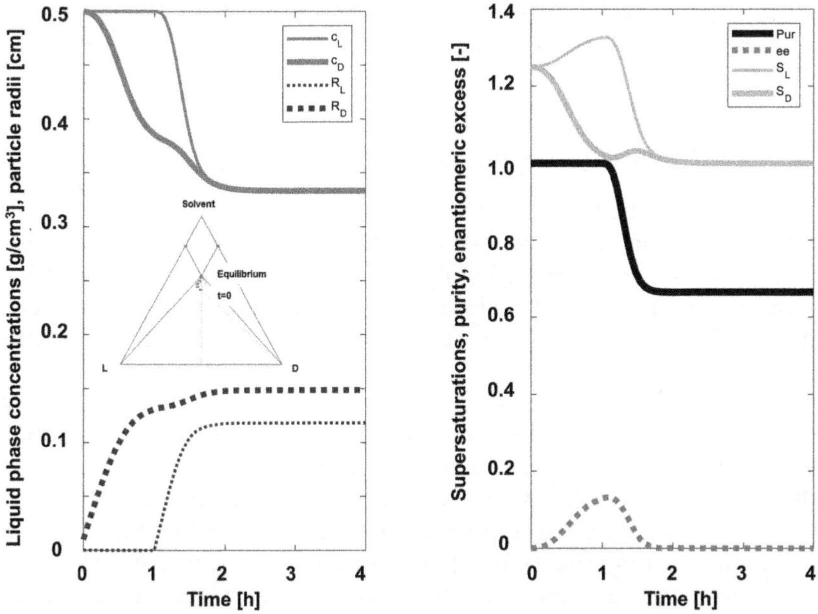

Figure 12.7 Schematic illustration of the transient behavior of a successful batch-wise operated PC (D-target) as predicted by the short-cut model, eqn (12.1)–(12.6). The predicted trajectories of the four relevant states (left) are complemented by the dynamically changing supersaturations (S_i), the solid phase purity (Pur) and the enantiomeric excess (ee) (right). The inset in the left figure shows the process trajectory in the corresponding triangular phase diagram. The following model parameters were used: $R_D^0 = 10^{-2}$ cm, $N^0_D = 100, R_L^0 = 10^{-7}$ cm, $N_L^0 = 200$, $V_{Liq} = 10\ cm^3$, $\rho_{Liq} = 1\ g\ cm^{-3}$, $\rho_{Sol} = 1.2\ g\ cm^{-3}$, $k_G^{eff} = 1$ g (cm^{-2} h^{-1}), $c_i^0 = 0.5$ g cm^{-3} (or $x_{i,0} = 0.25$ at $t = 0$), $x^0_{sat,D,L} = x^{DL}_{sat,D,L} = 0.2$, $t_{induction} = 1$ h.

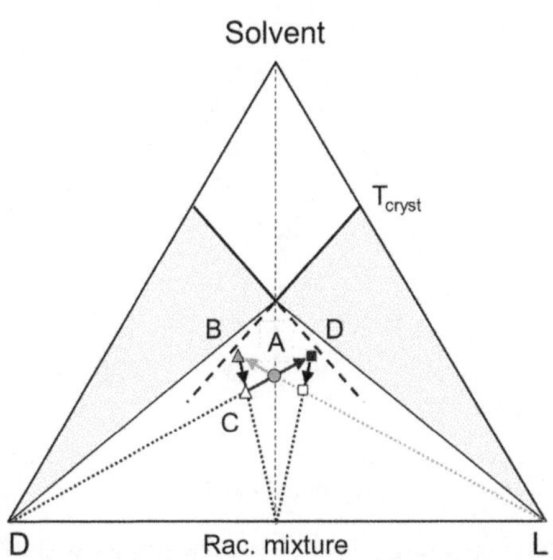

Figure 12.8 Trajectories of cyclic PC consecutively producing both enantiomers from a racemic mixture (starting with L crystallization, grey line) with solid–liquid separation and racemic feed addition (black downward lines) in-between. (See also Figure 12.6a).

of pure crystalline enantiomers over several cycles instead of driving the process to the limit of metastability to reduce the risk of primary nucleation.

12.4.1.2 Coupled Batch Preferential Crystallization (CPC)

In conventional PC, the concentration of the unseeded enantiomer stays constant in the liquid phase, but its liquid phase mass fraction increases due to selective crystallization of the seeded enantiomer. An objective of alternative strategies is a simultaneous reduction of the supersaturation of both enantiomers by crystallizing them in two spatially separated subsystems.[61]

The CPC process consists of two preferential batch crystallizations starting from identical racemic saturated solutions (Figure 12.6b). Upon subcooling to the identical temperature or initiation of the same cooling strategy, both solutions are seeded with homochiral crystals of the respective enantiomer. Both processes are coupled with each other *via* an appropriate pipe installation, allowing the exchange of solid-free liquid phases. The enrichment of the respective unseeded enantiomer can be completely counterbalanced if both crystallizations are working precisely in parallel and the rate of mother liquor transfer is sufficiently large (grey dashed arrows in Figure 12.9). Ideally, both liquid phases are then permanently at racemic composition over the whole process time until supersaturation is depleted (black solid arrow in Figure 12.9). This creates a robust process, which can be operated efficiently even close to saturation conditions. The process yield is in this manner only defined by the difference between the solubility at

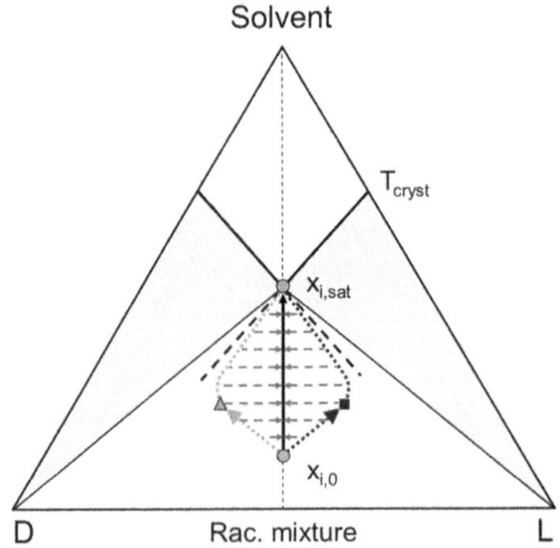

Figure 12.9 Transient composition of two simultaneously operated PCs (dotted curves) and the resulting coupled process if both liquid phases are exchanged during separation (CPC, black solid arrow). (See also Figure 12.6b).

initial and final temperature. The productivity can be increased by applying substance-specific optimized cooling profiles.[62] Furthermore, this process concept can be adapted also to systems forming racemic compounds if a previously enriched feed solution is available.[63,64]

12.4.1.3 Coupled Preferential Crystallization and Selective Dissolution (CPC-D)

The principle of coupled preferential crystallization and selective dissolution is also based on the operation of two coupled crystallizers initially filled with identical racemic saturated solution which, unlike in CPC, work at different, but constant temperatures (Figure 12.6c, CPC-D).[65,66] One crystallizer is operated at crystallization temperature within the metastable region (T_{cryst}, crystallization tank in Figure 12.6c), while the other is kept at saturation temperature (T_{sat}, dissolution tank in Figure 12.6c).

After the operation temperature is reached, a simple preferential crystallization is initiated in the low temperature crystallizer by adding seeds of the target enantiomer. The high temperature tank (dissolver) is fed with a given mass of racemic solid material to maintain a racemic saturated solution at T_{sat} during the whole process ($x_{i,0}$ in Figure 12.10).

The seeded enantiomer is hence depleted from the liquid phase during selective entrainment in the crystallizer. If this mother liquor is transported to the dissolver, held at higher temperature, a transient undersaturation with respect to the seeded enantiomer occurs. In order to reach the saturation composition at T_{sat} again, selective dissolution of this enantiomer from the racemic solid phase takes place. Therefore, the selective depletion in the crystallizer will be counterbalanced, provided the liquid exchange

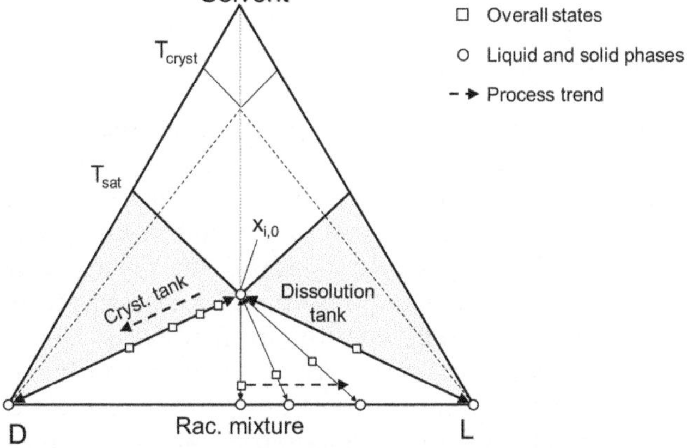

Figure 12.10 Evolution of the overall states of two coupled crystallizers during the preferential crystallization and selective dissolution process (CPC-D). (See also Figure 12.6c). Reprinted with permission from ref. 67. Copyright 2019 American Chemical Society.

rate is sufficiently large, and primary nucleation of the counter-species is prevented. As a result, the liquid phase in both reactors is kept racemic and the solid mass of the preferred species in the crystallizer increases steadily while the mass of the excess material reduces at the same rate in the dissolver (black dashed arrows in Figure 12.10). Finally, if the process is interrupted at the right time, only the seeded enantiomer is present as solid in the crystallizer, while in the dissolver only the unseeded enantiomer is left from the initially racemic solid material. The situation, in this way, reflects a CPC process with different temperatures. Further crystallization beyond this point leads to an enrichment of the unseeded enantiomer and, similarly to the conventional PC, increases the risk of primary nucleation. The main advantage of CPC-D in comparison to CPC is the need of only one pure enantiomer for the initial seeding while both enantiomers are produced. This process was demonstrated at first for resolving the enantiomers of the amino acid threonine.[65] Application to other conglomerate systems was shown more recently.[67]

12.4.2 Continuous Processes of PC

As described in other chapters of this book, there are numerous possibilities to perform crystallization processes in a continuous manner. Larger productivities and more constant product qualities are typically the targets compared to batch operation. As mentioned above, successful enantioseparation must deliver at first a high purity of the solid phase leaving the crystallizer. Due to the difficulty in achieving this requirement, so far only a few continuous operation concepts have been reported to be applicable.[17,68] Below we will introduce two concepts studied in our group. The applicability of these concepts is subsequently illustrated in two case studies.

12.4.2.1 MSPMR Concept

The continuously-operated, mixed-suspension, mixed-product-removal concept (MSMPR) has been intensively studied over several decades.[69] It is the first attractive concept to be applied for the separation of enantiomers. In order to exploit the broad knowledge regarding the MSMPR principle, it is convenient to directly adapt the CPC principle by using two connected crystallizers in a continuous operation. However, specific challenging requirements have to be met for chiral systems. A corresponding conventional PC process works by nature off the racemic composition, which increases the risk of primary nucleation of the counter-enantiomer and therefore complicates the implementation of the basic MSMPR concept.

To start the process, both crystallizers are fed continuously with a saturated racemic liquid phase at a certain initial temperature ($x_{i,0}$, T_{sat}, Figure 12.11 right). The tanks should be well mixed in order to meet the

MSMPR assumption that the withdrawn product suspension is characterized by the same crystal size distribution and liquid phase composition as the suspension inside the vessel (x_i at T_{cryst} in Figure 12.11 left and right). Enantioselective PC involves an initial provision of seed crystals to the system. Subsequently, the MSMPR operation can be maintained by secondary nucleation, which generates the required crystal surface of the desired target enantiomer. During operation, the liquid phases in both tanks evolve in the direction of the unseeded species (triangle and square in Figure 12.11 left). The exact position is determined by the specific crystallization kinetics (light and dark grey arrows in Figure 12.11 left) and the feed addition rate (black dotted arrows in Figure 12.11 left). To adapt the CPC principle, the tanks are coupled by an appropriate liquid phase exchange (black solid arrows in Figure 12.11 left), which ideally fully counterbalances the enrichment and therefore minimizes the risk of primary nucleation.

A key issue is the production of a sufficient amount of secondary nuclei that is required for the MSMPR process. Two options are possible: (a) the operation with a large driving force or (b) high suspension densities and an appropriate mechanical energy input. However, both favor nucleation of the counter-enantiomer due to the higher degree of supersaturation or the larger heterogeneous surface. A precise continuous seeding strategy is more desirable, as this increases robustness and predictability during the challenging enantioselective continuous crystallization.

Seed crystals can be provided either continuously, *e.g.* by a well-controlled wet milling, or periodically in portions.[68] The selection has to be made on a case-by-case basis, depending on the process scale, growth kinetics and plant set-up, to keep the mother liquor in the metastable zone throughout the process on the one hand and to yield the maximal productivity on the other.

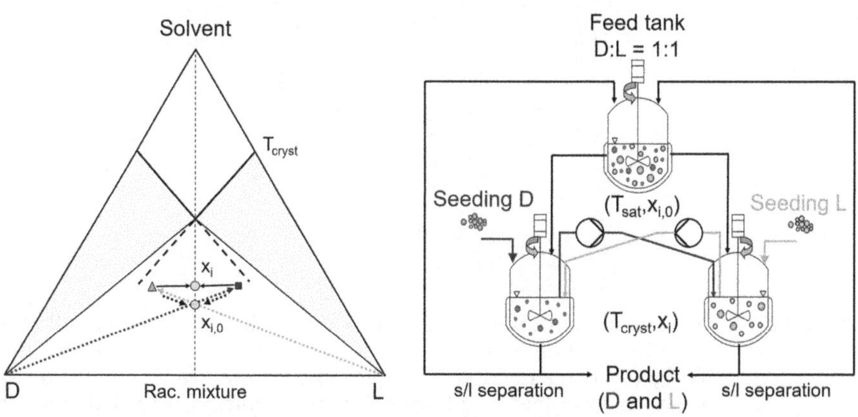

Figure 12.11 Schematic representation of the continuously-operated coupled preferential crystallization MSMPR process *via* its trajectories in the ternary phase diagram (left) and the corresponding process setup (right).

12.4.2.2 Continuous Enantioseparation in Fluidized Bed Crystallizers

A further promising variant for continuous preferential crystallization is using tubular reactors. In designing the tubes with a conical bottom part and operating them in a bottom-to-top flow regime, the fluidized bed–type crystallizers provide an improved separation performance with regard to supersaturation control, mixing, particle uniformity and size classification as well as process robustness.

A generic scheme for the continuous preferential crystallization setup in two parallel connected fluidized bed crystallizers (FBCs) is depicted in Figure 12.12. Racemic solution, introduced in a feed tank and saturated at a certain temperature ($x_{i,0}$, T_{sat}), is circulated through two FBCs from bottom to the top. The crystallizers are operated at a lower temperature

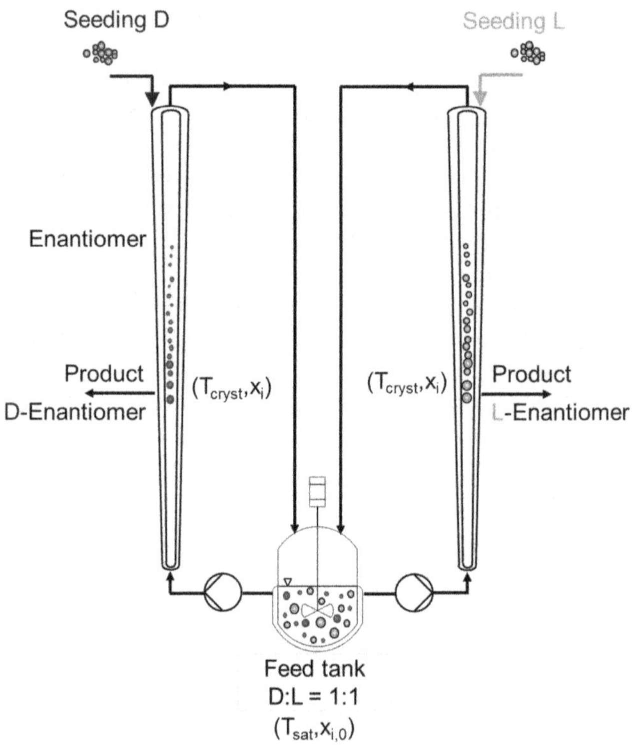

Figure 12.12 Scheme of a continuous PC setup with two fluidized bed crystallizers (one for each enantiomer) coupled *via* the feed (recycle) tank. The feed tank is operated at a higher temperature, where continuously solid racemic feed is dissolved in the depleted, recycled mother liquor to steadily provide supersaturated solution for continuous crystallization of the two enantiomers in the FBCs. In contrast to the MSMPR concept shown in Figure 12.11, there is no internal liquid phase exchange between the crystallizers.

(T_{cryst}) to provide the supersaturation as driving force for crystallization. To initiate preferential crystallization, seed crystals of both pure enantiomers are added to the respective FBC only once. The mother liquor leaving the crystallizers is sent back to the shared feed tank. Since the crystallization driving force for both enantiomers is equal if both columns are operated symmetrically, their ratio in the feed solution remains racemic throughout the process. An excess of solid racemic feed present in the feed tanks guarantees for re-saturation of the mother liquor and thus for a constant supersaturation within operation time.

The process trajectories of an FBC are close to those of the MSMPR variant (Figure 12.11). However, because of the lack of the internal coupling of the crystallizers *via* liquid phase exchange, there is no counterbalance of the enantiomer enrichment in the crystallizers leading to a slight enantiomeric excess of the respective counter-species. As a result of the continuous flow of fresh, supersaturated feed solution at a high rate through the crystallizer tubes and the comparatively small amount of crystallized material per time, this excess is almost negligible as shown in Section 12.5.2. This facilitates the robustness of the separation processes.

Successful preparative-scale enantiomer resolution in a crystallization setup of this type (with periodic batch-wise product removal) was demonstrated by the Merck Research Laboratories for Methyldopa[70] and a derivative of a drug candidate (3-fluoro-D-alanine-2-d).[71]

Compared to classical MSMPR crystallizers, the fluidized bed process variant presents four main benefits. First, the large feed flow rates ensure a constantly high and easily-controllable supersaturation level, which creates a robust process and offers relatively large crystal growth rates. Second, because of the minor mechanical strain on the growing crystals under fluidized bed conditions (low attrition), rather uniform particle sizes can be produced.[72] Third, and of high significance in preferential crystallization, the risk of product contamination by nucleation and crystal growth of the counter-enantiomer is reduced as respective nuclei or small crystals leave the crystallizers on top straight away with the mother liquor flow and can be dissolved in the feed tank. This enables a robust process operation, an important prerequisite for industrial application of a separation concept. Fourth, within certain limits, the particle size distribution of the product crystals is adjustable by the feed flow rate and the crystallizer dimensions (geometry and design of the conical part).

12.5 Case Studies

In this section we will demonstrate the results of applying some of the concepts of PC in two specific case studies. Regarding the separation of the enantiomers of DL-threonine, we will consider the cyclic batch operation, the coupled batch operation using the CPC-D mode, and finally the continuous operation in two coupled MSPMRs. For the separation of the enantiomers of asparagine monohydrate, we will essentially present results exploiting two coupled continuously operated fluidized bed crystallizers (FBC).

12.5.1 Resolution of DL-Threonine

L-Threonine belongs to the proteinogenic amino acids. It contains an uncharged polar side chain and is one of the nine essential amino acids that must be provided by proper nutrition, since humans and most other mammals are not able to synthesize them.[73] For this reason, it is often added to human food and animal feed as a dietary supplement and to improve the nutritional value.[74–76]

Because threonine contains two chiral centers, four stereoisomers occur, forming two isomeric pairs, namely L-threonine and D-threonine as well as L-*allo*-threonine and D-*allo*-threonine. The L-/D-threonine pair (L-/D-Thr) (Figure 12.13) was used in our crystallization studies.

12.5.1.1 Solubility Data for the DL-Threonine System

The threonine enantiomers form a racemic conglomerate, which makes the system a suitable candidate for PC based enantioseparation. The enantiomers D- and L-Thr were well characterized regarding their solubility limits in water.[77–79] Figure 12.14 presents solubility data determined in our group for the D-/L-Thr/water system for different temperatures in a ternary phase diagram. It is given in a three-dimensional representation using a triangular prism with the system components forming the apexes of the equilateral triangle of the prism's base and the temperature on the axis perpendicular to this triangle. The solubility isotherms depicted represent the ternary phase diagram of the system at the respective temperature as shown and discussed in Section 12.2.

12.5.1.2 Metastable Zone Width and Crystallization Kinetics

For this system, there are various data sets available regarding the metastable zone widths.[62,80] Furthermore, results of induction time measurements for primary nucleation were reported.[62] The results shown in Figure 12.15 indicate that for this system there is an attractive, sufficiently broad metastable

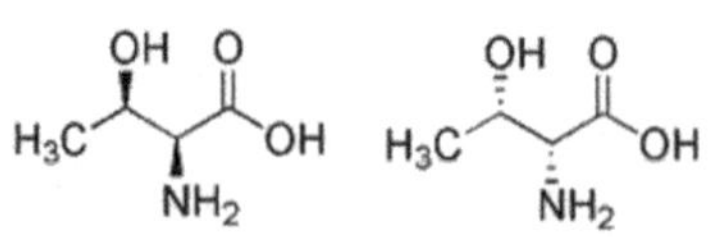

Figure 12.13 Structures of L-(2*S*,3*R*)-(−)-threonine and D-(2*R*,3*S*)-(+)-threonine (L-Thr and D-Thr).

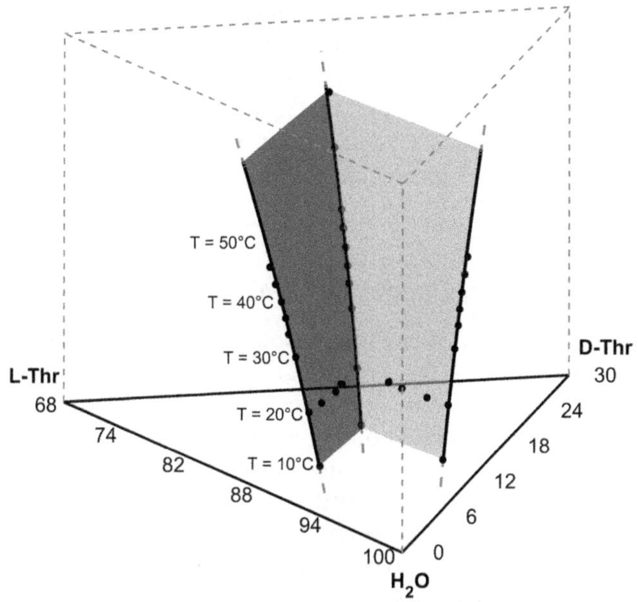

Figure 12.14 Ternary solubility phase diagram of the enantiomers of L- and D-threonine in water as a function of temperature (data measured in our group, in particular ref. 78). Axes are given in mass fractions x_i·100 (wt%); only a 30% section of the entire phase diagram is shown for clarity. Dark and light grey colors refer to L and D, respectively.

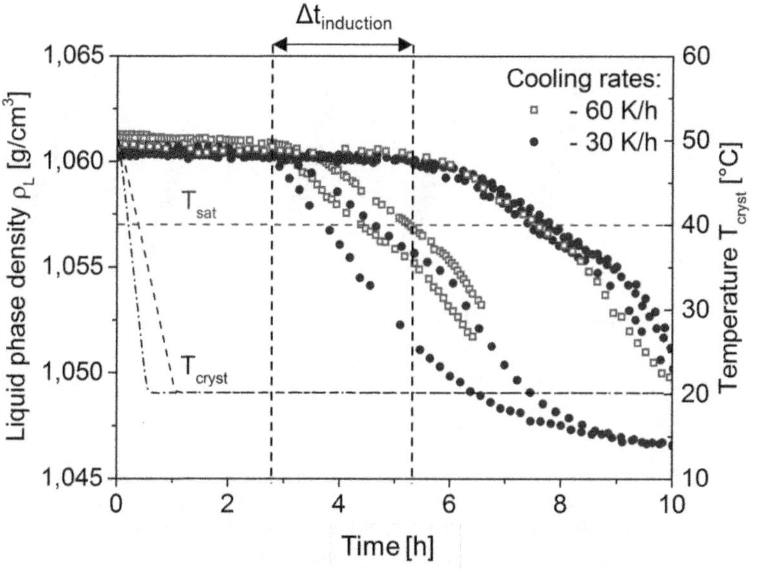

Figure 12.15 Illustration of the order of magnitude of exploitable induction times of primary nucleation for the system DL-threonine in water.[62] Experiments were performed at 0.5 L scale with racemic threonine solution and supersaturations of 1.37 for each enantiomer.

zone available to carry out PC. As seen, induction times of more than two hours can be exploited independent of the initial cooling rate. Additionally, the stochastic character of nucleation becomes evident from the large variations of the results in Figure 12.15. Therefore, these induction time measurements serve just as an orientation for subsequent process design and cannot be taken as a precisely-defined limit of exploitable supersaturation levels.

12.5.1.3 Cyclic Batch Operation of PC

Regarding the separation of the enantiomers of DL-threonine, we will first present results of applying the cyclic batch operation illustrated in Section 12.4.1.1. Figure 12.16 shows the corresponding polarimeter signals recorded for a series of five cycles carried out in a 1 L laboratory scale, initially starting from racemic composition and alternating periodically between compositions A, B, C and D (see also Figure 12.8). For the experiments reported, the auto-seeded polythermal process mode was used where only once (for the first half cycle) seeds have to be added. For the successive cycles, seeds are self-generated from the finely-ground racemic feed material *via*

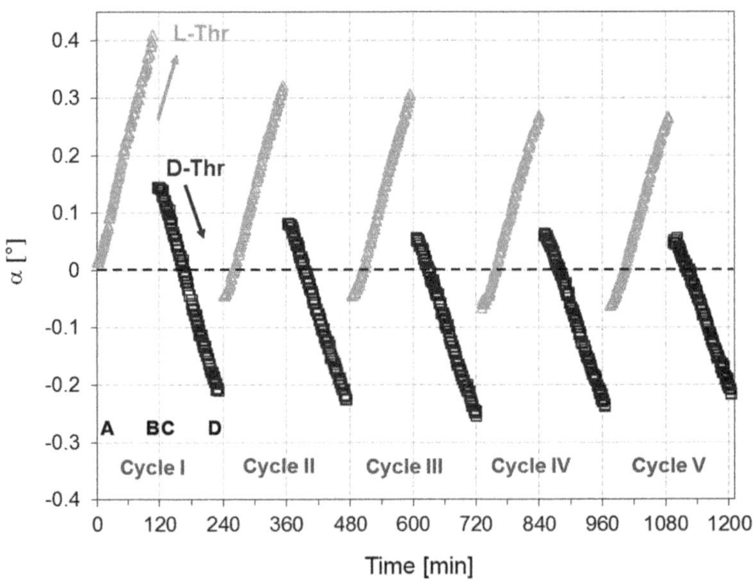

Figure 12.16 Transients of the optical rotation angle α recorded over five separation cycles in auto-seeded polythermal preferential crystallization of threonine depicting alternate crystallizations of L- and D-Thr. The solutions were prepared exploiting the solubility data; crystallization conditions: 1 kg solution, cooling from 39 °C to 4 °C at a rate of 20 K h^{-1}.[21,35] Reproduced from ref. 21 with permission from John Wiley and Sons, Copyright © 2014 WILEY-VCH Verlag GmbH & Co. KGaA, Weinheim.

Table 12.1 Results of the cyclic process shown in Figure 12.16 with crystallization of L-Thr and D-Thr (uneven and even half cycle numbers, respectively). The productivity (eqn (12.7)) is defined as the mass of pure product gained per hour and liter of the initially saturated clear solution with the time comprising preparation (Ø 1.3 hours) and the crystallization periods (Ø 1.8 hours) of each half cycle.

Cycle	Half cycle	Purity [%]	Productivity [g (h^{-1} L^{-1})]
I	1	99.2	14.7
	2	98.0	14.8
II	3	99.7	13.7
	4	99.9	12.1
III	5	99.5	13.0
	6	99.1	12.7
IV	7	99.4	11.4
	8	98.6	13.2
V	9	99.1	12.3
	10	98.4	11.5
Average		99.1	12.9

temperature-programmed selective dissolution of the respective counter-enantiomer. Table 12.1 gives an overview about the purity achieved in these stable and reproducible PC runs and provides the corresponding productivities as well. The purity of both enantiomers was on average 99.1%, which shows that there is still potential for further purification by intensifying the washing procedure. The productivities obtained are remarkably high compared to other process options given below (see Table 12.3).

12.5.1.4 Batch PC Coupled with Selective Dissolution (CPC-D)

Since the concept of coupling two crystallizers is demonstrated in the next section for the continuous mode, we will illustrate here a successful resolution exploiting a coupled batch operation using two tanks in which simultaneously two different sub-processes are carried out. The principle of the CPC-D process was introduced in Section 12.4.1.3 (Figure 12.10).

Figure 12.17 illustrates the transient mass fractions and supersaturations observed in a DL-threonine separation experiment carried out in a crystallization tank connected with a dissolution tank. Both vessels are initially fed with racemic liquid phase saturated at 45 °C. The temperature is lowered in the crystallization tank to 36 °C, which defines the driving force in this vessel, and seed crystals of pure L-Thr are introduced to initiate the entrainment. Simultaneously, racemic solid threonine is fed to the dissolution tank, which is kept at saturation temperature during the whole process. The entrainment in the crystallization tank should be counterbalanced by selective dissolution of L-Thr in the dissolution tank and the exchange of the liquid phases between both vessels. Two characteristic times are essential for a successful purification (t_{diss} and $t_{induction}$ marked in Figure 12.17).

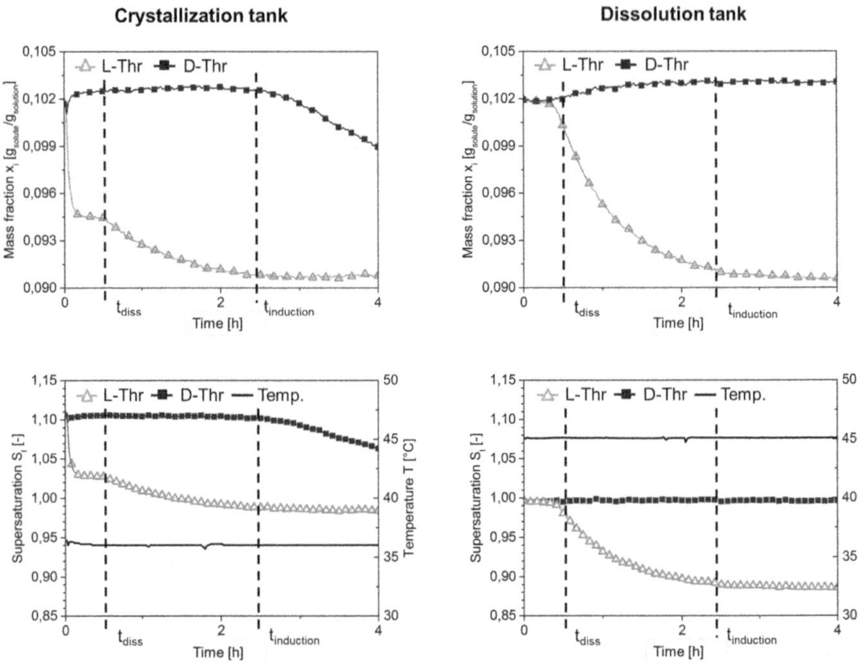

Figure 12.17 Mass fractions and supersaturations of the L- and D-threonine enantiomers in the crystallizer and dissolution tanks operated at different temperatures (36 °C and 45 °C).[62,66] Times t_{diss} and $t_{induction}$ mark the times of complete dissolution of the preferred enantiomer in the dissolution tank and nucleation of the counter-enantiomer in the crystallization tank, respectively (adapted from ref. 62).

The relatively large growth kinetics of L-Thr lead to a fast depletion of the supersaturation in the crystallization tank in the first few minutes. Afterwards, the driving force reaches a plateau at 1.03 until the marked time t_{diss} is reached. The mass fraction in the dissolution tank is constant at the same time indicating counterbalancing of crystallization of the preferred enantiomer by the selective dissolution. The difference between both vessels is related to the liquid exchange rate, which is not sufficient in this scenario. After the time t_{diss} is reached, the complete L-Thr is dissolved from the initially racemic excess material and only D-Thr remains in the dissolution tank. The further entrainment of L-Thr is now maintained solely by the excess that is still dissolved in the liquid phase in the crystallization tank. This results in an ongoing decrease of L-Thr in both vessels until an equilibrium state is reached after approximately 3 h.

The change of the transient mass fraction of the counter-enantiomer in the crystallization tank at $t_{induction}$ indicates nucleation of D-Thr. Subsequent growth of the nucleated antipode results in a decrease of the mass fraction of D-Thr in the crystallization tank. This enantiomer is dissolved in the dissolution tank after $t_{induction}$ from the remaining solid in the same way as L-Thr was dissolved from the racemate in the beginning of the process. Therefore, the crystallization must be stopped before the nucleation of the counter-enantiomer occurs, *i.e.* just before $t_{induction}$, to obtain pure product in the crystallization tank with maximal productivity. If pure crystals of both enantiomers are desired, one has to ensure that L-Thr is completely dissolved before D-Thr nucleates in the crystallization tank, *i.e.* after t_{diss}. The time between t_{diss} and $t_{induction}$ is therefore the operation window for the production of pure substances, which can be exploited for further process optimization. It should be mentioned again that the CPC-D concept is in particular attractive if seeds are available only for one of the enantiomers and if repetitive solid–liquid separations of the cyclic PC should be avoided.

12.5.1.5 PC in Continuously Operated Coupled MSPMR

Regarding the separation of the enantiomers of DL-threonine we present here finally a summary of acquired experimental results of continuous MSMPR operation (see Section 12.4.2.1). The MSMPR principle was recently successfully applied in two different experimental setups involving first only one MSMPR and then two MSMPR crystallizers, which were simultaneously operated and coupled *via* an exchange of their liquid phases.

Table 12.2 gives an overview about the implemented operating conditions. They were selected based on experience gained during performing the various above-mentioned batch separations and also on available theoretical results provided by a population balance-based process model.[57]

Results of exploiting the coupled MSPMR concept are summarized in Figure 12.18. The product's mass flows obtained at the outlet of the two tanks over a period of 18 hours are shown together with the purities, which are very satisfying.

Both product streams were found to be relatively stable. Only a few outliers are present between 13 and 14 h of operation due to pump malfunctions. Steady state operation was achieved quickly after approximately 3 h. Both enantiomers could be recovered simultaneously with high purities (≥99%) over the entire process time. The obtained product's mass flows as well as the high product purity are also a result of the beneficial effect from the continuous exchange of the mother liquors. The coupled

Table 12.2 Operating parameters of MSMPR runs carried out in two coupled crystallizers (Tanks 1 and 2) to resolve the enantiomers of DL-threonine.[68,81]

Operation parameter	Value	Unit
Liquid phase volume (each crystallizer)	0.45	L
Feed concentration	0.202	$g_{rac}\, g_{sol}^{-1}$
Crystallization temperature	38	°C
Mean residence time	50	min
Mean seed introduction	1	$g\, h^{-1}$
Exchange flow rate	30	$ml\, min^{-1}$

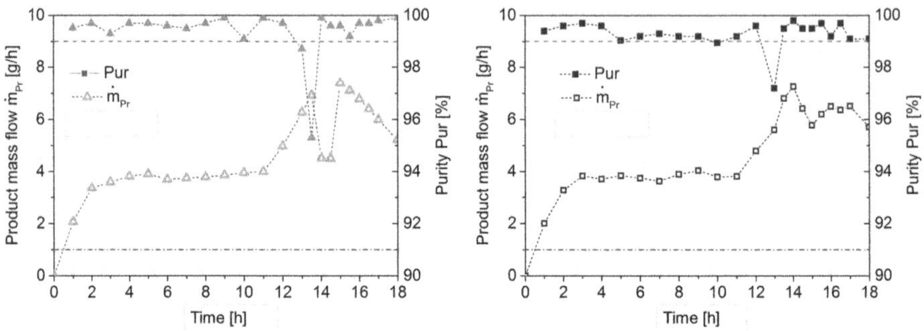

Figure 12.18 Product mass flows ($\dot{m}_{Pr}$) and purities (Pur) in Tank 1 seeded with L-Thr (left) and Tank 2 seeded with D-Thr (right). Conditions and parameters as given in Table 12.2 (adapted from ref. 68).

process was found to be more stable and robust compared to the runs using just a single vessel due to the decrease in the concentration of the counter-enantiomer in the respective vessel. Further results including a comparison with theoretical predictions using a detailed population balance are given in ref. 57.

12.5.1.6 Comparison of Different Process Options

Table 12.3 provides a short comparison of the results achieved for the separation of the enantiomers of DL-threonine using the different concepts discussed above. All process options were able to yield pure components if the operation parameters were chosen appropriately. Productivity is determined using eqn (12.7), with m_{product} being the total product mass as obtained, Pur – the purity of the product, m_{seeds} – the invested seed mass, V_{solution} – the solution volume, t_{process} – the actual process time period and t_{dead} – the batch dead time (*i.e.* cleaning and batch preparation time).

$$\text{Pr} = \frac{m_{\text{product}} \cdot \text{Pur} - m_{\text{seeds}}}{V_{\text{solution}}\left(t_{\text{process}} + t_{\text{dead}}\right)} \tag{12.7}$$

The lowest productivity eqn (12.7) is achieved, as expected, with the conventional PC connected with the low robustness of the process principle. The advanced batch PC options applying a second crystallizer or a periodic operation outperform the basic PC for this example by a factor 3 to 10. This demonstrates the large potential of more advanced variants for industrial applications. High productivities of the cyclic PC experiments (almost 10 times that of the single conventional batch run) result particularly from the broad temperature range covered with high cooling rates in the polythermal process mode, the comparatively short time periods needed for preparation of the next half cycle (including *in situ* seed formation), and the large surface of the *in situ* generated seed material. Comparing the processes carried out in the same plant and scale (conventional

Table 12.3 Purities and productivities (eqn (12.7)) of different batch and continuous processes carried out for the substance threonine[62,68] (first number: L-Thr, second number: D-Thr); productivity in gram pure product/volume of solution and dead (2 h) + process time except for the cyclic (dead time 1.3 h) and the continuous operation (no dead time).

	Process	Purity [%]	Productivity [g (h^{-1} L^{-1})]
Batch	Conventional PC (polythermal)	99.5	1.3
	Cyclic PC (auto-seeded polythermal)	99.4/98.8	13.0/12.9
	CPC (polythermal)	99.4/99.6	5.3/5.2
	CPC-D	95.3/98.8	5.9/3.1
Continuous	single MSMPR	99.1	8.4
	coupled MSMPR	99.6/99.1	6.6/6.6

PC, CPC, CPC-D and continuous MSMPR), the single continuous crystallization is the most productive. However, it is operated off the racemic state of the liquid phase, which increases the risk of product contamination by nucleation of the counter-enantiomer over time. Hence, for the long-term operation desirable in industry, the coupled MSMPR crystallization process is seen to be advantageous due to the increased process robustness caused by the liquid phase exchange together with comparable high productivities.

When considering the above case-specific statements, it should be noted that each of the compared options is not yet fully optimized. A more rigorous optimization could cause still different trends. Here, we also do not provide process yields, which are of course another important process characteristic. Respective numbers can be found, however, in the cited references.

12.5.2 Resolution of Racemic Asparagine Monohydrate

Regarding the separation of the second system, namely the enantiomers of asparagine monohydrate, results from the application of two coupled and continuously operated fluidized bed crystallizers (FBC) are presented here, as introduced in Section 12.4.2.2. A brief comparison with other batch process variants is given at the end.

DL-asparagine monohydrate is known to be a conglomerate. L-asparagine is one of the nonessential proteinogenic amino acids used in the human body for nitrogen transport and toxin excretion. Its structure is shown in Figure 12.19. Medically, it is applied in cases of low blood pressure, kidney problems, or liver damage. D-asparagine, included in the peptide fullicin, is an example of a variation in the biological activity of a snail protein. Fullicin, which contains a D-asparagine residue, shows higher bioactivity than fullicin containing the L-species.[82] From water, L- and D-asparagine crystallize as monohydrates, *i.e.* L-(−)-Asn·H_2O and D-(+)-Asn·H_2O. The racemate, DL-Asn·H_2O, is a mechanical mixture of the latter.

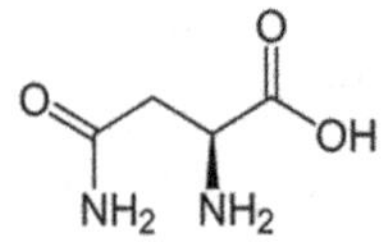

Figure 12.19 Structure of L-asparagine (L-Asn).

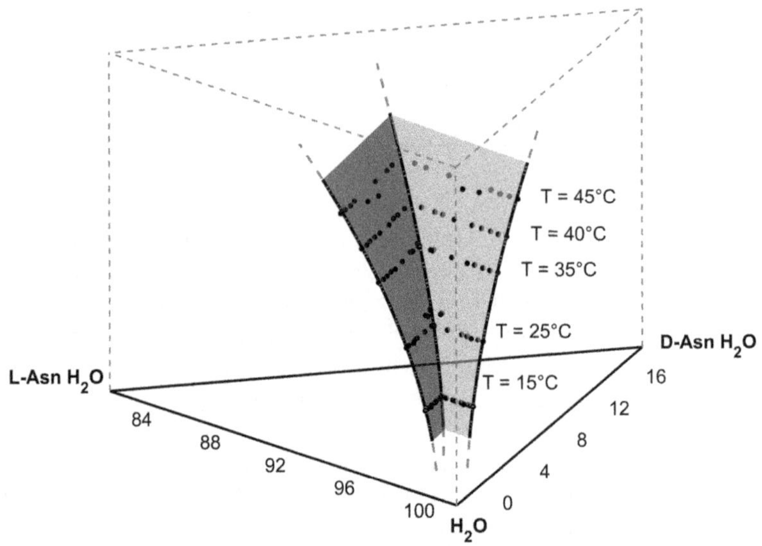

Figure 12.20 Ternary solubility phase diagram of the asparagine monohydrate enantiomers in water as a function of temperature. Axes are given in mass fractions x_i·100 (wt%); only a 20% section of the entire phase diagram is shown for clarity. Dark and light grey colors refer to L and D, respectively. Reprinted with permission from ref. 85. Copyright 2019 American Chemical Society.

12.5.2.1 Solubility Data for Asparagine Monohydrate

Solubilities of the asparagine monohydrate enantiomers and the racemate in water as solvent have been determined in previous works of our group in a broader temperature region (10–60 °C), showing the expected equality of the L- and D-enantiomers' solubility and in good agreement with literature sources.[83] The ternary phase diagram of the respective system with additional data points for enantiomer mixtures of various compositions is shown in Figure 12.20 (15–45 °C). As seen, solubility increases with temperature and the system's components exhibit conglomerate behavior. Within the studied conditions the solid phases in equilibrium are always the monohydrates of the asparagine enantiomers; no other phases have been observed. The ratio of the solubility of the racemic mixture to the solubility of the enantiomers is close to two, *i.e.* the solubility isotherms run almost parallel to the opposite triangle sides, meaning that the solubility of one enantiomer is not affected by the other one (ideal behavior). Moreover, linearizing the enantiomer's and racemate's solubility data in a van't Hoff plot[84] provides perfect linearization ($R^2 > 0.99$) and similar dissolution enthalpies (32.1 *vs.* 32.6 kJ mol^{-1}).

12.5.2.2 Metastable Zone Width and Crystallization Kinetics

Metastable zone width with respect to temperature was measured classically in cooling runs of multiple heating-cooling (thus dissolution-nucleation) cycles in a multi-reactor system in 1 mL volume scale. It was found to be between 9 and 11 K in a saturation temperature region between 25 and 40 °C. Since in continuous crystallization supersaturation is supposed to be constant, the time the supersaturated solution remains in that state without the occurrence of primary nucleation (induction period or nucleation time) is of additional significance. Further, as mentioned before (see Section 12.3), several variables may have an impact on MSZW. This motivated us to conduct nucleation time measurements directly in the coupled FBC setup starting with racemic asparagine solutions of different supersaturations. The results are shown in Table 12.4. At high supersaturation levels (ΔT 11 and 15 K corresponding to a supersaturation of 1.55 and 1.88 respectively) nucleation occurred within relatively short times (1 or 3 hours respectively). Applying a ΔT of 8 K, corresponding to a supersaturation of 1.37 at a crystallization temperature of 27 °C, no nucleation shower was observed within 24 hours of operation. It has to be noted that a non-significant number of nuclei could have been formed at an earlier point than given, but as a result of the continuous upward flow of solution in the crystallizers these nuclei were carried out of the columns and did not lead to a detectable concentration change. This is a favorable property of the FBC that also applies to unwanted nuclei of the counter-enantiomer during the actual separation process. Based on these results, first continuous preferential crystallization trials were planned to be conducted at a 35 °C saturation and a 27 °C crystallization temperature, respectively.[83]

For quantification of crystallization kinetics in the chiral asparagine–water system a short-cut method developed within the PhD thesis of one of the authors[87] was used. It is based on analysing the evolution of the crystal size distribution (CSD) during a few well-designed polythermal heating

Table 12.4 Nucleation time measurements (t_{nucl}) in the coupled FBC facility (T_{sat}, saturation temperature of the racemic asparagine solution; T_{cryst}, crystallizer temperature).[86]

T_{sat} [°C]	T_{cryst} [°C]	Supersaturation [–]	t_{nucl} [h]
35	20	1.88	1
35	24	1.55	3
35	27	1.37	>24

– cooling experiments. The desired information is subsequently derived from a characteristic part of the particle collective which enables a sequential quantification of the corresponding nucleation, growth and dissolution kinetics.[88,89] Experiments were performed in a 2.5 L batch crystallizer under temperature and concentration *i.e.*, the supersaturation conditions of the continuous preferential crystallization trials. Liquid phase changes were followed *via* inline density monitoring (total solution concentration) and chiral HPLC analyses after periodic sampling (solution composition). The solid phase evolution with time was recorded by an online microscope (QicPic, Sympatec) connected to the crystallizer *via* a sampling line. Up to 25 images per second were taken and subsequently, automatically evaluated with respect to CSD *via* an image processing toolbox programmed in Matlab and a modified algorithm regarding single crystals.[90] In addition, at the end of each batch after solid–liquid separation, the crystallized material was subjected to a sieve analysis for the final CSD and each sieve fraction to HPLC enantiomeric composition evaluation. Crystallization kinetics were determined for single L-/D-asparagine enantiomers in both racemic and enantiopure solutions.[85]

As a result, it could be shown that nucleation and growth kinetics differ significantly in presence of the counter-enantiomer. In Figure 12.21 exemplarily growth rate data of L-asparagine monohydrate, ascertained in racemic and L-asparagine solution, are presented in a 3D-plot as a function of crystallization temperature and applied supersaturation.

As expected, growth rates increase with both temperature and supersaturation, while in the range of interest the impact of supersaturation is considerably higher than that of temperature. Comparing growth kinetics in racemic

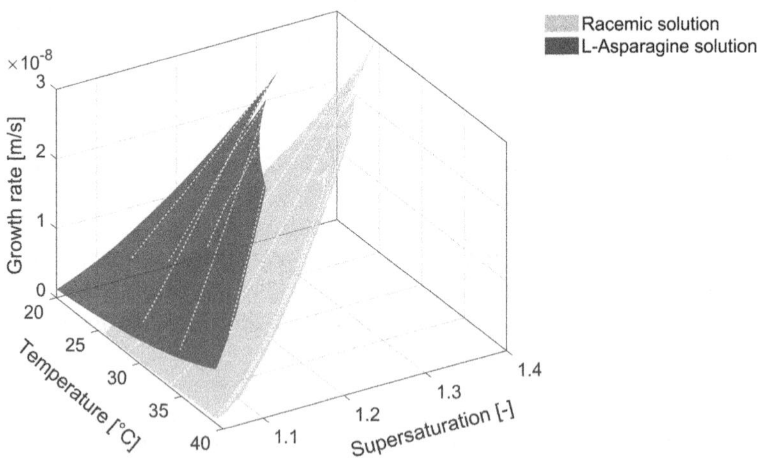

Figure 12.21 Growth rate of L-asparagine monohydrate in a racemic (light grey) and L-asparagine (grey) solution as a function of crystallization temperature and supersaturation. Dotted lines mark the experimentally studied parameter field. Reprinted with permission from ref. 85. Copyright 2019 American Chemical Society.

and pure L-asparagine solution, it becomes clear that the growth rates in enantiopure solution exceed those in racemic solution in the entire region of crystallization conditions covered by the experiments. Moreover, in presence of the counter-enantiomer, the dependence of growth rate on supersaturation and temperature becomes weaker. Thus, in contrast to solubility studies which revealed ideal solution behaviour of the enantiomers in aqueous solution, a significant effect of the counter-enantiomer on crystallization kinetics of the target enantiomer is observed. A decreasing growth rate in the presence of an impurity in general is often attributed to adsorption of the impurity molecules on the growing surface of the target crystals, thereby inhibiting surface diffusion and incorporation of the target molecules into the crystal lattice while decelerating crystal growth. The experiments also verified identical growth rates of L- and D-asparagine monohydrate in racemic solutions, which is a favourable property of the system for simultaneous preferential crystallization to be performed. The depressed growth rate of L-asparagine monohydrate in the presence of the D-species has also been confirmed in a recent article.[91]

12.5.2.3 Implementation of Coupled Continuously Operated Fluidized Bed Crystallizers

A pilot plant setup was constructed and improved within the last few years to study the performance of preferential crystallization in combination with fluidized bed reactors. A scheme of the plant (based on the principal setup shown in Figure 12.12) is depicted in Figure 12.22. Core parts of the plant are two identically constructed double jacketed glass-made tubular FBCs of 0.5 L volume having a conical bottom part. A feed tank of 6 L volume steadily supplies saturated racemic solution which is fed *via* filter frits (g) into the bottom of both crystallizers and afterwards recycled back *via* heated tubes (a) into the feed tank. Leaving mother liquor is analyzed inline with respect to optical rotation and concentration by means of polarimeters (α) and ultrasound devices (c). Several Pt100 sensors (T) and flowmeters (not shown) control temperature and solution flow continuously. To initiate crystallization, seeds of the respective enantiomer are added once to the corresponding crystallizer from the top. Large crystals or crystal agglomerates accumulating during continuous operation at the FBCs bottom are continuously extracted as suspension, fragmented in a targeted way (using ultrasound or a wet mill (z)), and afterwards, mixed with the main inlet flow, re-introduced as seed material into the FBCs bottom. This seed generation loop ensures the continuous operation of the separation process, providing the required seeds and preventing clogging of the crystallizer inlet. Product crystals of the two enantiomers were collected periodically after reaching a certain bed height *via* product outlets at both FBCs (b). Suspension removal is triggered automatically by optical sensors and realized *via* an electromagnetic valve. More details are provided in ref. 86.

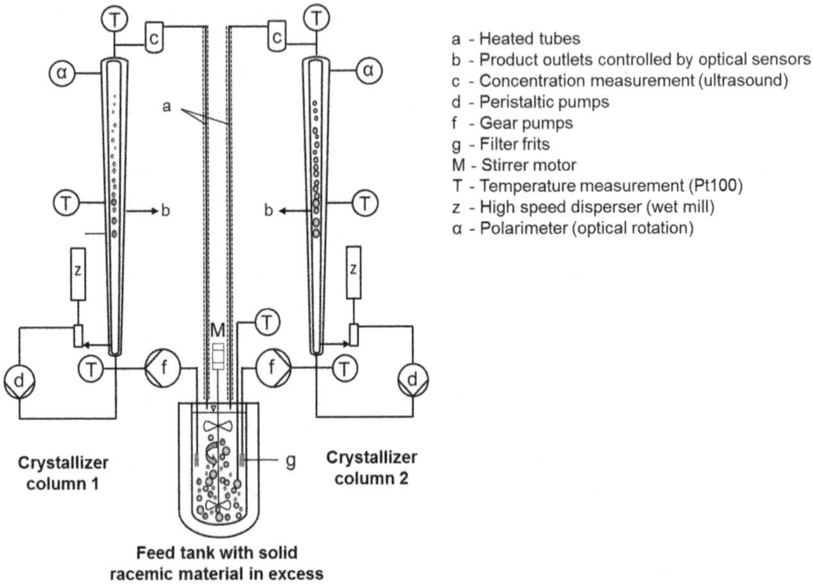

Figure 12.22 Scheme of the realized continuous enantiomer resolution setup with coupled fluidized bed crystallizers, a shared racemate feed tank, periodic product removal ports, continuous seed generation loops and process monitoring (temperature, optical sensors for detection of (product) bed height, densitometers and polarimeters for concentration and enantiomer composition observation).

12.5.2.4 Application of Coupled Continuously Operated Fluidized Bed Crystallizers

The feasibility of providing both pure asparagine enantiomers in crystalline form through continuous preferential crystallization in coupled FBCs has been reported by the authors' research group recently.[86] Typical process trajectories showing the periods of start-up, steady-state operation and shut-off are depicted in Figure 12.23. Illustrated is the progress of solution concentration and relative optical rotation as a function of time for both crystallizers, producing L- and D-asparagine monohydrate (marked in dark and light grey, respectively). Soon after initiation of crystallization by addition of seeds, steady state operation is observed indicated by constant and almost identical solution concentration at the crystallizer outlets. Further, the optical rotation of the mother liquors remains constant at values close to zero exhibiting a slight enantiomeric excess of opposite sign. Thus, crystallizing L-asparagine monohydrate generates a very small excess of the counter-D-(+)-enantiomer in the solution which is reversed for the D-asparagine monohydrate. The solution composition thus remains close to racemic composition, a beneficial behavior in comparison to the classical batch preferential crystallization

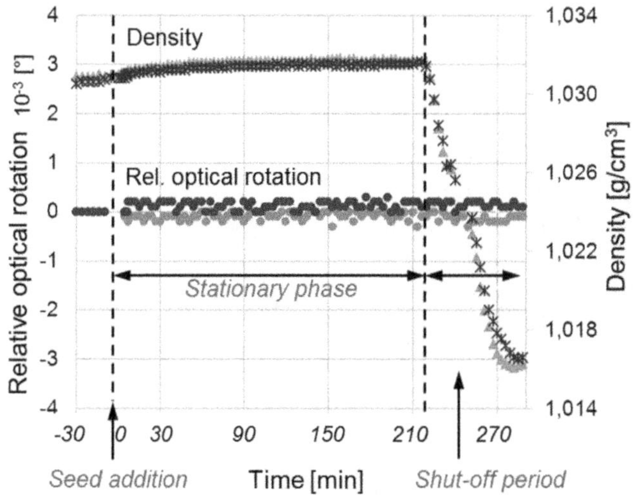

Figure 12.23 Trajectories of the solution concentration (as density) and relative optical rotation for a preferential crystallization experiment in the FBCs setup. Dark and light grey specify crystallizers for L- and D-asparagine. Experimental conditions were as follows: T_{sat} = 35 °C, T_{cryst} = 27 °C, mass and mean particle size of the seeds: 4 g and 55 μm respectively, feed flow rate = 9 L h^{-1}, excess of solid racemate in feed tank = 80 g (adapted from ref. 83).

processes. After ~220 minutes, the excess of solid racemic feed from the feed tank is consumed, finishing the steady-state run and leading to depletion of supersaturation (*i.e.* concentration) afterwards. During the shut-off period, preferential crystallization of the desired enantiomer in both crystallizers further proceeds, as indicated by the unchanged values of optical rotation. The crystallization process was stopped after ~4.7 hours and the last portion of product crystals was removed. Within the steady-state phase 46.2 g L- and 49 g D-asparagine monohydrate of purity ≥97% (dry, without washing) were produced; the shut-off period provided an additional 76 g of each enantiomer. As seen from the crystal size distribution of the steady-state products (Figure 12.24), uniform product crystals with a mean size of ~180 μm were generated for both enantiomers. The product amounts, CSDs and achieved purities prove that the coupled FBC setup is capable of continuously resolving racemic asparagine into its single enantiomers at high purity and uniform CSD in a robust mirror-image process operation.

Recent work was focused on a systematic study of the parameters influencing the separation performance with respect to product purity and particle size as well as productivity and yield. Main parameters varied were crystallization temperature, supersaturation and feed solution flow rate.

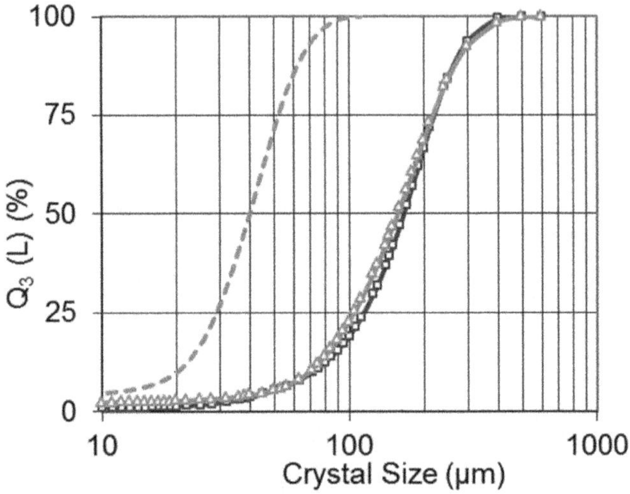

Figure 12.24 Seed (dashed line) and steady-state product crystal size distribution (light grey triangles: D-Asn·H_2O, dark grey squares: L-Asn·H_2O) as result from the separation shown in Figure 12.23 (adapted from ref. 83).

The steady-state behavior and reproducibility of product and process characteristics with regard to the same steady-state conditions has also been investigated.

Figure 12.25 shows exemplarily the evolution of process and product parameters for an experiment performed at constant process conditions for a period of nine hours. At $t = 0$ seeds were added to initiate crystallization. Periodically after detecting the bed height limit, product suspension was removed, filtered *via* a vacuum filter and the filter residue was washed according a preassigned protocol. After each product removal, 20 g of solid racemate was added to the feed tank. Shown are the mean product crystal size (μ_{product}), purity (Pur) and productivity of the respective period (Pr). As seen, the product properties remain constant within certain limits: the mean crystal size varies between 315 and 335 µm (standard deviation $\sigma = \pm$ 58–60 µm) while holding a purity mainly exceeding 99%. It should be noted that all the continuous resolution runs performed at stable crystallization conditions provided L- and D-asparagine monohydrate with purities higher than 99% (for the entire product mass obtained).

In Figure 12.26 the influence of the feed flow rate on crystal size distribution (CSD) of both enantiomers is demonstrated. Several experimental runs have been conducted for each of the selected flow rates, providing a remarkable, near perfect reproducibility. The CSDs received are monomodal, specifying the width of distribution between ~150 and 450 µm. With increasing flow rate, the mean crystal size (μ) significantly increases, because at higher

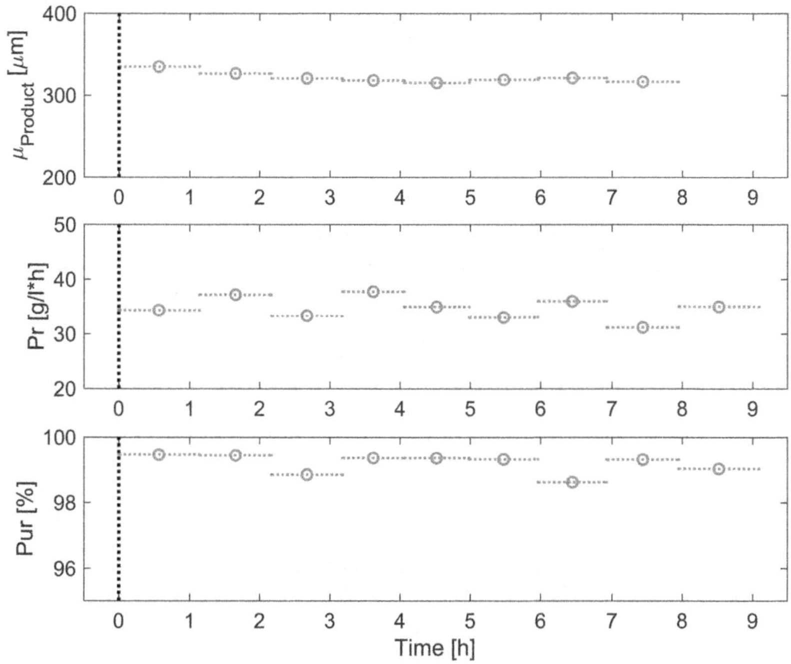

Figure 12.25 Performance of a continuous preferential crystallization experiment over time illustrated on evolution of mean crystal size ($\mu_{product}$), enantiomer purity (Pur) and productivity of the respective period (Pr). Points refer to the periodic automatic product suspension removal. Improved experimental conditions were as follows: T_{sat} = 35 °C, T_{cryst} = 30 °C, feed flow rate = 12 L h^{-1}. Shown are data for crystallization of D-Asn·H_2O.

fluid flow velocities bigger particles settle at the position of the product outlet. Moreover, the different operating points (10, 12 and 14 L h^{-1}) can be clearly distinguished according to the moderate standard deviation of the data, particularly for the L-Asn·H_2O case. Thus, definite crystal sizes (CSDs) can be target-oriented manufactured by adjusting the feed flow rates in the coupled FBC setup. The mean crystal sizes of D- and L-Asn·H_2O obtained differ steadily by about 10 μm with the bigger crystals for D-Asn·H_2O. This can be attributed to slight differences in the crystallizers' construction and the related flow characteristics, which cannot be discussed in detail here. Nevertheless, the CSDs for the two enantiomers are very similar and verify the mirror-shape symmetry of the process concept. It should be noted that Figure 12.26 depicts 16 different products from three operation points, which were reached in two separate experiments. The products of the same flow rates nearly superimpose each other and the products of Exp. 1 & 2 at 12 L h^{-1} are very close together, proving the high reproducibility of the FBC process.

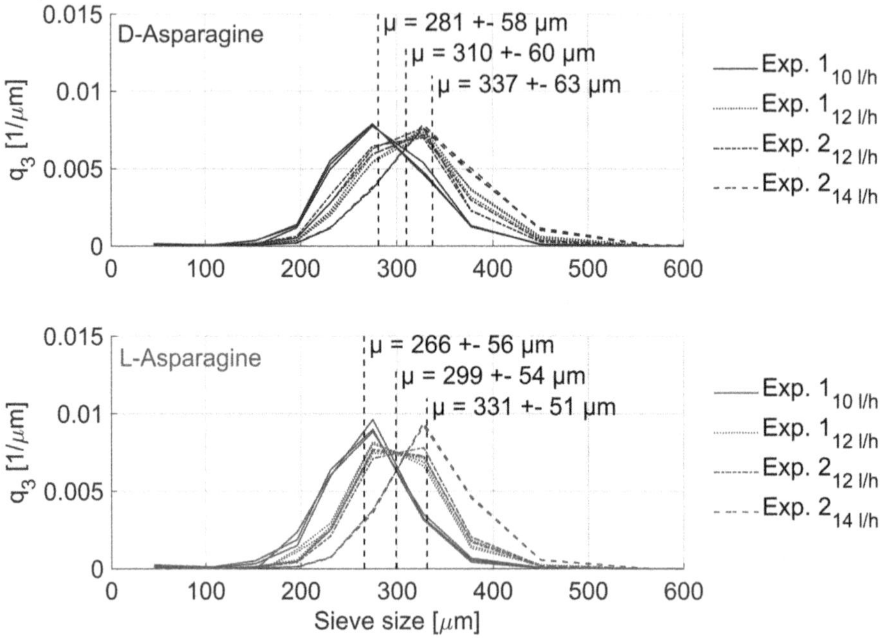

Figure 12.26 Influence of the feed solution flow rate on steady-state crystal size distribution in both crystallizers (top: D-Asn·H_2O, bottom: L-Asn·H_2O). Different line types in the graphic mark separate continuous resolution runs and numbers denominate mean size and standard deviation.

The effect of the crystallization temperature on productivity and yield is presented in Figure 12.27. The saturation temperature of the feed solution has been adapted in these experiments in order to ensure constant supersaturation. When working at the same saturation temperature but different crystallization temperatures, both supersaturation and crystallization temperature vary. For such experiments, a significant increase of productivity with increasing ΔT was observed as a result of the predominant influence of the supersaturation. Figure 12.27 proves that an increase in the crystallization temperature leads to elevated values for productivity and yield, which is caused by faster growth kinetics. A drop in crystallization temperature from 30 to 20 °C caused a decrease of the productivity by ~75%. The larger values of L-Asn·H_2O are again attributed to small differences in the experimental set-up regarding, for example, the pump outputs and the manufacturing tolerances of the crystallizer columns, which demonstrates the sensitivity of the FBC process to the fluid dynamics. It is therefore essential for the successful operation that the crystallization and the flow regime are well-balanced with each other.

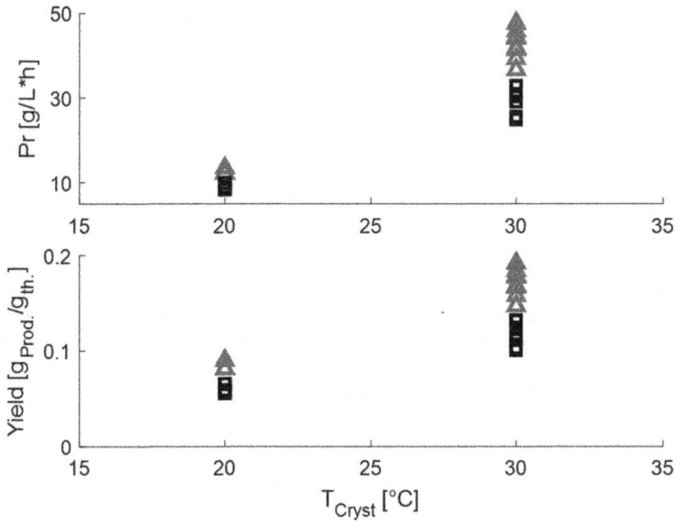

Figure 12.27 Influence of crystallization temperature on productivity (Pr) and yield at constant supersaturation ($S_{D/L}$ = 1.23, feed flow rate = 12 L h^{-1}, triangles: L-Asn·H_2O, squares: D-Asn·H_2O).

In parallel to the experimental investigations, complementary modeling work substantially supported understanding of (1) the flow characteristics in the fluidized bed crystallizer using CFD-DEM simulations[92] and (2) the relations of geometrical effects in combination with other operation parameters.[93] The process model developed[94] was utilized to identify the key operation and design parameters for maximizing the mass of product crystals and shaping their size distribution. It consists of a population balance for the particles in the crystallizer, a mass balance for the solute in the liquid phase, and a population balance for the particles in the fragmentation cycle. The model could, for example, reproduce an observed increase in productivity and crystal size with an increase in the flow rate as shown above. It was also shown that for the feed flow rate there is an optimum value where the productivity reaches a maximum.[93]

The successful applicability of the concept of coupled continuously-operated fluidized bed crystallizers could further be demonstrated in independent work for separation of two positional isomers, namely *ortho*- and *para*-aminobenzoic acid (OABA, PABA).[83,86]

12.5.2.5 Comparison with Batchwise Operated PC

As shown above regarding the threonine system (Section 12.5.1.6), we also present here a short summary comparing the productivities and purities of three batch variants of preferential crystallization with the coupled

Table 12.5 Purities and productivities (eqn (12.7)) of different batch and continuous processes carried out for the substance asparagine monohydrate (first number: L-Asn, second number: D-Asn; productivity in gram pure product/volume of solution and dead (2 h) + process time except for the continuous FBC (no dead time).

	Process	Purity [%]	Productivity $[g\ (h^{-1}\ L^{-1})]$
Batch	Conventional PC (polythermal)	99.7	1.6
	CPC (isothermal)	94.7/98.5	2.6/2.8
	CPC-D	91.7/99.3	3.7/2.5
Cont.	Coupled fluidized bed process (FBCs)	>97/>97[a,b]	19.7/20.9[b]

[a]Dry, without washing
[b]Later experiments gave even better purities and productivities (Figure 12.25)

fluidized bed process for the asparagine monohydrate system (Table 12.5). Similar trends for the different batch options were found. However, in this case the advantage of the continuous operation is more clearly pronounced. Again, all processes considered were not fully optimized.

12.6 Conclusions and Outlook

Preferential crystallization (PC) is a very attractive technique capable of directly resolving racemic mixtures of conglomerate forming chiral systems. We illustrated in this chapter of the book that there are several applicable process options available, performing the resolution batch-wise or in a continuous manner. A rational selection and a fair quantitative comparison between the rival options is not trivial and needs to be done considering compound-specific thermodynamic and kinetic data. Making these data available, short-cut and more detailed models can be of large help for selecting the best process mode and optimizing it. In particular, Preferential Crystallization in coupled fluidized beds appears to be promising, but also requires further investigation.

There are currently major efforts being made in order to enlarge the potential of PC by identifying metastable conglomerates of otherwise racemic compound-forming systems or derivatives possessing conglomerate behavior.[34,50,95,96] Also, there are attractive options available regarding the application of PC for processing enriched feed solutions of racemic compound-forming systems, *e.g.* through periodic addition of seed crystals of the pure enantiomer and the racemic compound.[35,36,97] Research has also intensified to include racemization of the unwanted enantiomer in the process chain in order to install recycling loops with the benefit of increased overall yields.[98] Furthermore, development and application of system-based workflows supporting a quick design of robust crystallization processes providing distinct product properties is suggested in the frame of pharmaceutical manufacturing.[99,100] Finally, a rational process selection requires evaluating crystallization-related alternatives beyond PC, such as *e.g.* Viedma Ripening,[101–103] processes applying grinding[104] and temperature cycling,[105] or other enantioselective separation techniques, such as chromatography using chiral stationary phases or liquid–liquid extraction using soluble chiral selectors.[21]

Abbreviations

CFD-DEM	computational fluid dynamics/discrete element method (model)
CPC	coupled batch preferential crystallization
CPC-D	coupled preferential crystallization and selective dissolution
CSD	crystal size distribution
D/L/DL	L-enantiomer/D-enantiomer/racemic composition
D-Asn/L-Asn	D- or L-asparagine
D-Thr/L-Thr	D- or L-threonine
FBC	fluidized bed crystallizer
HPLC	high performance liquid chromatography
l	liquid phase
MSMPR	mixed-suspension, mixed-product-removal concept
MSZW	metastable zone width
OABA/PABA	*ortho-/para*-aminobenzoic acid
PC	preferential crystallization
s	solid phase
SLE	solid–liquid phase equilibria

Roman Symbols

A	[cm^2]	surface
B	[$kg\ s^{-1}$]	mass depletion due to nucleation
c	[$g\ cm^{-3}$]	concentration
ee	[–]	enantiomeric excess
G	[$kg\ s^{-1}$]	mass depletion due to growth
k	[–]	kinetic constant
m	[g]	mass
N	[#]	crystal number
n	[–]	kinetic exponent
Pr	[$g\ L^{-1}\ h^{-1}$]	productivity
Pur	[–; %]	purity
q_3	[μm^{-1}]	mass related particle size distribution
Q_3	[%]	mass related cumulative particle size distribution
R	[cm]	crystal radius
S	[–]	relative supersaturation
T	[°C]	temperature
t	[h, min, s]	time
V	[cm^3]	volume
x	[–]	mass fraction

Greek Symbols

Δ	[–]	difference
μ	[μm]	mean particle size
ρ	[g cm^{-3}]	density
σ	[μm]	standard deviation

Superscripts

eff	effective variable
DL	racemic composition
L/D	pure enantiomer
0	initial model value

Subscripts

cryst	crystallization
eff	effective variable
diss	dissolution
G	growth
i	component i = A, B, solvent
L/D	pure enantiomer
Liq	liquid phase
N/Nuc	nucleation
sat	saturation
Sol/Solid	solid phase
0	conditions at the start of the process

Acknowledgements

The support by the European Union through the Project "Continuous Resolution (CORE)", the German Science Foundation within the Priority Program SPP-1679, the International Max Planck Research School "ProEng" and our former and current colleagues Linzhu Gou, Daniel Polenske, Kamila Galan, Matthias Eicke, Gonzalo Cambero Chaves and Jonathan Gänsch are gratefully acknowledged.

References

1. E. L. Eliel, S. Wilen and M. Doyle, *Basic Organic Stereochemistry*, Wiley-Interscience, New York, 2001.
2. U. Meierhenrich, *Amino Acids and the Asymmetry of Life*, Springer, Berlin, 2008.
3. *Origin of Life – Chemical Approach*, ed. P. Herdewijn and M. V. Kisakürek, Verlag Helvetica Chimica Acta, Zürich, 2008.

4. (a) P. Cintas, *Angew. Chem. Int. Ed.*, 2002, **41**, 1139; (b) P. Cintas, *Angew. Chem.*, 2002, **114**, 1187.
5. D. P. Glavin, J. E. Elisa, A. S. Burton, M. P. Callahan, J. P. Dworkin, R. W. Hilts and C. D. K. Herd, *Meteorit. Planet. Sci.*, 2012, **47**, 1347.
6. A. M. Rouhi, *Chem. Eng. News*, 2003, **81**(18), 56.
7. R. A. Sheldon, *Chirotechnology*, Marcel Dekker, New York, 1993.
8. *Chiral Drugs: Chemistry and Biological Action*, ed. G.-Q. Lin, Q.-D. You and J.-F. Cheng, John Wiley & Sons, Hoboken/NJ, 2011.
9. S. C. Stinson, *Chem. Eng. News*, 2001, **79**(20), 45.
10. A. M. Rouhi, *Chem. Eng. News*, 2003, **81**(18), 45.
11. J. Blumenstein, in *Chirality in Industry II. Developments in the Manufacture and Applications of Optically Active Compounds*, ed. A. N. Collins, G. N. Sheldrake and J. Crosby, John Wiley & Sons, Chichester, 1997, p. 11.
12. J. Crosby, in *Chirality in Industry. The Commercial Manufacture and Applications of Optically Active Compounds*, ed. A. Collins, G. Sheldrake and J. Crosby, John Wiley & Sons, Chichester, 1992, p. 1.
13. J. Crosby, in *Chirality in Industry II. Developments in the Manufacture and Applications of Optically Active Compounds*, ed. A. Collins, G. Sheldrake and J. Crosby, John Wiley & Sons, Chichester, 1997, p. 1.
14. B. A. Asteleford and L. O. Weigel, in *Chirality in Industry II. Developments in the Manufacture and Applications of Optically Active Compounds*, ed. A. Collins, G. Sheldrake and J. Crosby, John Wiley & Sons, Chichester, 1997, p. 119.
15. A. Collet, *Enantiomer*, 1999, **4**, 157.
16. J. Jacques, A. Collet and S. H. Wilen, *Enantiomers, Racemates and Resolutions*, Krieger, Malabar, 1994.
17. H.-H. Tung, E. L. Paul, M. Midler and J. A. McCauley, *Crystallization of Organic Compounds: An Industrial Perspective*, Wiley & Sons, Hoboken/NJ, 2009.
18. L. Synoradzki, H. Hajmowicz, J. Wisialski, A. Mizerski and T. Rowicki, *Org. Process Res. Dev.*, 2008, **12**, 1238.
19. M. Leeman, F. Querniard, T. R. Vries, B. Kaptein and R. M. Kellogg, *Org. Process Res. Dev.*, 2009, **13**, 1379.
20. A. Collet, M.-J. Brienne and J. Jacques, *Chem. Rev.*, 1980, **80**, 215.
21. H. Lorenz and A. Seidel-Morgenstern, *Angew. Chem., Int. Ed.*, 2014, **53**, 1218.
22. G. Coquerel, in *Novel Optical Resolution Technologies*, ed. K. Sakai, N. Hirayama and R. Tamura, Topics in Current Chemistry, 2007, vol. 269, p. 1.
23. M. Matsuoka, in *Advances in Industrial Crystallization*, ed. J. Garside, R. J. Davey and A. G. Jones, Butterworth-Heinemann, Oxford, 1991, p. 229.
24. L. Renou, T. Morelli, S. Coste, M.-N. Petit, B. Berton, J.-J. Malandain and G. Coquerel, *Cryst. Growth Des.*, 2007, **7**, 1599.
25. N. Taratin, H. Lorenz, E. Kotelnikova, A. Glikin, A. Galland, V. Dupray, G. Coquerel and A. Seidel-Morgenstern, *Cryst. Growth Des.*, 2012, **12**, 5882.

26. A. I. Isakov, E. N. Kotelnikova, L. Yu. Kryuchkova and H. Lorenz, in *BIWIC 2013-20th International Workshop on Industrial Crystallization*, ed. H. Qu, J. Rantanen and C. Malwade, University of Southern Denmark, Odense, Denmark, 2013, p. 395.
27. A. A. Bredikhin, D. V. Zakharychev, A. T. Gubaidullin, R. R. Fayzullin, A. V. Pashagin and Z. A. Bredikhina, *Cryst. Growth Des.*, 2014, **14**(4), 1676.
28. A. I. Isakov, E. N. Kotelnikova, S. Münzberg, S. N. Bocharov and H. Lorenz, *Cryst. Growth Des.*, 2016, **16**, 137.
29. R. R. Fayzullin, D. V. Zakharychev, A. T. Gubaidullin, O. A. Antonovich, D. B. Krivolapov, Z. A. Bredikhina and A. A. Bredikhin, *Cryst. Growth Des.*, 2017, **17**(1), 271.
30. T. Rekis, A. Berzins, L. Orola, T. Holczbauer, A. Actins, A. Seidel-Morgenstern and H. Lorenz, *Cryst. Growth Des.*, 2017, **17**, 1411.
31. T. Rekis, A. Berzins, I. Sarcevica, A. Kons, M. Balodis, L. Orola, H. Lorenz and A. Actins, *Cryst. Growth Des.*, 2018, **18**, 264.
32. A.-K. Kort, H. Lorenz and A. Seidel-Morgenstern, *J. Chem. Eng. Data*, 2017, **62**, 4411.
33. E. N. Kotelnikova, A. I. Isakov and H. Lorenz, *CrystEngComm*, 2017, **19**, 1851.
34. European ITN project CORE, https://www.coreitn.eu/home.htm, last accessed November 2019.
35. D. Polenske, Bewertung des Potentials der Bevorzugten Kristallisation zur Enantiomerentrennung, PhD thesis, Otto von Guericke University, Magdeburg, 2010.
36. (a) A. Seidel-Morgenstern, H. Lorenz and D. Polenske, Max-Planck-Gesellschaft zur Förderung der Wissenschaften e.V., *US Pat.*, US 7, 820,860 B2, 26 October 2010; (b) A. Seidel-Morgenstern, H. Lorenz and D. Polenske, Max-Planck-Gesellschaft zur Förderung der Wissenschaften e.V., *EP Pat.*, EP 1 924 540 B1, 03 July 2013.
37. J. Nývlt, O. Söhnel, M. Matuchová and M. Boul, *The Kinetics of Industrial Crystallization*, Elsevier, Amsterdam, 1985.
38. K. Sangwal, *Additives and Crystallization*, Wiley, New York, 2006.
39. J. Nývlt and J. Ulrich, *Admixtures in Crystallization*, VCH, Weinheim, 1995.
40. I. Weissbuch, L. Addadi, M. Lahav and L. Leiserowitz, *Science*, 1991, **253**(5020), 637.
41. K. A. Burrill, *J. Cryst. Growth*, 1972, **12**, 239.
42. B. Z. Kuvadia and M. F. Doherty, *Cryst. Growth Des.*, 2013, **13**, 1412.
43. N. Kubota, *Cryst. Res. Technol.*, 2001, **36**, 749.
44. H. Eisenschmidt, A. Voigt and K. Sundmacher, *Cryst. Growth Des.*, 2015, **15**, 219.

45. N. Zaitseva, L. Carman, I. Smolsky, R. Torres and M. Yan, *J. Cryst. Growth*, 1999, **204**, 512.
46. L. N. Rashkovich and N. V. Kronsky, *J. Cryst. Growth*, 1997, **182**, 434.
47. M. Lahav and L. Leiserowitz, *J. Phys. D: Appl. Phys.*, 1993, B22.
48. L. Klukas, H. Lorenz and A. Seidel-Morgenstern, *Cryst. Growth Des.*, 2012, **12**, 5197.
49. F. Czapla, D. Polenske, L. Klukas, H. Lorenz and A. Seidel-Morgenstern, *Chem. Eng. Process.*, 2010, **49**, 22.
50. G. Levilain and G. Coquerel, *CrystEngComm*, 2010, **12**, 1983.
51. D. Zbaida, M. Lahav, K. Drauz, G. Knaup and M. Kottenhahn, *Tetrahedron*, 2000, **56**, 6645.
52. N. Doki, M. Yokota, S. Sasaki and N. Kubota, *Cryst. Growth Des.*, 2004, **4**(6), 1359.
53. S. E. Gilks and R. J. Davey, *Chem. - Eur. J.*, 2016, **22**, 74.
54. J. W. Mullin and J. Nývlt, *Chem. Eng. Sci.*, 1971, **26**, 369.
55. F. Czapla, Modeling of polythermal preferential crystallization, PhD thesis, Otto von Guericke University, Magdeburg, 2010.
56. M. P. Elsner, G. Ziomek and A. Seidel-Morgenstern, *AIChE J.*, 2009, **55**, 640.
57. S. Qamar, K. Galan, M. P. Elsner, I. Hussain and A. Seidel-Morgernstern, *Chem. Eng. Sci.*, 2013, **98**, 25.
58. A. Majumder and Z. K. Nagy, *Pharmaceutics*, 2017, **9**, 55.
59. T. Köllges and T. Vetter, *Cryst. Growth Des.*, 2018, **18**, 1686.
60. T. Carneiro, S. Bhandari, E. Temmel, H. Lorenz and A. Seidel-Morgenstern, *Cryst. Growth Des.*, 2019, **19**, 3148.
61. M. P. Elsner, G. Ziomek and A. Seidel-Morgenstern, *Chem. Eng. Sci.*, 2011, **66**, 1269.
62. M. J. Eicke, Process Strategies for Batch Preferential Crystallization, PhD thesis, Otto von Guericke University, Magdeburg, Shaker, Aachen, 2016.
63. L. Gou, Erweiterung des Anwendungspotenzials der Bevorzugten Kristallisation zur Enantiomerentrennung, PhD thesis, Otto von Guericke University, Magdeburg, Dr Hut, München, 2016.
64. L. Klukas, D. Polenske, H. Lorenz and A. Seidel-Morgenstern, in *BIWIC 2009-16th International Workshop on Industrial Crystallization*, ed. M. Louhi-Kultanen and H. Hatakka, Lappeenranta University of Technology, Lappeenranta, Finland, 2009, p. 218.
65. G. Levilain, M. J. Eicke and A. Seidel-Morgenstern, *Cryst. Growth Des.*, 2012, **12**, 5396.
66. M. J. Eicke, G. Levilain and A. Seidel-Morgenstern, *Cryst. Growth Des.*, 2013, **13**, 1638.

67. E. Temmel, M. J. Eicke, F. Cascella, A. Seidel-Morgenstern and H. Lorenz, *Cryst. Growth Des.*, 2019, **19**, 3148.
68. K. Galan, Continuous Preferential Crystallization of Enantiomers: Simulation, Analysis, Process Design and Experimental Validation, PhD thesis, Otto von Guericke University, Magdeburg, Docupoint, Barleben, 2016.
69. D. R. Randolph and M. A. Larson, *Theory of Particulate Processes*, Academic Press Inc., San Diego, 1988, vol. 2.
70. (a) M. Midler Jr, *US Pat.*, 3,892,539, 1975; (b) M. Midler Jr, *US Pat.*, 3,996,018, 1976.
71. E. J. J. Grabowski, *Chirality*, 2005, **17**, 249.
72. D. Binev, A. Seidel-Morgenstern and H. Lorenz, *Chem. Eng. Sci.*, 2015, **133**, 116.
73. *Löffler/Petrides Biochemie und Pathobiochemie*, ed. P. C. Heinrich, M. Müller and L. Graeve, Springer, Berlin, 2014.
74. M. A. Bonato, N. K. Sakomura, J. C. Siqueira, J. B. K. Fernandes and R. M. Gous, *S. Afr. J. Anim. Sci.*, 2011, **41**, 209.
75. M. T. Kidd and B. J. Kerr, *J. Appl. Poult. Res.*, 1996, **5**, 358.
76. M. Xie, L. Zhang, Z. G. Wen, J. Tang, W. Huang and S. S. Hou, *Br. Poult. Sci.*, 2014, **55**, 553.
77. V. M. Profir and M. Matsuoka, *Colloids Surf., A*, 2000, **164**, 314.
78. D. Sapoundjiev, H. Lorenz and A. Seidel-Morgenstern, *J. Chem. Eng. Data*, 2006, **51**, 1562.
79. D. Binev, N. Taratin, E. Kotelnikova, A. Seidel-Morgenstern and H. Lorenz, *Cryst. Growth Des.*, 2014, **14**, 367.
80. H. Lorenz, A. Perlberg, D. Sapoundjiev, M. P. Elsner and A. Seidel-Morgenstern, *Chem. Eng. Process.*, 2006, **45**, 863.
81. K. Galan, M. Eicke, M. P. Elsner, H. Lorenz and A. Seidel-Morgenstern, *Cryst. Growth Des.*, 2015, **15**, 1808.
82. U. Meierhenrich, *Amino Acids and the Asymmetry of Life*, Springer–Verlag, Berlin Heidelberg, 2008.
83. D. Binev, Continuous fluidized bed crystallization, PhD thesis, Otto von Guericke University, Magdeburg, 2014.
84. H. Lorenz, in *Crystallization: Basic Concepts and Industrial Applications*, ed. W. Beckmann, Wiley-VCH, Weinheim, 2013, p. 35.
85. E. Temmel, J. Gänsch, H. Lorenz and A. Seidel-Morgenstern, *Cryst. Growth Des.*, 2018, **18**, 7504.
86. D. Binev, A. Seidel-Morgenstern and H. Lorenz, *Cryst. Growth Des.*, 2016, **16**, 1409.
87. E. Temmel, Design of Continuous Crystallization Processes, PhD thesis, Otto von Guericke University, Magdeburg, Shaker, Aachen, 2016.

88. E. Temmel, H. Eisenschmidt, H. Lorenz, K. Sundmacher and A. Seidel-Morgenstern, *Cryst. Growth Des.*, 2016, **16**, 6743.
89. E. Temmel, M. Eicke, H. Lorenz and A. Seidel-Morgenstern, *Cryst. Growth Des.*, 2016, **16**, 6756.
90. C. Borchert, Topics in Crystal Shape Dynamics, PhD thesis, Otto von Guericke University, Magdeburg, Shaker, Aachen, 2012.
91. P. Kongsamai, A. Maneedaeng, C. Flood, J. H. ter Horst and A. E. Flood, *Eur. Phys. J.: Spec. Top.*, 2017, **226**, 823.
92. K. Kerst, C. Roloff, L. Medeiros de Souza, A. Seidel-Morgenstern, D. Thévenin and G. Janiga, *Chem. Eng. Sci.*, 2017, **165**, 1.
93. M. Mangold, D. Khlopov, E. Temmel, H. Lorenz and A. Seidel-Morgenstern, *Chem. Eng. Sci.*, 2017, **160**, 281.
94. M. Mangold, L. Feng, D. Khlopov, S. Palis, P. Benner, D. Binev and A. Seidel-Morgenstern, *J. Comput. Appl. Math.*, 2015, **289**, 253.
95. J. von Langermann, E. Temmel, A. Seidel-Morgenstern and H. Lorenz, *J. Chem. Eng. Data*, 2015, **60**, 721.
96. C. Brandel, Y. Amharar, J. M. Rollinger, U. J. Griesser, Y. Cartigny, S. Petit and G. Coquerel, *Mol. Pharmaceutics*, 2013, **10**, 3850.
97. (a) H. Lorenz, D. Polenske, L. Klukas and A. Seidel-Morgenstern, Max-Planck-Gesellschaft zur Förderung der Wissenschaften e.V., *EP Pat.*, EP 2 292 306 A1, 09 March 2011; (b) H. Lorenz, D. Polenske, L. Klukas and A. Seidel-Morgenstern, Max-Planck-Gesellschaft zur Förderung der Wissenschaften e.V., *US Pat.*, US 8,822,721 B2, 02 September 2014; (c) H. Lorenz, D. Polenske, L. Klukas and A. Seidel-Morgenstern, Max-Planck-Gesellschaft zur Förderung der Wissenschaften e.V., *JP Pat.*, JP5513621 B2, 04 June 2014.
98. K. Petrusevska–Seebach, Overcoming yield limitations when resolving racemates by combination of crystallization and/or chromatographywith racemization, PhD thesis, Otto von Guericke University, Magdeburg, 2012.
99. European Collaboration Project INTENANT 2008–2011, Training package Crystallization.
100. C. J. Brown, T. McGlone, S. Yerdelen, V. Srirambhatla, F. Mabbott, R. Gurung, M. L. Briuglia, B. Ahmed, H. Polyzois, J. McGinty, F. Perciballi, D. Fysikopoulos, P. MacFhionnghaile, H. Siddique, V. Raval, T. S. Harrington, A. D. Vassileiou, M. Robertson, E. Prasad, A. Johnston, B. Johnston, A. Nordon, J. S. Srai, G. Halbert, J. H. ter Horst, C. J. Price, C. D. Rielly, J. Sefcika and A. J. Florence, *Mol. Syst. Des. Eng.*, 2018, **3**, 518.
101. C. Viedma, *Phys. Rev. Lett.*, 2005, **94**, 065504.
102. M. W. van der Meijden, M. Leeman, E. Gelens, W. L. Noorduin, H. Meekes, W. J. P. van Enckevort, B. Kaptein, E. Vlieg and R. M. Kellog, *Org. Process Res. Dev.*, 2009, **13**, 1195.

103. W. L. Noorduin, P. van der Asdonk, A. A. C. Bode, H. Meekes, W. J. P. van Enckevort, E. Vlieg, B. Kaptein, M. W. van der Meijden, R. M. Kellog and G. Deroover, *Org. Process Res. Dev.*, 2010, **14**, 908.
104. L. Spix, W. J. P. van Enckevort, L. J. M. van der Wal, H. Meekes and E. Vlieg, *CrystEngComm*, 2016, **18**, 9252.
105. K. Suwannasang, A. E. Flood, C. Rougeot and G. Coquerel, *Org. Process Res. Dev.*, 2017, **17**, 623.

CHAPTER 13

Continuous Isolation of Active Pharmaceutical Ingredients

C. J. PRICE*[a], A. BARTON[b] AND S. J. COLEMAN[b]

[a]University of Strathclyde, Department of Chemical and Process Engineering, Montrose Street, Glasgow, G1 1XQ, Scotland, UK; [b]Alconbury Weston Ltd, Oldfield House, Galveston Grove, Stoke-on-trent, Staffordshire, England, ST4 3PE, UK
*E-mail: chris.price@strath.ac.uk

13.1 Introduction

Continuous isolation is well established in the commodity chemical industry however it is not deployed routinely in the pharmaceutical industry. Potential reasons for this include; the predominantly batch mode of the upstream synthesis and crystallization processes, the lack of appropriately small scaled cGMP compliant equipment and the wide scale adoption of batch agitated filter driers with the accompanying confidence process developers have in transferring processes into agitated filter driers. This situation is now being challenged by the uptake of continuous synthesis and the adoption of continuous secondary processing which demands a feed of highly consistent particles with good and highly reproducible powder flow properties.

This chapter begins by summarising the essential underlying science and engineering of filtration, washing and drying. It then reviews the equipment variants used in associated industry sectors, and considers how these could

The Handbook of Continuous Crystallization
Edited by Nima Yazdanpanah and Zoltan K. Nagy

Published by the Royal Society of Chemistry, www.rsc.org

fit into a pharmaceutical context. The chapter concludes by examining some common isolation process problems, their potential causes and solution strategies.

13.2 Underlying Science and Engineering

Classically the isolation steps which follow crystallization are filtration, washing and drying, which together deliver an isolated crystalline product which is chemically pure and has bulk powder attributes consistent with the intended final use. In the case of pharmaceuticals, the final product is the active pharmaceutical ingredient (API) which is formulated into the drug product, most commonly *via* blending with excipients and compression into tablets. The position of filtration, washing and drying in the overall process context is illustrated in Figure 13.1.

The upstream crystallization step may be designed for two purposes; to purify the API and simultaneously to engineer crystals of a specific form and size distribution. Alternatively, as a recrystallization, it may be designed to engineer particles with the required form and with exquisite control of other attributes. This difference appears significant for the isolation process; in the first instance washing is a critical task which must be designed to remove dissolved impurities of synthesis to deliver the latent purification achieved by the selectivity of crystal growth. In the second case the role of washing is to remove mother liquors which contain only dissolved API and to minimise changes to the size distribution

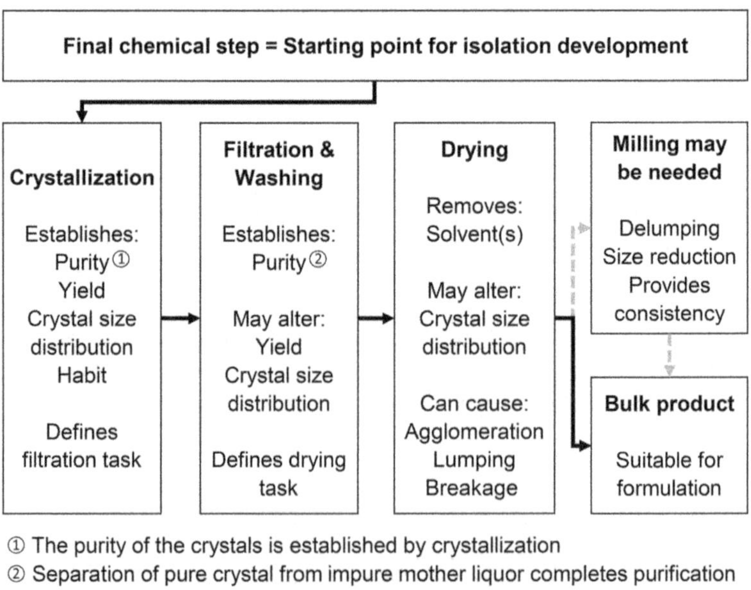

Figure 13.1 Overall context for filtration washing and drying.

arising from API deposition during the washing and drying steps. In fact, the requirement to prevent uncontrolled deposition of product dissolved in residual solvent during washing and drying means that the same approach to washing applies in either case. The drying process requires energy to provide the latent heat of vaporisation of the solvent which is evaporated. Delivery of this energy to the crystal surfaces from which solvent is to be evaporated without causing damage to the product crystals is one of the main challenges encountered in drying potentially heat sensitive APIs.

13.3 Filtration

A simple definition of filtration is the collection of particulate solids from suspension in a fluid as the fluid is driven through a porous medium. In this pharmaceutical application the valuable component is the crystalline drug substance and the fluid is the mother liquor solution from which the product has been crystallized. Classically the pharmaceutical industry employs dead end filtration in which the entire liquid volume is forced through the filter allowing all the particles above a collection threshold size to be retained. This collection threshold size is determined initially by the filter medium pore size but as a filter cake forms the collection threshold will be modified by the inter-particle path dimensions in the filter cake. This is determined by the product crystal size distribution and the filter cake structure. The process of dead end filtration is described in the equation first formalised by Darcy[2] which, in its most compact and widely used form, is shown in eqn (13.1).

$$\frac{t}{V} = \frac{\mu\alpha}{2\Delta P A^2}V + \frac{\mu R_m}{A\Delta P} \tag{13.1}$$

Here the duration of the filtration is expressed as the time, t, required to collect a volume of filtrate, V. Increasing values of t/V correspond to slower filtration rates, hence the viscosity of the solution being filtered, μ, the specific cake resistance, α, and the filter medium resistance, R_m are all in the numerator terms. The denominator terms which serve to increase the filtration rate are the pressure driving force, ΔP, and the filter area A. Filtration rate data can be collected relatively easily using a filter funnel, a calibrated filtrate receiver, a timer and a pressure controller. A suitable experimental arrangement for early development when API is in short supply is shown in Figure 13.2.

This is based on a Biotage Flash vac system in which the tube rack in the vacuum chamber has been modified to accommodate 50 mL measuring cylinders. Pressure control is provided by a Buchi vacuum controller. This design allows the experimentalist to charge a representative suspension sample to the filter tube, which is a syringe body, in the base of which a filter frit of known pore size has been positioned. When the desired pressure has been set in the vacuum chamber, filtration is started by opening a PTFE valve

(a) (b)

Figure 13.2 Small scale laboratory filtration unit adapted from a Biotage flash chromatography system; (a) filtration; (b) washing.

between the filter tube and the vacuum chamber, the experimentalist then records the time taken to collect predetermined volumes of filtrate. The data is plotted to generate a t/V *vs.* V plot in accordance with eqn (13.1); the value of the specific cake resistance, α, is derived from the gradient, and the filter medium resistance, R_m from the intercept. In principle a simple experiment along these lines provides the data needed to predict the duration of filtration at larger scale. However, there are some other important aspects must be considered.

13.3.1 Filter Medium and Medium Resistance

A number of key characteristics of the filter medium need to be considered. It must be chemically compatible with the intended process stream, for example, polymeric media must not be subject to leaching of plasticisers by the process solvents; metallic media should not be vulnerable to acid or base chemistry. The physical strength and rigidity of the filter medium must also be considered; the filter medium must not shed particles into the product. As the filter surface area is increased to provide sufficient capacity to deliver commercial quantities of product it becomes necessary to strengthen and to support the medium to prevent flexing. This dimensional stability is especially important when the product is to be removed from the filter by mechanical means and where close clearances are desirable, but the medium must not be damaged by mechanical contact. The filter pore size should be consistent across the entire medium surface and the filter structure should be such that the surface is not damaged easily. These requirements combined with commercial preference for filter media with a low resistance and ease of cleaning lead to a range of media types being available. The most significant classification is between depth filters and surface filters. Depth filters are relatively thick and form a cake by collecting product crystals within the structure of the filter medium. This makes depth filters excellent for recovering modest amounts of material from large process streams where the objective is to deliver a particle free solution stream, however, it difficult to recover all the product crystals from the filter at the end of the process and as a result depth filters are prone to particle blinding and tend to be single use devices. Where the desired product is the solid the preferred approach is to create a surface filter, typically these are multi-layer constructs in which the pore size is finest at the filtration surface but coarser on the exit side of the medium. The coarse structure is designed to support the fine structure providing the desired strength and rigidity whilst being freely draining and so not contributing to the medium resistance. Clearly it is desirable to use the intended media or an equivalent with similar characteristics when gathering laboratory data with the objective of predicting filtration scale up.

13.3.2 Specific Cake Resistance

The specific cake resistance is a characteristic which depends on the size distribution and shape of the crystals being isolated. Together these factors establish the cross-sectional area and tortuosity of the pores through which the filtrate must flow. It is essential that laboratory measurements are made on a particle suspension which is representative of the ultimate product. Indeed, in a manufacturing context where the crystallization is not well controlled and there is no *in situ* particle size measurement, the filtration rate is often the point where the variation in particle size is first identified. In order to correctly size a filter, it is essential to gather filtration rate data which spans the likely range of product crystal size distributions and solids loadings. Small particles have a disproportionate effect in increasing the specific cake resistance as does increasing the breadth of crystal size distribution, this is in part due to the increased efficiency of particle packing resulting in a lower void fraction and narrower pores through which the filtrate must pass. Fine particles tend to be carried through the cake towards the filter media where they are collected and contribute to increased resistance to filtrate flow. Increasing the pressure driving force for filtration is an obvious course of action to accelerate filtration. However, the specific cake resistance can increase with increasing pressure driving force; this phenomenon is known as cake compressibility and is typically linked with a reduction in filter cake volume with increasing pressure as the pore structure of the cake is compressed. Measuring the specific cake resistance at three different pressure driving forces allows the extent of cake compressibility to be determined and from this it is possible to determine the extent to which increasing pressure driving force will be useful in accelerating the filtration.

13.3.3 Mother Liquor Viscosity

The viscosity of the liquid which must pass through the network of pores which comprises the filter cake is obviously an important factor in determining the filtration rate. Viscosity increases with decreasing temperature, it also increases as the quantity of material dissolved in the solution increases. It is therefore important to evaluate the trade-off between isolating the product at a low temperature where solubility is usually lowest but viscosity may increase or alternatively sacrificing some yield and filtering at a higher temperature to achieve a faster filtration. A very rough rule of thumb is that solvent viscosity increases by between 10% and 20% for every 10 °C of cooling between 20 °C and 0 °C. This contrasts with "Black's rule",[1] typically, the solubility of organic compounds doubles every 20 °C the temperature increases, this would suggest it is advantage to maximise yield by isolating at the lowest practical temperature consistent with maintaining product purity.

Darcy's equation describes the flow of a fluid through a filter medium and through a filter cake as it forms. Reflecting on the equation indicates appropriate paths to take when selecting filters or designing filtration processes.

13.4 Washing

At the precise moment when the filter cake surface becomes exposed the cake is said to be saturated; that is all of the voids between the crystals are completely filled with the mother liquor from which the product was crystallized. If such a cake is then drained (usually termed deliquored) the solution in the large pores is removed. The mother liquor which is retained at the points of contact between particles is termed pendular solution. Depending on the contact angle of the mother liquor on specific crystal faces many of the crystal faces may also retain a thin film of mother liquor. This situation is illustrated in schematically in Figure 13.3. The relative volume of mother liquor which drains from large pores to that retained in the deliquored cake varies significantly with crystal habit and crystal size distribution and with solvent properties, especially viscosity.

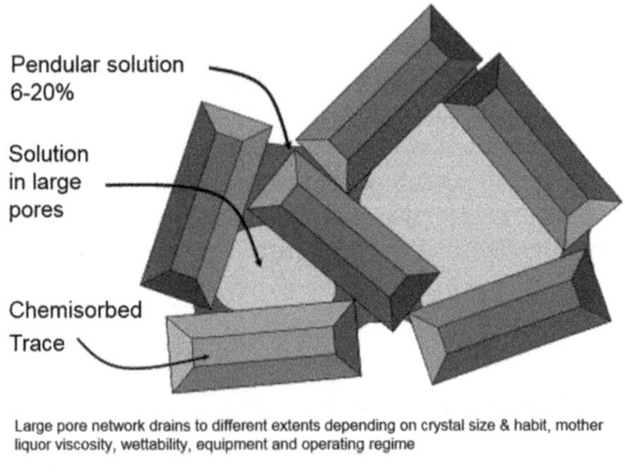

Figure 13.3 Schematic illustration of mother liquor hold up in a saturated filter cake.

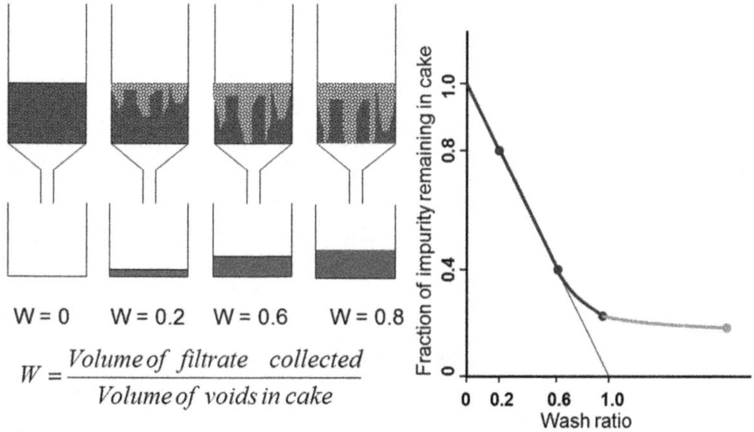

Figure 13.4 Deliquoring a filter cake and the effect of cracking.

If a filter cake is not washed but just drained (deliquored), all the product and impurities left in solution in the pendular mother liquor which remains in the cake will be deposited in the cake on drying, potentially leading to a lumpy and impure product. This is illustrated in Figure 13.4 in which the schematic image on the left indicates the passage of air in a laboratory setting or nitrogen in a plant context, through the cake. Once the gas penetrates the cake, bubbles can be seen on the lower surface of the filter medium. This is known as the bubble point, from this point on the rate of drainage of mother liquor from the cake tails off as the majority of the gas flow passes through the open pores rather than actively displacing mother liquor. The plot on the right of the figure indicates the typical fraction of the mother liquor and associated impurities remaining in the cake as the deliquoring progresses.

Three washing strategies are available to address this problem of incomplete removal of mother liquor; displacement washing of a fully saturated filter cake, washing an already deliquored cake, and resuspension washing a deliquored cake.

13.4.1 Displacement Washing

The wash liquid, either a pure solvent or solvent mixture, is forced through the already formed cake with the aim of displacing the mother liquor. Successful displacement washing requires that the wash solvent is applied to the filter cake without disturbing the cake surface. Delivering the wash solvent rapidly onto the saturated cake surface re-suspends product crystals and creates a thin point in the cake making a preferential path through which wash can pass more rapidly than through the remainder of the cake. This form of resuspension also causes the mother liquor held between the re-suspended particles to become mixed throughout the wash solvent volume making the wash less effective. If instead the wash is dispersed across the cake surface evenly, for example by atomisation, it is possible to establish a relatively flat wash front as the wash enters the cake. During the passage of the wash through the cake this wash front becomes more ragged and the boundary between mother liquor and wash becomes more diffuse as passage through the larger pores is faster than through the smaller ones. Thus to be reasonably effective the wash volume should exceed the mother liquor volume held up in the voids in the cake at the start of the wash. This is illustrated in Figure 13.5.

13.4.2 Deliquored Cake Washing

When a filter cake is deliquored the large pours are drained as illustrated in Figure 13.4. This is usually accompanied either by cake cracking or by cake shrinkage which often opens up a pathway between the filter vessel walls and the sides of the cake. The extent of cracking or shrinkage depends on the crystal size distribution, the crystal habit and the physical dimensions of the cake. In general cake cracking becomes more severe as the particle size becomes smaller and the cake becomes larger. The tendency for cake shrinkage is often associated with elongated, especially needle like particles which tend to form a three dimensional mat which rather than cracking shrinks from the walls of the filter body. Driving a wash solvent rapidly through a deliquored cake is accompanied by the risk that most of the wash solvent passes through the cracks and is not particularly effective in displacing the mother liquor from the un-cracked regions of the cake. Rather than forcing the wash through the cake in this situation it is preferable to maximise the wash efficiency by increasing the contact time – sometimes termed soak washing.

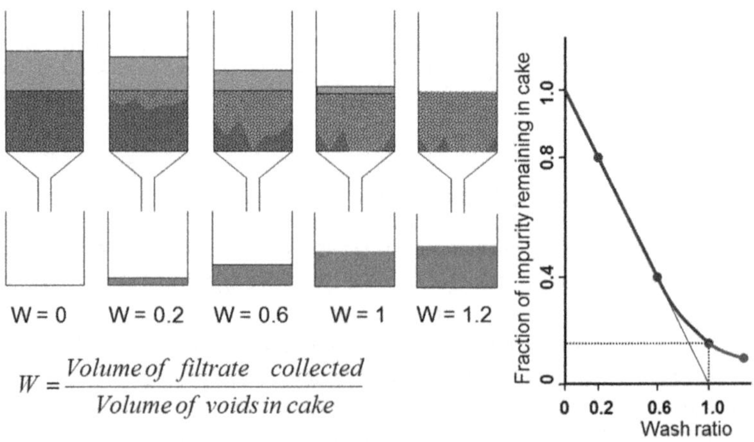

Figure 13.5 Role of wash solvent in displacing mother liquors, figure shows the effect of using 1.2 cake void volumes of wash solvent.

13.4.3 Resuspension Washing

This is an alternative approach to addressing the problem of washing a deliquored filter cake. The act of re-suspending the crystals in clean wash solvent results in the residual mother liquor becoming evenly diluted throughout the combined solution volume. This approach is very effective displacing material trapped at points of contact between particles but it requires at least three cake void volumes of solvent and risks damage to the crystals especially if the cake is re-suspended by mechanical agitation.

13.4.4 Wash Solvent Selection – Washing to Purify

It is essential to appreciate that filter cake washing is not just a physical process of forcing a fluid through a packed bed of particles to displace impure mother liquors. Depending on the quantity of API and impurities dissolved in the mother liquor there is the possibility of driving material out of solution as the wash passes through the filter cake. This may occur when the product or impurities are less soluble in the wash solvent than in the mother liquor, or mixtures of the two. Conversely there is the possibility of loss of product yield due to dissolution if the product is soluble in the wash solvent or mixtures of wash solvent and mother liquor. For these reasons it is important to quantify the solubility of the API along with the associated impurities of synthesis, residual reagents, by-products and degradants in the crystallization solvent and potential wash solvents and combinations thereof.

This is illustrated in Figure 13.6 for paracetamol in ethanol – water mixtures across a range of proportions at different temperatures using data

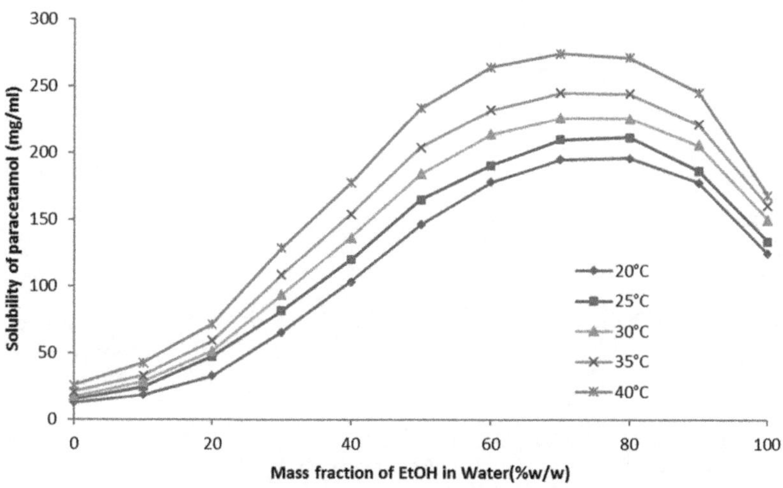

Figure 13.6 Solubility dependence of paracetamol in ethanol water mixtures across a range of temperatures. (Figure plotted from data reported by Jiménez and Martínez.[4])

collected by Jiménez and Martínez.[4] Here ethanol is the potential crystallization solvent and water the potential wash solvent; the solubility of paracetamol reaches a maximum at around 70% ethanol and 30% water by mass. Thus, using water to conduct a displacement wash of ethanol wet crystals would result in product dissolution as the water and ethanol mix. If the crystallization solvent comprised the 70% ethanol 30% water composition corresponding to the solubility maximum then washing with water could be effective; the decline in solubility expressed in mg per mL of solution as the composition changes with added water broadly keeps pace with the increase in solvent quantity. This solubility maximum occurring in solvent mixtures is not unusual, another example is found in the work by Jozwiakowski *et al.*[5] and concerns the solubility of the HIV product Lamivudine in a range of common solvents.

In addition to gathering data on the solubility of the API in appropriate crystallization and washing solvents it is essential also to consider the solubility of any by-products, major impurities of synthesis and any degradants in the same solvents. For example, Figure 13.7 shows the solubility of acetanilide and metacetamol in ethanol for comparison with paracetamol; in this case the impurities are more soluble than paracetamol and their relatively low abundance in reaction mixtures makes it unlikely that they will crystallize directly from solution in ethanol but they could be driven out of solution by addition of a wash solvent acting as an anti-solvent.

It is also important to consider how impurities modify the solubility of the API, the relationship between structural similarities of impurities and their API solubility modifying effect is not readily predicted and so

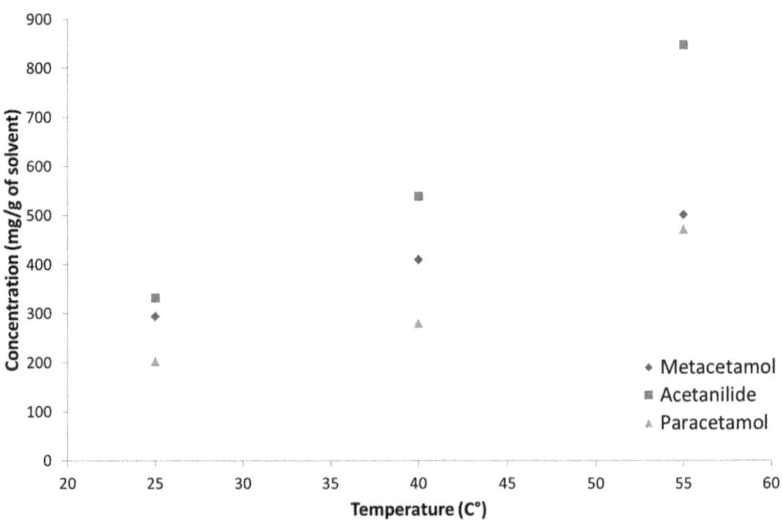

Figure 13.7 Solubility of metacetamol and acetanilide in comparison to the solubility of paracetamol in ethanol. (Data collected by Layla Mir-Bruce University of Strathclyde).

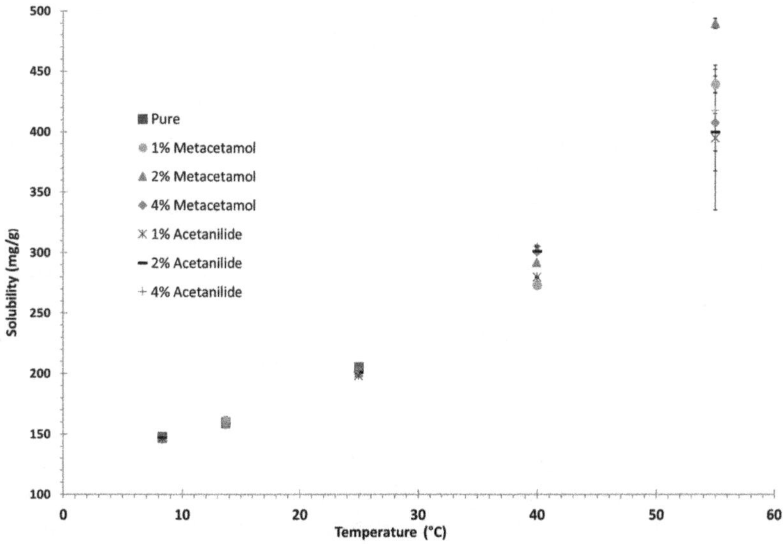

Figure 13.8 The effect of increasing quantities of the structurally similar impurities metacetamol and acetanilide on the solubility of paracetamol in ethanol. (Data collected by Layla Mir-Bruce University of Strathclyde).

must be measured experimentally. Figure 13.8 shows the effect of increasing amounts of the two impurities acetanilide and metacetamol on the solubility of paracetamol in ethanol. In this case the effect is modest and is not easily resolved from the noise in the experimental data at lower temperatures.

This approach, which already requires a significant body of experimental data, only addresses the equilibrium solubility arising from the addition of a wash solvent to a saturated solution; it does not consider the complexities arising when the mixing of mother liquor and wash solvent takes place as the wash solvent is driven through the inter crystal pore space in a saturated filter cake.

Mother liquor is a very useful resource for developing a filter cake washing procedure. Small scale array experiments allow alternative wash solvents to be assessed. A rapid practical assessment of the risk of driving dissolved impurities or API out of solution during washing can be made to aid wash solvent selection. The procedure is to take around 0.5 mL of mother liquor at the intended isolation temperature in an HPLC vial and to add 0.25 mL of the proposed wash solvent to form a 2 : 1 mother liquor to wash solvent ratio solution. The vial is shaken and then left standing for at least 10 minutes and up to the anticipated duration of the washing step. If nothing precipitates a further 0.25 mL aliquot of the potential wash solvent is added to form a 1 : 1 solvent ratio, and the shaking and waiting is repeated. If no crystals are observed a further 0.5 mL aliquot of the candidate wash solvent is added to form a mixture with a 2 : 1 ratio of wash solvent to solvent, and the shaking and waiting is repeated before assessing whether anything has nucleated. If crystals are found at any stage it is useful to leave the vial for several hours to allow crystallization to be completed and then to recover them, and to quantify how much material has been driven out of solution. It is useful to assay the solids to determine the purity of the deposited material. If nothing crystallizes over this range of mother liquor/potential wash solvent compositions, then the wash solvent may be useful providing it does not solubilise the API. This final point can be checked by adding a few mg of the API and determining whether this dissolves on mixing with the solution at the end point composition.

Once a potential wash solvent has been selected in this way its performance should be tested. A small scale procedure appropriate for early process development employing around 50 mL of suspension per test is illustrated in Figure 13.9.

The equipment is designed to process fully saturated filter cakes produced in the modified Biotage set up described earlier. The lid for the filter tube filter has connections to two syringes; the first syringe is used to deliver the pressure driving force for the filtration, the second syringe is charged with a predetermined volume of wash solvent which can be run down the side of the filter tube. This is operated manually, the wash being delivered based on visual inspection to achieve an addition rate such that the filter cake remains

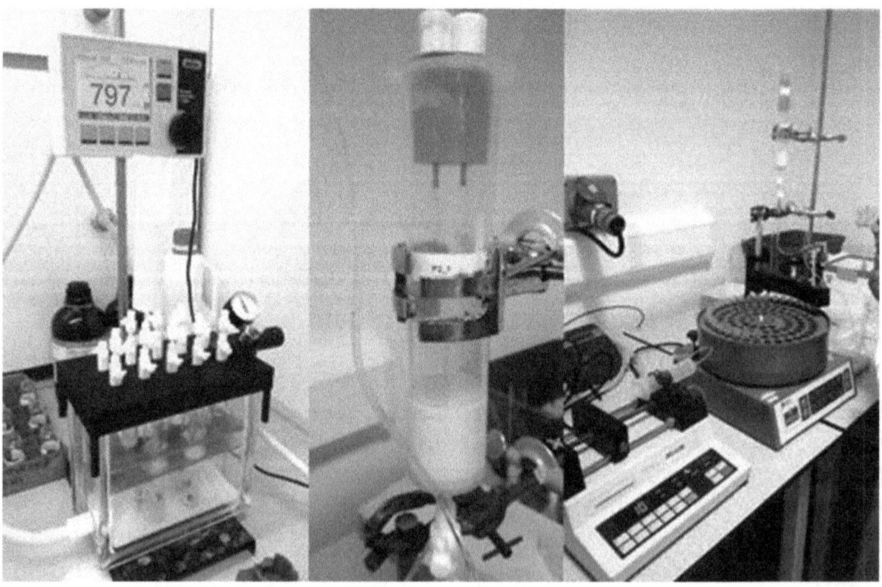

Figure 13.9 Use of a fraction collector to evaluate wash performance in detail by separating collected filtrate and wash solutions into small aliquots. (left) filtration system used to form a fully saturated filter cake; (centre) filter cake in Biotage filter tube with dropwise wash addition system in place, (right) system including pump and fraction collector. (Procedure developed by Chris Price & Sara Ottoboni University of Strathclyde).

Figure 13.10 Sequence of 10 drop aliquots of mother liquor/wash solvent displaced during filter cake washing 4 of the 46 samples (numbers 38 to 41) contain suspended crystals. (Data collected by Chris Price & Sara Ottoboni University of Strathclyde).

saturated until all the wash has been added. The outlet of the filter is connected to a fraction collector which allows sequential small aliquots, *e.g.* 10 or 20 drops of filtrate/wash solution, to be collected and checked for product/impurity crystallization by visual inspection and the product loss in washing to be quantified by analysis. An example of the resulting aliquots is shown in Figure 13.10.

The appearance of crystals in aliquots 38 to 41 indicates that there is a point in this washing procedure where the combination of mother liquor and wash solvent results in material being forced out of solution. If the washing process had been slower, for example using a thicker cake or targeting a longer contact time between the wash solvent and mother liquors, some of this crystallization could have taken place in the filter cake resulting in the crystalline bridges forming between particles resulting in granules forming in the product. An alternative wash solvent or solvent composition should be sought to avoid this.

13.4.5 Washing to Avoid Granule Formation During Drying

As well as the mechanism of granule formation due to product being forced out of solution during filter cake washing described in the preceding text it also very common to form granules during drying. The reason for this is the solubility of the API in the solvent which is to be removed during drying. Any solvent not removed during the final deliquoring stage of filtration is removed by evaporation and whatever material was dissolved in that solvent is deposited in the filter cake as additional solid. Knowing the solubility of the API in the final wash solvent and the loss of mass on drying allows the mass of material deposited to be estimated. Clearly it is desirable to leave the crystals wet with a non-solvent to avoid this undesired effect. Hydrocarbons, in particular *n*-heptane, are used to achieve this objective since the solubility of typical drug molecules in hydrocarbons is negligible. However, this requires an additional washing step to displace the preceding wash solvent, since washing the primary filter cake directly with a hydrocarbon almost invariably results in both product and impurity deposition leading to the formation of agglomerates and a failure to meet the target purity. The approach described in the previous section can be used to identify solvent ratios at which product or impurity deposition is minimised or eliminated.

In summary, the essential characteristics of a wash solvent are:

- Full miscibility with the primary solvent at the isolation temperature.
- The product should exhibit appropriately low solubility in the wash solvent and declining solubility in mixtures of the primary solvent and wash solvent as the proportion of wash solvent increases.
- Product should not be precipitated when contacting a saturated solvent with the proposed wash solvent.
- Solvents used in the later stages of API manufacturing are usually selected from category 3 in ICH guideline Q3C (R6).[3]
- It may be necessary to displace a good wash solvent with a suitable non-solvent to avoid granule formation and to facilitate drying.

13.4.6 Deliquoring the Washed Cake Prior to Drying

The final processing step prior to drying is deliquoring. Obviously it is desirable to minimise the drying duty by removing as much solvent as possible from the wet cake prior to drying. In traditional batch filtration this is achieved by blowing nitrogen through the wet cake. This is an area where a centrifuge offers some advantages over conventional pressure filtration.

13.5 Drying

From a process engineer's perspective drying is dominated by the evaporative mass transfer through which the solvent is removed and the associated heat transfer required to drive this process. From a material scientist's perspective, the focus is on the changes drying brings about in the API particle properties. These two perspectives need to be combined to achieve the desired goal of efficient drying to deliver a solvent free non-granulated API with good powder flow properties.

The key step in understanding the impact of drying on API particle attributes is to realise that most evaporation takes place at the surface of the wet cake. The simplest way to illustrate this is to take a sample of solvent wet filter cake which contains a low level of a soluble coloured impurity and leave it to dry without agitation. During drying the vast majority of the coloured impurity is transported to the surface of the wet powder resulting in a highly coloured lump of material which when broken apart reveals a much lighter core. The impurity is transported with the solvent to the surface through capillary forces acting on the film of residual liquid covering particle surfaces. A second consequence of this solvent transport mechanism is the tendency to form hardened surface layers because the vast majority of solvent evaporation takes place from the surface particle layers and any API dissolved in the residual solvent is deposited on the particles' surfaces and especially at points of contact. Both problems can be minimised by very effective washing such that virtually no impurity remains to be transported and by leaving the filter cake wet with a non-solvent such that there is nothing to deposit as the final wash solvent is evaporated; this is illustrated in Figure 13.11.

An alternative strategy is to agitate the filter cake during drying such that any impurities dissolved in the residual solvent are distributed evenly, and few strong inter particle bonds form as dissolved API is deposited by evaporation. Agitation also serves to aid heat transfer if drying is accomplished *via* contact with a heated surface. However, caution is needed as agitating the wet cake can promote granulation and lead to breakage of crystals with vulnerable crystal habits, especially needles.

13.5.1 Determining the Thermal Energy Required for Drying

Eqn (13.2) describes the energy needed to remove the solvent; if drying is isothermal then the second and third terms are zero.

$$E = \frac{\mathrm{LOD}_{S1} \cdot \Delta H_{\mathrm{vap}_{S1}}}{\mathrm{MW}_{S1}} + \frac{\Delta T \cdot M_{\mathrm{API}} \cdot \Delta C_{\mathrm{p}_{\mathrm{API}}}}{\mathrm{MW}_{\mathrm{API}}} + \frac{\Delta T \cdot \mathrm{LOD}_{S1} \cdot \Delta C_{\mathrm{p}_{S1}}}{\mathrm{MW}_{S1}} \tag{13.2}$$

(a) (b)

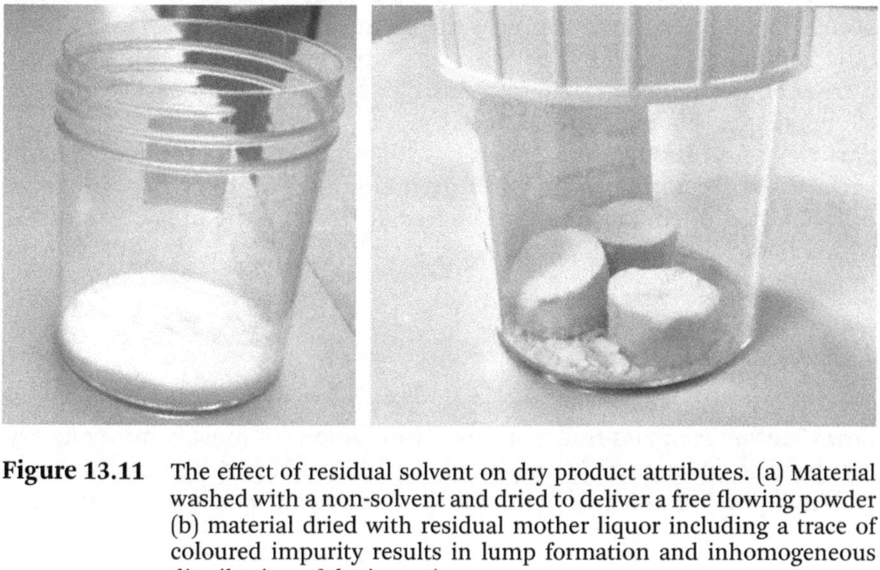

Figure 13.11 The effect of residual solvent on dry product attributes. (a) Material washed with a non-solvent and dried to deliver a free flowing powder (b) material dried with residual mother liquor including a trace of coloured impurity results in lump formation and inhomogeneous distribution of the impurity.

where E is the energy required to achieve complete drying (kJ); $\Delta H_{vap\,S1}$, latent heat of vaporisation of solvent 1 (kJ mol^{-1}); LOD_{S1}, mass of solvent 1 evaporated (g); MW_{S1}, molecular weight of solvent 1; ΔC_{pS1}, heat capacity of solvent 1 (kJ mol^{-1} K^{-1}); ΔT, temperature change from input temp and drier operating temperature (K); M_{API}, mass of API; ΔC_{pAPI}, heat capacity of API (kJ mol^{-1} K^{-1}); MW_{API}, molecular mass of API.

Once the heating duty has been determined it is necessary to determine how the heat will be delivered. The two most widely used mechanisms of delivering heat during drying are provision of a heat transfer surface and appropriate agitation to ensure temperature uniformity, or flowing heated nitrogen through the drier. Obviously as the drier becomes larger the available surface area to volume ratio becomes less favourable, the extent to which this becomes a problem then depends on the selected equipment geometry. The heat capacity of dry nitrogen close to room temperature is 1.04 kJ kg^{-1} K^{-1} which limits the energy which can be delivered. To put this into context, the drying duty associated with the removal of typical organic solvents ranges from 500 to 1000 kJ kg^{-1} of solvent evaporated. Presuming an inlet nitrogen temperature of 60 °C and almost complete heat transfer from the nitrogen to the solvent in the filter cake, the volume of nitrogen required to remove 1 kg of solvent would be between 10 and 20 m^3.

13.5.2 Agitation

Agitation of the wet filter cake can cause "balling-up" which ultimately results in large agglomerates forming and may result in uneven impurity distribution. This problem is especially severe if agitation is applied around the critical solvent content (sometimes referred to as the sticky point); this point cannot be predicted but must be determined experimentally and is influenced by particle size distribution and solvent choice. A further consequence of agitation during drying is the potential for particle breakage; this may both reduce the mean particle size of the primary particles and produce fine particles which increase the product specific surface area resulting in impaired particle flow. There is also the possibility of introducing amorphous character through the breakage process which has the potential to reduce the stability of both the API and the final drug product.

13.5.3 Drying Kinetics

The drying process can be regarded as controlling temperature and heat flow to influence mass transport and evaporation. In most continuous dryers, particle movement is employed, *e.g.* fluidised bed, internal agitator/paddle/flights, to assist in maintaining a uniform product temperature, so minimising the heat transfer resistance. For a more detailed review of drying see the paper by Kemp.[6]

13.6 Application of These Principles to Continuous Isolation

The principles and methodologies outlined above are useful for gathering data to select isolation conditions using very small quantities of material typically available in early development. The data collected can be used to select processing conditions for batch or continuous processes. Indeed, the conditions experienced by the material being isolated are for the most part comparable whether the processes deployed are batch or continuous.

Combining all these factors to deliver continuous filtration washing and drying at a scale suitable for early pharmaceutical development and extending into small manufacturing scale of a few kilos per hour represents a significant challenge. The classical options for continuous filtration are either to implement a moving filter medium, *i.e.* a belt filter or a drum filter, or to adopt sequential semi batch dead end filtration. A key distinction between typical batch isolation in a pressure Nutsche filter or filter dryer and continuous filtration on a drum or belt filter is the cake thickness which is usually rather thinner on the drum or belt filter, typically ranging from less than 10 mm to 50 mm but in some cases reaching 100 mm. Batch filtration often operates with cake thickness from 100 to 500 mm with 300 mm being typical. Thus the approach outlined above, based on collecting filtration and washing data with around 50 mL of suspension and collecting 10 to 30 mm of cake, provides data which is consistent with the operation of both moving media continuous filter types and allows the duration of cake formation, washing time and deliquoring time to be estimated and the rate of motion of the belt or drum to be set appropriately and hence the likely throughput to be assessed.

13.6.1 Drum Filtration

Large scale commercial drum filters divide the internal volume of the drum such that the filter operates as a series of sectors allowing the pressure driving force to be adjusted for different sectors of the operation *via* a rotary valve; cake formation, initial deliquoring, wash application, deliquoring and discharge; however this is challenging to do at small scales *e.g.* $< 1 g\ s^{-1}$ dry mass. It should be noted that the discharged product will be a damp powder which will need subsequent drying. It is almost inevitable that the wash is applied to a partially deliquored cake, as applying wash to a fully saturated cake will tend to wash the cake off the drum. This is likely to impact wash efficiency since the deliquored cake will contain some low resistance pathways for the wash to take. Furthermore, the opportunity for deliquoring to remove the wash is very limited. Uniform delivery of the wash to the cake is also potentially challenging since the rate of wash delivery *via* an atomiser/spray bar is likely to differ along the axis of the drum. Containment of vacuum drum filters is relatively straight forwards and operation can be automated.

13.6.2 Belt Filtration

There are some similarities in the challenges faced by belt and drum filters. The problem of cake cracking in this case is linked to the movement of the belt; minimising flexing of the cake is essential to the efficiency of washing which may be impaired both by having to wash a partially deliquored cake and through cake cracking associated with flexing of the cake. Optimising the belt velocity and suspension feed rate allows the cake thickness to be controlled; the optimum distance from the suspension feed to the wash feed can be set based on the degree of deliquoring which is deemed acceptable. The wash feed rate and wash zone length can then be adjusted based on the level of purification required. There is the possibility of reducing wash solvent consumption by adopting counter current washing in which fresh wash solvent is added at the end of the wash zone, the collected slightly contaminated wash then being fed back to an earlier point in the wash sequence, collected and applied even earlier in washing. If this approach is adopted, then the procedure outlined above to investigate appropriate wash compositions can be adapted to address the compositions likely to be encountered in such a counter current washing process. The length of the active washing area along the belt can be greater than for drum filters which may be another advantage, and it is possible to introduce multiple washing steps including wash solvent switches, again something which is very challenging to accomplish on a small drum filter. The final advantage of belt filters is that it is relatively easy to wash the belt after product discharge. Disadvantages include relatively complex mechanisms associated with the moving belt which make cleaning verification for product switchover challenging. In addition, there is greater technical complexity associated with containment. A useful concept which facilitates efficient operation is the Indexing Belt Filter.

13.6.3 Semi Continuous (Sequential Batch Filtration)

In principle this could be represented by a pair of filter dryers operating sequentially in parallel but the resulting relatively infrequent discharge may not be accepted as a continuous process. However, if the number of filter units is increased and the operation automated to deliver a substantially increased frequency of discharge this becomes a continuous process. This approach can combine the flexibility of conventional batch filtration, washing and drying with the advantage of processing thinner filter cakes resulting in improved wash performance, shorter cycle times and potentially delivery of a largely dry product compared to the drum or belt filter.

13.7 Commercially Available Filtration and Drying Technologies

There are several types of commercially available fully continuous and semi-continuous solid liquid separation devices. Here, a selection of equipment has been identified which can operate at industrial scale and within an end to end continuous processing system. The technologies have been evaluated based on their known system attributes and use for separating and retaining solids as the product.

The following technologies have been selected as potential semi-continuous and fully continuous isolation, washing and drying devices for inclusion in an end-to-end continuous processing system.

- Rotary Drum Vacuum Filters
- Rotary Pressure Filter/Dryer
- Indexing Belt Filter
- Carousel Filter/Dryer
- Multiple-Stage Batch Pressure Filtration and Drying

13.7.1 Rotary Drum Vacuum Filters (RDVF)

The Filtration Services Ltd rotary drum vacuum filter is a continuously operating vacuum filter, pictured in Figure 13.12. RDVFs have been used for industrial filtration during the last 3 centuries. The technology was

Figure 13.12 Rotary drum vacuum filter (RDVF). Image courtesy of Filtration Services Ltd.

patented in 1872 and is one of the oldest commercially available solid–liquid separation devices. Although the science behind perfecting the operation of these systems still requires a great deal of experience and understanding, the operation principle is beautifully simple. Whilst the fundamental principles of operation have not changed over the decades, the level of automation, reliability, safety and containment have evolved by multiple factors.

A schematic of an RDVF is shown in Figure 13.13. The slurry is pumped into the open tank and the drum is immersed axially within. The solids are kept in suspension by means of a reciprocating agitator arm within the tank. This 'beam' agitator is quite effective in keeping the solids suspended due to the low volume void between the drum and trough. The tank has an overflow and recirculation route to the feed tank. This ensures that the slurry level is constant, and the drum is consistently covered in cake. The constant layer of cake and thickness of cake dictates the ultimate vacuum level divided over the area of the drum.

The drum is wrapped with a filter medium and the separation is driven by vacuum being applied to the internal of the drum. Solids accumulate on the surface of the drum and the filtrate is collected *via* a dip tube within the drum.

Cake thickness is variable by altering the drum rotation speed. Cake washing is achieved by a spray bar mounted above the cake. The wash is limited due to the limited available surface area between the filtration

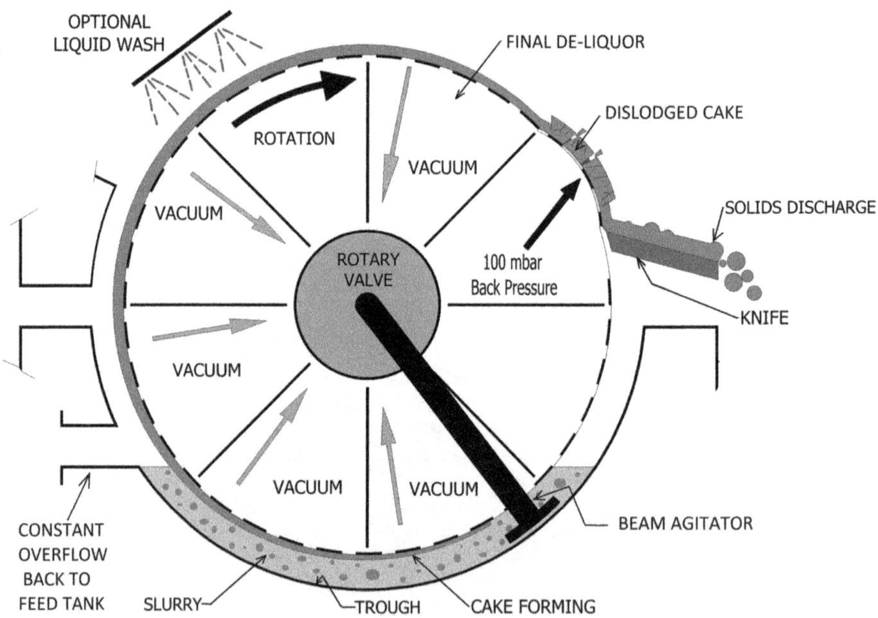

Figure 13.13 Schematic of a rotary drum vacuum filter. Image courtesy of Filtration Services Ltd.

zone and the discharge zone. The wash can be easily applied whereas the wash liquid must then be removed without compromising the required ultimate moisture removal. Washing can only commence when the cake is sufficiently dry otherwise the cake will be washed from the drum. The diagram shows the different zones and principle of operation. Reducing the rotation speed to increase the wash window has a direct reduction in throughput.

The cake removal is assisted by applying back pressure from the underside of the filter medium within the discharge zone. The typical back pressure applied is 100 mbar which is sufficient to dislodge the cake. The discharge knife acts as a weir causing the disrupted cake to be dislodged and fall into the discharge route. Successful filtration is all about building a cake, maintaining a good level of vacuum, setting the drum speed and removing sufficient moisture to allow the cake to be completely discharged.

The ability to filter continuously relies on the complete cake being discharged before a new cake is formed within the filtration zone. An additional wash can be introduced between the discharge and filtration zones to clean the filtration medium however, this reduces the product yield unless recycled to the start of the process. Although effective, this adds complexity to the process and requires an additional separate isolation process to recover the solids.

RDVFs are used within the pharmaceutical, food, cosmetic, chemical, water and other industries. Alongside belt filters they are have the widest and most diverse span over industrial multiple industrial sectors. Maintaining RVDFs is low cost and consists of remeshing the filter medium and replacing several bearings and seals. Operating costs and the comparable cost of ownership is relatively low. Typical operating parameters for an RDVF are displayed in Table 13.1.

13.7.2 Rotary Pressure Filter/Dryer (RPF)

The BHS rotary pressure filter (type RPF), pictured in Figure 13.14 is a continuously operating filter for pressure filtration. It facilitates gas-tight cake handling in a series of separate process steps. Cake discharge can take place at atmospheric pressure.

The rotary pressure filter supports a broad range of process steps during a single rotation of the drum which takes typically 3 minutes. Further processes can be carried out after separating the slurry into filtrate and filter cake:

- Single- or multi-stage cake washing
- Displacement washing, counter-current washing, closed-cycle washing
- Reslurrying
- Solvent exchange
- Steaming, extraction
- Mechanical or thermal cake dehumidification

Table 13.1 Rotary Drum Vacuum Filter – operating parameters and statistics.

Rotary Drum Vacuum Filters (RDVF)		
Commercial introduction late 18th century	MIN	MAX
Slurry volume required for lab-scale optimisation (20% solids content)	500 mL	5000 mL
Pilot/Production scale equipment slurry throughput (20% solids content)	5 L h^{-1}	120 L h^{-1}
Solids content	5%	50%
Filtration area	0.05 m^2	60 m^2
Max pressure differential	—	−0.7 bar
Typical moisture content post filtration	25%	35%
Residence time from slurry to filtered product	60 s	300 s
Cake thickness (typical)	10 mm	50 mm
Wash regime (optional)	—	1
Dead volume or heel (typical)	—	5 mm
Containment methods	Physical cover to improve containment	
Number of cGMP qualified processes (most processes are used to remove solids as the waste stream)	0	
Status of cGMP qualification (in this build form)	Not current	
Cost of typical initial 2 years strategic spare parts	€10 000	€20 000

Figure 13.14 BHS Rotary Pressure Filter. Image courtesy of BHS Sonthofen Inc.

All process steps are performed within the filter's compartmentalised zones, making it possible to segregate and separately discharge the filtrate or other substances. The operation is a fully encapsulated unit that can be integrated into production processes in a gas-tight manner. This means that the Rotary Pressure Filter is also suitable for use in operating scenarios with increased safety or hygienic requirements.

The Rotary Pressure Filter operates at a pressure of up to 7 bar (g) and generally within a temperature range of −20° to 160° Celsius. Pressure is generated hydraulically rather than pneumatically by means of gas stratification during all process steps. This enables individual optimization of each and every process step. Each process step is sealed off into a segment zone by means of static and dynamic pressure-tight separation elements inside of the rotary pressure filter.

The individual process steps are represented in the diagram in Figure 13.15.

Slurry is pumped into the feed port filtration area. The filtration is driven by the pump pressure creating back pressure against the formed cake. Back pressure and the ultimate filtration pressure vary depending on the pump flow rate, cake resistance and cake height. This relatively high-pressure capability makes the RPF the highest pressure rated continuous filter.

Cake wash is pumped into the wash segments to remove impurities. The wash flow and pressure can be varied to optimise the wash efficiency. Multiple solvents can be used by switching the flow from one supply to the other automatically during the wash regime. The constant stream of wash is advantageous when compared to batch dosing as the cake is not de-liquored, cake cracking is minimised, and contact time can be increased. Solvent exchanges can take place automatically without additional processing time.

Drying is achieved by means of convection drying in the form of heated nitrogen being blown through the de-liquored cake. The drying rate is a function of gas flow through the cake, cake porosity and gas temperature. Convection drying is limited to porous cakes as flow is required to remove the moisture. The RPF rotates at a speed of 3 rpm limiting the active drying time to approx. 45 s. A separate dedicated dryer is used for applications requiring extended drying times.

The internal workings of the equipment are manufactured to a very high and precise standard to achieve such a high level of compartmental pressure integrity. Complex internals within process equipment normally comes with a trade off with problematic cleaning to satisfy cGMP manufacturing cleaning regimes. The RPF has strategically positioned clean in place nozzles to provide excellent coverage and validated cleaning processes.

The ability to filter continuously relies on the complete cake being discharged before a new cake is formed within the feed and filtration zone. An

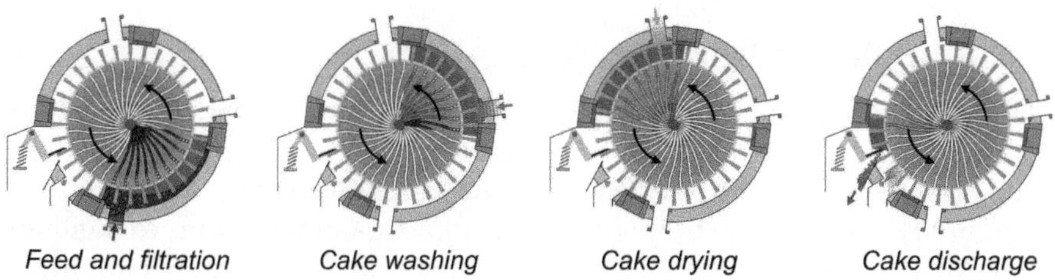

Figure 13.15 Diagram of the process steps of a BHS rotary pressure filter. Image courtesy of BHS Sonthofen Inc.

Table 13.2 Rotary Pressure Filter – operating parameters and statistics.

Rotary Pressure Filter		
Commercial introduction 1950s	MIN	MAX
Slurry volume required for lab-scale optimisation (20% solids content)	—	10 L
Pilot/Production scale equipment slurry throughput (20% solids content)	1 L h^{-1}	1500 L h^{-1}
Solids content	10%	60%
Filtration area	0.18 m^2	8.6 m^2
Max pressure differential (pressure version)	—	6.0 bar
Typical moisture content post filtration and drying	8%	10%
Residence time from slurry to discharge <8% moisture content	—	3 minutes
Cake thickness	5 mm	150 mm
Wash regime. Multi-wash and solvent exchange	1	2
Dead volume or heel	—	20 mm
Containment methods	Nitrogen purging operation in fume cupboard containment in glove box isolator	
Number of cGMP qualified processes	>10s	—
Status of cGMP qualification for API production	Current	
Cost of typical initial 2 years strategic spare parts	€20 000	€30 000

additional wash can be introduced between the discharge and feed filtration zones however, this reduces the product yield unless a separate isolation process is introduced to reclaim the solids.

RPFs are used within the pharmaceutical, food, cosmetic, chemical, bioprocessing and other industries. The cost of maintaining RPFs is relatively low considering the precise manufacturing procedures and level of complexity. The absence of linear sealing bellows and mechanical seals brings the cost of strategic spare parts dramatically down when compared to Agitated Nutsche Filters.

The BHS rotary pressure filter was invented by Dr Fest at BHS in the 1950s. BHS Sonthofen have further improved and optimised the rotary filter over the decades to provide a versatile pressure filtration and drying system. The details of the rotary pressure filter and typical operating parameters are shown in Table 13.2.

13.7.3 Indexing Belt Filter (BF)

The BHS indexing belt filter (type BF) is a continuously operating, horizontal vacuum filter used for the efficient and also gentle separation of sedimenting solids from suspensions. An image of a typical indexing belt filter is shown in Figure 13.16. The system is split into compartmental sections to facilitate the individual process steps. These steps or stations act as individual filters and discharge is semi-continuous.

Figure 13.16 BHS indexing belt filter. Image courtesy of BHS Sonthofen Inc.

A typical process setup is configured as follows:

- Steps 1 and 2 accept slurry and separate the solids from the liquid
- Step 3 wash 1
- Step 4 wash 2
- Steps 5 and 6 de-liquoring
- Steps 7 to 10 convection drying

Filtrates can be individually recovered and processed. The BHS belt filter thus offers very high process versatility.

Filtration medium is formed into an endless belt and stepped to multi positions aligning with the appropriate sequential process step. Slurry feeding is usually continuous, and the vacuum trays are fixed in place on the frame. The filter cloth with the suspension moves in steps over the trays. Each time the cloth stops, the filtrate is sucked downwards. Then the vacuum is switched off. This releases the filter cloth, which can be advanced again.

The filter cake forms on top of the belt and can undergo further treatment:

- Washing (co-current or counter-current), re-slurrying
- Steaming, extraction
- Vacuum drying or pressing

The wash filtrates can be recovered individually from each vacuum tray and further processed without any cross-contamination. The filter cake is discharged at the discharge roller. The cloth is cleaned as the belt returns.

Besides the standard open-frame design a range of the following options are available:

- Closed hood over the entire filter
- Closed box design
- Gas- and pressure-tight enclosure
- Pharmaceutical version

Table 13.3 Indexing Belt Filter – operating parameters and statistics.

Indexing Belt Filter		
Commercial introduction 1960s	MIN	MAX
Slurry volume required for lab-scale optimisation (20% solids content)	10 L	50 L
Pilot/Production scale equipment slurry throughput (20% solids content)	100 L h^{-1}	1000s L h^{-1}
Solids content	10%	50%
Filtration area	0.1 m^2	90 m^2
Max pressure differential	—	0.5 bar
Typical moisture content post filtration	20%	30%
Residence time	5 minutes	30 minutes
Cake thickness	5 mm	50 mm
Wash regime. Multi-wash and solvent exchange	1	3
Dead volume or heel (adjustable)	—	5 mm
Containment methods	Complete cover operation in pressurised purged enclosure	
Number of cGMP qualified processes	>10s	
Status of cGMP qualification for API production	Intermediates/Current	
Cost of typical initial 2 years strategic spare parts	€10 000	€20 000

The range of operating parameters for an indexing belt filter is displayed in Table 13.3.

13.7.4 Carousel Vacuum and Pressure Filter/Dryer

The AWL continuous feed carousel filter dryer (type CCFD) is a continuous or batch fed optimisation and production system. Throughput is aimed at small to medium scale pharmaceutical production capacities, in line with several key API manufacturers future expectations of 300 kg to 4000 kg per annum for new drug production. Variations of this design facilitate throughputs in line with the batch production volumes of 1 m^2 to 2 m^2 filter dryers. The carousel is a series of vertical cylinders moved around a central axis to fixed processing stations. Port diameters are limited to <100 mm to avoid cake cracking and wall effects during de-liquoring and drying thereby negating the need for agitation. Key advantages include the mobile pilot system fitting within a walk-in fume cupboard and it can be validated for a number of processes rather than the manufacturer adopting the traditional route of validating a fixed asset for a specific process. The ability to deploy validated, mobile isolation and drying systems to meet specific geographical demands gives the manufacture new levels of supply chain versatility and flexibility. An image of the AWL CCFD is shown in Figure 13.17.

The operating principles are aligned with Nutsche filtration whereby material is processed in a series of mini batches with a constant

Figure 13.17 AWL continuous carousel filter dryer (CCFD).

semi-continuous output in the form of a rotary Nutsche. The technology choice was a conscious decision by the manufactures to ease the transition from batch processing without the unnecessary complication of extensive evaluation, testing, validation and acceptance of a completely new processing philosophy.

The carousel pressure filter supports a broad range of process steps during a single rotation of the carousel. Multiple process steps can be performed simultaneously in 'production mode':

- Vacuum and pressure filtration
- Single- or multi-stage cake washing
- Solvent exchange
- Thermal cake dehumidification

All process steps are performed within the filter's segregated vertical chambers and zones. This makes it possible to operate the system as a filtration and drying 'optimisation' platform as well as a continuous 'production' system.

A schematic of the process sequence of the AWL CCFD is illustrated in Figure 13.18. Slurry is fed into the system by means of a semi continuous dosing system. Feed volume, transfer intervals and filtration pressures are all managed through the operator interface and recipe configuration. Initial operation is conducted in 'optimisation mode'. A 50 mL aliquot of slurry is 'jogged' through the process stages including isolation, washing and drying. De-liquoring is managed automatically by a vision system to detect 'dry land' and to prevent the cake being de-liquored between isolation and subsequent wash steps. The control software uses the vision system to

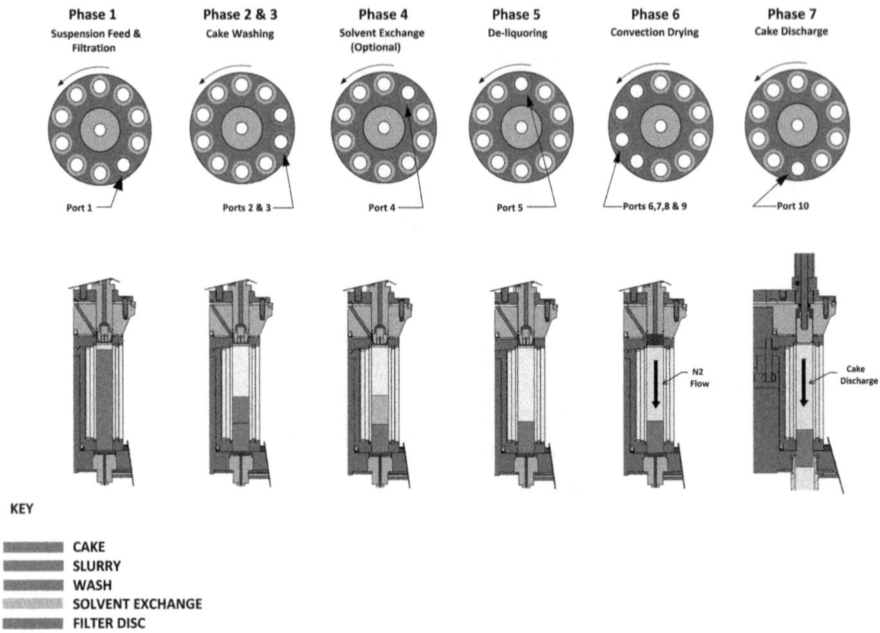

Figure 13.18 Schematic of the process sequence of the AWL Continuous Carousel Filter Dryer.

manage the wash contact times and solvent exchanges. The optimisation process can be paused to allow the wash liquid to be sampled offline during each step significantly reducing the overall quantity of material required to optimise a process. Convection drying is used to de-humidify the cake before being discharged. These optimisation parameters can be saved and recalled for use in 'production mode'. Production mode automatically isolates, washes, dries and discharges product in a fully automatic sequential process.

The carousel pressure filter operates at a pressure of up to 2 bar (g) generally within a temperature range of 0° to 100° Celsius. The filtration driving force can be vacuum or pressure. The effectiveness of convection drying is limited to products which allow gas flow through the cake and the process solvent choice. When moderate gas flow cannot be achieved or a solvent with a relatively high boiling point is being used an additional drying step will be required. A summary of the operating parameters and statistics of the AWL continuous carousel filter dryer is shown in Table 13.4.

The AWL carousel filter was invented by Mr A. Barton in 2012. The AWL development team, Strathclyde University and several end users have contributed greatly to improving and expanding the uses of the equipment over recent years. The development of this technology is summarised in the patent by A. Barton[7] and in the paper by Ottoboni *et al.*[8]

Table 13.4 AWL continuous carousel filter dryer – operating parameters and statistics.

Commercial introduction 2017	MIN	MAX
Slurry volume required for lab-scale optimisation (20% solids content)	50 mL	5000 mL
Pilot/production scale equipment slurry throughput (20% solids content)	5 L h^{-1}	100 L h^{-1}
Solids content (multi dosing available to build cake heights)	5%	50%
Filtration area	0.05 m^2	60 m^2
Max pressure differential (vacuum version)	—	900 mbar
Max pressure differential (pressure version)	—	2.5 bar
Typical moisture content post filtration	10%	15%
Residence time from slurry to dry API exiting the system <1% moisture content	600 s	1800 s
Cake thickness (including multi dose accumulative cake building)	10 mm	40 mm
Wash regime. Multi-wash and solvent exchange	1	3
Dead volume or heel (adjustable)	0.1 mm	1 mm
Containment methods	Nitrogen purging operation in fume cupboard containment in glove box isolator	
Number of cGMP qualified processes	0	
Status of cGMP qualification for cGMP production	Ongoing, estimated availability 2018	
Cost of typical initial 2 years strategic spare parts	€5000	€10 000

13.7.5 Agitated Nutsche Filter Dryers (ANFDs)

The agitated Nutsche filter and filter dryer is an industry standard batch processing unit. It is the most widely used and accepted technology within the pharmaceutical industry due to its complete inert containment, versatile processing capabilities and combined isolation and drying. Figure 13.19 shows an image of a typical agitated Nutsche filter dryer.

In theory, 2 units can be set up in tandem to accept slurry in a semi-continuous configuration. Units can be sized at a minimum of 50% of the total slurry feed production rate (excluding contingency and potential downtime). The key to using 2 independent batch units as a semi continuous system relies on the potential availability of a suitable free flowing crevice free heated diverter valve and a common discharge collection system.

Large scale vacuum filters were widely used in the chemical and pharmaceutical industry in the early 19th century. These versions were based on open vats containing a filter medium. Cake cracking over large diameters was inevitable and manual practices more commonly associated with grape and cheese pressing were employed. As plant safety demands increased in

Figure 13.19 Agitated Nutsche filter dryer. Image courtesy of Charles Thompson Ltd.

the 1960s, agitators and electric drives were introduced to smooth and re-slurry the cake. Process containment improved by providing covers to reduce solvent evaporation, improve manufacturing procedures and operator safety. Positive pressure filtration vessels were introduced in the mid 1960s and this saw the exponential uptake of the technology for the wider chemical and pharmaceutical industries.

Combined filter dryer versions appeared in the early 1980s resulting in an exponential increase in installations worldwide. Combined systems were attractive when compared to separate filters and dryers due to reduced footprint, capital cost and elimination of transfer steps. The downside is that a filter dryer processing cycle is between 4 and 12 hours and typically 8 hours. Separating the two processes allows the overall plant throughput to be increased. Combined filter dryer units grew to an enormous capacity in the 1990s, systems reaching up to 16 m^2 filtration area or 8 m^3 of dry product discharged every batch. Typical machine sizes used today for new processes in the pharmaceutical industry is 1 m^2 to 4 m^2 due to increased drug potencies.

Multiple processes can be performed sequentially:

- Isolation, cake smoothing
- Single or multi-stage cake washing
- Displacement washing, re-slurry washing (active re-slurry agitation)
- Solvent exchange
- Drying (active agitation)

Both cake smoothing and the discharge of the product can be carried out by rotating the agitator in different directions. Figure 13.20 illustrates how

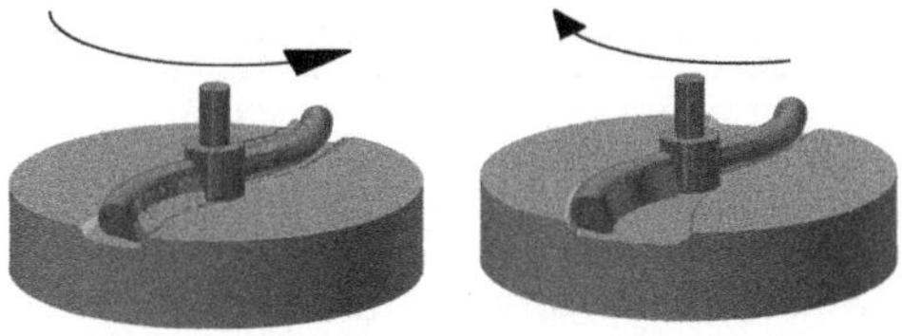

Figure 13.20 Deployment of the agitator in an agitated filter drier (left) the agitator used as a plough to discharge product and (right) rotation in the opposite direction to smooth the cake.

the agitator in an agitated filter dryer can be used as a 'plough' by rotating in one direction and can be used to smooth the cake by rotating in the other direction.

Filter dryer vessels are heated typically *via* a jacket to supply the energy required to evaporate the solvent. It is essential to heat all the dryer surfaces to prevent condensation forming as drops of solvent falling into the dry product powder bed lead to agglomerate/granule formation. The most effective filter dryers have heated agitators to increase the overall drying capability and to reduce the processing time. Although the typical surface area of an agitator relative to the total available heat exchange surface area is 15%, the effective overall heat contribution of the agitator is nearer 30% to 40% of the total, due to the agitator being in direct contact with the product and the product being in motion. The agitator has a wall and base clearance in the order of 10 mm due to fabrication tolerances and safety requirements. This often allows a layer of product to form on these non-agitated areas (vessel base and wall) which acts as an insulator effectively reducing the heat transfer. If this is allowed to remain between batches this "heel" of residual product reduces the filtration rate.

Two significant unresolved issues remain with the current design of Nutsche filter dryers. Non-representative sampling and the 'heel' arising from the dead area beneath the agitator caused by the necessity to have clearances between the agitator, vessel wall and filter base. Sampling is required to allow offline analysis to confirm the batch is dry. The majority of the sampling devices are non-intrusive to prevent accidental engagement with the rotating agitator, instead product samples are pushed into the ball or piston type samplers. The first material to enter the sampling receptor is generally not a true representation of the complete batch. This material is in direct contact with the vessel wall heating jacket and is often drier than the bulk and not disturbed by the agitator.

The 'heel' can vary from 6 mm to 20 mm depending on the size of the vessel. Larger vessels have larger machining and fabrication tolerances. After each batch is discharged the 'heel' remains in the vessel. Invariably the heel becomes consolidated and the voids become packed with fine particles increasing the resistance to filtrate flow. If this material is then discharged as part of the product batch, it manifests itself as large hard lumps impacting product performance. It is accepted practice to fill a

Table 13.5 Agitated Nutsche filter dryer – operating parameters and statistics.

Commercial introduction 1960s	MIN	MAX
Slurry volume required for lab-scale optimisation (20% solids content)	0.5 L	20 L
Pilot/production scale equipment slurry throughput (20% solids content)	$1\ L\ h^{-1}$	$3000\ L\ h^{-1}$
Solids content	10%	40%
Filtration area (typical and current demand)	$0.002\ m^2$	$6\ m^2$
Max pressure differential (typical)	0.5 bar	2.0 bar
Typical moisture content post filtration	20%	25%
Residence time from slurry to dry API	4 h	12 h
Cake thickness (typical)	100 mm	300 mm
Wash regime. Multi-wash and solvent exchange	1	3
Dead volume or heel (adjustable)	5 mm	20 mm
Containment methods	Nitrogen purging, discharge contained in glove discharge isolator or discharge valve	
Number of cGMP qualified processes	>1000	
Status of cGMP qualification for API production	Current	
Cost of typical initial 2 years strategic spare parts	€50 000	€150 000

new batch on top of the existing cake then reclaim the heel after multiple batches. Because this material is different in character to the bulk of the product the heel is typically recovered by dissolving the heel and reprocessing but this adds additional complexity to the process especially if it's a cGMP system.

These points aside, agitated Nutsche filter dryers are the dominant technology used within the pharmaceutical and fine chemical industries. They are widely accepted for cGMP manufacturing and likely to remain the chosen technology for large scale production volumes for the foreseeable future. Maintaining ANFDs can be relatively costly. These machines are often bespoke or only a small number of specific designs are repeated meaning spare parts are often made to order and command a high price. The statistics and operating parameters of a typical ANFD is shown in Table 13.5.

13.8 General Guidance and Troubleshooting

13.8.1 Cake Formation

Aim for an appropriate cake thickness; wash efficiency typically peaks somewhere between 200 and 1000 particle layers depending on particle shape, size distribution and contact time; this represents a 20–100 mm cake for 100 μm particles. If the solids loading is very low then, depending on the equipment selected, it may be necessary to dose a portion of the product suspension, filter it partially and then dose an additional aliquot of additional material, filter again and repeat until an appropriately thick cake has been built. In this sequence of operations if is desirable that the cake is not deliquored but is kept saturated to avoid cracking.

It is essential to avoid inducing crystallization in the filter medium and cake due to generating supersaturation either by drowning out during washing or solvent evaporation during drying. Remember materials continue to crystallize (or dissolve) during filtration and washing if they encounter supersaturation or undersaturation. For similar reasons beware of solvent evaporation on the vacuum side of the filter plate where product and impurities may gum out blocking the medium.

13.8.2 Cake Cracking

The best strategy is not to allow cracks to form in the cake, in addition to maintaining the cake in a saturated state as far as practicable. It is advisable to aim for smooth pressure transitions to avoid flexing of the filter medium. A common mistake with pressure filtration is to halt the filtration by venting the filter above the filter cake. This results in a situation where the pressure below the cake exceeds the pressure above and the cake either cracks or lifts from the filter medium and subsequently cracks.

Avoid disturbing the filter cake for example generating a thin spot by feeding the wash solvent too rapidly at a single point. It is much better to "atomise" the wash and add it slowly such that the cake is washed effectively, therefore focus on gentle wash solvent delivery. It is recommended that the user should be cautious about large temperature differences between mother liquor and wash liquor.

The underlying message is that it is much better not to crack the cake rather than to attempt to smooth it after the event.

13.8.3 Isolating Large Crystals/Agglomerates

Large and agglomerated particles lead to coarse structured cakes with big open channels for drainage making it hard to wash internal space within agglomerates due to the short contact time. Artificially extending wash contact time may be very helpful. Thus a strategy to improve filtration by agglomeration may be detrimental when it comes to washing. This is why it is important to take a whole process view and to pilot the process from end to end to avoid arriving at a collection of local optima which do not sum to become the overall process optimum.

13.8.4 Reasonable Washing Expectations

Estimating that 80% of mother liquor can be removed in a "good" displacement wash of 1 cake volume is a reasonable assumption. A 3 cake volume displacement wash followed by deliquoring could be expected to lower the mother liquor based impurity carry over to 0.4% of that in a saturated cake. Whilst this rule of thumb is useful the eventual outcome is highly dominated by the particle properties.

13.8.5 Drying

A number of processes may accompany the desired process of solvent removal. Frequently a dryer also functions as an agglomerator/granulator. In all instances where the product has solubility in the solvent(s) removed during drying the dryer functions as an evaporative crystallizer. The mass of material deposited during drying corresponds broadly to the solubility of the product in the mass of solvent removed at the drying temperature. However, this deposition is not uniform throughout the process volume as drying is a surface phenomenon, thus the migration of solvent to the evaporative surface provides a mass transport mechanism for dissolved product and impurities. Whilst the consequences of this uneven deposition can be mitigated by agitation during drying, an agitated drier also operates as a crystal breaker, especially for needle-like particles.

It is useful both to tray-dry some material from laboratory experiments and to subject a subsample of the same batch to intense agitated drying. These two materials typically represent the extremes of the attainable product attributes, thus evaluating the differences in performance between these two materials is very informative. It is recommended to think holistically across the whole process and identify the important parameters which should be investigated.

13.8.6 Further Troubleshooting Strategies

When addressing an isolation problem it is obvious that a researcher should start with filtration, washing and drying but it is important to be aware that this may point back to the crystallization step, for example, a change in filtration rate, cake bulk density, drying time and product assay may be linked to a change in CSD from the crystallization process for example linked to the presence of additional fines. It is important to be able to distinguish between this and for example issues arising from incomplete washing of a cracked filter cake.

Dealing with chemical impurities it is essential to both identify and quantify impurities. The identity may suggest the source and may possibly indicate a mechanism of incorporation. The quantity is useful in understanding the magnitude of the challenge, a stoichiometric quantity of a by-product is a different challenge to a smaller quantity of a structurally similar impurity.

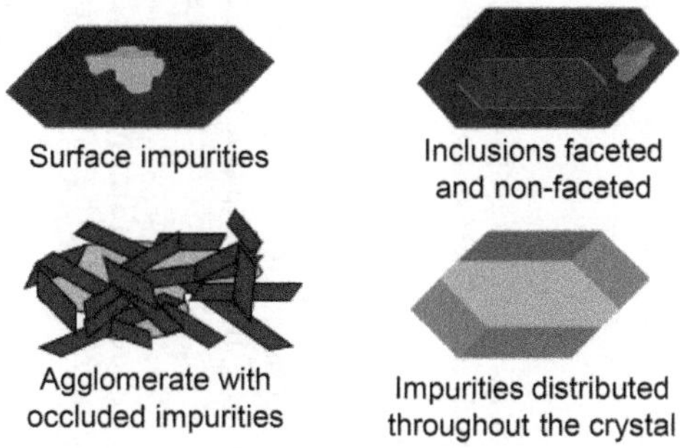

Figure 13.21 Potential locations of impurities in isolated product.

Measuring relative segregation coefficients of impurities is helpful since the differences in impurity incorporation relative to the other impurities in the mother liquor may indicate the mechanism by which impurities are retained in the product. Consider inclusions as an example.

Where impurities are a problem, determining their likely location as shown in Figure 13.21 may indicate a solution strategy.

13.9 Solutions to Issues Observed in Isolation Systems

This chapter has demonstrated that there are numerous factors that can affect the filtration and washing performance in isolation systems. It is therefore important to have a good understanding of the effect of these factors and to incorporate a systematic troubleshooting strategy in order to locate the cause of an issue and find a potential solution.

There are four main issues that can be observed in isolation systems which may indicate a problem that is potentially inhibiting the filtration and washing performance:

- Extended filtration or washing cycle time, resulting in low or no filtrate flow
- Excessive impurities remain in the cake after filtration and washing
- Lower yield than expected
- Product passing through the filter medium

Table 13.6 describes the potential causes for each of these problems along with suggested solutions. It can be used as a guide to further understand the potential challenges of isolation systems and includes potential routes towards finding a solution.

Table 13.6 Solutions to challenges observed in isolation systems.

Issues observed	Reasons for issues	Possible solutions	
Extended filtration or washing cycle time – low or no filtrate flow	Overall crystal size is too small	Investigate ways of increasing the crystal size in upstream crystallization process	Application of seeding and supersaturation control Investigate post crystallisation process to increase particle size
	Large cake depth	Reduce cake height by using larger filter surface area	Select filter based on achieving a cake thickness no greater than 1000 particle dimensions
		Measure filtration rate as a function of solids loading, cake thickness and pressure driving force	Determine the best compromise between solids throughput and filtrate flowrate
	Inappropriate pressure driving force	Insufficient pressure differential	If filtrate flow increases with increased pressure driving force, then the issue may be resolved by increasing the pressure driving force. If there is no measurable change in rate on increasing pressure driving force the filter plate could be blinded or the cake is compressed. Repeat with a clean filter plate/medium and compare filtration rate with previous measurements.
		Excessive pressure differential	Determine cake compressibility, if the cake is found to be compressible then use reduced pressure driving force in subsequent filtrations.
	Blinding of filter plate due to excessive fines, these may be carried by the crystallisation solvent to the filter medium and lead to blinding of the pores in the medium.	Dissolve away heel on the filter medium	Determine the appropriate frequency of filter medium cleaning required to prevent blinding reducing the filtration rate such that it impacts cycle time.
		Back flush the filtration membrane	This approach is only possible when the filtration medium is robustly fabricated in such a way as to withstand back pressure. Note typically filter media are designed for uni-directional use, with the pressure applied upstream the most fragile parts of the filter are protected supported by the sub structure which provide both strength and facilitates drainage.
		Ultrasonic cleaning – typically off line	Generally deployed as an offline technique for multi-layer industrial filter mesh

Excessive impurities remain in cake after filtration and washing	Cake cracking and 'rat holing' causing inefficient washing due to wash solvent taking a lower resistance path through cracks/voids in the cake.	It is desirable to keep the cake saturated at the end of filtration and then apply the wash such that cracks in the cake do not form. Some filter dryer designs allow the filter cake surface to be smoothed with the agitator. Whilst this can be useful it is preferable to avoid this situation by careful and where necessary coordinated operation.	A useful indication that smoothing has alleviated cake cracking is if the filtration rate and the flowrate of the gas delivering pressurisation reduce back to a normal level. Cake smoothing will not resolve the effects of 'rat holing' but may make a modest contribution to minimise the consequences of cake cracking. 'Dry land detection' equipment may be fitted to provide an automatic means of detecting when the mother liquor level advances towards or reaches the top of the cake. Maintaining a saturated cake minimises the extent of cake cracking and allows the mother liquor to be 'chased out' by the subsequent wash or washes (commonly referred to as displacement washing).
	Insufficient washing	Insufficient wash volume	Increase wash volume, relate wash volumes to cake volumes, determine the product loss and impurity reduction as a function of wash volume. Divide wash into several distinct wash volumes and quantify the effect of increasing wash volume on purity.
		Insufficient wash contact time between wash and particles	Increase contact time to allow diffusional mixing to occur between wash solvent and trapped mother liquor in fine pores within the cake. Typically wash contact time is increased by reducing the pressure driving force. If none of the suggestions above bring sufficient improvement, consider re-slurry wash if the filter is equipped with an agitator. Note: This may have an adverse effect on particle size increasing breakage, also the volume of solvent needed for a resuspension wash is several times the cake volume so significantly more than would be needed for conventional washing.
	Impurities precipitated during washing	Inappropriate wash solvent selection	Select wash solvent based on minimising product loss on washing but ensuring impurities remain in solution during washing. Use solubility as a criterion for conducting wash solvent trails.

(*continued*)

Table 13.6 (*continued*)

Issues observed	Reasons for issues	Possible solutions	
Lower yield than expected	Loss of material due to size of heel (material not discharged from the filter)	Reduce the clearance between the agitator and the filter medium	This clearance must be in line with the manufacturer's recommendations
		Use a smaller filter noting reduced throughput	Generally smaller filters have reduced heel heights
		Consider alternative filter designs	Some technologies have virtually no heel
	High solubility of product in wash solvent	Wash solvent selection and quantity	Choose a solvent in which the product is less soluble and use less wash solvent (whilst still meeting the purification objective)
Product passing through the filter medium – evidenced by translucent appearance of filtrate or solids present in filtrate receiver	Crystal size smaller than the filter medium aperture	If overall crystal size is too small, investigate ways of increasing crystal size	Consider seeding the crystallizer and controlling supersaturation during crystallization
		Particle attrition due to high shear, causing abrasion and fragmentation of crystals and excessive secondary nucleation	Decrease agitator rotation speed Change to agitator blade type with lower shear
		Confirm that the crystals are passing the filter medium and not forming in the receiver	Potential cause would be contact between wash solvent and mother liquor in the receiver. (This might be confounded with reduced filtration rate and impurity precipitation)
	Damage to the filter medium of seal – visible holes in the surface of the cake are an indication	Check leakage of seals between the filter medium and body and the filter body	Repair or replace components as needed
		Check the filter medium for damage	See below – Repair or replace filter medium if necessary
	Filter membrane damage	Contact with the agitator	Excessive load on agitator shaft causing flexing leading to the blade touching the filtration membrane. Reduce load as necessary. Less likely on small diameter filters. Clearance between the agitator and the filtration membrane too small
		Back pressure under the filter medium causing damage.	Investigate cause of high back pressure to prevent the issue re-occurring

References

1. S. Black and F. Muller, *Org. Process Res. Dev.*, 2010, **14**(3), 661–665.
2. H. Darcy, *Les Fontaines Publiques de la Ville de Dijon*, Dalmont, Paris, 1856.
3. ICH guideline Q3C (R6) on impurities: guideline for residual solvents, European Medicines Agency Committee for Human Medicinal Products EMA/CHMP/ICH/82260/2006 Issued 6 December 2016.
4. J. Jiménez and F. Martínez, *Rev. Acad. Colomb. Cienc. Exactas, Fis. Nat.*, 2006, **30**(114), 87–89.
5. M. J. Jozwiakowski, N. A. Nguyen, J. M. Sisco and C. W. Spancake, *J. Pharm. Sci.*, 1996, **85**(2), 193–199.
6. I. C. Kemp, *Drying Technol.*, 2017, **35**(8), 918–924.
7. A. Barton, *WO Pat.*, 033117 A1, 2015.
8. S. Ottoboni, C. J. Price, C. Steven, E. Meehan, A. Barton, P. Firth, A. Mitchell and F. Tahir, *J. Pharm. Sci.*, 2019, **108**(1), 372–381.

CHAPTER 14

Continuous Eutectic Freeze Crystallization

JEMITIAS CHIVAVAVA, DEBBIE JOOSTE, BENITA ASPELING, EDWARD PETERS, DERECK NDORO, HILTON HEYDENRYCH, MARCOS RODRIGUEZ PASCUAL AND ALISON LEWIS*

Crystallization and Precipitation Research Unit, Department of Chemical Engineering, University of Cape Town, Private Bag, Rondebosch, 7701, Cape Town, South Africa
*E-mail: Alison.Lewis@uct.ac.za

14.1 Introduction

14.1.1 What Is Eutectic Freeze Crystallization (EFC)?

Eutectic freeze crystallization (EFC) is a technology for treating saline streams in order to recover both water and dissolved salts (or acids). EFC works by cooling a saline stream to the eutectic temperature, at which point both ice and salt (or acid) will crystallize out of solution. The ice, being less dense than water, will float, and the salt, because it is denser than water, will sink – thus effecting a gravity separation. Theoretically, the process can result in zero liquid discharge (ZLD)[1] and has the significant added advantage of producing valuable products from a waste stream. It also changes the focus from cost to value by introducing a separation process that converts waste into potentially useful raw materials in an energy efficient process.[2]

The Handbook of Continuous Crystallization
Edited by Nima Yazdanpanah and Zoltan K. Nagy

Published by the Royal Society of Chemistry, www.rsc.org

EFC has several advantages over conventional separation techniques such as evaporative crystallization (EC). Since the latent heat of fusion of ice is about six times lower than the latent heat of vaporization of water, EFC is up to 85% more energy efficient than evaporative crystallization.[3] van der Ham[4] reported a 30 and 65% reduction in energy requirement upon using EFC in place of the conventional multi-step EC process for separation of sodium nitrate and copper sulfate from aqueous solutions, respectively. In addition to this, EFC has a smaller land footprint compared to evaporative ponds which are in wide use for storage of saline streams.

EFC also offers a number of other advantages: it generates products that have very high purity levels; it can be applied to temperature sensitive products; it requires cheaper materials of construction due to minimal corrosion at the low temperatures characteristic of EFC processes;[5,6] it offers safer separations since solid/liquid separations are generally safer than vapor/liquid separation; and it avoids undesirable side effects such as the production of fumes and the necessity to introduce additional reagents.

Theoretically, a 100% yield can be achieved from binary systems and this makes the method superior to cooling crystallization whose salt recovery is limited by residual solubility at the final operating temperatures.[7,8]

14.1.2 EFC Compared With Other Separation Technologies

Based on the driving force, separation technology can be broadly divided into four categories: *e.g.*, thermal (based on temperature differences), pressure, electrical potential, and chemical potential-driven processes.[9] Depending on the types of the aqueous solutions, the most commonly used technologies for separation, and their principles and disadvantages are listed in Table 14.1. As can be seen from the table, EFC has many advantages and can potentially be combined with other separation technologies to optimise separation processes.

EFC may be economically unviable for aqueous solutions with very low eutectic temperatures since the surface area required for heat transfer is very large. As a new technology, EFC equipment costs are still high, but these will decrease with time as better designs are developed.[11] The major limitation in EFC is ice scale layer formation on the heat exchanger surface. Ice scale layer fouling severely reduces the heat transfer coefficient and hence the heat flux, which reduces the rate of crystallization. The ice layer can further damage mechanical equipment inside the crystallizer and lead to unplanned shutdowns.

Initial studies into EFC are always carried out in batch mode, which is essential for testing the proof of concept, as well as for initial feasibility testing. However, in practice, the volumes of effluent generated by industrial operations are sometimes large and cannot be viably treated using a batch process. This is because of high labour and equipment costs. In addition, batch processes often produce inconsistent product quality due to variations from batch to batch, as well as batch variations due to the transient process.

Table 14.1 Conventional separation technologies.[10]

Technology	Separation principle	Disadvantages
Removal of low solute concentrations		
Solvent extraction	A selective solute from the aqueous phase is separated out by dissolving in an additional solvent	Additional chemicals Re-separation requirements
Reverse osmosis	The solution is pressurized against a selective membrane, which only allows the pure solvent to pass	Sensitive to fouling and scaling
Ion exchange/adsorption	To remove dissolved ions from solution and replace them with other similarly charged ions/to adhere the dissolved solute to a surface of the adsorbent.	Fouling from iron and/or calcium sulfate Organic contamination from the resin Thermal/design problems of adsorbent
Removal of high solute concentrations		
Evaporative crystallization	Solutes are crystallized by evaporating the solvent until the required supersaturation is reached	High energy consumption Operation at high working temperature
Cooling crystallization	Solutes are crystallized by cooling the solution below its saturation temperature	A positive and steep solubility–temperature line Limited yield
Anti-solvent crystallization	An anti-solvent that reduces the salt solubility is added to the solvent and thus crystallization occurs	Extra chemicals Recovery of anti-solvent
Precipitation	A reagent that will form a sparingly soluble product is added to the existing solution and thus precipitation is effected.	Extra chemicals Formation of unexpected solid phase
Membrane crystallization	The solvent is removed from the solution by a membrane which creates or enhances the generation of crystals	Scale formation on the membrane surface
Removal of both low and high solute concentrations		
Freeze crystallization	Water is removed by ice formation	Not suitable for eutectic systems

A continuous EFC process is required in order to cost effectively treat large volumes of brines, since it has the capability to generate products of a consistent quality and is cheaper in terms of equipment.

14.1.3 Theoretical Basis – Binary Phase Diagrams

EFC achieves simultaneous recovery of salts and highly pure water from brine solutions by operating at the eutectic point of the salt-water system in consideration. The eutectic point is defined as the composition and temperature below which both solvent and solute crystallize out of solution. Different salt-water systems have different eutectic compositions and temperatures.

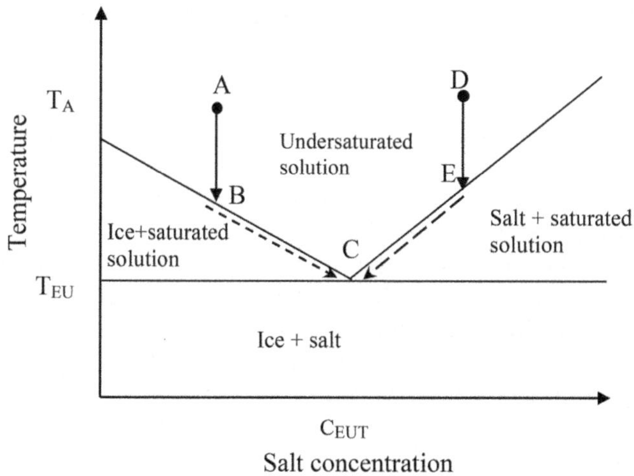

Figure 14.1 Modified phase equilibrium diagram of a binary system. Adapted from the MSc report of Willem Van der Tempel (TU Delft) with permission from Professor Geert-Jan Witkamp.

The principle of operation of EFC is best explained using a phase equilibrium diagram shown in Figure 14.1.

The freezing point of water decreases with increasing solute concentration. Cooling a hypoeutectic aqueous solution (*i.e.* more dilute than the eutectic) from point A to sub-freezing temperatures results in the crystallization of ice. The crystallization of ice causes the salt concentration to increase, thus depressing the freezing point. Further cooling results in the crystallization of more ice at progressively lower temperatures, moving the system along the line BC, until the system is saturated with the salt at the eutectic temperature. Further reduction of the temperature to sub-eutectic values causes the simultaneous crystallization of ice and salt. Thermodynamically, this occurs at a constant composition until the entire solution is recovered as separate solids.

Likewise, cooling an undersaturated hypereutectic mixture (starting at point D) to temperatures below the solubility line will result in a solution that is supersaturated in salt. The crystallization of salt dilutes the solution and the system moves along the line DC if cooling is continued. The point identifying the system characteristics on the phase diagram moves towards the eutectic point and the system is once again saturated in both ice and salt once this point is reached.

This example has illustrated a binary system. However, real brines are usually multicomponent in nature. These systems are more complicated than binary systems since multiple solutes interact with each other[12] and the solvent, thus altering the thermodynamic behaviour of the system. The solubility and activity of species, as well as eutectic parameters of systems often change in the presence of other species.

14.1.4 Theoretical Basis – Ternary and Quaternary Phase Diagrams

Figure 14.2 shows a three-dimensional representation of a three component (ternary) system, A, B and C, which gives an indication of the complexity brought to a phase diagram upon addition of another species.

Each new salt introduces an additional dimension to the phase diagram. At constant pressure, the phase diagram becomes three dimensional with composition on the sides of the basal triangle and temperature on the vertical axis. The region above the three concave surfaces represents an undersaturated solution while the surfaces, crossed with isotherms in Figure 14.2, represent the saturation values at different temperatures. The surfaces are separated by boundary curves which represent a series of pseudo-binary eutectic conditions where two solid phases on either side of the curves are at equilibrium with the liquid of the composition determined by the temperature. The concentration of the third component, in aqueous solution, increases along these curves and the curves converge on the ternary eutectic point.

At the ternary eutectic point, all three solid phases, A, B and C, are in equilibrium with a liquid solution containing constituents of the three solids. Ternary phase diagrams can be constructed by considering either the

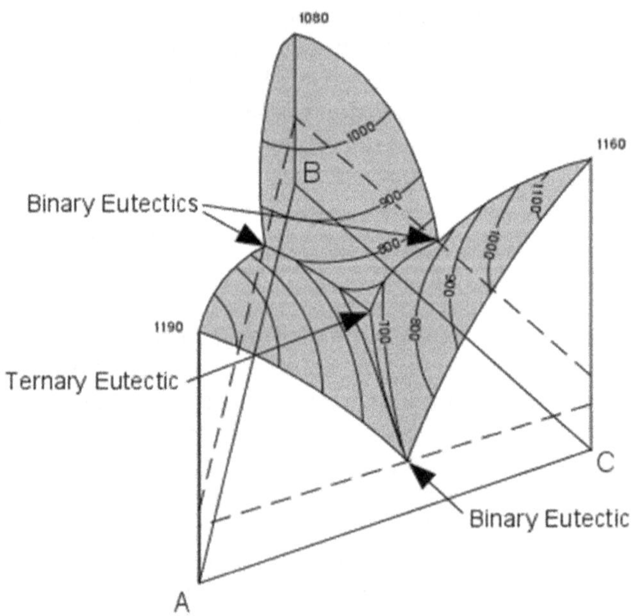

Figure 14.2 3D ternary phase diagram. Reproduced from ref. 13 with permission from Professor Stephen A. Nelson.

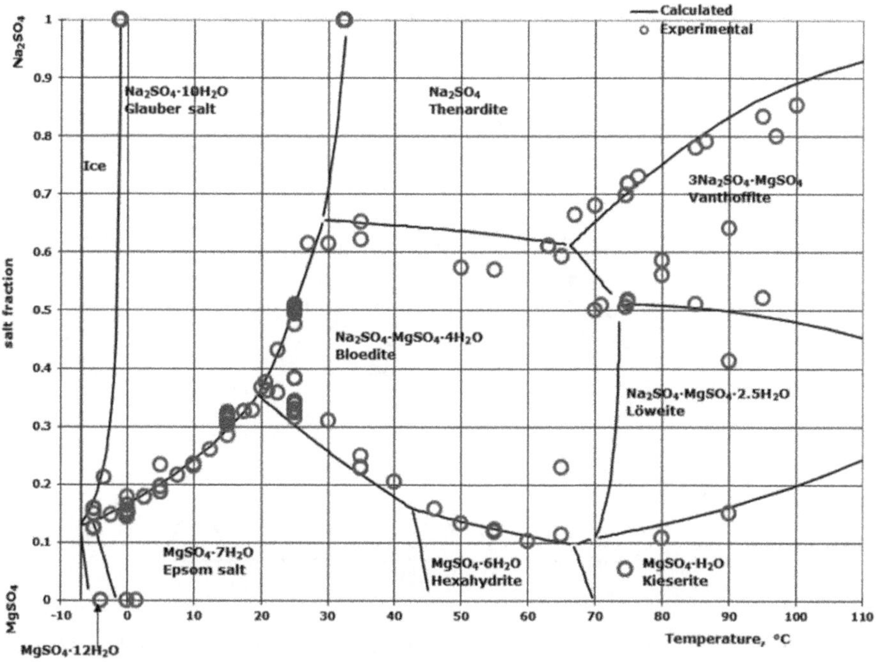

Figure 14.3 Phase diagram for Na_2SO_4–$MgSO_4$–H_2O system. Reproduced from ref. 14 with permission from Professor Kaj Thomsen.

cations or anions only in ternary salt-salt-water systems with common ions. The relative concentrations of two species are then plotted on one axis and temperature on the other axis, as shown for the Na_2SO_4 and $MgSO_4$ system in Figure 14.3.

Alternatively, ternary phase diagrams show concentrations of the three components on an equilateral triangle and leave out the temperature. An example of such a ternary phase diagram is shown in Figure 14.4, where composition is represented along the sides of the triangle. The dotted lines represent isotherms while pseudo-binary eutectic points are represented by the solid lines extending inward. The pseudo-binary eutectic lines meet at the ternary eutectic point, C.

The crystallization of the first salt and ice from a multicomponent system, under EFC conditions, increases the concentration of non-crystallizing aqueous species in the residual solution. Cooling a solution with composition D results in the crystallization of salt A and ice, which increases the concentration of B in the residual solution. The increased concentration of B depresses the eutectic temperatures resulting in salt A and ice crystallizing at progressively lower temperatures (along line DC) if cooling is continued. The system eventually reaches the ternary eutectic point (C) where it becomes saturated with the second salt.

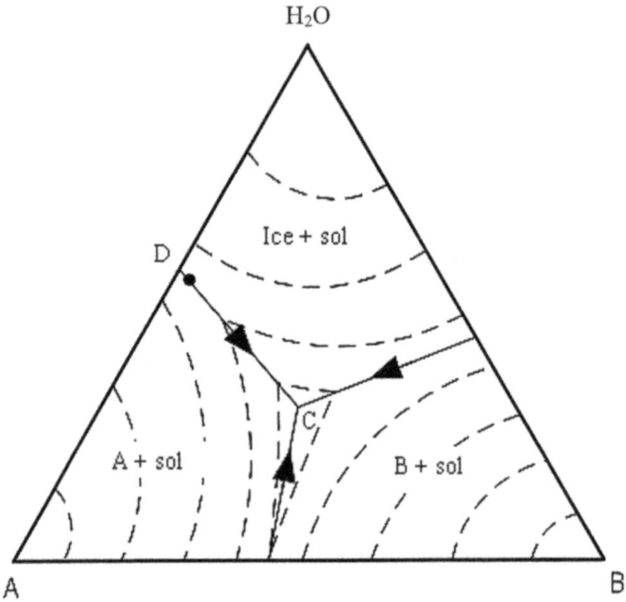

Figure 14.4 Typical phase diagram for a ternary eutectic system. Reproduced from ref. 15 with permission from Professor Stephen A. Nelson.

Further cooling of the solution below this temperature causes the crystallization of the second salt (B), thus resulting in the simultaneous crystallization of all three components. As the two salts and ice crystallize out, the liquid composition remains invariant until all the liquid solution has turned into the three solids.

The situation is complex in higher order systems. In this case the concentrations of non-crystallizing species increase and phase diagrams describing such systems become complicated. However, a point will be reached where all the various salts in solution will crystallize out at the eutectic composition and temperature of the specific multicomponent system. The removal of water in EFC, through ice crystallization, also concentrates salts whose solubilities are weak functions of temperature, such as $CaCO_3$ and $CaSO_4$. These salts will therefore eventually precipitate out and this makes predicting the crystallizing salts, nucleation temperatures and the sequence of crystallization non-trivial.

14.2 Thermodynamic Modelling of EFC for Saline Streams

In order to deal with the complexity of many of these brines, accurate predictions using thermodynamic modelling software are an essential tool for process design. It is important to select the appropriate thermodynamic modelling software in order to be able to accurately predict the eutectic temperature, eutectic composition, nucleation temperatures of ice and salt components, ice and salt recoveries and order of crystallization for crystal products.[16]

The following software packages have potential to be used in this application:

- Aspen Plus V10
- FactSage V7.2
- HSC Chemistry ® 5.1
- MINTEQ A2
- OLI Stream Analyzer 9.5
- PHREEQC

In order to assess whether the software is appropriate for modelling hypersaline streams, it needs to be able to fulfil the following criteria:

i. Ability to model solid–liquid equilibria for multi-component, mixed-salt electrolytes
ii. Robust concentration range
iii. Robust temperature range
iv. Ability to reconcile electro-neutrality of aqueous streams
v. Comprehensive databank
vi. Ability to be used as a predictive tool for:
  - Eutectic temperatures and compositions
  - Nucleation temperatures
  - Salt/ice recoveries
  - Order of salt/ice crystallization
vii. Convergence
viii. User friendly interface and debugger
ix. Ease of input

14.2.1 ASPEN Plus V10

Aspen Plus allows the building and running of process simulations in order to predict the behavior of processes by using basic engineering relationships, such as mass and energy balances and chemical and phase equilibria. Aspen Plus calculates liquid phase properties from the Electrolyte-NRTL activity coefficient model. The model also incorporates the Henry's Law parameters for the calculation of partial pressures and the Meissner equation for the calculation of the water activities. The vapor phase properties are calculated from the Redlich–Kwong Equation of State. Aspen Plus also has a built-in model that includes ice formation at temperatures down to −263 °C. However, there are limitations to the database, especially when modelling solid species. Furthermore, many of the hydrated salts that are present at the sub-zero temperature ranges are not included in the database. It is only recently that Aspen Plus has started to further develop its electrolyte database. Aspen Plus is available with the more comprehensive OLI Stream Analyzer database, since we evaluated OLI Stream Analyzer on its own, Aspen Plus was not deemed suitable as a thermodynamic package for EFC systems.

14.2.2 FactSage V7.2

The FactSage software is the fusion of two well-known software packages in the field of computational thermo-chemistry: FACT-Win (formerly F*A*C*T) and ChemSage (formerly SOLGASMIX).[17] This software package allows for the calculation and manipulation of phase diagrams although FactSage's database is limited at the lower temperature ranges. It does have the facility to import databanks from the OLI Stream Analyzer, but unfortunately, this databank can only be exported to the equilibrium model of the FactSage software and not the phase diagram section. Therefore, FactSage was not further investigated.

14.2.3 HSC Chemistry V5.1

The HSC Chemistry software has been designed to model various chemical reactions as well as equilibria. It contains fourteen calculation modules, of which the equilibrium module is of interest for eutectic freeze crystallization. This enables the calculation of multi-component equilibrium compositions in heterogeneous systems. Calculations are based on the GIBBS or SOLGASMIX solvers which use the Gibbs energy minimization method.[18] HSC Chemistry uses Criss–Cobble extrapolation for the heat capacity of aqueous species at elevated temperature (>25 °C), up to 300 °C. The activity coefficients are calculated using the Raoultian activity scale.[18] A limitation of the software is that it is essential for the user to input all defined phases into the system. Since, in simulating EFC processes, it is a requirement that the model predicts the crystallization of salts, this software was found to be inapplicable for EFC simulation.

14.2.4 MINTEQ V3.1

MINTEQ is a chemical equilibrium model that has an extensive thermodynamic database that allows for the calculation of speciation, solubility, and equilibrium of solid and dissolved phases of minerals in an aqueous solution.[19] However, it is not able to model streams at temperatures less than 0 °C.[20] Therefore, this software package is not suitable for EFC processes.

14.2.5 PHREEQC V3

PHREEQC is a general geochemical program and is used extensively for hydro-geochemical environments. The software uses ion-association and Debye Hückel expressions to account for the non-ideality of aqueous solutions. However, it is not applicable at high ionic strengths and also suffers from lack of internal consistency of the data in the databases. The databases that are distributed with the code have different origins and no systematic attempt has been made to ensure that the databases are congruent. Therefore PHREEQC was not investigated as a software package for EFC processes.[21]

14.2.6 OLI Stream Analyzer 9.5

The OLI Stream Analyzer software uses a speciation-based model that employs both accepted equations of state and experimental data to predict thermodynamic properties of multi-electrolyte aqueous solutions over a wide range of concentrations, temperatures, and pressures.[22] In addition, the software has in-built aqueous models which it applies to electrolyte systems in water, organic or mixed solvents.[23] The thermodynamic framework has been designed by combining the excess Gibbs energy model with detailed speciation calculations. The excess Gibbs energy model consists of a long-range interaction contribution represented by the Pitzer–Debye–Hückel expression, a short-range term expressed by the UNIQUAC model and a second-virial-coefficient term for specific ion interactions. The Helgeson–Kirkham–Flowers equation of state is incorporated to predict the equilibrium constants. The built-in model also integrates the Bromley–Meissner, Pitzer, Helgeson and Bromley–Zematis equations for calculations of the activity coefficients.[24] OLI software also includes a mixed solvent electrolyte (MSE) database, which uses the Helgeson direct method, and has the capability of successfully modelling temperatures as low as −200 °C.

The parameters of the model are determined using thermodynamic data of various types including vapor–liquid equilibria, activity and osmotic coefficients, solubility of salts in water or other solvents, amongst others.[23] The ability of the model to reproduce solubilities in multi-component systems together with accurately predicting the correct solid phases in multi-component systems means that the software is applicable to EFC processes.

14.2.7 Summary of Thermodynamic Software Packages

Presented in Table 14.2 is a comparison of the criteria used to evaluate the various thermodynamic models. As can be seen, OLI Stream Analyzer fulfils all the criteria set out.

Table 14.2 Comparison of thermodynamic software used for modelling of hypersaline solutions.

No.	Software model	Aspen plus	Fact Sage	HSC chemistry	MINTEQ	PHREEQC	OLI stream analyzer
i	Solid–liquid equilibria	Yes	Yes	Yes	Yes	Yes	Yes
ii	Concentration range	6M	6M?	6M?	1M (dilute aqueous streams)	Dilute	6M
iii	Temperature range	−263 °C to 200 °C	>0 °C	−269 °C to 300 °C	0 °C to 99.99 °C	25 °C	−200 °C to 225 °C
iv	Comprehensive databank (low temp. Hydrates)	No	Yes	Yes	No	No	Yes
v	Use as an electro-neutrality reconciliation tool	No	Yes	No	Yes	No	Yes
vi	Use as a predictive tool	Yes	No	No	Yes	No	Yes
vii	Convergence	Depends on complexity of stream	Fast	Depends on complexity	Fast	Depends on complexity	Fast
viii	User friendly interface/ debugger	Yes	Yes	No	No	Yes	Yes
ix	Ease of input	Complex	Simple	Complex	Complex	Simple	Simple

14.3 Understanding EFC from a Melt Crystallization Point of View

It is useful to understand eutectic freeze crystallization from a melt crystallization point of view, since melt crystallization forms an important "subclass" of crystallization that has a number of specialized applications. There is no real theoretical basis for the distinction between solution crystallization and melt crystallization, but historically, the very different industrial techniques used for melt crystallization have justified it being considered as a separate class of crystallization method.[25]

A melt is the common name given to a liquid or a liquid mixture at a temperature near its freezing point. Eutectic freeze crystallization is classified as a melt crystallization for two main reasons: firstly, the crystallizing phase is the solvent itself and secondly, (and this follows from the first reason) the operating temperature is close to the melting temperature of the main component.

Since eutectic freeze crystallization is dominated by ice crystallization, it has been classified as a melt crystallization problem. However, because of the simultaneous crystallization of salt, EFC also exhibits solution crystallization characteristics. This will be explained in more detail in a later part of this chapter, under Section 14.6: Coupled heat and mass transfer problem.

14.4 Defining Supersaturation in Eutectic Freeze Crystallization

Table 14.3 summarizes some expressions used to describe supersaturation for different crystallization methods.

Table 14.3 Summary of expressions for supersaturation, as well as expressions used in practice for cooling, melt and evaporative crystallization. Note that practical expressions usually are not consistent with the thermodynamic ones. c is the actual concentration in the solution and c_{eq} the equilibrium concentration (kg solute/kg solution or kg solute/kg solvent or kg solute/kg mixed solvent in the case of anti-solvent crystallization); similarly, T and T^* (K) are the actual and the equilibrium temperatures.[26]

Thermodynamic expression	Practical expressions	Crystallization method	Restriction
$(\Delta h_f/T^*)(T - T^*)$ (J mol^{-1})	$\Delta T = T^* - T$(K)	Cooling, melt	P constant, low supersaturation
RT ln S_{cm} RT ln S_c (J mol^{-1})	$\Delta c = c - c_{eq}$ (g solute /g solution) or (g solute/g solvent)	Evaporative	T, P constant, low supersaturation. T, P constant, low supersaturation, single solute

In order to calculate the supersaturation of the ice phase in Eutectic Freeze Crystallization, the expression in (14.1) should be used, since the activity of water remains relatively constant throughout the crystallization process:

$$\Delta T = T^{*} - T \tag{14.1}$$

Where ΔT = supersaturation of ice [K]
T = operating temperature [K]
T^{*} = equilibrium temperature [K]
However, the supersaturation of the salt phase in the crystallization of electrolytic aqueous solutions is estimated from the expression in (14.2):

$$\Delta c = c - c_{\text{eq}} \tag{14.2}$$

Where Δc = supersaturation of salt [(kg solute/kg solution) or (kg solute/kg solvent)]
c = actual concentration of salt [(kg solute/kg solution) or (kg solute/kg solvent)]
c_{eq} = equilibrium salt concentration [(kg solute/kg solution) or (kg solute/kg solvent)]

The use of this expression is justified if the solute concentration in solution is low and when the difference in the operating and equilibrium concentrations is small.[27]

It will be clear from these two equations that two different expressions for supersaturation, each with different units, will be generated. Both expressions are approximations to the real value of the supersaturation, or the *real* driving force for crystallization. These simplifications have also been applied under eutectic freeze crystallization conditions (*i.e.* simultaneous crystallization of both ice and salt), where the supersaturation of ice is considered to only be dependent on ΔT and the supersaturation of salt dependent on ΔC, taken as equilibrium points from the extrapolated solubility lines.[28] The value estimated from (14.2) is expected to be reasonably accurate especially in indirectly cooled eutectic systems, where low supersaturations are typically used.

14.5 Mechanisms

In any crystallization system, the processes of nucleation and growth are influenced not only by the thermodynamic properties of the system, but also, to a very large extent, by its kinetics.

14.5.1 Metastable Zone Width

The first important kinetic property of any crystallization process is the metastable zone width. The metastable zone comprises an area on the phase diagram below the equilibrium line, in which only primary heterogeneous and secondary nucleation occur.[29] The metastable zones for ice and salt overlap at sub-eutectic temperatures in eutectic freeze crystallization processes and both products can crystallize from solutions in this region.

14.5.2 Nucleation

14.5.2.1 Primary Nucleation

Homogeneous primary nucleation refers to the formation of new crystalline particles within the bulk solution while heterogeneous primary nucleation refers to the formation of new crystalline particles on a solid surface within the crystallizer, such as the walls of the vessel or on the mixing equipment.[26,30]

14.5.2.2 Secondary Nucleation

Secondary nucleation refers to the formation of new crystalline particles induced by the presence of particles of same crystalline nature.[31] The addition of such similar existing particles to the crystallizer, in order to initiate nucleation within the metastable zone, is called seeding and is a method used to control crystal size and morphology.

The formation of new crystals in continuous crystallization processes is dominated by secondary nucleation. This occurs through initial/dust breeding, needle/dendritic breeding, contact nucleation, shear and attrition breeding.[12]

Genceli[2] indicated that the secondary nucleation of ice and salts during EFC operations is proportional to the specific cooling surface area and solids fraction. This emphasizes the significance of crystal–wall and crystal–crystal collisions for both ice and salt nucleation mechanisms near the scraped, cooled walls. An analysis by Kane[32] revealed that the rate of secondary nucleation of ice in continuous mixed suspension mixed product removal crystallizers increases with the ice fraction. Crystal–crystal collisions were found to contribute 25% towards the total nucleation rate at ice fractions of around 10 wt.%. However, the contribution of this mechanism is negligible at low ice fractions.

The importance of ice solids fraction was also reported by Margolis[33] in the freezing of ice from a sodium chloride aqueous solution. The secondary nucleation rate of ice was correlated with the total surface area of the ice crystals and there is general agreement in literature that crystals contribute to secondary nucleation after attaining a certain size.[34]

Although needle breeding is expected at high supersaturations, Huige[35] noted that needles and thin platelets, which promote this mechanism, are formed at subcoolings around 0.5 °C in ice crystallization. This mechanism is unlikely to contribute significantly to the nucleation rate in the suspension, where the degree of subcooling is very low, but may dominate the zone near the cooled wall where supersaturations are high.

The authors acknowledge that secondary nucleation is still poorly understood yet it plays a critical role in the control of crystal size distribution. As in primary nucleation, it requires supersaturation, although it occurs even at quite low supersaturations.

14.5.3 Growth

Crystal growth can be described simply as the deposition of solute (or solvent) molecules onto an existing crystal surface. Measuring the growth rate of crystals growing in suspension can be complex, as each crystalline face exhibits its own linear growth rate. The overall linear growth rate is usually described as a combined function of the increase in specific crystal mass and the individual growth rates in the direction normal to each crystalline face.

The description of the growth rate is simplified in the case of growth on a solid surface such as a crystallizer wall, in which case it can be described as planar growth in the direction normal to the wall.[36] Planar growth of ice on a cooled surface from a dilute electrolyte solution involves three steps. Since the ice lattice excludes foreign ions/molecules at slow growth rates, solutes must diffuse away from the growing ice front while water molecules must be incorporated into the crystal lattice and the heat of crystallization must be removed so as to maintain the supersaturation.[37] These processes are graphically shown in Figure 14.5, where δ_d is the thickness of the diffusion boundary layer and δ_{ice} is the thickness of the ice layer.

14.5.4 Ice Growth

Depending on the process characteristics, any one of the three steps involved in ice layer growth could be the rate limiting step. For ice crystallization in EFC, the removal of heat from the growing crystal front can be the limiting process.[26] The heat of crystallization can be transferred either to the bulk solution or to the existing crystal solid. In the case of ice layer growth on the crystallizer walls, as illustrated in Figure 14.5, it can be assumed that the majority of the heat transfer happens through the existing ice layer since the surface is in contact with the coolant. Heat transfer is, therefore, conductive and the growth rate for a thin layer can be described by eqn (14.3):[36]

$$G = \frac{\gamma_{ice}}{\rho_{ice}\delta_{ice}\Delta h_f}(T_{int} - T_w) \tag{14.3}$$

where γ_{ice}, ρ_{ice} and δ_{ice} are the thermal conductivity, density and thickness of ice respectively, Δh_f is the heat of fusion of water, T_{int} is the solid–liquid interface temperature and T_w is the temperature of the cooled surface.

In the case of ice layer growth from concentrated solutions, the growth rate is limited by the diffusion of solute molecules away from the growing front and can be modelled as a transport process through a semi-permeable membrane according to eqn (14.4):[12]

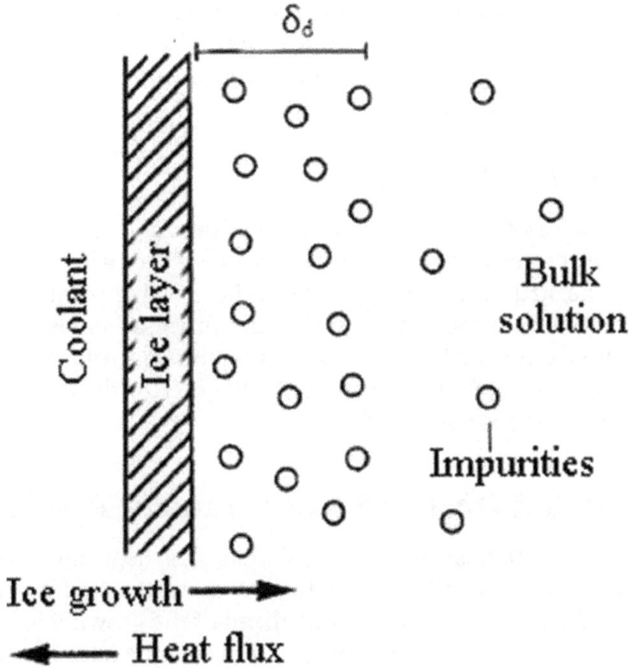

Figure 14.5 Ice layer growth on a cooled surface. Reproduced from ref. 37 with permission from Elsevier, Copyright 1990.

$$G = \frac{k_d}{x_b}\frac{\rho_{liq}}{\rho_{ice}}\frac{MM_{H_2O}}{MM_{liq}}(x_{int} - x_b) \quad (14.4)$$

where k_d is the mass transfer coefficient, x_b is the solute mole fraction in the bulk liquid, ρ_{ice} and ρ_{liq} is the density of the ice and liquid respectively, MM_{H_2O} and MM_{liq} are the molar mass of water and the liquid respectively and x_{int} is the solute mole fraction at the solid–liquid interface.

The rate of incorporation of water molecules into the crystal lattice is a property specific to each system. When crystal growth is limited by the surface integration kinetics, the growth rate can be described by eqn (14.5):[5]

$$G = 2.7 \times 10^{-3}\,(T^{*}x_{int} - T_{int})^{1.55} \quad (14.5)$$

where T^* is the equilibrium temperature and x_{int} and T_{int} is the solute mole fraction and the temperature respectively at the solid–liquid interface.

In the case of dilute electrolyte systems, the supersaturation causing the mass transfer limitation can be expressed in terms of temperature, which allows the combination of eqn (14.3)–(14.5) to give the overall growth rate of ice crystals on a cooled surface:[38]

$$G = 2.7\times10^{-3}\left(T^{*}x_b - T_w + G\left(\frac{x_b}{k_d}\frac{\rho_{ice}}{\rho_{liq}}\frac{MM_{liq}}{MM_{H_2O}}\left(\frac{dT^{*}}{dx}\right)_b - \frac{\rho_{ice}\delta_{ice}\Delta h_f}{\gamma_{ice}}\right)\right)^{1.55} \quad (14.6)$$

From eqn (14.1)–(14.6) it is clear that the rate of ice layer growth is strongly dependent on the local supersaturation at the solid–liquid interface and any change in process variables, which affect the local supersaturation, will have a major effect on the growth rate.

14.5.5 Salt Crystal Growth

During the growth of salt crystals from solution, molecules or ions diffuse either towards or away from the growing crystal surface where they are incorporated into the structure or rejected from the growing front. The diffusion of these species limits the growth rate if the process is mass transfer controlled. Otherwise the surface incorporation step controls the growth rate. The latter was reported for the growth of $MgSO_4 \cdot 11H_2O$ during EFC of a magnesium sulfate aqueous solution.[2]

14.6 Coupled Heat and Mass Transfer Problem

Eutectic freeze crystallization is a classic coupled heat and mass transfer problem. During ice crystallization, which is an exothermic process, heat is released, and if the rate of heat removal ahead of the growing ice front is not sufficiently fast, the crystallization process will become heat transfer limited. However, slow diffusion of rejected solutes from the growing interface may also limit the growth rate of ice. In EFC of binary systems, the rejected solutes are transported to the salt crystal growing interface where they are integrated into the crystal lattice.

Since these two processes happen simultaneously, a coupled heat and mass transfer problem is created. The two processes influence and are influenced by each other as follows.

As the ice crystallizes, the concentration of salt in solution increases, causing the driving force for salt crystallization to increase and the driving force for ice crystallization to decrease. However, at the same time, the crystallization of salt causes the solution to become more dilute, causing the driving force for ice crystallization to increase and the driving force for salt crystallization to decrease.

At the same time, ice crystallization releases heat (as does, to a lesser extent, salt crystallization). In classical theories, it is assumed that all the heat of crystallization is transferred to the colder side, that is, the solid phase.[12,36] However, later experimental studies[39,40] showed that the released heat diffuses to both the solid and liquid sides of the developing interface. This was predicted by using heat and mass transfer equations based on the irreversible thermodynamics of Onsager relations presented by Kjelstrup and Bedeaux.[41] Following this point of view, the release of heat to the liquid side is coupled with the rejection of solutes when ice grows from aqueous salt solutions.[42] This coupled effect results in the formation of a thermal and mass diffusion boundary layer in front of the growing ice layer.[37] Because of these coupled temperature and concentration gradients in the liquid phase in front of the growing ice interface, heat and mass diffuse into the bulk liquid. These cause the temperature of the solution to rise, and thus a decrease in the driving force for both ice and salt crystallization in an adiabatic operation.

All of these effects occur simultaneously in a dynamic and interactive way. The rates of heat and mass transfer are responsible for local supersaturation and therefore affect all the crystallization processes of nucleation, growth and morphology of ice. The rate of these transport processes is governed by the magnitude of the thermal and concentration gradients as well as the thermal and molecular/ion diffusivities. The driving forces are functions of the ice growth rate and rejection rate of the solutes as the water molecules integrate into the ice crystal lattice.[40]

The heat and mass transfer coefficients also depend on the hydrodynamics, *i.e.* flow regime and the thickness of the boundary layer in front of the growing interface. Heat and solutes rejected from the growing, suspended ice crystals are transported into the subcooled bulk solution. The heat is eventually removed from the liquid through heat exchange with the crystallizer wall, which is in thermal contact with a refrigerant. In the EFC of binary systems, the rejected solutes are transported, through the bulk liquid, to the surface of growing salt crystals where they are incorporated into the crystal lattice.

The transport of lattice ions from the bulk solution (or ice growing front) to the salt crystal surface or their integration into the salt lattice can limit the salt crystallization rate. Essentially any of the steps described here can be rate limiting. The simultaneous crystallization of the two species allows continual replenishment of supersaturation, thus allowing EFC processes to achieve high yields.

14.7 Heat Transfer

EFC is achieved through the removal of sensible heat of the feed and heat released by the formation of ice and salt products. In agitated suspension crystallizers, the heat of crystallization is transported from the growing interfaces of suspended crystals to the cooled wall through forced convection in the subcooled bulk solution. The heat is then conducted to the refrigerant through the walls in indirectly cooled systems and the refrigerant absorbs the heat and rejects it to the heat sink for the system such as the condenser.

The amount of sensible heat entering the system can be minimized through sound design principles, rendering the rate of heat removal proportional to the rate of crystallization.[43] It is, therefore, critical to promote a high rate of heat transfer in EFC processes. The level of turbulence in the suspension influences the rate of heat removal from the crystal–solution interfaces since it affects the thickness of the thermal and concentration boundary layers around the suspended crystals.

14.8 Why Continuous EFC?

A continuous EFC process is appropriate for the treatment of large volumes of saline streams since it generates products of a consistent quality, allows good control and reduces waste.

14.8.1 Continuous EFC Process Flow

The most basic version of a continuous EFC process consists of feed tank, a pre-cooler, EFC crystallizer-separator, filtration and washing units, as well as product tanks for the residual mother liquor, ice and salt (Figure 14.6). Pre-cooling removes sensible heat and maintains feed temperatures close to the crystallizer operating temperature to ease crystallizer operation and prevent scale formation. A plate type heat exchanger was employed by Genceli for this purpose.[44]

The pre-cooled feed is introduced into the crystallizer at temperatures 2 to 3 °C warmer than the crystallizer operating temperature. Residual sensible heat is removed and simultaneous crystallization of ice and salts occur in the crystallizer. Various crystallizer designs have been used, including several versions of scraped wall cooled crystallizers (SWCC),[46–48] cooled disk column crystallizers (CDCC)[2,44,49] and non-scraped agitated crystallizers.[50,51] The crystallizers allow simultaneous crystallization and initial solid–solid–liquid separation of the products, thus acting as hybrid crystallizer-separator equipment.

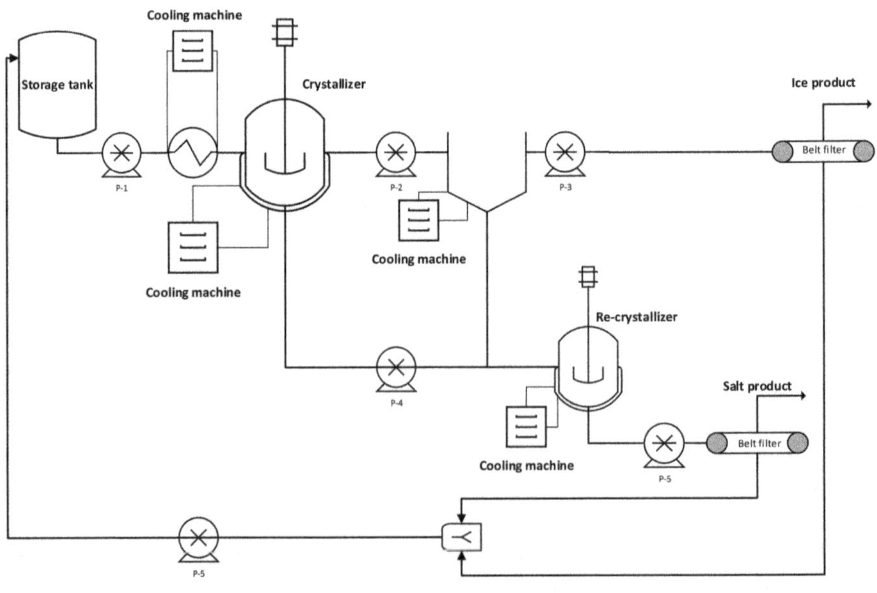

Figure 14.6 Continuous EFC set up.[45]

Eutectic crystallizers are cooled using jackets or cooled disks and are agitated using scrapers, which primarily prevent ice build-up on the cooled walls. The crystallizers have provisions for product separation zones in the top and bottom sections.

The efficiency of solid–solid–liquid separation and the downstream separation steps depend on the quality of the ice and salt crystals. These crystal characteristics are shaped by the operating conditions in the crystallizer, with the degree of supercooling governing the rates of nucleation, growth and agglomeration, which eventually determine the particle size distribution and morphologies. Disk-shaped ice crystals[28,44] were produced in several continuous EFC processes (Figure 14.7) while salt crystals of different morphologies varying from rhombohedral $CuSO_4 \cdot 5H_2O$,[7] hexagonal prisms ($MgSO_4 \cdot 12H_2O$)[44] and monoclinic $Na_2SO_4 \cdot 10H_2O$[51] have also been produced.

Residence times varying from 20 minutes to 2.5 hours[46] were employed in the various continuous EFC tests and average heat transfer driving forces of up to 4.5 K are widely reported in the different tests. Mean crystal sizes of 100 to 300 μm[28,43] were obtained in scraped eutectic crystallizers while particles as large as 1000 μm were reported by Peters.[51]

The products are withdrawn from the crystallizer separation compartments as salt (underflow) and ice (overflow) slurries from the bottom and top of the crystallizer, respectively. Ice and salt (s) are then filtered from the overflow and underflow slurries using vacuum belt filters. The products are often washed to remove impurities from the surfaces of the crystals. These impurities are usually entrained due to mother liquor adhesion onto the crystal surfaces since ice crystals themselves are pure due to rejection of impurities from the lattice.

It is important to note that, although high yields are achievable with binary systems, current designs allow partial crystallization in a single pass and the residual liquid is useful as a medium in the gravitational separation

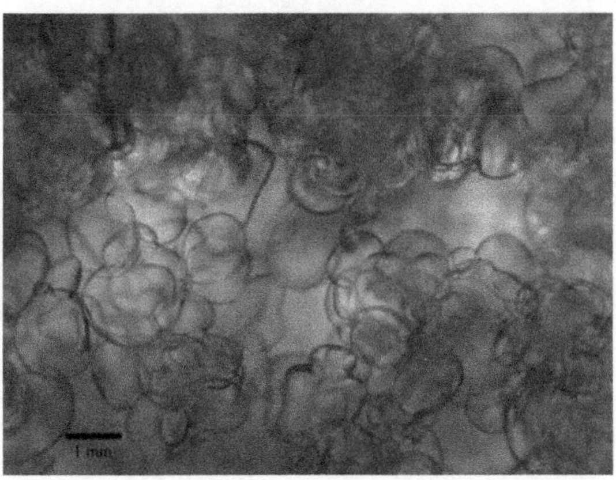

Figure 14.7 Example of ice produced from an EFC process.

step. From mass integration perspective, the filtrate from the filtration of both ice and salt can be recycled back to the feed stream and fed into the EFC crystallizer. However, the stream should be treated in further crystallization stages operating at lower temperatures in the case of a multicomponent system.

From a heat integration perspective, the feed can be precooled using the ice that is produced in the EFC crystallizer. This has the added advantage of ensuring that the temperature difference between the incoming feed stream and the EFC crystallizer is small. This small temperature difference ensures ease of operability and minimizes the problem of ice scaling on the crystallizer walls.

14.9 Stages in Continuous Eutectic Freeze Crystallization

Continuous crystallization involves a start-up stage, ideally followed by a long period of steady state operation and eventually shut-down. Following initial bulk nucleation is an unsteady period during which the magma density adapts to its steady state value, in accordance with the rate of heat removal. After nucleation, the growing crystals consume supersaturation and induce secondary nucleation. Nucleation, however, is very fast in comparison to growth and the lag between the two processes results in periods of high supersaturation dominated by nucleation and periods of low supersaturation dominated by growth.[31]

Depending on the process conditions, this stage of unsteady, periodic behaviour can be short and damped, as shown in Figure 14.8a or long and

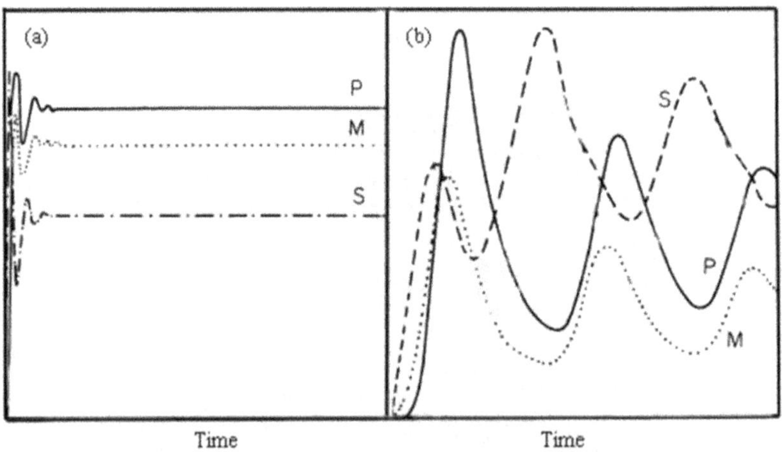

Figure 14.8 Dynamic behavior during start-up of a continuous crystallization process (a) short and damped, (b) long and profound. Reproduced from ref. 31 with permission from Butterworth-Heinemann, Copyright 2001.

profound, as shown in Figure 14.8b, where *P* denotes production rate, *M*, magma density and *S*, supersaturation ratio.

In crystallization operations, steady state is usually defined in terms of macroscopic variables which can be measured with ease, such as temperature in the bulk solution and production rates. When steady state is disrupted due to a step change in process conditions, similar periodic behavior as described for the start-up period will prevail until a new steady state is reached. Conditions which are not conducive to steady state will eventually lead to process failure.

The temperature profile in Figure 14.9 summarizes the stages in a continuous EFC process. During start-up, a solution is gradually cooled to temperatures below the eutectic temperature. Controlled nucleation of ice or salt can be achieved at small supersaturations through the introduction of seed crystals. Heat of crystallization is released and this raises the temperature of the solution (see Figure 14.9). After the initial spike, the temperature of the suspension drops slightly and remains constant since the system is constantly cooled. During this latter period, the heat removed by the coolant is equivalent to the sum of sensible heat and latent heat from the crystallization of ice and salts for systems with good insulation and negligible heat losses.

Although the temperature stabilizes quickly in most eutectic systems, other macroscopic properties of the system such as crystal size distribution, morphology and magma density may take longer to reach steady state. The evolution of these properties in continuous EFC is complex due to the simultaneous crystallization of products and the subsequent interdependence of supersaturation as described by Himawan.[52] Steady state may not be achieved due to scale formation in systems where large production rates are achieved by operating at large heat transfer driving forces.

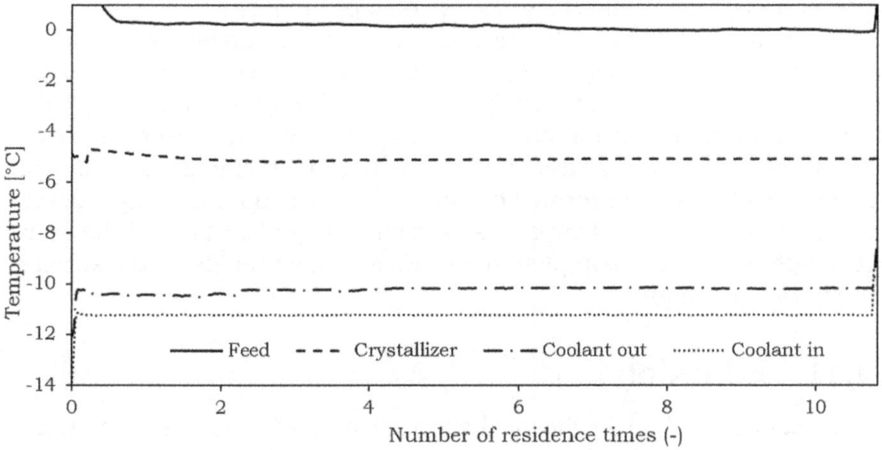

Figure 14.9 Temperature profile during continuous eutectic freeze crystallization.

14.10 Scaling

During suspension crystallization from aqueous solutions, in indirectly cooled systems, conditions that favour the formation of an ice scale layer on cooled heat exchanger surfaces are created. The formation of the ice layer is undesirable because conductivity of ice is several times smaller than that of materials typically used for the construction of crystallization vessels. The ice layer, therefore, increases thermal resistance and decreases heat removal rate from the suspension, adversely affecting the production rates of ice and salt in an EFC process.

The formation of the ice scale layer can be particularly severe due to the surface that is provided for heterogeneous nucleation, the temperature driving force that is applied across the surface and brines that typically consist of more than 80% water which is, in this case, the scalant. Furthermore, any mechanical equipment within the crystallizer serves as additional surfaces for adhesion and ice growth. Unfortunately, the mechanisms of ice scale formation in a crystallization environment are still poorly understood. Although heterogeneous nucleation is the most likely mechanism in scraped wall crystallizers, the adhesion of the ice crystals onto the cooled wall may initiate scale layer formation, especially in non-scraped crystallizers.[50]

14.10.1 Thermal Boundary Layer

Crystallizer walls experience the lowest temperature within the crystallizers since their externals are in thermal contact with the coolant flowing through the jacket. A thermal boundary layer develops between the wall and the bulk liquid, as depicted in Figure 14.10a. The solution adjacent to the wall surfaces is, therefore, at a higher supersaturation than the bulk. This high local supersaturation prompts heterogeneous nucleation and growth of an ice scale layer on the crystallizer surface as noted by Pronk.[53]

Two main crystallizer configurations have been researched with the aim of combatting scale formation. In a fluidised bed configuration, metal particles are used to mechanically remove ice from the crystallizer walls during fluidisation.[53] Alternatively, a scraped wall crystallizer has been used, where each scraper pass removes the thermal boundary layer as well as any crystals that have formed on the surface. The boundary layer develops again rapidly after each pass, as shown in Figure 14.11, and the average temperature over a complete scrape cycle remains colder at the surface than in the bulk liquid.

14.11 Adhesion

The crystallizer wall is a subcooled surface where adhesion of ice crystals, that formed in the bulk solution, can occur. Furthermore, in scraped surface crystallizers, the mechanical scrapers are subcooled themselves, due to the

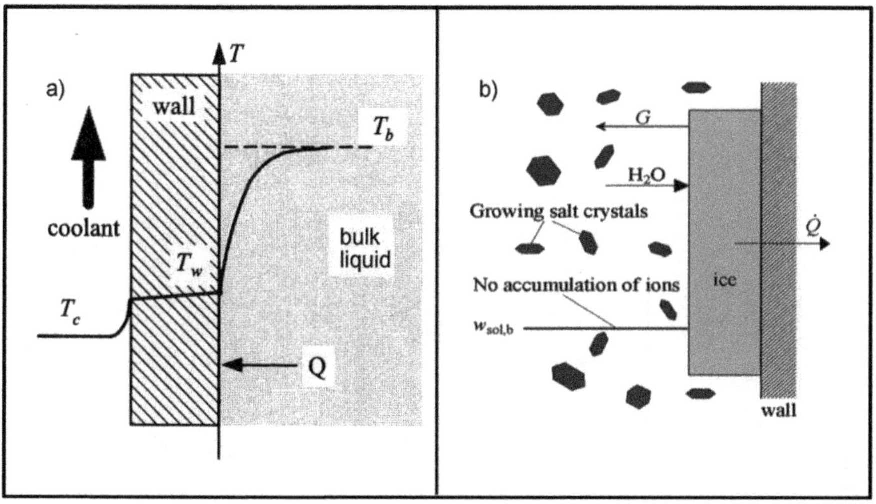

Figure 14.10 (a)Temperature profile promoting scale formation[54] (b) Scale formation during EFC. Reproduced from ref. 53 with permission from Dr Pepijn Pronk.

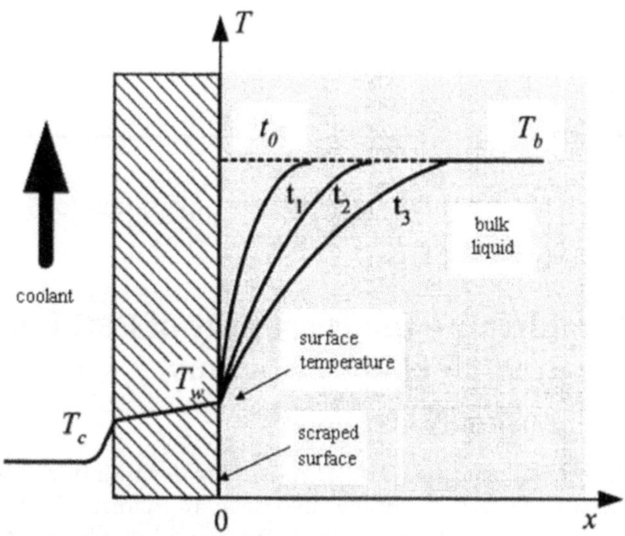

Figure 14.11 Development of the thermal boundary layer after each scraper pass. Reproduced from ref. 54 with permission from John Wiley and Sons, Copyright © 2003 American Institute of Chemical Engineers (AIChE).

continuous movement through the thermal boundary layer. Ice crystals that are scraped off the wall collide with the scraper blades and crystals which formed in the bulk solution also come into contact with the scrapers. The adhesive strength of ice on a solid surface is a function of temperature, as

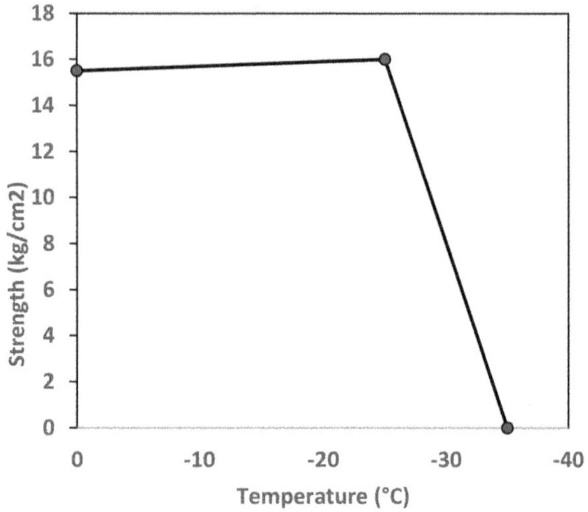

Figure 14.12 Shear strength of ice on a metal disc as a function of temperature.[55] Courtesy of Engineer Research and Development Center (US), Cold Regions Research and Engineering Laboratory (US).

Table 14.4 Salinity levels of various streams.

Fresh water	(<0.05%)
Brackish water	(0.05–3%)
Saline water	(3–5%)
Brine	(>5%)

shown in Figure 14.12. In addition to higher supersaturation, a colder wall also provides conditions for stronger adhesion of ice.

14.12 Establishing the Feasibility of EFC for Treatment of Saline Streams

14.12.1 What Is a Saline Stream?

With the increasing use of reverse osmosis for desalination and for treating industrial waste water, there has also been an increase in the volume of saline streams, brines and wastewaters produced.[56,57] The consequence of pure water recovery from a sea water or wastewater stream is the generation of a reject saline stream that is highly concentrated in ionic salts. The salinity levels of various streams are defined in Table 14.4.

Therefore, although desalination technologies can contribute to local water security, it is also critically important to find sustainable solutions to deal with associated brines and saline streams that are generated as waste products. This requires new, innovative approaches.[57] The current practice is to store this reject saline stream in evaporative ponds or, in the case of desalination plants, to pump it out to sea. However, eutectic freeze crystallization can be used as a subsequent step to reverse osmosis treatment in order to recover both water and salts from the reject saline stream. This is illustrated in Figure 14.13.

Table 14.5 summarizes the methods available for treating highly saline streams.

14.12.2 Options for Treatment of Highly Saline Streams

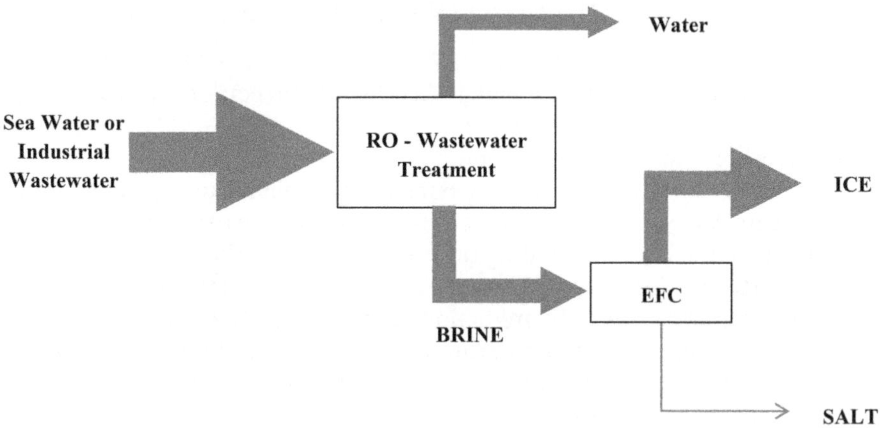

Figure 14.13 Reverse osmosis–eutectic freeze crystallization process for water and salt recovery.

Table 14.5 Comparison of saline brine treatment technologies available.

Crystallization method	Advantages	Disadvantages
Evaporative crystallization (EC)	High solute recovery[58]	Costly due to high energy costs from stream requirements as this is a high temperature process. Poor product quality.
Eutectic freeze crystallization (EFC)	Low energy as heat of fusion of ice is six times less than the heat of evaporation of water.[59] Can treat hypersaline brines. Possibility of higher conversion than other methods. Both crystallization and gravity separation occur in the same vessel. No additional chemicals needed to induce separation.[1]	Fouling on cooling surface reducing the heat transfer coefficient. High capital costs[28] Not applicable to solutions that have a reverse solubility.
Cooling crystallization (CC)	Lower energy costs when compared to EC[58]	Limited by the residual solute solubility at low temperature[60] Not applicable to solutions that have a reverse solubility.

14.13 Example of Thermodynamic Modelling of a Brine Stream Being Subjected to EFC

In order to establish the feasibility of EFC for treatment of a saline stream, the first step is to use a thermodynamic model to predict the nucleation temperatures and the salts that will be formed.

14.13.1 Modelling Using the OLI Stream Analyzer 9.5

The reverse osmosis concentrate stream specified in Table 14.6 is used as an example in thermodynamic modelling of a brine stream being subjected to EFC.

a. Creating a synthetic salt stream
   The first step involves carrying out a charge balance using OLI Systems Stream Analyzer.[61] Adjustment is also made for the pH by adding either an acid or a base, depending on the properties of the stream.
b. Prediction of salt crystallization at reduced temperatures
   Figure 14.14 shows the predictions of the successive crystallization of various salts as the temperature is lowered from ambient (from right to left on the *x* axis). Even at ambient temperatures, the RO retentate stream was supersaturated with $CaSO_4$, and this is reflected by the crystallization of $CaSO_4{\cdot}2H_2O$ across the entire temperature range. This would need to be taken into account in designing an appropriate EFC process.

Once the temperature reaches −1 °C, ice crystallizes out, followed by $Na_2SO_4{\cdot}10H_2O$ at −2 °C and $NaCl{\cdot}2H_2O$ at −23 °C. Trace quantities of CaF_2 and KNO_3 also crystallize out of the brine – these are species that will accumulate with increased brine volumes.

Table 14.6 Composition of stream to be modelled.

Species	Units	Concentration
Na^+	$mg\ l^{-1}$	5400
K^+	$mg\ l^{-1}$	300
Mg^{2+}	$mg\ l^{-1}$	41
Ca^{2+}	$mg\ l^{-1}$	390
Li^+	$mg\ l^{-1}$	0
NH_4^+	$mg\ l^{-1}$	0.5
Br^-	$mg\ l^{-1}$	<0.1
Cl^-	$mg\ l^{-1}$	4010
F^-	$mg\ l^{-1}$	13.86
NO_3^-	$mg\ l^{-1}$	29.6
SO_4^{2-}	$mg\ l^{-1}$	8690
pH		7.26
κ	$mS\ cm^{-1}$	>20
TDS	ppt	>10

Table 14.7 presents the predicted eutectic compositions and temperatures for the binary salt-water species found in the various streams obtained from the OLI Stream Analyzer. These are in good agreement with results from literature.

This table provides vital information for the evaluation of sequential EFC applications to selectively crystallize out different salts.

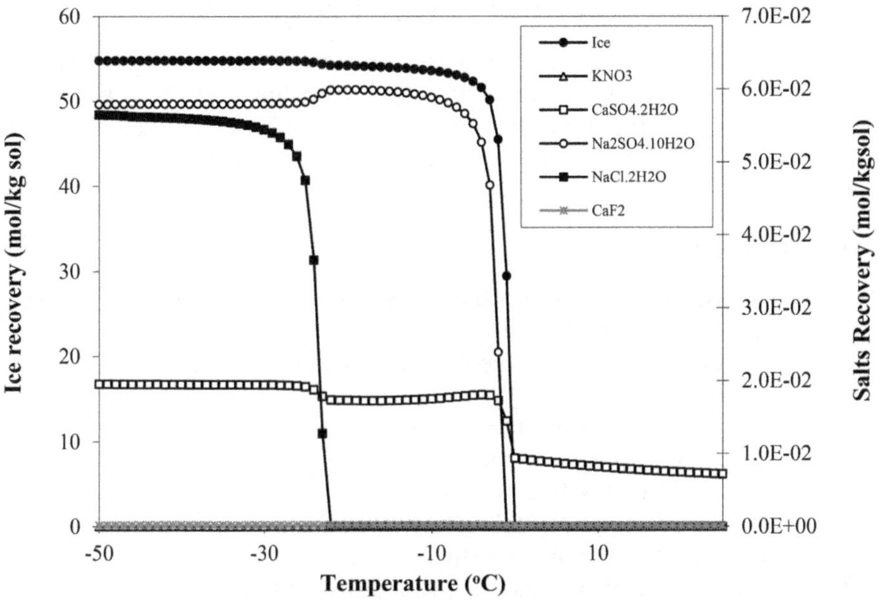

Figure 14.14 Salt and ice crystallization as a result of cooling for the brine.

Table 14.7 Predicted eutectic compositions and temperatures from OLI Stream Analyzer compared to those obtained from Pronk.[53]

Salt	Predicted eutectic composition (wt.%)	Literature eutectic composition (wt.%)	Predicted eutectic temperature (°C)	Literature eutectic temperature (°C)
CaF_2	0.001		−0.01	
$CaSO_4(2H_2O)$	0.17		−0.05	
$Na_2SO_4(10H_2O)$	3.97	3.8	−1.14	−1.2
$CuSO_4(5H_2O)$	11.90	11.9	−1.59	−1.5
NaF	3.33		−2.67	
KNO_3	9.79	10.4	−2.80	−2.9
$MgSO_4(12H_2O)$	16.93	18.0	−3.94	−3.9
$KCl(1H_2O)$	19.50	19.7	−10.61	−10.6
$NaCl(2H_2O)$	23.27	23.3	−21.02	−21.2

14.14 Scaling Up EFC

Scaling up of the EFC process to commercial scale EFC is at the early stages. There are currently a number of commercial plants that are in the process of being commissioned. The existence and development of these commercial units is extremely significant for EFC and for the future of water treatment globally.

14.15 Conclusions and Future Perspectives

Eutectic freeze crystallization (EFC) is a treatment technology for saline streams that can recover both water and dissolved salts. It has been shown to be successful on a laboratory scale in both batch and continuous mode and has the potential to be a feasible technology on a large scale.

Roman Symbols

c	actual solute concentration (kg solute per kg solvent) or (kg solute per kg solution)
c_{eq}	equilibrium solute concentration (kg solute per kg solvent) or (kg solute per kg solution)
C_{EUT}	eutectic concentration of the solute (kg solute per kg solvent) or (kg solute per kg solution)
G	growth rate (m s^{-1})
Δh_f	heat of fusion or crystallization (J mol^{-1})
k_d	mass transfer coefficient (m s^{-1})
MM_i	molar mass of species i, *i.e.*, H_2O or liquid (kg mol^{-1})
M	suspension or magma density (kg m^{-3})
P	pressure (Pa) or production rate (kg s^{-1})
Q	heat (W)
Q	heat flux (W m^{-2})
R	universal gas constant (J mol^{-1} K)
S_i	saturation ratio (–)
t_i	time (s)
T	temperature (K)
T_b	temperature of the bulk liquid (K)
T^*	freezing/melting temperature (K)
T_{int}	interfacial temperature (K)
T^*_{int}	equilibrium temperature of the interface (K)
T^*_b	equilibrium temperature of the bulk (K)
T_w	temperature of the cooled surface (K)
T_c	average temperature of the coolant (K)
ΔT	supersaturation/undercooling of ice (K)
x	solute mol fraction (–)

x_{int} solute mol fraction on the interface (–)
x_b solute mole fraction in the bulk (–)
$w_{sol,b}$ weight/mass fraction of the solute in the bulk (–)

Greek Symbols

κ electrical conductivity (mS cm^{-1})
δ_{ice} thickness of the ice layer (m)
δ_d thickness of the boundary layer (m)
γ_{ice} thermal conductivity of ice (W m^{-1} K)
ρ_{ice} density of ice (kg m^{-3})
ρ_{liquid} density of the liquid (kg m^{-3})

Abbreviations

CC cooling crystallization
EC evaporative crystallisation
EFC eutectic freeze crystallisation
TDS total dissolved solids (ppt)
RO reverse osmosis
$P-i$ pump

Acknowledgements

The support from the University of Cape Town, the Water Research Commission of South Africa, as well as our Sponsors, Eskom, Impala Platinum, Anglo Coal, Sasol, Coaltech, Johnson Matthey, Lonmin, South Africa's Department of Trade and Industry, Proxa and Prentec are acknowledged.

References

1. D. G. Randall, J. Nathoo and A. E. Lewis, A case study for treating a reverse osmosis brine using Eutectic Freeze Crystallization–Approaching a zero waste process, *Desalination*, 2011, **266**(1–3), 256–262.
2. F. E. Genceli, *Scaling Up Eutectic Freeze Crystallization*, Technical University of Delft, The Netherlands, 2008.
3. S. T. Reddy, A. E. Lewis, G. J. Witkamp, H. J. M. Kramer and J. van Spronsen, Recovery of $Na_2SO_4{\cdot}10H_2O$ from a reverse osmosis retentate by eutectic freeze crystallisation technology, *Chem. Eng. Res. Des.*, 2010, **88**(9), 1153–1157.
4. F. van der Ham, G. J. Witkamp, J. de Graauw and G. M. van Rosmalen, Eutectic freeze crystallization: Application to process streams and waste water purification, *Chem. Eng. Process.*, 1998, **37**(2), 207–213.

5. N. J. J. Huige and H. A. C. Thijssen, Production of large crystals by continuous ripening in a stirrer tank, *J. Cryst. Growth*, 1972, **13–14**(0), 483–487.
6. G. L. Stepakoff, D. Siegelman, R. Johnson and W. Gibson, Development of a eutectic freezing process for brine disposal, *Desalination*, 1974, **15**(1), 25–38.
7. F. Van der Ham, *Eutectic Freeze Crystallization*, Technical University of Delft, The Netherlands, 1999.
8. M. M. Seckler, D. Verdoes and G.-J. Witkamp, *Application of Eutectic Freeze Crystallization to Process Streams and Wastewater Purification*, Eutectic Freeze Crystallization EETK97129 Project, 2002, pp. 1–14.
9. T. K. A. Waly, Minimizing the Use of Chemicals to Control Scaling in Sea Water Reverse Osmosis: Improved Prediction of the Scaling Potential of Calcium Carbonate: UNESCO-IHE, PhD thesis, CRC Press, 2011.
10. X. Lu, *Novel Applications of Eutectic Freeze Crystallization*, TU Delft, Delft University of Technology, 2014.
11. A. Lewis, J. Nathoo, S. Reddy, D. Randall, L. Zibi and R. Jivanji, *Novel Technology for Recovery of Water and Solid Salts From Hypersaline Brines: Eutectic Freeze Crystallization*, Water Research Commission, South Africa, Progress Report, 2008, p. 2.
12. A. Mersmann, *Crystallization Technology Handbook*, Marcel Dekker, New York, 2nd edn, 2001.
13. S. A. Nelson, *Crystallization in Ternary Systems*, Tulane University, Petrology, 2011.
14. K. Thomsen, *Aqueous Electrolytes: Model Parameters and Process Simulation*, Technical University of Denmark, Lyngby, 1997.
15. S. A. Nelson, *Ternary Phase Diagrams*, 2011, Available from: http://www.tulane.edu/~sanelson/eens212/ternaryphdiag.pdf.
16. S. Lacour, R. P. Van Hille, K. Peterson and A. E. Lewis, Comparison of simulators for process and aqueous chemistry modeling, *AIChE J.*, 2005, **51**(8), 2358–2368.
17. I.-H. Jung, Overview of the applications of thermodynamic databases to steelmaking processes, *CALPHAD: Comput. Coupling Phase Diagrams Thermochem.*, 2010, **34**(3), 332–362.
18. A. Roine, *Outokumpu HSC Chemistry for Windows: Chemical Reaction and Equilibrium Software With Extensive Thermochemical Database*, Outokumpu research OY, Pori, 2002.
19. I. Çelen, J. R. Buchanan, R. T. Burns, R. B. Robinson and D. R. Raman, Using a chemical equilibrium model to predict amendments required to precipitate phosphorus as struvite in liquid swine manure, *Water Res.*, 2007, **41**(8), 1689–1696.
20. S. M. Serkiz, J. D. Allison, E. M. Perdue, H. E. Allen and D. S. Brown, Correcting errors in the thermodynamic database for the equilibrium speciation model MINTEQA2, *Water Res.*, 1996, **30**(8), 1930–1933.

21. D. L. Parkhurst and C. Appelo, *User's guide to PHREEQC (Version 2): A computer program for speciation, batch-reaction, one-dimensional transport, and inverse geochemical calculations*, 1999.
22. A. Rahardianto, J. Gao, C. J. Gabelich, M. D. Williams and Y. Cohen, High recovery membrane desalting of low-salinity brackish water: integration of accelerated precipitation softening with membrane RO, *J. Membr. Sci.*, 2007, **289**(1–2), 123–137.
23. Speciation and Phase Behavior in Mixed Solvent Electrolyte Solutions: Thermodynamic Modeling, in *Proceedings: 17th International Symposium on Industrial Crystallization*, ed. P. Wang, A. Anderko, R. Springer and M. Lencka, EFCE, Maastricht, The Netherlands, 2008.
24. M. Rafal, J. Berthold and D. Linz, *A Guide to Using OLI Studio Version 9.5*, OLI Systems, Morris Plains, USA, 2017.
25. T. Tähti, *Suspension Melt Crystallization in Tubular and Scraped Surface Heat Exchangers*, University Zentrum für Ingenieurwissenschaften, Halle, 2010.
26. A. E. Lewis, M. M. Seckler, H. Kramer and G. M. van Rosmalen, *Industrial Crystallization: Fundamentals and Applications*, University Press, Cambridge, 2015.
27. A. M. Schwartz and A. S. Myerson, 1-Solutions and solution properties, in *Handbook of Industrial Crystallization*, ed. S. M. Allan, Butterworth-Heinemann, Woburn, 2nd edn, 2002, pp. 201–230.
28. C. Himawan, H. J. M. Kramer and G. J. Witkamp, Study on the recovery of purified $MgSO_4 \cdot 7H_2O$ crystals from industrial solution by eutectic freezing, *Sep. Purif. Technol.*, 2006, **50**(2), 240–248.
29. J. Ulrich and M. Jones, *Heat and Mass Transfer Operations—Crystallization*, 2006.
30. D. Kashchiev, *Nucleation: Basic Theory with Applications*, Elsevier Science, 1st edn, 2000.
31. J. W. Mullin, *Crystallization*, Butterworth-Heinemann, 2001.
32. S. Kane, T. Evans, P. Brian and A. Sarofim, The kinetics of the secondary nucleation of ice: Implications to the operation of continuous crystallizers, *Desalination*, 1975, **17**(1), 17–30.
33. G. Margolis, T. K. Sherwood, P. T. Brian and A. F. Sarofim, The performance of a continuous well stirred ice crystallizer, *Ind. Eng. Chem. Fundam.*, 1971, **10**(3), 439–452.
34. L. Bauer, R. Rousseau and W. McCabe, Influence of crystal size on the rate of contact nucleation in stirred-tank crystallizers, *AIChE J.*, 1974, **20**(4), 653–659.
35. N. Huige, M. Senden and H. Thijssen, Nucleation and growth kinetics for the crystallization of ice from dextrose solutions in a continuous stirred tank crystallizer with supercooled feed, *Cryst. Res. Technol.*, 1973, **8**(7), 785–801.
36. P. Pronk, C. Infante Ferreira and G. Witkamp, Influence of solute type and concentration on ice scaling in fluidized bed ice crystallizers, *Chem. Eng. Sci.*, 2006, **61**(13), 4354–4362.

37. A. S. Myerson, *Handbook of Industrial Crystallization*, Butterworth-Heinemann, 2002.
38. R. De Goede and G. M. Van Rosmalen, Modelling of crystal growth kinetics: A simple but illustrative approach, *J. Cryst. Growth*, 1990, **104**(2), 392–398.
39. F. E. Genceli, M. Rodriguez Pascual, S. Kjelstrup and G.-J. Witkamp, Coupled Heat and Mass Transfer during Crystallization of $MgSO_4Â·7H_2O$ on a Cooled Surface, *Cryst. Growth Des.*, 2009, **9**(3), 1318–1326.
40. M. Kapembwa, M. Rodríguez-Pascual and A. E. Lewis, Heat and Mass Transfer Effects on Ice Growth Mechanisms in Pure Water and Aqueous Solutions, *Cryst. Growth Des.*, 2014, **14**(1), 389–395.
41. S. Kjelstrup and D. Bedeaux, *Non-equilibrium Thermodynamics of Heterogeneous Systems*, World Scientific, 2008.
42. P. Harriott, The growth of ice crystals in a stirred tank, *AIChE J.*, 1967, **13**(4), 755–759.
43. M. Rodriguez Pascual, Physical Aspects of Scraped Heat Exchanger Crystallizers: An Application in Eutectic Freeze Crystallization, PhD thesis, Technical University of Delft, The Netherlands, 2009.
44. F. E. Genceli, R. Gärtner and G. J. Witkamp, Eutectic freeze crystallization in a 2nd generation cooled disk column crystallizer for $MgSO_4·H_2O$ system,*J. Cryst. Growth*, 2005, **275**(1–2), e1369–e1372.
45. R. Jivanji, J. Nathoo, W. van der Merwe, A. Human and A. Lewis, Application of Eutectic Freeze Crystallization to the Treatment of Mining Wastewaters, *22nd World Mining Congress 11–16 September 2011*, Istanbul, 2011.
46. R. J. Vaessen, *Development of Scraped Eutectic Crystallizers*, Technical University of Delft, The Netherlands, 2003.
47. M. Rodriguez Pascual, F. Genceli, D. Trambitas, H. Evers, J. Van Spronsen and G. Witkamp, A novel scraped cooled wall crystallizer: Recovery of sodium carbonate and ice from an industrial aqueous solution by eutectic freeze crystallization, *Chem. Eng. Res. Des.*, 2010, **88**(9), 1252–1258.
48. J. Chivavava, M. Rodriguez-Pascual and A. E. Lewis, Effect of operating conditions on ice characteristics in continuous eutectic freeze crystallization, *Chem. Eng. Technol.*, 2014, **37**(8), 1314–1320.
49. F. van der Ham, M. M. Seckler and G. J. Witkamp, Eutectic freeze crystallization in a new apparatus: the cooled disk column crystallizer, *Chem. Eng. Process.*, 2004, **43**(2), 161–167.
50. M. Hasan, R. Filimonov, J. Chivavava, J. Sorvari, M. Louhi-Kultanen and A. E. Lewis, Ice growth on the cooling surface in a jacketed and stirred eutectic freeze crystallizer of aqueous Na2SO4 solutions, *Sep. Purif. Technol.*, 2017, **175**, 512–526.
51. E. M. Peters, J. Chivavava, M. Rodriguez-Pascual and A. E. Lewis, Effect of a Phosphonate Antiscalant during Eutectic Freeze Crystallization of a Sodium Sulfate Aqueous Stream, *Ind. Eng. Chem. Res.*, 2016, **55**(35), 9378–9386.

52. C. Himawan, R. J. C. Vaessen, H. J. M. Kramer, M. M. Seckler and G. J. Witkamp, Dynamic modeling and simulation of eutectic freeze crystallization, *J. Cryst. Growth*, 2002, **237–239**(Part 3), 2257–2263.
53. P. Pronk, Fluidised bed heat exchangers to prevent fouling in ice slurry systems and industrial crystallisers, PhD thesis, Technical University of Delft, The Netherlands, 2006.
54. F. G. Qin, X. D. Chen and A. B. Russell, Heat transfer at the subcooled-scraped surface with/without phase change, *AIChE J.*, 2003, **49**(8), 1947–1955.
55. V. F. Petrenko, *Electrical Properties of Ice*, Cold Regions Research and Engineering Lab, Hanover, NH, 1993.
56. A. Pérez-González, A. M. Urtiaga, R. Ibáñez and I. Ortiz, State of the art and review on the treatment technologies of water reverse osmosis concentrates, *Water Res.*, 2012, **46**(2), 267–283.
57. I. W. van der Merwe, A. Lourens and C. Waygood, *An Investigation of Innovative Approaches to Brine Handling*, Water Research Commission, 2010.
58. M. C. Georgiadis, J. R. Banga, E. N. Pistikopoulos and V. Dua, *Dynamic Process Modeling*, Wiley, 2013.
59. J. Nathoo, R. Jivanji and A. E. Lewis, Freezing your Brines off: Eutectic Freeze Crystallization for Brine Treatment, in *International Mine Water Conference, 19–21 October 2009*, ed. C. Taylor, Pretoria, 2009, pp. 431–7.
60. F. van der Ham, G. J. Witkamp, J. de Graauw and G. M. van Rosmalen, Eutectic freeze crystallization simultaneous formation and separation of two solid phases, *J. Cryst. Growth*, 1999, **198–199**(Part 1(0)), 744–748.
61. OLI Systems Inc, *OLI Studio 9.3, Stream Analyzer, (Revision 9.3.2) September 2016*, OLI Systems Inc, Morris Plains, New Jersey, 2016.

CHAPTER 15

Economic Analysis of Continuous Crystallisation†

SAMIR DIAB[a], HIKARU G. JOLLIFFE[b] AND DIMITRIOS I. GEROGIORGIS*[a]

[a]Institute for Materials and Processes (IMP), School of Engineering, University of Edinburgh, The King's Buildings, Edinburgh, EH9 3FB, UK;
[b]EPSRC Centre for Innovative Manufacturing in Continuous Manufacturing and Crystallisation, University of Strathclyde, Glasgow, G1 1RD, UK
*E-mail: D.Gerogiorgis@ed.ac.uk

15.1 Introduction

Crystallisation is an essential unit operation in several manufacturing sectors, ranging from the production of commodity to speciality chemical products.[1,2] Crystallisation processes are particularly important for pharmaceutical manufacturing, due to the industry's significant portion of solid form pharmaceutical products, *e.g.* tablets, dispersions, gels or topical treatments, with most small molecule drug products being manufactured and handled as solids.[3] Batch crystallisation techniques currently dominate industrial practice due to their ease of development and design from existing stirred tanks and ability to reach equilibrium during operation.[4] Batch methods have been widely studied and are well-understood, but imply issues of

†Electronic supplementary information (ESI) available. Colour versions of Figures 15.9 and 15. 10. See DOI: 10.1039/9781788013581.

The Handbook of Continuous Crystallization
Edited by Nima Yazdanpanah and Zoltan K. Nagy

Published by the Royal Society of Chemistry, www.rsc.org

consistency of product quality attributes (*i.e.*, purity, crystal size distribution, attained product polymorph);[5] achieving consistent product quality *via* robust crystallisation process design is essential in order to meet increasingly stringent regulations.[6]

Continuous manufacturing has received significant attention from both academia and industry due to its potential for significant operational and economic advantages. Implementation of continuous manufacturing implies smaller equipment footprints, significant reduction in material usage and waste and enhanced heat and mass transfer efficiencies.[7] Significant developments in continuous flow organic chemistry[8,9] and synthesis of active pharmaceutical ingredients (APIs)[10–13] have been made, with some examples of end-to-end continuous pharmaceutical manufacturing (CPM), from synthesis to product formulation, being demonstrated,[14–17] including integrated continuous crystallisation processes[18,19] and subsequent downstream operations. Continuous crystallisation processes allow higher reproducibility of crystal product quality attributes compared to batch crystallisations due to their operation at steady-state.

Ultimately, any process design option must demonstrate feasible operability and economic viability prior to expensive and laborious experimental campaigns and pilot plant studies.[20] Conceptual modelling and simulation studies are often implemented for comparative evaluation of process options.[21] Rapid screening of viable design alternatives allows investigation of design spaces and avoiding pitfalls early on in the development stages.[22] Comparative economic evaluation using established costing methodologies for pharmaceutical processes can elucidate cost optimal designs.[23]

This chapter describes methodologies and considerations for costing and comparative economic evaluation of continuous crystallisation processes. Total cost components, price and costing databases and additional resources are first presented as the foundation for economic evaluations. Different continuous crystalliser designs are then briefly discussed, including efforts towards their optimisation demonstrated in the literature. Formulation of nonlinear optimisation problems toward total cost minimisation of process designs, incorporating crystallisation kinetics, population balances, mass balances and plant costing are then presented. These methodologies are then combined in the demonstration of various economic analysis and optimisation case studies implemented by our group towards establishing cost optimal continuous crystallisation designs for various societally-important APIs. This work demonstrates the importance and utility of modelling and simulation approaches towards establishing economically viable continuous crystallisation designs.

15.2 Economic Analysis of Pharmaceutical Processes

Economic evaluation of pharmaceutical manufacturing processes requires methodologies for costing different candidate process designs for comparative analysis. In this section we describe different total cost components for the costing of different pharmaceutical manufacturing processes, including continuous crystallisation, as used in literature examples.

15.2.1 Capital Expenditure (CapEx)

Capital expenditure (CapEx) is the investment required for the acquirement and maintenance of fixed assets. CapEx estimation depends on a number of different essential considerations and assumptions. Whether a design is a green-field construction of a new, dedicated plant or if the design is to be constructed at an existing plant with essential auxiliary equipment already in place is important for CapEx component calculation; the number of production lines to be implemented at any site is another important consideration, amongst many more.

Equipment purchase costs should be taken from a variety of vendor sources to establish an average cost of specific unit process equipment; where possible, finding vendor data for equipment of similar capacity to that being designed is desirable in order to minimise errors in equipment cost estimation. Equipment costs must be scaled to account for varying capacity, design variations and year of purchase. When costing continuous processes, vendor prices for continuous process units should be sourced where possible; when prices are only available for batch equipment, a price premium can be applied to continuous unit prices relative to a batch unit of the same size to account for the increased process engineering and complexity required to operate continuously with feedback control, in comparison to batch processes, which are typically operated in open-loop.

Eqn (15.1) is a cost-capacity correlation widely implemented for the scaling of equipment with costs for different capacities to those available from vendors. P_j is the equipment purchase cost at capacity S_j, the units of which vary depending on the equipment in question. Where the reference purchase cost (P_A) is taken from the past, chemical engineering plant cost indices (CEPCIs) are used to calculate the corresponding purchase cost in the present day. Values for CEPCI are published monthly.

$$P_B = fP_A\left(\frac{S_B}{S_A}\right)^n \tag{15.1}$$

The sum of all inflation-adjusted equipment costs (P_B) gives the Free-on-Board (FOB) cost, which is the total cost of required processing equipment

Table 15.1 Reference equipment purchase costs and related capacities with corresponding values for parameter n and f (eqn (15.1)).[25]

Equipment	Year	Ref. Capacity, S_A	Units	Ref. Cost, P_A (GBP)	n	f(%)
Batch Crystalliser	2007	7.5	Volume (m^3)	192 000	0.68	10.33
Forced Circulation Crystalliser	2007	1	kg crystal s^{-1}	520 000	0.53	10.33
Draft Tube MSMPR	2007	1	kg crystal s^{-1}	538 000	0.63	10.33
Pumps	2015	—	—	958	—	—
Pipes[a]	2007	1	Length (m)	62	1.33	—

[a]Additional cost factors for alloying considerations may be required.[25]

excluding ancillaries, equipment delivery, electrical, engineering and piping expenses.[24] Parameter n scales costs for differences in capacity, while f describes differences in design and operation; values for n and f can be found in the literature.[25] Examples of reference crystallisation equipment costs and their capacities with their corresponding values of n and f are provided in Table 15.1; further values for specific equipment can be found in the references therein.[25]

Various methods exist to calculate the cost of delivered installed equipment costs. Wroth factors can be used to calculate the cost of delivered installed equipment;[26] additional costs are associated with the calculation of construction, process piping and instrumentation to give the battery limits installed costs (BLIC), a major component of CapEx. The Chilton method can be implemented for the estimation of BLIC for manufacturing processes.[23] Working capital (WC, *i.e.*, short term investments required to maintain operation) is calculated as a function of the costs of annual material requirements. There are additional factors for construction (typically 30% of BLIC[27]) and contingency to account for error in cost estimation (20% of BLIC[27]). Extra considerations for offsite capital are associated with grass-roots construction.[24] Total CapEx is taken as the sum of BLIC, WC, construction, contingency and offsite capital components.

15.2.2 Operating Expenditure (OpEx)

Total operating expenditure (OpEx) is the sum of materials, waste handling, utilities, labour and quality control.[27] Material costs are calculated from process mass balances required to meet a desired product capacity; material prices are estimated from available vendors and databases. Variation of material prices can be considered as a form of sensitivity analysis for comparative evaluation.[23] Waste handling and utilities costs vary with location, waste stream composition and extent of waste treatment required, but can be estimated as a function of material throughput and volumetric waste output.[24,27] Labour costs can be estimated if rates local to the plant location are known, but are typically lower than batch processes due to the automated nature of continuous manufacturing with respect to batch mode. Quality control for continuous processes is typically higher due to the lower state of development of online analytical methods (Process Analytical Technology, PAT) than those for batch that are well developed.

15.2.3 Prices and Costing Factor Databases

Various sources are available for equipment prices for scaling *via* eqn (15.1). Commonly implemented plant equipment typical of processes across a variety of manufacturing sectors can be estimated based on historical data and general rules of thumb implemented in the literature.[24,25] Specialised equipment (*e.g.* specific types of microreactor, pumping systems) can be found from specific vendors; comparison of prices across as wide a range of sources as possible is recommended to establish reliable averages for different types of equipment. The CEPCI values for calculation of inflation-adjusted equipment costs (eqn (15.1)) are published monthly.[28]

Material prices can be found from a variety of sources. Independent Chemical Information Services (ICIS) provide a range of prices from different countries for different quantities of bulk chemicals;[29] fine and speciality chemicals can be sourced from specific vendors. As described previously, consideration of a range of prices from a variety of vendors is recommended to allow a more reliable average for material cost calculations.

15.2.4 Costing of Continuous Processes

Here, we describe a methodology implemented by our group in various case studies[30–36] to estimate total cost components of different continuous manufacturing processes. The FOB costs are calculated as described previously (sum of all equipment purchase costs as calculated in eqn (15.1)). The Chilton method is used to calculate BLIC as follows.[24] The installed equipment cost (IEC), process piping and instrumentation (PPI) and total physical plant cost (TPPC) are calculated from eqn (15.2)–(15.4). A construction factor of 30% is added to TPPC to calculate the BLIC (eqn (15.5)).[27]

$$\mathrm{IEC} = 1.43\,\mathrm{FOB} \tag{15.2}$$

$$\mathrm{PPI} = 0.42\,\mathrm{IEC} \tag{15.3}$$

$$\mathrm{TPPC} = \mathrm{IEC} + \mathrm{PPI} \tag{15.4}$$

$$\mathrm{BLIC} = 1.3\,\mathrm{TPPC} \tag{15.5}$$

Working capital and contingency costs (WCC) are calculated as follows. Working capital (WC) costs are taken as 3.5% of annual material (mother liquor solvent) costs (MAT_{annual}). Contingency costs (CC) are calculated as 20% of BLIC. The sum of BLIC and WCC gives the total CapEx.

$$WC = 0.035\,MAT_{annual} \tag{15.6}$$

$$CC = 0.2\,BLIC \tag{15.7}$$

$$WCC = WC + CC \tag{15.8}$$

$$CapEx = BLIC + WCC \tag{15.9}$$

Annual utilities cost ($UTIL_{annual}$) is calculated as 0.96 GBP kg^{-1} of material input (Q_{plant}); the annual waste cost ($Waste_{annual}$) is 0.35 GBP L^{-1} of waste produced (Q_{waste}).[27] Annual operating expenditure ($OpEx_{annual}$) is calculated as the sum of annual material (MAT_{annual}), utilities ($UTIL_{annual}$) and waste disposal ($Waste_{annual}$).

$$UTIL_{annual} = 0.96 Q_{plant} \tag{15.10}$$

$$Waste_{annual} = 0.35 Q_{waste} \tag{15.11}$$

$$OpEx_{annual} = MAT_{annual} + UTIL_{annual} + Waste_{annual} \tag{15.12}$$

Operating expenditure can be normalised with respect to the attained yield or residence time if comparing continuous to batch designs to evaluate the relative productivities of each; however, comparison of absolute values for a set plant lifetime is a more valuable for absolute comparison of economic performance.

It is important to note that the presented costing equations for working capital, contingency and operating expenditure components presented here pertain specifically to pharmaceutical manufacturing processes, as implemented in various case studies by our group. Cost and estimation of these components requires specific data in order to gain an accurate estimation of various total cost components, *e.g.* costs of waste treatment specific to a process, local utility costs *etc.*

Total costs are calculated as the sum of CapEx and the sum of discounted $OpEx_{annual}$ over years $k = 1 - t$ (eqn (15.13)); plant operating lifetimes of $t = 20$ years and annual operation of 8000 hours per year are often considered. Alternatively, net present value (NPV) of designs can be calculated (eqn (15.14)) if reliable product sales prices are known in order to estimate annual revenues (Rev). The discount rate, y, considers the effect of time-value of money to express total costs in present day terms; a conservative estimate of $y = 5\%$ is often used. All CapEx is assumed to occur in year 0 and operation is assumed to begin in year 1.

$$\text{Total Cost} = \text{CapEx} + \sum_{k=1}^{t=20} \frac{\text{OpEx}_{\text{annual}}}{(1+y)^k} \tag{15.13}$$

$$\text{NPV} = -\text{CapEx} + \sum_{k=1}^{t=20} \frac{\text{Rev–OpEx}_{\text{annual}}}{(1+y)^k} \tag{15.14}$$

Comparison of continuous designs to their batch alternatives is essential to establish whether a continuous design may allow for economic advantages as per the implemented costing methodologies.[23,32] Comparison of total cost component savings of different candidate continuous designs with respect to batch modes *via* modelling and optimisation can systematically screen for the best process configuration. Furthermore, conducting various sensitivity analyses with respect to operating (*e.g.*, different assumptions of attainable solvent recovery[23]), and/or economic parameters (*e.g.*, key ingredient prices,[23] assumed interest rate[37]) is also essential to ensure robust technoeconomic evaluations.

15.3 Continuous Crystalliser Designs

In this section, we present different case studies implemented by our group using the costing methodologies described previously. Process and costing considerations differ between case studies due to the varying nature of each crystalliser design and approach of each case – specific details on the costing methodologies of each crystalliser design are briefly discussed, with more detailed modelling description and solution methods described in our recent publications.[33,38,39]

15.3.1 Mixed Suspension-mixed Product Removal Crystalliser (MSMPR)

Mixed suspension-mixed product removal (MSMPR) crystallisers are popular designs for implementing continuous crystallisation. MSMPR designs are easily adapted from existing jacketed continuous stirred tanks and are easy to maintain and operate; for these reasons, such designs have already been incorporated into end-to-end CPM plants.[14,19] They exhibit longer mean residence times compared to other continuous crystalliser designs and are best suited for systems with lower conversions and slower crystallisation kinetics.[40] Their drawbacks include longer start-up periods and difficulty in scale-up.[41] Implementation of MSMPRs in various configurations and novel applications have been demonstrated in various experimental and theoretical studies in the literature. MSMPRs are often used for elucidation of growth and nucleation kinetics parameters[18,42–45] of continuous crystallisation systems due to their longer mean residence times.[41] Implementation of novel operating regimes and techniques, control schemes,[46–58] specialised purifications and separations[59–62] and designs towards high polymorph selectivity[63,64] are implemented in MSMPRs.

15.3.2 Plug Flow Crystalliser (PFC)

Plug flow crystallisers (PFCs) are tubular designs best suited for systems with faster crystallisation kinetics and lower slurry densities to avoid equipment clogging. They attain narrower residence time distributions than MSMPR designs and are easier to scale-up; however, they are more prone to clogging due to the narrow inner diameters required to reduce radial concentration and temperature gradients and typically are more expensive and complex to operate and maintain.[41] Various PFC demonstrations in the literature show the promise of this technology for continuous crystallisation implementation, including studies into the effect of mother liquor seeding and nucleation control,[65–69] tuning mean product crystal size and shape,[70–73] the effect of multisegment designs with multiple antisolvent inlets[59,74] and mitigating crystalliser clogging and crust formation.[75]

15.3.3 Continuous Oscillatory Baffled Crystallisers (COBC)

Continuous oscillatory baffled crystallisers (COBCs) are an adaptation of oscillatory baffled reactors, comprised of a series of baffles in high aspect ratio tubes. Baffles in the crystalliser enhance heat and mass transfer efficiencies associated with mixing induced by baffles.[76,77] Implementing a reciprocating pump generates the required oscillatory flow; as the crystallisation magma is contacted with the baffles, eddies are generated which enhance mixing and therefore performance. COBC designs have been implemented for a wide variety of design objectives, including meeting a desired mean product size, purity and yield.[77] However, the baffled design of COBCs makes them more difficult to maintain.

15.4 Nonlinear Optimisation

Mathematical optimisation has been used in process systems engineering (PSE) applications,[78–80] including optimal reactor design,[81] life cycle assessments (LCAs),[82,83] separation process designs[62,74] and optimisation of pharmaceutical manufacturing campaigns.[84–87] The definition of a nonlinear optimisation problem is for the minimisation or maximisation of an objective function by varying certain decision variables; these are usually constrained to certain ranges representing mathematical, physical or operational limits. For the design and optimisation of manufacturing processes, decision variables are often process operating or design variables that significantly affect process performance and thus total cost components; these could be crystalliser operating temperatures or equipment volumes, which will likely have to respect limits reflecting available utilities and applicability to equipment materials and process streams. An example of an objective function for minimisation could be the total costs of a manufacturing process (or alternatively, the maximisation of NPV).

$$\text{min Total Costs} \tag{15.15}$$

$$\text{max NPV} \tag{15.16}$$

15.5 Economic Analysis and Optimisation Case Studies of Various Active Pharmaceutical Ingredients

This section describes case studies implementing the economic evaluation methodologies described in Section 15.2 and nonlinear optimisation towards total cost minimisation as described in Section 15.4 for the continuous crystallisation of various APIs. Investigations of design spaces and subsequent economic evaluation and cost optimisation of different processes allows rapid screening and elucidation of economically viable configurations, demonstrating the importance and utility of conceptual modelling approaches to facilitate the rapid transition to leaner manufacturing routes. Although the continuous crystallisation case studies presented here are specifically for the production of solid APIs (*i.e.*, pharmaceutical production), the overall methodological approaches are broadly applicable for similar configurations and designs, *e.g.*, cascades of MSMPRs, PFCs, COBCs *etc.*, with only specific model parameters requiring substitution given a similar process configuration.

15.5.1 Comparative Economic Evaluation of MSMPR Configurations: Cyclosporine

Cyclosporine is a World Health Organisation (WHO) Essential Medicine. The API is an immunosuppressant used primarily for organ transplant rejection prevention,[88] also used for treatment of psoriasis, rheumatoid arthritis and dermatitis.[89] Cyclosporine continuous cooling crystallisation in MSMPRs has been investigated *via* experimental and modelling studies. Measurement of crystal growth and nucleation under various operating conditions (varying residence times, operating temperatures, suspension densities and supersaturations) have elucidated crystallisation kinetic parameters.[42] Various process stream recycling options have also been considered for continuous cyclosporine crystallisation in MSMPRs. Investigated mother liquor recycle implementation showed enhanced yields, but issues of impurity accumulation rendered the process inferior to other configurations.[58] Cascades of MSMPR crystallisers without recycle have been investigated in the literature,[62] including options with incorporated solids recycle.[51] Here, we describe the technoeconomic evaluation of different recycle configurations for continuous cyclosporine crystallisation of cyclosporine. Steady-state process modelling is implemented to compare different designs.[33]

15.5.1.1 Steady-state MSMPR Crystallisation: With and Without Solids Recycle

Processes with and without solids recycle for continuous cyclosporine crystallisation in MSMPRs are based upon experimental demonstrations.[58,62] All designs presented here produce 100 kg of cyclosporine per annum, assuming steady-state production. The steady-state process model describing MSMPR crystallisation, including growth and nucleation kinetics[42] and temperature-dependent API solubility data,[58] population balances and mass balances is described in the literature.[51,62]

Figure 15.1 shows the process flowsheet for MSMPR crystallisation of cyclosporine without recycle. This configuration features a cascade of MSMPR crystallisers for continuous cyclosporine crystallisation with slurry transfer between crystallisers using peristaltic pumps. A clear mother liquor stream containing API dissolved in acetone at a feed concentration of 25% w/w. The product magma of one crystalliser is the feed stream to the subsequent crystalliser in the cascade; the product of the final crystalliser is the product stream of the cascade and is filtered for crystal removal.

Figure 15.2 shows the process flowsheet for MSMPR cyclosporine crystallisation with solids recycle.[51] Flowsheet configurations are similar to those without recycle, but have a gravity-driven separation column following the final crystalliser in the cascade recycling a concentrated slurry back to different locations in the crystalliser cascade. The extent of solids recycle to each crystalliser can be varied by altering the amount of clear liquor removed

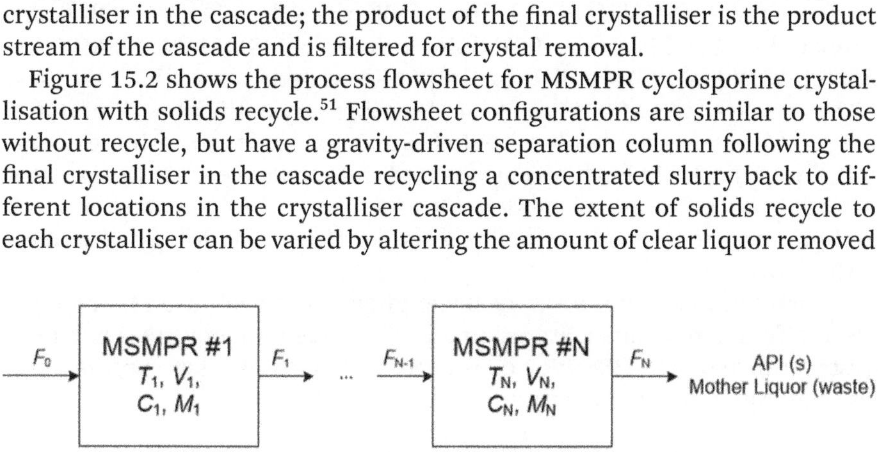

Figure 15.1 MSMPR crystallisation of cyclosporine without process recycle.[42,62] Reproduced from ref. 33, https://doi.org/10.1021/acs.oprd.7b00225, with permission from American Chemical Society, Copyright 2017.

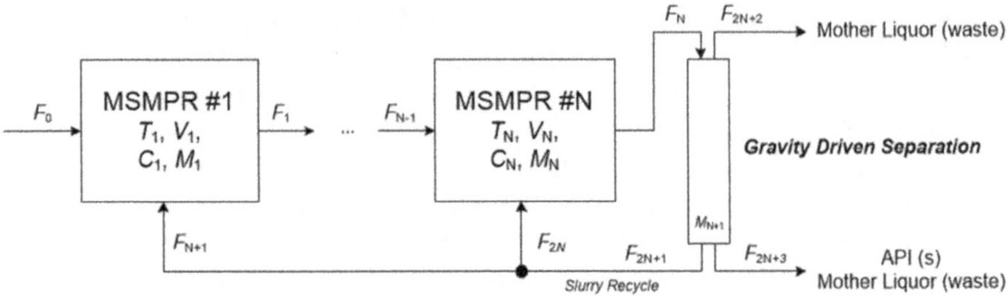

Figure 15.2 MSMPR crystallisation of cyclosporine with solids liquor recycle.[54] Reproduced from ref. 33, https://doi.org/10.1021/acs.oprd.7b00225, with permission from American Chemical Society, Copyright 2017.

from the top of the separation column (Figure 15.2). The removed mother liquor from the top of the column and that in the product magma is considered to be discarded as waste.

In all process designs, supersaturation is generated by cooling, without the need for any antisolvent or source of counter-ions for salt formation; stream impurities are not considered here. It is assumed that the feed stream of cyclosporine in acetone contains negligible amounts of impurities that would affect the attained product purity due to the sample composition variation present in real processes. In practice, impurity considerations are imperative to meet the strict requirements.[6]

For all configurations, the number of implementable crystallisers, $N = \{1, 2, 3\}$. For multiple MSMPRs, the operating temperature of the first crystalliser, $T_1 = \{10, 15, 20\}$ °C, with the final crystalliser temperature, $T_N = 0$ °C, is chosen, with linear temperature decrease from T_1 to T_N from the beginning to the end of the cascade, similar to experimental operation practices.[51] For processes with solids recycle, total solids recycle = {50, 70, 90} % are compared. The solids recycle feed point location = {MSMPR #1, MSMPR #2, MSMPR #3, Equal MSMPR distribution} is also varied.

The steady-state MSMPR model describes crystallisation kinetics,[42] temperature-dependent solubility behaviour for supersaturation calculation,[58] population balance equations and mass balances.[51,62] Design and operating variables at the model solution are used to calculate process configuration total cost components as per the methodology described in this work.

Crystallisation kinetics describe linear growth and nucleation of API crystals as a function of supersaturation and operating temperature. The linear crystal growth rate in MSMPR i, G_i, is described by a power law expression in supersaturation and temperature.

$$G_i = k_{g0} \exp\left(-\frac{E_{ag}}{RT_i}\right)\left(\frac{C_i}{C_i^{sat}} - 1\right)^g \tag{15.17}$$

k_{g0} is the pre-exponential growth factor, E_{ag} is the growth energy barrier, R is the universal gas constant, T_i is the operating temperature of MSMPR

i, C_i is the API concentration in the mother liquor within and discharged from MSMPR i and g is the crystal growth exponent. The API solubility concentration, $C^{sat}{}_i$, is a function of T_i, calculated *via* a surrogate polynomial regressed from published temperature-dependent saturation data for cyclosporine.[58]

The nucleation rate in MSMPR i, B_i, is described by the power law expression in supersaturation and temperature.[18]

$$B_i = k_{b0}\exp\left(-\frac{E_{ab}}{RT_i}\right)\left(\frac{C_i}{C_i^{sat}}-1\right)^b M_i^m \tag{15.18}$$

k_{b0} is the pre-exponential factor for nucleation, E_{ab} is the nucleation energy barrier, b is the crystal nucleation exponent, M_i is the slurry density in MSMPR i and m is the exponent of the slurry density. All crystallisation kinetic parameters are taken from previous work.[42,51]

The one-dimensional population balance model is described by a system of ordinary differential equations (ODEs).

$$G_1V_1\frac{dn_1}{dL} = F_{N+1}n_{N+1} - F_1n_1 \tag{15.19}$$

$$G_iV_i\frac{dn_i}{dL} = F_{i-1}n_{i-1} + F_{N+i}n_{N+i} - F_in_i \qquad i = 2 \ldots N. \tag{15.20}$$

F_{i-1} and F_i are the volumetric flowrates of streams entering and leaving MSMPR i, respectively, F_{N+i} is the recycle volumetric flowrate entering MSMPR i, N is the total number of crystallisers, n_i is the crystal population density function in MSMPR i and L is the characteristic length of the crystal. For the process without recycle, F_{N+i} terms are equal to zero. The system of ODEs formed by the population balance equations are satisfied by the boundary condition, $n_i^0 = n_i\,(L = 0) = B_i/G_i$, corresponding to the population density of nuclei.

The slurry density in MSMPR i, M_i, is calculated from the population density function as follows:

$$M_i = k_v\rho_{API}\int n_iL^3\,dL \tag{15.21}$$

k_v is the crystal volume shape factor (= π/6 for spherical crystals, assumed constant for linear crystal growth) and ρ_{API} is the crystal density. A value of $\rho_{API} = 1.3$ g cm^{-3} (average for solid APIs[88]) is assumed.

Steady-state mass balances assume no accumulation and account for volumetric changes due to API crystallisation. The general mass balance equations for processes are:

$$F_0C_0 + F_{N+1}\left(1-\frac{M_{N+1}}{\rho_{API}}\right)C_N + F_{N+1}M_{N+1} - F_1\left(1-\frac{M_1}{\rho_{API}}\right)C_1 - F_1M_1 = 0 \tag{15.22}$$

$$F_{i-1}\left(1-\frac{M_{i-1}}{\rho_{\text{API}}}\right)C_{i-1}+F_{i-1}M_{i-1}+F_{N+i}\left(1-\frac{M_{N+1}}{\rho_{\text{API}}}\right)C_N$$
$$+F_{N+i}M_{N+1}-F_i\left(1-\frac{M_i}{\rho_{\text{API}}}\right)C_i-F_iM_i=0 \quad i=2\ldots N. \tag{15.23}$$

F_0 and C_0 are the volumetric flowrate and mother liquor API concentration of the fresh feed stream, respectively. For the process without recycle, F_{N+i} terms are equal to zero. For all processes, an API balance across mother liquor and crystallised solid phases also gives the following expression for the slurry density from the process mass balances:

$$M_i = C_{i-1} - C_i \tag{15.24}$$

Solution of this process model describes MSMPR crystallisation of a solute; its solution is described in our work[33] as well as others',[44] as follows. For a single crystalliser, a guess of the outlet concentration of solute in the product mother liquor (C_i) must be made, which must lie between the solubility and the initial concentration ($C_0 > C_i > C_i^{\text{sat}}$). The slurry density calculated from the MSMPR mass balance (eqn (15.24)) is then calculated. Computation of crystallisation kinetics (eqn (15.17) and (15.18)) for set design and operating parameters (crystalliser volumes and temperatures, respectively) are then used for solution of population balance equations (eqn (15.19) and (15.20)). The slurry density calculated from solution of the population balance equations (eqn (15.21)) is then compared to that calculated from the process mass balances; if these values agree (*i.e.*, the difference between them is lower than a predefined tolerance) then the model is solved, otherwise a new guess of C_i must be made and iterated upon until solution.

15.5.1.2 Operational Performances of Different Process Configurations

Our work in continuous cyclosporine crystallisation showed that beyond certain residence times, there is no appreciable increase in yield for different designs.[33] Cascades with longer residence times will result in unnecessarily large volumes and increases in CapEx. Maximum residence times beyond which there is no appreciable increase in yield are selected for crystalliser volume calculation.[33]

Crystalliser volumes for the process without recycle are shown in Figure 15.3. Increasing the number of crystallisers decreases the total crystallisation volume due to enhanced yields attainable with multiple crystalliser usage; operating temperature has only a small effect on the total crystalliser volume due to its small effect on yields. When three crystallisers are used, operating the first crystalliser at 20 °C leads to a significant increase in total crystalliser volume (compared to operating at 10 and 15 °C) due to the longer

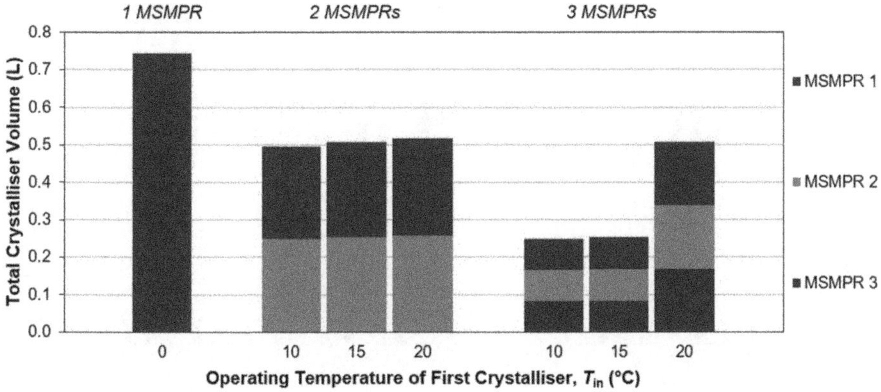

Figure 15.3 Required crystalliser volumes for continuous cyclosporine crystallisation without recycle.[33] Reproduced from ref. 33, https://doi.org/10.1021/acs.oprd.7b00225, with permission from American Chemical Society, Copyright 2017.

residence time, and thus total crystalliser volume, required to reach the maximum attainable yield.[33]

Crystalliser volumes for processes with solids recycle are shown in Figure 15.4. Increasing the number of crystallisers and varying operating temperatures show similar trends as for processes without recycle. Increasing the extent of recycle leads to larger crystalliser volumes. When the recycle stream is fed to the first crystalliser, all crystallisers have similar volumes as large throughputs are present throughout the whole cascade; results are similar when the recycle stream is equally distributed across all crystallisers. However, when the recycle stream is fed to the second or third crystallisers, only these crystallisers must be larger than the previous, to accommodate the higher total throughput.

15.5.1.3 Technoeconomic Comparative Evaluations

Crystalliser purchase costs are a significant portion of CapEx, and so correlating crystalliser volumes with cost components provides insight to differences in total costs between different process configurations.

Total cost components of different process configurations without recycle are shown in Figure 15.5. BLIC constitutes the most significant portion of CapEx; WCC contributes less due to the low material intensity of the process.[33,62] Implementing a single crystalliser at 0 °C has the lowest total costs due to the significantly reduced CapEx (BLIC and WCC) costs. Increasing the number of crystallisers and associated pumps and cooling equipment has a significant impact on total plant costs.[33] The dominant OpEx component is utilities and waste handling costs. Material costs for this process are relatively small due to the low process material intensity *i.e.* cooling crystallisation from mother liquor without the need for antisolvent addition for supersaturation generation.[62]

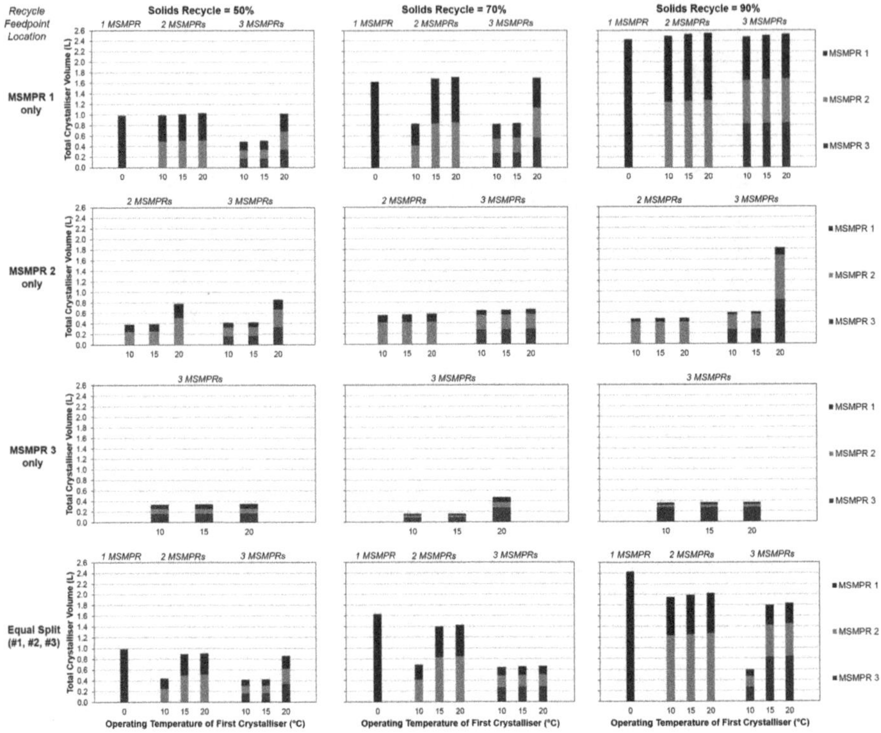

Figure 15.4 Required crystalliser volumes for continuous cyclosporine crystallisation with solids recycle.[33] Reproduced from ref. 33, https://doi.org/10.1021/acs.oprd.7b00225, with permission from American Chemical Society, Copyright 2017.

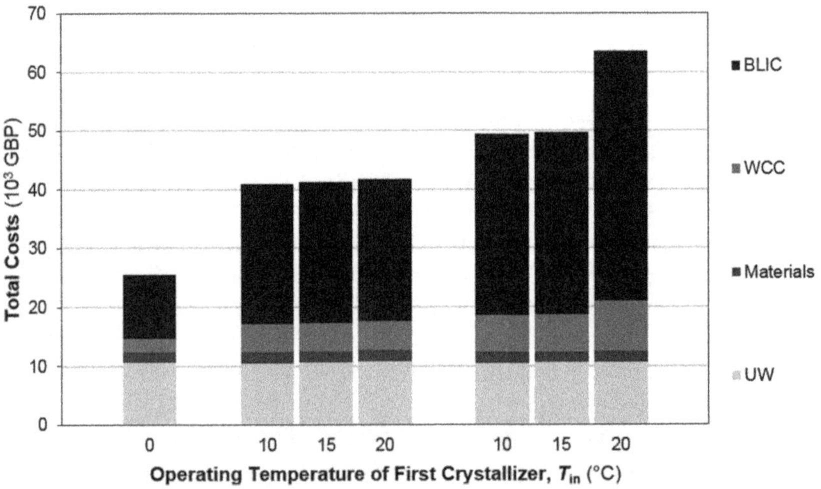

Figure 15.5 Cost components for the continuous cyclosporine crystallisation without recycle.[33] Reproduced from ref. 33, https://doi.org/10.1021/acs.oprd.7b00225, with permission from American Chemical Society, Copyright 2017.

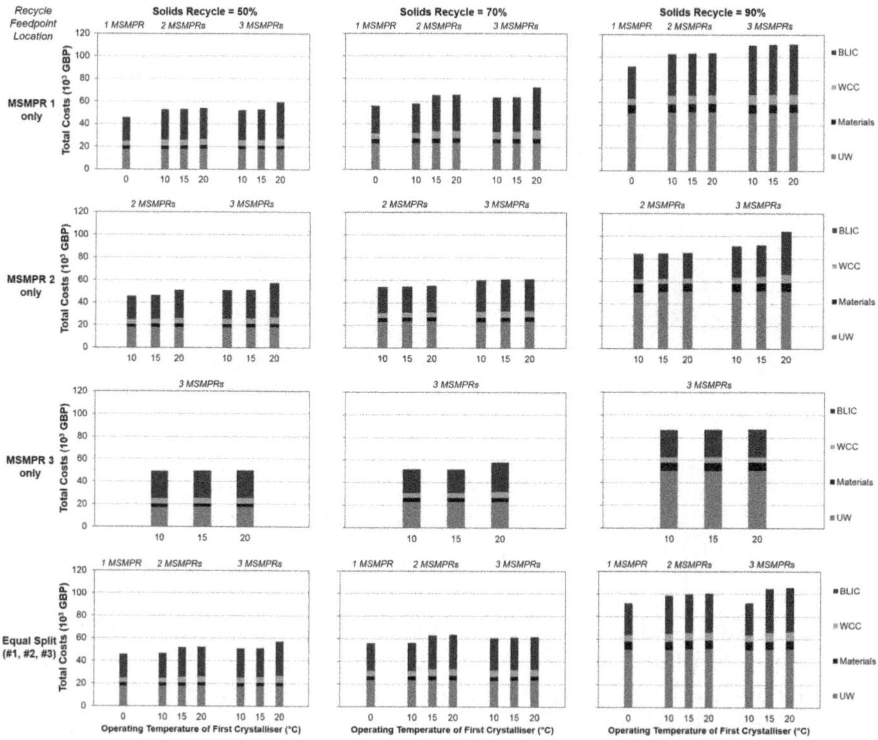

Figure 15.6 Cost components for the continuous cyclosporine crystallisation with solids recycle.[33] Reproduced from ref. 33, https://doi.org/10.1021/acs.oprd.7b00225, with permission from American Chemical Society, Copyright 2017.

Cost components of different solids recycle process configurations are shown in Figure 15.6. The best configuration (*i.e.* lowest cost) implements two crystallisers with all recycle fed to the first crystalliser, operating at 10 °C. As for configurations without recycle, increasing the number of crystallisers significantly increases CapEx. Increasing solids recycle and feeding the recycle to the first crystalliser leads to increased total costs due to the increased crystalliser volumes corresponding to increase throughputs at all stages in the cascade.[33] Highest total costs occur when recycle is evenly divided across the cascade due to the large vessel volumes caused by increased internal flowrates and poorer yields in comparison to when the recycle stream is fed to the final crystalliser. As for no recycle configurations, BLIC contributes a significant portion to CapEx; utilities and waste contributes a significant portion to OpEx.

For all process configurations, we have compared cost components with assumed total cascade residence times[33] and resulting crystalliser volumes presented in Figures 15.3 and 15.4, respectively. The selection of total

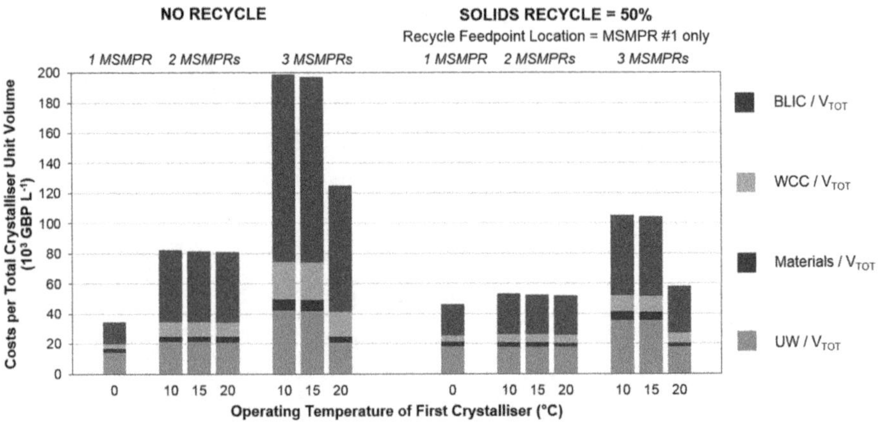

Figure 15.7 Cost components per total crystalliser unit volume for continuous cyclosporine crystallisation.[33] Reproduced from ref. 33, https://doi.org/10.1021/acs.oprd.7b00225, with permission from American Chemical Society, Copyright 2017.

cascade residence times is so that crystallisers are not over-sized and no unnecessary capital costs are incurred. This can lead to unfair cost comparisons due to the differences in crystalliser volumes, which significantly affects CapEx. Total cost components normalised with respect to the total cascade volume are presented to allow an alternative economic comparison of different configurations; these are shown in Figure 15.7 for selected designs. The results indicate that cost components are very sensitive to the selection of total cascade residence time for crystallisation process design. A more rigorous approach towards fair comparisons of design and operating variables for economic analyses of candidate configurations is required for such cascades.

15.5.2 Cost Optimisation of MSMPR Cascades: Cyclosporine, Paracetamol, Aliskiren

The previous section demonstrated the importance of rigorous selection of various design and operating variables for MSMPR cascades, as well as the economic disadvantages of operating MSMPRs with solids recycle due to the material losses in purge streams required to maintain the desired process mass balance. Technoeconomic performances of MSMPR cascade designs are very sensitive to crystalliser operating temperatures, cascade residence time and the number of implemented crystallisers; conducting MSMPR cascade design *via* nonlinear optimisation for total cost minimisation for cost optimal design and operating parameter selection allows systematic screening for the best process configurations. Here, we conduct nonlinear optimisation of MSMPR cascades without recycle for three APIs:[36] cyclosporine, paracetamol (the popular analgaesic) and aliskiren (the anticoagulant), whose continuous MSMPR crystallisation kinetics, solubility data and experimental performance have been demonstrated in the literature.[18,42,43,58]

15.5.2.1 Nonlinear Optimisation of MSMPR Configurations

Cyclosporine is crystallised from the same fresh mother liquor as described previously. Paracetamol is crystallised from a 4 : 1 mixture (volume basis) of isopropanol:water[43] and aliskiren is crystallised from a 1 : 1 mixture of ethyl acetate:ethanol (mass basis).[18] As before, the number of implementable crystallisers, N = 1–3. Additionally, we consider varying plant API capacities, $Q_{\text{API}} = \{10^2, 10^3, 10^4\}$ kg API per year, to investigate the effects of production scale.[23]

The total cost objective function of the nonlinear optimisation problem is the total cost, estimated as per the previously described methodology. Optimisation decision variables are the residence time and temperature of each crystalliser in the cascade. Crystallisation temperatures are constrained between −10 and 20 °C and the temperature of each crystalliser must be lower than or equal to the previous to distribute the cooling load across multiple crystallisers. Crystallisers of equal residence times are considered, as designing crystallisers of equal volume makes their purchase and acquisition less expensive. Additionally, the total cascade residence time is limited to 15 h in accordance with previous work.[33,62]

$$-10\ ^\circ\text{C} \leq T_N \leq \ldots \leq T_1 \leq 20\ ^\circ\text{C} \tag{15.25}$$

$$\tau_1 = \ldots = \tau_N \tag{15.26}$$

$$\sum_{i=1}^{N} \tau_i \leq 15h \tag{15.27}$$

The optimisation problem can be solved using various programming approaches; MATLAB is frequently implemented for such purposes, having many built-in nonlinear optimisation functions. The case study described here implemented the MATLAB built-in solver fmincon, a popular function for solving constrained nonlinear optimisation problems. For the case study described here, the optimisation problem was solved separately for all API choices {cyclosporine, paracetamol, aliskiren}, number of implemented crystallisers, $N = \{1, 2, 3\}$, and plant capacity, $Q_{\text{API}} = \{10^2, 10^3, 10^4\}$ kg API per annum.

15.5.2.2 Cost Optimal MSMPR Design and Operating Parameters

Figure 15.8 illustrates optimal design (crystalliser residence times and volumes) and operating variables (temperatures) for each API with different numbers of crystallisers and API capacities. Equipment volumes

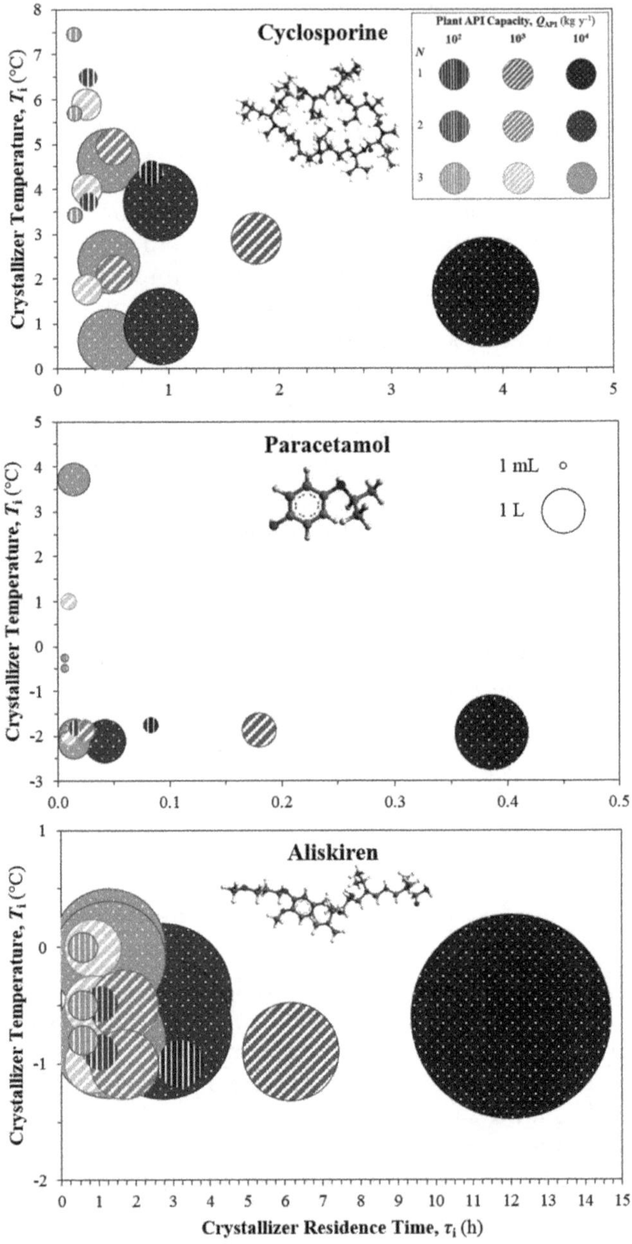

Figure 15.8 Crystalliser operating temperatures and residence times corresponding to total cost minima; bubble diameters are proportional to crystalliser volumes.[36] Reproduced from ref. 36 with permission from American Chemical Society, Copyright 2017.

increase with plant capacity to accommodate increased material throughputs, while increasing the number of crystallisers decreases the total crystallisation volume. MSMPR residence times for aliskiren are long, and thus, large crystalliser volumes are required, due to the slow crystallisation kinetics of the API.

Cyclosporine operating temperatures are above zero and decrease along the crystalliser cascade. Crystallisers for paracetamol and aliskiren also decrease in temperature along cascades, however operate at lower temperatures; as the number of implemented crystallisers is increased, operating temperatures increase and residence times decrease. Rigorous temperature control *via* PAT implementation is essential to maintain optimal operation; recent work illustrates its importance in crystallisation applications.[49]

15.5.2.3 Minimum Total Cost Components

Figure 15.9 shows minimum total cost components for cyclosporine, paracetamol and aliskiren. For all three APIs, total CapEx increases with capacity due to larger crystalliser volumes required for higher material

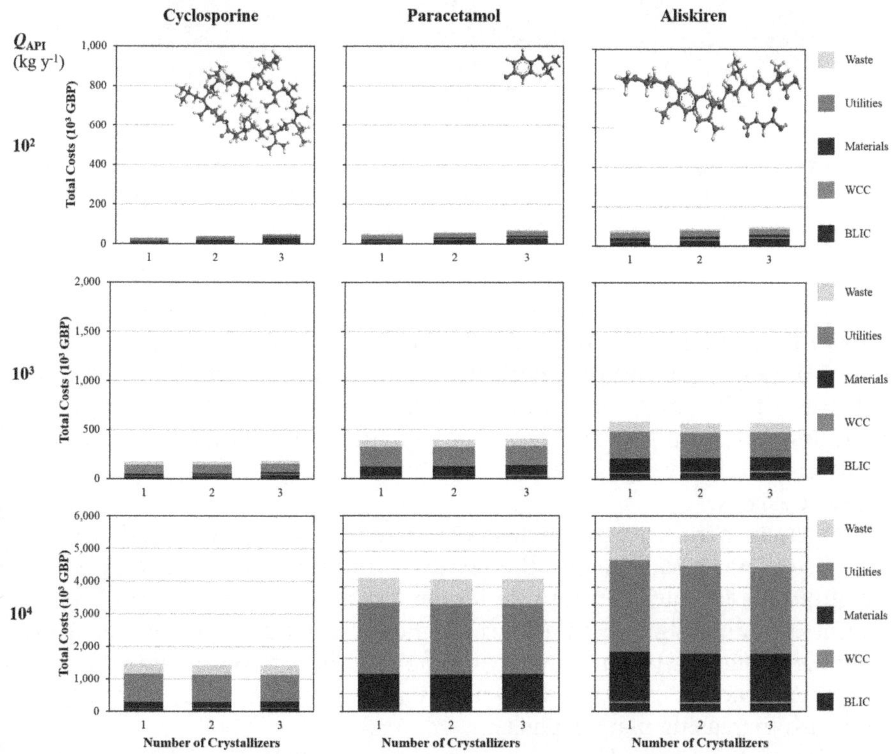

Figure 15.9 Minimum total cost components for each API at different plant capacities.[36] Reproduced from ref. 36 with permission from American Chemical Society, Copyright 2017.

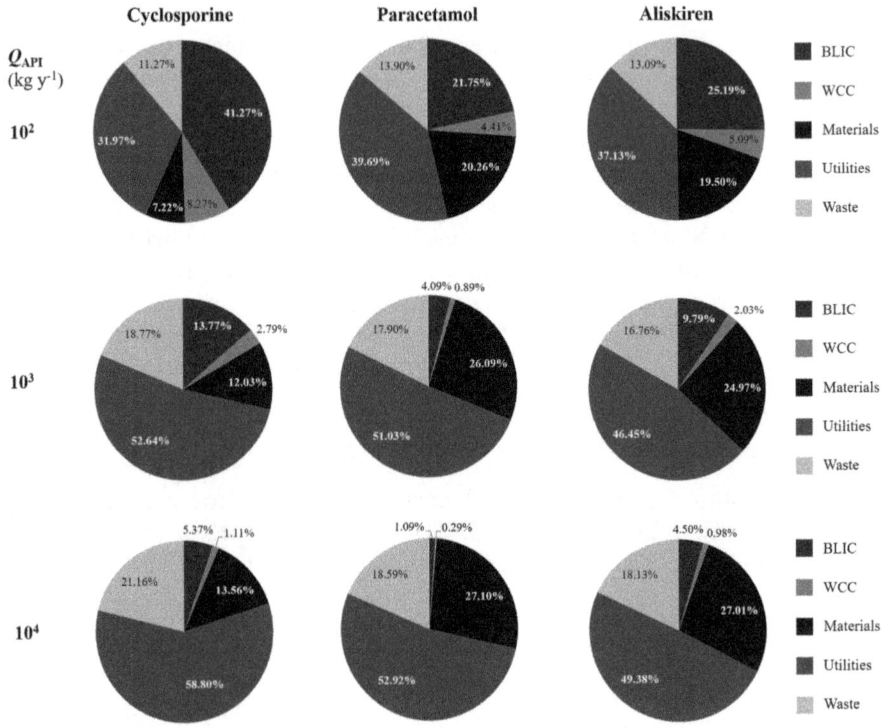

Figure 15.10 Component contributions towards total costs when implementing one crystallizer.[36] Reproduced from ref. 36 with permission from American Chemical Society, Copyright 2017.

throughputs. BLIC dominates CapEx contributions due to expensive crystallisation equipment;[36] WCC contributions increase as a linear function of material throughput. Both BLIC and WCC contributions increase with the number of implemented crystallisers, despite decreasing total crystallisation volumes and material requirements, due to the cost of additional auxiliary equipment (pumps and cooling) accompanying crystalliser designs.

Total OpEx increases with capacity due to higher throughputs and utilities and waste. Figure 15.10 shows the relative minimum total cost component contributions for each API for one implemented crystalliser. CapEx contributions are more significant at lower capacities, but as capacity increases, OpEx contributions become more dominant. Utilities dominate OpEx contributions, however materials and waste handling costs become more significant with increasing plant capacity.

Minimum total costs for cyclosporine are attained when implementing one crystalliser at 10^2 and 10^3 kg API year ^{1}. When capacity is increased to 10^4 kg API year^{-1}, lower total costs are attained when two crystallisers are implemented. For paracetamol, at capacities of 10^2 and 10^3 kg API year^{-1}, cost optimal designs also only require one crystalliser. Implementing two crystallisers is favourable at 10^4 kg API year^{-1}. Continuous crystallisation of aliskiren at a capacity of 10^2 kg API year^{-1} is cost optimal when implementing one crystalliser; implementing two crystallisers at 10^3 kg API year^{-1} and three crystallisers at 10^4 kg API year^{-1} is more economically favourable.

15.5.3 Design and Optimisation of COBCs: Paracetamol

Paracetamol continuous crystallisation has been demonstrated in a number of different studies and crystalliser designs. Continuous combined cooling-antisolvent paracetamol crystallisation in COBCs was experimentally demonstrated and crystallisation kinetics and population balance models elucidated;[89] recent work considered technoeconomic optimisation of COBC crystallisation of paracetamol *via* investigation of several different parameters in the design space.[35]

The general population balance for a COBC is described by

$$\frac{\delta n}{\delta t}+\frac{G\delta n}{\delta L}=0 \tag{15.28}$$

The API mass balance in the mother liquor stream is calculated from the population balance.

$$\frac{\mathrm{d}C}{\mathrm{d}t}=-3\rho_{\mathrm{API}}k_{\mathrm{v}}G\int_0^\infty L^2 n\,\mathrm{d}L \tag{15.29}$$

Population balances and API mass balances (eqn (15.28) and (15.29)) are transformed *via* the method of moments (eqn (15.30)), which give the following set of ODEs (eqn (15.31)–(15.33)).

$$\mu_j=\int_0^\infty L^j n(L,t)\,\mathrm{d}L \tag{15.30}$$

$$\frac{\mathrm{d}\mu_0}{\mathrm{d}t}=B \tag{15.31}$$

$$\frac{\mathrm{d}\mu_j}{\mathrm{d}t}=jG\mu_{j-1}+Br_0^j \quad \text{for } j=1,2,3 \tag{15.32}$$

$$\frac{\mathrm{d}C_{\mathrm{API}}}{\mathrm{d}t}=-\rho_{\mathrm{API}}k_{\mathrm{v}}\left(3G\mu_1+Br_0^3\right) \tag{15.33}$$

Here, μ_j is the jth moment of the population balance, L is the characteristic (one-dimensional) length of the crystal, r_0 is the average seed size (= 40 microns); $\rho_{\mathrm{paracetamol}}$ = 1.3 g cm^{-3}. The crystal growth rate is defined by the following semi-empirical correlation; Re_o is the oscillatory Reynolds' number.[39]

$$G \cdot 10^8 = (3.78 \cdot 10^{-12})(C - C^{sat})^{1.570} Re_o^{2.155} + 21.50 \quad (15.34)$$

15.5.3.1 COBC Design Space Investigation for Paracetamol Crystallisation

Figure 15.11 shows the evolution of yield, mother liquor API concentration (C_{API}) and API solubility (C_{API}^{sat}) as a function of COBC residence time for a seed size of 40 microns and outlet temperature of 5 °C; panels A and B assume inlet temperatures of 50 °C and panel C assumes an inlet temperature of 70 °C. Growth rates are initially slow, hence yields take time to develop. As material moves through the COBC, supersaturation develops, and thus yield increases. Eventually, the API concentration in mother liquor decreases, thus decreasing supersaturation and causing a plateau in the yield trajectory. The bottom plot shows the effect of desired yield on COBC volume (calculated from required volumetric throughput and residence time) and attained product crystal size. Achieving higher yields under the same conditions becomes increasingly difficult; thermodynamic constraints imposed on the system (*i.e.*, the solubility behaviour) incurred by assumed inlet temperatures significantly affects supersaturation profiles across COBCs for different target yields. One method of achieving higher yields for shorter residence times/by implementing smaller crystallisers would be to operate at a higher inlet temperature; this allows a greater supersaturation to be maintained, allowing high yields in a reasonable time and crystalliser size.

The response surface of product crystal size as a function of the optimisation decision variables, antisolvent feed rate (ASR) and seed loading (SL) is shown in Figure 15.12. Lower antisolvent amounts result in larger product crystals as lower supersaturation favours crystal growth. For lower antisolvent usage, the absolute mass of seed crystals added is less, therefore, the seed count will lower, meaning crystals must grow more to achieve the same yield.

15.5.3.2 Nonlinear Optimisation Problem Formulation

The objective function for minimisation is the CapEx of the COBC. Antisolvent usage and seed loading are decision variables in the optimisation: antisolvent usage (ASR) is constrained between 50 : 50 and 20 : 80 (mass basis of process solvent to antisolvent) and seed loading (SL) is constrained between 0.5% and 2.0% by weight. The inlet temperature is assumed to be

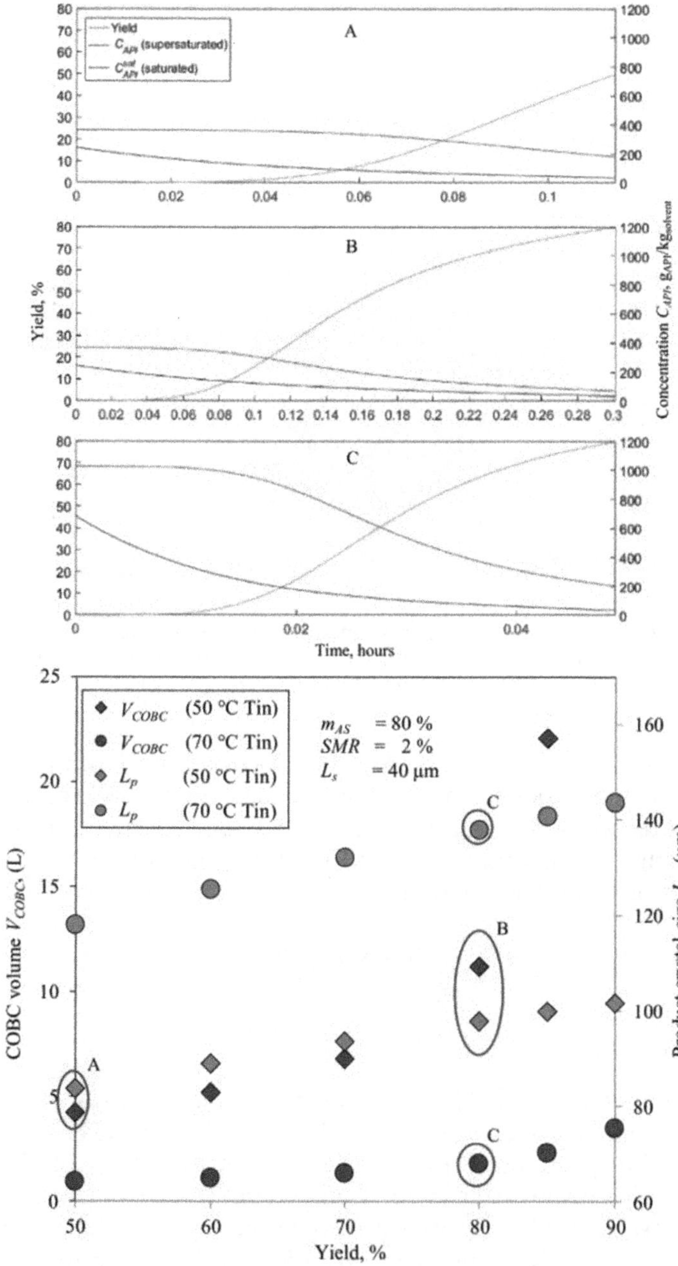

Figure 15.11 Trajectories of key process variables. In all cases the seed crystal size is 40 microns and the outlet temperature is 5 °C. (A) and (B): inlet temperature = 50 °C; (C): inlet temperature = 70 °C. Reproduced from ref. 35 with permission from Elsevier, Copyright 2019.

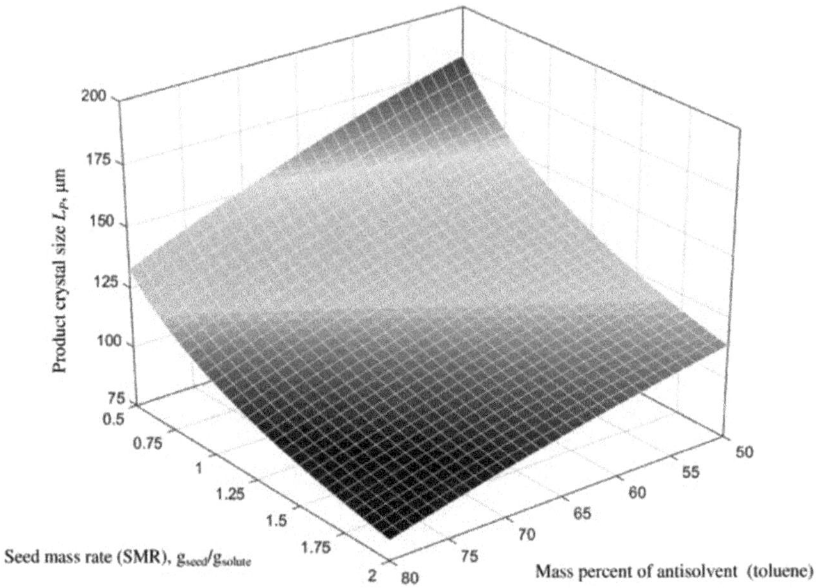

Figure 15.12 Product crystal size response surface for an inlet temperature of 50 °C and a seed crystal size of 40 microns. Reproduced from ref. 35 with permission from Elsevier, Copyright 2019.

50 °C and the mixture is cooled to 5 °C.[35] Seed crystal size was assumed as 40 microns, in order to ensure that the correlation used complies with the literature; only crystal sizes equal to or higher than this threshold have been investigated, to study the variation of optima. Solution of optimisation problems is implemented in MATLAB as described in recent work.[35]

$$50\% < \text{ASR} < 80\% \tag{15.35}$$

$$0.5\% < \text{SL} < 2.0\% \tag{15.36}$$

Given a desired yield of 50%, the total cost response surface for an inlet temperature of 50 °C and a seed crystal size of 40 microns is given in Figure 15.13. The optimal solution (minimum cost = 101 370 GBP) is pushed to bounds with respect to the decision variables. Antisolvent use affects the total cost more significantly than the seed mass loading, comparing the gradients of each of the decision variables. Greater rates of antisolvent use result in lower costs by reducing required residence times. More detailed design and cost estimations could encompass additional cost components other than CapEx. However, as the process being evaluated here is a single crystallisation operation, with no upstream or downstream operations, inclusion of additional cost detail as described above is complicated.

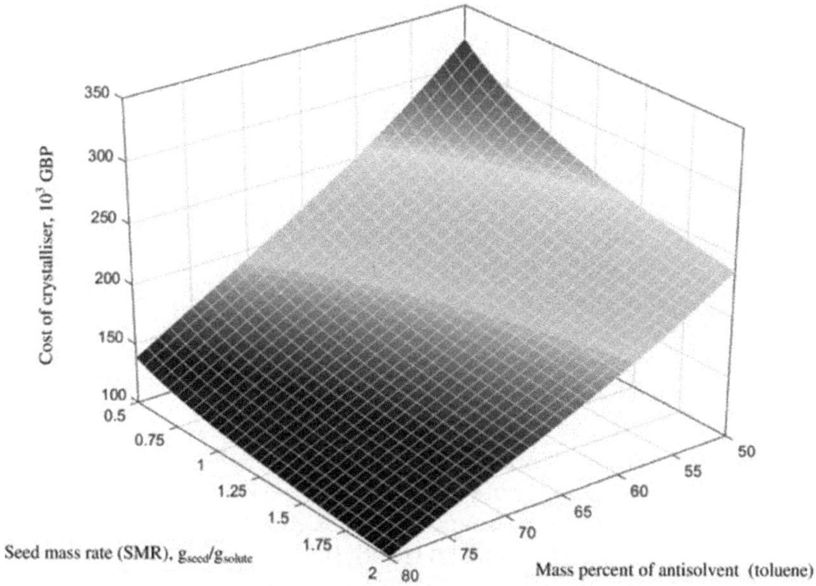

Figure 15.13 Total cost response surface for an inlet temperature of 50 °C and a seed crystal size of 40 microns. Reproduced from ref. 35 with permission from Elsevier, Copyright 2019.

15.6 Conclusions

Continuous pharmaceutical manufacturing (CPM) has demonstrated promise for both operational and economic benefits for a variety of APIs. Screening of candidate continuous crystallisation processes is essential for the rapid development of end-to-end continuous plants, in order to realise the full potential of CPM. Process modelling, simulation and optimisation allows comparative evaluation of technologically feasible and economically viable candidate process designs. Discussion of total cost component considerations, a description of an economic analysis methodology for continuous manufacturing processes, and their implementation in economic evaluation and nonlinear optimisation case studies of continuous pharmaceutical crystallisation design demonstrates the importance and utility of conceptual modelling approaches such as these to facilitate the rapid transition to leaner manufacturing routes.

Roman Symbols

ASR	antisolvent feed rate (mass ratio to feed mother liquor stream)
B	crystal nucleation rate (# m^{-3} s^{-1})

b	crystal nucleation exponent of eqn (15.18)
C_i	mother liquor solute concentration (kg m^{-3})
C_i^{sat}	solubility concentration of solute (kg m^{-3})
C_0	fresh feed mother liquor solute concentration (kg m^{-3})
CapEx	capital expenditure (GBP)
CC	contingency costs (GBP)
E_{ab}	crystal nucleation energy barrier (J mol^{-1})
E_{ag}	crystal growth energy barrier (J mol^{-1})
F_i	volumetric flowrate of stream i (m^3 s^{-1})
F_0	volumetric flowrate of the fresh feed stream (m^3 s^{-1})
f	coefficient in eqn (15.1)
FOB	free-on-board costs (GBP)
G	linear crystal growth rate (m s^{-1})
g	crystal growth exponent of eqn (15.17)
IEC	installed equipment costs (GBP)
k_{g0}	pre-exponent of eqn (15.17)
k_{v}	solute volume-shape factor
L	characteristic crystal length (m)
M_i	slurry density in MSMPR i (kg m^{-3})
m	slurry density exponent of eqn (15.18)
$\text{MAT}_{\text{annual}}$	annual material costs (GBP year^{-1})
N	total number of MSMPRs in crystallisation cascade
n	exponent in eqn (15.1)
n_{i}	crystal population density in crystalliser i (# m m^{-3})
NPV	net present value (GBP)
OpEx	total operating expenditure over plant lifetime (GBP)
$\text{OpEx}_{\text{annual}}$	annual operating expenditure (GBP year^{-1})
PPI	process, piping and instrumentation costs (GBP)
Q_{plant}	plant material throughput (kg year^{-1})
Q_{waste}	volumetric flow of waste production (L s^{-1})
P_j	purchase cost of equipment item at capacity j (GBP)
R	universal gas constant (8.3147 J mol^{-1} K^{-1})
Re_{o}	oscillatory Reynolds' number
Rev	annual revenue (GBP year^{-1})
r_0	average seed size (m)
S_j	capacity of equipment item (varying units)
SL	seed loading of COBC feed stream (% w/w seeds in feed mixture)
T_i	operating temperature of MSMPR i (K)
t	plant operating lifetime (year)
TPPC	total physical plant cost (GBP)
$\text{UTIL}_{\text{annual}}$	annual utilities costs (GBP year^{-1})
V_i	volume of crystalliser i (m^3)
$\text{Waste}_{\text{annual}}$	annual waste handling costs (GBP year^{-1})
WC	working capital costs (GBP)

WCC	working capital and construction costs (GBP)
y	interest rate (%)

Greek Symbols

ρ_{API}	solute density (kg m^{-3})
τ_i	residence time in MSMPR i (s)
μ_j	jth moment of crystal population balance (varying units)

Abbreviations

API	active pharmaceutical ingredient
CEPCI	chemical engineering plant cost index
COBC	continuous oscillatory baffled crystalliser
CPM	continuous pharmaceutical manufacturing
ICIS	Independent Chemical Information Services
LCA	life cycle assessment
MSMPR	mixed suspension, mixed product removal
ODE	ordinary differential equation
PAT	process analytical technology
PFC	plug flow crystalliser
PSE	process systems engineering
WHO	World Health Organisation

Acknowledgements

Mr Samir Diab gratefully acknowledges the financial support of the Engineering and Physical Sciences Research Council (EPSRC) *via* a Doctoral Training Partnership (DTP) PhD Fellowship (Grant # EP/N509644/1). Dr D.I. Gerogiorgis acknowledges a Royal Academy of Engineering (RAEng) Industrial Fellowship.

References

1. L. Yu and S. M. Reutzel-Edens, Crystallization – Basic Principles, *Encyclopedia of Food Sciences and Nutrition*, Elsevier, 2003, pp. 1697–1702.
2. A. S. Myerson, *Handbook of Industrial Crystallization*, Butterworth-Heinemann, 2002.
3. C. R. Gardner, C. T. Walsh and Ö. Almarsson, Drugs as materials: valuing physical form in drug discovery, *Nat. Rev. Drug Discovery*, 2004, **3**(11), 926–934.
4. O. Söhnel and J. Nýavlt, Induction time in batch crystallisation, *Krist. Tech.*, 1976, **11**(3), 239–244.
5. J. Chen, B. Sarma, J. M. B. Evans and A. S. Myerson, Pharmaceutical crystallization, *Cryst. Growth Des.*, 2011, **11**(4), 887–895.

6. L. X. Yu, G. Amidon, M. A. Khan, S. W. Hoag, J. Polli, G. K. Raju and J. Woodcock, Understanding pharmaceutical quality by design, *AAPS J.*, 2014, **16**(4), 771–783.
7. S. L. Lee, T. F. O'Connor, X. Yang, C. N. Cruz, S. Chatterjee, R. D. Madurawe, C. M. V. Moore, L. X. Yu and J. Woodcock, Modernizing pharmaceutical manufacturing: from batch to continuous production, *J. Pharm. Innovation*, 2015, **10**, 191–199.
8. R. Gérardy, N. Emmanuel, T. Toupy, V.-E. Kassin, N. N. Tshibalonza, M. Schmitz and J.-C. M. Monbaliu, Continuous flow organic chemistry: successes and pitfalls at the interface with current societal challenges, *Eur. J. Org. Chem.*, 2018, 2301–2351.
9. M. B. Plutschack, B. Pieber, K. Gilmore and P. H. Seeberger, The Hitchhiker's guide to flow chemistry, *Chem. Rev.*, 2017, **117**(18), 11796–11893.
10. L. Malet-Sanz and F. Susanne, Continuous flow synthesis. A pharma perspective,, *J. Med. Chem.*, 2012, **55**(9), 4062–4098.
11. M. Baumann and I. R. Baxendale, The synthesis of active pharmaceutical ingredients (APIs) using continuous flow chemistry, *Beilstein J. Org. Chem.*, 2015, **11**, 1194–1219.
12. R. Porta, M. Benaglia and A. Puglisi, Flow chemistry: recent developments in the synthesis of pharmaceutical products, *Org. Process Res. Dev.*, 2016, **20**, 2–25.
13. J. Britton and C. L. Raston, Multi-step continuous-flow synthesis, *Chem. Soc. Rev.*, 2017, **52**(5), 10159–10162.
14. S. Mascia, P. L. Heider, H. Zhang, R. Lakerveld, B. Benyahia, P. I. Barton, R. D. Braatz, C. L. Cooney, J. M. B. Evans, T. F. Jamison, K. F. Jensen, A. S. Myerson and B. L. Trout, End-to-end continuous manufacturing of pharmaceuticals: integrated synthesis, purification, and final dosage formation, *Angew. Chem., Int. Ed.*, 2013, **52**(47), 12359–12363.
15. A. Adamo, R. L. Beingessner, M. Behnam, J. Chen, T. F. Jamison, K. F. Jensen, J.-C. C. Monbaliu, A. S. Myerson, E. M. Revalor, D. R. Snead, T. Stelzer, N. Weeranoppanant, S. Y. Wong and P. Zhang, On-demand continuous-flow production of pharmaceuticals in a compact, reconfigurable system, *Science*, 2016, **352**(6281), 61–67.
16. S. E. Kuehn, Janssen embraces continuous manufacturing for Prezista, *Pharmaceutical Manufacturing*, 2015, Available at: http://www.pharmamanufacturing.com/articles/2015/janssen-embraces-continuous-manufacturing-for-prezista/.
17. E. Palmer, GSK commits to continuous processing, 2013, Available at, http://www.fiercepharmamanufacturing.com/story/gsk-commits-continuous-processing/2013-02-19?utm_medium=nl&utm_source=internal.
18. J. L. Quon, H. Zhang, A. Alvarez, J. Evans, A. S. Myerson and B. L. Trout, Continuous crystallization of aliskiren hemifumarate, *Cryst. Growth Des.*, 2012, **12**(6), 3036–3044.
19. J.-C. M. Monbaliu, T. Stelzer, E. Revalor, N. Weeranoppanant, K. F. Jensen and A. S. Myerson, Compact and integrated approach for

advanced end-to-end production, purification, and aqueous formulation of lidocaine hydrochloride, *Org. Process Res. Dev.*, 2016, **20**(7), 1347–1353.
20. A. Manipura, E. B. Martin, G. A. Montague, P. N. Sharratt and I. Houson, Risk-based decision making in early chemical process development of pharmaceutical and fine chemical industries, *Comput. Chem. Eng.*, 2013, **55**, 71–82.
21. S. K. Teoh, C. Rathi and P. Sharratt, Practical assessment methodology for converting fine chemicals processes from batch to continuous, *Org. Process Res. Dev.*, 2015, **20**(2), 414–431.
22. P. Bana, R. Örkényi, K. Lövei, Á. Lakó, G. I. Túrós, J. Éles, F. Faigl and I. Greiner, The route from problem to solution in multistep continuous flow synthesis of pharmaceutical compounds, *Bioorg. Med. Chem.*, 2017, **25**(23), 6180–6189.
23. H. G. Jolliffe and D. I. Gerogiorgis, Plantwide design and economic evaluation of two continuous pharmaceutical manufacturing (CPM) cases: ibuprofen and artemisinin, *Comput. Chem. Eng.*, 2016, **91**, 269–288.
24. J. R. Couper, *Process Engineering Economics*, CRC Press, 2003.
25. D. R. Woods, *Rules of Thumb in Engineering Practice*, Wiley, 2007.
26. T. Brown, *Engineering Economics and Economic Design for Process Engineers*, CRC Press, 2007.
27. H. G. Jolliffe and D. I. Gerogiorgis, Plantwide design and economic evaluation of two continuous pharmaceutical manufacturing (CPM) cases: ibuprofen and artemisinin, *Comput.-Aided Chem. Eng.*, 2015, **37**, 2213–2218.
28. Plant Cost Index Archives, Available at, http://www.chemengonline.com/site/plant-cost-index/.
29. ICIS: Indicative Chemical Prices, Available at, https://www.icis.com/chemicals/channel-info-chemicals-a-z/.
30. H. G. Jolliffe and D. I. Gerogiorgis, Technoeconomic optimization of a conceptual flowsheet for continuous separation of an analgaesic active pharmaceutical ingredient (API), *Ind. Eng. Chem. Res.*, 2017, **56**(15), 4357–4376.
31. H. G. Jolliffe and D. I. Gerogiorgis, Technoeconomic optimisation and comparative environmental impact evaluation of continuous crystallisation and antisolvent selection for artemisinin recovery, *Comput. Chem. Eng.*, 2017, **103**, 218–232.
32. S. Diab and D. I. Gerogiorgis, Process modeling, simulation, and technoeconomic evaluation of separation solvents for the continuous pharmaceutical manufacturing (CPM) of diphenhydramine, *Org. Process Res. Dev.*, 2017, **21**, 924–946.
33. S. Diab and D. I. Gerogiorgis, Technoeconomic evaluation of multiple mixed suspension-mixed product removal (MSMPR) crystallizer configurations for continuous cyclosporine crystallization, *Org. Process Res. Dev.*, 2017, **21**(7), 1571–1587.

34. S. Diab and D. I. Gerogiorgis, Process modelling, simulation and technoeconomic evaluation of crystallisation antisolvents for the continuous pharmaceutical manufacturing of rufinamide, *Comput. Chem. Eng.*, 2018, **111**, 102–114.
35. H. G. Jolliffe and D. I. Gerogiorgis, Process modelling, design and technoeconomic evaluation for continuous paracetamol crystallisation, *Comput. Chem. Eng.*, 2018, **118**, 224–235.
36. S. Diab and D. I. Gerogiorgis, Technoeconomic optimization of continuous crystallization for three active pharmaceutical ingredients (APIs): cyclosporine, paracetamol and aliskiren, *Ind. Eng. Chem. Res.*, 2018, **57**(29), 9489–9499.
37. S. Diab and D. I. Gerogiorgis, Plantwide technoeconomic analysis and separation solvent selection for continuous pharmaceutical manufacturing: ibuprofen, artemisinin and diphenhydramine, in *Process Systems Engineering for Pharmaceutical Manufacturing*, Elsevier, 2017, pp. 85–120.
38. A. Dafnomilis, S. Diab, A. D. Rodman, A. G. Boudouvis and D. I. Gerogiorgis, Multiobjective dynamic optimization of ampicillin batch crystallization: sensitivity analysis of attainable performance vs product quality constraints, *Ind. Eng. Chem. Res.*, 2019, **58**(40), 18756–18771.
39. H. G. Jolliffe and D. I. Gerogiorgis, Process modelling, design and technoeconomic evaluation for continuous paracetamol crystallisation, *Comput.-Aided Chem. Eng.*, 2018, **43**, 1637–1642.
40. G. Capellades, P. U. Joshi, K. Dam-Johansen, M. J. Mealy, T. V. Christensen and S. Kiil, Characterization of a multistage continuous MSMPR crystallization process assisted by image analysis of elongated crystals, *Cryst. Growth Des.*, 2018, **18**(11), 6455–6469
41. D. Zhang, S. Xu, S. Du, J. Wang and J. Gong, Progress of pharmaceutical continuous crystallization, *Engineering*, 2017, **3**(3), 354–364.
42. A. J. Alvarez, A. Singh and A. S. Myerson, Crystallization of cyclosporine in a multistage continuous MSMPR crystallizer, *Cryst. Growth Des.*, 2011, **11**(10), 4392–4400.
43. G. Power, G. Hou, V. K. Kamaraju, G. Morris, Y. Zhao and B. Glennon, Design and optimization of a multistage continuous cooling mixed suspension, mixed product removal crystallizer, *Chem. Eng. Sci.*, 2015, **133**, 125–139.
44. G. Morris, G. Power, S. Ferguson, M. Barrett, G. Hou and B. Glennon, Estimation of nucleation and growth kinetics of benzoic acid by population balance modeling of a continuous cooling mixed suspension, mixed product removal crystallizer, *Org. Process Res. Dev.*, 2015, **19**(12), 1891–1902.
45. J. M. Schall, J. S. Mandur, R. D. Braatz and A. S. Myerson, Nucleation and growth kinetics for combined cooling and antisolvent crystallization in a mixed-suspension, mixed-product removal system: Estimating solvent dependency, *Cryst. Growth Des.*, 2018, **18**(3), 1560–1570.

46. S. Ferguson, G. Morris, H. Hao, M. Barrett and B. Glennon, Characterization of the anti-solvent batch, plug flow and MSMPR crystallization of benzoic acid, *Chem. Eng. Sci.*, 2013, **104**, 44–54.
47. Q. Su, Z. K. Nagy and C. D. Rielly, Pharmaceutical crystallisation processes from batch to continuous operation using MSMPR stages: Modelling, design, and control, *Chem. Eng. Process.*, 2015, **89**, 41–53.
48. D. Acevedo, Y. Yang, D. J. Warnke and Z. K. Nagy, Model-based evaluation of direct nucleation control approaches for the continuous cooling crystallization of paracetamol in a mixed suspension mixed product removal system, *Cryst. Growth Des.*, 2017, **17**(10), 5377–5383.
49. X. Yang, D. Acevedo, A. Mohammad, N. Pavurala, H. Wu, A. L. Brayton, R. A. Shaw, M. J. Goldman, F. He, S. Li, R. J. Fisher, T. F. O'Connor and C. N. Cruz, Risk considerations on developing a continuous crystallization system for carbamazepine, *Org. Process Res. Dev.*, 2017, **21**(7), 1021–1033.
50. L. R. Agnew, T. McGlone, H. P. Wheatcroft, A. Robertson, A. R. Parsons and C. C. Wilson, Continuous crystallization of paracetamol (acetaminophen) form II: selective access to a metastable solid form, *Cryst. Growth Des.*, 2017, **17**(5), 2418–2427.
51. J. Li, B. L. Trout and A. S. Myerson, Multistage continuous mixed-suspension, mixed-product removal (MSMPR) crystallization with solids recycle, *Org. Process Res. Dev.*, 2016, **20**(2), 510–516.
52. K. Park, D. Y. Kim and D. R. Yang, Operating strategy for continuous multistage mixed suspension and mixed product removal (MSMPR) crystallization processes depending on crystallization kinetic parameters, *Ind. Eng. Chem. Res.*, 2016, **55**(26), 7142–7153.
53. H. Zhang, J. Quon, A. J. Alvarez, J. Evans, A. S. Myerson and B. Trout, Development of continuous anti-solvent/cooling crystallization process using cascaded mixed suspension, mixed product removal crystallizers, *Org. Process Res. Dev.*, 2012, **16**(5), 915–924.
54. H. Gruber-Woelfler, M. Escribà-Gelonch, T. Noël, M. C. Maier and V. Hessel, Effect of acetonitrile-based crystallization conditions on the crystal quality of vitamin D_3, *Chem. Eng. Technol.*, 2017, **40**(11), 2016–2024.
55. Y. Yang and Z. K. Nagy, Combined cooling and antisolvent crystallization in continuous mixed suspension, mixed product removal cascade crystallizers: steady-state and startup optimization, *Ind. Eng. Chem. Res.*, 2015, **54**(21), 5673–5682.
56. Y. Yang and Z. K. Nagy, Advanced control approaches for combined cooling/antisolvent crystallization in continuous mixed suspension mixed product removal cascade crystallizers, *Chem. Eng. Sci.*, 2015, **127**, 362–373.
57. T. Vetter, C. L. Burcham and M. F. Doherty, Regions of attainable particle sizes in continuous and batch crystallization processes, *Chem. Eng. Sci.*, 2014, **106**, 167–180.

58. S. Y. Wong, A. P. Tatusko, B. L. Trout and A. S. Myerson, Development of continuous crystallization processes using a single-stage mixed-suspension, mixed-product removal crystallizer with recycle, *Cryst. Growth Des.*, 2012, **12**(11), 5701–5707.
59. Q. Su, B. Benyahia, Z. K. Nagy and C. D. Rielly, Mathematical modeling, design, and optimization of a multisegment multiaddition plug-flow crystallizer for antisolvent crystallizations, *Org. Process Res. Dev.*, 2015, **19**(12), 1859–1870.
60. K. A. Powell, A. N. Saleemi, C. D. Rielly and Z. K. Nagy, Periodic steady-state flow crystallization of a pharmaceutical drug using MSMPR operation, *Chem. Eng. Process.*, 2015, **97**, 195–212.
61. K. Tahara, M. O'Mahony and A. S. Myerson, Continuous spherical crystallization of albuterol sulfate with solvent recycle system, *Cryst. Growth Des.*, 2015, **15**(10), 5155–5156.
62. J. Li, T. C. Lai, B. L. Trout and A. S. Myerson, Continuous crystallization of cyclosporine: the effect of operating conditions on yield and purity, *Cryst. Growth Des.*, 2017, **17**(3), 1000–1007.
63. T.-T. C. Lai, S. Ferguson, L. Palmer, B. L. Trout and A. S. Myerson, Continuous crystallization and polymorph dynamics in the l-glutamic acid system, *Org. Process Res. Dev.*, 2014, **18**(11), 1382–1390.
64. T.-T. C. Lai, J. Cornevin, S. Ferguson, N. Li, B. L. Trout and A. S. Myerson, Control of polymorphism in continuous crystallization *via* mixed suspension mixed product removal systems cascade design, *Cryst. Growth Des.*, 2015, **15**(7), 3374–3382.
65. Y. Cui, J. J. Jaramillo, T. Stelzer and A. S. Myerson, Statistical design of experiment on contact secondary nucleation as a means of creating seed crystals for continuous tubular crystallizers, *Org. Process Res. Dev.*, 2015, **19**(9), 1101–1108.
66. S. Y. Wong, Y. Cui and A. S. Myerson, Contact secondary nucleation as a means of creating seeds for continuous tubular crystallizers, *Cryst. Growth Des.*, 2013, **13**(6), 2514–2521.
67. R. J. P. Eder, S. Radl, E. Schmitt, S. Innerhofer, M. Maier, H. Gruber-Woelfler and J. G. Khinast, Continuously seeded, continuously operated tubular crystallizer for the production of active pharmaceutical ingredients, *Cryst. Growth Des.*, 2010, **10**(5), 2247–2257.
68. A. Majumder and Z. K. Nagy, Fines removal in a continuous plug flow crystallizer by optimal spatial temperature profiles with controlled dissolution, *AIChE J.*, 2013, **59**(12), 4582–4594.
69. R. J. P. Eder, E. K. Schmitt, J. Grill, S. Radl, H. Gruber-Woelfler and J. G. Khinast, Seed loading effects on the mean crystal size of acetylsalicylic acid in a continuous-flow crystallization device, *Cryst. Res. Technol.*, 2011, **46**(3), 227–237.
70. S. Ferguson, G. Morris, H. Hao, M. Barrett and B. Glennon, In-situ monitoring and characterization of plug flow crystallizers, *Chem. Eng. Sci.*, 2012, **77**, 105–111.

71. M. Jiang, Z. Zhu, E. Jimenez, C. D. Papageorgiou, J. Waetzig, A. Hardy, M. Langston and R. D. Braatz, Continuous-flow tubular crystallization in slugs spontaneously induced by hydrodynamics, *Cryst. Growth Des.*, 2014, **14**(2), 851–860.
72. D. Rossi, R. Jamshidi, N. Saffari, S. Kuhn, A. Gavriilidis and L. Mazzei, Continuous-flow sonocrystallization in droplet-based microfluidics, *Cryst. Growth Des.*, 2015, **15**(11), 5519–5529.
73. J. Sang-Il Kwon, M. Nayhouse, G. Orkoulas and P. D. Christofides, Crystal shape and size control using a plug flow crystallization configuration, *Chem. Eng. Sci.*, 2014, **119**, 30–39.
74. B. J. Ridder, A. Majumder and Z. K. Nagy, Population balance model-based Multiobjective optimization of a multisegment multiaddition (MSMA) continuous plug-flow antisolvent crystallizer, *Ind. Eng. Chem. Res.*, 2014, **53**(11), 4387–4397.
75. A. Majumder and Z. K. Nagy, Dynamic modeling of encrust formation and mitigation strategy in a continuous plug flow crystallizer, *Cryst. Growth Des.*, 2015, **15**(3), 1129–1140.
76. S. Lawton, G. Steele, P. Shering, L. Zhao, I. Laird and X.-W. Ni, Continuous crystallization of pharmaceuticals using a continuous oscillatory baffled crystallizer, *Org. Process Res. Dev.*, 2009, **13**(6), 1357–1363.
77. T. McGlone, N. E. B. Briggs, C. A. Clark, C. J. Brown, J. Sefcik and A. J. Florence, Oscillatory flow reactors (OFRs) for continuous manufacturing and crystallization, *Org. Process Res. Dev.*, 2015, **19**(9), 1186–1202.
78. M. S. Escotet-Espinoza, A. Rogers and M. G. Ierapetritou, Optimization methodologies for the production of pharmaceutical products, in *Process Simulation and Data Modeling in Solid Oral Drug Development and Manufacture*, Springer, 2016, pp. 281–309.
79. L. T. Biegler and I. E. Grossmann, Retrospective on optimization, *Comput. Chem. Eng.*, 2004, **28**, 1169–1192.
80. K. V. Gernaey, A. E. Cervera-Padrell and J. M. Woodley, A perspective on PSE in pharmaceutical process development and innovation, *Comput. Chem. Eng.*, 2012, **42**, 15–29.
81. M. Grom, G. Stavber, P. Drnovšek and B. Likozar, Modelling chemical kinetics of a complex reaction network of active pharmaceutical ingredient (API) synthesis with process optimization for benzazepine heterocyclic compound, *Chem. Eng. J.*, 2016, **283**, 703–716.
82. E. J. Cavanagh, M. J. Savelski and C. S. Slater, Optimization of environmental impact reduction and economic feasibility of solvent waste recovery using a new software tool, *Chem. Eng. Res. Des.*, 2014, **92**(10), 1942–1954.
83. D. Ott, S. Borukhova and V. Hessel, Life cycle assessment of multi-step rufinamide synthesis – from isolated reactions in batch to continuous microreactor networks, *Green Chem.*, 2016, **18**(4), 1096–1116.

84. R. Lakerveld, B. Benyahia, P. L. Heider, H. Zhang, A. Wolfe, C. Testa, S. Ogden, D. R. Hersey, S. Mascia, J. M. B. Evans, R. D. Braatz and P. I. Barton, The application of an automated plant-wide control strategy for a continuous pharmaceutical pilot plant, in *2014 American Control Conference*, IEEE, 2014, pp. 3512–3517.
85. F. Boukouvala and M. G. Ierapetritou, Surrogate-based optimization of expensive flowsheet modeling for continuous pharmaceutical manufacturing, *J. Pharm. Innovation*, 2013, **8**(2), 131–145.
86. R. B. Nussenblatt and A. G. Palestine, Cyclosporine: Immunology, pharmacology and therapeutic uses, *Surv. Ophthalmol.*, 1986, **31**(3), 159–169.
87. D. S. Groisser, C. E. Griffiths, C. N. Ellis and J. J. Voorhees, A review and update of the clinical uses of cyclosporine in dermatology, *Dermatol. Clin.*, 1991, **9**(4), 805–817.
88. J. L. Atwood, Separation of active pharmaceutical ingredients (APIs) from excipients in pharmaceutical formulations, *Cryst. Growth Des.*, 2015, **15**(6), 2874–2877.
89. C. J. Brown and X. Ni, Online evaluation of paracetamol antisolvent crystallization growth rate with video imaging in an oscillatory baffled crystallizer, *Cryst. Growth Des.*, 2011, **11**(3) ,719–725.

CHAPTER 16

Digital Design and Operation of Continuous Crystallization Processes via *Mechanistic Modelling Tools*

NIALL A. MITCHELL*[a], SEAN K. BERMINGHAM[a], CHRISTOPHER L. BURCHAM[b], CHRISTOPHER S. POLSTER, FURQAN TAHIR[c] AND JOHN MACK[c]

[a]Process Systems Enterprise (PSE) Ltd., Hammersmith, London, UK; [b]Eli Lilly & Co. Inc., Indianapolis, Indiana, USA; [c]Perceptive Engineering Ltd., Daresbury, UK
*E-mail: n.mitchell@psenterprise.com

16.1 Introduction

Process modelling is becoming more widely applied within academia and industry to assist in the digital design and operation of crystallization for both batch and continuous crystallization processes. The utilization of mechanistic modelling approaches *via* process or flowsheeting modelling tools allows for the virtual assessment of the design space for a continuous crystallization process and the attainable regions of the key critical quality attribute(s) (CQAs) of the process prior to experimental verification on the physical system, allowing for a more efficient workflow in terms of time

The Handbook of Continuous Crystallization
Edited by Nima Yazdanpanah and Zoltan K. Nagy

Published by the Royal Society of Chemistry, www.rsc.org

and materials. Mechanistic process modelling has several purposes in the design of a pharmaceutical continuous crystallization processes. A well-developed and validated mechanistic kinetic model can be utilized for the *in silico* determination of process sensitivity to parametric changes and process disturbances, aiding in the determination of proven acceptable ranges, establishing a design space and optimization of a process.

Mechanistic process models for crystallization processes use scientific understanding to formulate equations describing the process. The models include mass balances, energy balances, population balances, first-principles science, and kinetic theories for the active crystallization mechanisms. The primary value of a mechanistic model-based approach to process development as opposed to a more traditional approach using statistical analysis (*e.g.* design of experiments) is to reduce the number of experiments required to gain sufficient process understanding and more importantly to predict the crystallization behaviour over a wide range of process conditions with the ability to extrapolate beyond the regions probed experimentally (with caution). The mechanistic modelling approach allows for the prediction of the process behaviour at the same scale or at different scales, with the same or different equipment configurations. Examples of these scenarios might include:

1. Troubleshooting an existing process.
2. Reduction of batch time whilst satisfying particle size distribution (PSD) and/or product purity constraints.
3. Process scale-up.
4. Process technical transfer.

Utilization of process models for the *in silico* determination of process sensitivity and determination of proven acceptable ranges/design space for continuous crystallization has the obvious impact that is similar for most other unit operations and it allows the reduction of experiments needed to adequately validate a process. A mechanistic model for this purpose can be used to direct experimentation toward higher value experiments, or to evaluate the process under conditions at an extreme, which, with a properly benchmarked model, can be used as a surrogate for experimental data. Continuous process models are either steady-state or dynamic. Steady-state models describe the system after start-up, when nothing within the system is changing with respect to time. In contrast, dynamic models capture changes in the state of the system over time, describing start-up and shut-down behaviour, and propagation of disturbances throughout the system. Dynamic models can reach a steady-state, matching the results of the corresponding steady-state model. Many process models in the chemical industry describe only steady-state behaviour. However, capturing the process dynamics is important in continuous crystallization processes, as disturbance propagation and control is a key concern.

Table 16.1 Typical key inputs and outputs to crystallization process models in pharmaceutical manufacturing.

Inputs	Examples include the following:
Critical process parameters (CPPs)	Seed particle size distribution (PSD), solute liquid composition, residence time, temperature and/or composition set points for each stage
Equipment parameters	Impeller diameter(s), impeller type (*i.e.* power and pumping numbers)
Material and thermodynamic properties	Solubility, molecular weight, crystal stoichiometry, volume shape factors
Disturbances	Flow rates (feed & between stages – continuous, semi-continuous), seeding, fouling on surfaces and/or transfer lines
Outputs	Examples include the following:
Product critical quality attributes (CQAs)	PSD, particle shape, particle physical purity (form/polymorph), particle chemical purity (process impurities, solvents)
Process economics	Yield and productivity of process

Process models describe the quantitative relationships between inputs and outputs of a process. By correlating the design decisions, risk factors, and material attributes to measurable process and product outcomes, these models can be used as predictive tools for process design and operation. Table 16.1 shows typical key inputs and outputs of process models for pharmaceutical drug substance manufacturing.

The design of a continuous crystallization process is somewhat unique, so process design and/or optimization can benefit much more substantially. Figure 16.1 shows an example of an anti-solvent driven continuous crystallization process consisting of two mixed-suspension mixed-product removal (MSMPR) crystallizers, also known as a cascade continuous crystallizer. The solubility of the compound of interest (solute) is shown as a solid curved line, while the meta-stable zone limit for the process is shown by the dashed curved line. Assuming a feed concentration is fixed based on upstream process considerations and a final composition is chosen based on optimal yield and impurity rejection considerations, three process variables are left to choose to define the process (MSMPR1 anti-solvent concentration (x_1) and the mean residence time for both crystallizer stages (τ_1, and τ_2)).

A factorial experiment design consisting of three levels for each process variable would result in twenty-seven experiments to fully define the process behaviour. This experiment design corresponds to an optimistic estimate of a minimum of seven weeks for process development, including experimentation (based on completing four experiments per work week) and modelling elements. In addition, due to the proximity of feed concentrations to the process meta-stable limit, it may be wise

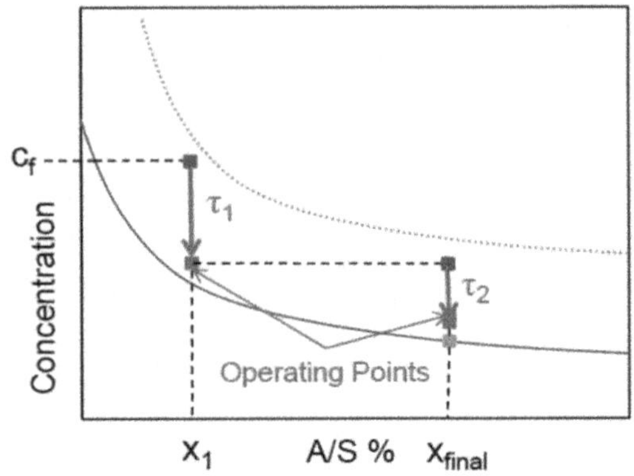

Figure 16.1 Conceptual design of a 2-stage MSMPR continuous crystallization process. An example of an isothermal anti-solvent driven crystallization is given.

to consider a higher number of stages for process development. This greatly expands the experimental space that needs to be explored; thus, the economy and potential benefits in terms of efficiency both in time and material of a mechanistic process modelling approach as opposed to a purely empirical method is readily apparent in this representative example.

The next section will describe a recommended workflow for the development of a continuous crystallization process and the utility of mechanistic process modelling tools in facilitating this workflow. In Section 16.3, we will focus on the lifecycle of a mechanistic crystallization model to describe continuous crystallization processes, including model configuration, validation and application stages. Subsequently, the utility of process mechanistic modelling to enable the digital design and control of continuous crystallization processes will be detailed *via* some industrial case studies in Sections 16.4 and 16.5, respectively.

16.2 Process Development Workflows for Continuous Crystallization

A number of process development workflows have been suggested in the literature for the development of crystallization processes. One suggested workflow for the development of a continuous cooling crystallization processes is given in Figure 16.2 below. These workflows aim to provide a clear sequence of steps to study a new crystallization system of interest with clear decision points following each stage. Please refer to 20 for further details on the focus of each workflow stage and more detail on the decision criteria.

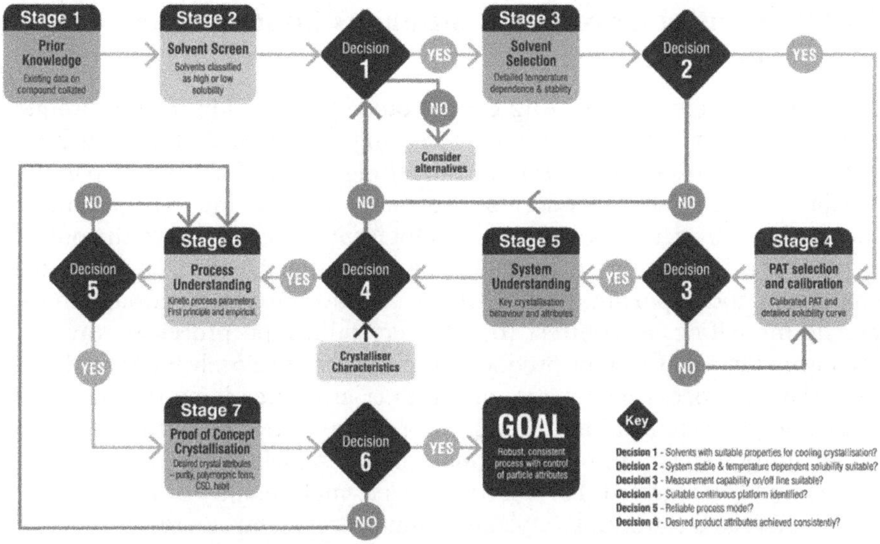

Figure 16.2 Workflow for the development of a cooling continuous crystallization process. Reproduced from ref. 9 with permission from the Royal Society of Chemistry.

16.3 Fundamentals of Mechanistic Process Modelling in Continuous Crystallization Processes

16.3.1 Purposes of Process Modelling in Pharmaceutical Applications

As demonstrated by the two case studies presented in Sections 16.4 and 16.5, process modelling can be used to meet a variety of objectives for continuous crystallization processes:

- Improve research and development efficiency and minimize the use of costly API at early stages of process development by reducing the number of experiments required to characterize and gain sufficient understanding of a crystallization process,
- Assess the robustness of the continuous crystallization process with respect to changes in operating conditions and/or sources of variability,
- Facilitate model transfer from batch to continuous crystallization by utilizing data rich batch laboratory-scale crystallization experiments to build an understanding that is transferable to continuous crystallization process,
- Enable model predictive control (MPC) by utilizing the mechanistic model on-line.

16.3.2 Considerations for Continuous Crystallization Processes

The rise in interest in utilizing continuous crystallization for the production of APIs has coincided with an increasing awareness on the application and utility of mechanistic process modelling tools. This may be partially attributed to a desire to modernize the pharmaceutical industry and implement quality-by-design (QbD) methodologies in processes from the outset. Beyond this connection, continuous crystallization processes pose unique challenges and opportunities that can be effectively explored through mechanistic modelling. In contrast to batch crystallization processes, the unit operations in a continuous production line are more closely integrated. The interplay of process parameters on product attributes is more non-linear. Continuous crystallization has the unfortunate attribute that there is less opportunity to test and reject the material after each "step" since the process is highly integrated. This all combines to make quality-by-testing much less feasible in a continuous crystallization process, and as a result, a deeper process understanding, sound control strategies, and risk mitigation methodologies are required.

16.3.3 Process Systems Engineering Tools

The formulation of a mechanistic model describing a process is only a part of the model lifecycle or workflow required to make use of the model in real industrial applications. The field of process systems engineering provides well-established tools and workflows for calibrating and validating mechanistic models and applying them to understand and improve the process in a rigorous manner.

16.3.4 Model Verification and Validation

A typical model development workflow starts with establishing a conceptual model on paper, or model formulation. Next is to transform the conceptual model to a computational model that evaluates the outputs of the process from a number of inputs. The relationships between the conceptual model, the computational model, and the real system are shown in Figure 16.3.[19]

After building a conceptualized model based on an analysis of the real system, the model is programmed to form a computerized model, which can be simulated to describe the process. Establishing trust that the computerized model represents reality requires three activities: model qualification, model verification, and model validation. Model verification is the confirmation that the computerized model as programmed is the correct implementation of the conceptual model, while model validation is the comparison of the computerized model to data from the real process. Model qualification involves an assessment of whether the conceptual model accurately describes reality.

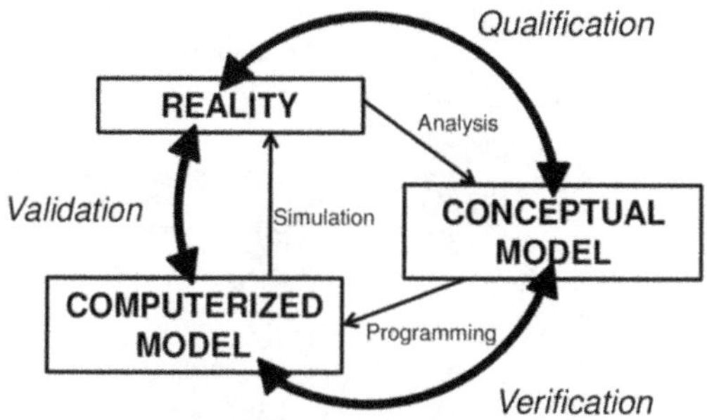

Figure 16.3 Phases of modelling and simulation and the role of verification and validation.[18] Reproduced from ref. 20 with permission from SAGE, Copyright 1979.

This terminology is accepted in the pharmaceutical industry as the standard. However, there is still debate within the industry, particularly because of the regulatory connotations of the word "validation."

The model validation step can be broken into two components: internal model validation, or calibration against experimental data, and external model validation, or blind testing against additional data not used to calibrate the model. Internal model validation is typically achieved through parameter estimation, or the tuning of kinetic and other unknown parameters within the model. Parameter estimation can be performed in a rigorous manner using optimization algorithms to minimize an objective function describing the difference between experimental data and model predictions. Statistical tests are utilized to assess the goodness-of-fit and identify bias in both the internal and external validation stages. A typical model lifecycle for a mechanistic crystallizer model encompasses model configuration, model validation and application, as described in Figure 16.4 below.

16.3.5 Uncertainty Analysis

Mechanistic modelling uses well-established methodologies that have proven success in ensuring quality, safety, and efficiency in other industries for decades. Though newer to the pharmaceutical industry, mechanistic models have seen increased use in R&D in recent years to improve the science (physics and chemistry) behind quality-by-design. A result of this is an increase in R&D efficiency since fewer experiments are needed to characterize a process, independent of the number of CPPs. This approach turns data into knowledge that can be extrapolated to other scales and/or operating conditions with known confidence and enables the construction of probabilistic design spaces to characterize robustness.

Once a model of a continuous crystallization process has been developed, it can be used to: (1) analyse the process performance under nominal conditions, (2) predict the ability of a process to cope with "abnormal"

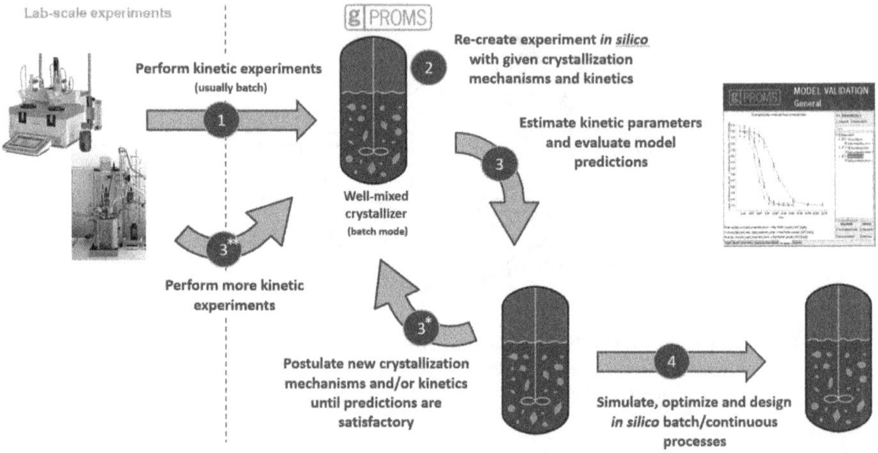

Figure 16.4 Mechanistic model lifecycle for a typical batch crystallization process.

conditions and disturbances (*e.g.* seeding loading), and (3) evaluate the effects of changes in the operating policy, or the design of individual equipment types. From the mathematical point of view, all of these correspond to process simulation calculations where the user specifies all the process inputs and the model equations are solved to determine all the relevant outputs, and in particular the key performance indicators (KPIs) of the individual steps and of the overall process. These single-point calculations can answer targeted questions about the process design space. Does an operating point yield a quality product, provided that the sources of variability are well controlled? What is the expected value of a quality metric for a given set of operating conditions?

In reality, the operating point can be variable, and the sources of common cause variability are not always negligible. As a result, "single-point" predictions of process KPIs (including critical quality attributes) computed *via* isolated process simulations are often of limited value. Instead, of more interest may be the probability distributions of these KPIs, and also the manner in which uncertainty in process KPIs can be attributed to the uncertainty and variability of individual inputs. The key question then becomes: What is the probability of product meeting quality and the desired process performance throughout the operational space of the continuous crystallization process? The same mechanistic model used for single-point calculations can be used to answer these questions through sensitivity and uncertainty analyses.

16.3.6 Risk Management through Sensitivity Analysis

The concept of the design space is central to current thinking regarding the degree of flexibility afforded to pharmaceutical process operation within a regulatory framework. In principle, process simulations could be used to determine whether given operating points fall within the design space. However, such single-point calculations may be unreliable in view of the process uncertainties: in reality, one can only determine the *probability* that a particular point is within the design space. Moreover, establishing the shape and size of the entire design space of any non-trivial process using isolated process simulations is impractical, and one may need to resort to more complex types of calculations.

Risk management is a key aspect of quality-by-design through systems-based approaches as mechanistic models are used to assess risk and evaluate mitigation strategies. Unlike one-factor-at-a-time approaches or statistical, data-driven techniques, integrated mechanistic models can be used to quantify and predict the combined effects of uncertainty and variability in process, product, and patient parameters on probability distributions of risk-related critical quality attributes, as shown in Figure 16.5 below. Further, risk is assessed as a probability distribution, or a likelihood of failure, rather than a point calculation that provides little insight beyond a single operating point. By globally assessing uncertainty and variability, risk mitigation strategies can be designed to target and reduce the highest-impact sources of variability.

Another dimension of complexity is added by the variability that is inherent in both process inputs (*e.g.* material flowrates) and in the underlying processing step models. Formal mathematical techniques can be used to perform *elementary effects analysis* and identify which of the input factors have

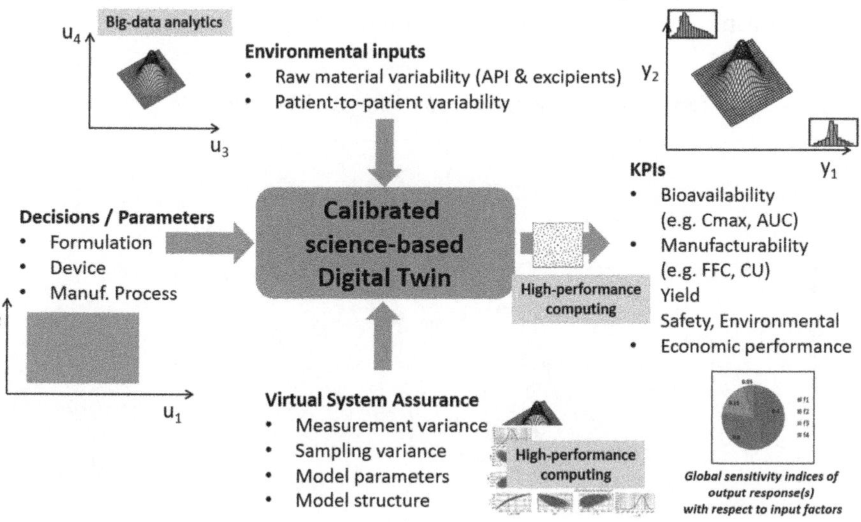

Figure 16.5 Global sensitivity and uncertainty analysis methodology wherein a mechanistic model describes the physical system of interest, and the KPIs are quantified as probabilistic distributions based on the multivariate combination of design/operational decisions, disturbances, and model uncertainty.

a non-negligible effect on at least one of the output responses in order to eliminate unimportant factors from further consideration at an early stage. *Uncertainty analysis* is next performed to determine the probability distribution of output responses in order to quantify the impact of uncertainty and variability. If the resulting probability distribution of the output response is unacceptable, then a model-based *global sensitivity analysis* (GSA) can be performed to compute global sensitivity indices of each response with respect to each factor and apportion variability of outputs to that of inputs and enable targeted risk management actions.

16.4 Digital Design Case Study – Batch to Continuous Workflow

This section describes the utilization of mechanistic process modelling tools to enable the efficient development and design of an industrial active pharmaceutical ingredient (API). In order to develop a model for a specific crystallization, many approaches can be taken. The example presented here utilized isothermal batch desupersaturation experiments, described in more detail elsewhere.[1] Figure 16.6 shows a phase diagram for the cooling crystallization of the solute studied in this case. The points (squares) in the figure show the seeding points which were utilized for the experiment design of each the isothermal desupersaturation condition. It was important to consider a broad range of temperatures and supersaturation levels in order to cover the applicable space for possible process optimization of the continuous crystallization. The broader experimental space covered

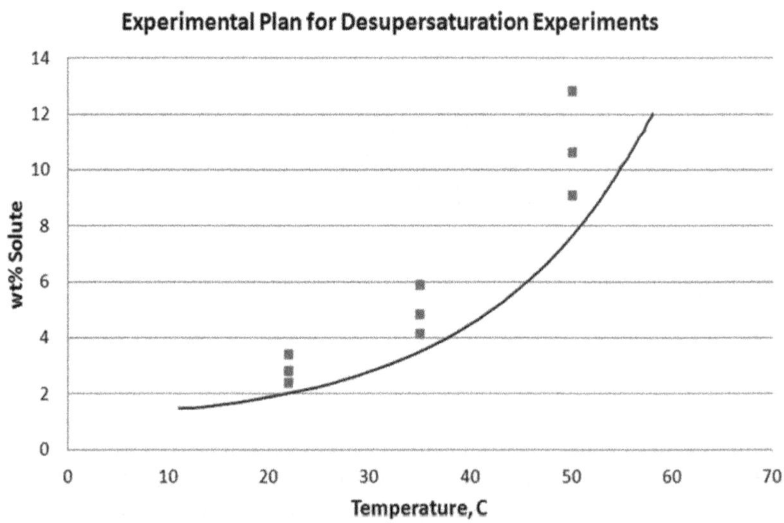

Figure 16.6 Identified experimental points (squares) and solubility (curve) for isothermal desupersaturation data discussed here.

allowed for minimization of model extrapolation as well as improving the significance of estimated kinetic model parameters. The experiments were conducted over the span of less than a week, with multiple experiments conducted each day.

Each experiment consisted of establishing the initial condition of a supersaturated solution of known composition and well-controlled temperature, then introducing a well-characterized and accurately weighed amount of seed crystals. *In situ* attenuated total reflectance infra-red spectroscopy (ATR-IR) was used for solute concentration measurements and focused beam reflectance measurements (FBRM) were used to monitor chord length distribution trends during desupersaturation, providing an indication of the evolution of the PSD. An offline sample was taken at the end of the experiment for determination of final particle size distribution *via* laser diffraction and evaluated visually by optical microscopy. Ultimately, the IR and laser diffraction data were employed for model validation, while the FBRM data was useful for understanding qualitative trends in the process data gathered. Figure 16.7 is a typical example of data collected during these experiments, showing the trends observed for solute concentration (*via* FTIR) and Chord Length Distribution (CLD – *via* FBRM).

Several approaches were considered for model development when considering the collected experimental data. A simple approach considered was an initial rate approximation, in which the initial mass deposition rate was estimated by extrapolating rate data to zero time, thus providing the growth rate on the initial seed surface area added to the solution (which was known accurately). This approach provided good estimates of growth kinetic model parameters but neglected a large amount of the collected data from which more detailed information could be gathered. For instance, if changes in particle size and distribution are considered in concert with changes in solute concentration, not only is there more growth rate kinetic information

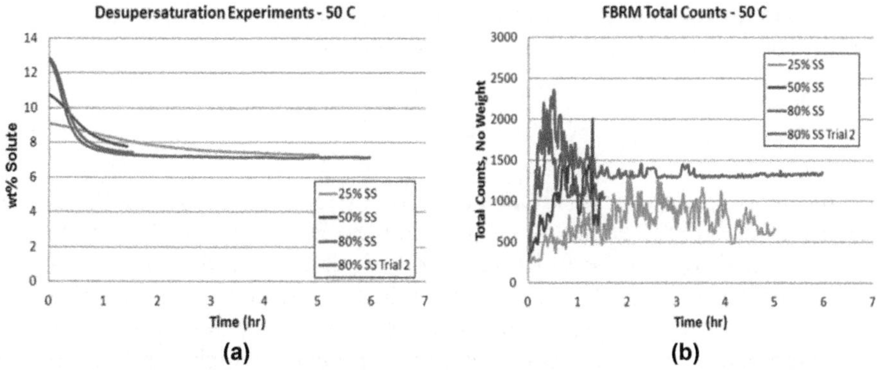

Figure 16.7 Typical data collected during desupersaturation experiments. On left (a) – concentration data as determined by ATR-IR for all experiments conducted at 50C. On right (b) – CLD total counts (non-weighted) for experiments conducted at 50C.

to extract, there is possible information about secondary nucleation rates, which are essential for an effective continuous crystallization design process. As evident from Figure 16.7, changes in particle count from the CLD data are greater in experiments with higher initial supersaturation levels. In order to maximize the information generated by these experiments a population balance process modelling approach was employed.

Selection of kinetic expressions for crystal growth and secondary nucleation can strongly influence the quality of the model fit achieved, so for this exercise multiple options were considered. The consideration of multiple rate expressions to describe a given mechanism, such as crystal growth, is commonly termed model discrimination. For growth rates, literature precedence is established well for an expression utilizing absolute supersaturation (eqn (16.1)).[2–4] This growth model will be referred to here as the Delta C model.

$$R_G = K_g \exp\left(\frac{-E_{A,g}}{RT}\right)\left(C - C^*\right)^g \tag{16.1}$$

Significant literature precedence is also established for an expression based on relative supersaturation (eqn (16.2)).[5–7] This growth model will be referred to as the Sigma model.

$$R_G = K_g \exp\left(\frac{-E_{A,g}}{RT}\right)\left(\frac{C - C^*}{C^*}\right)^g \tag{16.2}$$

Considering the CLD data in Figure 16.7, a mechanism for particle generation must be considered, namely secondary nucleation. For this work, the model chosen is from the work of Evans *et al.*,[8] which is a secondary nucleation rate expression considering crystal-impeller collisions as the main source of attrition based secondary nucleation (eqn (16.3)). In addition to this secondary nucleation model (referred to as Nuc Model 1), two variants were considered where the rate constant is temperature dependent (Nuc Model 2) and the nucleation rate is proportional to growth rate instead of simply the supersaturation level in solution (Nuc Model 3).

$$J_{\text{sec}} = K_n \sigma_n \frac{N_Q}{N_P} k_v \rho_c \varepsilon \int_{L_{\min}}^{\infty} n L^3 \, \mathrm{d}L \tag{16.3}$$

Combinations of the two chosen growth rate kinetic models and three secondary nucleation models led to six sets of kinetic models which must be discriminated to find the most suitable combination to describe the available experimental data. This activity is typically referred to as model discrimination. Each set of growth and nucleation kinetic model parameters was regressed against all ten experiments to help decide which combination of kinetic expressions described the data better. Figure 16.8 is a sampling of all of the data regression done, and clearly shows that the Sigma growth kinetic model significantly outperforms the Delta C model in fitting the experimental data for this system.

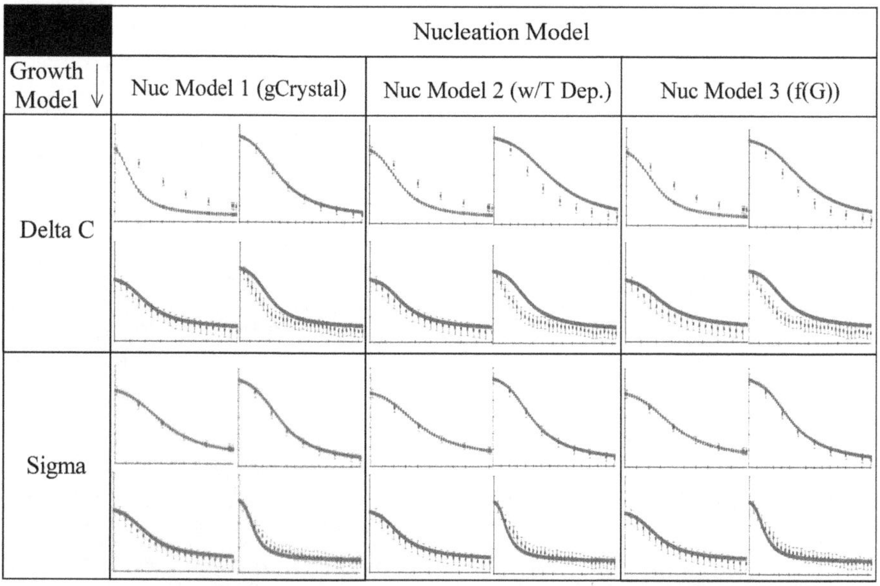

Figure 16.8 Sampling of results (concentration *versus* time) from parameter regressions during model discrimination testing. Plots shown are a subset of the ten experiments performed.

There is no visual way from this figure to distinguish the secondary nucleation kinetic models from one another. Since population balance modelling was utilized in this case, the product particle size quantiles were also considered in the parameter regression. Nuc Model 1 slightly outperformed both models 2 and 3 in terms of particle size fit to the data (Figure 16.9). Due to the small difference, Nuc Model 2 could not be eliminated as a viable model at this stage. Nuc Model 3 was eliminated based on the poor parameter fitting statistics. The data are not shown here, but Nuc Model 3 had one additional parameter that was not statistically significant, indicating an over-fitting of kinetic parameters in this case.

Batch isothermal desupersaturation experiments provided a very effective means to determine crystallization kinetics (for both crystal growth and secondary nucleation) over a wide range of conditions using a small number of experiments. Model discrimination was achieved mostly through data regression, but two candidates remained as possible best fits for the solution system. Further model benchmarking was done against two continuous crystallization experiments. The experiments used for benchmarking (Figure 16.10) were performed before data regression, thus the conditions were chosen heuristically. The conditions were chosen to exemplify the large number of variables available for experimentation.

The experiments were simulated with the Sigma growth kinetic model and using either Nuc Model 1 or 2. The metric used for final model

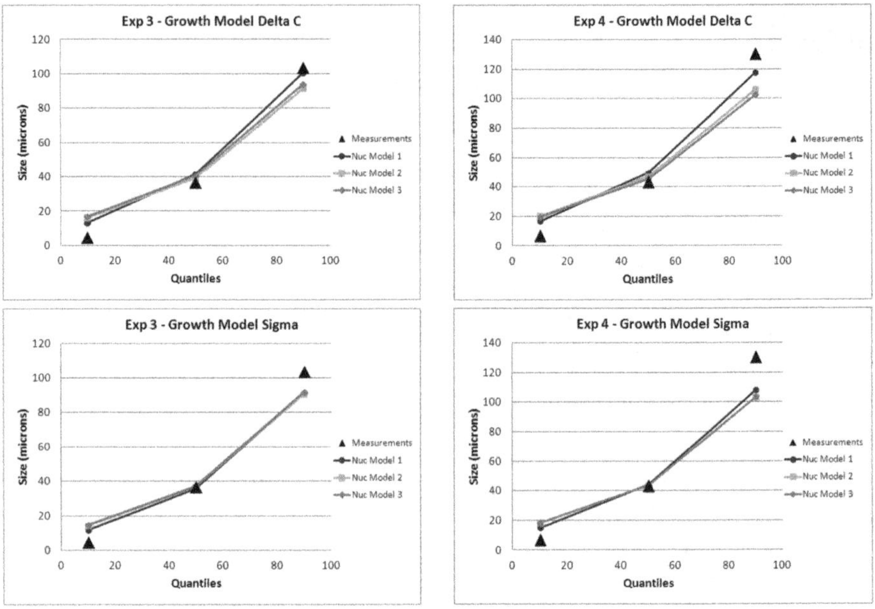

Figure 16.9 Sampling of results (ending particle size quantiles) from parameter regressions during model discrimination testing. Plots shown are a subset of the ten experiments performed.

discrimination was the steady-state solute concentrations measured in either experiment. Table 16.2 shows a summary of simulations of these experiments with both model sets. Model 1 (growth model Sigma with Nuc Model 1) performs better than Model 2, which tends to over-predict concentrations.

With a model combination selected and calibrated many activities can be performed, such as process optimization for the continuous crystallization process. The model can also be utilized in the determination of critical process parameters or design space determination, but those are not discussed in detail here. Process optimization for this case was done by considering a fixed process throughput (feed rate and solute concentration) with controls and constraints as dictated by product demands and practical process considerations (Figure 16.11). Product yield was optimized, while a small penalty was paid for additional crystallizer residence time (*i.e.* larger crystallizer size) to avoid the obvious solution of infinitely large crystallizer volume (*i.e.* operation at or near equilibrium to maximize yield). Constraints of maximum supersaturation (particularly in the first stage) were used as a means to minimize crystallizer fouling while increasing crystal production rates. Particle size constraints were used to manage product physical properties (*e.g.* volume-based quantile d10 of the PSD > 10 microns). Anti-solvent concentrations, flow rates and MSMPR volumes were allowed to vary within reasonable ranges as the process parameters.

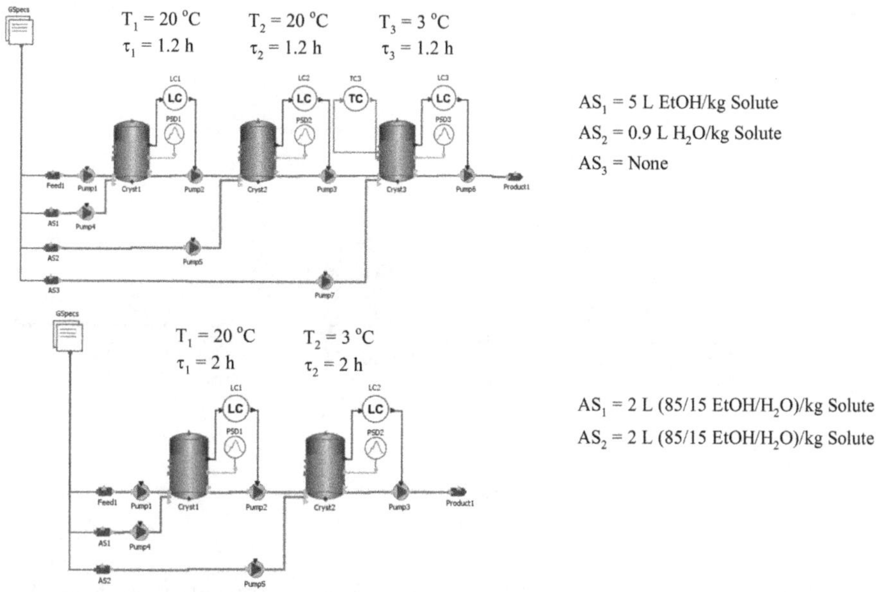

Figure 16.10 Continuous crystallization experiments used for final model discrimination and benchmarking. Top: Three stage MSMPR system with two unique anti-solvent streams added at two separate locations. Bottom: Two stage MSMPR system with the same anti-solvent mixture split over both stages.

Table 16.2 Steady state solute concentrations for all MSMPRs in either Exp 1 or Exp 2.

Value	Actual (wt%)	Model 1 (wt%)	Model 2 (wt%)
Exp 1 – MSMPR1 S–S	7.9	8.2	10.0
Exp 1 – MSMPR2 S–S	1.7	1.6	2.0
Exp 1 – MSMPR3 S–S	0.83	0.86	1.0
Exp 2 – MSMPR1 S–S	8.8	10.0	13.0
Exp 2 – MSMPR2 S–S	2.0	2.4	3.4

Optimization led to the determination of operating conditions which may have taken many experimental trials to reach heuristically (Table 16.3). This approach can also be used to identify Pareto optimal fronts, in which certain active constraints can be adjusted to understand trade-offs that must be made to achieve different results in terms of the CQAs of the process.

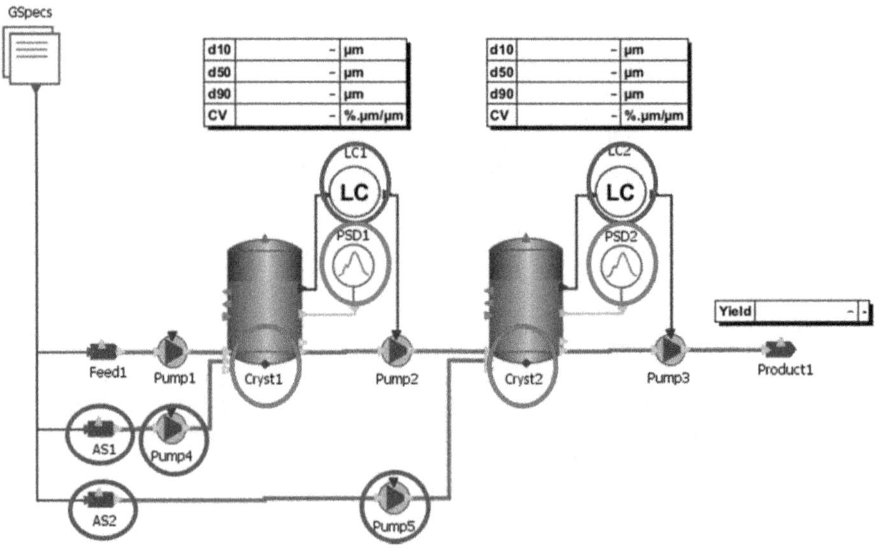

Figure 16.11 Process diagram (2 MSMPRs) emphasizing control variables (LC, AS, Pumps 4 and 5) and constraints (Cryst and PSD).

Table 16.3 Model prediction of the optimal conditions for 2 stage MSMPR system.

Process parameter	Value
Anti-solvent feed composition – Stage 1	88 (w/w%)
Anti-solvent feed composition – Stage 2	80% (w/w%) (lower bound)
Crystallizer operating volume – Stage 1	30 mL (lower bound)
Crystallizer operating volume – Stage 2	524 mL
Residence time – Stage 1	47 minutes
Residence – Stage 2	~6 hours
Anti-solvent feed flowrate – Stage 1	~4 L kg^{-1} feed
Anti-solvent feed flowrate – Stage 2	~8.5 L kg^{-1} feed
Steady-state yield	89.9%

16.5 Digital Operation Case Study: Utilizing Mechanistic Modelling for Development of a Model Predictive Controller (MPC)

In this section, we present the application of model predictive control (MPC) to a crystallization process. First, an introduction to MPC is presented followed by the development of a transferrable, data driven MPC scheme. Finally, a digital design approach to MPC for crystallization is also presented, where a mechanistic model is utilized as a virtual plant/digital twin.

16.5.1 Introduction to Model Predictive Control

MPC is a multivariable advanced control technique which uses a mathematical model to predict and optimise the future process behaviour.[9] The basic structure of MPC is shown in Figure 16.12.

The MPC algorithm computes a control sequence by solving online, at each sampling instance, a mathematical optimisation and the first control action in this sequence is then applied to the plant.[10] The optimisation problem typically involves a cost function of the following form:

$$J = \sum_{i=1}^{N} (y_{k+i} - y_{\text{des}})^T Q (y_{k+i} - y_{\text{des}}) + \Delta u_{k+i-1}^T R \Delta u_{k+i-1} \tag{16.4}$$

Subject to the constraints:

$$u_L \le u \le u_H \tag{16.5}$$

Where y_k is the controlled variable (CQA) at time k (*e.g.* concentration), y_{des} is the desired value, u_k is the control action (*e.g.* pump speed), N is the prediction horizon, and Q, R represent the cost function weights.

MPC offers several advantages over traditional control schemes, including the ability to explicitly take account of process constraints, as well as to handle

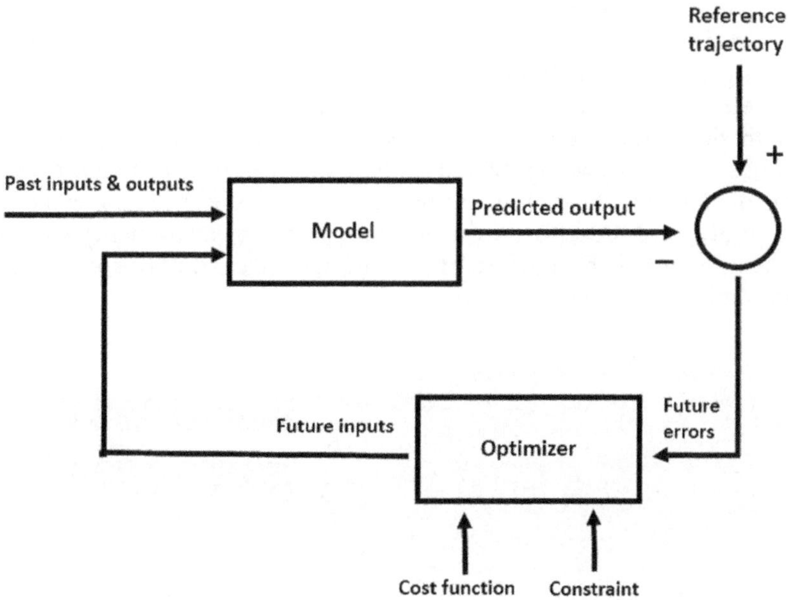

Figure 16.12 Structure of MPC.

multivariable, non-minimum phase and unstable processes. Due to these reasons, MPC has been widely implemented for advanced process control within various industry sectors.[11] Furthermore, over the past few decades, it has also been the subject of extensive academic research to improve robustness whilst reducing computational load and conservatism.[12,13] MPC technology is well suited to continuous crystallization processes as it involves multivariable interactions, long time delays and possible disturbances in the feed conditions. In this section, two approaches for advanced control of crystallization process are presented, namely: data driven MPC and a digital design approach. Additional details on control and MPC development can be found in Chapter 4.

16.5.2 Data Driven Approach to Advanced Control for Crystallization

This section presents the design and implementation of a transferrable, data driven MPC scheme for the control of two different continuous crystallization units. This work was carried out as part of the MOPPS project (Innovate UK, grant number: 101334) with the results published in ref. 14. The first continuous oscillatory baffled reactor (COBR) unit considered in this work is the Rattlesnake produced by Cambridge Reactor Design. As shown in Figure 16.13 (left), it consists of four cylindrical modules with a volume of 2760 mL. The temperature of each module is controlled through a water-filled jacket and the residence time within the unit can vary from 60 minutes upwards depending on the throughput.

The second considered COBR unit is the DN15 developed by Nitech. As shown in Figure 16.13 (right), it comprises of 31 tubes each consisting of baffled zones. The tubes were jacketed for temperature control and gave a total volume of 2500 mL. Peristaltic pumps were used for the seed and feed flows. Furthermore, an FBRM (focused beam reflectance measurement) was used for measurement of chord length distribution (CLD) and FTIR spectrometer provides inline spectra measurement.

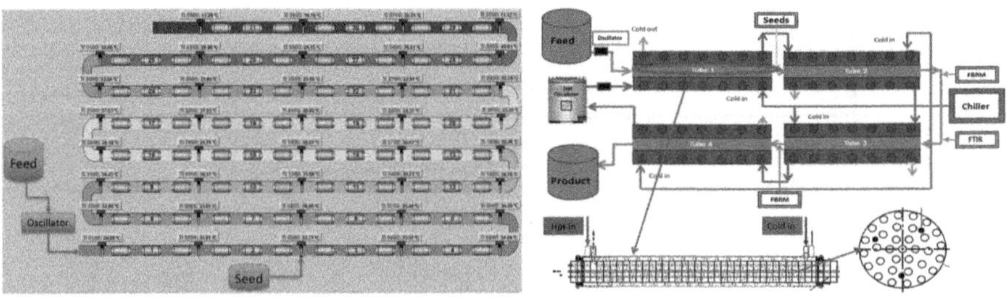

Figure 16.13 Rattlesnake (left) and DN15 continuous crystallization reactors.[14] Reproduced from ref. 14 with permission from Elsevier, Copyright 2017.

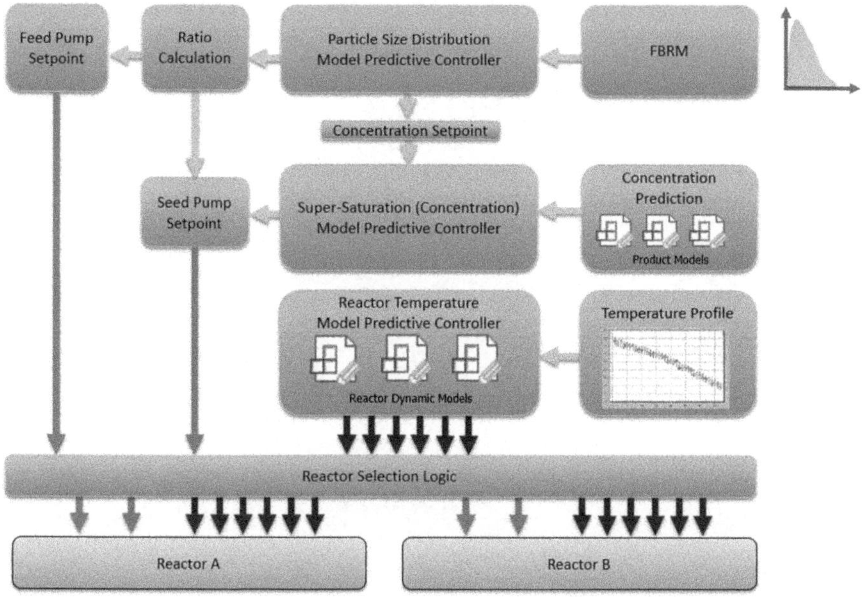

Figure 16.14 Transferrable MPC solution for continuous crystallization.[14] Reproduced from ref. 14 with permission from Elsevier, Copyright 2017.

A transferrable MPC solution was developed[17] so that it could be applied to both the Rattlesnake and DN15 reactors. The overall control scheme is shown in Figure 16.14 below.

As shown in Figure 16.14, this data driven advanced control scheme consisted of a calibration model to predict the API (Lactose) concentration based on the FTIR spectra. This prediction was then controlled to the setpoint by the concentration MPC controller manipulating the seed/feed pumps. The concentration setpoint was adjusted by the PSD MPC controller to drive the D50 to its desired setpoint. Finally, a temperature MPC controller manipulated the circulator temperatures to drive the reactor temperature profile to the desired setpoint.

The calibration model for the API was developed by preparing solutions of known concentration and then correlating this theoretical concentration with the corresponding inline FTIR spectra using PLS (partial least squares) regression. The training dataset for calibration model development is shown in Figure 16.15. It can be seen that the model predictions closely match the theoretical concentration.

To develop a data driven model for the concentration MPC controller, step tests were carried out on the feed and seed pump speeds, respectively (as shown in Figure 16.16). Then, RLS (recursive least squares) regression was applied to develop an output-error time series model.[14]

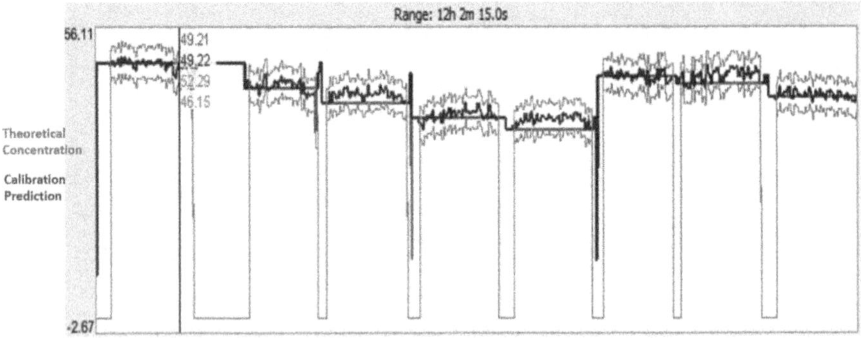

Figure 16.15 Concentration predictions by the PLS calibration model.[14] Reproduced from ref. 14 with permission from Elsevier, Copyright 2017.

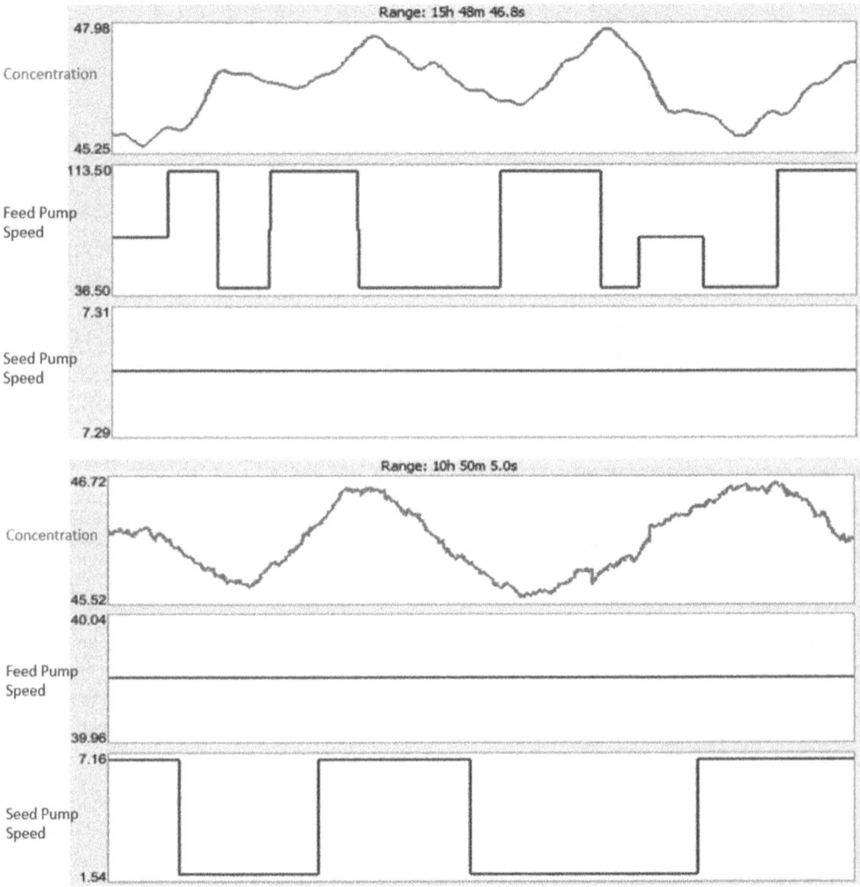

Figure 16.16 Step tests on the feed and seed pumps speeds.[14] Reproduced from ref. 14 with permission from Elsevier, Copyright 2017.

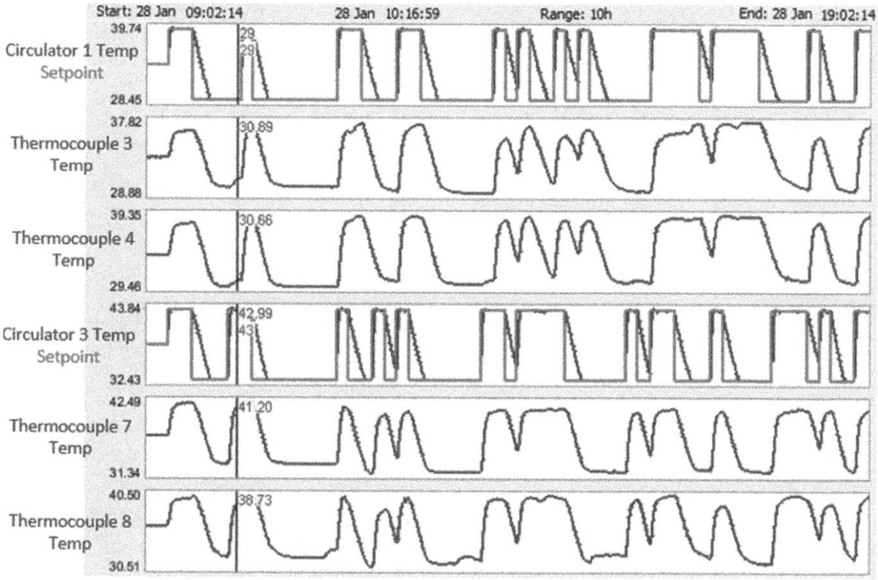

Figure 16.17 Step tests on the circulator temperatures.[14] Reproduced from ref. 14 with permission from Elsevier, Copyright 2017.

Similar to above, step tests were applied to the circulator temperature setpoints as shown in Figure 16.17 below. Then, the thermocouple temperature response was regressed with the circulator temperature to develop a data driven reactor temperature control model.

The results for the developed cascade MPC control strategy are shown in Figure 16.18. Note that the (master) particle size MPC controller is able to drive the D50 to the desired setpoint by manipulating the concentration setpoint. This concentration setpoint is then successfully tracked by the slave MPC through real-time manipulation of the feed/seed pump speeds. The proposed data driven control scheme is not only flexible enough to be applied to different continuous crystallization reactors but can also be extended to accommodate different APIs.[14]

16.5.3 Digital Design Approach to Advanced Control for Crystallization

In the previous section, a data driven approach was presented for advanced control of crystallization units. Whilst this approach is highly effective, its main drawback is the time and materials requirements associated with the experimental work for developing the control and calibration models. Therefore, a digital design approach has also been investigated for the development of hybrid control models for crystallization. The work presented in this section is based on.[15]

As part of the digital design approach, a mechanistic flowsheeting model was developed.[16] The flowsheet consisted of an MSMPR tank with a temperature

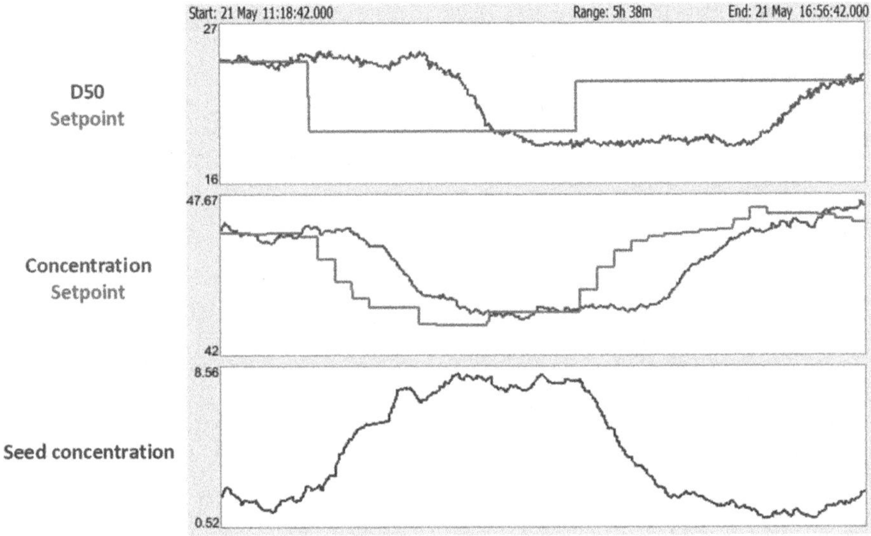

Figure 16.18 Control results on the DN15 COBR.[14] Reproduced from ref. 14 with permission from Elsevier, Copyright 2017.

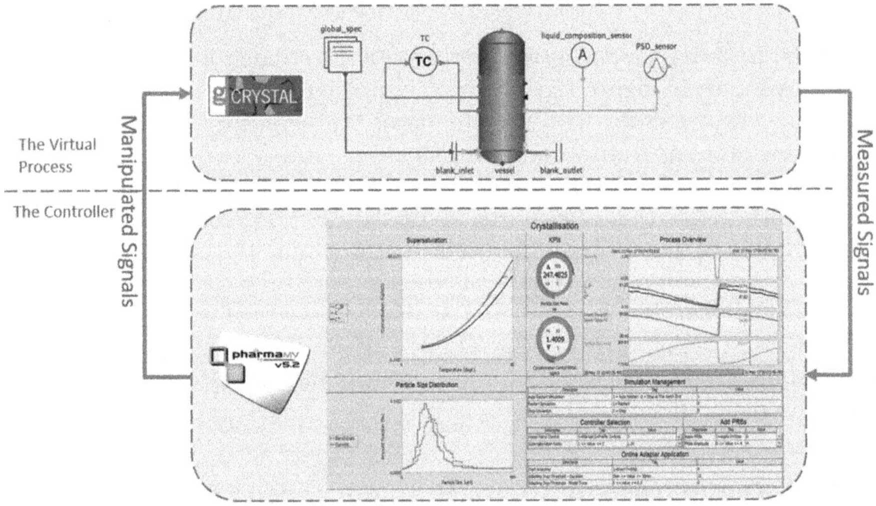

Figure 16.19 Digital design approach to MPC control of the crystallization process.[15]

controller, liquid composition sensor and a PSD sensor. The nucleation and growth crystallization kinetic models were parameterized through a small number of targeted laboratory batch desuperaturation crystallization experiments. As shown in Figure 16.19 below, this mechanistic model served as a virtual plant or digital twin for the development of an MPC scheme.[17]

To develop a suitable control model of the process, PRBS (pseudo random binary sequence) step tests were applied on the flowsheet MSMPR cooling rate (as shown in Figure 16.20).

Using the PRBS data, an empirical time series control model was developed. Then, the MPC controller was designed to adjust the MSMPR cooling rate to maintain the solute concentration to its desired supersaturation profile (C^*). The control results are shown in Figure 16.21. This shows that the controller, designed only using the flowsheet model, was able to effectively maintain the concentration, C on the desired supersaturation profile, C^* through the real-time adjustment of the vessel temperature.

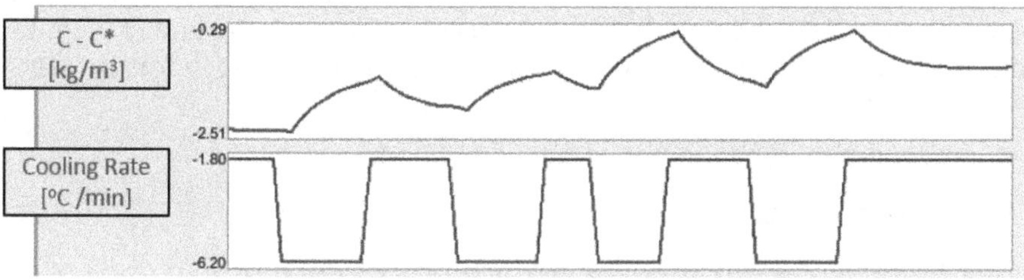

Figure 16.20 PRBS step tests on the flowsheet MSMPR cooling rate.[15]

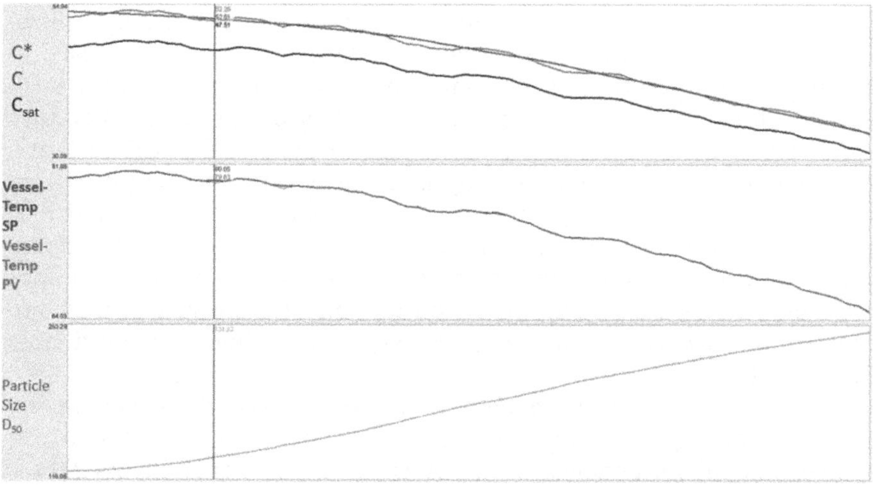

Figure 16.21 MPC control results using the digital design approach[15].

16.6 Conclusion

MPC is an optimisation-based control technique that is particularly suitable for the continuous crystallization process due to its ability to handle non-linear systems with multi-variable interactions as well as long time delays. In this chapter, two approaches for the MPC control of crystallization have been presented. A transferrable, data driven MPC scheme has been described for two commercial continuous crystallization reactors. This involved performing PRBS step tests on the crystallization platforms, building empirical models followed by commissioning and testing. Whilst this approach yields accurate control performance, it requires significant experimental effort in terms of time and materials. To address this, a digital design approach has also been presented. This involves using flowsheet mechanistic models, calibrated against significantly less data (often generated at smaller scales) to develop and test the MPC scheme before commissioning on the hardware platform. The benefits of this digital design approach include a significant reduction in experimental effort as well as material usage for the MPC development.

16.7 Summary

As outlined in this chapter, the application of digital design tools, specifically mechanistic models, is becoming more widespread in the pharmaceutical industry to enable more efficient development, design and control of API crystallization processes, both batch and continuous. Calibrated mechanistic models for crystallization can facilitate a more efficient workflow for studying and designing new continuous crystallization processes, by characterising the process kinetics using batch crystallization data. The calibrated crystallization kinetics can then be utilised to inform the configuration of the continuous crystallization process, such as the number of stages to employ and the set-points for temperature and/or composition to utilise in each stage to achieve the desired CQAs, in particular product PSD. This analysis can be performed *in silico via* simulation prior to performing any continuous crystallization runs. Furthermore, it was shown in the first case study that the mechanistic model of the continuous crystallization can subsequently be employed to optimise the process to desired objective functions and provide a means to understand the various processing trade-offs and map out the attainable regions of the process.

Finally, it was also shown *via* the second case study how these mechanistic models, that are typically applied in R&D steps only for process design, can be further applied to enable the more efficient development of online MPCs. In comparison to the data-driven approach, the mechanistic models have the potential to significantly reduce the experimental effort as well as material usage for MPC development.

References

1. J. Garside, A. Mersmann and J. Nyvlt, *Measurement of Crystal Growth and Nucleation Rates*, 2nd edn, 2002.
2. M. Aoun, *et al.*, *Chem. Eng. Sci.*, 1999, **54**, 1161–1180.
3. Z. Nagy, *et al.*, *Ind. Eng. Chem. Res.*, 2008, **47**, 1245–1252.
4. Z. Nancollas, *J. Cryst. Growth*, 1968, **3**(4), 335–339.
5. G. L. Zipp, *et al.*, *Int. J. Pharm.*, 1989, **51**, 147–156.
6. E. Kougoulos, *et al.*, *J. Cryst. Growth*, 2005, **273**, 520–528.
7. A. Mangood, *et al.*, *J. Cryst. Growth*, 2006, **290**, 565–570.
8. T. W. Evans, *et al.*, *AIChE J.*, 1974, **20**, 959–966.
9. J. B. Rawlings and D. Q. Mayne, *Model Predictive Control: Theory and Design, Nob Hill Publishing*, Wisconsin, Madison, 2012.
10. J. M. Maciejowski, *Predictive Control with Constraints*, Prentice Hall, Essex, England, 2002.
11. S. Joe Qin and T. A. Badgwell, A survey of industrial model predictive control technology, *Control Eng. Pract.*, 2003, **11**(7), 733–764.
12. D. Q. Mayne, J. B. Rawlings, C. V. Rao and P. O. M. Scokaert, Constrained model predictive control: Stability and optimality, *Automatica*, 2000, **36**(6), 789–814.
13. F. Tahir and I. M. Jaimoukha, Robust feedback model predictive control of constrained uncertain systems, *J. Process Control*, 2013, **23**, 189–200.
14. F. Tahir, K. Krzemieniewska-Nandwani, J. Mack, D. Lovett, H. Siddique, F. Mabbot, V. Raval, I. Houson and A. Florence, Advanced control of a continuous oscillatory flow crystalliser, *Control Eng. Pract.*, 2017, **67**, 64–75.
15. Y. Salman, C. Ma, J. Mack, T. Mahmud, N. Mitchell and K. Roberts, Application of mechanistic models for the online control of crystallization processes, *In 20th International Symposium of Industrial Crystallization (ISIC20)*, Ireland, Dublin, 2017.
16. Process Systems Enterprise, https://www.psenterprise.com/home.
17. Perceptive Engineering Limited, http://www.perceptiveapc.com.
18. W. L. Oberkampf and C. J. Roy, *Verification and Validation in Scientific Computing*, Cambridge Press, 2010, p. 23.
19. CMAC, Developing workflows for continuous crystallization, 2019, online accessible, https://www.cmac.ac.uk/workflows.htm.
20. S. Schlesinger, Terminology for model credibility, *Simulation*, 1979, **32**, 103–104.

Subject Index